SELECTED TABLES IN MATHEMATICAL STATISTICS

Volume I

This volume was prepared with the aid of:

Paul C. Cox, White Sands Missile Range, New Mexico
W. J. Dixon, University of California at Los Angeles
Z. Govindarajulu, University of Kentucky
S. S. Gupta, Purdue University
H. O. Hartley, Texas A&M University
W. H. Kruskal, University of Chicago
G. J. Lieberman, Stanford University
M. E. Muller, University of Wisconsin
G. P. Steck, Sandia Laboratory, Albuquerque, New Mexico
R. H. Wampler, National Bureau of Standards
Max A. Woodbury, Duke University

SELECTED TABLES IN MATHEMATICAL STATISTICS

Volume I

Edited by the Institute of Mathematical Statistics

Coeditors

H. L. Harter

Aerospace Research Laboratories

and

D. B. Owen

Southern Methodist University

AMERICAN MATHEMATICAL SOCIETY

PROVIDENCE, RHODE ISLAND

Second printing with revisions 1973
by
American Mathematical Society
for
Institute of Mathematical Statistics

International Standard Book Number 0-8218-1901-1
Library of Congress Catalog Card Number 71-111-981

Printed in United States of America

PREFACE

This volume of mathematical tables has been prepared under the aegis of the Institute of Mathematical Statistics. The Institute of Mathematical Statistics is a professional society for mathematically oriented statisticians. The purpose of the Institute is to encourage the development, dissemination, and application of mathematical statistics. The Committee on Mathematical Tables of the Institute of Mathematical Statistics is responsible for preparing and editing this series of tables. The Institute of Mathematical Statistics has entered into an agreement with the American Mathematical Society to jointly publish this series of volumes. At the time of this writing, Volume II is being prepared for publication, and submissions for Volume III are being solicited. No set number of volumes has been established for this series. As many volumes as are necessary to reach publication for meritorious material will be considered and every effort will be made to get them published.

Potential authors should consider the following rules when submitting material.

1. The manuscript must be prepared by the author in a form acceptable for photo-offset. This includes both the tables and introductory material. The author should assume that nothing will be set in type although the editors reserve the right to make editorial changes.

2. While there are no fixed upper and lower limits on the length of tables, authors should be aware that the purpose of this series is to provide an outlet for tables of high quality and utility which are too long to be accepted by a technical journal but too short for separate publication in book form.

3. The author must, wherever applicable, include in his introduction the following:

   (a) He should give the formula used in the calculation, and the computational procedure (or algorithm) used to generate his tables. Generally speaking, FORTRAN or ALGOL programs will not be included but the description of the algorithm used should be complete enough that such programs can be easily prepared.

   (b) A recommendation for interpolation in the tables should be given. The author should give the number of figures of accuracy which can be obtained with linear (and higher degree) interpolation.

   (c) Adequate references must be given.

   (d) The author should give the accuracy of the table and his method of rounding.

   (e) In considering possible formats for his tables, the author should attempt to give as much information as possible in as little space as possible. Generally speaking, critical values of a distribution convey more information than the distribution itself, but each case must be judged on its own merits.

   (f) The table should adequately cover the entire function. Asymptotic results should be given and tabulated if informative.

   (g) An example or examples of the use of the tables should be included.

4. The author should submit as accurate a tabulation as he can. The table will be checked before publication, and any excess of errors will be considered grounds for rejection. The manuscript introduction will be subjected to refereeing and an inadequate introduction may also lead to rejection.

5. Authors having tables they wish to submit should send two copies to:

Dr. H. Leon Harter, Co-editor
Aerospace Research Laboratories
Wright-Patterson Air Force Base
Ohio 45433

At the same time, a third copy should be sent to:

Dr. D. B. Owen, Co-editor
Department of Statistics
Southern Methodist University
Dallas, Texas 75275

Additional copies may be required, as needed for the editorial process. After the editorial process is complete, a camera–ready copy must be prepared for the publisher.

Authors should check several current issues of *The Institute of Mathematical Statistics Bulletin* and *The American Statistician* for any up-to-date announcements about submissions to this series.

The tables included in the present volume were checked at Wright–Patterson Air Force Base. Dr. H. L. Harter arranged for this checking which was done under the direction of Mr. Clem Grabner and Mr. James P. Hudson. The editors and the Institute of Mathematical Statistics wish to express their great appreciation for this invaluable assistance from Mr. Grabner, Mr. Hudson, and their group. So many other people have contributed to the instigation and preparation of this volume that it would be impossible to record their names here. To all of these people, who will remain anonymous, the editors and the Institute also wish to express their thanks.

TABLE OF CONTENTS

TABLES OF THE CUMULATIVE NON-CENTRAL CHI-SQUARE DISTRIBUTION

G. E. Haynam*
Z. Govindarajulu†
F. C. Leone‡

*** Currently at Vanderbilt University.**
† Currently at University of Kentucky.
‡Currently at University of Iowa, Iowa City.

This research, at Case Institute of Technology, Cleveland, Ohio, was supported in part by the United States Air Force through the Air Force Office of Scientific Research, under Research Grant AFOSR 62-72. Reproduction in whole or in part is permitted for any purpose of the United States Government.

SUMMARY

Two extensive tables related to the cumulative non-central chi-square distribution are presented. In Table I, the power of the chi-square test accurate to four decimal places is presented for

$\alpha = 0.001, 0.005, 0.01, 0.025, 0.05, 0.1$;

$\lambda = 0(0.1)\ 1.0\ (0.2)\ 3.0\ (0.5)\ 5\ (1)\ 40\ (2)\ 50\ (5)\ 100$; and

$\nu = 1\ (1)\ 30\ (2)\ 50\ (5)\ 100$, where α, λ, and ν respectively denote the level of significance, the non-centrality parameter, and the degrees of freedom. In Table II, the non-centrality parameter, namely λ, is presented accurate to three decimal places for $\alpha = 0.001, 0.005, 0.01, 0.025, 0.05, 0.1$, power $(1 - \beta) = 0.1\ (0.02)\ 0.7\ (0.01)\ 0.99$ and $\nu = 1\ (1)\ 30\ (2)\ 50\ (5)\ 100$.

INTRODUCTION

In testing statistical hypotheses, the performance of a statistical test procedure is usually judged by its power of detecting departures from the null hypothesis. Thus, it is necessary to know the distribution of the test statistic T not only under the null hypothesis but also under the admissible alternative hypotheses. In the case of the usual and well known test procedures using t, χ^2 and F, the evaluation of their power functions involves their non-central distributions. The non-central chi-square distribution arises naturally when one wishes to find the distribution of a sum of squares of independent normal variables, having common variance and non-zero means. Apart from power considerations, as remarked by Patnaik [3], non-central chi-square distributions are of considerable interest. Patnaik [3] also gave some applications of the non-central chi-square distribution. Patnaik [3], realizing the considerable labor involved in calculating the percentage points of the non-central chi-square distribution, considered some approximations to the non-central chi-square distribution. One of these approximations is a central chi-square distribution with modified degrees of freedom and a suitable scale parameter. Fix [1] tabulated the non-centrality parameter accurate to three decimal places for $\nu = 6\ (1)\ 20\ (2)\ 40\ (5)\ 60\ (10)\ 100$, $\alpha = 0.01, 0.05$ and power $= 0.1\ (0.1)\ 0.9$ where ν and α respectively denote the degrees of freedom and the level of significance. Recently, Ruben [4] found that the distribution function of a non-negative and non-homogeneous quadratic form of a finite number of correlated normal random variables can be expressed in terms of non-central chi-square distribution functions with arbitrary scale parameter. Hence, it is of interest and use to tabulate the non-central chi-square distribution. This distribution involves three parameters:

(i) the non-centrality parameter denoted by λ,
(ii) the degrees of freedom denoted by ν, and
(iii) the power (that is, the cumulative probability) denoted by $1 - \beta$.

COMPUTATIONAL ALGORITHM

If we let $y = \chi^2$ then the cumulative non-central χ^2 distribution as given in [3] is:

$$F(y,\lambda,\nu) = \int_0^y \frac{e^{-x/2}\ e^{-\lambda/2}}{2^{\nu/2}} \sum_{j=0}^{\infty} \frac{x^{(\nu/2)+j-1}\ \lambda^j}{\Gamma[(\nu/2)+j]\ 2^{2j}\ j!}\ dx. \tag{3.1}$$

A convenient computational formula for $F(y, \lambda, \nu)$ has been suggested by Johnson [2]. Let us consider an associated function $G(y, \lambda, \nu)$ defined by:

$$G(y,\lambda,\nu) = 1 - F(y,\lambda,\nu) = \int_y^\infty \frac{e^{-x/2}\ e^{-\lambda/2}}{2^{\nu/2}} \sum_{j=0}^{\infty} \frac{x^{(\nu/2)+j-1}\ \lambda^j}{\Gamma[(\nu/2)+j]\ 2^{2j}\, j!}\ dx. \tag{3.2}$$

Now, after interchanging the order of summation and integration, we can write (3.2) as:

$$G(y,\lambda,\nu) = \sum_{j=0}^{\infty} \frac{e^{-\lambda/2}\ (\lambda/2)^j}{j!} \left[\frac{1}{\Gamma[(\nu/2)+j]} \int_y^\infty (x/2)^{(\nu/2)+j-1}\ e^{-x/2}\ d(x/2)\right] \tag{3.3}$$

Consider the integral

$$I(p,y) = \frac{1}{\Gamma(p/2)} \int_y^\infty u^{(p/2)-1} e^{-u}\, du \tag{3.4}$$

where $p = \nu + 2j$ is an integer, $p \geq 1$. Let us first consider the case where p is even, then $(p/2)$ is an integer and (3.4) upon successive integration by parts reduces to:

$$I(p,y) = \left[\frac{e^{-y} y^{(p/2)-1}}{[(p/2)-1]!}\right] + \left[\frac{e^{-y} y^{(p/2)-2}}{[(p/2)-2]!}\right] + \ldots + \left[\frac{e^{-y} y^0}{0!}\right] = \sum_{i=0}^{(p/2)-1} P_i\,(y) \tag{3.5}$$

where $P_i\,(y) = e^{-y} y^i/(i!)$ is the Poisson density function. When p is an odd integer, $p/2$ is an odd multiple of ½; after performing repeated integration by parts we can write (3.4) as:

$$\begin{aligned} I(p,y) = {} & [e^{-y} y^{(p/2)-1}/\Gamma(p/2)] + [e^{-y} y^{(p/2)-2}/\Gamma[(p/2)-1)] + \ldots \\ & + [e^{-y} y^{½}/\Gamma(3/2)] + [\Gamma(½)]^{-1} \int_y^\infty e^{-x} x^{-(½)} dx \end{aligned} \tag{3.6}$$

where the last integral, on substitution of $u^2 = x$, reduces to the complementary error function:

$$2\pi^{-(½)} \int_{y^{½}}^\infty e^{-u^2}\, du.$$

(3.6) can now be rewritten as:

$$\begin{aligned} I(p,y) = {} & [e^{-y} y^{(p/2)-1}/\Gamma(p/2)] + \left[e^{-y} y^{(p/2)-2}/\Gamma[(p/2)-1]\right] + \ldots \\ & + [e^{-y} y^{½}/\Gamma(3/2)] + 2\pi^{-(½)} \int_{y^{½}}^\infty e^{-u^2}\, du = \sum_{i=0}^{(p/2)-(½)} Q_i\,(y), \end{aligned} \tag{3.7}$$

where $Q_i\,(y)$ is defined as

$$Q_i\,(y) = \begin{cases} e^{-y} y^{i-(½)}/\Gamma[i+(½)] & \text{for } i \geq 1, \\ 2\pi^{-(½)} \int_{y^{½}}^\infty e^{-u^2}\, du & \text{for } i = 0. \end{cases} \tag{3.8}$$

Combining (3.3), (3.5) and (3.7), one obtains

$$G(y,\lambda,\nu) = \begin{cases} \sum_{j=0}^\infty P_j\,(\lambda/2) \sum_{i=0}^{(\nu/2)+j-1} P_i\,(y/2) & \text{if } \nu \text{ is even,} \\ \sum_{j=0}^\infty P_j\,(\lambda/2) \sum_{i=0}^{(\nu/2)+j-(½)} Q_i\,(y/2) & \text{if } \nu \text{ is odd.} \end{cases} \tag{3.9}$$

The function $F(y,\lambda,\nu)$ can be computed from (3.2) and (3.9), which, together with

$$\sum_{j=0}^{\infty} P_j\,(\lambda/2) \sum_{i=0}^{\infty} P_i\,(y/2) = 1,$$

yield:

$$F(y,\lambda,\nu) = \begin{cases} \sum_{j=0}^{\infty} P_j\,(\lambda/2) \sum_{i=(\nu/2)+j}^{\infty} P_i\,(y/2) & (\nu \text{ even}), \\ \sum_{j=0}^{\infty} P_j\,(\lambda/2) \sum_{i=(\nu/2)+j+(\frac{1}{2})}^{\infty} Q_i\,(y/2) & (\nu \text{ odd}). \end{cases} \tag{3.10}$$

Now, if we interchange the order of summation, the series takes the form of

$$F(y,\lambda,\nu) = \begin{cases} \sum_{i=0}^{\infty} P_{i+(\nu/2)}\,(y/2) \sum_{j=0}^{i} P_j\,(\lambda/2) & (\nu \text{ even}), \\ \sum_{i=0}^{\infty} Q_{i+(\nu/2)+(\frac{1}{2})}\,(y/2) \sum_{j=0}^{i} P_j\,(\lambda/2) & (\nu \text{ odd}). \end{cases} \tag{3.11}$$

We define two vectors $S\,(y/2)$ and $T\,(\lambda/2)$ as follows:

$$T_k\,(\lambda/2) = \sum_{i=0}^{k} P_i\,(\lambda/2)\,,\ k \geq 0. \tag{3.12}$$

$$\left.\begin{array}{l} S_{2k}\,(y/2) = P_k\,(y/2) \\ S_{2k-1}\,(y/2) = Q_k\,(y/2) \end{array}\right\} k \geq 1\,. \tag{3.13}$$

Then, (3.11) can be rewritten as:

$$F(y,\lambda,\nu) = \sum_{i=0}^{\infty} S_{2i+\nu}\,(y/2)\ T_i(\lambda/2)\,. \tag{3.14}$$

Since the vectors $S\,(y/2)$ and $T(\lambda/2)$ do not depend upon ν, once these vectors are calculated, the function $F\,(y,\lambda,\nu)$ can be tabulated for all ν by using (3.14). The computation is terminated when the value of F no longer changes in the eighth significant figure. Also, it should be noted that the value of $T(\lambda/2)$ remains constant for all y and ν, and $S\,(y/2)$ remains constant for all λ and ν. This fact greatly reduces the amount of computation involved.

SUMMARY OF TABLES

Table I – Power of the χ^2 Test.

For various levels of significance α it was first necessary to compute the corresponding abscissa $x(\nu,\alpha)$ from the central χ^2 distribution:

$$a = \int_{x(\nu,a)}^{\infty} \frac{e^{-y/2}\, y^{(\nu/2)-1}}{2^{\nu/2}\,\Gamma(\nu/2)}\, dy; \tag{4.1}$$

then from these values of $x\ (\nu, a)$ we calculated the power of the non-central χ^2 distribution:

$$1-\beta(\nu,\lambda,a) = \int_0^{x(\nu,a)} \frac{e^{-y/2}\, e^{-\lambda/2}}{2^{\nu/2}} \sum_{j=0}^{\infty} \frac{y^{(\nu/2)+j-1}\,\lambda^j}{\Gamma[(\nu/2)+j]\, 2^{2j}\, j!}\, dy \tag{4.2}$$

for all combinations of the values of the parameters as follows:

a = 0.1, 0.05, 0.025, 0.01, 0.005, 0.001

λ = 0 (0.1) 1.0 (0.2) 3.0 (0.5) 5.0 (1.0) 40.0 (2.0) 50.0 (5.0) 100.0

ν = 1 (1) 30 (2) 50 (5) 100

where the final resulting power has been rounded to four decimal places. It should be noted that $x(\nu, a)$ was calculated correct to six decimal places.

Table II – Non-centrality Parameter of Non-central χ^2

The same values of $x(\nu, a)$ as used in preparing Table I were used in calculating this table of the non-centrality parameter $\lambda\ (\nu,\ \beta,\ a)$ from the equation:

$$1-\beta = \int_0^{x(\nu,a)} \frac{e^{-y/2}\, e^{-\lambda(\nu,\beta,a)/2}}{2^{\nu/2}} \sum_{j=0}^{\infty} \frac{y^{(\nu/2)+j-1}\,\lambda^j\,(\nu,\beta,a)}{\Gamma[(\nu/2)+j]\, 2^{2j}\, j!}\, dy \tag{4.3}$$

for all combinations of the values of the parameters as follows:

a = 0.1, 0.05, 0.025, 0.01, 0.005, 0.001

$1-\beta$ = 0.1 (0.02) 0.7 (0.01) 0.99

ν = 1 (1) 30 (2) 50 (5) 100.

The computational procedure used for this table is an iteration process.

The final resulting values of the non-centrality parameter are correctly rounded to three decimal places. Since this table contains Fix's [1] table as a subset, all of the values in Fix's tables were checked with perfect agreement. It should be noted that the computational time for this table was about 60 hours on Univac I.

DISCUSSION ABOUT ERRORS

The error in the computations of Tables I and II can arise from two sources: due to (i) the truncation of the series given by (3.14) and (ii) rounding of the pivotal values. Let us consider the error due to the truncation of the series. From (3.14) $F\ (y,\lambda,\nu)$ is given by:

$$F(y,\lambda,\nu) = \sum_{i=0}^{\infty} S_{2i+\nu}(y/2)\, T_i(\lambda/2)$$

where

$$T_i(u) = \sum_{j=0}^{i} P_j(u),$$

$$S_{2k}(u) = P_k(u), \quad S_{2k-1}(u) = Q_k(u),$$

$$P_k(u) = e^{-u} u^k/k!, \qquad k = 0,1,\ldots$$

and

$$Q_k(u) = e^{-u} u^{k-(1/2)}/\Gamma[k+(1/2)], \qquad k \geq 1,$$

$$= 2\pi^{-(1/2)} \int_u^{\infty} e^{-v^2}\, dv, \qquad k = 0.$$

Let q be the number of terms of the series in (3.14) considered. Then the error due to truncation of the series in (3.14) is given by:

$$\sum_{i=q+1}^{\infty} S_{2i+\nu}(y/2)\, T_i(\lambda/2) \leq \sum_{i=q+1}^{\infty} S_{2i+\nu}(y/2), \qquad \text{since } T_i \leq 1.$$

Also

$$\sum_{i=q+1}^{\infty} S_{2i+\nu}(y/2) = \begin{cases} \displaystyle\sum_{i=q+1}^{\infty} e^{-y/2}(y/2)^{2i+\nu}/(2i+\nu)!, & \text{if } \nu \text{ is even,} \\ \displaystyle\sum_{i=q+1}^{\infty} e^{-y/2}(y/2)^{2i+\nu-(1/2)}/\Gamma[2i+\nu+(1/2)], & \text{if } \nu \text{ is odd.} \end{cases}$$

Since the upper tail of a Poisson distribution can be represented as the lower tail of a gamma distribution, when ν is even, we obtain

$$\sum_{i=q+1}^{\infty} S_{2i+\nu}(y/2) = [(2q+\nu+1)!] \int_0^{y/2} e^{-u} u^{2q+\nu+1}\, du. \tag{5.1}$$

When ν is odd,

$$\sum_{i=q+1}^{\infty} S_{2i+\nu}(y/2) \leq \sum_{i=q+1}^{\infty} e^{-y/2}(y/2)^{2i+\nu-(1/2)}/(2i+\nu-1)!$$

$$= (y/2)^{1/2} \sum_{i=q+1}^{\infty} e^{-y/2}(y/2)^{2i+\nu-1}/(2i+\nu-1)! \tag{5.2}$$

$$= [(2q+\nu)!]\, (y/2)^{1/2} \int_0^{y/2} e^{-u} u^{2q+\nu}\, du.$$

From (5.1) and (5.2) it is clear that the error due to truncation will be small, especially when y is small and q is sufficiently large. Next, consider the error due to rounding of the pivotal values in the series given by (3.14). Let ϵ and ϵ' denote the absolute values of the rounding errors in S_k and T_k respectively. Then the maximum error involved in computing $F(y,\lambda,\nu)$ by considering only the first q terms of the series is given by

$$\sum_{i=0}^{q} [S_{2i+\nu}(y/2)+\epsilon]\,[T_i(\lambda/2)+\epsilon'] - \sum_{i=0}^{q} S_{2i+\nu}(y/2)\,T_i(\lambda/2)$$

$$= \epsilon \sum_{i=0}^{q} T_i(\lambda/2) + \epsilon' \sum_{i=0}^{q} S_{2i+\nu}(y/2) + \epsilon\epsilon' q$$

$$\leq q\,\epsilon\,(1+\epsilon') + \epsilon' \quad \text{or} \quad q\epsilon\,(1+\epsilon') + (\tfrac{1}{2}) y \epsilon'$$

according as ν is even or odd, since

$$\sum_{i=0}^{q} S_{2i+\nu}(y/2)$$

is bounded by unity when ν is even and is bounded by $y/2$ when ν is odd. Thus, the error due to rounding of the pivotal values is not serious if both $q\epsilon$ and $y\epsilon'$ are small.

REMARK.

Obviously, the number of terms considered, namely q, depends on y, and in fact q increases with y. Consequently, q differs from calculation to calculation. Also, it is programmed in such a way that the computer neglects those terms, the magnitude of which is less than one unit in the eighth significant figure. We believe that this kind of a rule ensures the accuracy claimed in the tables presented.

INTERPOLATION METHODS IN TABLES 1 AND 2.

It is of interest to explore some interpolation methods that could be used in Tables 1 and 2. First let us define the linear, harmonic and Lagrange interpolation formulae: Let $y = f(x)$ where x denotes the independent variable. Then, for $x_1 < x < x_2$, letting $y_i = f(x_i)$, we have:

(i) Linear interpolation gives

$$y = \frac{x - x_1}{x_2 - x_1}\, y_2 + \frac{x_2 - x}{x_2 - x_1}\, y_1 .$$

(ii) Harmonic interpolation yields:

(a) When x is discrete

$$y = \frac{x_1 y_1 (x_2 - x) + x_2 y_2 (x - x_1)}{x(x_2 - x_1)} .$$

(b) When y is discrete

$$y = \frac{y_1 y_2 (x_2 - x_1)}{y_1 (x - x_1) + y_2 (x_2 - x)} .$$

(iii) The Lagrange interpolation (six-point) yields (after taking $x_1 < x_2 < x_3 < x < x_4 < x_5 < x_6$):

$$y = \sum_{i=1}^{6} \prod_{\substack{j=1 \\ j \neq i}}^{6} \frac{(x-x_j)}{(x_i-x_j)} y_i \; .$$

In the following we choose specific combinations of the degrees of freedom ν and the non-centrality parameter λ and employ the various interpolation methods and study their approximations in the light of the exact values obtained from Tables 1 and 2.

Interpolation For Power In Table 1

Given non-centrality parameter (λ) = 5, interpolate power at degrees of freedom (ν) = 9 for α = .01 and .05:

	α = .05	α = .01
Linear	.2822	.1142
Harmonic	.2805	.1131
Lagrange	.2810	.1134
True value	.2811	.1135

Given non-centrality parameter (λ) = 50, interpolate power at ν = 85 for α = .01 and .05:

	α = .01	α =.05
Linear	.8058	.9311
Harmonic	.8048	.9308
Lagrange	.8058	.9312
True value	.8058	.9312

Given degrees of freedom (ν) = 10, interpolate power at non-centrality parameter (λ) = 4 for α = .01 and .05:

	α = .01	α = .05
Linear	.0776	.2154
Harmonic	.0792	.2185
Lagrange	.0770	.2149
True value	.0770	.2149

Given degrees of freedom (ν) = 90, interpolate power at non-centrality parameter (λ) = 50 for α = .01 and .05:

	α = .01	α = .05
Linear	.7851	.9203
Harmonic	.7878	.9216
Lagrange	.7884	.9230
True value	.7885	.9230

Notice that the Lagrange interpolation is superior either to the linear or harmonic interpolation. Hence only Lagrange interpolation is studied in Table 2.

Interpolation For Non-Centrality Parameter In Table 2

Given power = .10, interpolate non-centrality parameter at $\nu = 9$ for $\alpha = .01$ and .05:

	α = .05	α = .01
Lagrange	1.404	4.583
True value	1.404	4.583

Given power = .90, interpolate non-centrality parameter at $\nu = 85$ for $\alpha = .01$ and .05:

	α = .05	α =.01
Lagrange	46.042	58.371
True value	46.042	58.371

Given ν(d.f.) = 10, interpolate non-centrality parameter (λ) at power = .16 for $\alpha = .01$ and .05:

	α = .05	α = .01
Lagrange	2.884	6.586
True value	2.884	6.586

Given ν(d.f.) = 90, interpolate non-centrality parameter (λ) at power = .90 for $\alpha = .01$ and .05:

	α = .05	α = .01
Lagrange	47.156	59.744
True value	47.156	59.745

Thus, the Lagrange interpolation with three points above and three points below would certainly suffice.

Next it is of interest to interpolate for the required degrees of freedom for specified α, λ and the power using harmonic interpolation (b) in Table 1. We obtain the following:

a	λ	Power	ν	True ν
.05	42	.93	56	56
.05	13	.30	62	62
.01	30	.48	68	68

The true values are obtained from the authors' table giving the exact d.f. ν for specified a, λ and power. However, since the agreement between the true and the interpolated values is perfect, it is felt that the table is somewhat redundant. Thus, it is not presented here.

EXAMPLES ILLUSTRATING THE USE OF TABLES 1 AND 2.

Example 1.

Suppose that we have a multinomial sample $(r_1, \ldots r_s)$ based on a total sample of size $n = r_1 + \ldots r_s$. Under the null hypothesis outcome i has probability p_i^o (p_i^o specified) and we know that the test statistic

$$T = n \sum_1^s (r_i/n - p_i^o)^2/p_i^o \tag{7.1}$$

is asymptotically chi-square with $(s-1)$ degrees of freedom. Suppose that under the alternative hypothesis, the outcome i has probability p_i^*. Then it is well known that the T given in (7.1) has asymptotically a non-central chi-square with $(s-1)$ degrees of freedom and non-centrality parameter

$$\lambda = n \sum_1^s (p_i^* - p_i^o)^2/p_i^o \, . \tag{7.2}$$

To be specific let $s = 3$ and $n = 20$ and $p_1^o = p_2^o = p_3^o = 1/3$. Then, we reject H_o when $T \geq 5.991$ (at 5 per cent level) and when $T \geq 9.210$ (at one per cent level). Now, let us calculate the power of T when $p_1^* = .2$, $p_2^* = .3$ and $p_3^* = .5$ so that

$$\lambda = 20 \times 3 \; [(\frac{1}{5} - \frac{1}{3})^2 + (\frac{3}{10} - \frac{1}{3})^2 + (\frac{1}{2} - \frac{1}{3})^2]$$

$$= 60 \; [4/225 + 1/900 + 1/36] = 2.8.$$

Thus, from Table 1 we find that the asymptotic power is .3024 (at 5 per cent level) and .1281 (at one per cent level).

In order to illustrate the use of Table 2, let us consider the following example.

Example 2.

A pharmaceutical company is trying to evaluate the relative effectiveness of three drugs for lowering the cholesterol level in human blood. Each drug is administered to four men having the same cholesterol levels. At the conclusion of the trial, a blood sample is taken from each man. Each of the four technicians analyzes three blood samples, one from each drug group. Assume the following model:

$$y_{ij} = \theta + \delta_i + \nu_j + \epsilon_{ij}, \; i = 1, \ldots, 3, \; j = 1, \ldots, 4,$$

where y_{ij} = the level of cholesterol determined by the jth technician on a subject receiving the ith drug,

θ = mean level common to all determinations
δ_i = the effect of the ith drug
ν_j = the effect peculiar to j^{th} technician
ϵ_{ij} = uncorrected random errors which are distributed normally with zero mean and known variance σ^2.

Suppose we wish to test the hypothesis: H_0: $\delta_1 = \delta_2 = \delta_3$ against H_a: $\delta_i \neq \delta_{i'}$ for at least one pair (i, i'), $i = 1, \ldots, 3$ and $i' = 1, \ldots, 3$. The appropriate test statistic to be used would be

$$S = 4\sigma^2 \sum_{i=1}^{3} (y_{i.} - \bar{y}_{..})^2$$

where

$$y_{i.} = \sum_{j=1}^{4} y_{ij} /4 \text{ and } \bar{y}_{..} = \sum_i \sum_j y_{ij} /12.$$

It is well known that S is distributed as non-central chi-square with 2 degrees of freedom and non-centrality parameter

$$\lambda = 4\sigma^{-2} \sum_{i=1}^{3} (\delta_i - \bar{\delta})^2 ,$$

$$\bar{\delta} = \sum_{1}^{3} \delta_i /3.$$

Using this test statistic we would like to have power = 0.95 with $\alpha = .05$. Can one determine the configuration $(\delta_1, \delta_2, \delta_3)$ subject to the constraints: $\Sigma\delta_i = 0$ and $\delta_1 = 0$? Entering Table 2 with $\alpha = .05$, power = .95 and degrees of freedom = 2, we obtain $\lambda = 15.443$. Hence

$$\delta_2^2 + \delta_3^2 = 2\delta_2^2 = 15.443\, \sigma^2/4.$$

That is,

$$\delta_2^2 = 1.93\sigma^2, \quad \delta_2 \doteq 1.39\sigma$$

and

$$\delta_3 = -\, 1.39\sigma.$$

Acknowledgement.

We wish to thank Hardeo Sahai of the University of Kentucky for his assistance in numerically studying the various interpolation formulae appropriate for Tables 1 and 2.

REFERENCES

1. Fix, Evelyn (1949). "Tables of Non-Central χ^2," *University of California Publications in Statistics*, 1: No. 2, 15–19.

2. Johnson, N. L. (1959). "On an extension of the connection between Poisson and χ^2 distributions," *Biometrika*, 46: 352–363.

3. Patnaik, P. B. (1949). "The noncentral χ^2 and F distributions and their applications," *Biometrika*, 36: 202–232.

4. Ruben, H. (1962). "Probability content of regions under spherical normal distributions, IV: The distribution of homogeneous and non-homogeneous quadratic functions of normal variables," *Annals of Mathematical Statistics*, 33:542–570.

TABLE I. POWER OF THE CHI-SQUARE TEST ALPHA = .100

N	LAMBDA= .00	.10	.20	.30	.40	.50	.60	.70	.80	.90	1.00	1.20	1.40	1.60	1.80
1	0.1000	0.1169	0.1337	0.1505	0.1671	0.1835	0.1999	0.2160	0.2321	0.2479	0.2636	0.2944	0.3245	0.3538	0.3823
2	0.1000	0.1116	0.1232	0.1349	0.1466	0.1584	0.1702	0.1821	0.1939	0.2058	0.2177	0.2413	0.2649	0.2883	0.3115
3	0.1000	0.1092	0.1185	0.1279	0.1373	0.1469	0.1565	0.1662	0.1759	0.1858	0.1956	0.2154	0.2353	0.2552	0.2752
4	0.1000	0.1078	0.1157	0.1237	0.1318	0.1400	0.1483	0.1566	0.1650	0.1735	0.1821	0.1993	0.2168	0.2343	0.2520
5	0.1000	0.1069	0.1138	0.1209	0.1280	0.1353	0.1426	0.1500	0.1575	0.1651	0.1727	0.1881	0.2038	0.2196	0.2356
6	0.1000	0.1062	0.1125	0.1188	0.1253	0.1319	0.1385	0.1452	0.1520	0.1588	0.1658	0.1798	0.1941	0.2086	0.2232
7	0.1000	0.1057	0.1114	0.1173	0.1232	0.1292	0.1353	0.1415	0.1477	0.1540	0.1604	0.1733	0.1865	0.1999	0.2135
8	0.1000	0.1053	0.1106	0.1160	0.1215	0.1271	0.1327	0.1385	0.1443	0.1501	0.1560	0.1681	0.1804	0.1928	0.2055
9	0.1000	0.1049	0.1099	0.1150	0.1201	0.1253	0.1306	0.1360	0.1414	0.1469	0.1525	0.1638	0.1753	0.1870	0.1989
10	0.1000	0.1046	0.1093	0.1141	0.1190	0.1239	0.1289	0.1339	0.1390	0.1442	0.1494	0.1601	0.1710	0.1820	0.1933
11	0.1000	0.1044	0.1088	0.1134	0.1180	0.1226	0.1273	0.1321	0.1370	0.1419	0.1469	0.1570	0.1673	0.1778	0.1885
12	0.1000	0.1042	0.1084	0.1127	0.1171	0.1215	0.1260	0.1306	0.1352	0.1399	0.1446	0.1542	0.1640	0.1741	0.1843
13	0.1000	0.1040	0.1081	0.1122	0.1164	0.1206	0.1249	0.1292	0.1337	0.1381	0.1426	0.1518	0.1612	0.1708	0.1806
14	0.1000	0.1038	0.1077	0.1117	0.1157	0.1197	0.1239	0.1280	0.1323	0.1366	0.1409	0.1497	0.1587	0.1679	0.1773
15	0.1000	0.1037	0.1074	0.1112	0.1151	0.1190	0.1230	0.1270	0.1310	0.1352	0.1393	0.1478	0.1565	0.1653	0.1743
16	0.1000	0.1036	0.1072	0.1108	0.1146	0.1183	0.1221	0.1260	0.1299	0.1339	0.1379	0.1461	0.1544	0.1629	0.1716
17	0.1000	0.1034	0.1069	0.1105	0.1141	0.1177	0.1214	0.1251	0.1289	0.1328	0.1366	0.1445	0.1526	0.1608	0.1692
18	0.1000	0.1033	0.1067	0.1101	0.1136	0.1171	0.1207	0.1243	0.1280	0.1317	0.1355	0.1431	0.1509	0.1589	0.1670
19	0.1000	0.1032	0.1065	0.1098	0.1132	0.1166	0.1201	0.1236	0.1272	0.1308	0.1344	0.1418	0.1494	0.1571	0.1650
20	0.1000	0.1031	0.1063	0.1096	0.1128	0.1162	0.1195	0.1229	0.1264	0.1299	0.1334	0.1406	0.1479	0.1554	0.1631
21	0.1000	0.1031	0.1062	0.1093	0.1125	0.1157	0.1190	0.1223	0.1257	0.1291	0.1325	0.1395	0.1466	0.1539	0.1613
22	0.1000	0.1030	0.1060	0.1091	0.1122	0.1153	0.1185	0.1217	0.1250	0.1283	0.1317	0.1385	0.1454	0.1525	0.1597
23	0.1000	0.1029	0.1059	0.1089	0.1119	0.1150	0.1181	0.1212	0.1244	0.1276	0.1309	0.1375	0.1443	0.1512	0.1582
24	0.1000	0.1028	0.1057	0.1086	0.1116	0.1146	0.1176	0.1207	0.1238	0.1270	0.1301	0.1366	0.1432	0.1500	0.1568
25	0.1000	0.1028	0.1056	0.1085	0.1113	0.1143	0.1172	0.1202	0.1233	0.1263	0.1295	0.1358	0.1422	0.1488	0.1555
26	0.1000	0.1027	0.1055	0.1083	0.1111	0.1140	0.1169	0.1198	0.1228	0.1258	0.1288	0.1350	0.1413	0.1477	0.1543
27	0.1000	0.1027	0.1054	0.1081	0.1109	0.1137	0.1165	0.1194	0.1223	0.1252	0.1282	0.1343	0.1404	0.1467	0.1531
28	0.1000	0.1026	0.1053	0.1079	0.1107	0.1134	0.1162	0.1190	0.1218	0.1247	0.1276	0.1336	0.1396	0.1458	0.1520
29	0.1000	0.1026	0.1052	0.1078	0.1105	0.1131	0.1159	0.1186	0.1214	0.1242	0.1271	0.1329	0.1388	0.1449	0.1510
30	0.1000	0.1025	0.1051	0.1076	0.1103	0.1129	0.1156	0.1183	0.1210	0.1238	0.1266	0.1323	0.1381	0.1440	0.1500
32	0.1000	0.1024	0.1049	0.1074	0.1099	0.1124	0.1150	0.1176	0.1203	0.1229	0.1256	0.1311	0.1367	0.1424	0.1482
34	0.1000	0.1024	0.1047	0.1071	0.1096	0.1120	0.1145	0.1171	0.1196	0.1222	0.1248	0.1301	0.1355	0.1410	0.1466
36	0.1000	0.1023	0.1046	0.1069	0.1093	0.1117	0.1141	0.1165	0.1190	0.1215	0.1240	0.1291	0.1343	0.1397	0.1451
38	0.1000	0.1022	0.1045	0.1067	0.1090	0.1113	0.1137	0.1160	0.1184	0.1208	0.1233	0.1282	0.1333	0.1385	0.1437
40	0.1000	0.1022	0.1043	0.1065	0.1088	0.1110	0.1133	0.1156	0.1179	0.1203	0.1226	0.1274	0.1324	0.1374	0.1425
42	0.1000	0.1021	0.1042	0.1064	0.1085	0.1107	0.1129	0.1152	0.1174	0.1197	0.1220	0.1267	0.1315	0.1363	0.1413
44	0.1000	0.1020	0.1041	0.1062	0.1083	0.1104	0.1126	0.1148	0.1170	0.1192	0.1215	0.1260	0.1307	0.1354	0.1402
46	0.1000	0.1020	0.1040	0.1061	0.1081	0.1102	0.1123	0.1144	0.1166	0.1187	0.1209	0.1254	0.1299	0.1345	0.1392
48	0.1000	0.1020	0.1039	0.1059	0.1079	0.1100	0.1120	0.1141	0.1162	0.1183	0.1204	0.1248	0.1292	0.1337	0.1383
50	0.1000	0.1019	0.1038	0.1058	0.1078	0.1097	0.1118	0.1138	0.1158	0.1179	0.1200	0.1242	0.1285	0.1329	0.1374
55	0.1000	0.1018	0.1036	0.1055	0.1074	0.1093	0.1112	0.1131	0.1150	0.1170	0.1190	0.1230	0.1271	0.1312	0.1354
60	0.1000	0.1017	0.1035	0.1052	0.1070	0.1088	0.1106	0.1125	0.1143	0.1162	0.1181	0.1219	0.1258	0.1297	0.1337
65	0.1000	0.1017	0.1033	0.1050	0.1067	0.1084	0.1102	0.1119	0.1137	0.1155	0.1173	0.1209	0.1247	0.1284	0.1323
70	0.1000	0.1016	0.1032	0.1048	0.1065	0.1081	0.1098	0.1115	0.1132	0.1149	0.1166	0.1201	0.1237	0.1273	0.1309
75	0.1000	0.1015	0.1031	0.1047	0.1062	0.1078	0.1094	0.1110	0.1127	0.1143	0.1160	0.1194	0.1228	0.1262	0.1298
80	0.1000	0.1015	0.1030	0.1045	0.1060	0.1076	0.1091	0.1107	0.1122	0.1138	0.1154	0.1187	0.1220	0.1253	0.1287
85	0.1000	0.1014	0.1029	0.1043	0.1058	0.1073	0.1088	0.1103	0.1118	0.1134	0.1149	0.1181	0.1212	0.1245	0.1278
90	0.1000	0.1014	0.1028	0.1042	0.1056	0.1071	0.1085	0.1100	0.1115	0.1130	0.1145	0.1175	0.1206	0.1237	0.1269
95	0.1000	0.1014	0.1027	0.1041	0.1055	0.1069	0.1083	0.1097	0.1112	0.1126	0.1141	0.1170	0.1200	0.1230	0.1261
100	0.1000	0.1013	0.1026	0.1040	0.1053	0.1067	0.1081	0.1095	0.1108	0.1123	0.1137	0.1165	0.1194	0.1224	0.1254

POWER OF THE CHI-SQUARE TEST ALPHA = .100

N	LAMBDA= 2.00	2.20	2.40	2.60	2.80	3.00	3.50	4.00	4.50	5.00	6.00	7.00	8.00	9.00	10.00
1	0.4099	0.4367	0.4626	0.4876	0.5118	0.5351	0.5896	0.6389	0.6832	0.7229	0.7895	0.8416	0.8817	0.9123	0.9354
2	0.3344	0.3570	0.3793	0.4012	0.4227	0.4438	0.4945	0.5423	0.5869	0.6283	0.7017	0.7631	0.8136	0.8545	0.8873
3	0.2950	0.3149	0.3346	0.3541	0.3735	0.3926	0.4394	0.4844	0.5274	0.5681	0.6424	0.7070	0.7623	0.8087	0.8473
4	0.2697	0.2875	0.3053	0.3230	0.3407	0.3582	0.4016	0.4439	0.4848	0.5242	0.5975	0.6631	0.7206	0.7703	0.8127
5	0.2517	0.2679	0.2841	0.3004	0.3167	0.3330	0.3734	0.4132	0.4522	0.4900	0.5616	0.6269	0.6855	0.7371	0.7820
6	0.2380	0.2529	0.2680	0.2830	0.2982	0.3134	0.3513	0.3889	0.4260	0.4623	0.5319	0.5965	0.6552	0.7079	0.7545
7	0.2272	0.2411	0.2551	0.2692	0.2833	0.2976	0.3333	0.3690	0.4044	0.4393	0.5067	0.5702	0.6287	0.6820	0.7296
8	0.2184	0.2314	0.2445	0.2578	0.2711	0.2845	0.3184	0.3523	0.3861	0.4197	0.4851	0.5472	0.6053	0.6586	0.7070
9	0.2110	0.2233	0.2357	0.2482	0.2608	0.2735	0.3057	0.3381	0.3705	0.4028	0.4661	0.5269	0.5843	0.6375	0.6862
10	0.2048	0.2164	0.2281	0.2400	0.2520	0.2641	0.2947	0.3257	0.3569	0.3880	0.4494	0.5088	0.5653	0.6182	0.6671
11	0.1994	0.2104	0.2216	0.2329	0.2444	0.2559	0.2852	0.3149	0.3449	0.3749	0.4345	0.4925	0.5481	0.6006	0.6494
12	0.1947	0.2052	0.2159	0.2267	0.2377	0.2487	0.2768	0.3054	0.3342	0.3632	0.4211	0.4777	0.5324	0.5844	0.6330
13	0.1905	0.2006	0.2108	0.2212	0.2317	0.2423	0.2693	0.2968	0.3247	0.3528	0.4089	0.4643	0.5180	0.5694	0.6178
14	0.1868	0.1965	0.2063	0.2163	0.2264	0.2366	0.2626	0.2892	0.3161	0.3433	0.3979	0.4520	0.5047	0.5554	0.6035
15	0.1835	0.1923	0.2023	0.2119	0.2216	0.2315	0.2566	0.2822	0.3083	0.3347	0.3878	0.4407	0.4925	0.5425	0.5902
16	0.1805	0.1895	0.1986	0.2079	0.2173	0.2268	0.2511	0.2759	0.3012	0.3268	0.3785	0.4302	0.4811	0.5304	0.5776
17	0.1777	0.1864	0.1953	0.2042	0.2133	0.2225	0.2460	0.2701	0.2947	0.3196	0.3700	0.4205	0.4704	0.5191	0.5659
18	0.1753	0.1837	0.1922	0.2009	0.2097	0.2186	0.2414	0.2648	0.2887	0.3129	0.3621	0.4115	0.4605	0.5084	0.5547
19	0.1730	0.1811	0.1894	0.1978	0.2064	0.2151	0.2372	0.2599	0.2831	0.3067	0.3547	0.4031	0.4512	0.4985	0.5443
20	0.1709	0.1788	0.1868	0.1950	0.2033	0.2117	0.2333	0.2554	0.2780	0.3010	0.3478	0.3952	0.4425	0.4890	0.5343
21	0.1689	0.1766	0.1845	0.1924	0.2005	0.2087	0.2296	0.2511	0.2732	0.2956	0.3414	0.3879	0.4343	0.4802	0.5249
22	0.1671	0.1746	0.1822	0.1900	0.1978	0.2058	0.2262	0.2472	0.2687	0.2906	0.3354	0.3809	0.4266	0.4718	0.5160
23	0.1654	0.1727	0.1802	0.1877	0.1954	0.2032	0.2231	0.2435	0.2645	0.2859	0.3298	0.3744	0.4193	0.4638	0.5075
24	0.1638	0.1710	0.1782	0.1856	0.1931	0.2007	0.2201	0.2401	0.2606	0.2816	0.3245	0.3683	0.4124	0.4562	0.4994
25	0.1624	0.1693	0.1764	0.1836	0.1909	0.1983	0.2173	0.2369	0.2569	0.2774	0.3195	0.3624	0.4058	0.4490	0.4917
26	0.1610	0.1678	0.1747	0.1817	0.1889	0.1961	0.2147	0.2338	0.2534	0.2735	0.3147	0.3569	0.3996	0.4422	0.4843
27	0.1597	0.1663	0.1731	0.1800	0.1870	0.1941	0.2122	0.2309	0.2502	0.2698	0.3102	0.3517	0.3936	0.4357	0.4773
28	0.1584	0.1650	0.1716	0.1783	0.1852	0.1921	0.2099	0.2282	0.2471	0.2663	0.3060	0.3467	0.3880	0.4294	0.4705
29	0.1573	0.1637	0.1702	0.1768	0.1835	0.1903	0.2077	0.2256	0.2441	0.2630	0.3020	0.3420	0.3826	0.4235	0.4641
30	0.1562	0.1624	0.1688	0.1753	0.1818	0.1885	0.2056	0.2232	0.2413	0.2599	0.2981	0.3375	0.3775	0.4178	0.4579
32	0.1541	0.1602	0.1663	0.1725	0.1788	0.1853	0.2017	0.2187	0.2361	0.2540	0.2910	0.3290	0.3679	0.4071	0.4463
34	0.1523	0.1581	0.1640	0.1700	0.1761	0.1823	0.1982	0.2146	0.2314	0.2487	0.2844	0.3213	0.3591	0.3973	0.4355
36	0.1506	0.1562	0.1619	0.1677	0.1736	0.1796	0.1950	0.2108	0.2271	0.2438	0.2784	0.3143	0.3510	0.3882	0.4255
38	0.1491	0.1545	0.1600	0.1657	0.1714	0.1772	0.1920	0.2074	0.2232	0.2394	0.2729	0.3077	0.3434	0.3797	0.4163
40	0.1476	0.1529	0.1583	0.1637	0.1693	0.1749	0.1893	0.2042	0.2195	0.2353	0.2679	0.3017	0.3365	0.3719	0.4076
42	0.1463	0.1515	0.1567	0.1620	0.1674	0.1728	0.1868	0.2012	0.2161	0.2314	0.2631	0.2961	0.3300	0.3645	0.3995
44	0.1451	0.1501	0.1552	0.1603	0.1656	0.1709	0.1845	0.1985	0.2130	0.2279	0.2587	0.2908	0.3239	0.3577	0.3919
46	0.1440	0.1489	0.1538	0.1588	0.1639	0.1691	0.1823	0.1960	0.2101	0.2246	0.2547	0.2860	0.3182	0.3513	0.3848
48	0.1429	0.1477	0.1525	0.1574	0.1623	0.1674	0.1803	0.1936	0.2074	0.2215	0.2508	0.2814	0.3129	0.3452	0.3781
50	0.1420	0.1466	0.1513	0.1560	0.1609	0.1658	0.1784	0.1914	0.2048	0.2186	0.2472	0.2771	0.3079	0.3395	0.3717
55	0.1397	0.1441	0.1486	0.1531	0.1576	0.1623	0.1742	0.1864	0.1991	0.2121	0.2391	0.2674	0.2966	0.3267	0.3573
60	0.1378	0.1420	0.1462	0.1505	0.1548	0.1592	0.1705	0.1821	0.1941	0.2065	0.2321	0.2589	0.2868	0.3154	0.3447
65	0.1362	0.1401	0.1441	0.1482	0.1523	0.1565	0.1673	0.1784	0.1898	0.2016	0.2260	0.2515	0.2781	0.3055	0.3335
70	0.1347	0.1385	0.1423	0.1462	0.1502	0.1542	0.1644	0.1750	0.1860	0.1972	0.2205	0.2450	0.2704	0.2966	0.3235
75	0.1334	0.1370	0.1407	0.1444	0.1482	0.1521	0.1619	0.1721	0.1825	0.1933	0.2157	0.2391	0.2635	0.2887	0.3146
80	0.1322	0.1357	0.1392	0.1428	0.1465	0.1502	0.1596	0.1694	0.1795	0.1898	0.2113	0.2338	0.2572	0.2815	0.3065
85	0.1311	0.1345	0.1379	0.1414	0.1449	0.1485	0.1576	0.1670	0.1767	0.1866	0.2073	0.2290	0.2516	0.2750	0.2991
90	0.1301	0.1334	0.1367	0.1400	0.1434	0.1469	0.1557	0.1648	0.1741	0.1838	0.2037	0.2247	0.2465	0.2691	0.2924
95	0.1292	0.1324	0.1356	0.1388	0.1421	0.1455	0.1540	0.1628	0.1718	0.1811	0.2004	0.2207	0.2418	0.2636	0.2862
100	0.1284	0.1315	0.1346	0.1377	0.1409	0.1442	0.1524	0.1609	0.1697	0.1787	0.1974	0.2170	0.2374	0.2586	0.2805

POWER OF THE CHI-SQUARE TEST ALPHA = .100

N	LAMBDA= 11.00	12.00	13.00	14.00	15.00	16.00	17.00	18.00	19.00	20.00	21.00	22.00	23.00	24.00	25.00
1	0.9527	0.9656	0.9750	0.9820	0.9871	0.9907	0.9934	0.9953	0.9967	0.9977	0.9983	0.9988	0.9992	0.9994	0.9996
2	0.9133	0.9333	0.9497	0.9620	0.9714	0.9786	0.9841	0.9882	0.9913	0.9936	0.9953	0.9965	0.9975	0.9982	0.9987
3	0.8790	0.9048	0.9256	0.9421	0.9552	0.9655	0.9736	0.9799	0.9847	0.9885	0.9913	0.9935	0.9951	0.9964	0.9973
4	0.8484	0.8781	0.9026	0.9227	0.9390	0.9521	0.9625	0.9708	0.9774	0.9826	0.9866	0.9898	0.9922	0.9941	0.9955
5	0.8206	0.8533	0.8809	0.9038	0.9228	0.9384	0.9511	0.9613	0.9695	0.9761	0.9814	0.9855	0.9888	0.9913	0.9933
6	0.7952	0.8303	0.8603	0.8856	0.9070	0.9247	0.9394	0.9514	0.9613	0.9692	0.9757	0.9808	0.9850	0.9882	0.9908
7	0.7718	0.8087	0.8407	0.8681	0.8915	0.9111	0.9276	0.9413	0.9527	0.9620	0.9696	0.9758	0.9808	0.9848	0.9880
8	0.7503	0.7886	0.8222	0.8513	0.8764	0.8978	0.9159	0.9311	0.9439	0.9545	0.9632	0.9704	0.9763	0.9811	0.9849
9	0.7303	0.7697	0.8046	0.8352	0.8617	0.8846	0.9042	0.9209	0.9350	0.9468	0.9566	0.9648	0.9715	0.9771	0.9816
10	0.7117	0.7519	0.7879	0.8197	0.8475	0.8718	0.8927	0.9107	0.9260	0.9389	0.9498	0.9589	0.9665	0.9728	0.9780
11	0.6944	0.7352	0.7720	0.8048	0.8338	0.8592	0.8814	0.9005	0.9170	0.9310	0.9429	0.9529	0.9614	0.9684	0.9742
12	0.6781	0.7195	0.7569	0.7906	0.8205	0.8470	0.8702	0.8905	0.9080	0.9230	0.9359	0.9468	0.9560	0.9638	0.9703
13	0.6629	0.7045	0.7425	0.7769	0.8077	0.8351	0.8593	0.8805	0.8990	0.9150	0.9287	0.9405	0.9505	0.9590	0.9662
14	0.6486	0.6904	0.7288	0.7638	0.7953	0.8235	0.8486	0.8707	0.8901	0.9070	0.9216	0.9342	0.9450	0.9541	0.9619
15	0.6351	0.6771	0.7158	0.7512	0.7833	0.8123	0.8381	0.8611	0.8813	0.8990	0.9144	0.9278	0.9393	0.9491	0.9576
16	0.6224	0.6644	0.7033	0.7391	0.7718	0.8013	0.8279	0.8516	0.8726	0.8910	0.9072	0.9213	0.9335	0.9441	0.9531
17	0.6104	0.6523	0.6914	0.7275	0.7606	0.7907	0.8179	0.8422	0.8639	0.8832	0.9001	0.9149	0.9277	0.9389	0.9485
18	0.5990	0.6408	0.6800	0.7163	0.7498	0.7804	0.8081	0.8331	0.8555	0.8753	0.8929	0.9084	0.9219	0.9337	0.9439
19	0.5882	0.6299	0.6691	0.7056	0.7394	0.7704	0.7986	0.8241	0.8471	0.8676	0.8858	0.9019	0.9161	0.9284	0.9392
20	0.5779	0.6194	0.6586	0.6953	0.7293	0.7606	0.7893	0.8154	0.8389	0.8600	0.8788	0.8955	0.9102	0.9231	0.9344
21	0.5681	0.6094	0.6485	0.6853	0.7195	0.7512	0.7803	0.8068	0.8308	0.8524	0.8718	0.8890	0.9043	0.9178	0.9296
22	0.5588	0.5999	0.6389	0.6757	0.7101	0.7420	0.7714	0.7983	0.8228	0.8449	0.8648	0.8826	0.8984	0.9124	0.9248
23	0.5499	0.5907	0.6297	0.6664	0.7010	0.7331	0.7628	0.7901	0.8150	0.8376	0.8580	0.8763	0.8926	0.9071	0.9199
24	0.5414	0.5820	0.6208	0.6575	0.6921	0.7244	0.7544	0.7820	0.8073	0.8303	0.8512	0.8699	0.8868	0.9017	0.9150
25	0.5333	0.5736	0.6122	0.6489	0.6836	0.7160	0.7462	0.7741	0.7998	0.8232	0.8445	0.8637	0.8809	0.8964	0.9101
26	0.5255	0.5655	0.6039	0.6406	0.6753	0.7078	0.7382	0.7664	0.7924	0.8162	0.8378	0.8575	0.8752	0.8910	0.9052
27	0.5181	0.5578	0.5960	0.6325	0.6672	0.6999	0.7305	0.7589	0.7851	0.8093	0.8313	0.8513	0.8694	0.8857	0.9003
28	0.5109	0.5503	0.5883	0.6248	0.6594	0.6922	0.7229	0.7515	0.7780	0.8024	0.8248	0.8452	0.8637	0.8804	0.8954
29	0.5041	0.5431	0.5809	0.6172	0.6519	0.6846	0.7155	0.7443	0.7710	0.7958	0.8185	0.8392	0.8581	0.8751	0.8905
30	0.4975	0.5362	0.5738	0.6100	0.6445	0.6773	0.7082	0.7372	0.7642	0.7892	0.8122	0.8332	0.8525	0.8699	0.8856
32	0.4851	0.5231	0.5602	0.5961	0.6305	0.6633	0.6943	0.7235	0.7509	0.7763	0.7999	0.8215	0.8414	0.8595	0.8759
34	0.4735	0.5110	0.5475	0.5830	0.6172	0.6500	0.6811	0.7105	0.7381	0.7639	0.7879	0.8101	0.8305	0.8493	0.8663
36	0.4628	0.4996	0.5356	0.5708	0.6047	0.6373	0.6684	0.6979	0.7258	0.7519	0.7763	0.7990	0.8199	0.8392	0.8568
38	0.4527	0.4889	0.5245	0.5592	0.5928	0.6253	0.6564	0.6860	0.7140	0.7404	0.7651	0.7881	0.8095	0.8293	0.8474
40	0.4434	0.4789	0.5139	0.5482	0.5816	0.6139	0.6449	0.6745	0.7026	0.7292	0.7542	0.7776	0.7994	0.8196	0.8382
42	0.4346	0.4695	0.5040	0.5379	0.5709	0.6030	0.6339	0.6634	0.6916	0.7184	0.7436	0.7673	0.7894	0.8100	0.8291
44	0.4263	0.4606	0.4946	0.5281	0.5608	0.5926	0.6233	0.6529	0.6811	0.7080	0.7334	0.7573	0.7797	0.8007	0.8202
46	0.4185	0.4523	0.4857	0.5188	0.5512	0.5827	0.6133	0.6427	0.6710	0.6979	0.7235	0.7476	0.7703	0.7916	0.8114
48	0.4112	0.4443	0.4773	0.5099	0.5420	0.5732	0.6036	0.6330	0.6612	0.6882	0.7138	0.7382	0.7611	0.7826	0.8028
50	0.4042	0.4368	0.4693	0.5015	0.5332	0.5642	0.5944	0.6236	0.6518	0.6788	0.7045	0.7290	0.7521	0.7739	0.7943
55	0.3884	0.4197	0.4510	0.4822	0.5129	0.5432	0.5729	0.6017	0.6297	0.6566	0.6824	0.7071	0.7306	0.7529	0.7739
60	0.3745	0.4045	0.4347	0.4649	0.4948	0.5243	0.5534	0.5818	0.6094	0.6362	0.6620	0.6868	0.7105	0.7331	0.7546
65	0.3621	0.3910	0.4202	0.4494	0.4784	0.5072	0.5356	0.5635	0.5908	0.6173	0.6430	0.6678	0.6916	0.7144	0.7362
70	0.3510	0.3789	0.4071	0.4353	0.4636	0.4917	0.5194	0.5468	0.5736	0.5999	0.6254	0.6501	0.6739	0.6968	0.7188
75	0.3410	0.3680	0.3952	0.4226	0.4501	0.4774	0.5046	0.5314	0.5578	0.5837	0.6089	0.6335	0.6573	0.6802	0.7023
80	0.3320	0.3580	0.3844	0.4110	0.4377	0.4644	0.4909	0.5172	0.5431	0.5686	0.5936	0.6180	0.6416	0.6646	0.6867
85	0.3238	0.3490	0.3745	0.4003	0.4263	0.4523	0.4783	0.5040	0.5295	0.5546	0.5793	0.6034	0.6269	0.6498	0.6719
90	0.3162	0.3406	0.3654	0.3905	0.4158	0.4412	0.4665	0.4918	0.5168	0.5415	0.5658	0.5897	0.6130	0.6358	0.6578
95	0.3093	0.3330	0.3571	0.3815	0.4061	0.4309	0.4557	0.4804	0.5049	0.5292	0.5532	0.5768	0.5999	0.6225	0.6445
100	0.3029	0.3259	0.3494	0.3731	0.3971	0.4213	0.4455	0.4697	0.4938	0.5177	0.5414	0.5647	0.5876	0.6100	0.6318

POWER OF THE CHI-SQUARE TEST ALPHA = .100

N	LAMBDA= 26.00	27.00	28.00	29.00	30.00	31.00	32.00	33.00	34.00	35.00	36.00	37.00	38.00	39.00	40.00
1	0.9997	0.9998	0.9999	0.9999	0.9999	1.0000	1.0000	1.0000	1.0000	1.0000	1.0000	1.0000	1.0000	1.0000	1.0000
2	0.9990	0.9993	0.9995	0.9996	0.9997	0.9998	0.9999	0.9999	0.9999	1.0000	1.0000	1.0000	1.0000	1.0000	1.0000
3	0.9980	0.9985	0.9989	0.9992	0.9994	0.9996	0.9997	0.9998	0.9998	0.9999	0.9999	0.9999	1.0000	1.0000	1.0000
4	0.9966	0.9974	0.9981	0.9986	0.9989	0.9992	0.9994	0.9996	0.9997	0.9998	0.9998	0.9999	0.9999	0.9999	0.9999
5	0.9949	0.9961	0.9970	0.9977	0.9983	0.9987	0.9990	0.9993	0.9995	0.9996	0.9997	0.9998	0.9998	0.9999	0.9999
6	0.9929	0.9945	0.9958	0.9967	0.9975	0.9981	0.9985	0.9989	0.9992	0.9994	0.9995	0.9996	0.9997	0.9998	0.9998
7	0.9906	0.9926	0.9943	0.9955	0.9965	0.9973	0.9979	0.9984	0.9988	0.9991	0.9993	0.9995	0.9996	0.9997	0.9998
8	0.9881	0.9906	0.9926	0.9942	0.9954	0.9964	0.9972	0.9978	0.9983	0.9987	0.9990	0.9992	0.9994	0.9996	0.9997
9	0.9853	0.9883	0.9907	0.9926	0.9942	0.9954	0.9964	0.9972	0.9978	0.9983	0.9987	0.9990	0.9992	0.9994	0.9995
10	0.9823	0.9858	0.9886	0.9909	0.9928	0.9942	0.9954	0.9964	0.9972	0.9978	0.9983	0.9986	0.9989	0.9992	0.9994
11	0.9791	0.9831	0.9863	0.9890	0.9912	0.9930	0.9944	0.9955	0.9965	0.9972	0.9978	0.9983	0.9986	0.9989	0.9992
12	0.9757	0.9802	0.9839	0.9870	0.9895	0.9916	0.9932	0.9946	0.9957	0.9966	0.9973	0.9978	0.9983	0.9987	0.9989
13	0.9722	0.9772	0.9814	0.9848	0.9877	0.9900	0.9919	0.9935	0.9948	0.9958	0.9967	0.9974	0.9979	0.9983	0.9987
14	0.9685	0.9740	0.9787	0.9825	0.9857	0.9884	0.9906	0.9924	0.9938	0.9950	0.9960	0.9968	0.9974	0.9980	0.9984
15	0.9647	0.9708	0.9758	0.9801	0.9837	0.9866	0.9891	0.9911	0.9928	0.9942	0.9953	0.9962	0.9970	0.9976	0.9980
16	0.9608	0.9673	0.9729	0.9776	0.9815	0.9848	0.9875	0.9898	0.9917	0.9932	0.9945	0.9955	0.9964	0.9971	0.9977
17	0.9568	0.9633	0.9698	0.9749	0.9792	0.9828	0.9858	0.9884	0.9905	0.9922	0.9936	0.9948	0.9958	0.9966	0.9973
18	0.9527	0.9602	0.9667	0.9722	0.9768	0.9808	0.9841	0.9869	0.9892	0.9911	0.9927	0.9941	0.9952	0.9961	0.9968
19	0.9485	0.9565	0.9634	0.9693	0.9744	0.9786	0.9822	0.9853	0.9878	0.9900	0.9917	0.9932	0.9945	0.9955	0.9963
20	0.9442	0.9528	0.9601	0.9664	0.9718	0.9764	0.9803	0.9836	0.9864	0.9887	0.9907	0.9923	0.9937	0.9948	0.9958
21	0.9399	0.9489	0.9567	0.9634	0.9692	0.9741	0.9783	0.9819	0.9849	0.9875	0.9896	0.9914	0.9929	0.9942	0.9952
22	0.9356	0.9450	0.9532	0.9603	0.9665	0.9717	0.9762	0.9801	0.9833	0.9861	0.9884	0.9904	0.9921	0.9934	0.9946
23	0.9312	0.9411	0.9497	0.9572	0.9637	0.9693	0.9741	0.9782	0.9817	0.9847	0.9872	0.9894	0.9912	0.9927	0.9939
24	0.9268	0.9371	0.9461	0.9540	0.9608	0.9667	0.9719	0.9762	0.9800	0.9832	0.9859	0.9883	0.9902	0.9919	0.9933
25	0.9223	0.9330	0.9425	0.9507	0.9579	0.9642	0.9696	0.9742	0.9782	0.9817	0.9846	0.9871	0.9892	0.9910	0.9925
26	0.9178	0.9290	0.9388	0.9474	0.9550	0.9615	0.9672	0.9722	0.9764	0.9801	0.9832	0.9859	0.9882	0.9901	0.9917
27	0.9133	0.9249	0.9351	0.9441	0.9520	0.9589	0.9649	0.9701	0.9746	0.9785	0.9818	0.9846	0.9871	0.9892	0.9909
28	0.9088	0.9208	0.9313	0.9407	0.9489	0.9561	0.9624	0.9679	0.9727	0.9768	0.9803	0.9834	0.9860	0.9882	0.9901
29	0.9043	0.9166	0.9276	0.9373	0.9458	0.9533	0.9599	0.9657	0.9707	0.9750	0.9788	0.9820	0.9848	0.9872	0.9892
30	0.8998	0.9125	0.9238	0.9338	0.9427	0.9505	0.9574	0.9634	0.9687	0.9732	0.9772	0.9806	0.9836	0.9861	0.9883
32	0.8908	0.9042	0.9161	0.9268	0.9364	0.9448	0.9522	0.9588	0.9645	0.9695	0.9739	0.9777	0.9810	0.9839	0.9863
34	0.8818	0.8958	0.9085	0.9198	0.9299	0.9389	0.9469	0.9540	0.9602	0.9657	0.9705	0.9747	0.9783	0.9815	0.9842
36	0.8729	0.8875	0.9007	0.9127	0.9234	0.9329	0.9415	0.9490	0.9557	0.9617	0.9669	0.9714	0.9754	0.9789	0.9820
38	0.8641	0.8793	0.8930	0.9055	0.9168	0.9269	0.9359	0.9440	0.9512	0.9575	0.9632	0.9681	0.9725	0.9763	0.9796
40	0.8553	0.8710	0.8854	0.8984	0.9101	0.9207	0.9303	0.9388	0.9465	0.9533	0.9593	0.9646	0.9693	0.9735	0.9771
42	0.8467	0.8629	0.8777	0.8912	0.9035	0.9146	0.9246	0.9336	0.9417	0.9489	0.9553	0.9611	0.9661	0.9706	0.9745
44	0.8382	0.8543	0.8701	0.8841	0.8968	0.9084	0.9189	0.9283	0.9368	0.9445	0.9513	0.9574	0.9628	0.9676	0.9718
46	0.8298	0.8468	0.8625	0.8769	0.8901	0.9021	0.9131	0.9230	0.9319	0.9399	0.9471	0.9536	0.9593	0.9645	0.9690
48	0.8215	0.8390	0.8550	0.8699	0.8835	0.8959	0.9073	0.9176	0.9269	0.9353	0.9429	0.9497	0.9558	0.9613	0.9661
50	0.8134	0.8312	0.8476	0.8628	0.8768	0.8897	0.9014	0.9121	0.9219	0.9307	0.9386	0.9458	0.9522	0.9580	0.9631
55	0.7937	0.8122	0.8295	0.8455	0.8604	0.8742	0.8869	0.8985	0.9091	0.9188	0.9276	0.9357	0.9429	0.9494	0.9553
60	0.7748	0.7940	0.8119	0.8287	0.8443	0.8589	0.8724	0.8848	0.8963	0.9068	0.9164	0.9252	0.9332	0.9405	0.9471
65	0.7569	0.7765	0.7949	0.8123	0.8286	0.8438	0.8580	0.8712	0.8834	0.8947	0.9051	0.9146	0.9233	0.9313	0.9386
70	0.7398	0.7597	0.7786	0.7965	0.8133	0.8291	0.8439	0.8578	0.8706	0.8826	0.8936	0.9038	0.9132	0.9219	0.9298
75	0.7235	0.7437	0.7629	0.7812	0.7985	0.8148	0.8301	0.8445	0.8580	0.8705	0.8822	0.8930	0.9030	0.9123	0.9208
80	0.7080	0.7284	0.7479	0.7665	0.7842	0.8009	0.8167	0.8316	0.8455	0.8586	0.8708	0.8822	0.8928	0.9026	0.9117
85	0.6932	0.7138	0.7335	0.7523	0.7703	0.7874	0.8035	0.8189	0.8333	0.8468	0.8596	0.8715	0.8826	0.8929	0.9025
90	0.6792	0.6998	0.7197	0.7387	0.7569	0.7743	0.7908	0.8064	0.8213	0.8353	0.8484	0.8608	0.8724	0.8832	0.8933
95	0.6658	0.6865	0.7064	0.7256	0.7440	0.7616	0.7784	0.7944	0.8095	0.8239	0.8375	0.8503	0.8623	0.8735	0.8841
100	0.6531	0.6737	0.6937	0.7130	0.7315	0.7493	0.7663	0.7826	0.7981	0.8128	0.8267	0.8399	0.8523	0.8639	0.8749

ALPHA = .100

N	LAMBDA= 42.00	44.00	46.00	48.00	50.00	55.00	60.00	65.00	70.00	75.00	80.00	85.00	90.00	95.00	100.00
1	1.0000	1.0000	1.0000	1.0000	1.0000	1.0000	1.0000	1.0000	1.0000	1.0000	1.0000	1.0000	1.0000	1.0000	1.0000
2	1.0000	1.0000	1.0000	1.0000	1.0000	1.0000	1.0000	1.0000	1.0000	1.0000	1.0000	1.0000	1.0000	1.0000	1.0000
3	1.0000	1.0000	1.0000	1.0000	1.0000	1.0000	1.0000	1.0000	1.0000	1.0000	1.0000	1.0000	1.0000	1.0000	1.0000
4	1.0000	1.0000	1.0000	1.0000	1.0000	1.0000	1.0000	1.0000	1.0000	1.0000	1.0000	1.0000	1.0000	1.0000	1.0000
5	0.9999	1.0000	1.0000	1.0000	1.0000	1.0000	1.0000	1.0000	1.0000	1.0000	1.0000	1.0000	1.0000	1.0000	1.0000
6	0.9999	1.0000	1.0000	1.0000	1.0000	1.0000	1.0000	1.0000	1.0000	1.0000	1.0000	1.0000	1.0000	1.0000	1.0000
7	0.9999	0.9999	1.0000	1.0000	1.0000	1.0000	1.0000	1.0000	1.0000	1.0000	1.0000	1.0000	1.0000	1.0000	1.0000
8	0.9998	0.9999	0.9999	1.0000	1.0000	1.0000	1.0000	1.0000	1.0000	1.0000	1.0000	1.0000	1.0000	1.0000	1.0000
9	0.9997	0.9998	0.9999	0.9999	1.0000	1.0000	1.0000	1.0000	1.0000	1.0000	1.0000	1.0000	1.0000	1.0000	1.0000
10	0.9996	0.9998	0.9999	0.9999	1.0000	1.0000	1.0000	1.0000	1.0000	1.0000	1.0000	1.0000	1.0000	1.0000	1.0000
11	0.9995	0.9997	0.9998	0.9999	0.9999	1.0000	1.0000	1.0000	1.0000	1.0000	1.0000	1.0000	1.0000	1.0000	1.0000
12	0.9993	0.9996	0.9998	0.9999	0.9999	1.0000	1.0000	1.0000	1.0000	1.0000	1.0000	1.0000	1.0000	1.0000	1.0000
13	0.9992	0.9995	0.9997	0.9998	0.9999	1.0000	1.0000	1.0000	1.0000	1.0000	1.0000	1.0000	1.0000	1.0000	1.0000
14	0.9990	0.9994	0.9996	0.9998	0.9999	1.0000	1.0000	1.0000	1.0000	1.0000	1.0000	1.0000	1.0000	1.0000	1.0000
15	0.9988	0.9992	0.9995	0.9997	0.9998	0.9999	1.0000	1.0000	1.0000	1.0000	1.0000	1.0000	1.0000	1.0000	1.0000
16	0.9985	0.9991	0.9994	0.9996	0.9998	0.9999	1.0000	1.0000	1.0000	1.0000	1.0000	1.0000	1.0000	1.0000	1.0000
17	0.9982	0.9989	0.9993	0.9995	0.9997	0.9999	1.0000	1.0000	1.0000	1.0000	1.0000	1.0000	1.0000	1.0000	1.0000
18	0.9979	0.9987	0.9991	0.9994	0.9997	0.9999	1.0000	1.0000	1.0000	1.0000	1.0000	1.0000	1.0000	1.0000	1.0000
19	0.9976	0.9984	0.9990	0.9993	0.9996	0.9999	1.0000	1.0000	1.0000	1.0000	1.0000	1.0000	1.0000	1.0000	1.0000
20	0.9972	0.9982	0.9988	0.9992	0.9995	0.9998	1.0000	1.0000	1.0000	1.0000	1.0000	1.0000	1.0000	1.0000	1.0000
21	0.9968	0.9979	0.9986	0.9991	0.9994	0.9998	0.9999	1.0000	1.0000	1.0000	1.0000	1.0000	1.0000	1.0000	1.0000
22	0.9964	0.9976	0.9984	0.9989	0.9993	0.9998	0.9999	1.0000	1.0000	1.0000	1.0000	1.0000	1.0000	1.0000	1.0000
23	0.9959	0.9972	0.9982	0.9988	0.9992	0.9997	0.9999	1.0000	1.0000	1.0000	1.0000	1.0000	1.0000	1.0000	1.0000
24	0.9954	0.9969	0.9979	0.9986	0.9991	0.9997	0.9999	1.0000	1.0000	1.0000	1.0000	1.0000	1.0000	1.0000	1.0000
25	0.9949	0.9965	0.9976	0.9984	0.9989	0.9996	0.9999	1.0000	1.0000	1.0000	1.0000	1.0000	1.0000	1.0000	1.0000
26	0.9943	0.9961	0.9973	0.9982	0.9988	0.9996	0.9999	1.0000	1.0000	1.0000	1.0000	1.0000	1.0000	1.0000	1.0000
27	0.9937	0.9956	0.9970	0.9980	0.9986	0.9995	0.9998	0.9999	1.0000	1.0000	1.0000	1.0000	1.0000	1.0000	1.0000
28	0.9931	0.9952	0.9967	0.9977	0.9985	0.9994	0.9998	0.9999	1.0000	1.0000	1.0000	1.0000	1.0000	1.0000	1.0000
29	0.9924	0.9947	0.9963	0.9975	0.9983	0.9994	0.9998	0.9999	1.0000	1.0000	1.0000	1.0000	1.0000	1.0000	1.0000
30	0.9917	0.9942	0.9959	0.9972	0.9981	0.9993	0.9997	0.9999	1.0000	1.0000	1.0000	1.0000	1.0000	1.0000	1.0000
32	0.9902	0.9931	0.9951	0.9966	0.9976	0.9991	0.9997	0.9999	1.0000	1.0000	1.0000	1.0000	1.0000	1.0000	1.0000
34	0.9886	0.9918	0.9942	0.9959	0.9972	0.9989	0.9996	0.9998	0.9999	1.0000	1.0000	1.0000	1.0000	1.0000	1.0000
36	0.9869	0.9905	0.9932	0.9952	0.9966	0.9986	0.9995	0.9998	0.9999	1.0000	1.0000	1.0000	1.0000	1.0000	1.0000
38	0.9850	0.9891	0.9921	0.9944	0.9960	0.9983	0.9993	0.9998	0.9999	1.0000	1.0000	1.0000	1.0000	1.0000	1.0000
40	0.9831	0.9876	0.9910	0.9935	0.9953	0.9980	0.9992	0.9997	0.9999	1.0000	1.0000	1.0000	1.0000	1.0000	1.0000
42	0.9810	0.9860	0.9897	0.9925	0.9946	0.9977	0.9990	0.9996	0.9999	0.9999	1.0000	1.0000	1.0000	1.0000	1.0000
44	0.9788	0.9842	0.9884	0.9915	0.9938	0.9973	0.9989	0.9995	0.9998	0.9999	1.0000	1.0000	1.0000	1.0000	1.0000
46	0.9766	0.9824	0.9869	0.9904	0.9929	0.9969	0.9987	0.9994	0.9998	0.9999	1.0000	1.0000	1.0000	1.0000	1.0000
48	0.9742	0.9805	0.9854	0.9892	0.9920	0.9964	0.9984	0.9993	0.9997	0.9999	1.0000	1.0000	1.0000	1.0000	1.0000
50	0.9718	0.9786	0.9839	0.9879	0.9910	0.9959	0.9982	0.9992	0.9997	0.9999	1.0000	1.0000	1.0000	1.0000	1.0000
55	0.9653	0.9733	0.9796	0.9845	0.9883	0.9944	0.9974	0.9989	0.9995	0.9998	0.9999	1.0000	1.0000	1.0000	1.0000
60	0.9584	0.9676	0.9749	0.9807	0.9853	0.9927	0.9965	0.9984	0.9993	0.9997	0.9999	1.0000	1.0000	1.0000	1.0000
65	0.9512	0.9615	0.9698	0.9765	0.9819	0.9908	0.9955	0.9979	0.9990	0.9996	0.9998	0.9999	1.0000	1.0000	1.0000
70	0.9436	0.9550	0.9644	0.9720	0.9781	0.9886	0.9942	0.9972	0.9987	0.9994	0.9997	0.9999	1.0000	1.0000	1.0000
75	0.9358	0.9483	0.9587	0.9672	0.9741	0.9861	0.9928	0.9964	0.9983	0.9992	0.9996	0.9998	0.9999	1.0000	1.0000
80	0.9278	0.9414	0.9527	0.9621	0.9698	0.9834	0.9912	0.9955	0.9978	0.9989	0.9995	0.9998	0.9999	1.0000	1.0000
85	0.9197	0.9342	0.9465	0.9568	0.9653	0.9805	0.9895	0.9945	0.9972	0.9986	0.9994	0.9997	0.9999	0.9999	1.0000
90	0.9114	0.9269	0.9401	0.9513	0.9606	0.9774	0.9875	0.9933	0.9966	0.9983	0.9992	0.9996	0.9998	0.9999	1.0000
95	0.9031	0.9195	0.9336	0.9456	0.9556	0.9741	0.9854	0.9920	0.9958	0.9979	0.9989	0.9995	0.9998	0.9999	1.0000
100	0.8948	0.9120	0.9269	0.9397	0.9505	0.9706	0.9831	0.9906	0.9950	0.9974	0.9987	0.9993	0.9997	0.9999	0.9999

ALPHA = .050

N	LAMBDA= .00	.10	.20	.30	.40	.50	.60	.70	.80	.90	1.00	1.20	1.40	1.60	1.80
1	0.0500	0.0615	0.0732	0.0850	0.0969	0.1090	0.1211	0.1332	0.1455	0.1578	0.1701	0.1948	0.2195	0.2441	0.2687
2	0.0500	0.0576	0.0653	0.0733	0.0814	0.0896	0.0980	0.1065	0.1151	0.1239	0.1327	0.1507	0.1691	0.1877	0.2065
3	0.0500	0.0559	0.0620	0.0682	0.0746	0.0811	0.0878	0.0946	0.1015	0.1085	0.1157	0.1303	0.1453	0.1606	0.1763
4	0.0500	0.0550	0.0601	0.0653	0.0707	0.0762	0.0818	0.0876	0.0934	0.0994	0.1055	0.1180	0.1308	0.1441	0.1576
5	0.0500	0.0543	0.0588	0.0634	0.0681	0.0729	0.0778	0.0829	0.0880	0.0932	0.0986	0.1096	0.1209	0.1326	0.1447
6	0.0500	0.0539	0.0579	0.0620	0.0662	0.0705	0.0749	0.0794	0.0841	0.0888	0.0936	0.1034	0.1136	0.1242	0.1351
7	0.0500	0.0535	0.0572	0.0609	0.0648	0.0687	0.0727	0.0768	0.0810	0.0853	0.0897	0.0987	0.1080	0.1177	0.1276
8	0.0500	0.0533	0.0566	0.0601	0.0636	0.0673	0.0710	0.0748	0.0786	0.0826	0.0866	0.0949	0.1035	0.1124	0.1216
9	0.0500	0.0531	0.0562	0.0594	0.0627	0.0661	0.0695	0.0731	0.0767	0.0803	0.0841	0.0918	0.0998	0.1081	0.1167
10	0.0500	0.0529	0.0558	0.0588	0.0619	0.0651	0.0683	0.0716	0.0750	0.0785	0.0820	0.0892	0.0967	0.1045	0.1125
11	0.0500	0.0527	0.0555	0.0584	0.0613	0.0643	0.0673	0.0704	0.0736	0.0769	0.0802	0.0870	0.0941	0.1014	0.1090
12	0.0500	0.0526	0.0552	0.0579	0.0607	0.0635	0.0664	0.0694	0.0724	0.0755	0.0786	0.0851	0.0918	0.0987	0.1059
13	0.0500	0.0525	0.0550	0.0576	0.0602	0.0629	0.0657	0.0685	0.0714	0.0743	0.0773	0.0834	0.0898	0.0964	0.1032
14	0.0500	0.0524	0.0548	0.0572	0.0598	0.0624	0.0650	0.0677	0.0704	0.0732	0.0761	0.0820	0.0880	0.0943	0.1008
15	0.0500	0.0523	0.0546	0.0570	0.0594	0.0619	0.0644	0.0670	0.0696	0.0723	0.0750	0.0806	0.0865	0.0925	0.0987
16	0.0500	0.0522	0.0544	0.0567	0.0590	0.0614	0.0638	0.0663	0.0689	0.0714	0.0741	0.0795	0.0851	0.0908	0.0968
25	0.0500	0.0517	0.0534	0.0552	0.0570	0.0588	0.0607	0.0625	0.0645	0.0664	0.0684	0.0725	0.0767	0.0811	0.0856
26	0.0500	0.0517	0.0533	0.0551	0.0568	0.0586	0.0604	0.0623	0.0641	0.0661	0.0680	0.0720	0.0761	0.0803	0.0847
27	0.0500	0.0516	0.0533	0.0550	0.0567	0.0584	0.0602	0.0620	0.0638	0.0657	0.0676	0.0715	0.0755	0.0797	0.0839
28	0.0500	0.0516	0.0532	0.0549	0.0565	0.0582	0.0600	0.0617	0.0635	0.0654	0.0672	0.0710	0.0750	0.0790	0.0832
29	0.0500	0.0516	0.0531	0.0548	0.0564	0.0581	0.0598	0.0615	0.0633	0.0651	0.0669	0.0706	0.0744	0.0784	0.0825
30	0.0500	0.0515	0.0531	0.0547	0.0563	0.0579	0.0596	0.0613	0.0630	0.0648	0.0665	0.0702	0.0739	0.0778	0.0818
32	0.0500	0.0515	0.0530	0.0545	0.0560	0.0576	0.0592	0.0609	0.0625	0.0642	0.0659	0.0694	0.0730	0.0768	0.0806
34	0.0500	0.0514	0.0529	0.0543	0.0558	0.0574	0.0589	0.0605	0.0621	0.0637	0.0654	0.0687	0.0722	0.0758	0.0795
36	0.0500	0.0514	0.0528	0.0542	0.0557	0.0571	0.0586	0.0602	0.0617	0.0633	0.0649	0.0681	0.0715	0.0749	0.0785
38	0.0500	0.0513	0.0527	0.0541	0.0555	0.0569	0.0584	0.0598	0.0613	0.0629	0.0644	0.0675	0.0708	0.0741	0.0776
40	0.0500	0.0513	0.0526	0.0540	0.0553	0.0567	0.0581	0.0596	0.0610	0.0625	0.0640	0.0670	0.0702	0.0734	0.0767
42	0.0500	0.0513	0.0526	0.0539	0.0552	0.0565	0.0579	0.0593	0.0607	0.0621	0.0636	0.0665	0.0696	0.0727	0.0759
44	0.0500	0.0512	0.0525	0.0538	0.0551	0.0564	0.0577	0.0590	0.0604	0.0618	0.0632	0.0661	0.0691	0.0721	0.0752
46	0.0500	0.0512	0.0524	0.0537	0.0549	0.0562	0.0575	0.0588	0.0602	0.0615	0.0629	0.0657	0.0686	0.0715	0.0746
48	0.0500	0.0512	0.0524	0.0536	0.0548	0.0561	0.0573	0.0586	0.0599	0.0612	0.0626	0.0653	0.0681	0.0710	0.0739
50	0.0500	0.0511	0.0523	0.0535	0.0547	0.0559	0.0572	0.0584	0.0597	0.0610	0.0623	0.0649	0.0677	0.0705	0.0734
55	0.0500	0.0511	0.0522	0.0533	0.0545	0.0556	0.0568	0.0580	0.0592	0.0604	0.0616	0.0641	0.0667	0.0694	0.0721
60	0.0500	0.0510	0.0521	0.0532	0.0542	0.0553	0.0565	0.0576	0.0587	0.0599	0.0611	0.0634	0.0659	0.0684	0.0710
65	0.0500	0.0510	0.0520	0.0530	0.0541	0.0551	0.0562	0.0573	0.0583	0.0594	0.0606	0.0628	0.0652	0.0676	0.0700
70	0.0500	0.0510	0.0519	0.0529	0.0539	0.0549	0.0559	0.0570	0.0580	0.0591	0.0601	0.0623	0.0645	0.0668	0.0691
75	0.0500	0.0509	0.0519	0.0528	0.0538	0.0547	0.0557	0.0567	0.0577	0.0587	0.0597	0.0618	0.0640	0.0661	0.0684
80	0.0500	0.0509	0.0518	0.0527	0.0536	0.0546	0.0555	0.0565	0.0574	0.0584	0.0594	0.0614	0.0635	0.0656	0.0677
85	0.0500	0.0509	0.0517	0.0526	0.0535	0.0544	0.0553	0.0562	0.0572	0.0581	0.0591	0.0610	0.0630	0.0650	0.0671
90	0.0500	0.0508	0.0517	0.0525	0.0534	0.0543	0.0552	0.0560	0.0570	0.0579	0.0588	0.0607	0.0626	0.0645	0.0665
95	0.0500	0.0508	0.0516	0.0525	0.0533	0.0541	0.0550	0.0559	0.0567	0.0576	0.0585	0.0603	0.0622	0.0641	0.0660
100	0.0500	0.0508	0.0516	0.0524	0.0532	0.0540	0.0549	0.0557	0.0566	0.0574	0.0583	0.0600	0.0618	0.0637	0.0656

ALPHA = .050

N	LAMBDA= 2.00	2.20	2.40	2.60	2.80	3.00	3.50	4.00	4.50	5.00	6.00	7.00	8.00	9.00	10.00
1	0.2930	0.3171	0.3408	0.3643	0.3873	0.4100	0.4646	0.5160	0.5641	0.6088	0.6878	0.7536	0.8074	0.8508	0.8854
2	0.2255	0.2447	0.2639	0.2831	0.3024	0.3215	0.3690	0.4154	0.4604	0.5037	0.5840	0.6554	0.7176	0.7707	0.8154
3	0.1922	0.2084	0.2248	0.2413	0.2579	0.2746	0.3166	0.3585	0.3999	0.4405	0.5181	0.5896	0.6541	0.7113	0.7611
4	0.1715	0.1856	0.1999	0.2145	0.2293	0.2442	0.2819	0.3201	0.3583	0.3962	0.4701	0.5400	0.6047	0.6635	0.7160
5	0.1570	0.1696	0.1825	0.1955	0.2088	0.2223	0.2567	0.2918	0.3272	0.3627	0.4329	0.5006	0.5644	0.6236	0.6774
6	0.1462	0.1576	0.1693	0.1813	0.1934	0.2057	0.2373	0.2698	0.3028	0.3362	0.4028	0.4681	0.5307	0.5895	0.6438
7	0.1378	0.1483	0.1590	0.1700	0.1812	0.1926	0.2219	0.2522	0.2831	0.3145	0.3779	0.4408	0.5017	0.5598	0.6142
8	0.1311	0.1408	0.1507	0.1609	0.1713	0.1819	0.2093	0.2376	0.2667	0.2964	0.3568	0.4173	0.4766	0.5337	0.5877
9	0.1255	0.1345	0.1438	0.1533	0.1631	0.1730	0.1987	0.2254	0.2529	0.2811	0.3387	0.3969	0.4545	0.5104	0.5639
10	0.1208	0.1293	0.1380	0.1470	0.1561	0.1655	0.1897	0.2149	0.2410	0.2678	0.3229	0.3790	0.4349	0.4896	0.5424
11	0.1168	0.1248	0.1330	0.1415	0.1501	0.1590	0.1819	0.2058	0.2307	0.2562	0.3090	0.3631	0.4173	0.4708	0.5228
12	0.1133	0.1209	0.1287	0.1367	0.1449	0.1533	0.1751	0.1979	0.2216	0.2460	0.2967	0.3488	0.4015	0.4538	0.5049
13	0.1102	0.1175	0.1249	0.1325	0.1403	0.1483	0.1691	0.1909	0.2135	0.2369	0.2857	0.3360	0.3872	0.4382	0.4884
14	0.1075	0.1144	0.1215	0.1288	0.1363	0.1439	0.1638	0.1846	0.2063	0.2288	0.2757	0.3244	0.3741	0.4240	0.4732
15	0.1051	0.1117	0.1185	0.1255	0.1326	0.1400	0.1590	0.1790	0.1998	0.2215	0.2667	0.3139	0.3622	0.4108	0.4591
16	0.1030	0.1093	0.1158	0.1225	0.1294	0.1364	0.1547	0.1739	0.1940	0.2148	0.2585	0.3042	0.3512	0.3987	0.4461
17	0.1010	0.1071	0.1134	0.1198	0.1264	0.1332	0.1508	0.1693	0.1886	0.2088	0.2510	0.2953	0.3411	0.3875	0.4339
18	0.0992	0.1051	0.1112	0.1174	0.1237	0.1303	0.1472	0.1651	0.1838	0.2032	0.2441	0.2872	0.3317	0.3770	0.4225
19	0.0976	0.1033	0.1091	0.1151	0.1213	0.1276	0.1440	0.1612	0.1793	0.1981	0.2378	0.2796	0.3229	0.3673	0.4119
20	0.0961	0.1016	0.1073	0.1131	0.1190	0.1251	0.1409	0.1577	0.1752	0.1934	0.2319	0.2726	0.3148	0.3581	0.4019
21	0.0948	0.1001	0.1055	0.1112	0.1169	0.1228	0.1382	0.1544	0.1713	0.1891	0.2264	0.2660	0.3072	0.3496	0.3925
22	0.0935	0.0986	0.1040	0.1094	0.1150	0.1207	0.1356	0.1513	0.1678	0.1850	0.2213	0.2599	0.3001	0.3416	0.3837
23	0.0923	0.0973	0.1025	0.1078	0.1132	0.1187	0.1332	0.1485	0.1645	0.1812	0.2166	0.2542	0.2935	0.3340	0.3754
24	0.0912	0.0961	0.1011	0.1062	0.1115	0.1169	0.1310	0.1458	0.1614	0.1777	0.2121	0.2488	0.2872	0.3269	0.3675
25	0.0902	0.0949	0.0998	0.1048	0.1100	0.1152	0.1289	0.1434	0.1585	0.1744	0.2080	0.2438	0.2813	0.3202	0.3600
26	0.0892	0.0938	0.0986	0.1035	0.1085	0.1136	0.1270	0.1410	0.1558	0.1713	0.2041	0.2390	0.2757	0.3139	0.3529
27	0.0883	0.0928	0.0975	0.1022	0.1071	0.1121	0.1251	0.1389	0.1533	0.1684	0.2004	0.2345	0.2705	0.3078	0.3462
28	0.0875	0.0919	0.0964	0.1011	0.1058	0.1107	0.1234	0.1368	0.1509	0.1656	0.1969	0.2303	0.2655	0.3021	0.3398
29	0.0867	0.0910	0.0954	0.0999	0.1046	0.1094	0.1218	0.1349	0.1486	0.1630	0.1936	0.2263	0.2607	0.2967	0.3337
30	0.0859	0.0901	0.0945	0.0989	0.1034	0.1081	0.1202	0.1330	0.1465	0.1606	0.1904	0.2224	0.2562	0.2915	0.3279
32	0.0845	0.0886	0.0927	0.0970	0.1013	0.1058	0.1174	0.1297	0.1425	0.1560	0.1847	0.2154	0.2479	0.2819	0.3171
34	0.0833	0.0871	0.0911	0.0952	0.0994	0.1037	0.1148	0.1266	0.1390	0.1519	0.1794	0.2090	0.2403	0.2731	0.3071
36	0.0821	0.0859	0.0897	0.0936	0.0977	0.1018	0.1125	0.1238	0.1357	0.1482	0.1747	0.2031	0.2333	0.2650	0.2980
38	0.0811	0.0847	0.0884	0.0922	0.0961	0.1001	0.1104	0.1213	0.1328	0.1448	0.1703	0.1978	0.2270	0.2576	0.2896
40	0.0801	0.0836	0.0872	0.0909	0.0946	0.0985	0.1085	0.1190	0.1301	0.1416	0.1663	0.1928	0.2211	0.2508	0.2818
42	0.0792	0.0826	0.0861	0.0896	0.0933	0.0970	0.1067	0.1169	0.1276	0.1388	0.1626	0.1883	0.2157	0.2445	0.2746
44	0.0784	0.0817	0.0851	0.0885	0.0920	0.0957	0.1050	0.1149	0.1253	0.1361	0.1592	0.1841	0.2106	0.2386	0.2679
46	0.0777	0.0809	0.0841	0.0875	0.0909	0.0944	0.1035	0.1131	0.1231	0.1336	0.1560	0.1802	0.2059	0.2331	0.2616
48	0.0770	0.0801	0.0833	0.0865	0.0898	0.0932	0.1021	0.1114	0.1211	0.1313	0.1531	0.1765	0.2016	0.2280	0.2557
50	0.0763	0.0793	0.0824	0.0856	0.0888	0.0922	0.1007	0.1098	0.1193	0.1292	0.1504	0.1731	0.1975	0.2232	0.2503
55	0.0749	0.0777	0.0806	0.0836	0.0866	0.0897	0.0978	0.1063	0.1151	0.1244	0.1442	0.1655	0.1883	0.2125	0.2379
60	0.0736	0.0763	0.0790	0.0818	0.0847	0.0876	0.0952	0.1032	0.1116	0.1203	0.1389	0.1590	0.1805	0.2032	0.2272
65	0.0725	0.0750	0.0777	0.0803	0.0830	0.0858	0.0930	0.1006	0.1085	0.1168	0.1344	0.1533	0.1736	0.1951	0.2178
70	0.0715	0.0740	0.0765	0.0790	0.0816	0.0842	0.0911	0.0983	0.1058	0.1136	0.1303	0.1483	0.1676	0.1880	0.2096
75	0.0707	0.0730	0.0754	0.0778	0.0803	0.0828	0.0894	0.0962	0.1034	0.1109	0.1268	0.1439	0.1622	0.1817	0.2022
80	0.0699	0.0721	0.0744	0.0767	0.0791	0.0815	0.0878	0.0944	0.1012	0.1084	0.1236	0.1399	0.1574	0.1760	0.1957
85	0.0692	0.0714	0.0736	0.0758	0.0781	0.0804	0.0864	0.0927	0.0993	0.1062	0.1207	0.1364	0.1531	0.1709	0.1897
90	0.0686	0.0706	0.0728	0.0749	0.0771	0.0794	0.0852	0.0912	0.0975	0.1041	0.1181	0.1331	0.1492	0.1663	0.1844
95	0.0680	0.0700	0.0720	0.0741	0.0763	0.0784	0.0840	0.0899	0.0960	0.1023	0.1157	0.1302	0.1457	0.1621	0.1795
100	0.0675	0.0694	0.0714	0.0734	0.0755	0.0776	0.0830	0.0886	0.0945	0.1006	0.1136	0.1275	0.1424	0.1582	0.1750

ALPHA = .050

N	LAMBDA= 11.00	12.00	13.00	14.00	15.00	16.00	17.00	18.00	19.00	20.00	21.00	22.00	23.00	24.00	25.00
1	0.9126	0.9337	0.9501	0.9626	0.9721	0.9793	0.9847	0.9888	0.9918	0.9940	0.9956	0.9968	0.9977	0.9984	0.9988
2	0.8526	0.8832	0.9080	0.9280	0.9440	0.9567	0.9667	0.9745	0.9805	0.9852	0.9888	0.9916	0.9937	0.9953	0.9965
3	0.8039	0.8402	0.8707	0.8961	0.9170	0.9341	0.9479	0.9591	0.9680	0.9751	0.9807	0.9851	0.9885	0.9912	0.9933
4	0.7622	0.8024	0.8370	0.8664	0.8912	0.9119	0.9290	0.9431	0.9547	0.9640	0.9716	0.9776	0.9825	0.9863	0.9894
5	0.7258	0.7686	0.8061	0.8386	0.8665	0.8902	0.9102	0.9269	0.9408	0.9523	0.9618	0.9695	0.9757	0.9808	0.9848
6	0.6934	0.7380	0.7777	0.8126	0.8429	0.8692	0.8916	0.9107	0.9268	0.9403	0.9515	0.9608	0.9684	0.9747	0.9797
7	0.6644	0.7102	0.7514	0.7882	0.8206	0.8489	0.8735	0.8946	0.9126	0.9279	0.9408	0.9516	0.9606	0.9681	0.9742
8	0.6382	0.6847	0.7271	0.7652	0.7993	0.8294	0.8558	0.8787	0.8985	0.9155	0.9299	0.9422	0.9525	0.9611	0.9683
9	0.6144	0.6613	0.7045	0.7437	0.7791	0.8106	0.8386	0.8631	0.8845	0.9030	0.9189	0.9325	0.9440	0.9538	0.9620
10	0.5926	0.6397	0.6834	0.7234	0.7598	0.7926	0.8219	0.8478	0.8706	0.8905	0.9078	0.9226	0.9354	0.9463	0.9555
11	0.5726	0.6197	0.6636	0.7043	0.7416	0.7754	0.8058	0.8329	0.8570	0.8781	0.8966	0.9127	0.9266	0.9385	0.9487
12	0.5541	0.6010	0.6452	0.6863	0.7242	0.7588	0.7902	0.8184	0.8436	0.8659	0.8855	0.9027	0.9177	0.9306	0.9417
13	0.5371	0.5837	0.6278	0.6692	0.7076	0.7429	0.7751	0.8043	0.8304	0.8538	0.8745	0.8927	0.9087	0.9226	0.9346
14	0.5212	0.5675	0.6115	0.6531	0.6918	0.7277	0.7606	0.7905	0.8176	0.8419	0.8636	0.8827	0.8996	0.9145	0.9274
15	0.5065	0.5523	0.5962	0.6377	0.6768	0.7131	0.7465	0.7772	0.8051	0.8302	0.8527	0.8728	0.8906	0.9063	0.9200
16	0.4927	0.5381	0.5817	0.6232	0.6624	0.6990	0.7330	0.7643	0.7928	0.8187	0.8421	0.8630	0.8816	0.8981	0.9126
17	0.4798	0.5247	0.5680	0.6094	0.6487	0.6856	0.7199	0.7517	0.7809	0.8074	0.8315	0.8532	0.8726	0.8899	0.9051
18	0.4677	0.5120	0.5550	0.5963	0.6356	0.6726	0.7073	0.7395	0.7692	0.7964	0.8212	0.8435	0.8637	0.8816	0.8976
19	0.4564	0.5001	0.5427	0.5838	0.6230	0.6602	0.6951	0.7277	0.7579	0.7856	0.8110	0.8340	0.8548	0.8735	0.8901
20	0.4457	0.4888	0.5310	0.5719	0.6110	0.6483	0.6834	0.7163	0.7468	0.7751	0.8010	0.8246	0.8460	0.8653	0.8826
21	0.4356	0.4782	0.5199	0.5605	0.5995	0.6368	0.6720	0.7052	0.7361	0.7648	0.7911	0.8153	0.8373	0.8572	0.8751
22	0.4260	0.4681	0.5094	0.5496	0.5885	0.6257	0.6611	0.6944	0.7256	0.7547	0.7815	0.8062	0.8287	0.8492	0.8677
23	0.4170	0.4584	0.4993	0.5393	0.5779	0.6151	0.6505	0.6840	0.7155	0.7448	0.7721	0.7972	0.8202	0.8412	0.8602
24	0.4084	0.4493	0.4897	0.5293	0.5678	0.6048	0.6403	0.6739	0.7055	0.7352	0.7628	0.7883	0.8118	0.8333	0.8528
25	0.4003	0.4406	0.4806	0.5198	0.5580	0.5950	0.6304	0.6640	0.6959	0.7258	0.7537	0.7796	0.8035	0.8255	0.8455
26	0.3925	0.4323	0.4718	0.5107	0.5486	0.5854	0.6208	0.6545	0.6865	0.7166	0.7448	0.7711	0.7954	0.8177	0.8382
27	0.3852	0.4244	0.4634	0.5019	0.5396	0.5762	0.6115	0.6453	0.6774	0.7077	0.7361	0.7627	0.7873	0.8101	0.8310
28	0.3782	0.4168	0.4554	0.4935	0.5309	0.5674	0.6025	0.6363	0.6685	0.6989	0.7276	0.7545	0.7794	0.8026	0.8239
29	0.3715	0.4096	0.4477	0.4854	0.5226	0.5588	0.5939	0.6276	0.6598	0.6904	0.7193	0.7464	0.7716	0.7951	0.8168
30	0.3651	0.4026	0.4403	0.4777	0.5145	0.5505	0.5854	0.6191	0.6514	0.6821	0.7111	0.7384	0.7640	0.7878	0.8098
32	0.3531	0.3896	0.4264	0.4630	0.4992	0.5347	0.5694	0.6029	0.6352	0.6660	0.6953	0.7230	0.7490	0.7733	0.7960
34	0.3421	0.3776	0.4135	0.4493	0.4849	0.5200	0.5543	0.5876	0.6198	0.6507	0.6802	0.7081	0.7345	0.7594	0.7825
36	0.3319	0.3665	0.4015	0.4366	0.4716	0.5061	0.5401	0.5731	0.6052	0.6361	0.6657	0.6939	0.7206	0.7458	0.7694
38	0.3226	0.3563	0.3904	0.4248	0.4591	0.4931	0.5266	0.5594	0.5913	0.6221	0.6518	0.6801	0.7071	0.7326	0.7566
40	0.3139	0.3467	0.3801	0.4137	0.4474	0.4809	0.5140	0.5464	0.5781	0.6088	0.6384	0.6669	0.6940	0.7198	0.7442
42	0.3058	0.3378	0.3704	0.4033	0.4364	0.4693	0.5020	0.5341	0.5655	0.5961	0.6257	0.6542	0.6814	0.7074	0.7321
44	0.2982	0.3294	0.3613	0.3935	0.4260	0.4584	0.4906	0.5224	0.5535	0.5839	0.6134	0.6419	0.6693	0.6955	0.7203
46	0.2912	0.3216	0.3528	0.3844	0.4162	0.4481	0.4798	0.5112	0.5421	0.5723	0.6017	0.6301	0.6576	0.6838	0.7089
48	0.2846	0.3143	0.3447	0.3757	0.4070	0.4383	0.4696	0.5006	0.5312	0.5612	0.5904	0.6188	0.6462	0.6726	0.6978
50	0.2784	0.3074	0.3372	0.3675	0.3982	0.4291	0.4599	0.4905	0.5208	0.5505	0.5796	0.6079	0.6353	0.6617	0.6871
55	0.2644	0.2918	0.3201	0.3489	0.3782	0.4078	0.4375	0.4672	0.4966	0.5257	0.5543	0.5823	0.6096	0.6360	0.6615
60	0.2523	0.2783	0.3051	0.3326	0.3607	0.3891	0.4177	0.4464	0.4750	0.5034	0.5314	0.5589	0.5859	0.6122	0.6377
65	0.2416	0.2663	0.2919	0.3182	0.3450	0.3723	0.3999	0.4276	0.4554	0.4831	0.5105	0.5376	0.5642	0.5902	0.6156
70	0.2322	0.2558	0.2802	0.3053	0.3311	0.3573	0.3839	0.4107	0.4377	0.4646	0.4914	0.5179	0.5441	0.5698	0.5950
75	0.2238	0.2463	0.2697	0.2938	0.3185	0.3437	0.3694	0.3954	0.4215	0.4477	0.4738	0.4998	0.5256	0.5509	0.5759
80	0.2163	0.2378	0.2602	0.2834	0.3071	0.3315	0.3563	0.3814	0.4067	0.4322	0.4577	0.4831	0.5084	0.5334	0.5580
85	0.2095	0.2302	0.2517	0.2739	0.2968	0.3203	0.3442	0.3686	0.3932	0.4180	0.4428	0.4677	0.4925	0.5170	0.5413
90	0.2033	0.2232	0.2439	0.2653	0.2874	0.3101	0.3332	0.3568	0.3807	0.4048	0.4291	0.4534	0.4776	0.5017	0.5256
95	0.1977	0.2168	0.2368	0.2574	0.2788	0.3007	0.3231	0.3460	0.3692	0.3927	0.4163	0.4401	0.4638	0.4875	0.5110
100	0.1926	0.2110	0.2302	0.2502	0.2708	0.2920	0.3138	0.3360	0.3585	0.3814	0.4044	0.4276	0.4509	0.4741	0.4972

ALPHA = .050

N	LAMBDA= 42.00	44.00	46.00	48.00	50.00	55.00	60.00	65.00	70.00	75.00	80.00	85.00	90.00	95.00	100.00
1	1.0000	1.0000	1.0000	1.0000	1.0000	1.0000	1.0000	1.0000	1.0000	1.0000	1.0000	1.0000	1.0000	1.0000	1.0000
2	1.0000	1.0000	1.0000	1.0000	1.0000	1.0000	1.0000	1.0000	1.0000	1.0000	1.0000	1.0000	1.0000	1.0000	1.0000
3	1.0000	1.0000	1.0000	1.0000	1.0000	1.0000	1.0000	1.0000	1.0000	1.0000	1.0000	1.0000	1.0000	1.0000	1.0000
4	0.9999	0.9999	1.0000	1.0000	1.0000	1.0000	1.0000	1.0000	1.0000	1.0000	1.0000	1.0000	1.0000	1.0000	1.0000
5	0.9998	0.9999	0.9999	1.0000	1.0000	1.0000	1.0000	1.0000	1.0000	1.0000	1.0000	1.0000	1.0000	1.0000	1.0000
6	0.9997	0.9998	0.9999	0.9999	1.0000	1.0000	1.0000	1.0000	1.0000	1.0000	1.0000	1.0000	1.0000	1.0000	1.0000
7	0.9996	0.9998	0.9999	0.9999	1.0000	1.0000	1.0000	1.0000	1.0000	1.0000	1.0000	1.0000	1.0000	1.0000	1.0000
8	0.9994	0.9996	0.9998	0.9999	0.9999	1.0000	1.0000	1.0000	1.0000	1.0000	1.0000	1.0000	1.0000	1.0000	1.0000
9	0.9992	0.9995	0.9997	0.9998	0.9999	1.0000	1.0000	1.0000	1.0000	1.0000	1.0000	1.0000	1.0000	1.0000	1.0000
10	0.9989	0.9993	0.9996	0.9997	0.9998	1.0000	1.0000	1.0000	1.0000	1.0000	1.0000	1.0000	1.0000	1.0000	1.0000
11	0.9986	0.9991	0.9994	0.9997	0.9998	0.9999	1.0000	1.0000	1.0000	1.0000	1.0000	1.0000	1.0000	1.0000	1.0000
12	0.9982	0.9988	0.9993	0.9995	0.9997	0.9999	1.0000	1.0000	1.0000	1.0000	1.0000	1.0000	1.0000	1.0000	1.0000
13	0.9977	0.9985	0.9991	0.9994	0.9996	0.9999	1.0000	1.0000	1.0000	1.0000	1.0000	1.0000	1.0000	1.0000	1.0000
14	0.9972	0.9982	0.9989	0.9993	0.9995	0.9999	1.0000	1.0000	1.0000	1.0000	1.0000	1.0000	1.0000	1.0000	1.0000
15	0.9967	0.9978	0.9986	0.9991	0.9994	0.9998	0.9999	1.0000	1.0000	1.0000	1.0000	1.0000	1.0000	1.0000	1.0000
16	0.9961	0.9974	0.9983	0.9989	0.9993	0.9998	0.9999	1.0000	1.0000	1.0000	1.0000	1.0000	1.0000	1.0000	1.0000
17	0.9954	0.9969	0.9980	0.9987	0.9991	0.9997	0.9999	1.0000	1.0000	1.0000	1.0000	1.0000	1.0000	1.0000	1.0000
18	0.9946	0.9964	0.9976	0.9984	0.9990	0.9997	0.9999	1.0000	1.0000	1.0000	1.0000	1.0000	1.0000	1.0000	1.0000
19	0.9938	0.9958	0.9972	0.9981	0.9988	0.9996	0.9999	1.0000	1.0000	1.0000	1.0000	1.0000	1.0000	1.0000	1.0000
20	0.9930	0.9952	0.9968	0.9978	0.9986	0.9995	0.9998	0.9999	1.0000	1.0000	1.0000	1.0000	1.0000	1.0000	1.0000
21	0.9921	0.9945	0.9963	0.9975	0.9983	0.9994	0.9998	0.9999	1.0000	1.0000	1.0000	1.0000	1.0000	1.0000	1.0000
22	0.9911	0.9938	0.9958	0.9971	0.9980	0.9993	0.9998	0.9999	1.0000	1.0000	1.0000	1.0000	1.0000	1.0000	1.0000
23	0.9900	0.9930	0.9952	0.9967	0.9978	0.9992	0.9997	0.9999	1.0000	1.0000	1.0000	1.0000	1.0000	1.0000	1.0000
24	0.9889	0.9922	0.9946	0.9963	0.9974	0.9990	0.9997	0.9999	1.0000	1.0000	1.0000	1.0000	1.0000	1.0000	1.0000
25	0.9878	0.9913	0.9939	0.9958	0.9971	0.9989	0.9996	0.9999	1.0000	1.0000	1.0000	1.0000	1.0000	1.0000	1.0000
26	0.9865	0.9904	0.9933	0.9953	0.9967	0.9987	0.9995	0.9998	0.9999	1.0000	1.0000	1.0000	1.0000	1.0000	1.0000
27	0.9852	0.9895	0.9925	0.9948	0.9963	0.9986	0.9995	0.9998	0.9999	1.0000	1.0000	1.0000	1.0000	1.0000	1.0000
28	0.9839	0.9884	0.9918	0.9942	0.9959	0.9984	0.9994	0.9998	0.9999	1.0000	1.0000	1.0000	1.0000	1.0000	1.0000
29	0.9825	0.9874	0.9909	0.9936	0.9955	0.9982	0.9993	0.9997	0.9999	1.0000	1.0000	1.0000	1.0000	1.0000	1.0000
30	0.9811	0.9862	0.9901	0.9929	0.9950	0.9980	0.9992	0.9997	0.9999	1.0000	1.0000	1.0000	1.0000	1.0000	1.0000
32	0.9780	0.9839	0.9883	0.9915	0.9940	0.9975	0.9990	0.9996	0.9999	0.9999	1.0000	1.0000	1.0000	1.0000	1.0000
34	0.9748	0.9813	0.9863	0.9900	0.9928	0.9969	0.9987	0.9995	0.9998	0.9999	1.0000	1.0000	1.0000	1.0000	1.0000
36	0.9713	0.9786	0.9842	0.9884	0.9915	0.9963	0.9984	0.9994	0.9998	0.9999	1.0000	1.0000	1.0000	1.0000	1.0000
38	0.9677	0.9757	0.9819	0.9866	0.9901	0.9956	0.9981	0.9992	0.9997	0.9999	1.0000	1.0000	1.0000	1.0000	1.0000
40	0.9640	0.9727	0.9794	0.9847	0.9886	0.9948	0.9977	0.9991	0.9996	0.9999	0.9999	1.0000	1.0000	1.0000	1.0000
42	0.9601	0.9695	0.9769	0.9826	0.9870	0.9940	0.9973	0.9989	0.9995	0.9998	0.9999	1.0000	1.0000	1.0000	1.0000
44	0.9560	0.9661	0.9742	0.9804	0.9853	0.9931	0.9969	0.9986	0.9994	0.9998	0.9999	1.0000	1.0000	1.0000	1.0000
46	0.9518	0.9627	0.9713	0.9782	0.9835	0.9921	0.9963	0.9984	0.9993	0.9997	0.9999	1.0000	1.0000	1.0000	1.0000
48	0.9475	0.9591	0.9684	0.9758	0.9816	0.9910	0.9958	0.9981	0.9992	0.9997	0.9999	0.9999	1.0000	1.0000	1.0000
50	0.9431	0.9554	0.9653	0.9733	0.9795	0.9898	0.9952	0.9978	0.9990	0.9996	0.9998	0.9999	1.0000	1.0000	1.0000
55	0.9317	0.9457	0.9572	0.9665	0.9740	0.9866	0.9934	0.9969	0.9986	0.9994	0.9997	0.9999	1.0000	1.0000	1.0000
60	0.9198	0.9355	0.9485	0.9592	0.9679	0.9830	0.9913	0.9957	0.9980	0.9991	0.9996	0.9998	0.9999	1.0000	1.0000
65	0.9076	0.9249	0.9394	0.9514	0.9614	0.9789	0.9889	0.9944	0.9973	0.9987	0.9994	0.9997	0.9999	1.0000	1.0000
70	0.8952	0.9139	0.9298	0.9432	0.9544	0.9743	0.9861	0.9928	0.9964	0.9982	0.9992	0.9996	0.9998	1.0000	1.0000
75	0.8826	0.9027	0.9200	0.9346	0.9470	0.9694	0.9831	0.9910	0.9953	0.9977	0.9989	0.9995	0.9998	0.9999	1.0000
80	0.8700	0.8914	0.9099	0.9257	0.9392	0.9642	0.9797	0.9889	0.9941	0.9970	0.9985	0.9993	0.9997	0.9999	1.0000
85	0.8574	0.8800	0.8996	0.9166	0.9312	0.9586	0.9760	0.9866	0.9928	0.9962	0.9981	0.9991	0.9996	0.9998	0.9999
90	0.8449	0.8685	0.8892	0.9073	0.9230	0.9528	0.9721	0.9841	0.9912	0.9953	0.9976	0.9988	0.9994	0.9997	0.9999
95	0.8324	0.8570	0.8788	0.8979	0.9145	0.9466	0.9679	0.9813	0.9895	0.9943	0.9970	0.9985	0.9992	0.9996	0.9998
100	0.8201	0.8456	0.8683	0.8884	0.9059	0.9403	0.9635	0.9784	0.9876	0.9931	0.9963	0.9981	0.9990	0.9995	0.9998

ALPHA = .050

N	LAMBDA= 26.00	27.00	28.00	29.00	30.00	31.00	32.00	33.00	34.00	35.00	36.00	37.00	38.00	39.00	40.00
1	0.9992	0.9994	0.9996	0.9997	0.9998	0.9998	0.9999	0.9999	0.9999	1.0000	1.0000	1.0000	1.0000	1.0000	1.0000
2	0.9974	0.9981	0.9986	0.9989	0.9992	0.9994	0.9996	0.9997	0.9998	0.9998	0.9999	0.9999	0.9999	1.0000	1.0000
3	0.9949	0.9961	0.9971	0.9978	0.9983	0.9988	0.9991	0.9993	0.9995	0.9996	0.9997	0.9998	0.9998	0.9999	0.9999
4	0.9918	0.9936	0.9951	0.9962	0.9971	0.9978	0.9983	0.9987	0.9990	0.9993	0.9995	0.9996	0.9997	0.9998	0.9998
5	0.9881	0.9907	0.9927	0.9943	0.9956	0.9966	0.9974	0.9980	0.9984	0.9988	0.9991	0.9993	0.9995	0.9996	0.9997
6	0.9839	0.9872	0.9899	0.9920	0.9937	0.9951	0.9962	0.9970	0.9977	0.9982	0.9986	0.9989	0.9992	0.9994	0.9995
7	0.9792	0.9834	0.9867	0.9894	0.9916	0.9934	0.9948	0.9959	0.9968	0.9975	0.9980	0.9985	0.9988	0.9991	0.9993
8	0.9742	0.9791	0.9832	0.9865	0.9892	0.9914	0.9931	0.9945	0.9957	0.9966	0.9973	0.9979	0.9983	0.9987	0.9990
9	0.9689	0.9746	0.9793	0.9833	0.9865	0.9891	0.9913	0.9930	0.9944	0.9956	0.9965	0.9972	0.9978	0.9983	0.9986
10	0.9633	0.9698	0.9752	0.9798	0.9835	0.9866	0.9892	0.9913	0.9930	0.9944	0.9955	0.9964	0.9971	0.9977	0.9982
11	0.9574	0.9647	0.9709	0.9760	0.9804	0.9839	0.9869	0.9894	0.9914	0.9931	0.9944	0.9955	0.9964	0.9971	0.9977
12	0.9513	0.9594	0.9663	0.9721	0.9770	0.9811	0.9845	0.9873	0.9897	0.9916	0.9932	0.9945	0.9956	0.9964	0.9971
13	0.9450	0.9539	0.9615	0.9679	0.9734	0.9780	0.9818	0.9851	0.9878	0.9900	0.9918	0.9934	0.9946	0.9957	0.9965
14	0.9386	0.9482	0.9565	0.9636	0.9696	0.9747	0.9790	0.9827	0.9857	0.9883	0.9904	0.9921	0.9936	0.9948	0.9958
15	0.9320	0.9424	0.9514	0.9591	0.9657	0.9713	0.9761	0.9801	0.9835	0.9864	0.9888	0.9908	0.9925	0.9938	0.9950
16	0.9253	0.9364	0.9461	0.9544	0.9616	0.9677	0.9730	0.9774	0.9812	0.9844	0.9871	0.9893	0.9912	0.9928	0.9941
17	0.9186	0.9304	0.9407	0.9496	0.9574	0.9640	0.9697	0.9746	0.9788	0.9823	0.9853	0.9878	0.9899	0.9917	0.9931
18	0.9118	0.9243	0.9352	0.9447	0.9530	0.9602	0.9664	0.9717	0.9762	0.9801	0.9834	0.9861	0.9885	0.9905	0.9921
19	0.9050	0.9181	0.9296	0.9397	0.9486	0.9562	0.9629	0.9686	0.9735	0.9777	0.9813	0.9844	0.9870	0.9892	0.9910
20	0.8981	0.9118	0.9240	0.9346	0.9440	0.9522	0.9593	0.9654	0.9707	0.9753	0.9792	0.9825	0.9854	0.9878	0.9898
21	0.8912	0.9055	0.9182	0.9295	0.9393	0.9480	0.9556	0.9621	0.9678	0.9727	0.9770	0.9806	0.9837	0.9863	0.9886
22	0.8843	0.8992	0.9125	0.9242	0.9346	0.9438	0.9518	0.9588	0.9648	0.9701	0.9747	0.9786	0.9819	0.9848	0.9872
23	0.8774	0.8929	0.9067	0.9189	0.9298	0.9394	0.9479	0.9553	0.9618	0.9674	0.9722	0.9765	0.9801	0.9832	0.9858
24	0.8706	0.8865	0.9008	0.9136	0.9250	0.9351	0.9439	0.9518	0.9586	0.9646	0.9698	0.9743	0.9781	0.9815	0.9844
25	0.8637	0.8802	0.8950	0.9082	0.9201	0.9306	0.9399	0.9481	0.9554	0.9617	0.9672	0.9720	0.9761	0.9797	0.9828
26	0.8569	0.8738	0.8891	0.9028	0.9151	0.9261	0.9358	0.9445	0.9520	0.9587	0.9645	0.9696	0.9741	0.9779	0.9812
27	0.8501	0.8675	0.8832	0.8974	0.9102	0.9216	0.9317	0.9407	0.9487	0.9557	0.9618	0.9672	0.9719	0.9760	0.9795
28	0.8434	0.8612	0.8774	0.8920	0.9051	0.9170	0.9275	0.9369	0.9452	0.9526	0.9591	0.9647	0.9697	0.9740	0.9778
29	0.8367	0.8549	0.8715	0.8865	0.9001	0.9123	0.9233	0.9330	0.9417	0.9494	0.9562	0.9622	0.9674	0.9720	0.9760
30	0.8300	0.8487	0.8656	0.8811	0.8951	0.9077	0.9190	0.9291	0.9382	0.9462	0.9533	0.9596	0.9651	0.9699	0.9742
32	0.8169	0.8363	0.8540	0.8702	0.8849	0.8983	0.9104	0.9212	0.9309	0.9396	0.9473	0.9542	0.9603	0.9656	0.9703
34	0.8041	0.8241	0.8425	0.8594	0.8748	0.8889	0.9016	0.9132	0.9235	0.9329	0.9412	0.9486	0.9552	0.9611	0.9662
36	0.7915	0.8120	0.8310	0.8486	0.8647	0.8794	0.8928	0.9050	0.9160	0.9259	0.9349	0.9429	0.9500	0.9563	0.9620
38	0.7792	0.8002	0.8198	0.8379	0.8546	0.8699	0.8840	0.8968	0.9084	0.9189	0.9284	0.9369	0.9446	0.9514	0.9575
40	0.7671	0.7886	0.8087	0.8273	0.8446	0.8605	0.8751	0.8885	0.9007	0.9118	0.9218	0.9309	0.9391	0.9464	0.9529
42	0.7554	0.7773	0.7978	0.8169	0.8347	0.8511	0.8663	0.8802	0.8929	0.9046	0.9151	0.9247	0.9334	0.9412	0.9482
44	0.7439	0.7662	0.7871	0.8067	0.8249	0.8418	0.8575	0.8719	0.8852	0.8973	0.9084	0.9185	0.9276	0.9359	0.9433
46	0.7328	0.7553	0.7766	0.7965	0.8152	0.8326	0.8487	0.8636	0.8774	0.8900	0.9016	0.9121	0.9217	0.9304	0.9383
48	0.7219	0.7447	0.7663	0.7866	0.8056	0.8234	0.8400	0.8554	0.8696	0.8827	0.8947	0.9057	0.9158	0.9249	0.9332
50	0.7113	0.7343	0.7562	0.7768	0.7962	0.8144	0.8313	0.8471	0.8618	0.8753	0.8878	0.8992	0.9097	0.9193	0.9280
55	0.6860	0.7095	0.7319	0.7531	0.7733	0.7923	0.8101	0.8268	0.8425	0.8570	0.8705	0.8830	0.8945	0.9050	0.9147
60	0.6623	0.6861	0.7088	0.7306	0.7513	0.7709	0.7895	0.8070	0.8234	0.8388	0.8532	0.8666	0.8790	0.8905	0.9011
65	0.6402	0.6641	0.6871	0.7091	0.7303	0.7504	0.7696	0.7877	0.8048	0.8210	0.8361	0.8503	0.8635	0.8758	0.8873
70	0.6196	0.6434	0.6665	0.6888	0.7102	0.7308	0.7504	0.7690	0.7867	0.8035	0.8193	0.8342	0.8481	0.8612	0.8733
75	0.6003	0.6240	0.6472	0.6695	0.6912	0.7120	0.7319	0.7510	0.7691	0.7864	0.8028	0.8183	0.8329	0.8466	0.8594
80	0.5821	0.6058	0.6289	0.6513	0.6730	0.6940	0.7142	0.7336	0.7521	0.7698	0.7867	0.8027	0.8178	0.8321	0.8456
85	0.5652	0.5886	0.6116	0.6340	0.6557	0.6768	0.6972	0.7168	0.7357	0.7537	0.7710	0.7874	0.8030	0.8178	0.8318
90	0.5492	0.5725	0.5953	0.6176	0.6393	0.6605	0.6809	0.7007	0.7198	0.7382	0.7557	0.7725	0.7886	0.8038	0.8183
95	0.5343	0.5572	0.5799	0.6020	0.6237	0.6448	0.6654	0.6853	0.7045	0.7231	0.7409	0.7581	0.7744	0.7901	0.8049
100	0.5202	0.5429	0.5653	0.5873	0.6088	0.6299	0.6505	0.6704	0.6898	0.7085	0.7266	0.7440	0.7606	0.7766	0.7918

ALPHA = .025

N	LAMBDA= .00	.10	.20	.30	.40	.50	.60	.70	.80	.90	1.00	1.20	1.40	1.60	1.80
1	0.0250	0.0324	0.0400	0.0478	0.0558	0.0641	0.0725	0.0811	0.0899	0.0988	0.1078	0.1263	0.1453	0.1646	0.1843
2	0.0250	0.0297	0.0346	0.0397	0.0449	0.0504	0.0560	0.0618	0.0677	0.0738	0.0800	0.0929	0.1063	0.1202	0.1345
3	0.0250	0.0286	0.0324	0.0363	0.0404	0.0446	0.0490	0.0535	0.0581	0.0629	0.0678	0.0780	0.0886	0.0997	0.1113
4	0.0250	0.0280	0.0312	0.0344	0.0378	0.0413	0.0450	0.0487	0.0526	0.0566	0.0607	0.0692	0.0781	0.0875	0.0973
5	0.0250	0.0276	0.0303	0.0332	0.0361	0.0392	0.0423	0.0455	0.0489	0.0523	0.0559	0.0633	0.0711	0.0793	0.0878
6	0.0250	0.0273	0.0298	0.0323	0.0349	0.0376	0.0404	0.0433	0.0463	0.0493	0.0525	0.0590	0.0660	0.0732	0.0808
7	0.0250	0.0271	0.0293	0.0316	0.0340	0.0364	0.0389	0.0416	0.0442	0.0470	0.0499	0.0558	0.0621	0.0686	0.0755
8	0.0250	0.0270	0.0290	0.0311	0.0332	0.0355	0.0378	0.0402	0.0427	0.0452	0.0478	0.0532	0.0590	0.0650	0.0712
9	0.0250	0.0268	0.0287	0.0306	0.0327	0.0347	0.0369	0.0391	0.0414	0.0437	0.0461	0.0512	0.0564	0.0620	0.0678
10	0.0250	0.0267	0.0285	0.0303	0.0322	0.0341	0.0361	0.0382	0.0403	0.0425	0.0447	0.0494	0.0543	0.0595	0.0649
11	0.0250	0.0266	0.0283	0.0300	0.0318	0.0336	0.0355	0.0374	0.0394	0.0415	0.0436	0.0479	0.0526	0.0574	0.0625
12	0.0250	0.0265	0.0281	0.0297	0.0314	0.0331	0.0349	0.0367	0.0386	0.0406	0.0426	0.0467	0.0510	0.0556	0.0604
13	0.0250	0.0265	0.0280	0.0295	0.0311	0.0327	0.0344	0.0362	0.0380	0.0398	0.0417	0.0456	0.0497	0.0540	0.0585
14	0.0250	0.0264	0.0278	0.0293	0.0308	0.0324	0.0340	0.0357	0.0374	0.0391	0.0409	0.0446	0.0485	0.0526	0.0569
15	0.0250	0.0263	0.0277	0.0291	0.0306	0.0321	0.0336	0.0352	0.0368	0.0385	0.0402	0.0438	0.0475	0.0514	0.0555
16	0.0250	0.0263	0.0276	0.0290	0.0304	0.0318	0.0333	0.0348	0.0364	0.0380	0.0396	0.0430	0.0466	0.0503	0.0542
17	0.0250	0.0262	0.0275	0.0288	0.0302	0.0316	0.0330	0.0344	0.0359	0.0375	0.0391	0.0423	0.0457	0.0493	0.0531
18	0.0250	0.0262	0.0274	0.0287	0.0300	0.0313	0.0327	0.0341	0.0356	0.0370	0.0386	0.0417	0.0450	0.0484	0.0520
19	0.0250	0.0262	0.0274	0.0286	0.0298	0.0311	0.0325	0.0338	0.0352	0.0366	0.0381	0.0411	0.0443	0.0476	0.0511
20	0.0250	0.0261	0.0273	0.0285	0.0297	0.0309	0.0322	0.0335	0.0349	0.0363	0.0377	0.0406	0.0437	0.0469	0.0502
21	0.0250	0.0261	0.0272	0.0284	0.0296	0.0308	0.0320	0.0333	0.0346	0.0359	0.0373	0.0401	0.0431	0.0462	0.0494
22	0.0250	0.0261	0.0272	0.0283	0.0294	0.0306	0.0318	0.0331	0.0343	0.0356	0.0369	0.0397	0.0426	0.0456	0.0487
23	0.0250	0.0260	0.0271	0.0282	0.0293	0.0305	0.0316	0.0328	0.0341	0.0353	0.0366	0.0393	0.0421	0.0450	0.0480
24	0.0250	0.0260	0.0271	0.0281	0.0292	0.0303	0.0315	0.0326	0.0338	0.0351	0.0363	0.0389	0.0416	0.0444	0.0474
25	0.0250	0.0260	0.0270	0.0280	0.0291	0.0302	0.0313	0.0325	0.0336	0.0348	0.0360	0.0386	0.0412	0.0439	0.0468
26	0.0250	0.0260	0.0270	0.0280	0.0290	0.0301	0.0312	0.0323	0.0334	0.0346	0.0358	0.0382	0.0408	0.0435	0.0462
27	0.0250	0.0259	0.0269	0.0279	0.0289	0.0300	0.0310	0.0321	0.0332	0.0344	0.0355	0.0379	0.0404	0.0430	0.0457
28	0.0250	0.0259	0.0269	0.0278	0.0288	0.0299	0.0309	0.0320	0.0330	0.0342	0.0353	0.0376	0.0401	0.0426	0.0452
29	0.0250	0.0259	0.0268	0.0278	0.0288	0.0298	0.0308	0.0318	0.0329	0.0340	0.0351	0.0373	0.0397	0.0422	0.0448
30	0.0250	0.0259	0.0268	0.0277	0.0287	0.0297	0.0307	0.0317	0.0327	0.0338	0.0349	0.0371	0.0394	0.0418	0.0443
32	0.0250	0.0259	0.0267	0.0276	0.0285	0.0295	0.0304	0.0314	0.0324	0.0334	0.0345	0.0366	0.0388	0.0412	0.0436
34	0.0250	0.0258	0.0267	0.0275	0.0284	0.0293	0.0302	0.0312	0.0321	0.0331	0.0341	0.0362	0.0383	0.0405	0.0428
36	0.0250	0.0258	0.0266	0.0275	0.0283	0.0292	0.0301	0.0310	0.0319	0.0329	0.0338	0.0358	0.0379	0.0400	0.0422
38	0.0250	0.0258	0.0266	0.0274	0.0282	0.0291	0.0299	0.0308	0.0317	0.0326	0.0335	0.0354	0.0374	0.0395	0.0416
40	0.0250	0.0258	0.0265	0.0273	0.0281	0.0289	0.0298	0.0306	0.0315	0.0324	0.0333	0.0351	0.0370	0.0390	0.0411
42	0.0250	0.0257	0.0265	0.0272	0.0280	0.0288	0.0296	0.0305	0.0313	0.0322	0.0330	0.0348	0.0367	0.0386	0.0406
44	0.0250	0.0257	0.0264	0.0272	0.0279	0.0287	0.0295	0.0303	0.0311	0.0320	0.0328	0.0345	0.0364	0.0382	0.0401
46	0.0250	0.0257	0.0264	0.0271	0.0279	0.0286	0.0294	0.0302	0.0310	0.0318	0.0326	0.0343	0.0360	0.0379	0.0397
48	0.0250	0.0257	0.0264	0.0271	0.0278	0.0285	0.0293	0.0300	0.0308	0.0316	0.0324	0.0341	0.0358	0.0375	0.0393
50	0.0250	0.0257	0.0263	0.0270	0.0277	0.0285	0.0292	0.0299	0.0307	0.0315	0.0322	0.0338	0.0355	0.0372	0.0390
55	0.0250	0.0256	0.0263	0.0269	0.0276	0.0283	0.0290	0.0297	0.0304	0.0311	0.0318	0.0333	0.0349	0.0365	0.0382
60	0.0250	0.0256	0.0262	0.0268	0.0275	0.0281	0.0288	0.0294	0.0301	0.0308	0.0315	0.0329	0.0344	0.0359	0.0375
65	0.0250	0.0256	0.0262	0.0268	0.0274	0.0280	0.0286	0.0292	0.0299	0.0305	0.0312	0.0326	0.0340	0.0354	0.0369
70	0.0250	0.0256	0.0261	0.0267	0.0273	0.0279	0.0284	0.0291	0.0297	0.0303	0.0309	0.0322	0.0336	0.0349	0.0364
75	0.0250	0.0255	0.0261	0.0266	0.0272	0.0277	0.0283	0.0289	0.0295	0.0301	0.0307	0.0319	0.0332	0.0345	0.0359
80	0.0250	0.0255	0.0260	0.0266	0.0271	0.0276	0.0282	0.0288	0.0293	0.0299	0.0305	0.0317	0.0329	0.0342	0.0355
85	0.0250	0.0255	0.0260	0.0265	0.0270	0.0276	0.0281	0.0286	0.0292	0.0297	0.0303	0.0314	0.0326	0.0338	0.0351
90	0.0250	0.0255	0.0260	0.0265	0.0270	0.0275	0.0280	0.0285	0.0290	0.0296	0.0301	0.0312	0.0324	0.0335	0.0347
95	0.0250	0.0255	0.0259	0.0264	0.0269	0.0274	0.0279	0.0284	0.0289	0.0294	0.0300	0.0310	0.0321	0.0333	0.0344
100	0.0250	0.0255	0.0259	0.0264	0.0269	0.0273	0.0278	0.0283	0.0288	0.0293	0.0298	0.0309	0.0319	0.0330	0.0342

ALPHA = .025

N	LAMBDA= 2.00	2.20	2.40	2.60	2.80	3.00	3.50	4.00	4.50	5.00	6.00	7.00	8.00	9.00	10.00
1	0.2042	0.2243	0.2445	0.2647	0.2850	0.3053	0.3555	0.4046	0.4522	0.4979	0.5824	0.6570	0.7214	0.7760	0.8214
2	0.1493	0.1644	0.1798	0.1955	0.2114	0.2276	0.2686	0.3101	0.3517	0.3930	0.4730	0.5481	0.6167	0.6782	0.7324
3	0.1233	0.1356	0.1483	0.1613	0.1746	0.1882	0.2231	0.2592	0.2959	0.3330	0.4070	0.4788	0.5468	0.6098	0.6672
4	0.1075	0.1180	0.1289	0.1400	0.1515	0.1633	0.1939	0.2258	0.2587	0.2923	0.3606	0.4285	0.4944	0.5569	0.6152
5	0.0967	0.1059	0.1154	0.1253	0.1355	0.1459	0.1731	0.2018	0.2316	0.2623	0.3255	0.3896	0.4528	0.5140	0.5721
6	0.0887	0.0970	0.1055	0.1144	0.1235	0.1329	0.1575	0.1835	0.2108	0.2390	0.2978	0.3582	0.4188	0.4782	0.5354
7	0.0827	0.0901	0.0979	0.1059	0.1142	0.1227	0.1452	0.1691	0.1942	0.2204	0.2753	0.3324	0.3902	0.4477	0.5038
8	0.0778	0.0846	0.0917	0.0991	0.1067	0.1146	0.1353	0.1573	0.1807	0.2051	0.2565	0.3106	0.3659	0.4214	0.4760
9	0.0738	0.0802	0.0867	0.0935	0.1006	0.1079	0.1271	0.1476	0.1693	0.1922	0.2406	0.2919	0.3448	0.3983	0.4515
10	0.0705	0.0764	0.0825	0.0889	0.0954	0.1022	0.1202	0.1394	0.1598	0.1812	0.2270	0.2757	0.3263	0.3780	0.4297
11	0.0677	0.0732	0.0790	0.0849	0.0911	0.0974	0.1142	0.1323	0.1515	0.1718	0.2151	0.2615	0.3100	0.3599	0.4101
12	0.0653	0.0705	0.0759	0.0815	0.0873	0.0933	0.1091	0.1262	0.1443	0.1635	0.2047	0.2490	0.2955	0.3436	0.3924
13	0.0632	0.0681	0.0732	0.0785	0.0840	0.0897	0.1047	0.1208	0.1380	0.1562	0.1954	0.2378	0.2825	0.3290	0.3763
14	0.0614	0.0660	0.0709	0.0759	0.0811	0.0865	0.1007	0.1160	0.1324	0.1498	0.1872	0.2278	0.2708	0.3157	0.3616
15	0.0597	0.0642	0.0688	0.0735	0.0785	0.0836	0.0972	0.1118	0.1274	0.1440	0.1798	0.2187	0.2602	0.3036	0.3482
16	0.0583	0.0625	0.0669	0.0715	0.0762	0.0811	0.0940	0.1080	0.1229	0.1388	0.1731	0.2105	0.2505	0.2925	0.3358
17	0.0569	0.0610	0.0652	0.0696	0.0741	0.0788	0.0912	0.1046	0.1189	0.1341	0.1670	0.2030	0.2416	0.2823	0.3244
18	0.0558	0.0596	0.0637	0.0679	0.0722	0.0767	0.0886	0.1014	0.1152	0.1298	0.1615	0.1962	0.2335	0.2729	0.3139
19	0.0547	0.0584	0.0623	0.0663	0.0705	0.0748	0.0863	0.0986	0.1118	0.1259	0.1564	0.1899	0.2260	0.2642	0.3041
20	0.0537	0.0573	0.0610	0.0649	0.0689	0.0731	0.0841	0.0960	0.1087	0.1223	0.1517	0.1841	0.2190	0.2561	0.2949
21	0.0528	0.0562	0.0599	0.0636	0.0675	0.0715	0.0821	0.0936	0.1059	0.1189	0.1474	0.1787	0.2126	0.2486	0.2864
22	0.0519	0.0553	0.0588	0.0624	0.0662	0.0700	0.0803	0.0914	0.1032	0.1159	0.1434	0.1737	0.2066	0.2416	0.2784
23	0.0511	0.0544	0.0578	0.0613	0.0649	0.0687	0.0786	0.0893	0.1008	0.1130	0.1397	0.1691	0.2010	0.2351	0.2710
24	0.0504	0.0536	0.0569	0.0603	0.0638	0.0674	0.0771	0.0874	0.0985	0.1104	0.1362	0.1648	0.1958	0.2289	0.2639
25	0.0497	0.0528	0.0560	0.0593	0.0627	0.0663	0.0756	0.0857	0.0964	0.1079	0.1330	0.1607	0.1909	0.2232	0.2573
26	0.0491	0.0521	0.0552	0.0584	0.0617	0.0652	0.0742	0.0840	0.0945	0.1056	0.1300	0.1569	0.1863	0.2178	0.2511
27	0.0485	0.0514	0.0544	0.0576	0.0608	0.0641	0.0730	0.0825	0.0926	0.1035	0.1272	0.1534	0.1820	0.2127	0.2452
28	0.0480	0.0508	0.0537	0.0568	0.0599	0.0632	0.0718	0.0810	0.0909	0.1015	0.1245	0.1500	0.1779	0.2078	0.2397
29	0.0474	0.0502	0.0531	0.0560	0.0591	0.0623	0.0706	0.0796	0.0893	0.0996	0.1220	0.1469	0.1740	0.2033	0.2344
30	0.0469	0.0496	0.0524	0.0553	0.0583	0.0614	0.0696	0.0783	0.0877	0.0978	0.1196	0.1439	0.1704	0.1990	0.2294
32	0.0460	0.0486	0.0513	0.0541	0.0569	0.0598	0.0676	0.0760	0.0849	0.0945	0.1153	0.1384	0.1637	0.1910	0.2201
34	0.0452	0.0477	0.0503	0.0529	0.0556	0.0584	0.0659	0.0738	0.0824	0.0915	0.1114	0.1334	0.1576	0.1838	0.2117
36	0.0445	0.0469	0.0493	0.0519	0.0545	0.0572	0.0643	0.0719	0.0801	0.0888	0.1078	0.1290	0.1521	0.1772	0.2040
38	0.0438	0.0461	0.0485	0.0509	0.0534	0.0560	0.0628	0.0702	0.0780	0.0864	0.1046	0.1249	0.1471	0.1712	0.1970
40	0.0432	0.0454	0.0477	0.0500	0.0525	0.0550	0.0615	0.0686	0.0761	0.0842	0.1017	0.1211	0.1425	0.1657	0.1906
42	0.0427	0.0448	0.0470	0.0493	0.0516	0.0540	0.0603	0.0671	0.0744	0.0821	0.0990	0.1177	0.1383	0.1606	0.1847
44	0.0421	0.0442	0.0463	0.0485	0.0508	0.0531	0.0592	0.0658	0.0728	0.0802	0.0965	0.1146	0.1344	0.1560	0.1792
46	0.0417	0.0437	0.0457	0.0478	0.0500	0.0523	0.0582	0.0646	0.0713	0.0785	0.0942	0.1117	0.1308	0.1517	0.1741
48	0.0412	0.0432	0.0452	0.0472	0.0493	0.0515	0.0573	0.0634	0.0700	0.0769	0.0921	0.1090	0.1275	0.1477	0.1694
50	0.0408	0.0427	0.0446	0.0466	0.0487	0.0508	0.0564	0.0623	0.0687	0.0754	0.0901	0.1065	0.1244	0.1439	0.1650
55	0.0399	0.0416	0.0435	0.0453	0.0473	0.0492	0.0544	0.0600	0.0659	0.0721	0.0858	0.1009	0.1175	0.1356	0.1552
60	0.0391	0.0408	0.0425	0.0442	0.0460	0.0479	0.0528	0.0579	0.0635	0.0693	0.0820	0.0962	0.1117	0.1285	0.1468
65	0.0384	0.0400	0.0416	0.0433	0.0450	0.0467	0.0513	0.0562	0.0614	0.0669	0.0788	0.0921	0.1066	0.1224	0.1395
70	0.0378	0.0393	0.0408	0.0424	0.0440	0.0457	0.0500	0.0547	0.0596	0.0648	0.0760	0.0885	0.1022	0.1171	0.1332
75	0.0373	0.0387	0.0402	0.0417	0.0432	0.0448	0.0489	0.0533	0.0580	0.0629	0.0736	0.0854	0.0983	0.1124	0.1276
80	0.0368	0.0382	0.0396	0.0410	0.0425	0.0440	0.0479	0.0521	0.0566	0.0612	0.0714	0.0826	0.0948	0.1082	0.1226
85	0.0364	0.0377	0.0390	0.0404	0.0418	0.0433	0.0471	0.0511	0.0553	0.0598	0.0694	0.0801	0.0918	0.1045	0.1182
90	0.0360	0.0372	0.0385	0.0399	0.0412	0.0426	0.0462	0.0501	0.0541	0.0584	0.0677	0.0778	0.0890	0.1011	0.1142
95	0.0356	0.0368	0.0381	0.0394	0.0407	0.0420	0.0455	0.0492	0.0531	0.0572	0.0661	0.0758	0.0865	0.0981	0.1106
100	0.0353	0.0365	0.0377	0.0389	0.0402	0.0415	0.0448	0.0484	0.0522	0.0561	0.0646	0.0740	0.0842	0.0953	0.1073

ALPHA = .025

N	LAMBDA= 11.00	12.00	13.00	14.00	15.00	16.00	17.00	18.00	19.00	20.00	21.00	22.00	23.00	24.00	25.00
1	0.8589	0.8893	0.9137	0.9332	0.9486	0.9607	0.9701	0.9773	0.9829	0.9872	0.9904	0.9928	0.9947	0.9961	0.9971
2	0.7794	0.8196	0.8535	0.8819	0.9054	0.9247	0.9403	0.9530	0.9632	0.9713	0.9777	0.9828	0.9867	0.9898	0.9922
3	0.7185	0.7639	0.8034	0.8374	0.8665	0.8910	0.9115	0.9286	0.9427	0.9542	0.9636	0.9712	0.9773	0.9822	0.9861
4	0.6686	0.7163	0.7598	0.7978	0.8308	0.8593	0.8837	0.9044	0.9219	0.9364	0.9486	0.9585	0.9667	0.9734	0.9789
5	0.6262	0.6761	0.7213	0.7619	0.7979	0.8295	0.8570	0.8807	0.9011	0.9184	0.9330	0.9452	0.9554	0.9639	0.9708
6	0.5896	0.6402	0.6868	0.7292	0.7674	0.8014	0.8314	0.8577	0.8805	0.9002	0.9171	0.9314	0.9435	0.9536	0.9621
7	0.5575	0.6083	0.6556	0.6993	0.7390	0.7749	0.8070	0.8354	0.8604	0.8822	0.9010	0.9173	0.9311	0.9429	0.9529
8	0.5290	0.5796	0.6273	0.6717	0.7127	0.7500	0.7837	0.8139	0.8407	0.8643	0.8850	0.9030	0.9185	0.9319	0.9433
9	0.5036	0.5538	0.6015	0.6464	0.6881	0.7265	0.7615	0.7932	0.8215	0.8468	0.8691	0.8887	0.9057	0.9205	0.9333
10	0.4807	0.5302	0.5778	0.6228	0.6651	0.7043	0.7404	0.7732	0.8030	0.8296	0.8534	0.8744	0.8929	0.9090	0.9231
11	0.4600	0.5088	0.5560	0.6010	0.6436	0.6834	0.7202	0.7541	0.7850	0.8129	0.8379	0.8602	0.8800	0.8974	0.9127
12	0.4411	0.4891	0.5358	0.5807	0.6234	0.6636	0.7011	0.7358	0.7676	0.7965	0.8227	0.8462	0.8672	0.8858	0.9022
13	0.4239	0.4710	0.5171	0.5617	0.6044	0.6449	0.6828	0.7182	0.7508	0.7807	0.8078	0.8324	0.8544	0.8741	0.8916
14	0.4081	0.4543	0.4998	0.5440	0.5866	0.6271	0.6654	0.7013	0.7345	0.7652	0.7933	0.8188	0.8418	0.8625	0.8810
15	0.3935	0.4388	0.4836	0.5274	0.5698	0.6103	0.6488	0.6851	0.7189	0.7502	0.7791	0.8054	0.8294	0.8510	0.8703
16	0.3800	0.4244	0.4685	0.5118	0.5539	0.5944	0.6330	0.6695	0.7038	0.7357	0.7652	0.7923	0.8171	0.8395	0.8598
17	0.3675	0.4110	0.4544	0.4972	0.5389	0.5792	0.6179	0.6546	0.6893	0.7216	0.7517	0.7795	0.8050	0.8282	0.8492
18	0.3559	0.3985	0.4411	0.4833	0.5247	0.5648	0.6035	0.6403	0.6752	0.7080	0.7386	0.7670	0.7931	0.8170	0.8388
19	0.3451	0.3868	0.4287	0.4703	0.5112	0.5511	0.5897	0.6266	0.6617	0.6948	0.7258	0.7547	0.7814	0.8060	0.8284
20	0.3350	0.3758	0.4170	0.4580	0.4985	0.5381	0.5765	0.6134	0.6486	0.6820	0.7134	0.7427	0.7700	0.7951	0.8182
21	0.3255	0.3655	0.4059	0.4463	0.4864	0.5256	0.5638	0.6007	0.6360	0.6696	0.7013	0.7310	0.7587	0.7844	0.8081
22	0.3166	0.3558	0.3955	0.4353	0.4748	0.5138	0.5517	0.5885	0.6239	0.6576	0.6895	0.7196	0.7477	0.7739	0.7981
23	0.3083	0.3466	0.3856	0.4248	0.4639	0.5024	0.5402	0.5768	0.6121	0.6460	0.6781	0.7085	0.7370	0.7635	0.7882
24	0.3004	0.3380	0.3763	0.4148	0.4534	0.4916	0.5290	0.5655	0.6008	0.6347	0.6670	0.6976	0.7264	0.7534	0.7785
25	0.2930	0.3298	0.3674	0.4054	0.4434	0.4812	0.5184	0.5547	0.5899	0.6238	0.6562	0.6870	0.7161	0.7434	0.7689
26	0.2860	0.3221	0.3590	0.3963	0.4339	0.4712	0.5081	0.5442	0.5793	0.6132	0.6457	0.6767	0.7060	0.7336	0.7595
27	0.2793	0.3147	0.3509	0.3878	0.4248	0.4617	0.4983	0.5341	0.5691	0.6030	0.6355	0.6666	0.6962	0.7241	0.7503
28	0.2731	0.3077	0.3433	0.3795	0.4161	0.4526	0.4888	0.5244	0.5592	0.5930	0.6256	0.6568	0.6865	0.7147	0.7411
29	0.2671	0.3011	0.3360	0.3717	0.4077	0.4438	0.4797	0.5151	0.5497	0.5834	0.6160	0.6472	0.6771	0.7054	0.7322
30	0.2614	0.2947	0.3291	0.3642	0.3997	0.4354	0.4709	0.5060	0.5405	0.5741	0.6066	0.6379	0.6679	0.6964	0.7234
32	0.2509	0.2830	0.3162	0.3502	0.3847	0.4195	0.4543	0.4889	0.5229	0.5562	0.5886	0.6200	0.6501	0.6789	0.7063
34	0.2413	0.2722	0.3043	0.3373	0.3709	0.4048	0.4389	0.4728	0.5064	0.5394	0.5716	0.6029	0.6331	0.6621	0.6898
36	0.2325	0.2623	0.2934	0.3254	0.3580	0.3912	0.4245	0.4578	0.4909	0.5235	0.5555	0.5867	0.6169	0.6460	0.6739
38	0.2245	0.2533	0.2833	0.3143	0.3461	0.3785	0.4111	0.4438	0.4764	0.5086	0.5403	0.5713	0.6014	0.6306	0.6586
40	0.2171	0.2449	0.2740	0.3041	0.3351	0.3666	0.3985	0.4306	0.4626	0.4944	0.5258	0.5566	0.5866	0.6158	0.6439
42	0.2102	0.2372	0.2654	0.2947	0.3248	0.3555	0.3867	0.4182	0.4497	0.4811	0.5121	0.5426	0.5725	0.6016	0.6298
44	0.2039	0.2300	0.2574	0.2858	0.3151	0.3452	0.3757	0.4065	0.4375	0.4684	0.4990	0.5293	0.5590	0.5880	0.6162
46	0.1980	0.2234	0.2499	0.2776	0.3061	0.3354	0.3653	0.3955	0.4259	0.4564	0.4866	0.5166	0.5461	0.5749	0.6030
48	0.1926	0.2171	0.2429	0.2698	0.2977	0.3263	0.3555	0.3851	0.4150	0.4450	0.4748	0.5045	0.5337	0.5624	0.5904
50	0.1875	0.2113	0.2364	0.2626	0.2897	0.3177	0.3462	0.3753	0.4046	0.4341	0.4636	0.4929	0.5219	0.5504	0.5783
55	0.1761	0.1983	0.2218	0.2463	0.2718	0.2982	0.3253	0.3529	0.3810	0.4093	0.4377	0.4661	0.4944	0.5223	0.5499
60	0.1663	0.1871	0.2091	0.2322	0.2563	0.2812	0.3069	0.3332	0.3601	0.3872	0.4147	0.4422	0.4696	0.4969	0.5240
65	0.1579	0.1774	0.1981	0.2199	0.2426	0.2663	0.2907	0.3158	0.3415	0.3676	0.3940	0.4206	0.4472	0.4738	0.5003
70	0.1505	0.1689	0.1885	0.2091	0.2306	0.2531	0.2763	0.3003	0.3249	0.3499	0.3753	0.4010	0.4269	0.4528	0.4786
75	0.1439	0.1614	0.1799	0.1995	0.2199	0.2413	0.2635	0.2864	0.3099	0.3340	0.3585	0.3833	0.4084	0.4336	0.4588
80	0.1381	0.1547	0.1723	0.1909	0.2104	0.2308	0.2520	0.2739	0.2964	0.3196	0.3432	0.3672	0.3915	0.4159	0.4405
85	0.1329	0.1487	0.1655	0.1832	0.2018	0.2213	0.2415	0.2625	0.2842	0.3065	0.3292	0.3524	0.3759	0.3997	0.4236
90	0.1283	0.1433	0.1593	0.1762	0.1940	0.2126	0.2321	0.2522	0.2731	0.2945	0.3165	0.3389	0.3617	0.3848	0.4080
95	0.1240	0.1384	0.1537	0.1699	0.1869	0.2048	0.2235	0.2428	0.2629	0.2836	0.3048	0.3264	0.3485	0.3709	0.3936
100	0.1202	0.1339	0.1486	0.1641	0.1805	0.1976	0.2156	0.2342	0.2535	0.2735	0.2940	0.3150	0.3364	0.3581	0.3802

ALPHA = .025

N	LAMBDA= 26.00	27.00	28.00	29.00	30.00	31.00	32.00	33.00	34.00	35.00	36.00	37.00	38.00	39.00	40.00
1	0.9979	0.9984	0.9989	0.9992	0.9994	0.9996	0.9997	0.9998	0.9998	0.9999	0.9999	0.9999	1.0000	1.0000	1.0000
2	0.9941	0.9955	0.9966	0.9974	0.9981	0.9986	0.9989	0.9992	0.9994	0.9996	0.9997	0.9998	0.9998	0.9999	0.9999
3	0.9891	0.9916	0.9935	0.9950	0.9961	0.9970	0.9977	0.9983	0.9987	0.9990	0.9992	0.9994	0.9996	0.9997	0.9998
4	0.9832	0.9868	0.9896	0.9918	0.9936	0.9950	0.9961	0.9970	0.9977	0.9982	0.9986	0.9990	0.9992	0.9994	0.9995
5	0.9766	0.9812	0.9850	0.9881	0.9906	0.9925	0.9941	0.9954	0.9964	0.9972	0.9978	0.9983	0.9987	0.9990	0.9992
6	0.9692	0.9750	0.9798	0.9838	0.9870	0.9896	0.9917	0.9934	0.9948	0.9959	0.9968	0.9975	0.9980	0.9984	0.9988
7	0.9613	0.9683	0.9741	0.9790	0.9830	0.9863	0.9889	0.9911	0.9929	0.9943	0.9955	0.9964	0.9972	0.9978	0.9982
8	0.9529	0.9611	0.9680	0.9737	0.9785	0.9825	0.9858	0.9885	0.9907	0.9925	0.9940	0.9952	0.9962	0.9970	0.9976
9	0.9442	0.9535	0.9614	0.9681	0.9737	0.9784	0.9823	0.9856	0.9883	0.9905	0.9923	0.9938	0.9950	0.9960	0.9968
10	0.9352	0.9456	0.9545	0.9621	0.9686	0.9740	0.9786	0.9824	0.9856	0.9882	0.9904	0.9922	0.9937	0.9949	0.9959
11	0.9259	0.9374	0.9474	0.9558	0.9631	0.9693	0.9745	0.9789	0.9826	0.9857	0.9883	0.9904	0.9922	0.9936	0.9948
12	0.9165	0.9290	0.9399	0.9493	0.9574	0.9643	0.9702	0.9752	0.9794	0.9829	0.9859	0.9884	0.9905	0.9922	0.9936
13	0.9070	0.9205	0.9323	0.9425	0.9514	0.9590	0.9656	0.9712	0.9759	0.9800	0.9834	0.9863	0.9887	0.9907	0.9923
14	0.8973	0.9118	0.9245	0.9356	0.9452	0.9536	0.9608	0.9670	0.9723	0.9768	0.9807	0.9839	0.9867	0.9890	0.9909
15	0.8876	0.9030	0.9165	0.9284	0.9389	0.9480	0.9558	0.9626	0.9685	0.9735	0.9778	0.9814	0.9845	0.9871	0.9893
16	0.8779	0.8941	0.9085	0.9212	0.9324	0.9421	0.9507	0.9581	0.9645	0.9700	0.9747	0.9788	0.9822	0.9851	0.9876
17	0.8682	0.8852	0.9004	0.9138	0.9257	0.9362	0.9453	0.9533	0.9603	0.9663	0.9715	0.9759	0.9798	0.9830	0.9858
18	0.8585	0.8763	0.8922	0.9064	0.9189	0.9301	0.9399	0.9485	0.9560	0.9625	0.9681	0.9730	0.9772	0.9808	0.9838
19	0.8488	0.8673	0.8839	0.8988	0.9121	0.9239	0.9343	0.9434	0.9515	0.9585	0.9646	0.9699	0.9745	0.9784	0.9818
20	0.8393	0.8584	0.8757	0.8912	0.9052	0.9176	0.9286	0.9383	0.9469	0.9544	0.9609	0.9667	0.9716	0.9759	0.9796
21	0.8297	0.8495	0.8674	0.8836	0.8982	0.9112	0.9228	0.9331	0.9422	0.9502	0.9572	0.9633	0.9686	0.9733	0.9773
22	0.8203	0.8407	0.8592	0.8760	0.8911	0.9047	0.9169	0.9277	0.9373	0.9458	0.9533	0.9598	0.9656	0.9706	0.9749
23	0.8110	0.8319	0.8510	0.8683	0.8841	0.8982	0.9109	0.9223	0.9324	0.9414	0.9493	0.9563	0.9624	0.9677	0.9724
24	0.8017	0.8232	0.8428	0.8607	0.8770	0.8917	0.9049	0.9168	0.9274	0.9368	0.9452	0.9526	0.9591	0.9648	0.9698
25	0.7926	0.8145	0.8347	0.8531	0.8699	0.8851	0.8989	0.9112	0.9223	0.9322	0.9410	0.9488	0.9557	0.9618	0.9671
26	0.7836	0.8060	0.8266	0.8455	0.8628	0.8785	0.8928	0.9056	0.9172	0.9275	0.9368	0.9450	0.9522	0.9586	0.9643
27	0.7747	0.7975	0.8185	0.8379	0.8557	0.8719	0.8866	0.8999	0.9120	0.9228	0.9324	0.9410	0.9487	0.9554	0.9614
28	0.7660	0.7891	0.8106	0.8304	0.8486	0.8653	0.8805	0.8942	0.9067	0.9179	0.9280	0.9370	0.9450	0.9522	0.9585
29	0.7573	0.7808	0.8027	0.8229	0.8415	0.8587	0.8743	0.8885	0.9014	0.9130	0.9235	0.9329	0.9413	0.9488	0.9554
30	0.7488	0.7727	0.7949	0.8155	0.8345	0.8520	0.8681	0.8827	0.8961	0.9081	0.9190	0.9288	0.9375	0.9454	0.9523
32	0.7322	0.7566	0.7795	0.8008	0.8206	0.8389	0.8557	0.8712	0.8853	0.8981	0.9098	0.9203	0.9298	0.9383	0.9459
34	0.7161	0.7410	0.7644	0.7864	0.8068	0.8258	0.8434	0.8596	0.8745	0.8880	0.9004	0.9117	0.9218	0.9310	0.9393
36	0.7006	0.7258	0.7497	0.7722	0.7933	0.8129	0.8311	0.8480	0.8636	0.8779	0.8910	0.9029	0.9137	0.9235	0.9324
38	0.6855	0.7111	0.7354	0.7584	0.7799	0.8001	0.8190	0.8365	0.8527	0.8677	0.8814	0.8940	0.9054	0.9159	0.9253
40	0.6710	0.6968	0.7215	0.7448	0.7669	0.7876	0.8070	0.8251	0.8419	0.8574	0.8718	0.8850	0.8970	0.9081	0.9181
42	0.6569	0.6830	0.7079	0.7316	0.7540	0.7752	0.7951	0.8137	0.8311	0.8472	0.8621	0.8759	0.8886	0.9002	0.9108
44	0.6434	0.6696	0.6947	0.7187	0.7415	0.7631	0.7834	0.8025	0.8203	0.8370	0.8525	0.8668	0.8800	0.8922	0.9033
46	0.6303	0.6566	0.6819	0.7062	0.7292	0.7512	0.7719	0.7914	0.8097	0.8269	0.8428	0.8577	0.8714	0.8841	0.8958
48	0.6177	0.6441	0.6695	0.6939	0.7173	0.7395	0.7605	0.7805	0.7992	0.8168	0.8332	0.8486	0.8628	0.8760	0.8882
50	0.6055	0.6319	0.6574	0.6820	0.7056	0.7280	0.7494	0.7697	0.7888	0.8068	0.8237	0.8395	0.8542	0.8678	0.8805
55	0.5768	0.6032	0.6288	0.6536	0.6775	0.7005	0.7225	0.7435	0.7634	0.7823	0.8002	0.8169	0.8327	0.8474	0.8611
60	0.5506	0.5767	0.6023	0.6271	0.6512	0.6745	0.6969	0.7184	0.7390	0.7586	0.7772	0.7948	0.8114	0.8271	0.8418
65	0.5265	0.5523	0.5776	0.6024	0.6266	0.6500	0.6727	0.6946	0.7156	0.7357	0.7549	0.7732	0.7906	0.8070	0.8225
70	0.5043	0.5297	0.5548	0.5794	0.6035	0.6269	0.6497	0.6719	0.6932	0.7137	0.7334	0.7523	0.7702	0.7873	0.8035
75	0.4839	0.5089	0.5335	0.5579	0.5818	0.6052	0.6281	0.6503	0.6718	0.6927	0.7127	0.7320	0.7504	0.7681	0.7849
80	0.4650	0.4895	0.5138	0.5378	0.5615	0.5848	0.6076	0.6298	0.6515	0.6725	0.6928	0.7124	0.7313	0.7493	0.7666
85	0.4476	0.4716	0.4954	0.5191	0.5425	0.5656	0.5882	0.6105	0.6321	0.6532	0.6737	0.6936	0.7127	0.7312	0.7489
90	0.4314	0.4549	0.4783	0.5016	0.5247	0.5475	0.5700	0.5921	0.6137	0.6349	0.6555	0.6755	0.6948	0.7135	0.7316
95	0.4164	0.4393	0.4623	0.4851	0.5079	0.5304	0.5527	0.5747	0.5962	0.6173	0.6380	0.6581	0.6776	0.6965	0.7148
100	0.4024	0.4248	0.4473	0.4697	0.4921	0.5144	0.5364	0.5581	0.5796	0.6006	0.6212	0.6414	0.6610	0.6801	0.6986

ALPHA = .025

N	LAMBDA= 42.00	44.00	46.00	48.00	50.00	55.00	60.00	65.00	70.00	75.00	80.00	85.00	90.00	95.00	100.00
1	1.0000	1.0000	1.0000	1.0000	1.0000	1.0000	1.0000	1.0000	1.0000	1.0000	1.0000	1.0000	1.0000	1.0000	1.0000
2	0.9999	1.0000	1.0000	1.0000	1.0000	1.0000	1.0000	1.0000	1.0000	1.0000	1.0000	1.0000	1.0000	1.0000	1.0000
3	0.9999	0.9999	1.0000	1.0000	1.0000	1.0000	1.0000	1.0000	1.0000	1.0000	1.0000	1.0000	1.0000	1.0000	1.0000
4	0.9997	0.9998	0.9999	1.0000	1.0000	1.0000	1.0000	1.0000	1.0000	1.0000	1.0000	1.0000	1.0000	1.0000	1.0000
5	0.9995	0.9997	0.9998	0.9999	0.9999	1.0000	1.0000	1.0000	1.0000	1.0000	1.0000	1.0000	1.0000	1.0000	1.0000
6	0.9993	0.9996	0.9997	0.9998	0.9999	1.0000	1.0000	1.0000	1.0000	1.0000	1.0000	1.0000	1.0000	1.0000	1.0000
7	0.9989	0.9993	0.9996	0.9998	0.9999	1.0000	1.0000	1.0000	1.0000	1.0000	1.0000	1.0000	1.0000	1.0000	1.0000
8	0.9985	0.9991	0.9994	0.9996	0.9998	0.9999	1.0000	1.0000	1.0000	1.0000	1.0000	1.0000	1.0000	1.0000	1.0000
9	0.9980	0.9987	0.9992	0.9995	0.9997	0.9999	1.0000	1.0000	1.0000	1.0000	1.0000	1.0000	1.0000	1.0000	1.0000
10	0.9973	0.9983	0.9989	0.9993	0.9996	0.9999	1.0000	1.0000	1.0000	1.0000	1.0000	1.0000	1.0000	1.0000	1.0000
11	0.9966	0.9978	0.9986	0.9991	0.9994	0.9998	1.0000	1.0000	1.0000	1.0000	1.0000	1.0000	1.0000	1.0000	1.0000
12	0.9958	0.9972	0.9982	0.9989	0.9993	0.9998	0.9999	1.0000	1.0000	1.0000	1.0000	1.0000	1.0000	1.0000	1.0000
13	0.9949	0.9966	0.9978	0.9986	0.9991	0.9997	0.9999	1.0000	1.0000	1.0000	1.0000	1.0000	1.0000	1.0000	1.0000
14	0.9938	0.9959	0.9973	0.9982	0.9988	0.9996	0.9999	1.0000	1.0000	1.0000	1.0000	1.0000	1.0000	1.0000	1.0000
15	0.9927	0.9951	0.9967	0.9978	0.9986	0.9995	0.9998	1.0000	1.0000	1.0000	1.0000	1.0000	1.0000	1.0000	1.0000
16	0.9915	0.9942	0.9961	0.9974	0.9982	0.9994	0.9998	0.9999	1.0000	1.0000	1.0000	1.0000	1.0000	1.0000	1.0000
17	0.9901	0.9932	0.9954	0.9969	0.9979	0.9993	0.9997	0.9999	1.0000	1.0000	1.0000	1.0000	1.0000	1.0000	1.0000
18	0.9887	0.9921	0.9946	0.9963	0.9975	0.9991	0.9997	0.9999	1.0000	1.0000	1.0000	1.0000	1.0000	1.0000	1.0000
19	0.9871	0.9910	0.9938	0.9957	0.9971	0.9989	0.9996	0.9999	1.0000	1.0000	1.0000	1.0000	1.0000	1.0000	1.0000
20	0.9855	0.9898	0.9928	0.9950	0.9966	0.9987	0.9995	0.9998	0.9999	1.0000	1.0000	1.0000	1.0000	1.0000	1.0000
21	0.9837	0.9884	0.9919	0.9943	0.9961	0.9985	0.9994	0.9998	0.9999	1.0000	1.0000	1.0000	1.0000	1.0000	1.0000
22	0.9819	0.9870	0.9908	0.9936	0.9955	0.9982	0.9993	0.9998	0.9999	1.0000	1.0000	1.0000	1.0000	1.0000	1.0000
23	0.9799	0.9856	0.9897	0.9927	0.9949	0.9980	0.9992	0.9997	0.9999	1.0000	1.0000	1.0000	1.0000	1.0000	1.0000
24	0.9779	0.9840	0.9885	0.9918	0.9942	0.9977	0.9991	0.9997	0.9999	1.0000	1.0000	1.0000	1.0000	1.0000	1.0000
25	0.9758	0.9823	0.9873	0.9909	0.9935	0.9974	0.9990	0.9996	0.9999	1.0000	1.0000	1.0000	1.0000	1.0000	1.0000
26	0.9736	0.9806	0.9859	0.9899	0.9928	0.9970	0.9988	0.9995	0.9998	0.9999	1.0000	1.0000	1.0000	1.0000	1.0000
27	0.9713	0.9788	0.9845	0.9888	0.9920	0.9966	0.9986	0.9995	0.9998	0.9999	1.0000	1.0000	1.0000	1.0000	1.0000
28	0.9689	0.9770	0.9831	0.9877	0.9911	0.9962	0.9985	0.9994	0.9998	0.9999	1.0000	1.0000	1.0000	1.0000	1.0000
29	0.9665	0.9750	0.9816	0.9865	0.9902	0.9958	0.9983	0.9993	0.9997	0.9999	1.0000	1.0000	1.0000	1.0000	1.0000
30	0.9640	0.9730	0.9800	0.9853	0.9893	0.9953	0.9980	0.9992	0.9997	0.9999	1.0000	1.0000	1.0000	1.0000	1.0000
32	0.9587	0.9688	0.9766	0.9827	0.9872	0.9943	0.9975	0.9990	0.9996	0.9998	0.9999	1.0000	1.0000	1.0000	1.0000
34	0.9533	0.9644	0.9731	0.9798	0.9850	0.9931	0.9970	0.9987	0.9995	0.9998	0.9999	1.0000	1.0000	1.0000	1.0000
36	0.9475	0.9597	0.9693	0.9768	0.9826	0.9918	0.9963	0.9984	0.9994	0.9997	0.9999	1.0000	1.0000	1.0000	1.0000
38	0.9416	0.9548	0.9652	0.9735	0.9800	0.9904	0.9956	0.9981	0.9992	0.9997	0.9999	1.0000	1.0000	1.0000	1.0000
40	0.9355	0.9496	0.9610	0.9701	0.9772	0.9889	0.9948	0.9977	0.9990	0.9996	0.9998	0.9999	1.0000	1.0000	1.0000
42	0.9292	0.9444	0.9566	0.9665	0.9743	0.9872	0.9939	0.9972	0.9988	0.9995	0.9998	0.9999	1.0000	1.0000	1.0000
44	0.9228	0.9389	0.9521	0.9627	0.9712	0.9854	0.9930	0.9967	0.9985	0.9994	0.9997	0.9999	1.0000	1.0000	1.0000
46	0.9163	0.9333	0.9474	0.9588	0.9680	0.9835	0.9919	0.9962	0.9983	0.9992	0.9997	0.9999	0.9999	1.0000	1.0000
48	0.9096	0.9276	0.9425	0.9547	0.9646	0.9815	0.9907	0.9956	0.9980	0.9991	0.9996	0.9998	0.9999	1.0000	1.0000
50	0.9029	0.9218	0.9375	0.9505	0.9610	0.9793	0.9895	0.9949	0.9976	0.9989	0.9995	0.9998	0.9999	1.0000	1.0000
55	0.8857	0.9067	0.9245	0.9394	0.9517	0.9734	0.9860	0.9930	0.9966	0.9984	0.9993	0.9997	0.9999	0.9999	1.0000
60	0.8683	0.8913	0.9109	0.9276	0.9416	0.9669	0.9820	0.9906	0.9953	0.9977	0.9989	0.9995	0.9998	0.9999	1.0000
65	0.8508	0.8755	0.8969	0.9153	0.9309	0.9597	0.9775	0.9879	0.9937	0.9969	0.9985	0.9993	0.9997	0.9999	0.9999
70	0.8333	0.8596	0.8826	0.9026	0.9198	0.9520	0.9725	0.9848	0.9919	0.9958	0.9979	0.9990	0.9995	0.9998	0.9999
75	0.8159	0.8437	0.8682	0.8896	0.9082	0.9438	0.9670	0.9813	0.9898	0.9946	0.9973	0.9986	0.9993	0.9997	0.9999
80	0.7988	0.8278	0.8536	0.8765	0.8964	0.9352	0.9611	0.9775	0.9874	0.9932	0.9964	0.9982	0.9991	0.9996	0.9998
85	0.7820	0.8121	0.8391	0.8632	0.8844	0.9263	0.9548	0.9733	0.9847	0.9916	0.9955	0.9977	0.9988	0.9994	0.9997
90	0.7655	0.7965	0.8246	0.8498	0.8722	0.9170	0.9481	0.9687	0.9818	0.9897	0.9944	0.9970	0.9985	0.9992	0.9996
95	0.7494	0.7812	0.8103	0.8365	0.8600	0.9075	0.9411	0.9638	0.9785	0.9877	0.9931	0.9963	0.9980	0.9990	0.9995
100	0.7337	0.7662	0.7961	0.8232	0.8477	0.8978	0.9338	0.9587	0.9750	0.9854	0.9917	0.9954	0.9975	0.9987	0.9993

ALPHA = .010

N	LAMBDA= .00	.10	.20	.30	.40	.50	.60	.70	.80	.90	1.00	1.20	1.40	1.60	1.80
1	0.0100	0.0138	0.0179	0.0222	0.0267	0.0313	0.0362	0.0413	0.0466	0.0521	0.0577	0.0695	0.0820	0.0950	0.1086
2	0.0100	0.0124	0.0149	0.0176	0.0204	0.0234	0.0265	0.0298	0.0332	0.0367	0.0404	0.0482	0.0565	0.0653	0.0747
3	0.0100	0.0118	0.0137	0.0157	0.0179	0.0201	0.0225	0.0250	0.0276	0.0303	0.0331	0.0390	0.0454	0.0522	0.0595
4	0.0100	0.0115	0.0131	0.0147	0.0165	0.0183	0.0203	0.0223	0.0244	0.0266	0.0289	0.0338	0.0390	0.0446	0.0506
5	0.0100	0.0113	0.0126	0.0141	0.0156	0.0171	0.0188	0.0205	0.0223	0.0242	0.0262	0.0303	0.0348	0.0396	0.0447
6	0.0100	0.0111	0.0123	0.0136	0.0149	0.0163	0.0178	0.0193	0.0209	0.0225	0.0242	0.0279	0.0318	0.0360	0.0405
7	0.0100	0.0110	0.0121	0.0132	0.0144	0.0157	0.0170	0.0184	0.0198	0.0212	0.0228	0.0260	0.0295	0.0333	0.0373
8	0.0100	0.0109	0.0119	0.0130	0.0141	0.0152	0.0164	0.0176	0.0189	0.0203	0.0216	0.0246	0.0278	0.0311	0.0347
9	0.0100	0.0109	0.0118	0.0128	0.0138	0.0148	0.0159	0.0170	0.0182	0.0195	0.0207	0.0234	0.0263	0.0294	0.0327
10	0.0100	0.0108	0.0117	0.0126	0.0135	0.0145	0.0155	0.0166	0.0177	0.0188	0.0200	0.0225	0.0251	0.0280	0.0310
11	0.0100	0.0108	0.0116	0.0124	0.0133	0.0142	0.0152	0.0161	0.0172	0.0182	0.0193	0.0217	0.0242	0.0268	0.0296
12	0.0100	0.0107	0.0115	0.0123	0.0131	0.0140	0.0149	0.0158	0.0168	0.0178	0.0188	0.0210	0.0233	0.0258	0.0284
13	0.0100	0.0107	0.0114	0.0122	0.0130	0.0138	0.0146	0.0155	0.0164	0.0173	0.0183	0.0204	0.0226	0.0249	0.0274
14	0.0100	0.0107	0.0114	0.0121	0.0128	0.0136	0.0144	0.0152	0.0161	0.0170	0.0179	0.0199	0.0219	0.0241	0.0265
15	0.0100	0.0106	0.0113	0.0120	0.0127	0.0134	0.0142	0.0150	0.0158	0.0167	0.0176	0.0194	0.0214	0.0235	0.0257
16	0.0100	0.0106	0.0112	0.0119	0.0126	0.0133	0.0140	0.0148	0.0156	0.0164	0.0172	0.0190	0.0209	0.0229	0.0250
17	0.0100	0.0106	0.0112	0.0118	0.0125	0.0132	0.0139	0.0146	0.0154	0.0161	0.0169	0.0186	0.0204	0.0223	0.0243
18	0.0100	0.0106	0.0112	0.0118	0.0124	0.0131	0.0137	0.0144	0.0152	0.0159	0.0167	0.0183	0.0200	0.0218	0.0237
27	0.0100	0.0104	0.0109	0.0114	0.0119	0.0124	0.0129	0.0134	0.0140	0.0145	0.0151	0.0163	0.0176	0.0189	0.0203
28	0.0100	0.0104	0.0109	0.0113	0.0118	0.0123	0.0128	0.0133	0.0139	0.0144	0.0150	0.0162	0.0174	0.0187	0.0201
29	0.0100	0.0104	0.0109	0.0113	0.0118	0.0123	0.0128	0.0133	0.0138	0.0143	0.0149	0.0160	0.0172	0.0185	0.0198
30	0.0100	0.0104	0.0109	0.0113	0.0118	0.0122	0.0127	0.0132	0.0137	0.0142	0.0148	0.0159	0.0171	0.0183	0.0196
32	0.0100	0.0104	0.0108	0.0112	0.0117	0.0121	0.0126	0.0131	0.0136	0.0141	0.0146	0.0156	0.0168	0.0179	0.0192
34	0.0100	0.0104	0.0108	0.0112	0.0116	0.0121	0.0125	0.0130	0.0134	0.0139	0.0144	0.0154	0.0165	0.0176	0.0188
36	0.0100	0.0104	0.0108	0.0112	0.0116	0.0120	0.0124	0.0129	0.0133	0.0138	0.0142	0.0152	0.0163	0.0173	0.0185
38	0.0100	0.0104	0.0107	0.0111	0.0115	0.0119	0.0123	0.0128	0.0132	0.0136	0.0141	0.0150	0.0160	0.0171	0.0182
40	0.0100	0.0104	0.0107	0.0111	0.0115	0.0119	0.0123	0.0127	0.0131	0.0135	0.0140	0.0149	0.0158	0.0168	0.0179
42	0.0100	0.0103	0.0107	0.0111	0.0114	0.0118	0.0122	0.0126	0.0130	0.0134	0.0139	0.0147	0.0157	0.0166	0.0176
44	0.0100	0.0103	0.0107	0.0110	0.0114	0.0118	0.0121	0.0125	0.0129	0.0133	0.0137	0.0146	0.0155	0.0164	0.0174
46	0.0100	0.0103	0.0107	0.0110	0.0114	0.0117	0.0121	0.0125	0.0128	0.0132	0.0136	0.0145	0.0153	0.0162	0.0172
48	0.0100	0.0103	0.0106	0.0110	0.0113	0.0117	0.0120	0.0124	0.0128	0.0132	0.0135	0.0143	0.0152	0.0161	0.0170
50	0.0100	0.0103	0.0106	0.0110	0.0113	0.0116	0.0120	0.0123	0.0127	0.0131	0.0135	0.0142	0.0151	0.0159	0.0168
55	0.0100	0.0103	0.0106	0.0109	0.0112	0.0115	0.0119	0.0122	0.0125	0.0129	0.0133	0.0140	0.0148	0.0156	0.0164
60	0.0100	0.0103	0.0106	0.0109	0.0112	0.0115	0.0118	0.0121	0.0124	0.0127	0.0131	0.0138	0.0145	0.0153	0.0160
65	0.0100	0.0103	0.0105	0.0108	0.0111	0.0114	0.0117	0.0120	0.0123	0.0126	0.0129	0.0136	0.0143	0.0150	0.0157
70	0.0100	0.0103	0.0105	0.0108	0.0111	0.0113	0.0116	0.0119	0.0122	0.0125	0.0128	0.0134	0.0141	0.0148	0.0155
75	0.0100	0.0102	0.0105	0.0108	0.0110	0.0113	0.0116	0.0118	0.0121	0.0124	0.0127	0.0133	0.0139	0.0146	0.0152
80	0.0100	0.0102	0.0105	0.0107	0.0110	0.0112	0.0115	0.0118	0.0120	0.0123	0.0126	0.0132	0.0138	0.0144	0.0150
85	0.0100	0.0102	0.0105	0.0107	0.0109	0.0112	0.0114	0.0117	0.0120	0.0122	0.0125	0.0131	0.0136	0.0142	0.0148
90	0.0100	0.0102	0.0105	0.0107	0.0109	0.0112	0.0114	0.0116	0.0119	0.0122	0.0124	0.0130	0.0135	0.0141	0.0147
95	0.0100	0.0102	0.0104	0.0107	0.0109	0.0111	0.0114	0.0116	0.0118	0.0121	0.0123	0.0129	0.0134	0.0139	0.0145
100	0.0100	0.0102	0.0104	0.0106	0.0109	0.0111	0.0113	0.0116	0.0118	0.0120	0.0123	0.0128	0.0133	0.0138	0.0144

ALPHA = .010

N	LAMBDA= 2.00	2.20	2.40	2.60	2.80	3.00	3.50	4.00	4.50	5.00	6.00	7.00	8.00	9.00	10.00
1	0.1227	0.1373	0.1523	0.1677	0.1834	0.1994	0.2404	0.2824	0.3247	0.3670	0.4497	0.5279	0.5997	0.6643	0.7212
2	0.0845	0.0948	0.1055	0.1166	0.1281	0.1400	0.1711	0.2039	0.2382	0.2734	0.3453	0.4173	0.4871	0.5533	0.6148
3	0.0672	0.0752	0.0837	0.0925	0.1017	0.1113	0.1366	0.1639	0.1927	0.2228	0.2859	0.3512	0.4166	0.4806	0.5419
4	0.0569	0.0636	0.0707	0.0780	0.0858	0.0938	0.1152	0.1385	0.1634	0.1897	0.2458	0.3051	0.3660	0.4269	0.4865
5	0.0501	0.0559	0.0619	0.0682	0.0749	0.0818	0.1005	0.1208	0.1427	0.1661	0.2165	0.2707	0.3272	0.3848	0.4421
6	0.0452	0.0502	0.0555	0.0611	0.0670	0.0731	0.0896	0.1076	0.1272	0.1482	0.1939	0.2437	0.2963	0.3506	0.4055
7	0.0415	0.0460	0.0507	0.0557	0.0609	0.0664	0.0812	0.0974	0.1151	0.1342	0.1759	0.2219	0.2710	0.3223	0.3747
8	0.0386	0.0426	0.0469	0.0514	0.0561	0.0611	0.0745	0.0893	0.1054	0.1228	0.1613	0.2039	0.2499	0.2984	0.3484
9	0.0362	0.0399	0.0438	0.0479	0.0523	0.0568	0.0690	0.0826	0.0974	0.1135	0.1490	0.1888	0.2319	0.2778	0.3255
10	0.0343	0.0377	0.0413	0.0451	0.0490	0.0532	0.0645	0.0770	0.0907	0.1056	0.1387	0.1759	0.2165	0.2600	0.3056
11	0.0326	0.0358	0.0391	0.0426	0.0463	0.0502	0.0607	0.0723	0.0851	0.0989	0.1298	0.1647	0.2031	0.2444	0.2879
12	0.0312	0.0342	0.0373	0.0406	0.0440	0.0476	0.0574	0.0683	0.0802	0.0931	0.1221	0.1550	0.1913	0.2306	0.2722
13	0.0300	0.0328	0.0357	0.0388	0.0420	0.0454	0.0546	0.0647	0.0759	0.0881	0.1154	0.1465	0.1809	0.2183	0.2582
14	0.0289	0.0316	0.0343	0.0372	0.0403	0.0435	0.0521	0.0617	0.0722	0.0837	0.1095	0.1389	0.1716	0.2073	0.2455
15	0.0280	0.0305	0.0331	0.0358	0.0387	0.0417	0.0499	0.0589	0.0689	0.0798	0.1042	0.1321	0.1633	0.1974	0.2341
16	0.0272	0.0295	0.0320	0.0346	0.0373	0.0402	0.0479	0.0565	0.0660	0.0763	0.0995	0.1261	0.1558	0.1885	0.2237
17	0.0264	0.0287	0.0310	0.0335	0.0361	0.0388	0.0462	0.0544	0.0633	0.0731	0.0952	0.1206	0.1490	0.1803	0.2142
18	0.0258	0.0279	0.0302	0.0325	0.0350	0.0376	0.0446	0.0524	0.0610	0.0703	0.0914	0.1156	0.1428	0.1729	0.2055
19	0.0252	0.0272	0.0294	0.0316	0.0340	0.0365	0.0432	0.0506	0.0588	0.0677	0.0879	0.1111	0.1372	0.1661	0.1975
20	0.0246	0.0266	0.0287	0.0308	0.0331	0.0355	0.0419	0.0490	0.0568	0.0654	0.0847	0.1069	0.1320	0.1598	0.1901
21	0.0241	0.0260	0.0280	0.0301	0.0323	0.0345	0.0407	0.0475	0.0550	0.0632	0.0818	0.1031	0.1272	0.1540	0.1832
22	0.0237	0.0255	0.0274	0.0294	0.0315	0.0337	0.0396	0.0462	0.0534	0.0613	0.0791	0.0996	0.1228	0.1487	0.1769
23	0.0232	0.0250	0.0268	0.0288	0.0308	0.0329	0.0386	0.0449	0.0519	0.0595	0.0766	0.0964	0.1188	0.1437	0.1710
24	0.0228	0.0245	0.0263	0.0282	0.0301	0.0322	0.0377	0.0438	0.0505	0.0578	0.0743	0.0934	0.1150	0.1391	0.1655
25	0.0225	0.0241	0.0258	0.0276	0.0295	0.0315	0.0368	0.0427	0.0492	0.0562	0.0722	0.0906	0.1115	0.1348	0.1604
26	0.0221	0.0237	0.0254	0.0271	0.0290	0.0309	0.0360	0.0417	0.0480	0.0548	0.0702	0.0880	0.1082	0.1307	0.1556
27	0.0218	0.0234	0.0250	0.0267	0.0285	0.0303	0.0353	0.0408	0.0468	0.0534	0.0683	0.0855	0.1051	0.1270	0.1511
28	0.0215	0.0230	0.0246	0.0262	0.0280	0.0298	0.0346	0.0399	0.0458	0.0522	0.0666	0.0833	0.1022	0.1234	0.1468
29	0.0212	0.0227	0.0242	0.0258	0.0275	0.0292	0.0339	0.0391	0.0448	0.0510	0.0650	0.0811	0.0995	0.1201	0.1429
30	0.0210	0.0224	0.0239	0.0254	0.0271	0.0288	0.0333	0.0384	0.0439	0.0499	0.0634	0.0791	0.0970	0.1170	0.1391
32	0.0205	0.0218	0.0233	0.0247	0.0263	0.0279	0.0322	0.0370	0.0422	0.0479	0.0607	0.0755	0.0923	0.1112	0.1322
34	0.0200	0.0213	0.0227	0.0241	0.0256	0.0271	0.0312	0.0357	0.0407	0.0460	0.0582	0.0722	0.0882	0.1061	0.1260
36	0.0197	0.0209	0.0222	0.0235	0.0249	0.0264	0.0303	0.0346	0.0393	0.0444	0.0559	0.0693	0.0844	0.1015	0.1204
38	0.0193	0.0205	0.0217	0.0230	0.0244	0.0258	0.0295	0.0336	0.0381	0.0430	0.0539	0.0666	0.0811	0.0973	0.1153
40	0.0190	0.0201	0.0213	0.0226	0.0238	0.0252	0.0288	0.0327	0.0370	0.0416	0.0521	0.0642	0.0780	0.0935	0.1107
42	0.0187	0.0198	0.0209	0.0221	0.0234	0.0247	0.0281	0.0319	0.0360	0.0404	0.0505	0.0620	0.0752	0.0900	0.1065
44	0.0184	0.0195	0.0206	0.0217	0.0229	0.0242	0.0275	0.0311	0.0350	0.0393	0.0489	0.0600	0.0726	0.0869	0.1027
46	0.0182	0.0192	0.0203	0.0214	0.0225	0.0237	0.0269	0.0304	0.0342	0.0383	0.0475	0.0582	0.0703	0.0839	0.0991
48	0.0179	0.0189	0.0200	0.0210	0.0222	0.0233	0.0264	0.0298	0.0334	0.0374	0.0463	0.0565	0.0681	0.0812	0.0958
50	0.0177	0.0187	0.0197	0.0207	0.0218	0.0229	0.0259	0.0292	0.0327	0.0365	0.0451	0.0549	0.0661	0.0787	0.0928
55	0.0173	0.0181	0.0191	0.0200	0.0210	0.0221	0.0248	0.0278	0.0311	0.0346	0.0425	0.0515	0.0617	0.0733	0.0861
60	0.0168	0.0177	0.0186	0.0195	0.0204	0.0214	0.0239	0.0267	0.0297	0.0330	0.0403	0.0486	0.0580	0.0686	0.0805
65	0.0165	0.0173	0.0181	0.0190	0.0198	0.0207	0.0232	0.0258	0.0286	0.0316	0.0384	0.0461	0.0549	0.0647	0.0757
70	0.0162	0.0169	0.0177	0.0185	0.0193	0.0202	0.0225	0.0249	0.0276	0.0304	0.0368	0.0440	0.0521	0.0613	0.0715
75	0.0159	0.0166	0.0174	0.0181	0.0189	0.0197	0.0219	0.0242	0.0267	0.0294	0.0353	0.0421	0.0498	0.0584	0.0679
80	0.0157	0.0164	0.0171	0.0178	0.0185	0.0193	0.0214	0.0236	0.0259	0.0285	0.0341	0.0405	0.0477	0.0558	0.0647
85	0.0155	0.0161	0.0168	0.0175	0.0182	0.0189	0.0209	0.0230	0.0252	0.0276	0.0330	0.0390	0.0458	0.0534	0.0619
90	0.0153	0.0159	0.0165	0.0172	0.0179	0.0186	0.0205	0.0225	0.0246	0.0269	0.0320	0.0377	0.0442	0.0514	0.0594
95	0.0151	0.0157	0.0163	0.0170	0.0176	0.0183	0.0201	0.0220	0.0240	0.0262	0.0311	0.0365	0.0427	0.0495	0.0571
100	0.0149	0.0155	0.0161	0.0167	0.0174	0.0180	0.0197	0.0216	0.0235	0.0256	0.0303	0.0355	0.0413	0.0479	0.0551

ALPHA = .010

N	LAMBDA= 11.00	12.00	13.00	14.00	15.00	16.00	17.00	18.00	19.00	20.00	21.00	22.00	23.00	24.00	25.00
1	0.7706	0.8128	0.8484	0.8782	0.9027	0.9228	0.9391	0.9522	0.9627	0.9710	0.9776	0.9828	0.9868	0.9899	0.9923
2	0.6708	0.7211	0.7655	0.8043	0.8379	0.8666	0.8909	0.9112	0.9282	0.9423	0.9538	0.9632	0.9708	0.9769	0.9819
3	0.5996	0.6528	0.7014	0.7450	0.7838	0.8179	0.8476	0.8732	0.8951	0.9137	0.9293	0.9424	0.9533	0.9623	0.9697
4	0.5438	0.5979	0.6484	0.6947	0.7367	0.7745	0.8080	0.8375	0.8633	0.8856	0.9048	0.9211	0.9349	0.9466	0.9564
5	0.4982	0.5521	0.6032	0.6509	0.6950	0.7352	0.7715	0.8040	0.8329	0.8582	0.8803	0.8995	0.9160	0.9301	0.9421
6	0.4600	0.5131	0.5640	0.6124	0.6576	0.6994	0.7378	0.7726	0.8038	0.8317	0.8563	0.8780	0.8968	0.9132	0.9273
7	0.4273	0.4792	0.5297	0.5780	0.6238	0.6667	0.7065	0.7430	0.7762	0.8061	0.8329	0.8567	0.8776	0.8960	0.9120
8	0.3991	0.4496	0.4992	0.5472	0.5932	0.6367	0.6774	0.7152	0.7499	0.7815	0.8101	0.8357	0.8586	0.8788	0.8965
9	0.3743	0.4233	0.4719	0.5194	0.5653	0.6090	0.6504	0.6891	0.7249	0.7579	0.7880	0.8152	0.8397	0.8616	0.8809
10	0.3525	0.4000	0.4474	0.4942	0.5397	0.5835	0.6251	0.6645	0.7012	0.7353	0.7666	0.7952	0.8211	0.8445	0.8653
11	0.3330	0.3790	0.4253	0.4712	0.5162	0.5598	0.6016	0.6413	0.6788	0.7137	0.7461	0.7758	0.8030	0.8276	0.8498
12	0.3156	0.3601	0.4052	0.4501	0.4945	0.5378	0.5796	0.6195	0.6574	0.6930	0.7262	0.7569	0.7852	0.8110	0.8344
13	0.2999	0.3430	0.3868	0.4308	0.4745	0.5173	0.5589	0.5990	0.6371	0.6733	0.7071	0.7387	0.7679	0.7947	0.8191
14	0.2857	0.3274	0.3700	0.4130	0.4559	0.4982	0.5395	0.5795	0.6179	0.6544	0.6888	0.7210	0.7510	0.7787	0.8041
15	0.2728	0.3131	0.3545	0.3965	0.4386	0.4804	0.5213	0.5612	0.5996	0.6363	0.6711	0.7039	0.7346	0.7630	0.7893
16	0.2610	0.3000	0.3403	0.3813	0.4226	0.4637	0.5042	0.5438	0.5822	0.6190	0.6542	0.6874	0.7186	0.7478	0.7748
17	0.2502	0.2880	0.3271	0.3671	0.4076	0.4480	0.4881	0.5274	0.5656	0.6025	0.6379	0.6715	0.7031	0.7329	0.7606
18	0.2402	0.2769	0.3149	0.3539	0.3935	0.4333	0.4728	0.5118	0.5499	0.5867	0.6222	0.6561	0.6881	0.7184	0.7466
19	0.2311	0.2666	0.3035	0.3416	0.3804	0.4195	0.4584	0.4970	0.5348	0.5716	0.6072	0.6412	0.6736	0.7042	0.7330
20	0.2226	0.2570	0.2929	0.3301	0.3680	0.4064	0.4448	0.4830	0.5205	0.5572	0.5927	0.6268	0.6595	0.6905	0.7197
21	0.2147	0.2481	0.2831	0.3193	0.3564	0.3941	0.4319	0.4696	0.5069	0.5433	0.5788	0.6130	0.6458	0.6771	0.7067
22	0.2074	0.2398	0.2738	0.3092	0.3455	0.3825	0.4197	0.4569	0.4938	0.5300	0.5654	0.5996	0.6326	0.6641	0.6940
23	0.2005	0.2320	0.2651	0.2996	0.3352	0.3715	0.4081	0.4449	0.4813	0.5173	0.5525	0.5867	0.6198	0.6514	0.6816
24	0.1941	0.2247	0.2570	0.2907	0.3255	0.3610	0.3971	0.4333	0.4694	0.5051	0.5401	0.5743	0.6073	0.6391	0.6695
25	0.1882	0.2179	0.2493	0.2822	0.3163	0.3512	0.3866	0.4223	0.4580	0.4934	0.5282	0.5623	0.5953	0.6272	0.6578
26	0.1826	0.2115	0.2421	0.2742	0.3075	0.3418	0.3766	0.4118	0.4471	0.4822	0.5168	0.5507	0.5837	0.6156	0.6463
27	0.1773	0.2054	0.2353	0.2667	0.2993	0.3329	0.3671	0.4018	0.4367	0.4714	0.5057	0.5394	0.5724	0.6043	0.6351
28	0.1723	0.1997	0.2289	0.2595	0.2914	0.3244	0.3581	0.3922	0.4266	0.4610	0.4950	0.5286	0.5614	0.5934	0.6243
29	0.1676	0.1943	0.2228	0.2527	0.2840	0.3163	0.3494	0.3831	0.4170	0.4510	0.4848	0.5181	0.5508	0.5827	0.6137
30	0.1632	0.1892	0.2170	0.2463	0.2769	0.3086	0.3411	0.3743	0.4078	0.4414	0.4749	0.5080	0.5406	0.5724	0.6033
32	0.1551	0.1798	0.2063	0.2343	0.2637	0.2942	0.3256	0.3578	0.3904	0.4232	0.4561	0.4887	0.5210	0.5526	0.5835
34	0.1478	0.1713	0.1966	0.2234	0.2517	0.2810	0.3114	0.3426	0.3743	0.4064	0.4386	0.4707	0.5025	0.5339	0.5646
36	0.1411	0.1637	0.1878	0.2136	0.2407	0.2690	0.2984	0.3286	0.3594	0.3907	0.4223	0.4538	0.4852	0.5162	0.5468
38	0.1351	0.1567	0.1798	0.2045	0.2306	0.2579	0.2863	0.3156	0.3456	0.3761	0.4070	0.4379	0.4689	0.4995	0.5298
40	0.1297	0.1503	0.1725	0.1962	0.2213	0.2477	0.2752	0.3036	0.3328	0.3625	0.3927	0.4230	0.4534	0.4837	0.5137
42	0.1247	0.1444	0.1658	0.1886	0.2128	0.2382	0.2648	0.2924	0.3208	0.3498	0.3792	0.4090	0.4389	0.4688	0.4984
44	0.1201	0.1390	0.1596	0.1815	0.2049	0.2295	0.2552	0.2820	0.3096	0.3378	0.3666	0.3958	0.4252	0.4546	0.4838
46	0.1158	0.1341	0.1538	0.1750	0.1975	0.2213	0.2463	0.2722	0.2991	0.3266	0.3548	0.3833	0.4122	0.4411	0.4700
48	0.1119	0.1295	0.1485	0.1690	0.1907	0.2138	0.2379	0.2631	0.2892	0.3161	0.3436	0.3716	0.3999	0.4283	0.4568
50	0.1083	0.1252	0.1436	0.1633	0.1844	0.2067	0.2301	0.2546	0.2800	0.3062	0.3331	0.3605	0.3882	0.4162	0.4443
55	0.1003	0.1158	0.1327	0.1508	0.1703	0.1909	0.2127	0.2355	0.2593	0.2839	0.3092	0.3352	0.3616	0.3884	0.4154
60	0.0935	0.1079	0.1234	0.1402	0.1582	0.1774	0.1977	0.2190	0.2413	0.2645	0.2884	0.3130	0.3382	0.3638	0.3898
65	0.0878	0.1010	0.1155	0.1311	0.1479	0.1658	0.1847	0.2047	0.2256	0.2475	0.2701	0.2935	0.3174	0.3419	0.3668
70	0.0828	0.0952	0.1086	0.1232	0.1389	0.1556	0.1734	0.1922	0.2119	0.2325	0.2539	0.2761	0.2989	0.3223	0.3462
75	0.0784	0.0900	0.1026	0.1163	0.1310	0.1467	0.1634	0.1811	0.1997	0.2192	0.2395	0.2606	0.2824	0.3047	0.3276
80	0.0746	0.0855	0.0973	0.1102	0.1240	0.1388	0.1546	0.1713	0.1889	0.2074	0.2267	0.2467	0.2675	0.2888	0.3108
85	0.0712	0.0815	0.0926	0.1047	0.1178	0.1318	0.1467	0.1625	0.1792	0.1967	0.2151	0.2342	0.2540	0.2745	0.2955
90	0.0682	0.0779	0.0885	0.0999	0.1122	0.1255	0.1396	0.1546	0.1705	0.1872	0.2046	0.2229	0.2418	0.2614	0.2816
95	0.0655	0.0747	0.0847	0.0955	0.1072	0.1198	0.1332	0.1475	0.1626	0.1785	0.1952	0.2126	0.2307	0.2495	0.2689
100	0.0631	0.0718	0.0813	0.0916	0.1027	0.1147	0.1275	0.1410	0.1554	0.1706	0.1866	0.2032	0.2206	0.2386	0.2573

ALPHA = .010

N	LAMBDA= 26.00	27.00	28.00	29.00	30.00	31.00	32.00	33.00	34.00	35.00	36.00	37.00	38.00	39.00	40.00
1	0.9942	0.9956	0.9967	0.9975	0.9981	0.9986	0.9990	0.9992	0.9994	0.9996	0.9997	0.9998	0.9998	0.9999	0.9999
2	0.9858	0.9889	0.9914	0.9933	0.9948	0.9960	0.9969	0.9977	0.9982	0.9986	0.9990	0.9992	0.9994	0.9996	0.9997
3	0.9758	0.9807	0.9846	0.9878	0.9904	0.9925	0.9941	0.9954	0.9964	0.9972	0.9978	0.9983	0.9987	0.9990	0.9992
4	0.9645	0.9712	0.9767	0.9813	0.9850	0.9880	0.9905	0.9924	0.9940	0.9953	0.9963	0.9971	0.9977	0.9982	0.9986
5	0.9523	0.9608	0.9679	0.9738	0.9787	0.9828	0.9861	0.9888	0.9911	0.9929	0.9943	0.9955	0.9964	0.9972	0.9978
6	0.9393	0.9496	0.9583	0.9656	0.9717	0.9769	0.9811	0.9847	0.9876	0.9900	0.9919	0.9935	0.9948	0.9958	0.9967
7	0.9259	0.9378	0.9480	0.9567	0.9641	0.9703	0.9756	0.9799	0.9836	0.9866	0.9891	0.9912	0.9929	0.9942	0.9954
8	0.9121	0.9256	0.9373	0.9473	0.9559	0.9633	0.9695	0.9747	0.9791	0.9828	0.9859	0.9885	0.9906	0.9924	0.9938
9	0.8981	0.9131	0.9261	0.9375	0.9473	0.9557	0.9629	0.9691	0.9743	0.9787	0.9824	0.9855	0.9881	0.9902	0.9920
10	0.8839	0.9003	0.9147	0.9273	0.9383	0.9478	0.9560	0.9630	0.9690	0.9741	0.9785	0.9821	0.9852	0.9878	0.9900
11	0.8696	0.8873	0.9030	0.9168	0.9289	0.9395	0.9486	0.9566	0.9634	0.9692	0.9742	0.9785	0.9821	0.9851	0.9877
12	0.8554	0.8743	0.8912	0.9061	0.9193	0.9309	0.9410	0.9498	0.9574	0.9640	0.9697	0.9745	0.9787	0.9822	0.9852
13	0.8413	0.8613	0.8793	0.8953	0.9095	0.9221	0.9331	0.9428	0.9512	0.9585	0.9649	0.9703	0.9750	0.9790	0.9824
14	0.8273	0.8483	0.8673	0.8843	0.8995	0.9130	0.9250	0.9355	0.9447	0.9528	0.9598	0.9659	0.9711	0.9756	0.9795
15	0.8134	0.8354	0.8553	0.8733	0.8894	0.9038	0.9167	0.9280	0.9380	0.9468	0.9545	0.9612	0.9670	0.9720	0.9763
16	0.7997	0.8225	0.8433	0.8622	0.8792	0.8945	0.9082	0.9203	0.9311	0.9406	0.9489	0.9562	0.9626	0.9681	0.9729
17	0.7862	0.8098	0.8314	0.8511	0.8690	0.8851	0.8996	0.9125	0.9240	0.9342	0.9432	0.9511	0.9581	0.9641	0.9694
18	0.7729	0.7973	0.8196	0.8401	0.8587	0.8756	0.8909	0.9045	0.9168	0.9277	0.9373	0.9458	0.9533	0.9599	0.9656
19	0.7599	0.7849	0.8079	0.8291	0.8485	0.8661	0.8821	0.8965	0.9094	0.9210	0.9312	0.9404	0.9484	0.9555	0.9617
20	0.7471	0.7726	0.7963	0.8182	0.8383	0.8566	0.8732	0.8883	0.9019	0.9141	0.9250	0.9347	0.9434	0.9510	0.9577
21	0.7345	0.7606	0.7849	0.8073	0.8281	0.8470	0.8644	0.8801	0.8943	0.9072	0.9187	0.9290	0.9382	0.9463	0.9535
22	0.7222	0.7488	0.7736	0.7966	0.8179	0.8375	0.8555	0.8718	0.8867	0.9001	0.9122	0.9231	0.9328	0.9415	0.9491
23	0.7102	0.7372	0.7624	0.7860	0.8078	0.8280	0.8466	0.8635	0.8790	0.8930	0.9057	0.9171	0.9274	0.9365	0.9446
24	0.6984	0.7258	0.7515	0.7755	0.7978	0.8186	0.8377	0.8552	0.8712	0.8858	0.8991	0.9110	0.9218	0.9314	0.9400
25	0.6869	0.7146	0.7406	0.7651	0.7879	0.8092	0.8288	0.8469	0.8635	0.8786	0.8924	0.9049	0.9161	0.9263	0.9353
26	0.6757	0.7036	0.7300	0.7549	0.7782	0.7998	0.8200	0.8386	0.8557	0.8713	0.8856	0.8986	0.9104	0.9210	0.9305
27	0.6647	0.6928	0.7196	0.7448	0.7685	0.7906	0.8112	0.8303	0.8479	0.8640	0.8788	0.8923	0.9046	0.9156	0.9256
28	0.6539	0.6823	0.7093	0.7348	0.7589	0.7814	0.8025	0.8220	0.8401	0.8567	0.8720	0.8859	0.8987	0.9102	0.9206
29	0.6434	0.6720	0.6992	0.7251	0.7494	0.7724	0.7938	0.8138	0.8323	0.8494	0.8651	0.8795	0.8927	0.9047	0.9156
30	0.6332	0.6619	0.6893	0.7154	0.7401	0.7634	0.7852	0.8056	0.8245	0.8420	0.8582	0.8731	0.8867	0.8991	0.9104
32	0.6134	0.6423	0.6701	0.6966	0.7218	0.7457	0.7682	0.7893	0.8090	0.8274	0.8444	0.8601	0.8746	0.8878	0.8999
34	0.5946	0.6236	0.6516	0.6784	0.7041	0.7285	0.7516	0.7733	0.7937	0.8128	0.8306	0.8471	0.8623	0.8764	0.8893
36	0.5766	0.6057	0.6338	0.6609	0.6869	0.7117	0.7353	0.7576	0.7786	0.7984	0.8169	0.8341	0.8500	0.8648	0.8784
38	0.5595	0.5885	0.6167	0.6440	0.6702	0.6954	0.7194	0.7422	0.7638	0.7841	0.8032	0.8211	0.8377	0.8532	0.8675
40	0.5432	0.5721	0.6003	0.6277	0.6541	0.6796	0.7039	0.7271	0.7492	0.7700	0.7897	0.8082	0.8254	0.8415	0.8565
42	0.5277	0.5564	0.5846	0.6120	0.6386	0.6642	0.6888	0.7124	0.7348	0.7562	0.7763	0.7953	0.8132	0.8299	0.8454
44	0.5128	0.5414	0.5695	0.5969	0.6235	0.6493	0.6741	0.6980	0.7208	0.7425	0.7631	0.7826	0.8010	0.8182	0.8344
46	0.4987	0.5271	0.5550	0.5823	0.6090	0.6349	0.6599	0.6840	0.7071	0.7291	0.7502	0.7701	0.7889	0.8067	0.8233
48	0.4852	0.5133	0.5411	0.5683	0.5949	0.6209	0.6460	0.6703	0.6936	0.7160	0.7374	0.7577	0.7770	0.7952	0.8123
50	0.4723	0.5002	0.5277	0.5548	0.5814	0.6073	0.6326	0.6570	0.6805	0.7032	0.7248	0.7455	0.7652	0.7838	0.8014
55	0.4426	0.4696	0.4966	0.5233	0.5496	0.5754	0.6006	0.6252	0.6491	0.6722	0.6945	0.7159	0.7363	0.7558	0.7744
60	0.4159	0.4422	0.4684	0.4945	0.5204	0.5459	0.5711	0.5957	0.6197	0.6430	0.6657	0.6875	0.7086	0.7288	0.7481
65	0.3920	0.4174	0.4429	0.4683	0.4937	0.5188	0.5437	0.5682	0.5921	0.6156	0.6384	0.6606	0.6820	0.7027	0.7226
70	0.3704	0.3949	0.4196	0.4444	0.4692	0.4939	0.5184	0.5426	0.5664	0.5898	0.6127	0.6350	0.6567	0.6778	0.6981
75	0.3509	0.3746	0.3985	0.4226	0.4467	0.4709	0.4949	0.5187	0.5423	0.5656	0.5884	0.6108	0.6326	0.6539	0.6745
80	0.3332	0.3560	0.3791	0.4025	0.4260	0.4496	0.4731	0.4966	0.5198	0.5429	0.5656	0.5879	0.6098	0.6311	0.6520
85	0.3171	0.3391	0.3614	0.3841	0.4069	0.4299	0.4529	0.4759	0.4988	0.5216	0.5441	0.5662	0.5881	0.6095	0.6304
90	0.3024	0.3236	0.3452	0.3671	0.3893	0.4117	0.4342	0.4567	0.4792	0.5016	0.5238	0.5458	0.5675	0.5888	0.6098
95	0.2889	0.3093	0.3302	0.3515	0.3730	0.3948	0.4167	0.4388	0.4608	0.4828	0.5048	0.5265	0.5480	0.5692	0.5901
100	0.2765	0.2962	0.3164	0.3370	0.3579	0.3791	0.4005	0.4220	0.4436	0.4652	0.4868	0.5083	0.5296	0.5506	0.5714

ALPHA = .010

N	LAMBDA= 42.00	44.00	46.00	48.00	50.00	55.00	60.00	65.00	70.00	75.00	80.00	85.00	90.00	95.00	100.00
1	1.0000	1.0000	1.0000	1.0000	1.0000	1.0000	1.0000	1.0000	1.0000	1.0000	1.0000	1.0000	1.0000	1.0000	1.0000
2	0.9998	0.9999	0.9999	1.0000	1.0000	1.0000	1.0000	1.0000	1.0000	1.0000	1.0000	1.0000	1.0000	1.0000	1.0000
3	0.9996	0.9997	0.9999	0.9999	1.0000	1.0000	1.0000	1.0000	1.0000	1.0000	1.0000	1.0000	1.0000	1.0000	1.0000
4	0.9992	0.9995	0.9997	0.9998	0.9999	1.0000	1.0000	1.0000	1.0000	1.0000	1.0000	1.0000	1.0000	1.0000	1.0000
5	0.9986	0.9992	0.9995	0.9997	0.9998	1.0000	1.0000	1.0000	1.0000	1.0000	1.0000	1.0000	1.0000	1.0000	1.0000
6	0.9979	0.9987	0.9992	0.9995	0.9997	0.9999	1.0000	1.0000	1.0000	1.0000	1.0000	1.0000	1.0000	1.0000	1.0000
7	0.9970	0.9981	0.9988	0.9993	0.9995	0.9999	1.0000	1.0000	1.0000	1.0000	1.0000	1.0000	1.0000	1.0000	1.0000
8	0.9960	0.9974	0.9983	0.9989	0.9993	0.9998	0.9999	1.0000	1.0000	1.0000	1.0000	1.0000	1.0000	1.0000	1.0000
9	0.9947	0.9965	0.9977	0.9985	0.9991	0.9997	0.9999	1.0000	1.0000	1.0000	1.0000	1.0000	1.0000	1.0000	1.0000
10	0.9933	0.9955	0.9970	0.9981	0.9988	0.9996	0.9999	1.0000	1.0000	1.0000	1.0000	1.0000	1.0000	1.0000	1.0000
11	0.9916	0.9944	0.9962	0.9975	0.9984	0.9995	0.9998	0.9999	1.0000	1.0000	1.0000	1.0000	1.0000	1.0000	1.0000
12	0.9898	0.9930	0.9953	0.9969	0.9979	0.9993	0.9998	0.9999	1.0000	1.0000	1.0000	1.0000	1.0000	1.0000	1.0000
13	0.9878	0.9916	0.9943	0.9961	0.9974	0.9991	0.9997	0.9999	1.0000	1.0000	1.0000	1.0000	1.0000	1.0000	1.0000
14	0.9856	0.9900	0.9931	0.9953	0.9968	0.9988	0.9996	0.9999	1.0000	1.0000	1.0000	1.0000	1.0000	1.0000	1.0000
15	0.9832	0.9882	0.9918	0.9943	0.9961	0.9986	0.9995	0.9998	0.9999	1.0000	1.0000	1.0000	1.0000	1.0000	1.0000
16	0.9806	0.9863	0.9904	0.9933	0.9954	0.9982	0.9994	0.9998	0.9999	1.0000	1.0000	1.0000	1.0000	1.0000	1.0000
17	0.9779	0.9842	0.9888	0.9921	0.9945	0.9979	0.9992	0.9997	0.9999	1.0000	1.0000	1.0000	1.0000	1.0000	1.0000
18	0.9750	0.9820	0.9871	0.9909	0.9936	0.9975	0.9990	0.9997	0.9999	1.0000	1.0000	1.0000	1.0000	1.0000	1.0000
19	0.9719	0.9796	0.9853	0.9896	0.9926	0.9970	0.9989	0.9996	0.9999	0.9999	1.0000	1.0000	1.0000	1.0000	1.0000
20	0.9687	0.9771	0.9834	0.9881	0.9915	0.9965	0.9986	0.9995	0.9998	0.9999	1.0000	1.0000	1.0000	1.0000	1.0000
21	0.9654	0.9745	0.9814	0.9865	0.9904	0.9960	0.9984	0.9994	0.9998	0.9999	1.0000	1.0000	1.0000	1.0000	1.0000
22	0.9619	0.9717	0.9792	0.9849	0.9891	0.9954	0.9981	0.9993	0.9997	0.9999	1.0000	1.0000	1.0000	1.0000	1.0000
23	0.9583	0.9683	0.9770	0.9831	0.9878	0.9947	0.9978	0.9991	0.9997	0.9999	1.0000	1.0000	1.0000	1.0000	1.0000
24	0.9545	0.9658	0.9746	0.9813	0.9863	0.9940	0.9975	0.9990	0.9996	0.9999	0.9999	1.0000	1.0000	1.0000	1.0000
25	0.9507	0.9627	0.9721	0.9793	0.9848	0.9932	0.9971	0.9988	0.9995	0.9998	0.9999	1.0000	1.0000	1.0000	1.0000
26	0.9467	0.9595	0.9695	0.9773	0.9832	0.9924	0.9967	0.9987	0.9995	0.9998	0.9999	1.0000	1.0000	1.0000	1.0000
27	0.9426	0.9562	0.9669	0.9752	0.9816	0.9915	0.9963	0.9985	0.9994	0.9998	0.9999	1.0000	1.0000	1.0000	1.0000
28	0.9385	0.9528	0.9641	0.9729	0.9798	0.9906	0.9959	0.9982	0.9993	0.9997	0.9999	1.0000	1.0000	1.0000	1.0000
29	0.9342	0.9492	0.9612	0.9706	0.9780	0.9896	0.9954	0.9980	0.9992	0.9997	0.9999	1.0000	1.0000	1.0000	1.0000
30	0.9299	0.9456	0.9583	0.9682	0.9760	0.9886	0.9948	0.9978	0.9991	0.9996	0.9999	0.9999	1.0000	1.0000	1.0000
32	0.9210	0.9382	0.9521	0.9632	0.9720	0.9863	0.9937	0.9972	0.9988	0.9995	0.9998	0.9999	1.0000	1.0000	1.0000
34	0.9118	0.9304	0.9456	0.9579	0.9676	0.9839	0.9923	0.9965	0.9985	0.9994	0.9997	0.9999	1.0000	1.0000	1.0000
36	0.9024	0.9224	0.9389	0.9523	0.9630	0.9812	0.9909	0.9958	0.9981	0.9992	0.9997	0.9999	0.9999	1.0000	1.0000
38	0.8929	0.9142	0.9319	0.9464	0.9582	0.9783	0.9893	0.9949	0.9977	0.9990	0.9996	0.9998	0.9999	1.0000	1.0000
40	0.8832	0.9057	0.9246	0.9403	0.9530	0.9752	0.9875	0.9940	0.9972	0.9988	0.9995	0.9998	0.9999	1.0000	1.0000
42	0.8733	0.8971	0.9172	0.9339	0.9477	0.9719	0.9856	0.9929	0.9967	0.9985	0.9993	0.9997	0.9999	1.0000	1.0000
44	0.8634	0.8884	0.9096	0.9274	0.9421	0.9683	0.9835	0.9917	0.9960	0.9982	0.9992	0.9996	0.9999	0.9999	1.0000
46	0.8535	0.8795	0.9018	0.9207	0.9364	0.9647	0.9812	0.9905	0.9954	0.9978	0.9990	0.9996	0.9998	0.9999	1.0000
48	0.8435	0.8706	0.8940	0.9138	0.9305	0.9608	0.9789	0.9891	0.9946	0.9974	0.9988	0.9995	0.9998	0.9999	1.0000
50	0.8335	0.8616	0.8860	0.9068	0.9244	0.9567	0.9763	0.9876	0.9938	0.9970	0.9986	0.9994	0.9997	0.9999	1.0000
55	0.8086	0.8390	0.8656	0.8888	0.9086	0.9459	0.9694	0.9834	0.9914	0.9957	0.9979	0.9990	0.9996	0.9998	0.9999
60	0.7840	0.8163	0.8450	0.8703	0.8922	0.9343	0.9617	0.9786	0.9885	0.9940	0.9970	0.9985	0.9993	0.9997	0.9999
65	0.7600	0.7939	0.8243	0.8514	0.8753	0.9219	0.9532	0.9730	0.9851	0.9920	0.9959	0.9979	0.9990	0.9995	0.9998
70	0.7365	0.7717	0.8037	0.8325	0.8580	0.9089	0.9440	0.9669	0.9812	0.9897	0.9945	0.9972	0.9986	0.9993	0.9997
75	0.7138	0.7501	0.7834	0.8135	0.8406	0.8955	0.9342	0.9602	0.9768	0.9869	0.9929	0.9962	0.9981	0.9990	0.9995
80	0.6918	0.7290	0.7633	0.7947	0.8232	0.8817	0.9240	0.9529	0.9719	0.9838	0.9910	0.9951	0.9974	0.9987	0.9994
85	0.6706	0.7085	0.7437	0.7762	0.8058	0.8676	0.9132	0.9452	0.9666	0.9804	0.9888	0.9938	0.9967	0.9983	0.9991
90	0.6502	0.6886	0.7245	0.7579	0.7886	0.8534	0.9022	0.9370	0.9609	0.9765	0.9863	0.9923	0.9958	0.9978	0.9988
95	0.6307	0.6694	0.7058	0.7399	0.7715	0.8391	0.8908	0.9285	0.9548	0.9723	0.9836	0.9906	0.9947	0.9971	0.9985
100	0.6119	0.6503	0.6877	0.7224	0.7547	0.8247	0.8792	0.9196	0.9483	0.9678	0.9806	0.9886	0.9935	0.9964	0.9981

ALPHA = .005

N	LAMBDA= .00	.10	.20	.30	.40	.50	.60	.70	.80	.90	1.00	1.20	1.40	1.60	1.80
1	0.0050	0.0073	0.0097	0.0123	0.0151	0.0181	0.0212	0.0245	0.0280	0.0316	0.0354	0.0435	0.0522	0.0615	0.0714
2	0.0050	0.0064	0.0079	0.0095	0.0112	0.0130	0.0150	0.0170	0.0192	0.0215	0.0239	0.0290	0.0346	0.0407	0.0472
3	0.0050	0.0060	0.0072	0.0083	0.0096	0.0110	0.0124	0.0140	0.0156	0.0173	0.0191	0.0229	0.0271	0.0317	0.0366
4	0.0050	0.0058	0.0068	0.0077	0.0088	0.0099	0.0110	0.0123	0.0136	0.0150	0.0164	0.0195	0.0229	0.0266	0.0306
5	0.0050	0.0057	0.0065	0.0073	0.0082	0.0092	0.0101	0.0112	0.0123	0.0135	0.0147	0.0173	0.0201	0.0232	0.0266
6	0.0050	0.0056	0.0063	0.0071	0.0078	0.0087	0.0095	0.0104	0.0114	0.0124	0.0135	0.0157	0.0182	0.0209	0.0238
7	0.0050	0.0056	0.0062	0.0069	0.0075	0.0083	0.0090	0.0099	0.0107	0.0116	0.0125	0.0146	0.0167	0.0191	0.0217
8	0.0050	0.0055	0.0061	0.0067	0.0073	0.0080	0.0087	0.0094	0.0102	0.0110	0.0118	0.0136	0.0156	0.0177	0.0200
9	0.0050	0.0055	0.0060	0.0066	0.0071	0.0078	0.0084	0.0091	0.0098	0.0105	0.0113	0.0129	0.0147	0.0166	0.0187
10	0.0050	0.0055	0.0059	0.0065	0.0070	0.0076	0.0082	0.0088	0.0094	0.0101	0.0108	0.0123	0.0140	0.0157	0.0176
11	0.0050	0.0054	0.0059	0.0064	0.0069	0.0074	0.0080	0.0085	0.0091	0.0098	0.0104	0.0118	0.0133	0.0150	0.0167
12	0.0050	0.0054	0.0058	0.0063	0.0068	0.0073	0.0078	0.0083	0.0089	0.0095	0.0101	0.0114	0.0128	0.0143	0.0160
13	0.0050	0.0054	0.0058	0.0062	0.0067	0.0071	0.0076	0.0081	0.0087	0.0092	0.0098	0.0110	0.0124	0.0138	0.0153
14	0.0050	0.0054	0.0058	0.0062	0.0066	0.0070	0.0075	0.0080	0.0085	0.0090	0.0096	0.0107	0.0120	0.0133	0.0148
15	0.0050	0.0054	0.0057	0.0061	0.0065	0.0069	0.0074	0.0078	0.0083	0.0088	0.0093	0.0104	0.0116	0.0129	0.0143
16	0.0050	0.0053	0.0057	0.0061	0.0065	0.0069	0.0073	0.0077	0.0082	0.0087	0.0092	0.0102	0.0113	0.0125	0.0138
17	0.0050	0.0053	0.0057	0.0060	0.0064	0.0068	0.0072	0.0076	0.0081	0.0085	0.0090	0.0100	0.0110	0.0122	0.0134
18	0.0050	0.0053	0.0056	0.0060	0.0064	0.0067	0.0071	0.0075	0.0079	0.0084	0.0088	0.0098	0.0108	0.0119	0.0131
19	0.0050	0.0053	0.0056	0.0060	0.0063	0.0067	0.0070	0.0074	0.0078	0.0082	0.0087	0.0096	0.0106	0.0116	0.0127
28	0.0050	0.0052	0.0055	0.0058	0.0060	0.0063	0.0066	0.0069	0.0072	0.0075	0.0078	0.0085	0.0092	0.0100	0.0108
29	0.0050	0.0052	0.0055	0.0057	0.0060	0.0063	0.0065	0.0068	0.0071	0.0074	0.0078	0.0084	0.0091	0.0099	0.0107
30	0.0050	0.0052	0.0055	0.0057	0.0060	0.0062	0.0065	0.0068	0.0071	0.0074	0.0077	0.0083	0.0090	0.0098	0.0105
32	0.0050	0.0052	0.0055	0.0057	0.0059	0.0062	0.0065	0.0067	0.0070	0.0073	0.0076	0.0082	0.0089	0.0095	0.0103
34	0.0050	0.0052	0.0054	0.0057	0.0059	0.0061	0.0064	0.0067	0.0069	0.0072	0.0075	0.0081	0.0087	0.0094	0.0101
36	0.0050	0.0052	0.0054	0.0056	0.0059	0.0061	0.0063	0.0066	0.0069	0.0071	0.0074	0.0080	0.0086	0.0092	0.0099
38	0.0050	0.0052	0.0054	0.0056	0.0058	0.0061	0.0063	0.0065	0.0068	0.0070	0.0073	0.0079	0.0084	0.0090	0.0097
40	0.0050	0.0052	0.0054	0.0056	0.0058	0.0060	0.0063	0.0065	0.0067	0.0070	0.0072	0.0078	0.0083	0.0089	0.0095
42	0.0050	0.0052	0.0054	0.0056	0.0058	0.0060	0.0062	0.0065	0.0067	0.0069	0.0072	0.0077	0.0082	0.0088	0.0094
44	0.0050	0.0052	0.0054	0.0056	0.0058	0.0060	0.0062	0.0064	0.0066	0.0069	0.0071	0.0076	0.0081	0.0086	0.0092
46	0.0050	0.0052	0.0054	0.0056	0.0058	0.0060	0.0062	0.0064	0.0066	0.0068	0.0070	0.0075	0.0080	0.0085	0.0091
48	0.0050	0.0052	0.0054	0.0055	0.0057	0.0059	0.0061	0.0063	0.0065	0.0068	0.0070	0.0074	0.0079	0.0084	0.0090
50	0.0050	0.0052	0.0053	0.0055	0.0057	0.0059	0.0061	0.0063	0.0065	0.0067	0.0069	0.0074	0.0079	0.0083	0.0089
55	0.0050	0.0052	0.0053	0.0055	0.0057	0.0059	0.0060	0.0062	0.0064	0.0066	0.0068	0.0072	0.0077	0.0081	0.0086
60	0.0050	0.0052	0.0053	0.0055	0.0056	0.0058	0.0060	0.0062	0.0063	0.0065	0.0067	0.0071	0.0075	0.0080	0.0084
65	0.0050	0.0051	0.0053	0.0055	0.0056	0.0058	0.0059	0.0061	0.0063	0.0065	0.0066	0.0070	0.0074	0.0078	0.0082
70	0.0050	0.0051	0.0053	0.0054	0.0056	0.0057	0.0059	0.0061	0.0062	0.0064	0.0066	0.0069	0.0073	0.0077	0.0081
75	0.0050	0.0051	0.0053	0.0054	0.0056	0.0057	0.0059	0.0060	0.0062	0.0063	0.0065	0.0068	0.0072	0.0076	0.0079
80	0.0050	0.0051	0.0053	0.0054	0.0055	0.0057	0.0058	0.0060	0.0061	0.0063	0.0064	0.0068	0.0071	0.0075	0.0078
85	0.0050	0.0051	0.0053	0.0054	0.0055	0.0057	0.0058	0.0059	0.0061	0.0062	0.0064	0.0067	0.0070	0.0074	0.0077
90	0.0050	0.0051	0.0052	0.0054	0.0055	0.0056	0.0058	0.0059	0.0061	0.0062	0.0063	0.0066	0.0070	0.0073	0.0076
95	0.0050	0.0051	0.0052	0.0054	0.0055	0.0056	0.0058	0.0059	0.0060	0.0062	0.0063	0.0066	0.0069	0.0072	0.0075
100	0.0050	0.0051	0.0052	0.0054	0.0055	0.0056	0.0057	0.0059	0.0060	0.0061	0.0063	0.0065	0.0068	0.0071	0.0074

ALPHA = .005

N	LAMBDA= 2.00	2.20	2.40	2.60	2.80	3.00	3.50	4.00	4.50	5.00	6.00	7.00	8.00	9.00	10.00
1	0.0818	0.0928	0.1042	0.1161	0.1285	0.1412	0.1746	0.2098	0.2464	0.2840	0.3603	0.4359	0.5085	0.5765	0.6388
2	0.0541	0.0615	0.0693	0.0775	0.0861	0.0951	0.1192	0.1455	0.1735	0.2031	0.2658	0.3314	0.3978	0.4632	0.5261
3	0.0419	0.0475	0.0535	0.0599	0.0666	0.0736	0.0927	0.1137	0.1365	0.1610	0.2142	0.2715	0.3313	0.3921	0.4524
4	0.0348	0.0394	0.0443	0.0495	0.0550	0.0608	0.0765	0.0941	0.1134	0.1342	0.1802	0.2309	0.2850	0.3412	0.3981
5	0.0302	0.0341	0.0382	0.0426	0.0472	0.0521	0.0656	0.0806	0.0973	0.1154	0.1559	0.2012	0.2504	0.3022	0.3557
6	0.0269	0.0302	0.0338	0.0376	0.0417	0.0459	0.0576	0.0708	0.0854	0.1015	0.1375	0.1784	0.2233	0.2713	0.3214
7	0.0244	0.0274	0.0305	0.0339	0.0375	0.0412	0.0516	0.0633	0.0763	0.0907	0.1231	0.1603	0.2015	0.2460	0.2931
8	0.0225	0.0251	0.0280	0.0310	0.0342	0.0376	0.0468	0.0574	0.0691	0.0820	0.1115	0.1455	0.1835	0.2250	0.2692
9	0.0210	0.0234	0.0259	0.0286	0.0315	0.0346	0.0430	0.0525	0.0632	0.0750	0.1020	0.1333	0.1685	0.2072	0.2488
10	0.0197	0.0219	0.0242	0.0267	0.0294	0.0322	0.0399	0.0486	0.0583	0.0692	0.0940	0.1229	0.1557	0.1920	0.2312
11	0.0186	0.0207	0.0228	0.0251	0.0275	0.0301	0.0372	0.0452	0.0542	0.0642	0.0872	0.1141	0.1447	0.1788	0.2158
12	0.0177	0.0196	0.0216	0.0237	0.0260	0.0284	0.0349	0.0424	0.0507	0.0600	0.0814	0.1065	0.1352	0.1672	0.2022
13	0.0170	0.0187	0.0206	0.0226	0.0247	0.0269	0.0330	0.0399	0.0477	0.0564	0.0763	0.0998	0.1268	0.1570	0.1902
14	0.0163	0.0179	0.0197	0.0215	0.0235	0.0256	0.0313	0.0378	0.0451	0.0532	0.0719	0.0940	0.1194	0.1480	0.1795
15	0.0157	0.0173	0.0189	0.0207	0.0225	0.0245	0.0298	0.0359	0.0428	0.0504	0.0680	0.0888	0.1128	0.1399	0.1699
16	0.0152	0.0167	0.0182	0.0199	0.0216	0.0235	0.0285	0.0343	0.0407	0.0479	0.0645	0.0842	0.1069	0.1327	0.1613
17	0.0147	0.0161	0.0176	0.0192	0.0208	0.0226	0.0274	0.0328	0.0389	0.0457	0.0614	0.0800	0.1016	0.1261	0.1534
18	0.0143	0.0156	0.0170	0.0185	0.0201	0.0218	0.0263	0.0315	0.0373	0.0437	0.0586	0.0763	0.0968	0.1202	0.1463
19	0.0139	0.0152	0.0165	0.0180	0.0195	0.0210	0.0254	0.0303	0.0358	0.0419	0.0561	0.0729	0.0925	0.1148	0.1398
20	0.0136	0.0148	0.0161	0.0174	0.0189	0.0204	0.0245	0.0292	0.0344	0.0403	0.0538	0.0698	0.0885	0.1099	0.1338
21	0.0133	0.0144	0.0157	0.0170	0.0183	0.0198	0.0237	0.0282	0.0332	0.0388	0.0517	0.0670	0.0849	0.1054	0.1283
22	0.0130	0.0141	0.0153	0.0165	0.0178	0.0192	0.0230	0.0273	0.0321	0.0374	0.0498	0.0644	0.0816	0.1012	0.1232
23	0.0127	0.0138	0.0149	0.0161	0.0174	0.0187	0.0224	0.0265	0.0311	0.0362	0.0480	0.0621	0.0785	0.0974	0.1186
24	0.0125	0.0135	0.0146	0.0158	0.0170	0.0183	0.0218	0.0257	0.0301	0.0350	0.0464	0.0599	0.0757	0.0938	0.1142
25	0.0123	0.0133	0.0143	0.0154	0.0166	0.0178	0.0212	0.0250	0.0293	0.0340	0.0449	0.0579	0.0731	0.0905	0.1102
26	0.0120	0.0130	0.0140	0.0151	0.0162	0.0174	0.0207	0.0244	0.0284	0.0330	0.0435	0.0560	0.0706	0.0874	0.1064
27	0.0119	0.0128	0.0138	0.0148	0.0159	0.0171	0.0202	0.0238	0.0277	0.0321	0.0422	0.0543	0.0684	0.0846	0.1029
28	0.0117	0.0126	0.0135	0.0145	0.0156	0.0167	0.0198	0.0232	0.0270	0.0312	0.0410	0.0526	0.0662	0.0819	0.0996
29	0.0115	0.0124	0.0133	0.0143	0.0153	0.0164	0.0194	0.0227	0.0263	0.0304	0.0399	0.0511	0.0643	0.0794	0.0966
30	0.0113	0.0122	0.0131	0.0141	0.0151	0.0161	0.0190	0.0222	0.0257	0.0297	0.0388	0.0497	0.0624	0.0771	0.0937
32	0.0110	0.0119	0.0127	0.0136	0.0146	0.0156	0.0182	0.0213	0.0246	0.0283	0.0369	0.0471	0.0590	0.0728	0.0884
34	0.0108	0.0116	0.0124	0.0132	0.0141	0.0151	0.0176	0.0205	0.0236	0.0271	0.0352	0.0448	0.0560	0.0690	0.0837
36	0.0106	0.0113	0.0121	0.0129	0.0137	0.0146	0.0170	0.0197	0.0227	0.0261	0.0337	0.0428	0.0534	0.0656	0.0795
38	0.0103	0.0110	0.0118	0.0126	0.0134	0.0142	0.0165	0.0191	0.0220	0.0251	0.0323	0.0409	0.0510	0.0625	0.0757
40	0.0101	0.0108	0.0115	0.0123	0.0131	0.0139	0.0161	0.0185	0.0212	0.0242	0.0311	0.0393	0.0488	0.0598	0.0723
42	0.0100	0.0106	0.0113	0.0120	0.0128	0.0135	0.0157	0.0180	0.0206	0.0234	0.0300	0.0378	0.0468	0.0573	0.0692
44	0.0098	0.0104	0.0111	0.0118	0.0125	0.0132	0.0153	0.0175	0.0200	0.0227	0.0290	0.0364	0.0450	0.0550	0.0663
46	0.0097	0.0103	0.0109	0.0116	0.0123	0.0130	0.0149	0.0171	0.0195	0.0221	0.0281	0.0351	0.0434	0.0529	0.0637
48	0.0095	0.0101	0.0107	0.0114	0.0120	0.0127	0.0146	0.0167	0.0190	0.0215	0.0272	0.0340	0.0419	0.0510	0.0614
50	0.0094	0.0100	0.0106	0.0112	0.0118	0.0125	0.0143	0.0163	0.0185	0.0209	0.0264	0.0329	0.0405	0.0492	0.0592
55	0.0091	0.0096	0.0102	0.0108	0.0114	0.0120	0.0136	0.0155	0.0175	0.0197	0.0247	0.0306	0.0375	0.0454	0.0544
60	0.0089	0.0094	0.0099	0.0104	0.0110	0.0115	0.0131	0.0148	0.0166	0.0187	0.0233	0.0287	0.0350	0.0422	0.0504
65	0.0087	0.0091	0.0096	0.0101	0.0106	0.0112	0.0126	0.0142	0.0159	0.0178	0.0221	0.0271	0.0328	0.0395	0.0470
70	0.0085	0.0089	0.0094	0.0099	0.0103	0.0108	0.0122	0.0137	0.0153	0.0171	0.0210	0.0257	0.0310	0.0371	0.0441
75	0.0083	0.0088	0.0092	0.0096	0.0101	0.0106	0.0118	0.0132	0.0148	0.0164	0.0201	0.0244	0.0294	0.0351	0.0416
80	0.0082	0.0086	0.0090	0.0094	0.0099	0.0103	0.0115	0.0128	0.0143	0.0158	0.0193	0.0234	0.0280	0.0334	0.0394
85	0.0081	0.0084	0.0088	0.0092	0.0097	0.0101	0.0112	0.0125	0.0139	0.0153	0.0186	0.0224	0.0268	0.0318	0.0375
90	0.0080	0.0083	0.0087	0.0091	0.0095	0.0099	0.0110	0.0122	0.0135	0.0149	0.0180	0.0216	0.0257	0.0305	0.0358
95	0.0079	0.0082	0.0086	0.0089	0.0093	0.0097	0.0108	0.0119	0.0131	0.0145	0.0174	0.0209	0.0248	0.0292	0.0343
100	0.0078	0.0081	0.0084	0.0088	0.0092	0.0095	0.0106	0.0116	0.0128	0.0141	0.0169	0.0202	0.0239	0.0281	0.0329

ALPHA = .005

N	LAMBDA= 11.00	12.00	13.00	14.00	15.00	16.00	17.00	18.00	19.00	20.00	21.00	22.00	23.00	24.00	25.00
1	0.6948	0.7444	0.7877	0.8250	0.8568	0.8836	0.9059	0.9244	0.9397	0.9521	0.9621	0.9702	0.9766	0.9818	0.9858
2	0.5855	0.6406	0.6909	0.7362	0.7765	0.8119	0.8428	0.8693	0.8921	0.9113	0.9275	0.9411	0.9523	0.9616	0.9692
3	0.5111	0.5670	0.6196	0.6683	0.7128	0.7531	0.7891	0.8209	0.8488	0.8731	0.8941	0.9120	0.9273	0.9402	0.9510
4	0.4546	0.5097	0.5626	0.6126	0.6593	0.7024	0.7416	0.7771	0.8088	0.8369	0.8616	0.8832	0.9019	0.9180	0.9318
5	0.4096	0.4631	0.5154	0.5656	0.6132	0.6578	0.6992	0.7371	0.7716	0.8027	0.8304	0.8550	0.8766	0.8955	0.9119
6	0.3727	0.4243	0.4753	0.5250	0.5728	0.6182	0.6608	0.7005	0.7370	0.7703	0.8004	0.8275	0.8516	0.8730	0.8918
7	0.3417	0.3912	0.4407	0.4895	0.5370	0.5826	0.6260	0.6667	0.7047	0.7397	0.7718	0.8009	0.8272	0.8506	0.8715
8	0.3154	0.3627	0.4106	0.4583	0.5051	0.5506	0.5942	0.6356	0.6746	0.7109	0.7445	0.7753	0.8033	0.8286	0.8513
9	0.2926	0.3379	0.3841	0.4305	0.4765	0.5215	0.5651	0.6068	0.6465	0.6837	0.7185	0.7506	0.7801	0.8070	0.8314
10	0.2727	0.3161	0.3606	0.4057	0.4506	0.4950	0.5383	0.5802	0.6202	0.6581	0.6937	0.7270	0.7577	0.7860	0.8117
11	0.2553	0.2968	0.3396	0.3833	0.4272	0.4708	0.5137	0.5554	0.5956	0.6339	0.6702	0.7043	0.7360	0.7654	0.7925
12	0.2398	0.2795	0.3208	0.3631	0.4059	0.4486	0.4909	0.5323	0.5725	0.6111	0.6478	0.6826	0.7152	0.7455	0.7736
13	0.2261	0.2641	0.3038	0.3447	0.3863	0.4282	0.4698	0.5109	0.5508	0.5895	0.6266	0.6618	0.6950	0.7262	0.7552
14	0.2137	0.2501	0.2883	0.3279	0.3684	0.4094	0.4503	0.4908	0.5305	0.5691	0.6063	0.6419	0.6757	0.7075	0.7373
15	0.2026	0.2375	0.2743	0.3126	0.3520	0.3919	0.4321	0.4720	0.5114	0.5498	0.5871	0.6229	0.6570	0.6894	0.7198
16	0.1925	0.2260	0.2615	0.2985	0.3367	0.3757	0.4151	0.4544	0.4934	0.5316	0.5688	0.6047	0.6391	0.6719	0.7029
17	0.1833	0.2155	0.2497	0.2855	0.3227	0.3607	0.3993	0.4379	0.4764	0.5143	0.5514	0.5873	0.6219	0.6550	0.6865
18	0.1749	0.2059	0.2389	0.2736	0.3096	0.3467	0.3844	0.4224	0.4604	0.4979	0.5348	0.5707	0.6054	0.6387	0.6705
19	0.1672	0.1970	0.2289	0.2625	0.2975	0.3336	0.3706	0.4079	0.4453	0.4824	0.5190	0.5548	0.5895	0.6230	0.6551
20	0.1602	0.1889	0.2196	0.2521	0.2862	0.3214	0.3575	0.3941	0.4309	0.4676	0.5039	0.5396	0.5743	0.6078	0.6401
21	0.1537	0.1813	0.2110	0.2426	0.2756	0.3100	0.3453	0.3812	0.4174	0.4536	0.4896	0.5250	0.5596	0.5932	0.6256
22	0.1477	0.1743	0.2031	0.2336	0.2658	0.2992	0.3337	0.3689	0.4045	0.4403	0.4759	0.5110	0.5455	0.5791	0.6116
23	0.1421	0.1678	0.1956	0.2253	0.2565	0.2891	0.3228	0.3573	0.3924	0.4276	0.4628	0.4976	0.5319	0.5655	0.5980
24	0.1369	0.1618	0.1887	0.2174	0.2478	0.2796	0.3126	0.3464	0.3808	0.4155	0.4503	0.4848	0.5189	0.5523	0.5849
25	0.1321	0.1561	0.1822	0.2101	0.2397	0.2707	0.3029	0.3360	0.3698	0.4040	0.4383	0.4725	0.5064	0.5396	0.5721
26	0.1276	0.1509	0.1761	0.2032	0.2320	0.2622	0.2937	0.3261	0.3593	0.3930	0.4268	0.4607	0.4943	0.5274	0.5598
27	0.1234	0.1459	0.1704	0.1968	0.2248	0.2542	0.2850	0.3168	0.3494	0.3825	0.4159	0.4494	0.4827	0.5156	0.5479
28	0.1194	0.1413	0.1650	0.1906	0.2179	0.2467	0.2767	0.3079	0.3399	0.3724	0.4054	0.4385	0.4715	0.5042	0.5364
29	0.1157	0.1369	0.1600	0.1849	0.2115	0.2395	0.2689	0.2994	0.3308	0.3629	0.3954	0.4280	0.4607	0.4932	0.5252
30	0.1123	0.1328	0.1552	0.1794	0.2053	0.2327	0.2615	0.2913	0.3222	0.3537	0.3857	0.4180	0.4503	0.4825	0.5144
32	0.1059	0.1253	0.1465	0.1694	0.1940	0.2202	0.2476	0.2763	0.3060	0.3365	0.3676	0.3991	0.4307	0.4623	0.4938
34	0.1002	0.1185	0.1386	0.1604	0.1838	0.2088	0.2351	0.2626	0.2912	0.3207	0.3509	0.3815	0.4124	0.4435	0.4744
36	0.0951	0.1125	0.1315	0.1523	0.1746	0.1984	0.2236	0.2501	0.2776	0.3061	0.3354	0.3652	0.3954	0.4258	0.4563
38	0.0905	0.1070	0.1251	0.1449	0.1662	0.1890	0.2131	0.2386	0.2652	0.2927	0.3211	0.3501	0.3795	0.4093	0.4392
40	0.0864	0.1020	0.1193	0.1381	0.1585	0.1803	0.2035	0.2280	0.2536	0.2803	0.3078	0.3360	0.3647	0.3938	0.4231
42	0.0826	0.0975	0.1140	0.1320	0.1515	0.1724	0.1947	0.2183	0.2430	0.2687	0.2954	0.3228	0.3508	0.3793	0.4080
44	0.0791	0.0934	0.1091	0.1263	0.1450	0.1651	0.1865	0.2092	0.2331	0.2580	0.2839	0.3105	0.3378	0.3656	0.3938
46	0.0760	0.0896	0.1047	0.1212	0.1391	0.1584	0.1790	0.2009	0.2239	0.2480	0.2731	0.2990	0.3256	0.3527	0.3803
48	0.0731	0.0861	0.1005	0.1164	0.1336	0.1521	0.1720	0.1931	0.2154	0.2387	0.2630	0.2882	0.3141	0.3406	0.3676
50	0.0704	0.0829	0.0967	0.1119	0.1285	0.1463	0.1655	0.1859	0.2074	0.2300	0.2536	0.2781	0.3033	0.3292	0.3556
55	0.0645	0.0758	0.0884	0.1022	0.1173	0.1336	0.1511	0.1698	0.1897	0.2106	0.2325	0.2553	0.2790	0.3034	0.3283
60	0.0596	0.0700	0.0814	0.0941	0.1079	0.1228	0.1389	0.1562	0.1745	0.1940	0.2144	0.2357	0.2579	0.2808	0.3044
65	0.0555	0.0650	0.0755	0.0871	0.0998	0.1136	0.1285	0.1445	0.1615	0.1796	0.1986	0.2186	0.2394	0.2611	0.2834
70	0.0519	0.0607	0.0704	0.0812	0.0929	0.1057	0.1195	0.1344	0.1502	0.1671	0.1849	0.2036	0.2232	0.2436	0.2648
75	0.0489	0.0570	0.0661	0.0760	0.0869	0.0988	0.1117	0.1255	0.1403	0.1561	0.1728	0.1904	0.2089	0.2282	0.2482
80	0.0462	0.0538	0.0622	0.0715	0.0817	0.0928	0.1048	0.1178	0.1316	0.1464	0.1621	0.1787	0.1961	0.2144	0.2333
85	0.0438	0.0510	0.0588	0.0675	0.0771	0.0875	0.0987	0.1109	0.1239	0.1378	0.1526	0.1683	0.1848	0.2020	0.2200
90	0.0418	0.0484	0.0558	0.0640	0.0730	0.0827	0.0933	0.1048	0.1170	0.1302	0.1441	0.1589	0.1745	0.1909	0.2080
95	0.0399	0.0462	0.0532	0.0609	0.0693	0.0785	0.0885	0.0993	0.1109	0.1233	0.1365	0.1505	0.1653	0.1809	0.1972
100	0.0382	0.0442	0.0508	0.0580	0.0660	0.0747	0.0841	0.0944	0.1053	0.1171	0.1296	0.1429	0.1570	0.1718	0.1873

ALPHA = .005

N	LAMBDA= 26.00	27.00	28.00	29.00	30.00	31.00	32.00	33.00	34.00	35.00	36.00	37.00	38.00	39.00	40.00
1	0.9890	0.9916	0.9935	0.9950	0.9962	0.9971	0.9978	0.9983	0.9988	0.9991	0.9993	0.9995	0.9996	0.9997	0.9998
2	0.9754	0.9804	0.9845	0.9878	0.9904	0.9924	0.9941	0.9954	0.9964	0.9972	0.9979	0.9984	0.9987	0.9990	0.9993
3	0.9600	0.9675	0.9737	0.9788	0.9830	0.9864	0.9892	0.9914	0.9932	0.9946	0.9958	0.9967	0.9974	0.9980	0.9984
4	0.9435	0.9534	0.9617	0.9686	0.9744	0.9792	0.9832	0.9864	0.9891	0.9912	0.9930	0.9944	0.9956	0.9965	0.9972
5	0.9261	0.9382	0.9486	0.9574	0.9648	0.9710	0.9762	0.9806	0.9842	0.9872	0.9896	0.9916	0.9932	0.9946	0.9957
6	0.9082	0.9224	0.9347	0.9453	0.9543	0.9620	0.9685	0.9740	0.9786	0.9824	0.9856	0.9883	0.9905	0.9923	0.9937
7	0.8899	0.9061	0.9203	0.9325	0.9431	0.9522	0.9600	0.9667	0.9723	0.9771	0.9811	0.9844	0.9872	0.9895	0.9915
8	0.8716	0.8896	0.9054	0.9193	0.9314	0.9419	0.9510	0.9588	0.9655	0.9712	0.9760	0.9801	0.9835	0.9864	0.9888
9	0.8533	0.8729	0.8903	0.9057	0.9193	0.9311	0.9415	0.9504	0.9581	0.9648	0.9704	0.9753	0.9794	0.9829	0.9858
10	0.8351	0.8562	0.8751	0.8919	0.9068	0.9199	0.9315	0.9416	0.9503	0.9579	0.9645	0.9701	0.9749	0.9790	0.9825
11	0.8171	0.8395	0.8597	0.8779	0.8941	0.9085	0.9212	0.9323	0.9421	0.9507	0.9581	0.9645	0.9700	0.9747	0.9788
12	0.7994	0.8230	0.8444	0.8638	0.8812	0.8967	0.9106	0.9228	0.9336	0.9431	0.9513	0.9585	0.9648	0.9702	0.9748
13	0.7820	0.8067	0.8292	0.8497	0.8682	0.8848	0.8997	0.9130	0.9248	0.9352	0.9443	0.9523	0.9592	0.9653	0.9705
14	0.7650	0.7906	0.8141	0.8356	0.8552	0.8728	0.8887	0.9030	0.9157	0.9270	0.9369	0.9457	0.9534	0.9601	0.9660
15	0.7483	0.7747	0.7992	0.8216	0.8421	0.8608	0.8776	0.8928	0.9064	0.9186	0.9294	0.9389	0.9473	0.9547	0.9612
16	0.7320	0.7592	0.7844	0.8077	0.8291	0.8487	0.8664	0.8825	0.8970	0.9100	0.9215	0.9319	0.9410	0.9490	0.9561
17	0.7161	0.7440	0.7699	0.7940	0.8162	0.8366	0.8552	0.8721	0.8874	0.9012	0.9135	0.9246	0.9344	0.9431	0.9508
18	0.7007	0.7290	0.7556	0.7804	0.8033	0.8245	0.8439	0.8616	0.8777	0.8923	0.9054	0.9172	0.9277	0.9371	0.9454
19	0.6856	0.7145	0.7416	0.7670	0.7906	0.8125	0.8326	0.8511	0.8679	0.8832	0.8971	0.9096	0.9208	0.9308	0.9397
20	0.6710	0.7002	0.7279	0.7538	0.7781	0.8006	0.8214	0.8406	0.8581	0.8741	0.8887	0.9018	0.9137	0.9243	0.9338
21	0.6567	0.6863	0.7144	0.7409	0.7656	0.7888	0.8102	0.8301	0.8483	0.8650	0.8802	0.8940	0.9065	0.9177	0.9278
22	0.6429	0.6728	0.7012	0.7281	0.7534	0.7771	0.7991	0.8196	0.8384	0.8558	0.8716	0.8861	0.8992	0.9110	0.9217
23	0.6294	0.6596	0.6883	0.7156	0.7413	0.7655	0.7881	0.8091	0.8286	0.8465	0.8630	0.8780	0.8918	0.9042	0.9154
24	0.6164	0.6467	0.6757	0.7033	0.7295	0.7541	0.7772	0.7987	0.8188	0.8373	0.8543	0.8700	0.8842	0.8973	0.9090
25	0.6037	0.6342	0.6634	0.6913	0.7178	0.7428	0.7664	0.7884	0.8090	0.8280	0.8456	0.8618	0.8767	0.8902	0.9026
26	0.5914	0.6220	0.6514	0.6795	0.7063	0.7318	0.7557	0.7782	0.7992	0.8188	0.8369	0.8537	0.8690	0.8831	0.8960
27	0.5795	0.6101	0.6396	0.6680	0.6951	0.7208	0.7452	0.7681	0.7896	0.8096	0.8282	0.8455	0.8614	0.8760	0.8893
28	0.5679	0.5985	0.6282	0.6567	0.6840	0.7101	0.7348	0.7581	0.7800	0.8005	0.8196	0.8373	0.8537	0.8688	0.8826
29	0.5566	0.5873	0.6170	0.6457	0.6732	0.6995	0.7245	0.7481	0.7704	0.7914	0.8109	0.8291	0.8459	0.8615	0.8758
30	0.5457	0.5763	0.6061	0.6348	0.6625	0.6891	0.7144	0.7383	0.7610	0.7823	0.8023	0.8209	0.8382	0.8542	0.8690
32	0.5248	0.5553	0.5850	0.6140	0.6419	0.6688	0.6946	0.7191	0.7424	0.7644	0.7851	0.8046	0.8227	0.8396	0.8552
34	0.5051	0.5354	0.5650	0.5940	0.6221	0.6493	0.6754	0.7004	0.7242	0.7469	0.7682	0.7884	0.8073	0.8249	0.8414
36	0.4866	0.5165	0.5460	0.5749	0.6031	0.6305	0.6568	0.6822	0.7065	0.7296	0.7516	0.7724	0.7919	0.8103	0.8275
38	0.4690	0.4987	0.5280	0.5568	0.5849	0.6123	0.6389	0.6646	0.6892	0.7128	0.7353	0.7566	0.7768	0.7958	0.8136
40	0.4525	0.4818	0.5108	0.5394	0.5675	0.5949	0.6216	0.6475	0.6724	0.6963	0.7192	0.7411	0.7618	0.7814	0.7998
42	0.4369	0.4658	0.4945	0.5229	0.5508	0.5782	0.6050	0.6309	0.6561	0.6803	0.7036	0.7258	0.7470	0.7671	0.7861
44	0.4221	0.4506	0.4789	0.5070	0.5348	0.5622	0.5889	0.6150	0.6402	0.6647	0.6883	0.7108	0.7324	0.7530	0.7725
46	0.4082	0.4361	0.4641	0.4920	0.5195	0.5467	0.5734	0.5995	0.6249	0.6495	0.6733	0.6962	0.7181	0.7391	0.7591
48	0.3949	0.4225	0.4500	0.4776	0.5049	0.5319	0.5585	0.5846	0.6100	0.6348	0.6587	0.6819	0.7041	0.7254	0.7458
50	0.3824	0.4094	0.4366	0.4638	0.4909	0.5177	0.5441	0.5701	0.5956	0.6204	0.6445	0.6679	0.6904	0.7120	0.7327
55	0.3538	0.3797	0.4058	0.4320	0.4583	0.4845	0.5105	0.5362	0.5616	0.5864	0.6107	0.6344	0.6573	0.6795	0.7009
60	0.3286	0.3533	0.3783	0.4036	0.4290	0.4545	0.4799	0.5052	0.5302	0.5549	0.5792	0.6029	0.6261	0.6487	0.6705
65	0.3064	0.3299	0.3538	0.3781	0.4026	0.4273	0.4521	0.4768	0.5014	0.5258	0.5499	0.5735	0.5968	0.6195	0.6416
70	0.2866	0.3089	0.3318	0.3552	0.3788	0.4027	0.4267	0.4508	0.4749	0.4988	0.5226	0.5461	0.5692	0.5920	0.6142
75	0.2689	0.2902	0.3121	0.3344	0.3572	0.3803	0.4035	0.4270	0.4505	0.4739	0.4973	0.5205	0.5434	0.5661	0.5883
80	0.2530	0.2734	0.2943	0.3157	0.3376	0.3598	0.3823	0.4051	0.4279	0.4509	0.4738	0.4966	0.5193	0.5417	0.5639
85	0.2388	0.2582	0.2781	0.2987	0.3197	0.3411	0.3629	0.3849	0.4072	0.4295	0.4520	0.4744	0.4967	0.5189	0.5408
90	0.2259	0.2444	0.2635	0.2832	0.3034	0.3240	0.3450	0.3664	0.3880	0.4098	0.4316	0.4536	0.4755	0.4974	0.5191
95	0.2142	0.2318	0.2501	0.2690	0.2884	0.3083	0.3286	0.3493	0.3702	0.3914	0.4127	0.4342	0.4557	0.4772	0.4986
100	0.2035	0.2204	0.2379	0.2561	0.2747	0.2939	0.3134	0.3334	0.3537	0.3743	0.3951	0.4161	0.4371	0.4582	0.4793

ALPHA = .005

N	LAMBDA= 42.00	44.00	46.00	48.00	50.00	55.00	60.00	65.00	70.00	75.00	80.00	85.00	90.00	95.00	100.00
1	0.9999	0.9999	1.0000	1.0000	1.0000	1.0000	1.0000	1.0000	1.0000	1.0000	1.0000	1.0000	1.0000	1.0000	1.0000
2	0.9996	0.9998	0.9999	0.9999	1.0000	1.0000	1.0000	1.0000	1.0000	1.0000	1.0000	1.0000	1.0000	1.0000	1.0000
3	0.9990	0.9994	0.9997	0.9998	0.9999	1.0000	1.0000	1.0000	1.0000	1.0000	1.0000	1.0000	1.0000	1.0000	1.0000
4	0.9983	0.9989	0.9994	0.9996	0.9998	0.9999	1.0000	1.0000	1.0000	1.0000	1.0000	1.0000	1.0000	1.0000	1.0000
5	0.9972	0.9983	0.9989	0.9993	0.9996	0.9999	1.0000	1.0000	1.0000	1.0000	1.0000	1.0000	1.0000	1.0000	1.0000
6	0.9959	0.9974	0.9983	0.9990	0.9993	0.9998	0.9999	1.0000	1.0000	1.0000	1.0000	1.0000	1.0000	1.0000	1.0000
7	0.9944	0.9963	0.9976	0.9985	0.9990	0.9997	0.9999	1.0000	1.0000	1.0000	1.0000	1.0000	1.0000	1.0000	1.0000
8	0.9925	0.9950	0.9967	0.9979	0.9986	0.9996	0.9999	1.0000	1.0000	1.0000	1.0000	1.0000	1.0000	1.0000	1.0000
9	0.9903	0.9935	0.9956	0.9971	0.9981	0.9994	0.9998	0.9999	1.0000	1.0000	1.0000	1.0000	1.0000	1.0000	1.0000
10	0.9879	0.9917	0.9944	0.9962	0.9975	0.9991	0.9997	0.9999	1.0000	1.0000	1.0000	1.0000	1.0000	1.0000	1.0000
11	0.9852	0.9897	0.9930	0.9952	0.9968	0.9989	0.9996	0.9999	1.0000	1.0000	1.0000	1.0000	1.0000	1.0000	1.0000
12	0.9822	0.9875	0.9914	0.9941	0.9960	0.9985	0.9995	0.9998	0.9999	1.0000	1.0000	1.0000	1.0000	1.0000	1.0000
13	0.9789	0.9851	0.9896	0.9928	0.9950	0.9981	0.9993	0.9998	0.9999	1.0000	1.0000	1.0000	1.0000	1.0000	1.0000
14	0.9754	0.9825	0.9876	0.9913	0.9940	0.9977	0.9991	0.9997	0.9999	1.0000	1.0000	1.0000	1.0000	1.0000	1.0000
15	0.9717	0.9796	0.9854	0.9897	0.9928	0.9972	0.9989	0.9996	0.9999	1.0000	1.0000	1.0000	1.0000	1.0000	1.0000
16	0.9677	0.9765	0.9831	0.9879	0.9915	0.9966	0.9987	0.9995	0.9998	0.9999	1.0000	1.0000	1.0000	1.0000	1.0000
17	0.9636	0.9733	0.9806	0.9860	0.9901	0.9959	0.9984	0.9994	0.9998	0.9999	1.0000	1.0000	1.0000	1.0000	1.0000
18	0.9592	0.9698	0.9779	0.9840	0.9885	0.9952	0.9981	0.9993	0.9997	0.9999	1.0000	1.0000	1.0000	1.0000	1.0000
19	0.9546	0.9662	0.9751	0.9818	0.9868	0.9944	0.9977	0.9991	0.9997	0.9999	1.0000	1.0000	1.0000	1.0000	1.0000
20	0.9498	0.9624	0.9721	0.9794	0.9850	0.9935	0.9973	0.9989	0.9996	0.9999	0.9999	1.0000	1.0000	1.0000	1.0000
21	0.9449	0.9584	0.9689	0.9770	0.9831	0.9925	0.9968	0.9987	0.9995	0.9998	0.9999	1.0000	1.0000	1.0000	1.0000
22	0.9399	0.9543	0.9656	0.9743	0.9811	0.9915	0.9963	0.9985	0.9994	0.9998	0.9999	1.0000	1.0000	1.0000	1.0000
23	0.9346	0.9500	0.9621	0.9716	0.9789	0.9903	0.9958	0.9983	0.9993	0.9997	0.9999	1.0000	1.0000	1.0000	1.0000
24	0.9293	0.9456	0.9585	0.9687	0.9766	0.9891	0.9952	0.9980	0.9992	0.9997	0.9999	1.0000	1.0000	1.0000	1.0000
25	0.9238	0.9410	0.9548	0.9657	0.9742	0.9878	0.9945	0.9977	0.9990	0.9996	0.9999	0.9999	1.0000	1.0000	1.0000
26	0.9182	0.9364	0.9510	0.9626	0.9717	0.9865	0.9938	0.9973	0.9989	0.9996	0.9998	0.9999	1.0000	1.0000	1.0000
27	0.9126	0.9316	0.9470	0.9594	0.9691	0.9850	0.9931	0.9970	0.9987	0.9995	0.9998	0.9999	1.0000	1.0000	1.0000
28	0.9068	0.9267	0.9430	0.9560	0.9664	0.9835	0.9923	0.9966	0.9985	0.9994	0.9998	0.9999	1.0000	1.0000	1.0000
29	0.9009	0.9218	0.9388	0.9526	0.9636	0.9819	0.9914	0.9961	0.9983	0.9993	0.9997	0.9999	1.0000	1.0000	1.0000
30	0.8950	0.9167	0.9345	0.9490	0.9607	0.9802	0.9905	0.9957	0.9981	0.9992	0.9997	0.9999	1.0000	1.0000	1.0000
32	0.8830	0.9063	0.9257	0.9416	0.9546	0.9766	0.9885	0.9946	0.9976	0.9990	0.9996	0.9998	0.9999	1.0000	1.0000
34	0.8707	0.8957	0.9166	0.9339	0.9481	0.9727	0.9863	0.9934	0.9970	0.9987	0.9994	0.9998	0.9999	1.0000	1.0000
36	0.8584	0.8848	0.9072	0.9259	0.9413	0.9685	0.9839	0.9921	0.9963	0.9983	0.9993	0.9997	0.9999	1.0000	1.0000
38	0.8459	0.8738	0.8975	0.9176	0.9343	0.9640	0.9812	0.9906	0.9955	0.9979	0.9991	0.9996	0.9998	0.9999	1.0000
40	0.8334	0.8626	0.8877	0.9090	0.9269	0.9593	0.9784	0.9890	0.9946	0.9975	0.9989	0.9995	0.9998	0.9999	1.0000
42	0.8208	0.8513	0.8777	0.9003	0.9194	0.9543	0.9753	0.9872	0.9937	0.9970	0.9986	0.9994	0.9997	0.9999	1.0000
44	0.8083	0.8400	0.8676	0.8913	0.9116	0.9490	0.9720	0.9853	0.9926	0.9964	0.9983	0.9992	0.9997	0.9999	0.9999
46	0.7959	0.8286	0.8574	0.8823	0.9036	0.9436	0.9685	0.9832	0.9914	0.9958	0.9980	0.9991	0.9996	0.9998	0.9999
48	0.7835	0.8173	0.8471	0.8731	0.8955	0.9379	0.9648	0.9809	0.9901	0.9950	0.9976	0.9989	0.9995	0.9998	0.9999
50	0.7713	0.8059	0.8367	0.8638	0.8872	0.9321	0.9610	0.9785	0.9886	0.9942	0.9972	0.9987	0.9994	0.9997	0.9999
55	0.7411	0.7778	0.8108	0.8402	0.8661	0.9167	0.9505	0.9718	0.9846	0.9919	0.9959	0.9980	0.9990	0.9996	0.9998
60	0.7120	0.7502	0.7851	0.8165	0.8445	0.9005	0.9391	0.9643	0.9798	0.9890	0.9943	0.9971	0.9986	0.9993	0.9997
65	0.6839	0.7234	0.7598	0.7929	0.8228	0.8836	0.9268	0.9558	0.9744	0.9857	0.9923	0.9960	0.9980	0.9990	0.9995
70	0.6571	0.6974	0.7350	0.7695	0.8010	0.8663	0.9138	0.9467	0.9682	0.9818	0.9899	0.9946	0.9972	0.9986	0.9993
75	0.6314	0.6724	0.7108	0.7465	0.7794	0.8486	0.9002	0.9368	0.9615	0.9773	0.9871	0.9929	0.9962	0.9980	0.9990
80	0.6070	0.6483	0.6874	0.7241	0.7580	0.8307	0.8862	0.9264	0.9541	0.9724	0.9839	0.9909	0.9950	0.9974	0.9986
85	0.5838	0.6253	0.6649	0.7022	0.7371	0.8128	0.8717	0.9154	0.9462	0.9669	0.9803	0.9887	0.9937	0.9966	0.9982
90	0.5618	0.6033	0.6431	0.6810	0.7166	0.7949	0.8570	0.9040	0.9377	0.9610	0.9763	0.9861	0.9921	0.9956	0.9976
95	0.5409	0.5823	0.6222	0.6604	0.6966	0.7771	0.8421	0.8922	0.9288	0.9546	0.9719	0.9832	0.9902	0.9945	0.9970
100	0.5211	0.5622	0.6021	0.6406	0.6772	0.7595	0.8271	0.8801	0.9195	0.9478	0.9672	0.9800	0.9881	0.9932	0.9962

ALPHA = .001

N	LAMBDA= .00	.10	.20	.30	.40	.50	.60	.70	.80	.90	1.00	1.20	1.40	1.60	1.80
1	0.0010	0.0016	0.0023	0.0031	0.0040	0.0049	0.0060	0.0071	0.0083	0.0096	0.0110	0.0141	0.0175	0.0214	0.0257
2	0.0010	0.0014	0.0018	0.0022	0.0027	0.0033	0.0039	0.0046	0.0053	0.0060	0.0069	0.0087	0.0107	0.0130	0.0156
3	0.0010	0.0013	0.0016	0.0019	0.0023	0.0027	0.0031	0.0036	0.0041	0.0046	0.0052	0.0065	0.0080	0.0097	0.0115
4	0.0010	0.0012	0.0015	0.0017	0.0020	0.0023	0.0027	0.0030	0.0034	0.0039	0.0043	0.0053	0.0065	0.0078	0.0092
5	0.0010	0.0012	0.0014	0.0016	0.0019	0.0021	0.0024	0.0027	0.0030	0.0034	0.0038	0.0046	0.0055	0.0066	0.0078
6	0.0010	0.0012	0.0013	0.0015	0.0017	0.0020	0.0022	0.0025	0.0028	0.0031	0.0034	0.0041	0.0049	0.0058	0.0068
7	0.0010	0.0011	0.0013	0.0015	0.0017	0.0019	0.0021	0.0023	0.0026	0.0028	0.0031	0.0037	0.0044	0.0052	0.0060
8	0.0010	0.0011	0.0013	0.0014	0.0016	0.0018	0.0020	0.0022	0.0024	0.0026	0.0029	0.0034	0.0040	0.0047	0.0055
9	0.0010	0.0011	0.0013	0.0014	0.0016	0.0017	0.0019	0.0021	0.0023	0.0025	0.0027	0.0032	0.0038	0.0044	0.0050
10	0.0010	0.0011	0.0012	0.0014	0.0015	0.0017	0.0018	0.0020	0.0022	0.0024	0.0026	0.0030	0.0035	0.0041	0.0047
11	0.0010	0.0011	0.0012	0.0013	0.0015	0.0016	0.0018	0.0019	0.0021	0.0023	0.0025	0.0029	0.0033	0.0038	0.0044
12	0.0010	0.0011	0.0012	0.0013	0.0014	0.0016	0.0017	0.0019	0.0020	0.0022	0.0024	0.0027	0.0032	0.0036	0.0041
13	0.0010	0.0011	0.0012	0.0013	0.0014	0.0015	0.0017	0.0018	0.0020	0.0021	0.0023	0.0026	0.0030	0.0035	0.0039
14	0.0010	0.0011	0.0012	0.0013	0.0014	0.0015	0.0016	0.0018	0.0019	0.0021	0.0022	0.0025	0.0029	0.0033	0.0037
15	0.0010	0.0011	0.0012	0.0013	0.0014	0.0015	0.0016	0.0017	0.0019	0.0020	0.0021	0.0025	0.0028	0.0032	0.0036
16	0.0010	0.0011	0.0012	0.0013	0.0014	0.0015	0.0016	0.0017	0.0018	0.0020	0.0021	0.0024	0.0027	0.0031	0.0035
17	0.0010	0.0011	0.0012	0.0013	0.0014	0.0015	0.0016	0.0017	0.0018	0.0019	0.0020	0.0023	0.0026	0.0030	0.0033
18	0.0010	0.0011	0.0012	0.0012	0.0013	0.0014	0.0015	0.0016	0.0018	0.0019	0.0020	0.0023	0.0026	0.0029	0.0032
19	0.0010	0.0011	0.0012	0.0012	0.0013	0.0014	0.0015	0.0016	0.0017	0.0018	0.0020	0.0022	0.0025	0.0028	0.0031
20	0.0010	0.0011	0.0011	0.0012	0.0013	0.0014	0.0015	0.0016	0.0017	0.0018	0.0019	0.0022	0.0024	0.0027	0.0030
29	0.0010	0.0011	0.0011	0.0012	0.0012	0.0013	0.0014	0.0015	0.0015	0.0016	0.0017	0.0019	0.0021	0.0023	0.0025
30	0.0010	0.0011	0.0011	0.0012	0.0012	0.0013	0.0014	0.0015	0.0015	0.0016	0.0017	0.0019	0.0020	0.0022	0.0025
32	0.0010	0.0011	0.0011	0.0012	0.0012	0.0013	0.0014	0.0014	0.0015	0.0016	0.0017	0.0018	0.0020	0.0022	0.0024
34	0.0010	0.0011	0.0011	0.0012	0.0012	0.0013	0.0013	0.0014	0.0015	0.0016	0.0016	0.0018	0.0020	0.0021	0.0023
36	0.0010	0.0011	0.0011	0.0012	0.0012	0.0013	0.0013	0.0014	0.0015	0.0015	0.0016	0.0018	0.0019	0.0021	0.0023
38	0.0010	0.0010	0.0011	0.0012	0.0012	0.0013	0.0013	0.0014	0.0014	0.0015	0.0016	0.0017	0.0019	0.0020	0.0022
40	0.0010	0.0010	0.0011	0.0011	0.0012	0.0013	0.0013	0.0014	0.0014	0.0015	0.0016	0.0017	0.0018	0.0020	0.0022
42	0.0010	0.0010	0.0011	0.0011	0.0012	0.0012	0.0013	0.0014	0.0014	0.0015	0.0015	0.0017	0.0018	0.0020	0.0021
44	0.0010	0.0010	0.0011	0.0011	0.0012	0.0012	0.0013	0.0013	0.0014	0.0015	0.0015	0.0017	0.0018	0.0019	0.0021
46	0.0010	0.0010	0.0011	0.0011	0.0012	0.0012	0.0013	0.0013	0.0014	0.0015	0.0015	0.0016	0.0018	0.0019	0.0021
48	0.0010	0.0010	0.0011	0.0011	0.0012	0.0012	0.0013	0.0013	0.0014	0.0014	0.0015	0.0016	0.0017	0.0019	0.0020
50	0.0010	0.0010	0.0011	0.0011	0.0012	0.0012	0.0013	0.0013	0.0014	0.0014	0.0015	0.0016	0.0017	0.0019	0.0020
55	0.0010	0.0010	0.0011	0.0011	0.0012	0.0012	0.0013	0.0013	0.0013	0.0014	0.0015	0.0016	0.0017	0.0018	0.0019
60	0.0010	0.0010	0.0011	0.0011	0.0012	0.0012	0.0012	0.0013	0.0013	0.0014	0.0014	0.0015	0.0016	0.0017	0.0019
65	0.0010	0.0010	0.0011	0.0011	0.0011	0.0012	0.0012	0.0013	0.0013	0.0014	0.0014	0.0015	0.0016	0.0017	0.0018
70	0.0010	0.0010	0.0011	0.0011	0.0011	0.0012	0.0012	0.0013	0.0013	0.0013	0.0014	0.0015	0.0016	0.0017	0.0018
75	0.0010	0.0010	0.0011	0.0011	0.0011	0.0012	0.0012	0.0012	0.0013	0.0013	0.0014	0.0015	0.0015	0.0016	0.0017
80	0.0010	0.0010	0.0011	0.0011	0.0011	0.0012	0.0012	0.0012	0.0013	0.0013	0.0014	0.0014	0.0015	0.0016	0.0017
85	0.0010	0.0010	0.0011	0.0011	0.0011	0.0012	0.0012	0.0012	0.0013	0.0013	0.0013	0.0014	0.0015	0.0016	0.0017
90	0.0010	0.0010	0.0011	0.0011	0.0011	0.0012	0.0012	0.0012	0.0013	0.0013	0.0013	0.0014	0.0015	0.0016	0.0017
95	0.0010	0.0010	0.0011	0.0011	0.0011	0.0011	0.0012	0.0012	0.0012	0.0013	0.0013	0.0014	0.0015	0.0015	0.0016
100	0.0010	0.0010	0.0011	0.0011	0.0011	0.0011	0.0012	0.0012	0.0012	0.0013	0.0013	0.0014	0.0015	0.0015	0.0016

ALPHA = .001

N	LAMBDA= 2.00	2.20	2.40	2.60	2.80	3.00	3.50	4.00	4.50	5.00	6.00	7.00	8.00	9.00	10.00
1	0.0303	0.0354	0.0408	0.0467	0.0529	0.0596	0.0778	0.0984	0.1212	0.1458	0.2002	0.2595	0.3220	0.3857	0.4490
2	0.0185	0.0216	0.0250	0.0287	0.0326	0.0369	0.0489	0.0627	0.0784	0.0958	0.1357	0.1815	0.2320	0.2860	0.3421
3	0.0136	0.0158	0.0183	0.0210	0.0239	0.0270	0.0359	0.0463	0.0582	0.0717	0.1031	0.1403	0.1824	0.2288	0.2783
4	0.0108	0.0126	0.0145	0.0166	0.0189	0.0214	0.0284	0.0366	0.0462	0.0571	0.0829	0.1139	0.1499	0.1902	0.2341
5	0.0091	0.0105	0.0121	0.0138	0.0157	0.0177	0.0234	0.0303	0.0382	0.0473	0.0690	0.0955	0.1267	0.1621	0.2014
6	0.0079	0.0091	0.0104	0.0119	0.0134	0.0151	0.0200	0.0257	0.0325	0.0402	0.0589	0.0819	0.1093	0.1408	0.1760
7	0.0070	0.0080	0.0092	0.0104	0.0118	0.0132	0.0174	0.0224	0.0282	0.0349	0.0512	0.0715	0.0957	0.1239	0.1558
8	0.0063	0.0072	0.0082	0.0093	0.0105	0.0118	0.0155	0.0198	0.0249	0.0308	0.0452	0.0632	0.0849	0.1104	0.1394
9	0.0058	0.0066	0.0075	0.0085	0.0095	0.0106	0.0139	0.0178	0.0223	0.0276	0.0404	0.0565	0.0761	0.0992	0.1257
10	0.0054	0.0061	0.0069	0.0078	0.0087	0.0097	0.0126	0.0161	0.0202	0.0249	0.0365	0.0510	0.0688	0.0899	0.1142
11	0.0050	0.0057	0.0064	0.0072	0.0080	0.0090	0.0116	0.0148	0.0184	0.0227	0.0332	0.0464	0.0627	0.0820	0.1044
12	0.0047	0.0053	0.0060	0.0067	0.0075	0.0083	0.0107	0.0136	0.0170	0.0209	0.0304	0.0425	0.0574	0.0752	0.0960
13	0.0044	0.0050	0.0056	0.0063	0.0070	0.0078	0.0100	0.0126	0.0157	0.0193	0.0281	0.0392	0.0529	0.0694	0.0887
14	0.0042	0.0048	0.0053	0.0059	0.0066	0.0073	0.0094	0.0118	0.0146	0.0179	0.0260	0.0363	0.0490	0.0643	0.0823
15	0.0040	0.0045	0.0051	0.0056	0.0063	0.0069	0.0088	0.0111	0.0137	0.0168	0.0243	0.0338	0.0456	0.0599	0.0767
16	0.0039	0.0043	0.0048	0.0054	0.0059	0.0066	0.0083	0.0104	0.0129	0.0157	0.0227	0.0316	0.0426	0.0559	0.0717
17	0.0037	0.0042	0.0046	0.0051	0.0057	0.0063	0.0079	0.0099	0.0122	0.0148	0.0214	0.0297	0.0400	0.0524	0.0672
18	0.0036	0.0040	0.0044	0.0049	0.0054	0.0060	0.0075	0.0094	0.0115	0.0140	0.0201	0.0279	0.0376	0.0493	0.0632
19	0.0035	0.0039	0.0043	0.0047	0.0052	0.0057	0.0072	0.0090	0.0110	0.0133	0.0191	0.0264	0.0355	0.0465	0.0596
20	0.0034	0.0037	0.0041	0.0046	0.0050	0.0055	0.0069	0.0086	0.0105	0.0127	0.0181	0.0250	0.0336	0.0440	0.0564
21	0.0033	0.0036	0.0040	0.0044	0.0048	0.0053	0.0066	0.0082	0.0100	0.0121	0.0172	0.0237	0.0318	0.0417	0.0534
22	0.0032	0.0035	0.0039	0.0043	0.0047	0.0051	0.0064	0.0079	0.0096	0.0116	0.0164	0.0226	0.0303	0.0396	0.0508
23	0.0031	0.0034	0.0038	0.0041	0.0045	0.0050	0.0062	0.0076	0.0092	0.0111	0.0157	0.0216	0.0288	0.0377	0.0483
24	0.0030	0.0033	0.0037	0.0040	0.0044	0.0048	0.0060	0.0073	0.0089	0.0107	0.0150	0.0206	0.0275	0.0360	0.0461
25	0.0030	0.0033	0.0036	0.0039	0.0043	0.0047	0.0058	0.0071	0.0085	0.0103	0.0144	0.0197	0.0263	0.0344	0.0440
26	0.0029	0.0032	0.0035	0.0038	0.0042	0.0045	0.0056	0.0068	0.0083	0.0099	0.0139	0.0189	0.0252	0.0329	0.0421
27	0.0028	0.0031	0.0034	0.0037	0.0041	0.0044	0.0054	0.0066	0.0080	0.0095	0.0134	0.0182	0.0242	0.0316	0.0403
28	0.0028	0.0031	0.0033	0.0036	0.0040	0.0043	0.0053	0.0064	0.0077	0.0092	0.0129	0.0175	0.0233	0.0303	0.0387
29	0.0027	0.0030	0.0033	0.0036	0.0039	0.0042	0.0052	0.0062	0.0075	0.0089	0.0124	0.0169	0.0224	0.0291	0.0372
30	0.0027	0.0029	0.0032	0.0035	0.0038	0.0041	0.0050	0.0061	0.0073	0.0087	0.0120	0.0163	0.0216	0.0281	0.0358
32	0.0026	0.0028	0.0031	0.0034	0.0036	0.0039	0.0048	0.0058	0.0069	0.0082	0.0113	0.0152	0.0201	0.0261	0.0333
34	0.0025	0.0028	0.0030	0.0032	0.0035	0.0038	0.0046	0.0055	0.0066	0.0078	0.0107	0.0143	0.0189	0.0244	0.0310
36	0.0025	0.0027	0.0029	0.0031	0.0034	0.0037	0.0044	0.0053	0.0062	0.0074	0.0101	0.0135	0.0177	0.0229	0.0291
38	0.0024	0.0026	0.0028	0.0030	0.0033	0.0035	0.0042	0.0051	0.0060	0.0070	0.0096	0.0128	0.0167	0.0216	0.0273
40	0.0023	0.0025	0.0027	0.0030	0.0032	0.0034	0.0041	0.0049	0.0057	0.0067	0.0091	0.0121	0.0159	0.0204	0.0258
42	0.0023	0.0025	0.0027	0.0029	0.0031	0.0033	0.0040	0.0047	0.0055	0.0065	0.0087	0.0116	0.0151	0.0193	0.0244
44	0.0023	0.0024	0.0026	0.0028	0.0030	0.0032	0.0038	0.0045	0.0053	0.0062	0.0084	0.0110	0.0143	0.0183	0.0231
46	0.0022	0.0024	0.0026	0.0027	0.0029	0.0032	0.0037	0.0044	0.0051	0.0060	0.0080	0.0106	0.0137	0.0175	0.0220
48	0.0022	0.0023	0.0025	0.0027	0.0029	0.0031	0.0036	0.0043	0.0050	0.0058	0.0077	0.0101	0.0131	0.0167	0.0210
50	0.0021	0.0023	0.0025	0.0026	0.0028	0.0030	0.0035	0.0042	0.0048	0.0056	0.0075	0.0098	0.0126	0.0160	0.0200
55	0.0021	0.0022	0.0024	0.0025	0.0027	0.0029	0.0033	0.0039	0.0045	0.0052	0.0069	0.0089	0.0114	0.0144	0.0180
60	0.0020	0.0021	0.0023	0.0024	0.0026	0.0027	0.0032	0.0037	0.0042	0.0049	0.0064	0.0082	0.0105	0.0132	0.0164
65	0.0019	0.0021	0.0022	0.0023	0.0025	0.0026	0.0030	0.0035	0.0040	0.0046	0.0060	0.0076	0.0097	0.0121	0.0150
70	0.0019	0.0020	0.0021	0.0023	0.0024	0.0025	0.0029	0.0033	0.0038	0.0044	0.0056	0.0072	0.0090	0.0112	0.0139
75	0.0018	0.0020	0.0021	0.0022	0.0023	0.0025	0.0028	0.0032	0.0037	0.0042	0.0053	0.0067	0.0085	0.0105	0.0129
80	0.0018	0.0019	0.0020	0.0021	0.0023	0.0024	0.0027	0.0031	0.0035	0.0040	0.0051	0.0064	0.0080	0.0098	0.0121
85	0.0018	0.0019	0.0020	0.0021	0.0022	0.0023	0.0026	0.0030	0.0034	0.0038	0.0048	0.0061	0.0075	0.0093	0.0113
90	0.0017	0.0018	0.0019	0.0020	0.0021	0.0023	0.0026	0.0029	0.0033	0.0037	0.0046	0.0058	0.0072	0.0088	0.0107
95	0.0017	0.0018	0.0019	0.0020	0.0021	0.0022	0.0025	0.0028	0.0032	0.0036	0.0045	0.0055	0.0068	0.0084	0.0101
100	0.0017	0.0018	0.0019	0.0020	0.0021	0.0022	0.0024	0.0027	0.0031	0.0035	0.0043	0.0053	0.0065	0.0080	0.0096

ALPHA = .001

N	LAMBDA= 11.00	12.00	13.00	14.00	15.00	16.00	17.00	18.00	19.00	20.00	21.00	22.00	23.00	24.00	25.00
1	0.5104	0.5689	0.6236	0.6741	0.7199	0.7610	0.7975	0.8295	0.8573	0.8813	0.9018	0.9192	0.9339	0.9461	0.9563
2	0.3990	0.4556	0.5108	0.5638	0.6139	0.6607	0.7038	0.7431	0.7786	0.8103	0.8384	0.8631	0.8846	0.9032	0.9192
3	0.3299	0.3826	0.4355	0.4875	0.5381	0.5865	0.6322	0.6750	0.7145	0.7507	0.7835	0.8130	0.8394	0.8627	0.8832
4	0.2809	0.3295	0.3793	0.4293	0.4788	0.5270	0.5736	0.6178	0.6596	0.6985	0.7344	0.7673	0.7972	0.8241	0.8482
5	0.2438	0.2886	0.3352	0.3827	0.4304	0.4777	0.5240	0.5687	0.6115	0.6520	0.6899	0.7252	0.7577	0.7874	0.8143
6	0.2146	0.2560	0.2994	0.3443	0.3901	0.4359	0.4814	0.5259	0.5690	0.6103	0.6495	0.6863	0.7207	0.7525	0.7818
7	0.1911	0.2293	0.2698	0.3122	0.3558	0.4000	0.4443	0.4882	0.5311	0.5726	0.6125	0.6504	0.6861	0.7195	0.7506
8	0.1717	0.2070	0.2449	0.2849	0.3263	0.3688	0.4117	0.4546	0.4970	0.5384	0.5786	0.6171	0.6537	0.6883	0.7208
9	0.1555	0.1883	0.2237	0.2613	0.3007	0.3414	0.3829	0.4247	0.4663	0.5073	0.5474	0.5862	0.6234	0.6589	0.6923
10	0.1417	0.1722	0.2054	0.2409	0.2783	0.3172	0.3572	0.3978	0.4385	0.4789	0.5187	0.5575	0.5951	0.6310	0.6653
11	0.1299	0.1584	0.1895	0.2230	0.2585	0.2957	0.3342	0.3735	0.4132	0.4529	0.4922	0.5309	0.5685	0.6048	0.6396
12	0.1197	0.1463	0.1755	0.2072	0.2409	0.2765	0.3135	0.3515	0.3901	0.4290	0.4678	0.5060	0.5435	0.5800	0.6151
13	0.1108	0.1357	0.1632	0.1931	0.2252	0.2592	0.2948	0.3315	0.3690	0.4070	0.4451	0.4829	0.5202	0.5566	0.5919
14	0.1030	0.1263	0.1523	0.1806	0.2112	0.2436	0.2778	0.3132	0.3496	0.3867	0.4240	0.4613	0.4982	0.5345	0.5699
15	0.0960	0.1180	0.1425	0.1694	0.1985	0.2295	0.2623	0.2965	0.3318	0.3679	0.4044	0.4411	0.4776	0.5136	0.5489
16	0.0899	0.1106	0.1337	0.1592	0.1870	0.2167	0.2482	0.2812	0.3154	0.3505	0.3862	0.4222	0.4582	0.4939	0.5291
17	0.0844	0.1039	0.1258	0.1501	0.1765	0.2050	0.2352	0.2671	0.3002	0.3343	0.3692	0.4045	0.4400	0.4753	0.5102
18	0.0794	0.0979	0.1187	0.1418	0.1670	0.1943	0.2234	0.2541	0.2861	0.3193	0.3533	0.3879	0.4228	0.4576	0.4922
19	0.0749	0.0925	0.1122	0.1342	0.1583	0.1844	0.2124	0.2420	0.2731	0.3053	0.3385	0.3723	0.4065	0.4409	0.4751
20	0.0709	0.0875	0.1063	0.1273	0.1503	0.1754	0.2023	0.2309	0.2609	0.2923	0.3246	0.3577	0.3912	0.4251	0.4589
21	0.0672	0.0830	0.1009	0.1209	0.1430	0.1671	0.1929	0.2205	0.2496	0.2800	0.3115	0.3439	0.3768	0.4101	0.4435
22	0.0638	0.0789	0.0960	0.1151	0.1362	0.1593	0.1843	0.2109	0.2391	0.2686	0.2993	0.3308	0.3631	0.3958	0.4288
23	0.0607	0.0751	0.0914	0.1097	0.1300	0.1522	0.1762	0.2019	0.2292	0.2579	0.2877	0.3186	0.3502	0.3823	0.4148
24	0.0579	0.0716	0.0872	0.1048	0.1242	0.1456	0.1687	0.1936	0.2200	0.2478	0.2769	0.3070	0.3379	0.3695	0.4014
25	0.0553	0.0684	0.0833	0.1002	0.1188	0.1394	0.1617	0.1857	0.2113	0.2384	0.2667	0.2961	0.3263	0.3573	0.3887
26	0.0529	0.0654	0.0797	0.0959	0.1138	0.1336	0.1552	0.1784	0.2032	0.2295	0.2570	0.2857	0.3153	0.3457	0.3766
27	0.0507	0.0627	0.0764	0.0919	0.1092	0.1283	0.1491	0.1715	0.1956	0.2211	0.2479	0.2759	0.3049	0.3346	0.3650
28	0.0486	0.0601	0.0733	0.0882	0.1049	0.1232	0.1433	0.1651	0.1884	0.2132	0.2393	0.2666	0.2949	0.3241	0.3539
29	0.0467	0.0578	0.0704	0.0848	0.1008	0.1185	0.1380	0.1590	0.1816	0.2057	0.2311	0.2578	0.2855	0.3141	0.3433
30	0.0449	0.0555	0.0677	0.0815	0.0970	0.1141	0.1329	0.1533	0.1752	0.1987	0.2234	0.2494	0.2765	0.3045	0.3332
32	0.0417	0.0515	0.0628	0.0757	0.0901	0.1061	0.1237	0.1428	0.1635	0.1856	0.2091	0.2339	0.2598	0.2866	0.3143
34	0.0389	0.0480	0.0585	0.0705	0.0839	0.0989	0.1154	0.1335	0.1530	0.1739	0.1962	0.2198	0.2446	0.2703	0.2970
36	0.0364	0.0449	0.0547	0.0659	0.0785	0.0925	0.1081	0.1251	0.1435	0.1634	0.1846	0.2071	0.2307	0.2554	0.2811
38	0.0342	0.0421	0.0513	0.0618	0.0736	0.0868	0.1015	0.1175	0.1350	0.1538	0.1740	0.1954	0.2180	0.2417	0.2664
40	0.0322	0.0397	0.0483	0.0581	0.0692	0.0817	0.0955	0.1107	0.1272	0.1451	0.1643	0.1848	0.2064	0.2292	0.2529
42	0.0304	0.0374	0.0455	0.0548	0.0653	0.0771	0.0901	0.1045	0.1202	0.1372	0.1555	0.1750	0.1957	0.2176	0.2404
44	0.0288	0.0354	0.0431	0.0518	0.0617	0.0728	0.0852	0.0988	0.1138	0.1299	0.1474	0.1661	0.1859	0.2069	0.2289
46	0.0274	0.0336	0.0408	0.0491	0.0585	0.0690	0.0807	0.0937	0.1079	0.1233	0.1400	0.1578	0.1769	0.1970	0.2182
48	0.0260	0.0320	0.0388	0.0466	0.0555	0.0655	0.0767	0.0890	0.1025	0.1172	0.1331	0.1502	0.1685	0.1878	0.2082
50	0.0248	0.0305	0.0370	0.0444	0.0528	0.0623	0.0729	0.0847	0.0975	0.1116	0.1268	0.1432	0.1607	0.1793	0.1990
55	0.0223	0.0272	0.0330	0.0395	0.0470	0.0554	0.0648	0.0753	0.0868	0.0993	0.1130	0.1278	0.1436	0.1605	0.1784
60	0.0202	0.0246	0.0297	0.0356	0.0422	0.0497	0.0582	0.0675	0.0779	0.0892	0.1015	0.1149	0.1293	0.1447	0.1611
65	0.0184	0.0224	0.0270	0.0323	0.0383	0.0450	0.0526	0.0611	0.0704	0.0807	0.0919	0.1040	0.1172	0.1312	0.1463
70	0.0170	0.0206	0.0247	0.0295	0.0349	0.0411	0.0479	0.0556	0.0641	0.0734	0.0836	0.0948	0.1068	0.1197	0.1336
75	0.0157	0.0190	0.0228	0.0272	0.0321	0.0377	0.0440	0.0509	0.0587	0.0672	0.0766	0.0868	0.0978	0.1098	0.1225
80	0.0147	0.0177	0.0212	0.0251	0.0297	0.0348	0.0405	0.0469	0.0541	0.0619	0.0705	0.0799	0.0901	0.1011	0.1129
85	0.0137	0.0165	0.0197	0.0234	0.0276	0.0323	0.0376	0.0435	0.0500	0.0573	0.0652	0.0739	0.0833	0.0935	0.1045
90	0.0129	0.0155	0.0185	0.0219	0.0257	0.0301	0.0350	0.0404	0.0465	0.0532	0.0605	0.0686	0.0773	0.0868	0.0970
95	0.0122	0.0146	0.0174	0.0205	0.0241	0.0282	0.0327	0.0378	0.0434	0.0496	0.0564	0.0639	0.0721	0.0809	0.0904
100	0.0116	0.0138	0.0164	0.0193	0.0227	0.0264	0.0307	0.0354	0.0406	0.0464	0.0528	0.0598	0.0674	0.0756	0.0846

ALPHA = .001

N	LAMBDA= 26.00	27.00	28.00	29.00	30.00	31.00	32.00	33.00	34.00	35.00	36.00	37.00	38.00	39.00	40.00
1	0.9647	0.9717	0.9773	0.9819	0.9856	0.9886	0.9910	0.9929	0.9945	0.9957	0.9966	0.9974	0.9980	0.9984	0.9988
2	0.9329	0.9445	0.9543	0.9625	0.9694	0.9751	0.9798	0.9837	0.9868	0.9894	0.9915	0.9933	0.9946	0.9957	0.9966
3	0.9011	0.9166	0.9300	0.9415	0.9513	0.9596	0.9667	0.9726	0.9775	0.9816	0.9850	0.9878	0.9901	0.9920	0.9936
4	0.8696	0.8885	0.9050	0.9195	0.9320	0.9428	0.9521	0.9600	0.9667	0.9724	0.9772	0.9812	0.9846	0.9874	0.9897
5	0.8386	0.8603	0.8797	0.8967	0.9118	0.9249	0.9363	0.9462	0.9547	0.9620	0.9683	0.9735	0.9780	0.9818	0.9850
6	0.8084	0.8325	0.8543	0.8737	0.8910	0.9063	0.9197	0.9315	0.9418	0.9507	0.9583	0.9649	0.9706	0.9754	0.9795
7	0.7791	0.8053	0.8291	0.8506	0.8699	0.8872	0.9025	0.9161	0.9281	0.9385	0.9476	0.9555	0.9624	0.9683	0.9733
8	0.7509	0.7788	0.8043	0.8276	0.8488	0.8678	0.8849	0.9002	0.9137	0.9257	0.9362	0.9454	0.9535	0.9604	0.9665
9	0.7237	0.7530	0.7800	0.8049	0.8277	0.8484	0.8671	0.8839	0.8989	0.9123	0.9242	0.9347	0.9439	0.9520	0.9590
10	0.6977	0.7280	0.7564	0.7826	0.8068	0.8289	0.8491	0.8673	0.8838	0.8986	0.9118	0.9235	0.9339	0.9430	0.9511
11	0.6727	0.7040	0.7333	0.7607	0.7861	0.8096	0.8310	0.8506	0.8684	0.8845	0.8989	0.9119	0.9234	0.9336	0.9426
12	0.6488	0.6803	0.7110	0.7394	0.7659	0.7904	0.8131	0.8339	0.8529	0.8702	0.8858	0.8999	0.9125	0.9238	0.9338
13	0.6259	0.6585	0.6894	0.7186	0.7460	0.7716	0.7953	0.8172	0.8373	0.8557	0.8725	0.8876	0.9013	0.9136	0.9245
14	0.6041	0.6371	0.6685	0.6984	0.7266	0.7530	0.7777	0.8006	0.8217	0.8412	0.8590	0.8752	0.8898	0.9031	0.9150
15	0.5833	0.6165	0.6484	0.6788	0.7076	0.7348	0.7603	0.7841	0.8062	0.8266	0.8454	0.8625	0.8782	0.8924	0.9052
16	0.5634	0.5968	0.6289	0.6598	0.6892	0.7170	0.7432	0.7678	0.7908	0.8121	0.8317	0.8498	0.8663	0.8814	0.8951
17	0.5444	0.5778	0.6102	0.6414	0.6712	0.6996	0.7265	0.7518	0.7755	0.7976	0.8181	0.8370	0.8544	0.8703	0.8849
18	0.5263	0.5597	0.5922	0.6236	0.6538	0.6826	0.7100	0.7360	0.7603	0.7832	0.8044	0.8241	0.8424	0.8591	0.8744
19	0.5090	0.5423	0.5748	0.6064	0.6368	0.6661	0.6939	0.7204	0.7454	0.7689	0.7909	0.8113	0.8303	0.8478	0.8639
20	0.4925	0.5256	0.5581	0.5898	0.6204	0.6499	0.6782	0.7051	0.7307	0.7548	0.7774	0.7985	0.8182	0.8364	0.8532
21	0.4767	0.5097	0.5421	0.5737	0.6045	0.6343	0.6629	0.6902	0.7162	0.7408	0.7640	0.7858	0.8061	0.8250	0.8425
22	0.4617	0.4944	0.5266	0.5583	0.5891	0.6190	0.6479	0.6755	0.7020	0.7271	0.7508	0.7731	0.7941	0.8136	0.8318
23	0.4473	0.4797	0.5118	0.5433	0.5742	0.6042	0.6333	0.6612	0.6880	0.7135	0.7377	0.7606	0.7821	0.8022	0.8210
24	0.4335	0.4656	0.4975	0.5289	0.5597	0.5898	0.6190	0.6472	0.6743	0.7002	0.7248	0.7482	0.7702	0.7909	0.8102
25	0.4204	0.4521	0.4837	0.5150	0.5457	0.5758	0.6051	0.6335	0.6609	0.6871	0.7121	0.7359	0.7584	0.7796	0.7994
26	0.4078	0.4392	0.4705	0.5015	0.5322	0.5623	0.5916	0.6202	0.6477	0.6742	0.6996	0.7237	0.7466	0.7683	0.7887
27	0.3957	0.4267	0.4577	0.4886	0.5191	0.5491	0.5785	0.6071	0.6348	0.6616	0.6872	0.7117	0.7350	0.7571	0.7780
28	0.3842	0.4143	0.4455	0.4761	0.5064	0.5364	0.5657	0.5944	0.6223	0.6492	0.6751	0.6999	0.7236	0.7461	0.7674
29	0.3732	0.4033	0.4337	0.4640	0.4942	0.5240	0.5533	0.5820	0.6099	0.6370	0.6632	0.6882	0.7122	0.7351	0.7568
30	0.3626	0.3923	0.4223	0.4524	0.4823	0.5120	0.5412	0.5699	0.5979	0.6251	0.6514	0.6768	0.7011	0.7243	0.7463
32	0.3427	0.3716	0.4009	0.4303	0.4597	0.4891	0.5181	0.5467	0.5747	0.6021	0.6287	0.6544	0.6791	0.7029	0.7256
34	0.3244	0.3525	0.3809	0.4097	0.4386	0.4675	0.4962	0.5246	0.5526	0.5800	0.6068	0.6327	0.6579	0.6821	0.7053
36	0.3076	0.3347	0.3624	0.3905	0.4188	0.4472	0.4755	0.5037	0.5315	0.5589	0.5857	0.6119	0.6373	0.6618	0.6855
38	0.2920	0.3183	0.3452	0.3726	0.4003	0.4281	0.4560	0.4839	0.5115	0.5387	0.5655	0.5918	0.6173	0.6422	0.6662
40	0.2776	0.3030	0.3292	0.3558	0.3829	0.4102	0.4376	0.4651	0.4924	0.5195	0.5462	0.5724	0.5981	0.6231	0.6474
42	0.2642	0.2889	0.3142	0.3401	0.3665	0.3933	0.4202	0.4473	0.4742	0.5011	0.5276	0.5538	0.5795	0.6047	0.6291
44	0.2518	0.2757	0.3002	0.3255	0.3512	0.3773	0.4038	0.4304	0.4570	0.4835	0.5099	0.5360	0.5616	0.5868	0.6114
46	0.2403	0.2634	0.2872	0.3117	0.3368	0.3623	0.3882	0.4143	0.4406	0.4668	0.4929	0.5188	0.5444	0.5696	0.5943
48	0.2296	0.2519	0.2750	0.2988	0.3232	0.3482	0.3735	0.3992	0.4250	0.4509	0.4767	0.5024	0.5278	0.5530	0.5776
50	0.2196	0.2411	0.2635	0.2867	0.3105	0.3348	0.3596	0.3847	0.4101	0.4356	0.4612	0.4866	0.5119	0.5369	0.5616
55	0.1974	0.2172	0.2379	0.2594	0.2816	0.3045	0.3279	0.3518	0.3760	0.4005	0.4252	0.4500	0.4747	0.4993	0.5236
60	0.1785	0.1963	0.2159	0.2359	0.2567	0.2781	0.3002	0.3228	0.3459	0.3693	0.3930	0.4169	0.4409	0.4649	0.4889
65	0.1623	0.1792	0.1969	0.2155	0.2349	0.2550	0.2758	0.2972	0.3191	0.3414	0.3642	0.3872	0.4104	0.4337	0.4571
70	0.1483	0.1639	0.1804	0.1978	0.2159	0.2348	0.2543	0.2745	0.2953	0.3165	0.3382	0.3603	0.3827	0.4052	0.4280
75	0.1362	0.1507	0.1660	0.1822	0.1991	0.2168	0.2352	0.2543	0.2740	0.2942	0.3149	0.3360	0.3575	0.3793	0.4013
80	0.1256	0.1391	0.1534	0.1685	0.1843	0.2009	0.2183	0.2363	0.2549	0.2741	0.2939	0.3141	0.3347	0.3557	0.3769
85	0.1163	0.1288	0.1422	0.1563	0.1712	0.1868	0.2031	0.2201	0.2378	0.2560	0.2748	0.2942	0.3139	0.3341	0.3546
90	0.1080	0.1197	0.1322	0.1455	0.1594	0.1741	0.1896	0.2056	0.2224	0.2397	0.2576	0.2761	0.2950	0.3143	0.3341
95	0.1007	0.1117	0.1234	0.1358	0.1489	0.1628	0.1773	0.1926	0.2084	0.2249	0.2420	0.2596	0.2777	0.2963	0.3153
100	0.0942	0.1045	0.1155	0.1271	0.1395	0.1526	0.1663	0.1807	0.1958	0.2115	0.2277	0.2445	0.2619	0.2797	0.2979

ALPHA = .001

N	LAMBDA= 42.00	44.00	46.00	48.00	50.00	55.00	60.00	65.00	70.00	75.00	80.00	85.00	90.00	95.00	100.00
1	0.9993	0.9996	0.9998	0.9999	0.9999	1.0000	1.0000	1.0000	1.0000	1.0000	1.0000	1.0000	1.0000	1.0000	1.0000
2	0.9979	0.9987	0.9992	0.9995	0.9997	0.9999	1.0000	1.0000	1.0000	1.0000	1.0000	1.0000	1.0000	1.0000	1.0000
3	0.9959	0.9974	0.9984	0.9990	0.9994	0.9998	0.9999	1.0000	1.0000	1.0000	1.0000	1.0000	1.0000	1.0000	1.0000
4	0.9932	0.9956	0.9971	0.9982	0.9988	0.9996	0.9999	1.0000	1.0000	1.0000	1.0000	1.0000	1.0000	1.0000	1.0000
5	0.9899	0.9932	0.9955	0.9971	0.9981	0.9994	0.9998	0.9999	1.0000	1.0000	1.0000	1.0000	1.0000	1.0000	1.0000
6	0.9859	0.9904	0.9935	0.9957	0.9971	0.9990	0.9997	0.9999	1.0000	1.0000	1.0000	1.0000	1.0000	1.0000	1.0000
7	0.9813	0.9870	0.9911	0.9940	0.9959	0.9986	0.9995	0.9998	1.0000	1.0000	1.0000	1.0000	1.0000	1.0000	1.0000
8	0.9761	0.9832	0.9883	0.9919	0.9945	0.9980	0.9993	0.9998	0.9999	1.0000	1.0000	1.0000	1.0000	1.0000	1.0000
9	0.9704	0.9789	0.9851	0.9896	0.9928	0.9972	0.9990	0.9997	0.9999	1.0000	1.0000	1.0000	1.0000	1.0000	1.0000
10	0.9642	0.9741	0.9815	0.9869	0.9908	0.9964	0.9986	0.9995	0.9998	0.9999	1.0000	1.0000	1.0000	1.0000	1.0000
11	0.9576	0.9690	0.9775	0.9839	0.9886	0.9954	0.9982	0.9993	0.9998	0.9999	1.0000	1.0000	1.0000	1.0000	1.0000
12	0.9505	0.9634	0.9732	0.9806	0.9861	0.9942	0.9977	0.9991	0.9997	0.9999	1.0000	1.0000	1.0000	1.0000	1.0000
13	0.9430	0.9574	0.9685	0.9770	0.9834	0.9929	0.9971	0.9989	0.9996	0.9999	1.0000	1.0000	1.0000	1.0000	1.0000
14	0.9352	0.9511	0.9635	0.9731	0.9803	0.9914	0.9964	0.9986	0.9995	0.9998	0.9999	1.0000	1.0000	1.0000	1.0000
15	0.9271	0.9445	0.9582	0.9689	0.9771	0.9897	0.9956	0.9982	0.9993	0.9997	0.9999	1.0000	1.0000	1.0000	1.0000
16	0.9186	0.9376	0.9527	0.9644	0.9736	0.9879	0.9948	0.9978	0.9991	0.9997	0.9999	1.0000	1.0000	1.0000	1.0000
17	0.9100	0.9304	0.9468	0.9597	0.9698	0.9859	0.9938	0.9974	0.9989	0.9996	0.9999	0.9999	1.0000	1.0000	1.0000
18	0.9011	0.9230	0.9407	0.9548	0.9658	0.9837	0.9927	0.9969	0.9987	0.9995	0.9998	0.9999	1.0000	1.0000	1.0000
19	0.8921	0.9154	0.9343	0.9496	0.9616	0.9814	0.9915	0.9963	0.9985	0.9994	0.9998	0.9999	1.0000	1.0000	1.0000
20	0.8829	0.9075	0.9278	0.9441	0.9572	0.9789	0.9902	0.9956	0.9981	0.9992	0.9997	0.9999	1.0000	1.0000	1.0000
21	0.8735	0.8995	0.9210	0.9385	0.9526	0.9763	0.9887	0.9949	0.9978	0.9991	0.9996	0.9999	0.9999	1.0000	1.0000
22	0.8641	0.8914	0.9141	0.9327	0.9478	0.9735	0.9872	0.9941	0.9974	0.9989	0.9996	0.9998	0.9999	1.0000	1.0000
23	0.8546	0.8831	0.9070	0.9267	0.9428	0.9705	0.9856	0.9933	0.9970	0.9987	0.9995	0.9998	0.9999	1.0000	1.0000
24	0.8450	0.8747	0.8997	0.9206	0.9377	0.9674	0.9838	0.9923	0.9965	0.9985	0.9994	0.9998	0.9999	1.0000	1.0000
25	0.8353	0.8662	0.8924	0.9143	0.9324	0.9641	0.9819	0.9913	0.9960	0.9983	0.9993	0.9997	0.9999	1.0000	1.0000
26	0.8256	0.8576	0.8849	0.9079	0.9269	0.9607	0.9799	0.9903	0.9955	0.9980	0.9991	0.9996	0.9999	0.9999	1.0000
27	0.8159	0.8490	0.8773	0.9013	0.9213	0.9571	0.9778	0.9891	0.9949	0.9977	0.9990	0.9996	0.9998	0.9999	1.0000
28	0.8062	0.8403	0.8696	0.8946	0.9156	0.9534	0.9756	0.9879	0.9942	0.9974	0.9988	0.9995	0.9998	0.9999	1.0000
29	0.7966	0.8316	0.8619	0.8878	0.9098	0.9496	0.9733	0.9865	0.9935	0.9970	0.9987	0.9994	0.9998	0.9999	1.0000
30	0.7869	0.8228	0.8541	0.8810	0.9038	0.9457	0.9709	0.9851	0.9928	0.9966	0.9985	0.9993	0.9997	0.9999	1.0000
32	0.7677	0.8053	0.8383	0.8670	0.8916	0.9375	0.9657	0.9821	0.9911	0.9957	0.9980	0.9991	0.9996	0.9998	0.9999
34	0.7487	0.7878	0.8224	0.8528	0.8791	0.9288	0.9602	0.9788	0.9892	0.9947	0.9975	0.9989	0.9995	0.9998	0.9999
36	0.7300	0.7704	0.8065	0.8384	0.8663	0.9198	0.9542	0.9751	0.9871	0.9935	0.9969	0.9986	0.9994	0.9997	0.9999
38	0.7116	0.7531	0.7906	0.8239	0.8533	0.9104	0.9479	0.9712	0.9847	0.9922	0.9962	0.9982	0.9992	0.9996	0.9998
40	0.6935	0.7360	0.7747	0.8094	0.8401	0.9007	0.9413	0.9669	0.9821	0.9907	0.9954	0.9978	0.9990	0.9995	0.9998
42	0.6759	0.7192	0.7589	0.7948	0.8268	0.8907	0.9343	0.9623	0.9793	0.9891	0.9945	0.9973	0.9987	0.9994	0.9997
44	0.6586	0.7027	0.7433	0.7803	0.8135	0.8805	0.9271	0.9575	0.9763	0.9873	0.9935	0.9967	0.9984	0.9993	0.9997
46	0.6418	0.6865	0.7279	0.7658	0.8001	0.8701	0.9196	0.9524	0.9730	0.9853	0.9923	0.9961	0.9981	0.9991	0.9996
48	0.6254	0.6705	0.7127	0.7515	0.7868	0.8595	0.9118	0.9470	0.9695	0.9832	0.9911	0.9954	0.9977	0.9989	0.9995
50	0.6094	0.6550	0.6977	0.7372	0.7734	0.8489	0.9038	0.9415	0.9658	0.9809	0.9897	0.9946	0.9973	0.9987	0.9994
55	0.5715	0.6175	0.6613	0.7024	0.7405	0.8217	0.8830	0.9265	0.9557	0.9743	0.9857	0.9923	0.9960	0.9980	0.9990
60	0.5362	0.5824	0.6267	0.6688	0.7083	0.7944	0.8613	0.9103	0.9443	0.9667	0.9809	0.9894	0.9943	0.9970	0.9985
65	0.5036	0.5494	0.5939	0.6367	0.6772	0.7670	0.8389	0.8931	0.9319	0.9582	0.9752	0.9858	0.9922	0.9958	0.9978
70	0.4735	0.5187	0.5631	0.6060	0.6472	0.7400	0.8162	0.8752	0.9185	0.9487	0.9688	0.9817	0.9896	0.9943	0.9969
75	0.4457	0.4901	0.5340	0.5770	0.6184	0.7135	0.7933	0.8566	0.9042	0.9383	0.9616	0.9769	0.9866	0.9924	0.9958
80	0.4201	0.4635	0.5068	0.5495	0.5910	0.6876	0.7704	0.8376	0.8893	0.9272	0.9537	0.9715	0.9830	0.9902	0.9945
85	0.3964	0.4383	0.4813	0.5235	0.5649	0.6624	0.7477	0.8183	0.8738	0.9153	0.9451	0.9655	0.9790	0.9876	0.9929
90	0.3745	0.4158	0.4574	0.4990	0.5401	0.6381	0.7253	0.7989	0.8579	0.9029	0.9358	0.9589	0.9745	0.9846	0.9910
95	0.3543	0.3944	0.4351	0.4760	0.5166	0.6146	0.7033	0.7795	0.8416	0.8900	0.9260	0.9518	0.9695	0.9813	0.9888
100	0.3356	0.3745	0.4142	0.4543	0.4944	0.5920	0.6818	0.7601	0.8251	0.8766	0.9156	0.9441	0.9640	0.9775	0.9863

TABLE II. NON-CENTRALITY PARAMETER OF NON-CENTRAL CHI-SQUARE ALPHA = .100

N	POWER= .10	.12	.14	.16	.18	.20	.22	.24	.26	.28
1	0.000	0.118	0.237	0.357	0.479	0.601	0.725	0.850	0.977	1.106
2	0.000	0.173	0.344	0.513	0.682	0.851	1.020	1.189	1.358	1.529
3	0.000	0.216	0.428	0.636	0.841	1.045	1.246	1.447	1.648	1.849
4	0.000	0.254	0.500	0.740	0.976	1.208	1.437	1.664	1.890	2.115
5	0.000	0.288	0.564	0.833	1.095	1.352	1.605	1.855	2.103	2.349
6	0.000	0.318	0.623	0.917	1.203	1.482	1.756	2.027	2.294	2.560
7	0.000	0.346	0.676	0.994	1.302	1.602	1.896	2.185	2.470	2.753
8	0.000	0.373	0.727	1.066	1.394	1.713	2.025	2.332	2.634	2.932
9	0.000	0.398	0.774	1.134	1.481	1.818	2.147	2.469	2.787	3.101
10	0.000	0.421	0.819	1.198	1.563	1.917	2.262	2.600	2.932	3.260
11	0.000	0.444	0.862	1.259	1.642	2.012	2.372	2.724	3.070	3.412
12	0.000	0.465	0.903	1.318	1.717	2.102	2.477	2.843	3.202	3.556
13	0.000	0.486	0.942	1.374	1.789	2.189	2.577	2.956	3.329	3.695
14	0.000	0.506	0.980	1.428	1.858	2.272	2.674	3.066	3.450	3.828
15	0.000	0.525	1.016	1.481	1.925	2.352	2.767	3.172	3.568	3.957
16	0.000	0.544	1.052	1.531	1.989	2.430	2.858	3.274	3.681	4.082
17	0.000	0.562	1.086	1.581	2.052	2.506	2.945	3.373	3.791	4.202
18	0.000	0.580	1.119	1.628	2.113	2.579	3.030	3.469	3.898	4.320
19	0.000	0.597	1.152	1.675	2.173	2.651	3.113	3.563	4.002	4.434
20	0.000	0.614	1.184	1.720	2.230	2.720	3.194	3.654	4.103	4.545
21	0.000	0.630	1.214	1.764	2.287	2.788	3.272	3.743	4.202	4.653
22	0.000	0.646	1.245	1.807	2.342	2.854	3.349	3.830	4.299	4.759
23	0.000	0.662	1.274	1.849	2.396	2.919	3.424	3.915	4.393	4.862
24	0.000	0.677	1.303	1.891	2.448	2.983	3.498	3.998	4.485	4.963
25	0.000	0.692	1.331	1.931	2.500	3.045	3.570	4.079	4.576	5.062
26	0.000	0.707	1.359	1.971	2.551	3.106	3.641	4.159	4.665	5.160
27	0.000	0.721	1.386	2.010	2.601	3.165	3.710	4.237	4.752	5.255
28	0.000	0.735	1.413	2.048	2.649	3.224	3.778	4.314	4.837	5.348
29	0.000	0.749	1.440	2.086	2.697	3.282	3.845	4.390	4.921	5.440
30	0.000	0.763	1.465	2.123	2.744	3.339	3.910	4.464	5.003	5.531
32	0.000	0.790	1.516	2.195	2.837	3.449	4.039	4.609	5.164	5.707
34	0.000	0.816	1.565	2.265	2.926	3.557	4.163	4.750	5.320	5.878
36	0.000	0.841	1.612	2.332	3.012	3.661	4.284	4.886	5.472	6.044
38	0.000	0.865	1.659	2.399	3.097	3.762	4.401	5.019	5.619	6.205
40	0.000	0.889	1.704	2.463	3.179	3.861	4.516	5.148	5.763	6.363
42	0.000	0.912	1.748	2.526	3.259	3.957	4.627	5.274	5.903	6.516
44	0.000	0.935	1.791	2.587	3.338	4.052	4.736	5.398	6.040	6.666
46	0.000	0.957	1.833	2.647	3.414	4.144	4.843	5.518	6.173	6.813
48	0.000	0.979	1.874	2.706	3.489	4.234	4.947	5.636	6.304	6.956
50	0.000	1.000	1.914	2.763	3.563	4.322	5.050	5.752	6.433	7.096
55	0.000	1.052	2.012	2.903	3.740	4.535	5.297	6.031	6.742	7.436
60	0.000	1.101	2.105	3.035	3.910	4.739	5.533	6.298	7.038	7.760
65	0.000	1.149	2.194	3.163	4.072	4.935	5.760	6.554	7.323	8.071
70	0.000	1.194	2.280	3.286	4.229	5.123	5.978	6.800	7.596	8.371
75	0.000	1.238	2.363	3.404	4.380	5.305	6.188	7.038	7.860	8.659
80	0.000	1.281	2.444	3.519	4.527	5.481	6.392	7.268	8.115	8.939
85	0.000	1.322	2.522	3.630	4.668	5.651	6.589	7.491	8.363	9.210
90	0.000	1.362	2.597	3.738	4.806	5.817	6.781	7.707	8.603	9.473
95	0.000	1.401	2.671	3.843	4.940	5.978	6.967	7.918	8.837	9.729
100	0.000	1.439	2.743	3.946	5.071	6.135	7.149	8.123	9.064	9.979

ALPHA = .100

N	POWER= .30	.32	.34	.36	.38	.40	.42	.44	.46	.48
1	1.237	1.370	1.505	1.643	1.784	1.928	2.075	2.225	2.380	2.538
2	1.701	1.874	2.049	2.227	2.407	2.589	2.775	2.964	3.157	3.354
3	2.050	2.252	2.455	2.661	2.868	3.078	3.290	3.506	3.726	3.950
4	2.341	2.566	2.792	3.020	3.250	3.481	3.716	3.953	4.195	4.440
5	2.595	2.841	3.087	3.334	3.582	3.833	4.086	4.342	4.602	4.866
6	2.824	3.088	3.351	3.616	3.881	4.149	4.419	4.691	4.968	5.248
7	3.034	3.314	3.594	3.874	4.155	4.438	4.723	5.011	5.302	5.597
8	3.229	3.524	3.819	4.114	4.409	4.706	5.005	5.307	5.612	5.921
9	3.412	3.721	4.030	4.338	4.647	4.957	5.269	5.584	5.902	6.224
10	3.585	3.908	4.229	4.550	4.872	5.194	5.519	5.846	6.176	6.510
11	3.749	4.085	4.419	4.752	5.085	5.420	5.756	6.094	6.436	6.781
12	3.907	4.254	4.600	4.944	5.289	5.635	5.982	6.331	6.684	7.040
13	4.057	4.416	4.773	5.129	5.484	5.841	6.198	6.558	6.921	7.288
14	4.202	4.572	4.940	5.306	5.672	6.039	6.407	6.777	7.150	7.527
15	4.342	4.722	5.101	5.477	5.853	6.230	6.607	6.987	7.370	7.757
16	4.477	4.868	5.256	5.643	6.028	6.414	6.802	7.191	7.583	7.979
17	4.608	5.009	5.407	5.803	6.198	6.593	6.990	7.388	7.789	8.194
18	4.735	5.146	5.553	5.958	6.363	6.767	7.172	7.579	7.989	8.403
19	4.859	5.279	5.695	6.110	6.523	6.936	7.350	7.765	8.184	8.606
20	4.979	5.408	5.834	6.257	6.679	7.100	7.523	7.947	8.373	8.804
21	5.097	5.535	5.969	6.401	6.831	7.261	7.691	8.123	8.558	8.996
22	5.211	5.658	6.101	6.541	6.979	7.417	7.856	8.296	8.738	9.185
23	5.323	5.779	6.230	6.678	7.124	7.570	8.016	8.464	8.914	9.368
24	5.433	5.897	6.356	6.812	7.266	7.720	8.174	8.629	9.087	9.548
25	5.541	6.013	6.480	6.944	7.405	7.866	8.328	8.790	9.256	9.724
26	5.646	6.126	6.601	7.072	7.542	8.010	8.479	8.949	9.421	9.897
27	5.749	6.237	6.720	7.199	7.675	8.151	8.627	9.104	9.583	10.066
28	5.851	6.346	6.836	7.323	7.807	8.289	8.772	9.256	9.742	10.232
29	5.951	6.454	6.951	7.445	7.936	8.425	8.915	9.406	9.899	10.396
30	6.049	6.559	7.064	7.564	8.062	8.559	9.055	9.553	10.053	10.556
32	6.240	6.765	7.283	7.798	8.309	8.819	9.329	9.840	10.352	10.869
34	6.425	6.964	7.497	8.024	8.549	9.072	9.594	10.118	10.643	11.172
36	6.605	7.158	7.703	8.244	8.782	9.317	9.852	10.388	10.925	11.466
38	6.780	7.346	7.905	8.458	9.008	9.555	10.102	10.650	11.200	11.752
40	6.951	7.529	8.101	8.666	9.228	9.788	10.346	10.906	11.467	12.031
42	7.117	7.708	8.292	8.869	9.443	10.014	10.584	11.155	11.727	12.303
44	7.280	7.883	8.478	9.068	9.653	10.235	10.817	11.398	11.982	12.568
46	7.438	8.054	8.661	9.262	9.858	10.452	11.044	11.636	12.231	12.828
48	7.594	8.221	8.839	9.451	10.059	10.663	11.266	11.869	12.474	13.082
50	7.746	8.385	9.014	9.637	10.255	10.870	11.484	12.097	12.712	13.330
55	8.114	8.780	9.437	10.086	10.730	11.371	12.010	12.648	13.288	13.931
60	8.466	9.158	9.841	10.515	11.184	11.849	12.512	13.174	13.838	14.504
65	8.803	9.521	10.228	10.927	11.620	12.308	12.994	13.679	14.365	15.055
70	9.127	9.870	10.601	11.323	12.039	12.749	13.458	14.165	14.873	15.584
75	9.441	10.207	10.961	11.705	12.443	13.175	13.905	14.634	15.363	16.095
80	9.744	10.533	11.309	12.075	12.834	13.588	14.338	15.087	15.837	16.589
85	10.037	10.848	11.646	12.433	13.213	13.987	14.757	15.526	16.296	17.068
90	10.323	11.155	11.974	12.781	13.581	14.375	15.165	15.953	16.742	17.533
95	10.600	11.453	12.292	13.120	13.939	14.752	15.561	16.368	17.175	17.985
100	10.870	11.744	12.603	13.450	14.288	15.119	15.947	16.772	17.598	18.425

ALPHA = .100

N	POWER= .50	.52	.54	.56	.58	.60	.62	.64	.66	.68
1	2.701	2.869	3.043	3.222	3.408	3.601	3.802	4.012	4.231	4.462
2	3.556	3.763	3.975	4.194	4.420	4.654	4.896	5.148	5.411	5.686
3	4.178	4.412	4.652	4.898	5.152	5.414	5.684	5.965	6.258	6.563
4	4.690	4.946	5.207	5.476	5.752	6.036	6.330	6.634	6.950	7.280
5	5.135	5.409	5.689	5.977	6.272	6.575	6.889	7.213	7.550	7.901
6	5.533	5.824	6.121	6.425	6.737	7.058	7.388	7.731	8.086	8.455
7	5.898	6.203	6.515	6.834	7.162	7.498	7.845	8.203	8.575	8.961
8	6.235	6.555	6.881	7.214	7.555	7.906	8.267	8.640	9.027	9.429
9	6.551	6.883	7.222	7.569	7.923	8.287	8.662	9.049	9.450	9.867
10	6.849	7.194	7.545	7.903	8.270	8.647	9.035	9.435	9.849	10.279
11	7.132	7.488	7.851	8.221	8.600	8.988	9.388	9.800	10.227	10.670
12	7.402	7.769	8.142	8.524	8.914	9.314	9.725	10.149	10.588	11.043
13	7.660	8.038	8.422	8.814	9.214	9.625	10.047	10.483	10.933	11.400
14	7.909	8.296	8.691	9.093	9.503	9.925	10.357	10.803	11.264	11.743
15	8.148	8.545	8.949	9.361	9.782	10.213	10.656	11.112	11.584	12.073
16	8.380	8.786	9.200	9.621	10.051	10.492	10.945	11.411	11.893	12.392
17	8.604	9.019	9.442	9.872	10.312	10.762	11.224	11.700	12.191	12.701
18	8.822	9.246	9.677	10.116	10.565	11.024	11.495	11.980	12.481	13.001
19	9.033	9.466	9.905	10.353	10.810	11.278	11.758	12.253	12.763	13.292
20	9.239	9.680	10.128	10.584	11.050	11.526	12.015	12.518	13.037	13.575
21	9.440	9.889	10.345	10.809	11.283	11.767	12.265	12.776	13.305	13.852
22	9.636	10.093	10.557	11.029	11.510	12.003	12.509	13.029	13.565	14.121
23	9.827	10.292	10.763	11.243	11.733	12.233	12.747	13.275	13.820	14.385
24	10.014	10.487	10.966	11.453	11.950	12.459	12.980	13.516	14.069	14.642
25	10.198	10.677	11.164	11.659	12.163	12.679	13.208	13.752	14.313	14.894
26	10.378	10.864	11.358	11.860	12.372	12.895	13.432	13.984	14.553	15.141
27	10.554	11.048	11.548	12.058	12.577	13.107	13.651	14.210	14.787	15.384
28	10.727	11.227	11.735	12.251	12.778	13.315	13.866	14.433	15.017	15.621
29	10.897	11.404	11.919	12.442	12.975	13.519	14.078	14.651	15.243	15.855
30	11.064	11.578	12.099	12.629	13.169	13.720	14.285	14.866	15.465	16.084
32	11.390	11.917	12.451	12.994	13.547	14.112	14.690	15.285	15.898	16.532
34	11.705	12.245	12.792	13.347	13.913	14.491	15.083	15.691	16.317	16.965
36	12.012	12.563	13.122	13.690	14.269	14.859	15.464	16.085	16.724	17.385
38	12.310	12.873	13.444	14.024	14.614	15.217	15.834	16.467	17.120	17.794
40	12.600	13.175	13.758	14.349	14.951	15.566	16.195	16.840	17.505	18.193
42	12.883	13.469	14.063	14.666	15.279	15.905	16.546	17.204	17.881	18.581
44	13.160	13.757	14.361	14.975	15.600	16.237	16.889	17.559	18.248	18.960
46	13.430	14.037	14.653	15.278	15.913	16.562	17.225	17.906	18.606	19.330
48	13.694	14.312	14.938	15.573	16.220	16.879	17.553	18.245	18.957	19.692
50	13.953	14.581	15.217	15.863	16.520	17.189	17.874	18.577	19.300	20.047
55	14.578	15.231	15.892	16.563	17.245	17.940	18.651	19.380	20.130	20.904
60	15.175	15.852	16.537	17.231	17.937	18.657	19.392	20.146	20.922	21.722
65	15.748	16.448	17.155	17.872	18.601	19.344	20.103	20.881	21.681	22.507
70	16.299	17.020	17.750	18.489	19.240	20.005	20.787	21.588	22.412	23.261
75	16.831	17.573	18.324	19.084	19.856	20.643	21.447	22.270	23.117	23.989
80	17.345	18.108	18.878	19.659	20.452	21.260	22.085	22.930	23.798	24.694
85	17.844	18.626	19.416	20.217	21.030	21.858	22.703	23.569	24.459	25.376
90	18.328	19.129	19.939	20.758	21.591	22.438	23.304	24.190	25.100	26.038
95	18.798	19.618	20.446	21.285	22.136	23.003	23.888	24.793	25.724	26.682
100	19.257	20.095	20.941	21.798	22.668	23.553	24.456	25.381	26.331	27.310

ALPHA = .100

N	POWER= .70	.71	.72	.73	.74	.75	.76	.77	.78	.79
1	4.705	4.831	4.962	5.096	5.235	5.379	5.527	5.682	5.842	6.008
2	5.974	6.124	6.279	6.438	6.601	6.770	6.945	7.126	7.313	7.508
3	6.884	7.050	7.221	7.396	7.577	7.763	7.956	8.155	8.361	8.575
4	7.626	7.805	7.989	8.178	8.372	8.573	8.780	8.993	9.215	9.444
5	8.268	8.458	8.653	8.854	9.060	9.272	9.491	9.718	9.952	10.195
6	8.842	9.042	9.247	9.457	9.674	9.897	10.127	10.364	10.610	10.864
7	9.365	9.574	9.788	10.008	10.234	10.466	10.706	10.954	11.210	11.475
8	9.849	10.066	10.289	10.517	10.752	10.993	11.242	11.499	11.764	12.039
9	10.301	10.526	10.756	10.993	11.236	11.485	11.742	12.008	12.282	12.566
10	10.728	10.960	11.197	11.441	11.691	11.949	12.214	12.488	12.770	13.063
11	11.132	11.371	11.615	11.866	12.124	12.389	12.661	12.942	13.233	13.534
12	11.517	11.763	12.014	12.271	12.536	12.808	13.087	13.376	13.674	13.982
13	11.886	12.138	12.395	12.659	12.930	13.208	13.495	13.791	14.096	14.412
14	12.241	12.498	12.762	13.032	13.309	13.594	13.887	14.189	14.501	14.824
15	12.582	12.845	13.115	13.390	13.674	13.965	14.264	14.573	14.891	15.221
16	12.912	13.181	13.455	13.737	14.026	14.323	14.628	14.943	15.268	15.604
17	13.231	13.505	13.785	14.073	14.367	14.670	14.981	15.302	15.633	15.975
18	13.541	13.820	14.105	14.398	14.698	15.006	15.323	15.650	15.987	16.335
19	13.842	14.126	14.417	14.714	15.019	15.333	15.655	15.988	16.330	16.685
20	14.135	14.424	14.719	15.022	15.332	15.651	15.979	16.316	16.665	17.025
21	14.420	14.714	15.014	15.322	15.637	15.961	16.294	16.637	16.991	17.357
22	14.699	14.997	15.302	15.614	15.935	16.263	16.602	16.950	17.309	17.680
23	14.971	15.274	15.583	15.900	16.225	16.559	16.902	17.255	17.620	17.996
24	15.237	15.544	15.858	16.180	16.509	16.848	17.196	17.554	17.924	18.306
25	15.498	15.809	16.128	16.453	16.788	17.131	17.483	17.846	18.221	18.608
26	15.753	16.069	16.391	16.722	17.060	17.408	17.765	18.133	18.512	18.904
27	16.003	16.323	16.650	16.984	17.327	17.679	18.041	18.414	18.798	19.195
28	16.249	16.573	16.904	17.242	17.590	17.946	18.312	18.689	19.078	19.480
29	16.490	16.818	17.153	17.496	17.847	18.208	18.578	18.960	19.353	19.760
30	16.727	17.059	17.398	17.745	18.100	18.465	18.840	19.226	19.624	20.035
32	17.189	17.529	17.875	18.230	18.593	18.966	19.350	19.744	20.151	20.571
34	17.637	17.984	18.338	18.700	19.071	19.452	19.844	20.246	20.662	21.091
36	18.072	18.425	18.787	19.156	19.535	19.924	20.323	20.734	21.157	21.595
38	18.494	18.855	19.223	19.600	19.986	20.382	20.789	21.208	21.639	22.085
40	18.905	19.273	19.648	20.032	20.425	20.828	21.243	21.669	22.108	22.562
42	19.306	19.680	20.062	20.453	20.853	21.263	21.685	22.118	22.565	23.027
44	19.698	20.078	20.467	20.864	21.271	21.688	22.117	22.557	23.012	23.481
46	20.080	20.467	20.862	21.265	21.679	22.103	22.538	22.986	23.448	23.924
48	20.455	20.847	21.248	21.658	22.078	22.509	22.951	23.406	23.874	24.358
50	20.821	21.220	21.627	22.043	22.469	22.906	23.355	23.817	24.292	24.783
55	21.706	22.119	22.541	22.972	23.414	23.866	24.331	24.808	25.301	25.808
60	22.551	22.978	23.414	23.859	24.315	24.783	25.262	25.756	26.264	26.788
65	23.362	23.802	24.251	24.710	25.180	25.661	26.156	26.664	27.187	27.727
70	24.141	24.594	25.056	25.528	26.011	26.507	27.015	27.537	28.075	28.630
75	24.893	25.358	25.832	26.317	26.814	27.322	27.844	28.380	28.932	29.501
80	25.620	26.097	26.583	27.081	27.589	28.110	28.645	29.194	29.760	30.343
85	26.325	26.813	27.311	27.820	28.341	28.874	29.421	29.984	30.563	31.159
90	27.009	27.508	28.018	28.538	29.070	29.616	30.175	30.750	31.342	31.952
95	27.674	28.184	28.705	29.236	29.780	30.337	30.908	31.495	32.099	32.722
100	28.322	28.842	29.374	29.916	30.471	31.039	31.622	32.221	32.837	33.472

ALPHA = .100

N	POWER= .80	.81	.82	.83	.84	.85	.86	.87	.88	.89
1	6.182	6.364	6.554	6.755	6.966	7.189	7.426	7.680	7.951	8.245
2	7.711	7.922	8.144	8.376	8.620	8.879	9.152	9.444	9.756	10.093
3	8.798	9.030	9.273	9.527	9.794	10.076	10.375	10.693	11.033	11.400
4	9.683	9.931	10.191	10.463	10.749	11.050	11.369	11.708	12.071	12.460
5	10.447	10.710	10.984	11.271	11.572	11.890	12.226	12.583	12.965	13.375
6	11.129	11.404	11.691	11.992	12.307	12.639	12.990	13.364	13.762	14.190
7	11.750	12.037	12.335	12.648	12.976	13.321	13.686	14.074	14.488	14.932
8	12.324	12.621	12.931	13.255	13.594	13.952	14.330	14.731	15.159	15.618
9	12.861	13.168	13.487	13.821	14.172	14.541	14.930	15.344	15.785	16.258
10	13.367	13.682	14.011	14.355	14.716	15.095	15.496	15.921	16.374	16.861
11	13.846	14.170	14.508	14.861	15.231	15.621	16.032	16.468	16.933	17.432
12	14.302	14.635	14.981	15.343	15.722	16.121	16.542	16.989	17.465	17.975
13	14.739	15.079	15.434	15.804	16.192	16.600	17.031	17.487	17.974	18.496
14	15.158	15.506	15.868	16.247	16.643	17.060	17.500	17.966	18.462	18.995
15	15.562	15.917	16.287	16.673	17.077	17.503	17.951	18.426	18.933	19.476
16	15.953	16.314	16.691	17.085	17.497	17.930	18.387	18.871	19.387	19.940
17	16.330	16.699	17.083	17.483	17.903	18.344	18.809	19.302	19.827	20.389
18	16.696	17.072	17.462	17.870	18.297	18.745	19.218	19.719	20.253	20.825
19	17.052	17.434	17.831	18.245	18.679	19.135	19.616	20.125	20.667	21.248
20	17.399	17.786	18.189	18.610	19.051	19.514	20.002	20.519	21.070	21.659
21	17.736	18.129	18.539	18.966	19.414	19.884	20.379	20.904	21.462	22.060
22	18.065	18.464	18.880	19.314	19.767	20.244	20.747	21.279	21.845	22.452
23	18.387	18.792	19.213	19.653	20.113	20.596	21.106	21.645	22.219	22.834
24	18.701	19.112	19.539	19.984	20.451	20.940	21.457	22.003	22.585	23.207
25	19.009	19.425	19.858	20.309	20.781	21.277	21.800	22.354	22.942	23.573
26	19.310	19.732	20.170	20.627	21.105	21.607	22.137	22.697	23.293	23.931
27	19.606	20.032	20.476	20.939	21.423	21.931	22.466	23.033	23.636	24.282
28	19.896	20.327	20.776	21.245	21.734	22.248	22.790	23.364	23.973	24.626
29	20.181	20.617	21.071	21.545	22.040	22.560	23.108	23.688	24.304	24.964
30	20.460	20.902	21.361	21.840	22.341	22.866	23.420	24.006	24.629	25.296
32	21.006	21.457	21.926	22.415	22.927	23.463	24.029	24.627	25.263	25.944
34	21.534	21.995	22.473	22.972	23.494	24.042	24.618	25.228	25.877	26.571
36	22.047	22.516	23.004	23.513	24.045	24.603	25.190	25.812	26.473	27.179
38	22.546	23.024	23.521	24.039	24.580	25.148	25.746	26.379	27.052	27.771
40	23.031	23.517	24.023	24.550	25.101	25.679	26.287	26.931	27.615	28.346
42	23.504	23.999	24.513	25.049	25.609	26.196	26.815	27.469	28.164	28.907
44	23.966	24.468	24.991	25.536	26.105	26.702	27.330	27.994	28.700	29.455
46	24.417	24.927	25.458	26.011	26.589	27.195	27.833	28.507	29.224	29.990
48	24.858	25.376	25.915	26.476	27.063	27.678	28.325	29.009	29.736	30.513
50	25.290	25.816	26.363	26.932	27.527	28.150	28.807	29.501	30.238	31.026
55	26.333	26.878	27.443	28.032	28.647	29.291	29.970	30.687	31.449	32.262
60	27.330	27.891	28.474	29.082	29.716	30.381	31.080	31.819	32.604	33.443
65	28.285	28.863	29.463	30.088	30.741	31.425	32.145	32.905	33.712	34.575
70	29.203	29.797	30.414	31.056	31.727	32.429	33.169	33.949	34.778	35.663
75	30.089	30.699	31.331	31.990	32.678	33.398	34.156	34.956	35.805	36.712
80	30.946	31.570	32.218	32.893	33.597	34.335	35.110	35.929	36.799	37.727
85	31.776	32.415	33.078	33.767	34.488	35.242	36.035	36.873	37.761	38.710
90	32.582	33.234	33.912	34.617	35.352	36.123	36.933	37.788	38.696	39.664
95	33.365	34.031	34.723	35.442	36.193	36.979	37.806	38.678	39.604	40.592
100	34.128	34.808	35.513	36.246	37.012	37.813	38.656	39.545	40.488	41.495

ALPHA = .100

N	POWER= .90	.91	.92	.93	.94	.95	.96	.97	.98	.99
1	8.564	8.914	9.302	9.738	10.238	10.822	11.530	12.430	13.680	15.770
2	10.458	10.858	11.300	11.796	12.362	13.023	13.821	14.833	16.231	18.557
3	11.796	12.230	12.710	13.247	13.859	14.573	15.432	16.521	18.021	20.509
4	12.883	13.344	13.853	14.423	15.072	15.828	16.737	17.887	19.469	22.087
5	13.819	14.304	14.839	15.437	16.117	16.909	17.860	19.062	20.714	23.443
6	14.653	15.159	15.716	16.339	17.047	17.871	18.860	20.108	21.822	24.649
7	15.413	15.937	16.515	17.160	17.894	18.746	19.769	21.059	22.829	25.745
8	16.114	16.656	17.253	17.919	18.675	19.554	20.609	21.937	23.758	26.756
9	16.769	17.327	17.941	18.627	19.405	20.308	21.392	22.756	24.625	27.699
10	17.386	17.959	18.589	19.293	20.091	21.018	22.129	23.527	25.441	28.585
11	17.970	18.557	19.203	19.924	20.741	21.690	22.827	24.257	26.213	29.425
12	18.527	19.127	19.788	20.525	21.361	22.330	23.491	24.951	26.948	30.224
13	19.059	19.672	20.347	21.100	21.953	22.942	24.126	25.615	27.650	30.987
14	19.570	20.195	20.884	21.651	22.521	23.529	24.736	26.252	28.324	31.720
15	20.062	20.699	21.401	22.182	23.068	24.094	25.323	26.866	28.973	32.425
16	20.536	21.186	21.899	22.695	23.596	24.640	25.889	27.458	29.599	33.105
17	20.996	21.656	22.382	23.191	24.107	25.168	26.437	28.030	30.205	33.763
18	21.441	22.112	22.850	23.672	24.602	25.679	26.968	28.585	30.792	34.400
19	21.874	22.556	23.305	24.139	25.083	26.176	27.484	29.124	31.362	35.019
20	22.295	22.987	23.747	24.593	25.551	26.660	27.986	29.649	31.916	35.621
21	22.705	23.407	24.177	25.035	26.007	27.131	28.474	30.159	32.456	36.207
22	23.105	23.816	24.598	25.467	26.451	27.590	28.951	30.657	32.983	36.779
23	23.496	24.217	25.008	25.889	26.885	28.039	29.416	31.144	33.497	37.337
24	23.878	24.608	25.409	26.301	27.310	28.477	29.871	31.619	33.999	37.883
25	24.252	24.991	25.802	26.704	27.725	28.906	30.316	32.084	34.491	38.417
26	24.618	25.366	26.186	27.099	28.132	29.326	30.752	32.539	34.972	38.939
27	24.977	25.733	26.563	27.486	28.530	29.738	31.180	32.986	35.444	39.451
28	25.329	26.094	26.933	27.866	28.922	30.142	31.599	33.424	35.907	39.954
29	25.675	26.448	27.296	28.239	29.305	30.539	32.010	33.854	36.361	40.447
30	26.015	26.795	27.652	28.605	29.683	30.928	32.415	34.276	36.808	40.931
32	26.677	27.474	28.348	29.319	30.418	31.688	33.203	35.099	37.677	41.875
34	27.318	28.130	29.021	30.011	31.130	32.423	33.965	35.896	38.519	42.789
36	27.941	28.767	29.674	30.682	31.821	33.136	34.705	36.668	39.336	43.675
38	28.545	29.386	30.309	31.334	32.492	33.830	35.424	37.419	40.130	44.536
40	29.134	29.989	30.926	31.968	33.145	34.504	36.124	38.150	40.902	45.373
42	29.708	30.576	31.528	32.587	33.782	35.162	36.806	38.863	41.654	46.190
44	30.267	31.149	32.116	33.190	34.403	35.803	37.472	39.558	42.389	46.986
46	30.814	31.709	32.690	33.780	35.010	36.430	38.122	40.236	43.106	47.764
48	31.349	32.257	33.252	34.356	35.603	37.043	38.758	40.900	43.807	48.525
50	31.873	32.793	33.801	34.921	36.185	37.643	39.380	41.550	44.494	49.269
55	33.138	34.087	35.128	36.283	37.587	39.091	40.882	43.119	46.151	51.066
60	34.345	35.323	36.395	37.584	38.926	40.474	42.316	44.616	47.732	52.780
65	35.502	36.507	37.609	38.831	40.209	41.799	43.690	46.050	49.247	54.422
70	36.614	37.646	38.776	40.030	41.443	43.073	45.011	47.429	50.703	56.001
75	37.687	38.744	39.902	41.186	42.633	44.301	46.285	48.759	52.107	57.523
80	38.725	39.806	40.990	42.303	43.783	45.488	47.516	50.044	53.465	58.994
85	39.730	40.835	42.044	43.385	44.897	46.639	48.709	51.289	54.779	60.419
90	40.705	41.833	43.068	44.436	45.978	47.755	49.866	52.497	56.055	61.801
95	41.653	42.804	44.062	45.458	47.030	48.840	50.992	53.672	57.295	63.145
100	42.577	43.749	45.031	46.452	48.053	49.897	52.087	54.815	58.502	64.453

ALPHA = .050

N	POWER= .10	.12	.14	.16	.18	.20	.22	.24	.26	.28
1	0.426	0.591	0.755	0.918	1.080	1.242	1.404	1.566	1.729	1.893
2	0.624	0.856	1.081	1.302	1.518	1.731	1.942	2.151	2.360	2.567
3	0.779	1.060	1.330	1.592	1.847	2.096	2.342	2.585	2.825	3.064
4	0.910	1.232	1.539	1.835	2.121	2.401	2.675	2.944	3.211	3.475
5	1.026	1.384	1.723	2.048	2.362	2.667	2.966	3.259	3.547	3.833
6	1.131	1.521	1.889	2.241	2.579	2.907	3.228	3.541	3.850	4.155
7	1.228	1.648	2.042	2.418	2.779	3.128	3.468	3.801	4.128	4.450
8	1.319	1.766	2.185	2.582	2.964	3.333	3.691	4.042	4.385	4.724
9	1.404	1.876	2.318	2.737	3.138	3.525	3.901	4.268	4.627	4.981
10	1.485	1.981	2.445	2.884	3.303	3.707	4.099	4.481	4.856	5.224
11	1.562	2.081	2.566	3.023	3.460	3.880	4.287	4.684	5.073	5.455
12	1.636	2.177	2.681	3.156	3.610	4.045	4.467	4.878	5.280	5.675
13	1.706	2.269	2.791	3.284	3.753	4.204	4.640	5.064	5.479	5.886
14	1.775	2.357	2.898	3.407	3.891	4.357	4.806	5.243	5.670	6.089
15	1.840	2.443	3.001	3.526	4.025	4.504	4.966	5.416	5.855	6.285
16	1.904	2.525	3.100	3.641	4.154	4.646	5.121	5.583	6.033	6.475
17	1.966	2.605	3.197	3.752	4.279	4.784	5.272	5.744	6.206	6.658
18	2.026	2.683	3.291	3.860	4.401	4.918	5.417	5.902	6.374	6.836
19	2.085	2.759	3.382	3.966	4.519	5.049	5.559	6.054	6.537	7.010
20	2.142	2.833	3.471	4.068	4.634	5.176	5.698	6.203	6.696	7.179
21	2.197	2.905	3.558	4.168	4.747	5.300	5.832	6.349	6.851	7.343
22	2.252	2.976	3.642	4.266	4.857	5.421	5.964	6.490	7.003	7.504
23	2.305	3.044	3.725	4.362	4.964	5.539	6.093	6.629	7.151	7.661
24	2.357	3.112	3.806	4.455	5.069	5.655	6.219	6.765	7.296	7.815
25	2.408	3.178	3.886	4.547	5.172	5.769	6.342	6.898	7.438	7.966
26	2.458	3.243	3.964	4.637	5.273	5.880	6.463	7.028	7.577	8.114
27	2.507	3.306	4.040	4.725	5.372	5.989	6.582	7.156	7.713	8.258
28	2.555	3.369	4.115	4.812	5.469	6.096	6.699	7.281	7.847	8.401
29	2.602	3.430	4.189	4.897	5.565	6.202	6.813	7.404	7.979	8.540
30	2.649	3.490	4.262	4.981	5.659	6.305	6.926	7.526	8.109	8.678
32	2.740	3.608	4.403	5.144	5.842	6.507	7.146	7.762	8.361	8.946
34	2.828	3.722	4.541	5.302	6.020	6.703	7.359	7.991	8.606	9.205
36	2.914	3.833	4.674	5.456	6.193	6.893	7.565	8.214	8.844	9.458
38	2.997	3.941	4.804	5.606	6.361	7.079	7.767	8.431	9.075	9.703
40	3.078	4.046	4.930	5.751	6.524	7.259	7.963	8.642	9.300	9.942
42	3.158	4.149	5.054	5.894	6.684	7.435	8.154	8.847	9.520	10.175
44	3.235	4.249	5.174	6.033	6.840	7.607	8.341	9.048	9.735	10.403
46	3.311	4.347	5.292	6.169	6.992	7.775	8.523	9.245	9.944	10.626
48	3.385	4.443	5.407	6.302	7.142	7.939	8.702	9.437	10.150	10.844
50	3.458	4.537	5.520	6.432	7.288	8.100	8.877	9.626	10.351	11.057
55	3.633	4.764	5.793	6.747	7.641	8.489	9.300	10.081	10.837	11.573
60	3.801	4.981	6.055	7.048	7.979	8.861	9.705	10.516	11.301	12.065
65	3.962	5.190	6.305	7.336	8.303	9.218	10.093	10.933	11.747	12.538
70	4.117	5.391	6.546	7.614	8.615	9.562	10.466	11.335	12.176	12.993
75	4.267	5.584	6.779	7.883	8.916	9.894	10.826	11.723	12.590	13.432
80	4.412	5.772	7.004	8.142	9.207	10.214	11.175	12.098	12.990	13.857
85	4.552	5.953	7.223	8.394	9.490	10.525	11.513	12.462	13.378	14.269
90	4.689	6.130	7.435	8.638	9.764	10.827	11.841	12.815	13.755	14.668
95	4.821	6.302	7.641	8.876	10.031	11.121	12.161	13.159	14.122	15.057
100	4.951	6.469	7.842	9.108	10.291	11.408	12.472	13.493	14.479	15.436

ALPHA = .050

N	POWER= .30	.32	.34	.36	.38	.40	.42	.44	.46	.48
1	2.058	2.225	2.393	2.563	2.736	2.911	3.090	3.272	3.457	3.647
2	2.776	2.984	3.194	3.405	3.617	3.832	4.050	4.271	4.495	4.724
3	3.302	3.540	3.778	4.018	4.258	4.501	4.746	4.994	5.245	5.500
4	3.737	3.999	4.261	4.523	4.786	5.051	5.318	5.587	5.861	6.138
5	4.117	4.399	4.681	4.962	5.245	5.529	5.815	6.103	6.395	6.691
6	4.458	4.758	5.057	5.357	5.656	5.957	6.260	6.565	6.874	7.186
7	4.770	5.087	5.402	5.718	6.033	6.349	6.667	6.988	7.312	7.639
8	5.059	5.392	5.723	6.052	6.382	6.713	7.045	7.380	7.717	8.059
9	5.331	5.678	6.022	6.366	6.709	7.053	7.399	7.746	8.097	8.451
10	5.588	5.948	6.306	6.662	7.018	7.375	7.733	8.092	8.455	8.822
11	5.831	6.204	6.575	6.943	7.312	7.680	8.050	8.421	8.795	9.174
12	6.064	6.449	6.832	7.212	7.592	7.971	8.352	8.735	9.120	9.509
13	6.287	6.684	7.078	7.469	7.860	8.250	8.642	9.035	9.431	9.831
14	6.502	6.910	7.314	7.717	8.118	8.519	8.920	9.324	9.730	10.140
15	6.709	7.128	7.543	7.955	8.366	8.777	9.189	9.602	10.018	10.437
16	6.909	7.338	7.763	8.186	8.607	9.027	9.448	9.871	10.296	10.725
17	7.103	7.542	7.977	8.409	8.840	9.270	9.700	10.132	10.566	11.004
18	7.291	7.740	8.185	8.626	9.066	9.505	9.944	10.385	10.828	11.275
19	7.474	7.933	8.387	8.837	9.286	9.733	10.181	10.631	11.082	11.538
20	7.653	8.120	8.583	9.043	9.500	9.956	10.412	10.870	11.330	11.794
21	7.827	8.303	8.775	9.243	9.708	10.173	10.638	11.103	11.572	12.043
22	7.997	8.482	8.962	9.438	9.912	10.385	10.858	11.331	11.807	12.287
23	8.163	8.657	9.145	9.630	10.111	10.592	11.073	11.554	12.038	12.525
24	8.325	8.827	9.324	9.817	10.306	10.795	11.283	11.772	12.263	12.758
25	8.484	8.995	9.499	10.000	10.497	10.993	11.489	11.985	12.484	12.986
26	8.640	9.159	9.671	10.179	10.684	11.188	11.691	12.194	12.700	13.210
27	8.793	9.320	9.840	10.355	10.868	11.378	11.888	12.399	12.913	13.429
28	8.943	9.477	10.005	10.528	11.048	11.565	12.083	12.601	13.121	13.644
29	9.091	9.633	10.168	10.698	11.225	11.749	12.274	12.798	13.325	13.855
30	9.236	9.785	10.327	10.865	11.398	11.930	12.461	12.993	13.526	14.063
32	9.519	10.082	10.639	11.190	11.738	12.283	12.827	13.372	13.918	14.468
34	9.793	10.371	10.941	11.506	12.066	12.624	13.182	13.739	14.298	14.861
36	10.059	10.651	11.234	11.812	12.385	12.956	13.526	14.096	14.667	15.242
38	10.318	10.923	11.520	12.110	12.696	13.279	13.861	14.443	15.026	15.613
40	10.571	11.188	11.798	12.400	12.998	13.593	14.187	14.780	15.376	15.974
42	10.817	11.447	12.069	12.683	13.293	13.900	14.505	15.110	15.716	16.326
44	11.057	11.700	12.333	12.960	13.581	14.199	14.815	15.431	16.049	16.669
46	11.292	11.947	12.592	13.230	13.863	14.492	15.119	15.746	16.374	17.005
48	11.522	12.189	12.845	13.494	14.138	14.778	15.416	16.054	16.692	17.334
50	11.747	12.425	13.093	13.753	14.408	15.058	15.707	16.355	17.004	17.656
55	12.292	12.997	13.692	14.379	15.059	15.736	16.410	17.083	17.757	18.434
60	12.812	13.544	14.265	14.977	15.682	16.383	17.081	17.778	18.476	19.176
65	13.311	14.068	14.814	15.550	16.280	17.004	17.725	18.445	19.166	19.889
70	13.791	14.573	15.343	16.102	16.854	17.601	18.345	19.087	19.830	20.575
75	14.254	15.060	15.853	16.635	17.409	18.178	18.943	19.706	20.470	21.236
80	14.703	15.531	16.346	17.150	17.945	18.735	19.521	20.305	21.089	21.876
85	15.137	15.988	16.824	17.649	18.466	19.276	20.081	20.886	21.690	22.496
90	15.559	16.431	17.289	18.134	18.970	19.800	20.626	21.449	22.273	23.098
95	15.969	16.863	17.740	18.606	19.462	20.311	21.155	21.997	22.840	23.684
100	16.369	17.283	18.180	19.065	19.940	20.808	21.671	22.531	23.392	24.254

ALPHA = .050

N	POWER= .50	.52	.54	.56	.58	.60	.62	.64	.66	.68
1	3.841	4.040	4.245	4.456	4.673	4.899	5.132	5.375	5.628	5.893
2	4.957	5.195	5.439	5.689	5.947	6.213	6.487	6.772	7.069	7.378
3	5.760	6.026	6.297	6.575	6.860	7.154	7.457	7.771	8.097	8.437
4	6.419	6.707	7.000	7.300	7.607	7.924	8.250	8.588	8.938	9.302
5	6.991	7.297	7.609	7.928	8.255	8.591	8.937	9.294	9.665	10.050
6	7.503	7.826	8.154	8.490	8.834	9.187	9.551	9.926	10.315	10.719
7	7.971	8.309	8.653	9.004	9.363	9.732	10.111	10.503	10.908	11.329
8	8.405	8.756	9.114	9.479	9.853	10.236	10.630	11.037	11.458	11.894
9	8.810	9.175	9.546	9.924	10.311	10.708	11.116	11.536	11.971	12.422
10	9.193	9.570	9.953	10.344	10.743	11.153	11.573	12.007	12.455	12.920
11	9.556	9.945	10.340	10.742	11.153	11.575	12.008	12.454	12.914	13.392
12	9.903	10.302	10.708	11.122	11.544	11.977	12.422	12.879	13.352	13.842
13	10.235	10.645	11.061	11.486	11.919	12.363	12.818	13.287	13.771	14.273
14	10.554	10.974	11.401	11.835	12.279	12.733	13.199	13.679	14.174	14.687
15	10.861	11.291	11.728	12.172	12.626	13.090	13.566	14.056	14.562	15.086
16	11.159	11.598	12.044	12.497	12.961	13.434	13.920	14.421	14.937	15.472
17	11.446	11.895	12.350	12.813	13.285	13.768	14.264	14.774	15.300	15.845
18	11.726	12.183	12.647	13.119	13.600	14.092	14.597	15.116	15.652	16.207
19	11.997	12.463	12.935	13.416	13.906	14.407	14.921	15.449	15.994	16.558
20	12.262	12.736	13.216	13.705	14.204	14.713	15.236	15.773	16.327	16.901
21	12.519	13.001	13.490	13.987	14.494	15.012	15.543	16.089	16.652	17.234
22	12.771	13.261	13.758	14.263	14.777	15.304	15.843	16.397	16.968	17.560
23	13.017	13.514	14.019	14.532	15.054	15.588	16.136	16.698	17.278	17.878
24	13.257	13.762	14.274	14.795	15.325	15.867	16.422	16.993	17.581	18.189
25	13.493	14.005	14.525	15.052	15.590	16.140	16.703	17.281	17.877	18.493
26	13.723	14.243	14.770	15.305	15.850	16.407	16.978	17.563	18.167	18.792
27	13.950	14.476	15.010	15.553	16.105	16.669	17.247	17.841	18.452	19.084
28	14.172	14.705	15.246	15.796	16.355	16.926	17.511	18.112	18.731	19.372
29	14.390	14.930	15.478	16.034	16.601	17.179	17.771	18.379	19.006	19.653
30	14.604	15.151	15.706	16.269	16.842	17.427	18.026	18.642	19.275	19.931
32	15.023	15.583	16.150	16.726	17.313	17.911	18.524	19.153	19.801	20.471
34	15.428	16.000	16.581	17.169	17.769	18.381	19.007	19.649	20.311	20.994
36	15.821	16.406	16.998	17.600	18.212	18.836	19.475	20.130	20.805	21.502
38	16.204	16.801	17.405	18.018	18.642	19.279	19.930	20.598	21.286	21.996
40	16.576	17.185	17.800	18.425	19.061	19.710	20.373	21.053	21.754	22.477
42	16.939	17.559	18.186	18.823	19.470	20.130	20.805	21.498	22.210	22.946
44	17.294	17.925	18.563	19.211	19.869	20.540	21.227	21.931	22.656	23.404
46	17.641	18.282	18.931	19.590	20.259	20.941	21.639	22.355	23.091	23.851
48	17.980	18.632	19.292	19.960	20.641	21.334	22.043	22.769	23.517	24.289
50	18.312	18.974	19.644	20.324	21.014	21.718	22.438	23.175	23.934	24.717
55	19.115	19.802	20.496	21.201	21.917	22.646	23.392	24.156	24.941	25.752
60	19.881	20.592	21.311	22.039	22.779	23.533	24.303	25.092	25.903	26.740
65	20.616	21.350	22.091	22.842	23.605	24.383	25.177	25.990	26.826	27.688
70	21.324	22.079	22.842	23.616	24.401	25.201	26.017	26.854	27.713	28.600
75	22.006	22.783	23.567	24.361	25.168	25.990	26.828	27.687	28.569	29.479
80	22.666	23.463	24.268	25.083	25.910	26.753	27.612	28.493	29.397	30.329
85	23.306	24.123	24.947	25.782	26.629	27.492	28.372	29.274	30.199	31.153
90	23.927	24.763	25.607	26.461	27.328	28.210	29.110	30.032	30.978	31.953
95	24.532	25.386	26.248	27.121	28.007	28.908	29.828	30.769	31.736	32.731
100	25.120	25.992	26.873	27.764	28.668	29.588	30.527	31.487	32.473	33.489

ALPHA = .050

N	POWER= .70	.71	.72	.73	.74	.75	.76	.77	.78	.79
1	6.172	6.317	6.466	6.619	6.777	6.940	7.109	7.284	7.465	7.653
2	7.702	7.870	8.042	8.220	8.403	8.591	8.785	8.986	9.194	9.410
3	8.792	8.977	9.166	9.360	9.559	9.765	9.977	10.196	10.423	10.658
4	9.682	9.880	10.082	10.289	10.503	10.722	10.949	11.182	11.424	11.675
5	10.453	10.661	10.874	11.093	11.318	11.550	11.789	12.035	12.290	12.554
6	11.141	11.359	11.582	11.812	12.047	12.289	12.539	12.796	13.063	13.338
7	11.768	11.995	12.228	12.467	12.712	12.963	13.223	13.491	13.767	14.053
8	12.349	12.585	12.826	13.073	13.326	13.587	13.856	14.133	14.419	14.715
9	12.892	13.135	13.384	13.639	13.901	14.170	14.447	14.733	15.028	15.333
10	13.404	13.655	13.911	14.173	14.443	14.720	15.005	15.298	15.602	15.916
11	13.890	14.147	14.410	14.680	14.956	15.240	15.533	15.835	16.146	16.468
12	14.353	14.616	14.886	15.162	15.446	15.737	16.037	16.346	16.664	16.994
13	14.796	15.065	15.341	15.624	15.914	16.212	16.519	16.835	17.161	17.498
14	15.221	15.497	15.779	16.068	16.364	16.669	16.982	17.304	17.637	17.981
15	15.631	15.913	16.201	16.495	16.798	17.108	17.428	17.757	18.096	18.447
16	16.027	16.314	16.608	16.908	17.217	17.533	17.859	18.194	18.540	18.897
17	16.411	16.703	17.002	17.308	17.622	17.944	18.276	18.617	18.969	19.333
18	16.783	17.080	17.384	17.696	18.015	18.343	18.680	19.027	19.385	19.755
19	17.144	17.447	17.756	18.073	18.397	18.731	19.073	19.426	19.790	20.166
20	17.496	17.804	18.118	18.439	18.769	19.108	19.456	19.814	20.183	20.565
21	17.839	18.151	18.470	18.797	19.132	19.475	19.828	20.192	20.567	20.955
22	18.174	18.490	18.814	19.146	19.485	19.834	20.192	20.561	20.941	21.334
23	18.501	18.822	19.150	19.486	19.831	20.184	20.547	20.921	21.307	21.705
24	18.820	19.146	19.479	19.819	20.169	20.527	20.895	21.274	21.665	22.068
25	19.133	19.463	19.801	20.146	20.499	20.862	21.235	21.619	22.015	22.423
26	19.440	19.774	20.116	20.465	20.823	21.191	21.568	21.957	22.358	22.771
27	19.741	20.079	20.425	20.778	21.141	21.513	21.895	22.288	22.694	23.112
28	20.036	20.378	20.728	21.086	21.453	21.829	22.216	22.613	23.024	23.447
29	20.325	20.672	21.026	21.388	21.759	22.139	22.530	22.933	23.347	23.776
30	20.610	20.960	21.318	21.684	22.060	22.444	22.840	23.246	23.666	24.099
32	21.165	21.523	21.889	22.263	22.646	23.039	23.443	23.858	24.286	24.728
34	21.703	22.069	22.442	22.823	23.214	23.615	24.027	24.451	24.887	25.338
36	22.225	22.598	22.978	23.367	23.766	24.174	24.594	25.026	25.471	25.930
38	22.733	23.112	23.500	23.896	24.302	24.718	25.145	25.585	26.038	26.505
40	23.227	23.613	24.007	24.411	24.824	25.247	25.682	26.129	26.590	27.065
42	23.709	24.101	24.502	24.913	25.332	25.763	26.205	26.660	27.128	27.611
44	24.179	24.578	24.986	25.402	25.829	26.267	26.716	27.178	27.653	28.144
46	24.639	25.044	25.458	25.881	26.315	26.759	27.215	27.684	28.167	28.665
48	25.088	25.500	25.920	26.350	26.789	27.240	27.703	28.179	28.669	29.175
50	25.528	25.946	26.372	26.808	27.254	27.712	28.181	28.664	29.161	29.673
55	26.591	27.023	27.465	27.915	28.377	28.850	29.335	29.834	30.348	30.878
60	27.607	28.053	28.508	28.973	29.449	29.937	30.437	30.952	31.482	32.028
65	28.580	29.039	29.508	29.987	30.477	30.979	31.494	32.023	32.568	33.131
70	29.517	29.989	30.470	30.962	31.466	31.981	32.510	33.054	33.614	34.191
75	30.420	30.904	31.398	31.903	32.419	32.948	33.491	34.048	34.622	35.214
80	31.294	31.790	32.296	32.813	33.341	33.883	34.439	35.010	35.597	36.203
85	32.140	32.648	33.165	33.694	34.235	34.789	35.357	35.941	36.542	37.161
90	32.962	33.481	34.010	34.550	35.103	35.669	36.249	36.845	37.459	38.092
95	33.761	34.291	34.831	35.382	35.946	36.524	37.116	37.725	38.351	38.996
100	34.539	35.079	35.630	36.192	36.768	37.357	37.961	38.581	39.219	39.877

ALPHA = .050

N	POWER= .80	.81	.82	.83	.84	.85	.86	.87	.88	.89
1	7.849	8.053	8.268	8.492	8.729	8.978	9.243	9.526	9.828	10.154
2	9.635	9.869	10.113	10.370	10.639	10.923	11.224	11.544	11.887	12.255
3	10.903	11.157	11.423	11.701	11.993	12.301	12.627	12.973	13.343	13.741
4	11.935	12.206	12.489	12.785	13.095	13.422	13.768	14.136	14.528	14.949
5	12.828	13.113	13.410	13.721	14.047	14.391	14.754	15.139	15.550	15.992
6	13.624	13.922	14.232	14.556	14.897	15.255	15.633	16.034	16.462	16.922
7	14.351	14.659	14.981	15.318	15.671	16.042	16.434	16.850	17.293	17.769
8	15.022	15.341	15.674	16.022	16.386	16.770	17.174	17.604	18.062	18.552
9	15.650	15.979	16.322	16.680	17.055	17.450	17.866	18.308	18.779	19.284
10	16.241	16.579	16.932	17.299	17.685	18.090	18.518	18.972	19.455	19.973
11	16.802	17.149	17.510	17.887	18.282	18.697	19.136	19.600	20.095	20.626
12	17.336	17.691	18.061	18.447	18.851	19.276	19.724	20.199	20.705	21.247
13	17.847	18.210	18.588	18.982	19.395	19.829	20.287	20.772	21.289	21.842
14	18.338	18.709	19.094	19.497	19.918	20.361	20.828	21.322	21.849	22.414
15	18.811	19.189	19.582	19.992	20.421	20.873	21.348	21.852	22.389	22.964
16	19.268	19.652	20.053	20.470	20.908	21.367	21.851	22.364	22.910	23.495
17	19.710	20.101	20.509	20.933	21.378	21.845	22.338	22.859	23.414	24.009
18	20.139	20.537	20.951	21.383	21.835	22.309	22.810	23.340	23.903	24.507
19	20.555	20.960	21.380	21.819	22.278	22.760	23.268	23.806	24.378	24.991
20	20.961	21.371	21.798	22.243	22.709	23.199	23.714	24.260	24.841	25.462
21	21.356	21.772	22.205	22.657	23.130	23.626	24.149	24.702	25.291	25.921
22	21.741	22.164	22.603	23.061	23.540	24.043	24.573	25.134	25.730	26.369
23	22.118	22.546	22.991	23.455	23.940	24.450	24.987	25.555	26.160	26.806
24	22.486	22.920	23.370	23.840	24.332	24.848	25.392	25.967	26.579	27.234
25	22.847	23.285	23.742	24.218	24.716	25.238	25.788	26.371	26.990	27.652
26	23.200	23.644	24.106	24.587	25.091	25.620	26.177	26.766	27.392	28.062
27	23.546	23.995	24.463	24.950	25.459	25.994	26.557	27.153	27.787	28.464
28	23.885	24.340	24.813	25.305	25.821	26.361	26.931	27.533	28.173	28.858
29	24.219	24.679	25.156	25.655	26.175	26.722	27.297	27.906	28.553	29.246
30	24.547	25.011	25.494	25.998	26.524	27.076	27.658	28.273	28.927	29.626
32	25.186	25.660	26.153	26.666	27.203	27.767	28.360	28.987	29.654	30.367
34	25.804	26.288	26.790	27.314	27.862	28.436	29.040	29.679	30.359	31.085
36	26.405	26.898	27.409	27.943	28.500	29.085	29.700	30.351	31.043	31.782
38	26.989	27.490	28.011	28.554	29.121	29.716	30.342	31.004	31.707	32.459
40	27.557	28.067	28.597	29.149	29.725	30.330	30.966	31.639	32.354	33.118
42	28.111	28.629	29.168	29.728	30.314	30.929	31.575	32.259	32.985	33.761
44	28.652	29.178	29.725	30.294	30.889	31.513	32.169	32.863	33.600	34.388
46	29.181	29.715	30.269	30.847	31.451	32.084	32.750	33.454	34.201	35.000
48	29.698	30.239	30.802	31.388	32.001	32.642	33.318	34.032	34.790	35.599
50	30.204	30.753	31.324	31.918	32.539	33.189	33.874	34.597	35.365	36.186
55	31.426	31.993	32.583	33.197	33.838	34.509	35.216	35.963	36.755	37.602
60	32.593	33.178	33.785	34.418	35.078	35.770	36.498	37.267	38.083	38.954
65	33.711	34.313	34.938	35.588	36.267	36.978	37.726	38.516	39.355	40.250
70	34.787	35.405	36.046	36.714	37.410	38.140	38.908	39.718	40.578	41.496
75	35.825	36.458	37.115	37.799	38.513	39.261	40.047	40.877	41.758	42.698
80	36.829	37.477	38.149	38.849	39.580	40.345	41.149	41.998	42.899	43.860
85	37.801	38.463	39.151	39.866	40.613	41.394	42.216	43.084	44.004	44.986
90	38.745	39.421	40.123	40.853	41.616	42.414	43.252	44.137	45.077	46.079
95	39.663	40.353	41.069	41.813	42.591	43.404	44.260	45.162	46.120	47.141
100	40.556	41.260	41.989	42.748	43.540	44.369	45.240	46.160	47.135	48.176

ALPHA = .050

N	POWER= .90	.91	.92	.93	.94	.95	.96	.97	.98	.99
1	10.507	10.895	11.323	11.804	12.353	12.995	13.769	14.751	16.110	18.372
2	12.654	13.090	13.572	14.112	14.727	15.443	16.306	17.398	18.902	21.396
3	14.171	14.642	15.161	15.741	16.401	17.170	18.094	19.262	20.867	23.521
4	15.405	15.903	16.451	17.064	17.761	18.572	19.545	20.774	22.460	25.243
5	16.469	16.990	17.564	18.205	18.934	19.780	20.796	22.076	23.833	26.726
6	17.419	17.960	18.557	19.223	19.979	20.857	21.911	23.237	25.055	28.046
7	18.284	18.844	19.461	20.150	20.931	21.838	22.925	24.294	26.168	29.247
8	19.083	19.660	20.296	21.006	21.810	22.744	23.862	25.270	27.195	30.356
9	19.829	20.423	21.076	21.805	22.631	23.589	24.737	26.180	28.154	31.391
10	20.532	21.141	21.811	22.558	23.404	24.386	25.561	27.037	29.056	32.365
11	21.198	21.821	22.507	23.271	24.136	25.140	26.341	27.849	29.910	33.286
12	21.833	22.469	23.170	23.950	24.834	25.858	27.083	28.623	30.724	34.164
13	22.439	23.089	23.804	24.599	25.501	26.545	27.794	29.362	31.502	35.003
14	23.022	23.684	24.412	25.223	26.141	27.204	28.476	30.071	32.248	35.808
15	23.583	24.258	24.998	25.823	26.758	27.839	29.132	30.754	32.967	36.583
16	24.125	24.811	25.564	26.403	27.353	28.452	29.766	31.414	33.660	37.331
17	24.650	25.346	26.112	26.964	27.929	29.045	30.379	32.052	34.331	38.054
18	25.158	25.866	26.643	27.508	28.487	29.620	30.973	32.670	34.982	38.755
19	25.652	26.370	27.159	28.036	29.029	30.178	31.550	33.270	35.613	39.436
20	26.132	26.861	27.660	28.550	29.557	30.721	32.112	33.855	36.228	40.098
21	26.600	27.339	28.149	29.051	30.071	31.250	32.659	34.424	36.826	40.743
22	27.057	27.805	28.626	29.539	30.572	31.767	33.192	34.979	37.410	41.372
23	27.503	28.260	29.092	30.016	31.062	32.271	33.713	35.521	37.979	41.986
24	27.939	28.706	29.547	30.483	31.540	32.763	34.223	36.050	38.536	42.586
25	28.366	29.142	29.993	30.939	32.009	33.246	34.721	36.569	39.081	43.173
26	28.784	29.568	30.429	31.386	32.468	33.718	35.209	37.076	39.615	43.748
27	29.194	29.987	30.857	31.824	32.917	34.181	35.688	37.574	40.138	44.312
28	29.596	30.398	31.277	32.254	33.359	34.635	36.157	38.062	40.652	44.865
29	29.991	30.801	31.689	32.676	33.792	35.081	36.618	38.541	41.155	45.407
30	30.379	31.197	32.094	33.091	34.217	35.519	37.071	39.012	41.650	45.940
32	31.135	31.969	32.883	33.899	35.047	36.372	37.953	39.930	42.615	46.979
34	31.867	32.716	33.647	34.682	35.850	37.199	38.807	40.818	43.548	47.985
36	32.578	33.441	34.389	35.441	36.629	38.001	39.636	41.680	44.454	48.960
38	33.268	34.146	35.109	36.179	37.386	38.780	40.441	42.517	45.334	49.908
40	33.940	34.833	35.811	36.897	38.123	39.539	41.225	43.332	46.191	50.830
42	34.595	35.501	36.494	37.597	38.842	40.278	41.989	44.126	47.025	51.729
44	35.235	36.154	37.162	38.280	39.543	41.000	42.734	44.901	47.840	52.605
46	35.860	36.792	37.813	38.948	40.227	41.704	43.462	45.658	48.635	53.462
48	36.470	37.415	38.451	39.600	40.897	42.394	44.175	46.398	49.413	54.299
50	37.069	38.026	39.075	40.239	41.553	43.068	44.872	47.123	50.175	55.119
55	38.513	39.500	40.582	41.782	43.136	44.697	46.554	48.872	52.012	57.097
60	39.891	40.907	42.020	43.255	44.647	46.251	48.160	50.542	53.766	58.984
65	41.213	42.256	43.399	44.666	46.095	47.741	49.700	52.142	55.447	60.793
70	42.483	43.553	44.724	46.023	47.487	49.174	51.179	53.680	57.063	62.531
75	43.709	44.804	46.003	47.332	48.829	50.555	52.606	55.163	58.620	64.207
80	44.893	46.013	47.238	48.597	50.127	51.890	53.985	56.596	60.126	65.827
85	46.041	47.184	48.436	49.822	51.384	53.184	55.321	57.984	61.584	67.395
90	47.155	48.321	49.597	51.012	52.605	54.439	56.618	59.332	62.999	68.918
95	48.238	49.427	50.727	52.168	53.791	55.659	57.879	60.642	64.375	70.397
100	49.293	50.503	51.827	53.294	54.946	56.848	59.106	61.917	65.714	71.838

ALPHA = .025

N	POWER= .10	.12	.14	.16	.18	.20	.22	.24	.26	.28
1	0.914	1.132	1.345	1.552	1.757	1.958	2.158	2.356	2.553	2.750
2	1.307	1.597	1.875	2.142	2.403	2.657	2.906	3.152	3.396	3.638
3	1.604	1.946	2.270	2.580	2.880	3.171	3.456	3.736	4.011	4.284
4	1.854	2.238	2.599	2.944	3.276	3.597	3.911	4.218	4.520	4.818
5	2.073	2.493	2.888	3.263	3.622	3.970	4.308	4.638	4.963	5.283
6	2.271	2.724	3.148	3.550	3.934	4.305	4.665	5.017	5.361	5.701
7	2.454	2.937	3.387	3.813	4.220	4.612	4.993	5.363	5.726	6.084
8	2.624	3.134	3.610	4.058	4.486	4.898	5.297	5.685	6.065	6.439
9	2.784	3.320	3.819	4.288	4.736	5.166	5.582	5.987	6.383	6.772
10	2.935	3.496	4.016	4.506	4.972	5.419	5.852	6.272	6.683	7.086
11	3.079	3.663	4.204	4.713	5.197	5.660	6.108	6.543	6.969	7.385
12	3.217	3.823	4.384	4.911	5.411	5.890	6.353	6.802	7.241	7.671
13	3.349	3.976	4.556	5.100	5.617	6.111	6.588	7.051	7.502	7.944
14	3.476	4.124	4.722	5.283	5.815	6.324	6.814	7.290	7.753	8.208
15	3.599	4.267	4.883	5.459	6.006	6.528	7.032	7.520	7.996	8.462
16	3.718	4.405	5.038	5.630	6.191	6.727	7.243	7.743	8.230	8.707
17	3.834	4.539	5.188	5.795	6.370	6.919	7.447	7.959	8.457	8.945
18	3.946	4.669	5.334	5.956	6.544	7.105	7.646	8.169	8.678	9.176
19	4.055	4.795	5.476	6.112	6.713	7.287	7.839	8.373	8.893	9.401
20	4.161	4.919	5.615	6.264	6.878	7.464	8.027	8.572	9.102	9.620
21	4.265	5.039	5.750	6.413	7.039	7.636	8.210	8.766	9.305	9.833
22	4.367	5.157	5.882	6.558	7.196	7.805	8.389	8.955	9.504	10.041
23	4.466	5.272	6.011	6.700	7.350	7.970	8.565	9.140	9.699	10.245
24	4.563	5.384	6.137	6.839	7.500	8.131	8.736	9.321	9.889	10.444
25	4.658	5.495	6.261	6.975	7.648	8.289	8.904	9.498	10.076	10.640
26	4.751	5.603	6.382	7.108	7.792	8.443	9.069	9.672	10.259	10.831
27	4.843	5.709	6.501	7.239	7.934	8.595	9.230	9.843	10.438	11.019
28	4.933	5.813	6.618	7.367	8.073	8.745	9.389	10.010	10.614	11.203
29	5.021	5.915	6.733	7.493	8.210	8.891	9.544	10.175	10.787	11.384
30	5.108	6.016	6.846	7.618	8.344	9.035	9.697	10.337	10.957	11.562
32	5.278	6.213	7.066	7.860	8.606	9.316	9.996	10.652	11.289	11.909
34	5.442	6.403	7.280	8.095	8.861	9.589	10.286	10.958	11.610	12.246
36	5.602	6.588	7.488	8.323	9.108	9.853	10.567	11.256	11.923	12.573
38	5.757	6.768	7.690	8.545	9.348	10.111	10.841	11.545	12.227	12.891
40	5.909	6.944	7.887	8.761	9.582	10.362	11.108	11.826	12.523	13.201
42	6.057	7.115	8.079	8.972	9.811	10.607	11.368	12.101	12.811	13.503
44	6.201	7.282	8.266	9.178	10.034	10.846	11.622	12.369	13.093	13.798
46	6.343	7.446	8.450	9.380	10.252	11.079	11.870	12.632	13.369	14.087
48	6.481	7.606	8.629	9.577	10.466	11.308	12.113	12.888	13.639	14.369
50	6.616	7.763	8.805	9.770	10.675	11.532	12.352	13.140	13.903	14.645
55	6.944	8.142	9.231	10.237	11.180	12.074	12.927	13.748	14.542	15.314
60	7.257	8.505	9.637	10.683	11.664	12.591	13.477	14.328	15.152	15.952
65	7.557	8.853	10.027	11.112	12.127	13.088	14.005	14.886	15.737	16.565
70	7.847	9.187	10.403	11.524	12.574	13.566	14.513	15.422	16.301	17.155
75	8.126	9.511	10.765	11.922	13.004	14.027	15.003	15.940	16.845	17.724
80	8.396	9.824	11.116	12.307	13.421	14.474	15.477	16.440	17.371	18.274
85	8.658	10.127	11.456	12.681	13.825	14.907	15.937	16.926	17.881	18.808
90	8.913	10.422	11.786	13.043	14.218	15.327	16.384	17.397	18.376	19.326
95	9.160	10.708	12.107	13.396	14.600	15.736	16.818	17.856	18.858	19.830
100	9.402	10.987	12.420	13.740	14.972	16.134	17.241	18.303	19.327	20.321

ALPHA = .025

N	POWER= .30	.32	.34	.36	.38	.40	.42	.44	.46	.48
1	2.948	3.146	3.345	3.545	3.748	3.952	4.160	4.370	4.584	4.802
2	3.878	4.118	4.359	4.600	4.842	5.086	5.332	5.581	5.834	6.090
3	4.555	4.825	5.094	5.363	5.633	5.905	6.179	6.455	6.734	7.017
4	5.113	5.407	5.700	5.992	6.284	6.578	6.874	7.172	7.473	7.778
5	5.600	5.914	6.226	6.538	6.851	7.164	7.478	7.795	8.115	8.439
6	6.036	6.369	6.699	7.029	7.359	7.689	8.020	8.354	8.690	9.031
7	6.436	6.785	7.132	7.478	7.823	8.169	8.516	8.865	9.216	9.572
8	6.807	7.172	7.534	7.894	8.254	8.614	8.975	9.338	9.704	10.073
9	7.155	7.534	7.910	8.284	8.658	9.031	9.406	9.782	10.161	10.543
10	7.484	7.876	8.266	8.653	9.039	9.425	9.812	10.200	10.591	10.986
11	7.796	8.201	8.603	9.003	9.401	9.799	10.197	10.598	11.000	11.407
12	8.094	8.512	8.925	9.337	9.746	10.156	10.565	10.977	11.391	11.808
13	8.380	8.809	9.234	9.657	10.077	10.498	10.918	11.340	11.765	12.193
14	8.654	9.095	9.531	9.964	10.396	10.826	11.257	11.689	12.124	12.562
15	8.919	9.371	9.818	10.261	10.703	11.143	11.584	12.026	12.471	12.919
16	9.176	9.638	10.095	10.548	10.999	11.450	11.900	12.352	12.806	13.263
17	9.424	9.896	10.363	10.826	11.287	11.747	12.206	12.667	13.130	13.597
18	9.665	10.147	10.623	11.096	11.566	12.035	12.503	12.973	13.445	13.920
19	9.900	10.391	10.877	11.358	11.837	12.315	12.792	13.271	13.751	14.235
20	10.128	10.629	11.123	11.614	12.102	12.588	13.074	13.560	14.049	14.541
21	10.351	10.860	11.364	11.863	12.359	12.854	13.348	13.843	14.340	14.840
22	10.568	11.087	11.599	12.106	12.611	13.114	13.616	14.119	14.624	15.132
23	10.781	11.308	11.828	12.344	12.857	13.367	13.877	14.388	14.901	15.417
24	10.989	11.524	12.053	12.577	13.097	13.616	14.133	14.652	15.172	15.696
25	11.192	11.736	12.273	12.805	13.333	13.859	14.384	14.910	15.438	15.969
26	11.392	11.944	12.489	13.028	13.564	14.097	14.630	15.163	15.698	16.236
27	11.588	12.148	12.700	13.247	13.790	14.331	14.871	15.411	15.953	16.498
28	11.780	12.348	12.908	13.462	14.013	14.561	15.108	15.655	16.204	16.756
29	11.969	12.544	13.112	13.673	14.231	14.786	15.340	15.894	16.450	17.009
30	12.155	12.737	13.312	13.881	14.446	15.007	15.568	16.129	16.692	17.258
32	12.517	13.114	13.703	14.286	14.864	15.440	16.014	16.588	17.164	17.743
34	12.868	13.480	14.082	14.679	15.270	15.859	16.446	17.033	17.621	18.213
36	13.209	13.834	14.450	15.060	15.664	16.265	16.865	17.464	18.065	18.669
38	13.541	14.179	14.808	15.431	16.047	16.661	17.273	17.884	18.497	19.112
40	13.864	14.515	15.157	15.792	16.421	17.046	17.670	18.293	18.917	19.545
42	14.179	14.843	15.497	16.144	16.785	17.422	18.057	18.692	19.327	19.966
44	14.487	15.163	15.830	16.488	17.140	17.789	18.435	19.081	19.728	20.377
46	14.788	15.476	16.154	16.824	17.488	18.148	18.805	19.462	20.119	20.780
48	15.082	15.783	16.472	17.153	17.828	18.498	19.167	19.834	20.502	21.173
50	15.371	16.083	16.783	17.475	18.161	18.842	19.521	20.199	20.877	21.558
55	16.068	16.807	17.535	18.254	18.966	19.673	20.377	21.080	21.783	22.489
60	16.734	17.500	18.254	18.998	19.735	20.466	21.194	21.921	22.649	23.379
65	17.373	18.164	18.943	19.712	20.472	21.227	21.979	22.729	23.479	24.232
70	17.988	18.804	19.606	20.398	21.182	21.960	22.734	23.506	24.278	25.053
75	18.581	19.421	20.247	21.061	21.867	22.666	23.462	24.255	25.049	25.845
80	19.155	20.018	20.866	21.702	22.529	23.350	24.166	24.980	25.795	26.611
85	19.712	20.597	21.466	22.323	23.171	24.012	24.849	25.683	26.517	27.353
90	20.252	21.158	22.049	22.927	23.795	24.656	25.512	26.366	27.219	28.074
95	20.778	21.705	22.616	23.514	24.401	25.282	26.157	27.029	27.901	28.775
100	21.290	22.237	23.168	24.086	24.992	25.891	26.785	27.676	28.566	29.458

ALPHA = .025

N	POWER= .50	.52	.54	.56	.58	.60	.62	.64	.66	.68
1	5.024	5.251	5.484	5.723	5.970	6.224	6.487	6.759	7.043	7.339
2	6.351	6.617	6.889	7.167	7.453	7.747	8.051	8.365	8.691	9.030
3	7.305	7.598	7.897	8.203	8.516	8.838	9.170	9.513	9.869	10.239
4	8.088	8.402	8.723	9.051	9.387	9.732	10.086	10.453	10.832	11.227
5	8.767	9.101	9.440	9.787	10.142	10.506	10.881	11.267	11.667	12.083
6	9.376	9.726	10.082	10.446	10.818	11.199	11.592	11.996	12.414	12.849
7	9.932	10.297	10.669	11.048	11.436	11.833	12.241	12.661	13.096	13.548
8	10.447	10.827	11.213	11.606	12.008	12.419	12.842	13.278	13.728	14.195
9	10.930	11.323	11.721	12.128	12.543	12.968	13.405	13.854	14.319	14.800
10	11.385	11.790	12.201	12.620	13.048	13.486	13.935	14.398	14.876	15.371
11	11.818	12.234	12.657	13.088	13.527	13.977	14.439	14.914	15.404	15.913
12	12.230	12.657	13.092	13.533	13.984	14.445	14.919	15.406	15.908	16.429
13	12.625	13.063	13.508	13.960	14.422	14.894	15.378	15.877	16.391	16.923
14	13.005	13.453	13.908	14.371	14.843	15.325	15.820	16.329	16.854	17.398
15	13.371	13.829	14.293	14.766	15.248	15.741	16.246	16.765	17.301	17.856
16	13.725	14.192	14.666	15.148	15.640	16.142	16.657	17.187	17.733	18.298
17	14.067	14.544	15.027	15.518	16.019	16.531	17.056	17.595	18.151	18.727
18	14.400	14.885	15.377	15.878	16.387	16.908	17.442	17.991	18.557	19.142
19	14.723	15.217	15.718	16.227	16.745	17.275	17.818	18.376	18.951	19.546
20	15.038	15.540	16.049	16.567	17.094	17.632	18.184	18.750	19.335	19.939
21	15.345	15.855	16.372	16.898	17.433	17.980	18.540	19.116	19.709	20.322
22	15.644	16.163	16.688	17.221	17.765	18.320	18.888	19.472	20.074	20.696
23	15.937	16.463	16.996	17.537	18.089	18.652	19.228	19.820	20.430	21.061
24	16.223	16.757	17.297	17.846	18.406	18.976	19.561	20.161	20.779	21.418
25	16.504	17.045	17.593	18.149	18.716	19.294	19.886	20.494	21.121	21.768
26	16.779	17.327	17.882	18.446	19.020	19.606	20.205	20.821	21.455	22.110
27	17.048	17.603	18.166	18.737	19.318	19.911	20.518	21.141	21.783	22.446
28	17.312	17.875	18.444	19.022	19.610	20.211	20.825	21.456	22.105	22.776
29	17.572	18.141	18.717	19.302	19.898	20.505	21.127	21.765	22.421	23.100
30	17.828	18.403	18.986	19.578	20.180	20.794	21.423	22.068	22.732	23.418
32	18.326	18.914	19.510	20.115	20.731	21.359	22.001	22.660	23.338	24.039
34	18.808	19.410	20.018	20.636	21.264	21.905	22.561	23.233	23.925	24.640
36	19.277	19.890	20.512	21.142	21.783	22.436	23.104	23.790	24.495	25.224
38	19.732	20.358	20.991	21.633	22.286	22.952	23.633	24.331	25.049	25.791
40	20.176	20.813	21.458	22.112	22.777	23.455	24.148	24.858	25.589	26.343
42	20.609	21.257	21.913	22.579	23.255	23.945	24.649	25.372	26.115	26.882
44	21.031	21.691	22.358	23.034	23.722	24.423	25.139	25.873	26.629	27.408
46	21.444	22.115	22.793	23.480	24.178	24.890	25.618	26.364	27.131	27.922
48	21.848	22.529	23.218	23.916	24.625	25.348	26.086	26.843	27.622	28.425
50	22.244	22.935	23.634	24.342	25.062	25.796	26.545	27.313	28.102	28.917
55	23.200	23.916	24.640	25.373	26.118	26.877	27.653	28.447	29.264	30.106
60	24.113	24.853	25.600	26.358	27.127	27.911	28.711	29.531	30.373	31.241
65	24.989	25.751	26.522	27.302	28.095	28.902	29.726	30.570	31.436	32.330
70	25.831	26.616	27.408	28.211	29.026	29.855	30.702	31.569	32.460	33.378
75	26.644	27.450	28.264	29.088	29.924	30.775	31.644	32.533	33.447	34.388
80	27.431	28.257	29.091	29.935	30.792	31.665	32.555	33.466	34.401	35.365
85	28.193	29.039	29.893	30.757	31.634	32.527	33.437	34.369	35.326	36.312
90	28.933	29.798	30.671	31.555	32.451	33.364	34.294	35.247	36.224	37.231
95	29.653	30.536	31.428	32.331	33.246	34.177	35.128	36.100	37.098	38.125
100	30.354	31.255	32.165	33.086	34.020	34.970	35.939	36.931	37.948	38.996

ALPHA = .025

N	POWER= .70	.71	.72	.73	.74	.75	.76	.77	.78	.79
1	7.650	7.811	7.976	8.147	8.322	8.502	8.689	8.882	9.082	9.289
2	9.386	9.570	9.759	9.952	10.152	10.357	10.569	10.788	11.014	11.249
3	10.625	10.825	11.030	11.241	11.457	11.680	11.909	12.146	12.391	12.645
4	11.638	11.851	12.070	12.293	12.523	12.760	13.004	13.255	13.515	13.784
5	12.516	12.740	12.970	13.205	13.447	13.695	13.951	14.215	14.488	14.770
6	13.301	13.535	13.774	14.020	14.272	14.531	14.798	15.073	15.358	15.652
7	14.018	14.260	14.509	14.764	15.025	15.294	15.571	15.856	16.151	16.456
8	14.681	14.932	15.189	15.452	15.723	16.000	16.286	16.581	16.885	17.200
9	15.301	15.560	15.825	16.096	16.375	16.661	16.955	17.259	17.572	17.896
10	15.886	16.152	16.425	16.703	16.989	17.283	17.586	17.897	18.219	18.551
11	16.441	16.714	16.993	17.279	17.573	17.874	18.184	18.503	18.832	19.173
12	16.970	17.250	17.535	17.828	18.128	18.437	18.754	19.080	19.417	19.766
13	17.477	17.762	18.055	18.354	18.660	18.975	19.299	19.633	19.977	20.333
14	17.963	18.255	18.553	18.859	19.172	19.493	19.824	20.164	20.515	20.878
15	18.432	18.730	19.034	19.345	19.664	19.992	20.329	20.675	21.033	21.403
16	18.885	19.188	19.498	19.815	20.140	20.474	20.817	21.170	21.534	21.910
17	19.324	19.632	19.947	20.270	20.601	20.940	21.289	21.648	22.018	22.401
18	19.750	20.063	20.383	20.711	21.048	21.392	21.747	22.112	22.488	22.877
19	20.163	20.482	20.807	21.140	21.482	21.832	22.192	22.563	22.945	23.340
20	20.566	20.889	21.220	21.558	21.904	22.260	22.626	23.002	23.390	23.790
21	20.958	21.286	21.622	21.965	22.316	22.677	23.048	23.429	23.823	24.229
22	21.341	21.674	22.014	22.362	22.718	23.084	23.460	23.847	24.246	24.657
23	21.715	22.052	22.397	22.750	23.111	23.482	23.863	24.255	24.659	25.076
24	22.081	22.423	22.772	23.129	23.495	23.871	24.256	24.653	25.062	25.485
25	22.439	22.785	23.139	23.501	23.871	24.251	24.642	25.044	25.458	25.885
26	22.790	23.141	23.498	23.865	24.240	24.624	25.019	25.426	25.845	26.278
27	23.134	23.489	23.851	24.221	24.601	24.990	25.390	25.801	26.225	26.662
28	23.472	23.831	24.197	24.572	24.955	25.349	25.753	26.169	26.597	27.040
29	23.804	24.166	24.537	24.915	25.303	25.701	26.110	26.530	26.963	27.411
30	24.130	24.496	24.870	25.253	25.645	26.047	26.460	26.885	27.323	27.775
32	24.765	25.139	25.521	25.912	26.312	26.723	27.144	27.577	28.024	28.485
34	25.381	25.762	26.152	26.550	26.958	27.377	27.806	28.248	28.703	29.173
36	25.978	26.367	26.764	27.170	27.586	28.012	28.449	28.899	29.362	29.841
38	26.559	26.955	27.359	27.772	28.195	28.629	29.074	29.532	30.003	30.490
40	27.125	27.528	27.939	28.359	28.789	29.230	29.682	30.148	30.627	31.122
42	27.677	28.086	28.503	28.930	29.367	29.815	30.275	30.748	31.235	31.738
44	28.215	28.631	29.055	29.488	29.932	30.387	30.854	31.334	31.829	32.339
46	28.741	29.163	29.594	30.034	30.484	30.946	31.420	31.907	32.409	32.926
48	29.256	29.684	30.121	30.567	31.024	31.493	31.973	32.468	32.976	33.501
50	29.760	30.194	30.637	31.090	31.553	32.028	32.515	33.016	33.532	34.064
55	30.977	31.426	31.884	32.351	32.830	33.321	33.824	34.341	34.874	35.423
60	32.140	32.602	33.074	33.556	34.050	34.555	35.074	35.607	36.155	36.721
65	33.255	33.730	34.216	34.712	35.219	35.739	36.272	36.820	37.384	37.965
70	34.327	34.816	35.314	35.823	36.344	36.877	37.424	37.987	38.565	39.162
75	35.362	35.862	36.373	36.895	37.429	37.975	38.536	39.112	39.705	40.316
80	36.362	36.875	37.397	37.932	38.478	39.037	39.611	40.200	40.807	41.432
85	37.331	37.855	38.390	38.936	39.494	40.066	40.653	41.255	41.875	42.513
90	38.272	38.808	39.354	39.911	40.481	41.065	41.664	42.279	42.911	43.563
95	39.188	39.734	40.291	40.859	41.441	42.036	42.647	43.274	43.919	44.584
100	40.079	40.635	41.203	41.783	42.375	42.982	43.605	44.244	44.901	45.578

ALPHA = .025

N	POWER= .80	.81	.82	.83	.84	.85	.86	.87	.88	.89
1	9.505	9.730	9.965	10.212	10.471	10.744	11.034	11.342	11.672	12.027
2	11.493	11.747	12.012	12.290	12.581	12.888	13.213	13.559	13.928	14.324
3	12.908	13.183	13.469	13.768	14.082	14.413	14.763	15.134	15.530	15.956
4	14.064	14.354	14.657	14.974	15.307	15.656	16.026	16.418	16.837	17.286
5	15.063	15.368	15.686	16.018	16.366	16.732	17.118	17.529	17.966	18.435
6	15.957	16.274	16.605	16.950	17.312	17.692	18.094	18.520	18.975	19.462
7	16.772	17.100	17.443	17.800	18.175	18.568	18.984	19.425	19.895	20.398
8	17.526	17.865	18.218	18.587	18.973	19.379	19.807	20.261	20.745	21.264
9	18.231	18.580	18.943	19.322	19.719	20.136	20.577	21.043	21.540	22.073
10	18.896	19.254	19.626	20.015	20.422	20.850	21.302	21.780	22.289	22.835
11	19.526	19.892	20.274	20.672	21.089	21.527	21.989	22.478	22.999	23.558
12	20.127	20.501	20.891	21.298	21.724	22.172	22.644	23.144	23.676	24.246
13	20.702	21.084	21.482	21.898	22.333	22.789	23.271	23.781	24.324	24.905
14	21.254	21.644	22.050	22.473	22.917	23.382	23.873	24.392	24.945	25.537
15	21.786	22.183	22.597	23.028	23.479	23.953	24.453	24.982	25.544	26.147
16	22.300	22.704	23.125	23.564	24.023	24.505	25.013	25.551	26.123	26.735
17	22.797	23.208	23.636	24.082	24.549	25.039	25.555	26.101	26.683	27.305
18	23.280	23.697	24.132	24.585	25.059	25.556	26.081	26.635	27.226	27.857
19	23.749	24.173	24.614	25.074	25.555	26.059	26.591	27.154	27.753	28.394
20	24.205	24.635	25.083	25.549	26.037	26.549	27.089	27.659	28.266	28.916
21	24.650	25.086	25.539	26.012	26.507	27.026	27.573	28.151	28.767	29.425
22	25.084	25.526	25.985	26.465	26.966	27.492	28.045	28.632	29.255	29.921
23	25.508	25.955	26.421	26.906	27.414	27.946	28.507	29.100	29.731	30.406
24	25.922	26.375	26.847	27.338	27.852	28.391	28.959	29.559	30.197	30.880
25	26.328	26.787	27.264	27.761	28.281	28.826	29.400	30.008	30.654	31.344
26	26.725	27.190	27.672	28.175	28.701	29.252	29.833	30.448	31.101	31.799
27	27.115	27.585	28.073	28.581	29.113	29.670	30.258	30.879	31.539	32.244
28	27.498	27.972	28.466	28.980	29.517	30.081	30.674	31.302	31.969	32.682
29	27.873	28.353	28.851	29.371	29.914	30.483	31.083	31.717	32.391	33.111
30	28.242	28.727	29.230	29.755	30.304	30.879	31.484	32.125	32.805	33.533
32	28.962	29.456	29.970	30.505	31.064	31.650	32.268	32.920	33.614	34.355
34	29.659	30.162	30.686	31.231	31.800	32.398	33.026	33.691	34.397	35.151
36	30.335	30.848	31.380	31.935	32.515	33.123	33.762	34.438	35.157	35.924
38	30.993	31.514	32.056	32.620	33.210	33.827	34.478	35.165	35.895	36.675
40	31.633	32.163	32.713	33.287	33.886	34.514	35.174	35.873	36.614	37.406
42	32.257	32.795	33.354	33.937	34.545	35.183	35.853	36.562	37.315	38.119
44	32.866	33.413	33.980	34.571	35.188	35.835	36.516	37.235	37.999	38.815
46	33.462	34.016	34.592	35.191	35.817	36.473	37.164	37.893	38.667	39.495
48	34.044	34.606	35.190	35.797	36.432	37.097	37.797	38.536	39.321	40.159
50	34.614	35.184	35.775	36.391	37.034	37.708	38.417	39.166	39.961	40.810
55	35.991	36.579	37.189	37.824	38.488	39.183	39.914	40.687	41.506	42.382
60	37.305	37.911	38.539	39.193	39.876	40.592	41.344	42.138	42.982	43.882
65	38.566	39.188	39.833	40.506	41.207	41.942	42.714	43.530	44.396	45.320
70	39.778	40.416	41.078	41.768	42.487	43.240	44.032	44.869	45.756	46.703
75	40.947	41.601	42.279	42.985	43.721	44.493	45.304	46.160	47.068	48.037
80	42.078	42.746	43.440	44.162	44.915	45.704	46.533	47.408	48.336	49.326
85	43.173	43.856	44.565	45.302	46.071	46.877	47.723	48.617	49.564	50.575
90	44.237	44.934	45.657	46.409	47.194	48.016	48.879	49.790	50.757	51.788
95	45.271	45.981	46.719	47.485	48.286	49.123	50.003	50.932	51.916	52.967
100	46.278	47.002	47.753	48.534	49.348	50.201	51.097	52.043	53.046	54.115

ALPHA = .025

N	POWER= .90	.91	.92	.93	.94	.95	.96	.97	.98	.99
1	12.411	12.832	13.297	13.818	14.411	15.103	15.937	16.993	18.448	20.864
2	14.753	15.222	15.739	16.318	16.976	17.741	18.662	19.825	21.423	24.066
3	16.416	16.918	17.472	18.091	18.794	19.611	20.592	21.830	23.528	26.329
4	17.771	18.301	18.884	19.535	20.274	21.133	22.163	23.461	25.240	28.169
5	18.943	19.495	20.104	20.783	21.553	22.448	23.520	24.870	26.718	29.756
6	19.988	20.562	21.193	21.897	22.695	23.621	24.731	26.127	28.037	31.172
7	20.942	21.534	22.185	22.912	23.735	24.690	25.834	27.272	29.237	32.461
8	21.824	22.433	23.103	23.850	24.697	25.679	26.854	28.330	30.347	33.652
9	22.648	23.273	23.961	24.727	25.596	26.602	27.806	29.318	31.383	34.764
10	23.424	24.064	24.769	25.553	26.442	27.471	28.703	30.249	32.359	35.811
11	24.160	24.815	25.534	26.336	27.244	28.295	29.553	31.131	33.283	36.803
12	24.861	25.529	26.264	27.082	28.008	29.080	30.362	31.970	34.163	37.747
13	25.532	26.213	26.962	27.796	28.739	29.831	31.137	32.774	35.005	38.650
14	26.176	26.870	27.632	28.481	29.441	30.552	31.880	33.545	35.814	39.517
15	26.796	27.502	28.278	29.141	30.117	31.247	32.596	34.288	36.592	40.352
16	27.395	28.113	28.901	29.778	30.770	31.917	33.287	35.004	37.343	41.157
17	27.975	28.704	29.504	30.394	31.401	32.566	33.956	35.698	38.070	41.936
18	28.538	29.277	30.089	30.992	32.014	33.195	34.605	36.371	38.774	42.692
19	29.084	29.834	30.658	31.573	32.609	33.806	35.235	37.024	39.459	43.425
20	29.616	30.376	31.210	32.138	33.187	34.400	35.847	37.659	40.124	44.139
21	30.134	30.904	31.749	32.689	33.751	34.979	36.444	38.279	40.773	44.834
22	30.639	31.419	32.275	33.226	34.301	35.544	37.027	38.882	41.405	45.512
23	31.133	31.922	32.788	33.751	34.839	36.096	37.596	39.472	42.023	46.174
24	31.615	32.414	33.290	34.264	35.364	36.635	38.152	40.049	42.627	46.821
25	32.088	32.895	33.781	34.766	35.878	37.163	38.696	40.613	43.218	47.455
26	32.551	33.367	34.262	35.257	36.382	37.680	39.229	41.166	43.797	48.075
27	33.004	33.829	34.734	35.740	36.875	38.187	39.751	41.707	44.364	48.682
28	33.449	34.283	35.197	36.212	37.360	38.685	40.264	42.239	44.920	49.279
29	33.886	34.728	35.651	36.677	37.835	39.173	40.767	42.760	45.467	49.864
30	34.316	35.166	36.098	37.133	38.302	39.652	41.261	43.273	46.003	50.439
32	35.153	36.019	36.968	38.022	39.213	40.587	42.225	44.272	47.049	51.559
34	35.963	36.845	37.811	38.884	40.095	41.493	43.158	45.239	48.062	52.644
36	36.750	37.646	38.628	39.719	40.950	42.371	44.063	46.177	49.044	53.696
38	37.515	38.425	39.423	40.531	41.782	43.225	44.943	47.089	49.999	54.719
40	38.259	39.184	40.197	41.322	42.591	44.056	45.799	47.976	50.928	55.714
42	38.984	39.923	40.951	42.092	43.380	44.865	46.634	48.841	51.834	56.683
44	39.692	40.644	41.687	42.844	44.150	45.656	47.448	49.685	52.717	57.629
46	40.384	41.349	42.406	43.579	44.902	46.428	48.244	50.510	53.580	58.553
48	41.061	42.038	43.109	44.297	45.638	47.183	49.022	51.316	54.424	59.457
50	41.723	42.713	43.798	45.001	46.358	47.922	49.783	52.105	55.250	60.341
55	43.323	44.343	45.460	46.699	48.096	49.707	51.622	54.011	57.245	62.476
60	44.849	45.898	47.047	48.320	49.756	51.410	53.377	55.829	59.148	64.513
65	46.313	47.389	48.567	49.874	51.346	53.042	55.059	57.572	60.971	66.465
70	47.720	48.823	50.030	51.368	52.875	54.612	56.676	59.247	62.725	68.341
75	49.078	50.206	51.440	52.809	54.350	56.126	58.235	60.863	64.415	70.150
80	50.390	51.543	52.804	54.201	55.776	57.589	59.742	62.424	66.049	71.899
85	51.661	52.838	54.125	55.551	57.157	59.006	61.202	63.937	67.632	73.592
90	52.895	54.095	55.407	56.860	58.497	60.382	62.619	65.405	69.168	75.235
95	54.095	55.317	56.653	58.134	59.801	61.719	63.997	66.832	70.661	76.833
100	55.263	56.507	57.867	59.373	61.070	63.021	65.338	68.222	72.114	78.388

ALPHA = .010

N	POWER= .10	.12	.14	.16	.18	.20	.22	.24	.26	.28
1	1.674	1.962	2.236	2.501	2.757	3.007	3.253	3.495	3.734	3.972
2	2.299	2.660	3.001	3.325	3.638	3.941	4.237	4.526	4.811	5.093
3	2.763	3.176	3.563	3.931	4.283	4.624	4.955	5.278	5.595	5.908
4	3.149	3.605	4.031	4.433	4.818	5.189	5.548	5.899	6.243	6.581
5	3.488	3.981	4.440	4.873	5.285	5.682	6.067	6.441	6.808	7.168
6	3.794	4.320	4.809	5.268	5.706	6.126	6.533	6.928	7.315	7.695
7	4.075	4.632	5.147	5.631	6.092	6.534	6.960	7.375	7.780	8.178
8	4.337	4.921	5.462	5.969	6.451	6.912	7.358	7.790	8.212	8.626
9	4.583	5.194	5.757	6.286	6.787	7.267	7.730	8.179	8.617	9.047
10	4.816	5.451	6.037	6.586	7.106	7.603	8.082	8.547	9.000	9.444
11	5.038	5.696	6.303	6.871	7.408	7.922	8.417	8.897	9.364	9.821
12	5.250	5.930	6.557	7.143	7.697	8.227	8.737	9.231	9.712	10.182
13	5.453	6.155	6.801	7.404	7.975	8.520	9.044	9.551	10.045	10.528
14	5.649	6.372	7.036	7.656	8.242	8.801	9.339	9.859	10.365	10.860
25	7.466	8.378	9.211	9.985	10.714	11.406	12.069	12.708	13.329	13.934
26	7.609	8.537	9.383	10.169	10.908	11.611	12.284	12.933	13.562	14.176
27	7.750	8.692	9.551	10.349	11.099	11.812	12.495	13.153	13.791	14.413
28	7.888	8.845	9.717	10.526	11.287	12.010	12.702	13.369	14.016	14.646
29	8.024	8.995	9.879	10.700	11.472	12.204	12.905	13.581	14.237	14.875
30	8.158	9.142	10.039	10.871	11.653	12.395	13.106	13.790	14.454	15.100
32	8.419	9.430	10.351	11.205	12.007	12.768	13.496	14.198	14.877	15.539
34	8.671	9.709	10.653	11.528	12.350	13.129	13.875	14.592	15.288	15.965
36	8.917	9.980	10.947	11.842	12.683	13.480	14.242	14.976	15.687	16.378
38	9.156	10.244	11.233	12.148	13.007	13.821	14.600	15.349	16.075	16.780
40	9.389	10.501	11.511	12.446	13.323	14.154	14.948	15.713	16.452	17.172
42	9.617	10.752	11.783	12.737	13.631	14.479	15.288	16.067	16.821	17.554
44	9.839	10.997	12.048	13.021	13.932	14.796	15.620	16.414	17.181	17.928
46	10.056	11.236	12.308	13.298	14.227	15.106	15.945	16.752	17.533	18.292
48	10.269	11.471	12.562	13.570	14.515	15.409	16.263	17.084	17.878	18.650
50	10.477	11.700	12.811	13.836	14.797	15.706	16.574	17.408	18.215	18.999
55	10.980	12.256	13.412	14.480	15.479	16.424	17.326	18.193	19.030	19.844
60	11.462	12.787	13.987	15.095	16.131	17.111	18.045	18.943	19.810	20.652
65	11.924	13.296	14.539	15.685	16.757	17.769	18.735	19.662	20.558	21.427
70	12.369	13.787	15.070	16.253	17.359	18.403	19.399	20.355	21.277	22.173
75	12.799	14.261	15.583	16.802	17.940	19.015	20.040	21.023	21.972	22.893
80	13.214	14.719	16.079	17.332	18.503	19.608	20.660	21.669	22.644	23.590
85	13.617	15.163	16.560	17.847	19.048	20.182	21.261	22.296	23.295	24.265
90	14.009	15.595	17.028	18.347	19.578	20.739	21.845	22.905	23.928	24.920
95	14.390	16.015	17.482	18.833	20.093	21.281	22.413	23.497	24.544	25.558
100	14.761	16.424	17.925	19.306	20.595	21.810	22.966	24.074	25.143	26.179

ALPHA = .010

N	POWER= .30	.32	.34	.36	.38	.40	.42	.44	.46	.48
1	4.208	4.444	4.680	4.917	5.154	5.394	5.636	5.880	6.128	6.379
2	5.372	5.650	5.927	6.203	6.480	6.758	7.039	7.321	7.607	7.896
3	6.218	6.525	6.830	7.135	7.440	7.745	8.052	8.362	8.674	8.990
4	6.915	7.246	7.575	7.902	8.230	8.558	8.887	9.218	9.552	9.890
5	7.523	7.874	8.223	8.571	8.917	9.265	9.613	9.963	10.316	10.672
6	8.069	8.439	8.806	9.171	9.535	9.899	10.264	10.631	11.001	11.374
7	8.569	8.956	9.339	9.720	10.100	10.480	10.860	11.242	11.627	12.015
8	9.033	9.435	9.834	10.230	10.624	11.019	11.413	11.810	12.208	12.611
9	9.469	9.885	10.298	10.708	11.116	11.524	11.932	12.341	12.753	13.168
10	9.880	10.310	10.736	11.159	11.580	12.000	12.421	12.843	13.267	13.694
11	10.271	10.714	11.152	11.588	12.021	12.453	12.885	13.319	13.755	14.194
12	10.644	11.099	11.550	11.997	12.442	12.885	13.329	13.774	14.221	14.671
13	11.002	11.469	11.931	12.389	12.845	13.300	13.754	14.210	14.667	15.128
14	11.346	11.824	12.298	12.767	13.233	13.698	14.163	14.629	15.096	15.567
15	11.678	12.167	12.651	13.130	13.607	14.082	14.557	15.033	15.510	15.991
16	11.999	12.499	12.993	13.482	13.969	14.454	14.938	15.423	15.910	16.401
17	12.310	12.820	13.324	13.823	14.319	14.814	15.307	15.802	16.298	16.797
18	12.612	13.132	13.646	14.154	14.660	15.163	15.666	16.169	16.674	17.182
19	12.906	13.435	13.958	14.476	14.990	15.503	16.014	16.526	17.040	17.557
20	13.192	13.731	14.263	14.790	15.313	15.833	16.353	16.874	17.396	17.921
21	13.471	14.019	14.560	15.095	15.627	16.156	16.684	17.213	17.743	18.277
22	13.744	14.300	14.850	15.394	15.934	16.471	17.007	17.544	18.082	18.624
23	14.010	14.576	15.133	15.685	16.233	16.779	17.323	17.868	18.414	18.963
24	14.271	14.845	15.411	15.971	16.527	17.080	17.632	18.184	18.738	19.295
25	14.526	15.108	15.682	16.250	16.814	17.375	17.935	18.494	19.055	19.620
26	14.776	15.367	15.949	16.524	17.096	17.664	18.231	18.798	19.367	19.938
27	15.022	15.620	16.210	16.793	17.372	17.948	18.522	19.096	19.672	20.250
28	15.263	15.869	16.466	17.057	17.643	18.226	18.807	19.389	19.971	20.557
29	15.500	16.113	16.718	17.316	17.909	18.499	19.088	19.676	20.266	20.858
30	15.733	16.354	16.966	17.571	18.171	18.768	19.363	19.958	20.555	21.154
32	16.187	16.822	17.449	18.068	18.682	19.293	19.901	20.510	21.119	21.731
34	16.627	17.277	17.917	18.550	19.177	19.801	20.422	21.044	21.666	22.291
36	17.055	17.718	18.372	19.018	19.658	20.294	20.928	21.562	22.197	22.834
38	17.470	18.147	18.814	19.473	20.126	20.774	21.421	22.066	22.713	23.362
40	17.876	18.566	19.245	19.916	20.581	21.242	21.900	22.558	23.216	23.877
42	18.271	18.974	19.665	20.349	21.025	21.698	22.368	23.037	23.706	24.378
44	18.657	19.372	20.076	20.771	21.459	22.143	22.824	23.504	24.185	24.868
46	19.034	19.761	20.477	21.184	21.883	22.578	23.270	23.962	24.653	25.347
48	19.403	20.142	20.870	21.587	22.298	23.004	23.707	24.409	25.111	25.816
50	19.765	20.516	21.254	21.983	22.705	23.421	24.135	24.847	25.560	26.275
55	20.639	21.418	22.183	22.939	23.687	24.429	25.168	25.906	26.644	27.384
60	21.474	22.280	23.071	23.852	24.625	25.392	26.156	26.917	27.679	28.443
65	22.276	23.106	23.923	24.729	25.526	26.316	27.103	27.888	28.672	29.459
70	23.047	23.902	24.743	25.572	26.392	27.205	28.014	28.821	29.628	30.437
75	23.791	24.670	25.534	26.386	27.228	28.063	28.894	29.722	30.550	31.380
80	24.511	25.413	26.299	27.173	28.037	28.893	29.745	30.594	31.442	32.292
85	25.209	26.133	27.041	27.936	28.820	29.697	30.569	31.438	32.307	33.177
90	25.887	26.833	27.762	28.677	29.582	30.478	31.370	32.258	33.146	34.036
95	26.546	27.513	28.463	29.398	30.322	31.238	32.149	33.056	33.963	34.871
100	27.189	28.176	29.145	30.100	31.043	31.978	32.907	33.833	34.758	35.684

ALPHA = .010

N	POWER= .50	.52	.54	.56	.58	.60	.62	.64	.66	.68
1	6.635	6.896	7.162	7.435	7.716	8.004	8.302	8.610	8.930	9.263
2	8.190	8.488	8.793	9.105	9.424	9.752	10.089	10.438	10.799	11.175
3	9.310	9.636	9.968	10.306	10.653	11.008	11.374	11.751	12.142	12.547
4	10.232	10.579	10.932	11.293	11.662	12.040	12.428	12.829	13.243	13.673
5	11.033	11.399	11.771	12.150	12.538	12.935	13.344	13.764	14.199	14.650
6	11.751	12.134	12.523	12.919	13.324	13.738	14.164	14.602	15.055	15.525
7	12.408	12.806	13.210	13.622	14.042	14.472	14.914	15.369	15.838	16.325
8	13.017	13.429	13.847	14.273	14.708	15.153	15.609	16.079	16.563	17.066
9	13.588	14.013	14.444	14.883	15.331	15.790	16.260	16.743	17.242	17.759
10	14.126	14.564	15.007	15.459	15.919	16.390	16.873	17.370	17.883	18.414
11	14.638	15.087	15.542	16.005	16.478	16.961	17.456	17.965	18.491	19.034
12	15.126	15.586	16.052	16.527	17.011	17.505	18.012	18.533	19.071	19.627
13	15.593	16.064	16.541	17.027	17.521	18.027	18.545	19.077	19.626	20.194
14	16.043	16.524	17.011	17.507	18.012	18.528	19.056	19.600	20.160	20.739
15	16.476	16.967	17.464	17.970	18.485	19.011	19.550	20.103	20.674	21.264
16	16.895	17.395	17.902	18.417	18.942	19.478	20.027	20.590	21.171	21.772
17	17.301	17.810	18.326	18.851	19.385	19.930	20.488	21.062	21.653	22.264
18	17.695	18.213	18.738	19.271	19.814	20.369	20.936	21.519	22.120	22.741
19	18.078	18.605	19.138	19.680	20.232	20.795	21.372	21.964	22.574	23.205
20	18.451	18.986	19.528	20.078	20.639	21.211	21.796	22.397	23.016	23.656
21	18.814	19.358	19.908	20.466	21.035	21.616	22.210	22.819	23.447	24.096
22	19.169	19.721	20.279	20.845	21.422	22.011	22.613	23.231	23.868	24.526
23	19.516	20.075	20.641	21.216	21.800	22.397	23.007	23.634	24.279	24.945
24	19.856	20.422	20.996	21.578	22.170	22.775	23.393	24.028	24.681	25.356
25	20.188	20.762	21.343	21.933	22.533	23.145	23.771	24.413	25.075	25.758
26	20.514	21.095	21.683	22.280	22.887	23.507	24.141	24.791	25.460	26.151
27	20.833	21.421	22.017	22.621	23.236	23.863	24.504	25.162	25.839	26.538
28	21.146	21.742	22.344	22.956	23.577	24.211	24.860	25.525	26.210	26.917
29	21.454	22.057	22.666	23.284	23.913	24.554	25.210	25.882	26.574	27.289
30	21.757	22.366	22.982	23.607	24.243	24.891	25.554	26.233	26.933	27.655
32	22.347	22.970	23.599	24.237	24.886	25.548	26.224	26.918	27.631	28.368
34	22.920	23.554	24.196	24.848	25.509	26.184	26.874	27.581	28.309	29.059
36	23.475	24.122	24.777	25.440	26.115	26.802	27.505	28.225	28.966	29.730
38	24.015	24.674	25.341	26.017	26.703	27.403	28.118	28.851	29.605	30.383
40	24.542	25.212	25.890	26.578	27.276	27.988	28.715	29.461	30.227	31.018
42	25.055	25.737	26.426	27.125	27.835	28.559	29.298	30.055	30.834	31.637
44	25.556	26.249	26.949	27.659	28.381	29.116	29.866	30.636	31.426	32.242
46	26.045	26.749	27.461	28.182	28.914	29.660	30.422	31.203	32.005	32.833
48	26.525	27.239	27.961	28.693	29.436	30.193	30.966	31.758	32.572	33.411
50	26.994	27.719	28.451	29.193	29.947	30.715	31.498	32.301	33.127	33.977
55	28.128	28.877	29.635	30.402	31.181	31.975	32.785	33.614	34.466	35.345
60	29.211	29.984	30.766	31.557	32.360	33.178	34.013	34.868	35.746	36.651
65	30.250	31.046	31.850	32.665	33.491	34.333	35.191	36.070	36.973	37.903
70	31.249	32.068	32.894	33.730	34.579	35.443	36.325	37.227	38.154	39.108
75	32.214	33.053	33.901	34.759	35.629	36.515	37.419	38.343	39.293	40.271
80	33.147	34.007	34.875	35.753	36.644	37.551	38.476	39.423	40.394	41.395
85	34.051	34.930	35.818	36.717	37.628	38.556	39.501	40.469	41.462	42.484
90	34.929	35.828	36.735	37.653	38.584	39.531	40.496	41.484	42.498	43.542
95	35.782	36.700	37.626	38.563	39.513	40.479	41.464	42.472	43.506	44.571
100	36.614	37.550	38.494	39.449	40.418	41.403	42.407	43.434	44.488	45.572

ALPHA = .010

N	POWER= .70	.71	.72	.73	.74	.75	.76	.77	.78	.79
1	9.611	9.792	9.977	10.167	10.363	10.565	10.772	10.987	11.209	11.440
2	11.567	11.770	11.978	12.191	12.411	12.636	12.869	13.109	13.357	13.614
3	12.970	13.189	13.413	13.643	13.879	14.121	14.371	14.629	14.896	15.171
4	14.121	14.352	14.589	14.833	15.082	15.339	15.603	15.875	16.156	16.447
5	15.119	15.362	15.610	15.865	16.126	16.395	16.671	16.956	17.250	17.554
6	16.014	16.266	16.525	16.789	17.061	17.340	17.627	17.923	18.228	18.544
7	16.831	17.092	17.360	17.634	17.915	18.204	18.500	18.806	19.122	19.449
8	17.588	17.858	18.134	18.416	18.706	19.003	19.310	19.625	19.950	20.287
9	18.297	18.574	18.858	19.148	19.446	19.752	20.067	20.391	20.725	21.071
10	18.965	19.250	19.541	19.839	20.144	20.458	20.781	21.113	21.455	21.810
11	19.599	19.891	20.189	20.494	20.807	21.128	21.458	21.798	22.149	22.511
12	20.204	20.502	20.807	21.119	21.439	21.767	22.104	22.451	22.810	23.180
13	20.784	21.088	21.399	21.717	22.044	22.378	22.723	23.077	23.442	23.820
14	21.341	21.651	21.968	22.292	22.625	22.966	23.317	23.678	24.050	24.435
15	21.877	22.193	22.516	22.847	23.185	23.533	23.890	24.257	24.636	25.028
16	22.396	22.717	23.046	23.382	23.727	24.080	24.443	24.817	25.203	25.601
17	22.898	23.225	23.559	23.901	24.251	24.610	24.979	25.359	25.751	26.155
18	23.385	23.717	24.056	24.404	24.759	25.124	25.499	25.885	26.283	26.693
19	23.859	24.196	24.540	24.893	25.254	25.624	26.004	26.396	26.799	27.216
20	24.320	24.662	25.011	25.369	25.735	26.110	26.496	26.893	27.303	27.725
21	24.769	25.116	25.470	25.833	26.204	26.585	26.976	27.378	27.793	28.221
22	25.208	25.559	25.918	26.285	26.662	27.047	27.444	27.851	28.271	28.705
23	25.636	25.992	26.356	26.728	27.109	27.500	27.901	28.314	28.739	29.178
24	26.055	26.416	26.784	27.161	27.546	27.942	28.348	28.766	29.196	29.641
25	26.466	26.831	27.203	27.584	27.975	28.375	28.786	29.208	29.644	30.093
26	26.868	27.237	27.614	27.999	28.394	28.799	29.215	29.642	30.082	30.537
27	27.262	27.635	28.017	28.406	28.806	29.215	29.635	30.067	30.513	30.972
28	27.649	28.027	28.412	28.806	29.209	29.623	30.048	30.485	30.935	31.399
29	28.029	28.411	28.800	29.198	29.606	30.024	30.453	30.894	31.349	31.818
30	28.403	28.788	29.181	29.584	29.996	30.418	30.851	31.297	31.756	32.230
32	29.131	29.524	29.925	30.336	30.756	31.186	31.628	32.082	32.550	33.033
34	29.837	30.237	30.646	31.064	31.492	31.930	32.380	32.843	33.320	33.812
36	30.522	30.930	31.346	31.771	32.206	32.653	33.111	33.582	34.067	34.567
38	31.188	31.603	32.026	32.458	32.901	33.355	33.821	34.300	34.793	35.301
40	31.837	32.258	32.688	33.128	33.578	34.039	34.512	34.999	35.499	36.016
42	32.469	32.897	33.334	33.780	34.237	34.706	35.186	35.680	36.189	36.713
44	33.086	33.521	33.964	34.417	34.881	35.356	35.844	36.345	36.861	37.394
46	33.690	34.130	34.580	35.040	35.510	35.993	36.487	36.995	37.519	38.059
48	34.280	34.727	35.183	35.649	36.126	36.615	37.116	37.632	38.162	38.709
50	34.858	35.311	35.773	36.246	36.729	37.224	37.732	38.254	38.792	39.346
55	36.254	36.721	37.198	37.686	38.185	38.696	39.220	39.758	40.312	40.884
60	37.587	38.069	38.560	39.062	39.575	40.101	40.640	41.195	41.765	42.353
65	38.866	39.360	39.865	40.381	40.908	41.448	42.003	42.572	43.157	43.761
70	40.096	40.603	41.121	41.650	42.191	42.745	43.313	43.896	44.497	45.116
75	41.282	41.802	42.332	42.874	43.428	43.995	44.577	45.174	45.789	46.423
80	42.430	42.961	43.504	44.058	44.624	45.204	45.799	46.410	47.038	47.686
85	43.542	44.085	44.639	45.205	45.783	46.376	46.983	47.607	48.249	48.910
90	44.621	45.176	45.741	46.319	46.909	47.513	48.133	48.770	49.424	50.099
95	45.671	46.236	46.813	47.401	48.003	48.620	49.251	49.900	50.567	51.255
100	46.693	47.269	47.857	48.456	49.069	49.697	50.340	51.001	51.680	52.380

ALPHA = .010

N	POWER= .80	.81	.82	.83	.84	.85	.86	.87	.88	.89
1	11.679	11.928	12.188	12.461	12.747	13.048	13.367	13.706	14.069	14.458
2	13.881	14.158	14.448	14.751	15.068	15.403	15.756	16.131	16.532	16.962
3	15.458	15.755	16.065	16.390	16.730	17.087	17.465	17.866	18.293	18.752
4	16.749	17.063	17.390	17.732	18.090	18.466	18.864	19.285	19.735	20.216
5	17.869	18.197	18.538	18.895	19.269	19.661	20.076	20.515	20.984	21.486
6	18.872	19.212	19.566	19.936	20.324	20.731	21.161	21.616	22.101	22.621
7	19.787	20.139	20.505	20.887	21.287	21.707	22.151	22.621	23.121	23.657
8	20.635	20.997	21.374	21.767	22.179	22.611	23.068	23.551	24.065	24.616
9	21.429	21.800	22.187	22.591	23.013	23.457	23.925	24.421	24.948	25.513
10	22.177	22.557	22.954	23.367	23.800	24.254	24.733	25.241	25.781	26.359
11	22.886	23.276	23.681	24.104	24.546	25.011	25.500	26.019	26.570	27.161
12	23.563	23.961	24.374	24.806	25.258	25.732	26.231	26.760	27.323	27.926
13	24.211	24.616	25.038	25.478	25.939	26.422	26.931	27.470	28.044	28.657
14	24.833	25.246	25.676	26.124	26.593	27.085	27.604	28.152	28.736	29.360
15	25.433	25.854	26.291	26.747	27.224	27.724	28.251	28.809	29.403	30.038
16	26.013	26.440	26.885	27.348	27.833	28.341	28.877	29.444	30.047	30.692
17	26.574	27.008	27.460	27.930	28.422	28.939	29.483	30.059	30.671	31.325
18	27.118	27.559	28.017	28.495	28.995	29.519	30.071	30.655	31.276	31.940
19	27.647	28.094	28.559	29.044	29.550	30.082	30.642	31.234	31.863	32.537
20	28.162	28.616	29.087	29.578	30.092	30.630	31.198	31.798	32.435	33.118
21	28.664	29.123	29.601	30.099	30.619	31.165	31.739	32.347	32.993	33.684
22	29.154	29.619	30.103	30.607	31.133	31.686	32.268	32.883	33.537	34.236
23	29.632	30.103	30.593	31.103	31.636	32.195	32.784	33.407	34.068	34.776
24	30.100	30.577	31.072	31.588	32.128	32.693	33.289	33.919	34.588	35.303
25	30.558	31.041	31.541	32.063	32.609	33.181	33.783	34.420	35.097	35.820
26	31.007	31.495	32.001	32.529	33.080	33.659	34.268	34.911	35.595	36.326
27	31.447	31.940	32.452	32.985	33.543	34.127	34.742	35.393	36.084	36.822
28	31.879	32.377	32.894	33.433	33.996	34.587	35.208	35.865	36.563	37.309
29	32.303	32.806	33.329	33.873	34.442	35.038	35.666	36.329	37.034	37.787
30	32.720	33.228	33.756	34.305	34.880	35.482	36.115	36.785	37.496	38.256
32	33.533	34.050	34.588	35.148	35.733	36.346	36.992	37.674	38.398	39.172
34	34.320	34.847	35.394	35.964	36.560	37.184	37.841	38.535	39.272	40.059
36	35.084	35.620	36.177	36.757	37.362	37.997	38.665	39.370	40.120	40.920
38	35.827	36.372	36.938	37.527	38.142	38.787	39.466	40.183	40.944	41.757
40	36.550	37.104	37.678	38.277	38.902	39.557	40.245	40.973	41.746	42.571
42	37.256	37.817	38.401	39.008	39.642	40.307	41.006	41.744	42.528	43.365
44	37.944	38.514	39.105	39.721	40.365	41.039	41.748	42.496	43.291	44.140
46	38.617	39.194	39.794	40.419	41.071	41.754	42.473	43.232	44.037	44.898
48	39.275	39.860	40.468	41.101	41.762	42.454	43.182	43.951	44.767	45.638
50	39.919	40.512	41.128	41.769	42.438	43.139	43.876	44.655	45.481	46.364
55	41.475	42.086	42.721	43.382	44.071	44.794	45.553	46.355	47.206	48.115
60	42.960	43.590	44.242	44.922	45.631	46.374	47.154	47.979	48.853	49.787
65	44.385	45.031	45.701	46.398	47.126	47.888	48.689	49.535	50.432	51.389
70	45.755	46.417	47.104	47.819	48.564	49.345	50.166	51.032	51.950	52.931
75	47.077	47.754	48.457	49.188	49.951	50.750	51.589	52.475	53.415	54.417
80	48.355	49.047	49.766	50.513	51.292	52.109	52.966	53.871	54.831	55.855
85	49.593	50.300	51.033	51.796	52.592	53.425	54.300	55.223	56.203	57.247
90	50.796	51.516	52.264	53.042	53.853	54.703	55.595	56.536	57.534	58.599
95	51.965	52.699	53.461	54.253	55.080	55.945	56.854	57.813	58.829	59.913
100	53.103	53.851	54.627	55.433	56.275	57.155	58.080	59.056	60.090	61.193

ALPHA = .010

N	POWER= .90	.91	.92	.93	.94	.95	.96	.97	.98	.99
1	14.879	15.340	15.848	16.416	17.062	17.814	18.719	19.861	21.433	24.031
2	17.427	17.934	18.493	19.117	19.826	20.650	21.639	22.886	24.596	27.415
3	19.247	19.788	20.382	21.046	21.800	22.674	23.723	25.043	26.852	29.826
4	20.737	21.304	21.928	22.624	23.413	24.329	25.426	26.806	28.695	31.794
5	22.028	22.618	23.267	23.990	24.811	25.762	26.901	28.333	30.289	33.498
6	23.182	23.792	24.464	25.212	26.060	27.043	28.219	29.697	31.715	35.020
7	24.235	24.865	25.557	26.327	27.201	28.212	29.422	30.941	33.015	36.408
8	25.211	25.857	26.568	27.359	28.256	29.294	30.535	32.092	34.217	37.692
9	26.123	26.785	27.514	28.324	29.242	30.305	31.575	33.168	35.341	38.891
10	26.982	27.660	28.405	29.234	30.172	31.258	32.555	34.182	36.399	40.021
11	27.798	28.490	29.250	30.096	31.053	32.161	33.484	35.143	37.403	41.092
12	28.575	29.281	30.056	30.918	31.894	33.022	34.370	36.059	38.360	42.112
13	29.319	30.038	30.827	31.705	32.698	33.846	35.218	36.936	39.275	43.088
14	30.033	30.765	31.567	32.460	33.470	34.638	36.031	37.777	40.153	44.026
15	30.722	31.465	32.281	33.188	34.214	35.400	36.815	38.588	40.999	44.928
16	31.387	32.142	32.970	33.891	34.932	36.136	37.573	39.371	41.817	45.800
17	32.031	32.797	33.637	34.572	35.628	36.849	38.305	40.128	42.607	46.643
18	32.655	33.432	34.284	35.232	36.302	37.540	39.016	40.863	43.374	47.460
19	33.262	34.049	34.913	35.873	36.958	38.211	39.706	41.577	44.119	48.255
20	33.852	34.650	35.525	36.497	37.595	38.865	40.378	42.271	44.844	49.027
21	34.427	35.235	36.121	37.105	38.217	39.501	41.033	42.948	45.550	49.780
22	34.989	35.806	36.702	37.698	38.823	40.122	41.671	43.608	46.239	50.514
23	35.537	36.364	37.270	38.278	39.415	40.729	42.295	44.253	46.911	51.231
24	36.073	36.909	37.826	38.844	39.994	41.322	42.905	44.883	47.569	51.932
25	36.598	37.443	38.370	39.399	40.561	41.903	43.501	45.500	48.213	52.619
26	37.113	37.966	38.902	39.942	41.116	42.471	44.086	46.104	48.843	53.290
27	37.617	38.479	39.425	40.475	41.660	43.029	44.659	46.697	49.461	53.949
28	38.111	38.932	39.937	40.997	42.194	43.576	45.222	47.278	50.068	54.595
29	38.597	39.476	40.440	41.511	42.719	44.113	45.774	47.848	50.663	55.230
30	39.074	39.962	40.935	42.015	43.234	44.641	46.316	48.409	51.248	55.853
32	40.005	40.908	41.899	42.998	44.238	45.669	47.373	49.502	52.388	57.067
34	40.906	41.825	42.832	43.950	45.211	46.666	48.397	50.560	53.492	58.243
36	41.781	42.715	43.738	44.873	46.154	47.632	49.391	51.587	54.562	59.384
38	42.631	43.579	44.618	45.771	47.072	48.572	50.357	52.585	55.603	60.493
40	43.459	44.421	45.475	46.645	47.965	49.486	51.297	53.556	56.616	61.572
42	44.266	45.242	46.311	47.497	48.835	50.378	52.213	54.502	57.603	62.623
44	45.053	46.043	47.126	48.329	49.684	51.248	53.107	55.426	58.567	63.649
46	45.822	46.825	47.923	49.141	50.514	52.098	53.980	56.329	59.508	64.652
48	46.575	47.591	48.702	49.936	51.326	52.929	54.835	57.211	60.428	65.632
50	47.312	48.340	49.465	50.714	52.121	53.743	55.671	58.075	61.329	66.591
55	49.091	50.149	51.308	52.592	54.039	55.707	57.690	60.161	63.504	68.907
60	50.790	51.877	53.066	54.385	55.871	57.583	59.617	62.152	65.580	71.118
65	52.418	53.532	54.752	56.104	57.627	59.381	61.464	64.061	67.570	73.236
70	53.984	55.125	56.373	57.757	59.315	61.109	63.241	65.895	69.483	75.272
75	55.494	56.661	57.937	59.351	60.943	62.776	64.953	67.664	71.327	77.235
80	56.954	58.145	59.448	60.892	62.517	64.388	66.609	69.375	73.110	79.132
85	58.369	59.584	60.913	62.384	64.042	65.949	68.213	71.031	74.836	80.970
90	59.742	60.980	62.334	63.833	65.522	67.464	69.770	72.639	76.512	82.753
95	61.078	62.338	63.716	65.242	66.960	68.937	71.283	74.202	78.141	84.487
100	62.378	63.660	65.062	66.614	68.361	70.372	72.757	75.724	79.728	86.175

ALPHA = .005

N	POWER= .10	.12	.14	.16	.18	.20	.22	.24	.26	.28
1	2.327	2.664	2.981	3.285	3.578	3.863	4.141	4.413	4.682	4.947
2	3.105	3.515	3.899	4.263	4.612	4.949	5.276	5.596	5.909	6.219
3	3.680	4.142	4.573	4.980	5.368	5.742	6.105	6.458	6.804	7.144
4	4.158	4.663	5.133	5.574	5.995	6.400	6.791	7.171	7.544	7.909
5	4.578	5.120	5.622	6.094	6.543	6.974	7.390	7.794	8.189	8.576
6	4.956	5.532	6.064	6.563	7.037	7.491	7.929	8.355	8.770	9.177
7	5.304	5.910	6.469	6.993	7.490	7.965	8.424	8.868	9.302	9.727
8	5.628	6.262	6.846	7.393	7.911	8.406	8.883	9.346	9.796	10.238
9	5.932	6.592	7.200	7.768	8.306	8.820	9.314	9.794	10.260	10.717
10	6.220	6.905	7.535	8.123	8.680	9.211	9.722	10.217	10.699	11.170
11	6.494	7.202	7.853	8.461	9.035	9.583	10.110	10.619	11.115	11.600
12	6.756	7.487	8.158	8.784	9.375	9.938	10.480	11.004	11.514	12.011
13	7.007	7.760	8.450	9.093	9.700	10.279	10.835	11.373	11.895	12.406
14	7.249	8.023	8.731	9.391	10.014	10.607	11.177	11.728	12.263	12.785
15	7.483	8.276	9.003	9.679	10.317	10.924	11.507	12.070	12.617	13.151
16	7.710	8.522	9.265	9.957	10.610	11.230	11.826	12.401	12.960	13.505
17	7.929	8.760	9.520	10.227	10.893	11.527	12.135	12.722	13.292	13.848
18	8.142	8.992	9.768	10.489	11.169	11.816	12.435	13.034	13.615	14.181
19	8.350	9.217	10.009	10.744	11.437	12.096	12.728	13.337	13.929	14.505
20	8.552	9.436	10.243	10.993	11.699	12.370	13.012	13.633	14.234	14.821
21	8.750	9.650	10.472	11.236	11.954	12.636	13.290	13.921	14.532	15.129
22	8.943	9.859	10.696	11.472	12.203	12.897	13.561	14.202	14.824	15.429
23	9.131	10.064	10.915	11.704	12.446	13.151	13.826	14.477	15.108	15.723
24	9.316	10.264	11.129	11.931	12.685	13.401	14.086	14.747	15.387	16.011
25	9.497	10.460	11.338	12.153	12.918	13.645	14.340	15.010	15.660	16.293
26	9.674	10.653	11.544	12.371	13.147	13.884	14.589	15.269	15.928	16.569
27	9.848	10.841	11.746	12.584	13.372	14.119	14.834	15.523	16.190	16.840
28	10.019	11.027	11.944	12.794	13.592	14.350	15.074	15.772	16.448	17.106
29	10.187	11.209	12.139	13.000	13.809	14.576	15.310	16.016	16.701	17.367
30	10.353	11.388	12.330	13.203	14.022	14.799	15.542	16.257	16.950	17.624
32	10.675	11.738	12.704	13.599	14.438	15.234	15.994	16.727	17.436	18.126
34	10.988	12.077	13.066	13.982	14.841	15.655	16.433	17.182	17.907	18.612
36	11.292	12.406	13.418	14.355	15.233	16.065	16.859	17.624	18.364	19.084
38	11.588	12.727	13.761	14.717	15.614	16.463	17.274	18.054	18.809	19.543
40	11.876	13.039	14.095	15.071	15.985	16.851	17.678	18.473	19.242	19.990
42	12.158	13.344	14.421	15.416	16.348	17.230	18.072	18.882	19.665	20.427
44	12.432	13.642	14.739	15.752	16.701	17.600	18.457	19.281	20.078	20.853
46	12.701	13.933	15.050	16.082	17.047	17.961	18.834	19.672	20.482	21.270
48	12.964	14.218	15.354	16.404	17.386	18.315	19.202	20.054	20.877	21.677
50	13.222	14.497	15.653	16.720	17.718	18.662	19.563	20.428	21.264	22.077
55	13.845	15.172	16.374	17.483	18.520	19.500	20.435	21.333	22.200	23.042
60	14.441	15.817	17.064	18.212	19.286	20.301	21.269	22.197	23.094	23.965
65	15.013	16.437	17.725	18.913	20.022	21.070	22.069	23.027	23.952	24.850
70	15.563	17.033	18.362	19.587	20.730	21.810	22.839	23.825	24.778	25.702
75	16.095	17.609	18.977	20.237	21.414	22.524	23.582	24.596	25.575	26.525
80	16.609	18.166	19.573	20.867	22.075	23.215	24.301	25.342	26.346	27.320
85	17.108	18.706	20.150	21.477	22.717	23.886	24.998	26.065	27.094	28.091
90	17.593	19.231	20.710	22.070	23.339	24.536	25.675	26.767	27.820	28.841
95	18.064	19.742	21.255	22.647	23.945	25.169	26.334	27.450	28.526	29.569
100	18.524	20.239	21.787	23.209	24.536	25.786	26.976	28.115	29.214	30.279

ALPHA = .005

N	POWER= .30	.32	.34	.36	.38	.40	.42	.44	.46	.48
1	5.210	5.472	5.734	5.996	6.258	6.521	6.787	7.055	7.326	7.600
2	6.524	6.828	7.129	7.430	7.731	8.033	8.337	8.642	8.951	9.263
3	7.480	7.812	8.143	8.472	8.801	9.130	9.460	9.792	10.127	10.466
4	8.269	8.626	8.979	9.331	9.682	10.034	10.386	10.740	11.096	11.457
5	8.958	9.335	9.709	10.080	10.451	10.821	11.193	11.565	11.941	12.319
6	9.577	9.972	10.364	10.754	11.141	11.529	11.917	12.307	12.699	13.094
7	10.144	10.557	10.965	11.370	11.774	12.177	12.580	12.985	13.392	13.803
8	10.671	11.099	11.522	11.943	12.361	12.778	13.196	13.615	14.036	14.461
9	11.165	11.608	12.045	12.479	12.911	13.342	13.773	14.205	14.640	15.077
10	11.632	12.088	12.539	12.986	13.431	13.875	14.318	14.763	15.209	15.659
11	12.076	12.545	13.008	13.468	13.925	14.380	14.836	15.292	15.751	16.212
12	12.500	12.981	13.457	13.928	14.396	14.863	15.330	15.798	16.267	16.740
13	12.906	13.399	13.886	14.369	14.849	15.327	15.804	16.282	16.762	17.246
14	13.297	13.802	14.300	14.793	15.283	15.772	16.260	16.748	17.239	17.732
15	13.675	14.190	14.698	15.202	15.703	16.201	16.699	17.198	17.698	18.201
16	14.039	14.565	15.084	15.598	16.108	16.617	17.124	17.632	18.142	18.655
17	14.393	14.929	15.458	15.981	16.501	17.019	17.536	18.053	18.572	19.094
18	14.736	15.282	15.821	16.354	16.883	17.410	17.936	18.462	18.990	19.520
19	15.070	15.626	16.173	16.716	17.254	17.790	18.324	18.859	19.396	19.935
20	15.395	15.960	16.517	17.068	17.615	18.160	18.703	19.246	19.791	20.339
21	15.713	16.286	16.852	17.412	17.968	18.521	19.072	19.624	20.177	20.733
22	16.022	16.605	17.180	17.748	18.312	18.873	19.432	19.992	20.553	21.117
23	16.325	16.917	17.500	18.076	18.648	19.217	19.785	20.352	20.921	21.493
24	16.622	17.221	17.813	18.397	18.977	19.554	20.129	20.705	21.281	21.860
25	16.912	17.520	18.119	18.712	19.300	19.884	20.467	21.050	21.634	22.220
26	17.196	17.813	18.420	19.020	19.616	20.208	20.798	21.388	21.979	22.573
27	17.476	18.100	18.715	19.323	19.926	20.525	21.123	21.720	22.318	22.919
28	17.750	18.382	19.004	19.620	20.230	20.836	21.441	22.046	22.651	23.259
29	18.019	18.659	19.289	19.911	20.529	21.142	21.754	22.365	22.978	23.593
30	18.284	18.931	19.568	20.198	20.823	21.443	22.062	22.680	23.299	23.921
32	18.800	19.462	20.114	20.758	21.396	22.030	22.662	23.293	23.926	24.561
34	19.301	19.977	20.642	21.300	21.951	22.599	23.244	23.888	24.533	25.181
36	19.787	20.477	21.156	21.827	22.491	23.151	23.809	24.465	25.123	25.783
38	20.260	20.963	21.655	22.339	23.016	23.688	24.358	25.027	25.697	26.369
40	20.721	21.437	22.142	22.838	23.527	24.212	24.893	25.574	26.255	26.939
42	21.170	21.899	22.617	23.325	24.026	24.722	25.415	26.108	26.800	27.495
44	21.609	22.351	23.080	23.800	24.513	25.221	25.925	26.629	27.332	28.038
46	22.038	22.792	23.533	24.265	24.989	25.708	26.423	27.138	27.853	28.569
48	22.458	23.224	23.977	24.719	25.455	26.185	26.911	27.636	28.362	29.089
50	22.870	23.647	24.411	25.165	25.911	26.652	27.389	28.124	28.860	29.598
55	23.864	24.669	25.460	26.241	27.013	27.780	28.543	29.303	30.064	30.827
60	24.814	25.646	26.463	27.270	28.067	28.858	29.645	30.430	31.215	32.002
65	25.726	26.583	27.426	28.256	29.078	29.893	30.703	31.511	32.319	33.129
70	26.603	27.485	28.352	29.206	30.050	30.888	31.721	32.551	33.381	34.213
75	27.450	28.356	29.246	30.122	30.989	31.849	32.703	33.555	34.406	35.259
80	28.269	29.198	30.110	31.009	31.897	32.778	33.653	34.526	35.398	36.271
85	29.064	30.015	30.948	31.868	32.777	33.678	34.574	35.467	36.359	37.252
90	29.835	30.807	31.762	32.703	33.632	34.553	35.469	36.381	37.292	38.204
95	30.585	31.579	32.554	33.514	34.464	35.404	36.338	37.270	38.200	39.131
100	31.316	32.330	33.325	34.305	35.273	36.233	37.186	38.135	39.084	40.033

ALPHA = .005

N	POWER= .50	.52	.54	.56	.58	.60	.62	.64	.66	.68
1	7.879	8.164	8.453	8.750	9.054	9.366	9.688	10.020	10.365	10.724
2	9.579	9.900	10.228	10.562	10.904	11.255	11.616	11.989	12.374	12.775
3	10.808	11.156	11.510	11.871	12.241	12.619	13.008	13.409	13.823	14.254
4	11.821	12.191	12.566	12.949	13.340	13.741	14.153	14.577	15.015	15.469
5	12.702	13.091	13.485	13.887	14.297	14.717	15.148	15.592	16.051	16.526
6	13.494	13.898	14.310	14.728	15.156	15.593	16.042	16.503	16.980	17.474
7	14.218	14.638	15.064	15.498	15.941	16.394	16.859	17.337	17.830	18.341
8	14.889	15.324	15.764	16.212	16.669	17.137	17.616	18.109	18.618	19.144
9	15.519	15.966	16.420	16.881	17.352	17.833	18.326	18.833	19.356	19.897
10	16.113	16.573	17.039	17.513	17.996	18.490	18.996	19.516	20.052	20.607
11	16.678	17.149	17.627	18.113	18.608	19.113	19.631	20.164	20.713	21.281
12	17.217	17.699	18.188	18.685	19.191	19.708	20.238	20.783	21.344	21.924
13	17.733	18.226	18.726	19.233	19.751	20.279	20.820	21.375	21.948	22.540
14	18.230	18.733	19.243	19.761	20.288	20.827	21.379	21.945	22.529	23.132
15	18.709	19.222	19.741	20.269	20.807	21.356	21.917	22.494	23.089	23.703
16	19.172	19.694	20.223	20.761	21.308	21.866	22.438	23.025	23.630	24.255
17	19.620	20.152	20.690	21.237	21.793	22.361	22.943	23.540	24.154	24.790
18	20.055	20.596	21.143	21.699	22.264	22.842	23.432	24.039	24.663	25.308
19	20.479	21.028	21.584	22.148	22.723	23.309	23.908	24.524	25.158	25.812
20	20.891	21.448	22.013	22.586	23.169	23.763	24.372	24.996	25.639	26.303
21	21.293	21.858	22.431	23.012	23.603	24.207	24.824	25.457	26.109	26.782
22	21.685	22.259	22.839	23.429	24.028	24.639	25.265	25.906	26.567	27.249
23	22.069	22.650	23.238	23.835	24.443	25.062	25.696	26.346	27.015	27.705
24	22.444	23.033	23.629	24.234	24.849	25.476	26.117	26.776	27.453	28.152
25	22.811	23.408	24.011	24.623	25.246	25.881	26.530	27.196	27.882	28.589
26	23.171	23.775	24.386	25.005	25.636	26.278	26.935	27.609	28.302	29.017
27	23.525	24.135	24.753	25.380	26.018	26.667	27.332	28.013	28.714	29.437
28	23.871	24.489	25.114	25.748	26.393	27.050	27.721	28.410	29.119	29.850
29	24.212	24.837	25.468	26.109	26.761	27.425	28.104	28.800	29.516	30.255
30	24.547	25.178	25.817	26.464	27.123	27.794	28.480	29.183	29.906	30.653
32	25.200	25.844	26.496	27.157	27.829	28.513	29.213	29.930	30.668	31.429
34	25.832	26.490	27.154	27.828	28.513	29.211	29.924	30.655	31.406	32.181
36	26.447	27.117	27.794	28.480	29.177	29.888	30.614	31.358	32.122	32.911
38	27.045	27.726	28.415	29.114	29.823	30.546	31.285	32.041	32.819	33.621
40	27.627	28.320	29.021	29.731	30.452	31.187	31.938	32.707	33.497	34.313
42	28.194	28.899	29.611	30.333	31.066	31.812	32.575	33.356	34.159	34.987
44	28.749	29.464	30.188	30.921	31.665	32.423	33.197	33.990	34.805	35.645
46	29.290	30.017	30.751	31.495	32.250	33.020	33.805	34.609	35.436	36.288
48	29.821	30.558	31.303	32.057	32.823	33.603	34.400	35.216	36.054	36.918
50	30.340	31.087	31.843	32.608	33.384	34.175	34.982	35.809	36.658	37.534
55	31.594	32.367	33.147	33.937	34.740	35.556	36.390	37.243	38.119	39.023
60	32.793	33.589	34.394	35.208	36.034	36.875	37.734	38.613	39.515	40.445
65	33.942	34.762	35.589	36.426	37.276	38.141	39.023	39.926	40.853	41.808
70	35.048	35.890	36.739	37.599	38.471	39.358	40.263	41.190	42.141	43.120
75	36.116	36.978	37.849	38.730	39.624	40.533	41.460	42.409	43.383	44.386
80	37.148	38.031	38.922	39.824	40.739	41.669	42.618	43.588	44.584	45.610
85	38.149	39.052	39.963	40.885	41.819	42.770	43.739	44.731	45.748	46.796
90	39.121	40.043	40.973	41.914	42.868	43.839	44.828	45.840	46.879	47.948
95	40.066	41.007	41.956	42.916	43.889	44.879	45.888	46.919	47.978	49.068
100	40.986	41.945	42.913	43.891	44.883	45.891	46.919	47.970	49.049	50.159

ALPHA = .005

N	POWER= .70	.71	.72	.73	.74	.75	.76	.77	.78	.79
1	11.098	11.292	11.491	11.695	11.905	12.121	12.344	12.573	12.811	13.057
2	13.192	13.408	13.630	13.856	14.089	14.329	14.576	14.831	15.094	15.366
3	14.702	14.933	15.170	15.414	15.663	15.920	16.184	16.456	16.737	17.028
4	15.942	16.187	16.437	16.693	16.956	17.226	17.504	17.791	18.087	18.393
5	17.021	17.276	17.537	17.805	18.080	18.362	18.652	18.951	19.260	19.579
6	17.988	18.253	18.524	18.802	19.087	19.379	19.680	19.991	20.311	20.642
7	18.872	19.146	19.426	19.713	20.008	20.310	20.621	20.941	21.271	21.613
8	19.691	19.974	20.262	20.558	20.861	21.172	21.492	21.822	22.162	22.513
9	20.459	20.749	21.045	21.349	21.660	21.980	22.308	22.646	22.995	23.356
10	21.183	21.480	21.784	22.095	22.414	22.741	23.078	23.424	23.782	24.151
11	21.871	22.175	22.486	22.804	23.130	23.465	23.809	24.163	24.528	24.905
12	22.527	22.837	23.155	23.480	23.813	24.154	24.506	24.867	25.240	25.625
13	23.155	23.472	23.796	24.127	24.467	24.815	25.173	25.542	25.922	26.314
14	23.759	24.082	24.412	24.749	25.095	25.450	25.815	26.190	26.577	26.976
15	24.341	24.670	25.005	25.349	25.701	26.062	26.433	26.815	27.208	27.615
16	24.904	25.238	25.579	25.929	26.286	26.654	27.031	27.419	27.819	28.232
17	25.449	25.788	26.135	26.490	26.854	27.226	27.609	28.004	28.410	28.829
18	25.977	26.322	26.674	27.035	27.404	27.782	28.171	28.571	28.983	29.409
19	26.491	26.841	27.199	27.564	27.939	28.322	28.717	29.123	29.541	29.972
20	26.992	27.347	27.709	28.080	28.459	28.848	29.248	29.659	30.083	30.521
21	27.480	27.839	28.207	28.582	28.967	29.361	29.766	30.183	30.612	31.055
22	27.956	28.320	28.692	29.073	29.462	29.862	30.272	30.694	31.128	31.577
23	28.421	28.790	29.167	29.552	29.946	30.350	30.766	31.193	31.633	32.087
24	28.877	29.250	29.631	30.021	30.420	30.829	31.249	31.681	32.126	32.585
25	29.322	29.700	30.085	30.479	30.883	31.297	31.722	32.159	32.609	33.073
26	29.759	30.141	30.531	30.929	31.337	31.756	32.185	32.627	33.082	33.552
27	30.187	30.573	30.967	31.370	31.783	32.206	32.640	33.086	33.546	34.021
28	30.608	30.998	31.396	31.803	32.220	32.647	33.086	33.537	34.002	34.481
29	31.020	31.414	31.817	32.228	32.649	33.081	33.524	33.980	34.449	34.933
30	31.426	31.824	32.230	32.646	33.071	33.507	33.955	34.415	34.888	35.377
32	32.217	32.623	33.037	33.461	33.894	34.338	34.794	35.263	35.746	36.244
34	32.984	33.397	33.819	34.250	34.691	35.144	35.608	36.085	36.576	37.083
36	33.728	34.149	34.578	35.016	35.465	35.925	36.397	36.883	37.382	37.898
38	34.452	34.879	35.316	35.761	36.218	36.685	37.165	37.658	38.166	38.690
40	35.157	35.591	36.034	36.487	36.950	37.425	37.913	38.413	38.929	39.461
42	35.844	36.285	36.735	37.194	37.665	38.147	38.642	39.150	39.673	40.213
44	36.515	36.962	37.418	37.885	38.362	38.851	39.353	39.869	40.400	40.947
46	37.170	37.624	38.087	38.560	39.044	39.540	40.049	40.571	41.109	41.664
48	37.812	38.272	38.741	39.220	39.711	40.213	40.729	41.259	41.804	42.366
50	38.440	38.906	39.381	39.867	40.364	40.873	41.395	41.932	42.484	43.053
55	39.957	40.438	40.928	41.429	41.941	42.466	43.004	43.557	44.126	44.713
60	41.406	41.901	42.405	42.920	43.447	43.987	44.541	45.109	45.695	46.298
65	42.796	43.304	43.822	44.351	44.892	45.446	46.014	46.598	47.198	47.818
70	44.133	44.654	45.185	45.727	46.281	46.849	47.432	48.030	48.645	49.279
75	45.423	45.956	46.499	47.054	47.622	48.203	48.799	49.411	50.041	50.690
80	46.670	47.215	47.771	48.338	48.918	49.512	50.121	50.747	51.390	52.053
85	47.879	48.435	49.003	49.582	50.174	50.781	51.403	52.041	52.698	53.375
90	49.052	49.620	50.199	50.790	51.394	52.013	52.647	53.298	53.967	54.657
95	50.194	50.772	51.362	51.964	52.580	53.210	53.856	54.520	55.202	55.905
100	51.305	51.894	52.495	53.108	53.735	54.377	55.034	55.710	56.404	57.120

ALPHA = .005

N	POWER= .80	.81	.82	.83	.84	.85	.86	.87	.88	.89
1	13.313	13.579	13.856	14.147	14.451	14.772	15.112	15.472	15.856	16.270
2	15.649	15.942	16.249	16.569	16.905	17.258	17.631	18.027	18.449	18.902
3	17.330	17.643	17.970	18.311	18.669	19.045	19.443	19.864	20.313	20.794
4	18.710	19.040	19.384	19.742	20.118	20.513	20.929	21.371	21.842	22.346
5	19.910	20.253	20.611	20.985	21.376	21.787	22.220	22.680	23.169	23.694
6	20.984	21.340	21.711	22.098	22.503	22.928	23.377	23.852	24.359	24.901
7	21.967	22.334	22.716	23.115	23.533	23.971	24.434	24.924	25.445	26.003
8	22.877	23.255	23.648	24.058	24.487	24.938	25.413	25.916	26.451	27.025
9	23.729	24.116	24.520	24.940	25.380	25.842	26.329	26.845	27.393	27.980
10	24.533	24.929	25.342	25.772	26.222	26.695	27.193	27.720	28.281	28.882
11	25.296	25.701	26.122	26.562	27.022	27.505	28.013	28.552	29.124	29.737
12	26.023	26.437	26.867	27.315	27.784	28.277	28.795	29.344	29.928	30.553
13	26.720	27.142	27.580	28.037	28.514	29.016	29.544	30.103	30.697	31.333
14	27.390	27.819	28.265	28.730	29.216	29.726	30.264	30.832	31.437	32.084
15	28.035	28.472	28.925	29.398	29.892	30.411	30.957	31.535	32.149	32.807
16	28.659	29.102	29.563	30.043	30.545	31.072	31.627	32.214	32.838	33.505
17	29.263	29.713	30.181	30.668	31.178	31.713	32.276	32.871	33.504	34.181
18	29.849	30.306	30.780	31.275	31.792	32.334	32.905	33.509	34.151	34.837
19	30.419	30.882	31.363	31.864	32.388	32.938	33.517	34.129	34.779	35.475
20	30.973	31.442	31.930	32.438	32.969	33.526	34.112	34.732	35.391	36.095
21	31.514	31.989	32.482	32.997	33.535	34.099	34.692	35.320	35.987	36.700
22	32.041	32.522	33.022	33.543	34.087	34.658	35.259	35.894	36.569	37.290
23	32.556	33.043	33.549	34.076	34.627	35.204	35.812	36.454	37.137	37.867
24	33.060	33.553	34.064	34.597	35.154	35.738	36.353	37.003	37.693	38.431
25	33.554	34.052	34.569	35.108	35.671	36.261	36.883	37.539	38.237	38.983
26	34.037	34.541	35.064	35.608	36.177	36.774	37.402	38.065	38.770	39.523
27	34.512	35.020	35.548	36.099	36.674	37.276	37.910	38.581	39.293	40.054
28	34.977	35.491	36.024	36.580	37.161	37.769	38.410	39.087	39.806	40.574
29	35.434	35.953	36.491	37.053	37.639	38.254	38.900	39.584	40.309	41.085
30	35.883	36.407	36.951	37.517	38.109	38.729	39.382	40.072	40.804	41.587
32	36.759	37.292	37.846	38.423	39.026	39.657	40.322	41.024	41.769	42.566
34	37.607	38.150	38.714	39.301	39.914	40.556	41.232	41.946	42.704	43.514
36	38.430	38.982	39.555	40.152	40.775	41.428	42.115	42.841	43.611	44.434
38	39.231	39.792	40.374	40.980	41.613	42.277	42.974	43.711	44.493	45.329
40	40.011	40.580	41.171	41.786	42.429	43.102	43.810	44.558	45.352	46.200
42	40.771	41.348	41.948	42.572	43.224	43.907	44.626	45.384	46.189	47.049
44	41.513	42.098	42.707	43.340	44.001	44.693	45.421	46.190	47.006	47.878
46	42.238	42.832	43.448	44.090	44.759	45.461	46.199	46.978	47.805	48.688
48	42.947	43.549	44.173	44.823	45.502	46.212	46.960	47.749	48.586	49.480
50	43.642	44.251	44.883	45.541	46.228	46.948	47.704	48.503	49.351	50.256
55	45.319	45.947	46.598	47.276	47.983	48.724	49.503	50.325	51.197	52.128
60	46.921	47.566	48.236	48.932	49.659	50.420	51.220	52.065	52.961	53.917
65	48.457	49.119	49.806	50.520	51.266	52.046	52.867	53.733	54.651	55.631
70	49.934	50.612	51.316	52.048	52.811	53.611	54.451	55.337	56.277	57.280
75	51.360	52.053	52.773	53.521	54.302	55.119	55.978	56.885	57.846	58.871
80	52.738	53.446	54.181	54.946	55.743	56.578	57.455	58.381	59.362	60.409
85	54.073	54.796	55.546	56.326	57.140	57.992	58.886	59.830	60.831	61.899
90	55.370	56.107	56.871	57.667	58.496	59.364	60.276	61.237	62.257	63.345
95	56.630	57.381	58.160	58.970	59.814	60.698	61.626	62.606	63.644	64.751
100	57.858	58.622	59.415	60.239	61.098	61.997	62.942	63.938	64.994	66.120

ALPHA = .005

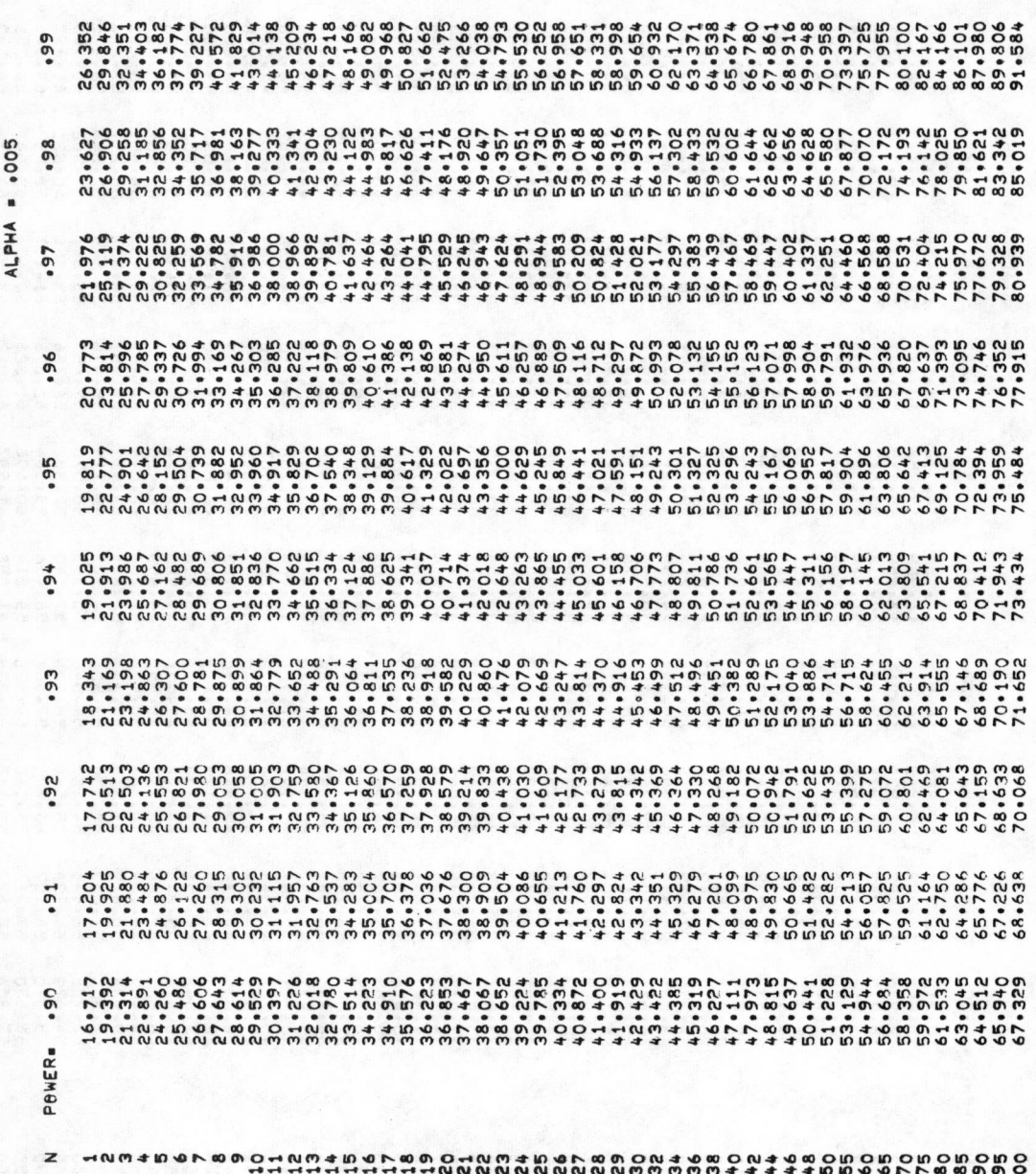

N	POWER= .90	.91	.92	.93	.94	.95	.96	.97	.98	.99
1	16.717	17.204	17.742	18.343	19.025	19.819	20.773	21.976	23.627	26.352
2	19.392	19.925	20.513	21.169	21.913	22.777	23.814	25.119	26.906	29.846
3	21.314	21.880	22.503	23.198	23.986	24.901	25.996	27.374	29.258	32.351
4	22.891	23.484	24.136	24.863	25.687	26.642	27.785	29.222	31.185	34.403
5	24.260	24.876	25.553	26.307	27.162	28.152	29.337	30.825	32.856	36.182
6	25.486	26.122	26.821	27.600	28.482	29.504	30.726	32.259	34.352	37.774
7	26.606	27.260	27.980	28.781	29.689	30.739	31.994	33.569	35.717	39.227
8	27.643	28.315	29.053	29.875	30.806	31.882	33.169	34.782	36.981	40.572
9	28.614	29.302	30.058	30.899	31.851	32.952	34.267	35.916	38.163	41.829
10	29.529	30.232	31.005	31.864	32.836	33.960	35.303	36.986	39.277	43.014
11	30.397	31.115	31.903	32.779	33.770	34.917	36.285	38.000	40.333	44.138
12	31.226	31.957	32.759	33.652	34.662	35.829	37.222	38.966	41.341	45.209
13	32.018	32.763	33.580	34.488	35.515	36.702	38.118	39.892	42.304	46.234
14	32.780	33.537	34.367	35.291	36.334	37.540	38.979	40.781	43.230	47.218
15	33.514	34.283	35.126	36.064	37.124	38.348	39.809	41.637	44.122	48.166
16	34.223	35.004	35.860	36.811	37.886	39.129	40.610	42.464	44.983	49.082
17	34.910	35.702	36.570	37.535	38.625	39.884	41.386	43.264	45.817	49.968
18	35.576	36.378	37.259	38.236	39.341	40.617	42.138	44.041	46.626	50.827
19	36.223	37.036	37.928	38.918	40.037	41.329	42.869	44.795	47.411	51.662
20	36.853	37.676	38.579	39.582	40.714	42.022	43.581	45.529	48.176	52.475
21	37.467	38.300	39.214	40.229	41.374	42.697	44.274	46.245	48.920	53.266
22	38.067	38.909	39.833	40.860	42.018	43.356	44.950	46.943	49.647	54.038
23	38.652	39.504	40.438	41.476	42.648	44.000	45.611	47.624	50.357	54.793
24	39.224	40.086	41.030	42.079	43.263	44.629	46.257	48.291	51.051	55.530
25	39.785	40.655	41.609	42.669	43.865	45.245	46.889	48.944	51.730	56.252
26	40.334	41.213	42.177	43.247	44.455	45.849	47.509	49.583	52.395	56.958
27	40.872	41.760	42.733	43.814	45.033	46.441	48.116	50.209	53.048	57.651
28	41.400	42.297	43.279	44.370	45.601	47.021	48.712	50.824	53.688	58.331
29	41.919	42.824	43.815	44.916	46.158	47.591	49.297	51.428	54.316	58.998
30	42.429	43.342	44.342	45.453	46.706	48.151	49.872	52.021	54.933	59.654
32	43.422	44.351	45.369	46.499	47.773	49.243	50.993	53.177	56.137	60.932
34	44.385	45.329	46.364	47.512	48.807	50.301	52.078	54.297	57.302	62.170
36	45.319	46.279	47.330	48.496	49.811	51.327	53.132	55.383	58.433	63.371
38	46.227	47.201	48.268	49.451	50.786	52.325	54.155	56.439	59.532	64.538
40	47.111	48.099	49.182	50.382	51.736	53.296	55.152	57.467	60.602	65.674
42	47.973	48.975	50.072	51.289	52.661	54.243	56.123	58.469	61.644	66.780
44	48.815	49.830	50.942	52.175	53.565	55.167	57.071	59.447	62.662	67.861
46	49.637	50.665	51.791	53.040	54.447	56.069	57.998	60.402	63.656	68.916
48	50.441	51.482	52.622	53.886	55.311	56.952	58.904	61.337	64.628	69.948
50	51.228	52.282	53.435	54.714	56.156	57.817	59.791	62.251	65.580	70.958
55	53.129	54.213	55.399	56.715	58.197	59.904	61.932	64.460	67.877	73.397
60	54.944	56.057	57.275	58.624	60.145	61.896	63.976	66.568	70.070	75.725
65	56.684	57.825	59.072	60.455	62.013	63.806	65.936	68.588	72.172	77.955
70	58.358	59.525	60.801	62.216	63.809	65.642	67.820	70.531	74.193	80.100
75	59.972	61.164	62.469	63.914	65.541	67.413	69.637	72.404	76.142	82.167
80	61.533	62.750	64.081	65.555	67.215	69.125	71.393	74.215	78.025	84.166
85	63.045	64.286	65.643	67.146	68.837	70.784	73.095	75.970	79.850	86.101
90	64.512	65.776	67.159	68.689	70.412	72.394	74.746	77.672	81.621	87.980
95	65.940	67.226	68.633	70.190	71.943	73.959	76.352	79.328	83.342	89.806
100	67.329	68.638	70.068	71.652	73.434	75.484	77.915	80.939	85.019	91.584

ALPHA = .001

N	POWER= .10	.12	.14	.16	.18	.20	.22	.24	.26	.28
1	4.036	4.476	4.885	5.272	5.641	5.997	6.342	6.678	7.008	7.332
2	5.113	5.625	6.100	6.545	6.970	7.376	7.770	8.152	8.525	8.892
3	5.909	6.473	6.994	7.482	7.945	8.389	8.817	9.232	9.637	10.034
4	6.571	7.178	7.737	8.261	8.756	9.230	9.686	10.129	10.560	10.982
5	7.153	7.797	8.389	8.943	9.466	9.966	10.447	10.914	11.367	11.811
6	7.678	8.355	8.977	9.558	10.106	10.630	11.133	11.621	12.095	12.558
7	8.160	8.868	9.517	10.123	10.695	11.240	11.763	12.270	12.762	13.243
8	8.610	9.346	10.020	10.649	11.242	11.807	12.349	12.873	13.383	13.881
9	9.032	9.795	10.493	11.143	11.756	12.339	12.899	13.440	13.966	14.479
10	9.432	10.220	10.940	11.610	12.242	12.843	13.419	13.976	14.517	15.044
11	9.813	10.624	11.366	12.055	12.704	13.322	13.914	14.486	15.041	15.582
12	10.178	11.011	11.773	12.480	13.147	13.780	14.387	14.973	15.542	16.096
13	10.527	11.383	12.163	12.889	13.571	14.220	14.841	15.441	16.022	16.589
14	10.864	11.740	12.539	13.282	13.980	14.643	15.278	15.891	16.485	17.064
15	11.189	12.085	12.903	13.661	14.374	15.051	15.700	16.325	16.931	17.522
16	11.504	12.420	13.254	14.028	14.756	15.446	16.108	16.745	17.363	17.965
17	11.810	12.744	13.595	14.384	15.126	15.830	16.503	17.152	17.782	18.395
18	12.106	13.059	13.926	14.730	15.485	16.202	16.888	17.548	18.188	18.812
19	12.395	13.365	14.248	15.067	15.835	16.564	17.261	17.933	18.584	19.218
20	12.677	13.664	14.562	15.395	16.176	16.917	17.626	18.308	18.969	19.613
21	12.951	13.955	14.869	15.715	16.509	17.261	17.981	18.674	19.345	19.999
22	13.220	14.240	15.168	16.027	16.834	17.598	18.328	19.032	19.713	20.376
23	13.483	14.519	15.461	16.333	17.151	17.927	18.668	19.381	20.072	20.744
24	13.740	14.792	15.748	16.633	17.462	18.249	19.000	19.723	20.423	21.104
25	13.992	15.059	16.029	16.926	17.767	18.564	19.326	20.058	20.768	21.458
26	14.239	15.321	16.304	17.213	18.066	18.873	19.645	20.387	21.105	21.804
27	14.481	15.578	16.574	17.496	18.359	19.177	19.958	20.710	21.437	22.144
28	14.719	15.831	16.840	17.773	18.647	19.475	20.266	21.026	21.762	22.477
29	14.953	16.079	17.101	18.045	18.930	19.768	20.568	21.337	22.081	22.805
30	15.183	16.323	17.357	18.313	19.208	20.056	20.865	21.643	22.396	23.127
32	15.633	16.799	17.858	18.836	19.751	20.618	21.445	22.240	23.009	23.756
34	16.068	17.262	18.343	19.342	20.278	21.163	22.007	22.819	23.603	24.366
36	16.492	17.711	18.815	19.835	20.789	21.692	22.553	23.381	24.181	24.958
38	16.904	18.148	19.274	20.314	21.287	22.207	23.085	23.928	24.743	25.534
40	17.306	18.573	19.722	20.781	21.772	22.709	23.603	24.461	25.290	26.096
42	17.698	18.989	20.158	21.237	22.245	23.199	24.108	24.981	25.824	26.644
44	18.081	19.395	20.585	21.682	22.708	23.677	24.602	25.489	26.346	27.179
46	18.455	19.792	21.002	22.117	23.160	24.145	25.084	25.986	26.856	27.702
48	18.822	20.181	21.410	22.543	23.602	24.603	25.557	26.472	27.356	28.214
50	19.181	20.562	21.810	22.961	24.036	25.052	26.019	26.948	27.845	28.715
55	20.050	21.482	22.777	23.970	25.084	26.136	27.138	28.099	29.027	29.927
60	20.880	22.363	23.702	24.935	26.086	27.173	28.207	29.199	30.157	31.086
65	21.678	23.208	24.589	25.861	27.048	28.167	29.233	30.255	31.241	32.198
70	22.446	24.021	25.444	26.753	27.974	29.125	30.221	31.272	32.285	33.268
75	23.187	24.807	26.269	27.613	28.867	30.050	31.175	32.253	33.293	34.301
80	23.905	25.567	27.067	28.446	29.732	30.944	32.098	33.202	34.268	35.300
85	24.600	26.305	27.841	29.254	30.571	31.812	32.992	34.123	35.213	36.269
90	25.276	27.021	28.593	30.039	31.385	32.654	33.861	35.017	36.131	37.210
95	25.934	27.718	29.325	30.802	32.178	33.474	34.707	35.887	37.024	38.126
100	26.575	28.396	30.038	31.546	32.950	34.273	35.530	36.734	37.894	39.018

ALPHA = .001

N	POWER= .30	.32	.34	.36	.38	.40	.42	.44	.46	.48
1	7.651	7.968	8.283	8.597	8.910	9.224	9.540	9.857	10.177	10.500
2	9.253	9.610	9.964	10.316	10.667	11.018	11.370	11.723	12.079	12.439
3	10.425	10.811	11.193	11.572	11.951	12.328	12.707	13.086	13.469	13.854
4	11.397	11.806	12.212	12.614	13.014	13.414	13.814	14.215	14.619	15.025
5	12.247	12.677	13.102	13.524	13.944	14.363	14.781	15.201	15.623	16.048
6	13.013	13.461	13.904	14.344	14.781	15.216	15.652	16.089	16.527	16.969
7	13.715	14.180	14.640	15.095	15.548	15.999	16.450	16.902	17.356	17.813
8	14.369	14.849	15.324	15.794	16.261	16.727	17.192	17.658	18.126	18.597
9	14.982	15.477	15.965	16.450	16.931	17.410	17.888	18.367	18.848	19.332
10	15.561	16.070	16.572	17.069	17.563	18.055	18.546	19.037	19.531	20.027
11	16.113	16.634	17.149	17.659	18.165	18.668	19.171	19.675	20.180	20.687
12	16.639	17.173	17.700	18.222	18.739	19.254	19.769	20.283	20.799	21.318
13	17.145	17.690	18.229	18.762	19.290	19.816	20.341	20.867	21.393	21.923
14	17.631	18.188	18.737	19.281	19.820	20.357	20.893	21.428	21.965	22.505
15	18.100	18.668	19.228	19.782	20.332	20.879	21.424	21.970	22.517	23.066
16	18.554	19.133	19.703	20.267	20.827	21.383	21.939	22.494	23.050	23.609
17	18.994	19.583	20.163	20.737	21.307	21.873	22.437	23.001	23.567	24.135
18	19.422	20.020	20.610	21.194	21.772	22.348	22.921	23.494	24.069	24.646
19	19.837	20.446	21.045	21.638	22.225	22.809	23.392	23.974	24.557	25.143
20	20.242	20.860	21.469	22.070	22.667	23.259	23.850	24.441	25.032	25.627
21	20.637	21.264	21.882	22.492	23.097	23.698	24.298	24.896	25.496	26.099
22	21.023	21.659	22.285	22.904	23.518	24.127	24.734	25.341	25.949	26.559
23	21.401	22.045	22.680	23.307	23.929	24.546	25.161	25.776	26.392	27.010
24	21.770	22.423	23.066	23.701	24.331	24.956	25.579	26.202	26.825	27.451
25	22.132	22.793	23.444	24.087	24.725	25.358	25.988	26.618	27.249	27.882
26	22.486	23.156	23.815	24.466	25.111	25.752	26.390	27.027	27.665	28.306
27	22.834	23.512	24.179	24.837	25.490	26.138	26.783	27.428	28.073	28.721
28	23.176	23.861	24.536	25.202	25.862	26.517	27.170	27.821	28.474	29.128
29	23.512	24.205	24.887	25.560	26.227	26.890	27.549	28.208	28.867	29.529
30	23.842	24.542	25.232	25.912	26.586	27.256	27.922	28.588	29.254	29.922
32	24.486	25.201	25.905	26.600	27.288	27.971	28.651	29.329	30.009	30.690
34	25.110	25.840	26.558	27.266	27.967	28.663	29.356	30.048	30.740	31.435
36	25.717	26.460	27.191	27.913	28.627	29.336	30.042	30.746	31.451	32.158
38	26.307	27.064	27.808	28.543	29.269	29.991	30.709	31.425	32.142	32.861
40	26.882	27.652	28.409	29.156	29.895	30.628	31.358	32.087	32.815	33.546
42	27.443	28.225	28.995	29.754	30.505	31.250	31.992	32.732	33.472	34.214
44	27.990	28.786	29.567	30.338	31.101	31.858	32.611	33.362	34.113	34.866
46	28.526	29.333	30.127	30.910	31.684	32.452	33.216	33.978	34.740	35.504
48	29.050	29.870	30.675	31.469	32.254	33.033	33.808	34.581	35.354	36.128
50	29.564	30.395	31.211	32.016	32.813	33.602	34.388	35.171	35.954	36.739
55	30.805	31.664	32.508	33.340	34.162	34.978	35.789	36.598	37.406	38.216
60	31.991	32.877	33.747	34.604	35.452	36.293	37.128	37.961	38.794	39.628
65	33.129	34.041	34.936	35.818	36.690	37.554	38.413	39.269	40.125	40.982
70	34.225	35.161	36.081	36.986	37.881	38.768	39.650	40.528	41.406	42.285
75	35.282	36.243	37.185	38.114	39.031	39.940	40.843	41.743	42.642	43.543
80	36.306	37.289	38.254	39.204	40.143	41.073	41.998	42.919	43.838	44.759
85	37.298	38.303	39.290	40.261	41.221	42.172	43.117	44.058	44.998	45.939
90	38.261	39.288	40.296	41.288	42.268	43.239	44.204	45.164	46.123	47.084
95	39.198	40.246	41.275	42.287	43.287	44.277	45.261	46.240	47.218	48.197
100	40.111	41.180	42.228	43.260	44.279	45.288	46.291	47.289	48.285	49.283

ALPHA = .001

N	POWER= .50	.52	.54	.56	.58	.60	.62	.64	.66	.68
1	10.828	11.160	11.499	11.844	12.197	12.559	12.931	13.315	13.712	14.124
2	12.802	13.171	13.546	13.928	14.317	14.717	15.127	15.549	15.986	16.438
3	14.243	14.638	15.039	15.447	15.863	16.290	16.727	17.177	17.642	18.123
4	15.436	15.852	16.274	16.704	17.142	17.590	18.050	18.523	19.011	19.516
5	16.478	16.912	17.353	17.801	18.258	18.726	19.205	19.697	20.205	20.731
6	17.415	17.866	18.323	18.788	19.262	19.747	20.243	20.753	21.279	21.823
7	18.274	18.740	19.212	19.693	20.182	20.682	21.194	21.721	22.263	22.824
8	19.071	19.552	20.038	20.533	21.036	21.551	22.078	22.619	23.177	23.753
9	19.820	20.313	20.813	21.320	21.837	22.365	22.906	23.461	24.033	24.624
10	20.527	21.032	21.545	22.065	22.594	23.135	23.688	24.257	24.842	25.447
11	21.199	21.716	22.240	22.772	23.313	23.866	24.432	25.012	25.611	26.229
12	21.841	22.369	22.904	23.447	24.000	24.564	25.142	25.734	26.344	26.975
13	22.456	22.995	23.541	24.095	24.659	25.234	25.822	26.426	27.048	27.690
14	23.049	23.598	24.154	24.718	25.292	25.877	26.477	27.091	27.724	28.378
15	23.620	24.179	24.744	25.319	25.903	26.498	27.108	27.733	28.377	29.041
16	24.172	24.740	25.316	25.899	26.493	27.099	27.718	28.354	29.007	29.683
17	24.707	25.285	25.869	26.462	27.066	27.681	28.310	28.955	29.619	30.304
18	25.227	25.814	26.407	27.009	27.621	28.245	28.884	29.538	30.212	30.907
19	25.732	26.328	26.930	27.540	28.161	28.795	29.442	30.106	30.789	31.494
20	26.225	26.828	27.439	28.058	28.688	29.330	29.986	30.658	31.350	32.065
21	26.705	27.317	27.935	28.563	29.201	29.851	30.516	31.197	31.898	32.622
22	27.174	27.793	28.420	29.056	29.702	30.360	31.033	31.723	32.433	33.165
23	27.632	28.260	28.894	29.538	30.192	30.858	31.539	32.238	32.956	33.697
24	28.080	28.716	29.358	30.009	30.671	31.345	32.034	32.741	33.467	34.217
25	28.520	29.162	29.812	30.471	31.140	31.822	32.519	33.234	33.968	34.726
26	28.950	29.600	30.257	30.923	31.600	32.290	32.994	33.717	34.459	35.225
27	29.372	30.030	30.694	31.367	32.051	32.748	33.460	34.190	34.940	35.714
28	29.787	30.451	31.123	31.803	32.494	33.199	33.918	34.655	35.413	36.195
29	30.194	30.865	31.544	32.231	32.929	33.641	34.367	35.112	35.877	36.667
30	30.595	31.272	31.958	32.652	33.357	34.075	34.809	35.561	36.333	37.130
32	31.376	32.067	32.765	33.473	34.191	34.923	35.671	36.437	37.224	38.035
34	32.133	32.837	33.548	34.268	35.000	35.745	36.506	37.285	38.086	38.912
36	32.868	33.584	34.308	35.041	35.785	36.543	37.317	38.109	38.924	39.763
38	33.583	34.312	35.047	35.792	36.549	37.319	38.106	38.911	39.738	40.591
40	34.280	35.020	35.767	36.524	37.293	38.075	38.874	39.692	40.532	41.398
42	34.959	35.711	36.470	37.238	38.018	38.812	39.623	40.453	41.305	42.184
44	35.623	36.385	37.156	37.935	38.727	39.532	40.354	41.196	42.061	42.952
46	36.272	37.045	37.826	38.617	39.419	40.236	41.070	41.923	42.799	43.702
48	36.906	37.690	38.482	39.283	40.097	40.925	41.769	42.634	43.522	44.436
50	37.528	38.323	39.125	39.937	40.760	41.599	42.454	43.330	44.229	45.156
55	39.030	39.850	40.677	41.514	42.364	43.228	44.110	45.012	45.939	46.893
60	40.466	41.309	42.161	43.022	43.896	44.785	45.692	46.619	47.572	48.553
65	41.843	42.709	43.584	44.468	45.365	46.278	47.209	48.161	49.138	50.144
70	43.168	44.056	44.953	45.860	46.780	47.715	48.669	49.644	50.645	51.676
75	44.447	45.357	46.274	47.203	48.144	49.101	50.077	51.075	52.099	53.154
80	45.684	46.614	47.553	48.502	49.464	50.442	51.440	52.460	53.506	54.583
85	46.883	47.833	48.791	49.761	50.743	51.742	52.760	53.801	54.869	55.968
90	48.047	49.017	49.995	50.983	51.985	53.004	54.042	55.104	56.193	57.313
95	49.180	50.168	51.165	52.172	53.194	54.232	55.289	56.371	57.480	58.621
100	50.283	51.290	52.305	53.331	54.370	55.427	56.504	57.605	58.734	59.895

ALPHA = .001

N	POWER= .70	.71	.72	.73	.74	.75	.76	.77	.78	.79
1	14.554	14.776	15.003	15.236	15.475	15.721	15.975	16.236	16.506	16.785
2	16.909	17.152	17.401	17.656	17.917	18.186	18.463	18.748	19.042	19.347
3	18.624	18.882	19.146	19.417	19.695	19.980	20.274	20.576	20.888	21.211
4	20.041	20.312	20.589	20.873	21.164	21.463	21.770	22.087	22.413	22.751
5	21.278	21.559	21.848	22.143	22.445	22.756	23.075	23.404	23.743	24.093
6	22.389	22.680	22.978	23.284	23.596	23.917	24.247	24.587	24.937	25.299
7	23.407	23.707	24.014	24.329	24.651	24.981	25.321	25.671	26.031	26.404
8	24.352	24.661	24.976	25.299	25.630	25.969	26.318	26.677	27.047	27.429
9	25.238	25.554	25.877	26.208	26.547	26.895	27.252	27.620	27.999	28.390
10	26.075	26.398	26.729	27.067	27.414	27.769	28.134	28.510	28.897	29.297
11	26.870	27.200	27.538	27.883	28.237	28.600	28.972	29.356	29.751	30.159
12	27.629	27.966	28.310	28.662	29.022	29.392	29.772	30.163	30.566	30.982
13	28.356	28.699	29.050	29.408	29.775	30.152	30.539	30.937	31.347	31.770
14	29.056	29.405	29.761	30.126	30.500	30.883	31.276	31.681	32.098	32.528
15	29.730	30.085	30.448	30.818	31.198	31.587	31.987	32.398	32.822	33.259
16	30.382	30.743	31.111	31.487	31.873	32.268	32.674	33.091	33.521	33.965
17	31.014	31.380	31.754	32.136	32.527	32.928	33.340	33.763	34.199	34.650
18	31.628	31.999	32.378	32.765	33.162	33.568	33.986	34.415	34.857	35.314
19	32.224	32.600	32.984	33.377	33.779	34.191	34.614	35.049	35.497	35.960
20	32.805	33.186	33.575	33.973	34.380	34.797	35.226	35.666	36.120	36.588
21	33.371	33.757	34.151	34.554	34.966	35.388	35.822	36.268	36.728	37.202
22	33.924	34.314	34.713	35.121	35.538	35.966	36.404	36.856	37.321	37.800
23	34.464	34.859	35.263	35.675	36.097	36.530	36.973	37.430	37.900	38.385
24	34.993	35.392	35.800	36.217	36.644	37.081	37.530	37.992	38.467	38.957
25	35.511	35.914	36.327	36.748	37.180	37.622	38.075	38.542	39.022	39.517
26	36.018	36.426	36.843	37.269	37.704	38.151	38.609	39.081	39.566	40.066
27	36.516	36.928	37.349	37.779	38.219	38.670	39.133	39.609	40.099	40.605
28	37.004	37.420	37.846	38.280	38.725	39.180	39.648	40.128	40.623	41.133
29	37.484	37.904	38.333	38.772	39.221	39.681	40.153	40.638	41.137	41.652
30	37.955	38.380	38.813	39.256	39.709	40.173	40.649	41.139	41.643	42.163
32	38.875	39.307	39.748	40.199	40.660	41.133	41.617	42.116	42.629	43.158
34	39.767	40.206	40.655	41.113	41.582	42.063	42.556	43.063	43.584	44.122
36	40.632	41.079	41.535	42.001	42.477	42.966	43.467	43.982	44.512	45.058
38	41.474	41.928	42.391	42.864	43.348	43.844	44.353	44.875	45.413	45.968
40	42.293	42.754	43.224	43.704	44.196	44.699	45.215	45.746	46.292	46.855
42	43.093	43.560	44.037	44.524	45.023	45.533	46.057	46.595	47.148	47.719
44	43.873	44.347	44.831	45.325	45.830	46.347	46.878	47.423	47.984	48.563
46	44.636	45.116	45.606	46.107	46.619	47.143	47.681	48.233	48.802	49.388
48	45.383	45.869	46.365	46.872	47.391	47.922	48.466	49.026	49.601	50.195
50	46.114	46.606	47.109	47.622	48.147	48.684	49.235	49.802	50.384	50.985
55	47.879	48.387	48.904	49.432	49.973	50.526	51.093	51.676	52.276	52.894
60	49.567	50.088	50.619	51.162	51.717	52.285	52.868	53.467	54.082	54.717
65	51.185	51.719	52.264	52.821	53.390	53.973	54.570	55.184	55.815	56.465
70	52.741	53.289	53.847	54.417	54.999	55.596	56.208	56.836	57.482	58.147
75	54.243	54.803	55.374	55.957	56.552	57.162	57.787	58.429	59.090	59.770
80	55.696	56.268	56.851	57.446	58.054	58.677	59.315	59.971	60.645	61.339
85	57.104	57.687	58.282	58.889	59.509	60.145	60.796	61.464	62.152	62.860
90	58.471	59.065	59.671	60.290	60.923	61.570	62.233	62.915	63.615	64.337
95	59.800	60.405	61.023	61.653	62.297	62.956	63.631	64.325	65.038	65.773
100	61.095	61.711	62.339	62.980	63.635	64.306	64.993	65.699	66.424	67.171

ALPHA = .001

N	POWER= .80	.81	.82	.83	.84	.85	.86	.87	.88	.89
1	17.075	17.376	17.690	18.017	18.361	18.723	19.104	19.509	19.941	20.404
2	19.662	19.990	20.332	20.688	21.062	21.454	21.868	22.307	22.775	23.276
3	21.545	21.892	22.253	22.631	23.026	23.440	23.878	24.342	24.835	25.364
4	23.100	23.463	23.841	24.235	24.647	25.081	25.538	26.021	26.536	27.088
5	24.456	24.833	25.224	25.633	26.061	26.510	26.984	27.485	28.019	28.590
6	25.674	26.063	26.468	26.890	27.331	27.795	28.283	28.800	29.350	29.939
7	26.790	27.190	27.606	28.040	28.494	28.970	29.472	30.004	30.569	31.174
8	27.825	28.235	28.662	29.107	29.573	30.061	30.576	31.121	31.700	32.319
9	28.795	29.215	29.652	30.108	30.584	31.084	31.610	32.167	32.759	33.393
10	29.711	30.140	30.587	31.052	31.538	32.049	32.586	33.155	33.759	34.406
11	30.581	31.019	31.475	31.949	32.445	32.965	33.513	34.093	34.709	35.368
12	31.412	31.858	32.322	32.805	33.310	33.840	34.398	34.988	35.616	36.286
13	32.208	32.662	33.134	33.625	34.139	34.678	35.246	35.846	36.484	37.166
14	32.973	33.435	33.914	34.414	34.937	35.484	36.061	36.671	37.319	38.012
15	33.711	34.180	34.667	35.175	35.705	36.261	36.847	37.466	38.124	38.827
16	34.424	34.900	35.395	35.910	36.448	37.012	37.606	38.235	38.902	39.615
17	35.115	35.598	36.099	36.622	37.168	37.740	38.342	38.979	39.655	40.378
18	35.786	36.275	36.783	37.313	37.866	38.446	39.056	39.701	40.387	41.119
19	36.438	36.933	37.448	37.985	38.545	39.132	39.750	40.404	41.098	41.839
20	37.073	37.574	38.096	38.639	39.206	39.800	40.426	41.087	41.790	42.540
21	37.692	38.199	38.727	39.276	39.850	40.452	41.085	41.754	42.464	43.223
22	38.296	38.810	39.343	39.899	40.479	41.088	41.728	42.404	43.123	43.890
23	38.886	39.406	39.945	40.507	41.094	41.709	42.356	43.040	43.766	44.542
24	39.464	39.989	40.534	41.102	41.695	42.317	42.971	43.662	44.396	45.180
25	40.030	40.560	41.111	41.685	42.284	42.912	43.573	44.271	45.012	45.804
26	40.584	41.120	41.676	42.256	42.861	43.495	44.163	44.868	45.616	46.415
27	41.127	41.669	42.231	42.816	43.427	44.067	44.741	45.453	46.208	47.015
28	41.661	42.207	42.775	43.365	43.982	44.629	45.309	46.027	46.790	47.604
29	42.185	42.736	43.309	43.905	44.528	45.180	45.866	46.591	47.360	48.182
30	42.700	43.256	43.834	44.436	45.064	45.722	46.414	47.145	47.921	48.750
32	43.704	44.271	44.858	45.470	46.109	46.779	47.483	48.226	49.015	49.858
34	44.678	45.253	45.851	46.473	47.122	47.803	48.518	49.274	50.075	50.931
36	45.623	46.207	46.814	47.446	48.106	48.797	49.523	50.290	51.104	51.973
38	46.542	47.135	47.751	48.393	49.062	49.763	50.500	51.279	52.105	52.987
40	47.436	48.038	48.664	49.314	49.993	50.704	51.452	52.241	53.079	53.973
42	48.309	48.919	49.553	50.213	50.901	51.622	52.380	53.180	54.029	54.935
44	49.161	49.779	50.422	51.090	51.787	52.518	53.285	54.096	54.956	55.874
46	49.993	50.620	51.270	51.947	52.654	53.393	54.171	54.991	55.862	56.791
48	50.808	51.442	52.101	52.786	53.501	54.250	55.037	55.867	56.749	57.689
50	51.605	52.248	52.914	53.607	54.331	55.089	55.885	56.725	57.617	58.568
55	53.532	54.193	54.878	55.591	56.335	57.114	57.933	58.797	59.713	60.691
60	55.372	56.050	56.754	57.486	58.250	59.049	59.889	60.775	61.715	62.718
65	57.137	57.832	58.553	59.303	60.085	60.904	61.765	62.672	63.635	64.662
70	58.835	59.546	60.284	61.051	61.851	62.689	63.569	64.497	65.482	66.532
75	60.473	61.199	61.953	62.737	63.555	64.411	65.309	66.258	67.263	68.335
80	62.056	62.798	63.568	64.368	65.202	66.075	66.992	67.960	68.986	70.079
85	63.591	64.348	65.132	65.948	66.798	67.689	68.623	69.609	70.655	71.769
90	65.082	65.852	66.651	67.482	68.348	69.255	70.207	71.211	72.275	73.410
95	66.531	67.315	68.128	68.974	69.855	70.778	71.746	72.768	73.851	75.005
100	67.942	68.740	69.567	70.427	71.323	72.261	73.246	74.284	75.385	76.558

ALPHA = .001

N	POWER= .90	.91	.92	.93	.94	.95	.96	.97	.98	.99
1	20.904	21.449	22.049	22.718	23.477	24.358	25.414	26.743	28.561	31.549
2	23.817	24.406	25.054	25.776	26.594	27.543	28.678	30.104	32.053	35.247
3	25.935	26.555	27.238	27.998	28.859	29.856	31.048	32.545	34.588	37.931
4	27.683	28.330	29.040	29.832	30.727	31.764	33.004	34.559	36.679	40.145
5	29.206	29.875	30.611	31.429	32.355	33.427	34.707	36.313	38.500	42.072
6	30.573	31.263	32.021	32.863	33.817	34.919	36.236	37.887	40.134	43.801
7	31.825	32.534	33.311	34.176	35.154	36.285	37.635	39.327	41.629	45.382
8	32.987	33.712	34.508	35.394	36.395	37.552	38.933	40.662	43.015	46.848
9	34.075	34.816	35.630	36.534	37.557	38.738	40.148	41.913	44.313	48.221
10	35.102	35.859	36.689	37.611	38.653	39.858	41.295	43.093	45.538	49.517
11	36.078	36.848	37.694	38.633	39.695	40.921	42.384	44.214	46.701	50.746
12	37.009	37.793	38.653	39.609	40.688	41.936	43.423	45.283	47.810	51.919
13	37.900	38.698	39.572	40.543	41.640	42.907	44.418	46.307	48.872	53.042
14	38.758	39.567	40.455	41.441	42.555	43.841	45.374	47.291	49.893	54.121
15	39.584	40.406	41.306	42.307	43.437	44.741	46.296	48.239	50.877	55.161
16	40.383	41.216	42.129	43.144	44.289	45.611	47.187	49.155	51.828	56.166
17	41.156	42.001	42.926	43.954	45.114	46.454	48.049	50.043	52.748	57.139
18	41.907	42.762	43.700	44.740	45.915	47.271	48.886	50.904	53.641	58.083
19	42.637	43.503	44.451	45.505	46.694	48.066	49.700	51.741	54.509	59.000
20	43.347	44.223	45.183	46.249	47.451	48.839	50.492	52.555	55.354	59.893
21	44.040	44.926	45.897	46.974	48.190	49.593	51.263	53.349	56.178	60.763
22	44.716	45.611	46.593	47.682	48.911	50.329	52.017	54.124	56.981	61.612
23	45.376	46.282	47.273	48.374	49.615	51.048	52.753	54.881	57.766	62.442
24	46.023	46.937	47.939	49.050	50.304	51.751	53.473	55.622	58.534	63.253
25	46.655	47.579	48.590	49.713	50.979	52.439	54.177	56.346	59.286	64.047
26	47.275	48.207	49.228	50.362	51.640	53.114	54.868	57.056	60.022	64.825
27	47.883	48.824	49.854	50.998	52.288	53.775	55.545	57.753	60.744	65.587
28	48.479	49.429	50.469	51.623	52.924	54.424	56.209	58.436	61.452	66.336
29	49.065	50.023	51.072	52.236	53.548	55.061	56.861	59.107	62.148	67.070
30	49.641	50.607	51.665	52.839	54.162	55.688	57.502	59.766	62.832	67.792
32	50.764	51.746	52.821	54.014	55.359	56.909	58.753	61.052	64.164	69.200
34	51.852	52.849	53.941	55.153	56.518	58.092	59.964	62.297	65.456	70.564
36	52.908	53.920	55.029	56.258	57.644	59.240	61.139	63.506	66.708	71.887
38	53.934	54.961	56.086	57.332	58.738	60.357	62.281	64.681	67.926	73.173
40	54.934	55.975	57.115	58.379	59.803	61.444	63.394	65.824	69.112	74.424
42	55.909	56.964	58.118	59.399	60.841	62.503	64.478	66.939	70.268	75.645
44	56.860	57.928	59.098	60.394	61.855	63.537	65.537	68.028	71.396	76.836
46	57.790	58.871	60.055	61.367	62.845	64.548	66.571	69.091	72.498	77.999
48	58.699	59.794	60.992	62.319	63.814	65.537	67.583	70.131	73.576	79.137
50	59.590	60.697	61.909	63.251	64.763	66.505	68.573	71.150	74.632	80.252
55	61.741	62.878	64.123	65.501	67.054	68.842	70.965	73.609	77.180	82.942
60	63.795	64.961	66.237	67.651	69.242	71.074	73.249	75.957	79.613	85.510
65	65.765	66.959	68.264	69.711	71.339	73.214	75.438	78.207	81.946	87.971
70	67.659	68.880	70.214	71.693	73.357	75.271	77.544	80.372	84.188	90.337
75	69.486	70.732	72.095	73.604	75.302	77.256	79.575	82.459	86.351	92.619
80	71.253	72.524	73.914	75.452	77.183	79.175	81.538	84.477	88.442	94.825
85	72.965	74.260	75.676	77.243	79.006	81.034	83.440	86.432	90.468	96.962
90	74.627	75.945	77.386	78.981	80.775	82.839	85.286	88.330	92.434	99.037
95	76.243	77.584	79.049	80.671	82.495	84.594	87.082	90.175	94.345	101.053
100	77.817	79.180	80.669	82.317	84.171	86.302	88.830	91.971	96.206	103.016

TABLES OF THE EXACT SAMPLING DISTRIBUTION OF THE TWO-SAMPLE KOLMOGOROV-SMIRNOV CRITERION, D_{mn}, $m \leq n$

P. J. Kim*
R. I. Jennrich†

***University of Oregon Medical School, Portland.**
†University of California, Los Angeles.

The preparation of the tables was supported in part by U.S. Public Health Service Research Grant No. MH-08667, U. S. Public Health Service General Research Support Grant No. 1-S01-FR-05632-01, and National Institutes of Health Grant FR-3.

INTRODUCTION

As a nonparametric procedure for two-sample problems, the Kolmogorov-Smirnov criterion $D_{mn} = \sup_x |S_m(x) - S_n(x)|$, where $S_m(x)$ and $S_n(x)$ are the empirical cumulative distributions of the two samples of sizes m and n, possesses simplicity and certain theoretical attractiveness. The exact sampling distribution of the criterion D_{mn}, except for $m = n$, is quite involved even with the aid of high-speed digital computers. Our aim is to extend the basic results of Massey (1951) to the case for $m \neq n$ and to present the tables in a form convenient for those engaged in theoretical as well as practical statistical inference problems.

In view of the volume involved, we present the tables in two parts. In Part I, the tables cover the upper α point, $.001 \leq \alpha \leq .10$, along with exact probabilities for $m \leq n = 1(1)25$. In Part II, we present the upper six α points only, $\alpha = .10, .05, .025, .01, .005, .001$, for $m \leq n = 1(1)100$. We hope that the user finds the tables of Part II easy to use. Even with these restraints, the task of compiling the tables was rather involved and it is our pleasure to gratefully acknowledge our indebtedness to the staff of the Socio-Behavioral Laboratory, Pacific State Hospital, Pomona, California and the Health Science Computing Facility, University of California, Los Angeles for the preparation of the tables.

Definitions and Notation

Suppose

$$x_1, x_2, \ldots, x_m; x'_1, x'_2, \ldots, x'_n$$

are $m + n$ mutually independent random variables with a common (but unknown) continuous distribution, $F(x)$ (without loss of generality, $m \leq n$). If $S_m(x)$ and $S_n(x)$ are the empirical distributions of the two samples, the Kolmogorov-Smirnov criterion is $D_{mn} = \sup_x |S_m(x) - S_n(x)|$, and its exact null distribution is $P(D_{mn} \leq \frac{c}{mn})$, $c = 1(1)mn$. The

large-sample null distribution of $S = [mn/(m+n)]^{1/2} D_{mn}$, $m/n > 0$

is $G(s) = P(S \leq s) = 1 - 2 \sum_{r=1}^{\infty} (-1)^{r-1} \exp\{-2r^2 s^2\}$.

We shall call $G(s)$ the Smirnov distribution and approximations based on $G(s)$ the Smirnov approximation.

When the size of one sample, n, is large in comparison with that of the other sample, m, we find that the approximation based on the exact null distribution of the one-sample Kolmogorov-Smirnov criterion D_m is better than the Smirnov approximation, where $D_m = \sup_x |S_m(x) - F(x)|$ and the underlying distribution $F(x)$ is assumed to be continuous. We shall call approximations based on the exact null distribution of D_m the Kolmogorov approximation. The large sample null distribution of D_m is of course the same as $G(s)$ with $S = \sqrt{m}D_m$.

The criterion D_{mn} was first introduced by Smirnov (1939, 1948) for two-sample problems. By showing that $G(s)$ is the limiting distribution of D_{mn} and giving tables of $G(s)$, Smirnov indicated the possibility of large sample approximation. Subsequently, Feller (1948) gave a much simpler derivation of $G(s)$. Treating the criterion D_{mn} as a rank test Steck (1969) derived the exact distribution of D_{mn} which is expressed in a determinantal form. Computationwise, however, the evaluation of the exact distribution through his expression appears rather cumbersome. As to other aspects such as exact power and efficiency of D_{mn}, there seem to be only few results (Massey, 1950) so far. For summaries of results on D_{mn} and related topics, the reader may be referred to Darling (1957) and more recently to Barton and Mallows (1965).

Despite the fact that the exact distribution of D_{mn} is rather involved, several efficient algorithms do exist which have been used to tabulate the exact distribution of D_{mn}. The basic algorithm seems to be the successive recursion relation Massey (1951) introduced to tabulate $P(D_{mn} \leq \frac{c}{mn})$, $m = n = 1(1)40$. Marliss and Zayachkowski (1962) used the "inside" method of Hodges (1957) to tabulate $P(D_{mn} \leq \frac{c}{mn})$, $m \leq n = 1(1)20$. The "inside" method is essentially a graphical variant of Massey's algorithm ($m = n$) with slightly less restrictive boundary conditions, i.e., $n - m = 1$. The algorithm we derived for the compilation of the tables may be considered as an extension of the successive recursion relation of Massey with a set of very general boundary conditions, i.e., $m \neq n$, details of which are included in the Appendix along with a FORTRAN (IV) function subroutine.

Construction of the Tables

Consideration of volume led us to present the tables in two parts.

Part I. $m \leq n = 1(1)25$.

The values of c and $P(D_{mn} \geq \frac{c}{mn})$ constitute the main entry at the $(n,m)^{\text{th}}$ row. Due to the discreteness of the distribution of D_{mn}, the values of P range slightly below and above α, $.001 \leq \alpha \leq .10$; more precisely, the critical region A is defined as:

$$A = \Big\{ c \,|\, c_1 \leq c'_1 \leq c \leq c'_2 \leq c_2 \,;$$

$$P(D_{mn} \geq \tfrac{c_1}{mn}) \geq .10,\; P(D_{mn} \geq \tfrac{c'_1}{mn}) \leq .10, \text{ and } P(\tfrac{c_1}{mn} < D_{mn} < \tfrac{c'_1}{mn}) = 0;$$

$$P(D_{mn} \geq \tfrac{c'_2}{mn}) \geq .001,\; P(D_{mn} \geq \tfrac{c_2}{mn}) \leq .001, \text{ and } P(\tfrac{c'_2}{mn} < D_{mn} < \tfrac{c_2}{mn}) = 0 \Big\}$$

The tabled critical region A is considerably smaller due to the fact that for all combinations of values of m and n for which $\binom{m+n}{m} \leq 2000$, the exact probabilities (the right-tail area) exceed the .1 percent level even when the value of c reaches the upper limit, mn, since $P(D_{mn} = 1) = 2/\binom{m+n}{m}$ for all m and n. The upper bound of c, namely c_2, in the definition of the critical region A is, therefore, to be taken as mn whenever $\binom{m+n}{m} \leq 2000$.

Part II. $m \leq n = 1(1)100$.

The main entry at the $(n,m)^{\text{th}}$ row is the smallest value of c for which $P(D_{mn} \geq \frac{c}{mn}) \leq \alpha$, $\alpha = .10, .05, .025, .01, .005, .001$. The critical region B for Part II is,

$$B = \Big\{ \text{smallest } c \,|\, P(D_{mn} \geq \tfrac{c}{mn}) \leq \alpha \Big\}.$$

Due to the discreteness of the distribution of D_{mn}, at least for small to moderate sizes of m and n, the compilation of Part II of the tables is slightly complicated. We use the cases $(n,m) = (13,7)$ and $(20,2)$ to illustrate the manner in which the critical region B is constructed.

Table Illustrating the Construction of the Critical Region B

n	m	c	$P(D_{mn} \geq \frac{c}{mn})$	α	smallest c
13	7	49	.111275		
		50	.094118	.10	50
		51	.072859		
		52	.054076		
		56	.045820	.05	56
		57	.036378		
		58	.024768	.025	58
		63	.016641		
		64	.011997		
		65	.006837	.01	65
		70	.004902	.005	70
		71	.002915		
		77	.001213		
		78	.000516	.001	78
20	2	32	.129870		
		34	.086580	.10	34
		36	.051948		
		38	.025974	.05	38
				.025	(40)
		40	.008658	.01	40
				.005	–

In the critical region B, we denote by "–" the impossible event. When the distribution is less smooth, there may not be a value of c for the specified level of α, in which case the value of c corresponding to the next level of α (if it exists) is entered in parentheses. In the example for $(n,m) = (20,2)$ we have the impossible event for the nominal level $\alpha =$ .005 (or less) since $P(D_{2,20} = 1) = 2/\binom{20+2}{2} = .008658 > .005$. For the nominal level $\alpha = .025$ there is no value of c which satisfies the definition of the critical region B. On the other hand, there is one for $\alpha = .01$ so that the value of $c = 40$ for $\alpha = .01$ also serves as the critical value for $\alpha = .025$, which we enter in parentheses.

It is noted that these complications occur when n is an integer multiple of m and/or $\binom{m+n}{m} < 2000$. When $\binom{m+n}{m} \geq 2000$, n is not an integer multiple of m, or the sizes of m and n becomes moderate to large, the distribution of D_{mn} becomes smooth enough so that we have few of these complications as in the case of $(n,m) = (13,7)$.

Throughout the construction of the critical regions A and B, the probabilities were computed with thirteen decimal place accuracy. Numerical checking indicates that the error is at most one unit in the 12th decimal place. The probabilities for the Critical region A were rounded off at the seventh decimal place. The computation was carried out in part at the Socio-Behavioral Laboratory, Pacific State Hospital, Pomona, California using Honeywell 1200 and in part at the Health Science Computing Facility, University of California, Los Angeles with IBM 360/91.

The critical region B in contrast to the critical region A is somewhat conservative due to the discreteness of the distribution of D_{mn} and the way it is constructed in that the actual size of Type I error is strictly less than or (rarely) equal to the nominal size of α although this effect does not appear serious for n as small as 50 and certainly negligible when $n = 100$.

Use of the Tables

Example 1.

Two samples of the random normal deviates of sizes 7 and 13 are drawn from *A Million Random Digits with 100,000 Normal Deviates,* by the Rand Corporation, (New York: The Free Press, 1966). The first sample is taken from the 5th column, page 1; the second from the 7th column, page 200.

X: .723, .382, –.854, –1.318, .847, –.410, –.111.

X': .558, –1.327, 1.778, .516, –1.751, –.135, –1.292, .846, –1.260, –1.943, –.128, 1.124, –1.813.

The null hypothesis to be tested is that the two samples are drawn from a common distribution (against the alternative that they are not). The evaluation of the Kolmogorov-Smirnov criterion D_{mn} may be carried out in the following manner.

(1) Pool and order the two samples:

$X + X'$: –1.943, –1.813, –1.751, –1.327, –1.318, –1.292, –1.260, –.854, –.410, –.135, –.128, –.111, .382, .516, .558, .723, .846, .847, 1.124, 1.778

(2) Evaluate $S_7(x)$, $S_{13}(x)$, and $7 \cdot 13|S_7(x) - S_{13}(x)|$:

$X + X'$	$S_7(x)$	$S_{13}(x)$	$91\|S_7(x)-S_{13}(x)\|$
–1.943	0	$\frac{1}{13}$	7
–1.813		$\frac{2}{13}$	14
–1.751		$\frac{3}{13}$	21
–1.327		$\frac{4}{13}$	28
–1.318	$\frac{1}{7}$		15
–1.292		$\frac{5}{13}$	24

$X+X'$	$S_7(x)$	$S_{13}(x)$	$91\|S_7(x)-S_{13}(x)\|$	$X+X'$	$S_7(x)$	$S_{13}(x)$	$91\|S_7(x)-S_{13}(x)\|$
−1.260		$\frac{6}{13}$	<u>29</u>	.516		$\frac{9}{13}$	2
− .854	$\frac{2}{7}$		16	.558		$\frac{10}{13}$	5
− .410	$\frac{3}{7}$		3	.723	$\frac{6}{7}$		8
− .135		$\frac{7}{13}$	10	.846		$\frac{11}{13}$	1
− .128		$\frac{8}{13}$	17	.847	1		14
− .111	$\frac{4}{7}$		4	1.124		$\frac{12}{13}$	7
.382	$\frac{5}{7}$		9	1.778		1	0

(3) Enter the tables at $(n,m;c) = (13, 7; 29)$.

Since the tables start with $c = 49$, and the observed c is less than this value, $P(D_{7,13} \geq \frac{29}{91}) > .10$ and, as expected, the null hypothesis would not be rejected at the usual levels of Type I error.

Example 2.

Now suppose the measurements of one sample seem larger than those of the other so that we would be now testing the same null hypothesis as in Example 1 but against a location shift alternative (in the parametric case). We add a value of 1.282 to the first sample of Example 1 to accomplish such a shift, equivalent to a shift of the unit normal distribution $\Phi(x)$ by an amount of 40 per cent of its probability mass. In terms of the maximum difference alternatives Δ of Massey (1950), where $\Delta = \sup_x |F(x) - F'(x)|$ and the underlying distributions $F(x)$ and $F'(x)$ of the two samples are assumed to be continuous, this amounts to specifying $\Delta = \sup_x |\Phi(x - 1.282)| = 2\Phi(\frac{1.282}{2}) - 1 = .478$ so that we would be testing $H_0 : \Delta = 0$ versus $H_1 : \Delta = .478$ (a non-parametric formulation of the problem).

X: <u>2.005</u>, <u>1.664</u>, <u>.428</u>, −<u>.036</u>, <u>2.129</u>, <u>.872</u>, <u>1.171</u>.

When pooled and ordered with the second sample of Example 1, we have

$X+X'$: −1.943, −1.813, −1.751, −1.327, −1.292, −1.260, −.135, −.128, −<u>.036</u>, <u>.428</u>, .516, .558, .846, <u>.872</u>, 1.124, <u>1.171</u>, <u>1.664</u>, 1.778, <u>2.005</u>, <u>2.129</u>.

We proceed as in Example 1 to evaluate $S_7(x)$, $S_{13}(x)$, and $7 \cdot 13\,|S_7(x) - S_{13}(x)|$:

$X+X'$	$S_7(x)$	$S_{13}(x)$	$91\|S_7(x)-S_{13}(x)\|$	$X+X'$	$S_7(x)$	$S_{13}(x)$	$91\|S_7(x)-S_{13}(x)\|$
−1.943	0	$\frac{1}{13}$	7	−1.292		$\frac{5}{13}$	35
−1.813		$\frac{2}{13}$	14	−1.260		$\frac{6}{13}$	42
−1.751		$\frac{3}{13}$	21	− .135		$\frac{7}{13}$	49
−1.327		$\frac{4}{13}$	28	− .128		$\frac{8}{13}$	<u>56</u>

$X+X'$	$S_7(x)$	$S_{13}(x)$	$91\,\lvert S_7(x)-S_{13}(x)\rvert$	$X+X'$	$S_7(x)$	$S_{13}(x)$	$91\,\lvert S_7(x)-S_{13}(x)\rvert$
$-.036$	$\frac{1}{7}$		43	1.124		$\frac{12}{13}$	45
.428	$\frac{2}{7}$		30	1.171	$\frac{4}{7}$		32
.516		$\frac{9}{13}$	37	1.664	$\frac{5}{7}$		19
.558		$\frac{10}{13}$	44	1.778		1	26
.846		$\frac{11}{13}$	51	2.005	$\frac{6}{7}$		13
.872	$\frac{3}{7}$		38	2.129	1		0

The observed c is 56 and the entry at $(n,m;c) = (13,7;56)$ of the tables reads $P(D_{7,13} \geq \frac{56}{91}) = .045820$ so that the null hypothesis would be rejected at the five per cent level.

As with any test criterion which makes use of rank-order statistics, the problem of how to treat ties is an inherent one. Three methods have been discussed in the literature. These are (1) discard the tied observations if the number of ties is small, with possible loss of power; (2) order tied observations at random with possible loss of efficiency; and (3) assign the average rank to each of the tied observations, in which case the resulting distribution may be a slightly different one, i.e., the variance may be affected.

Unfortunately little investigation has been done on the relative merits of these methods relevant to D_{mn}. With the Wilcoxon test, at least, the asymptotic relative efficiency is less for the random ordering method than the average ranking method. Until further information is available, we may regard the average ranking method as preferable.

Approximations

The investigation of the adequacy of and possible improvement over the Smirnov approximation appears spotty until recently. Hodges (1957) examined the adequacy for the case of m/n being 1 or nearly so and suggested a continuity correction to correct the overestimating effect (to the left-tail area) of the Smirnov approximation for this case. More recently a detailed study on various aspects of large sample approximations was carried out by one of the authors (see Kim, 1969), in which (1) the adequacy of the Smirnov approximation to the exact probability integral as well as exact moments was examined and (2) relative merits of the Smirnov, the Kolmogorov, and a normal approximation were compared. It appears that not only the sample sizes m and n, but also the relation between m and n, i.e., whether n is or is not an integer multiple of m affects appreciably the range of m and n over which these large sample approximations may apply.

In what follows, we present those results relevant for the use of the tables, omitting details.

(1) The Smirnov approximation:

Evaluate the observed $S = [mn/(m+n)]^{(1/2)} D_{mn}$ and enter the Smirnov tables (1948). The Smirnov approximation is reasonable for $n \geq 100$, $m \geq [.10n]$, and $\alpha \leq .10$. It is noted that a continuity correction (in the S scale) improves the Smirnov approximation markedly. This amounts to adding $1/(2\sqrt{n})$ to the observed value of S and the corrected value of $S_c = [mn/(m+n)]^{(1/2)} D_{mn} + 1/(2\sqrt{n})$ is then referred to the Smirnov tables.

(2) The Kolmogorov approximation:

For $m \leq [.10\,n]$, $n \geq 50$, and $\alpha \leq .10$, the Kolmogorov approximation is superior to the Smirnov approximation. For the approximation, one treats the observed D_{mn} as D_m and enters the Miller tables (1956) at the mth row. As in the

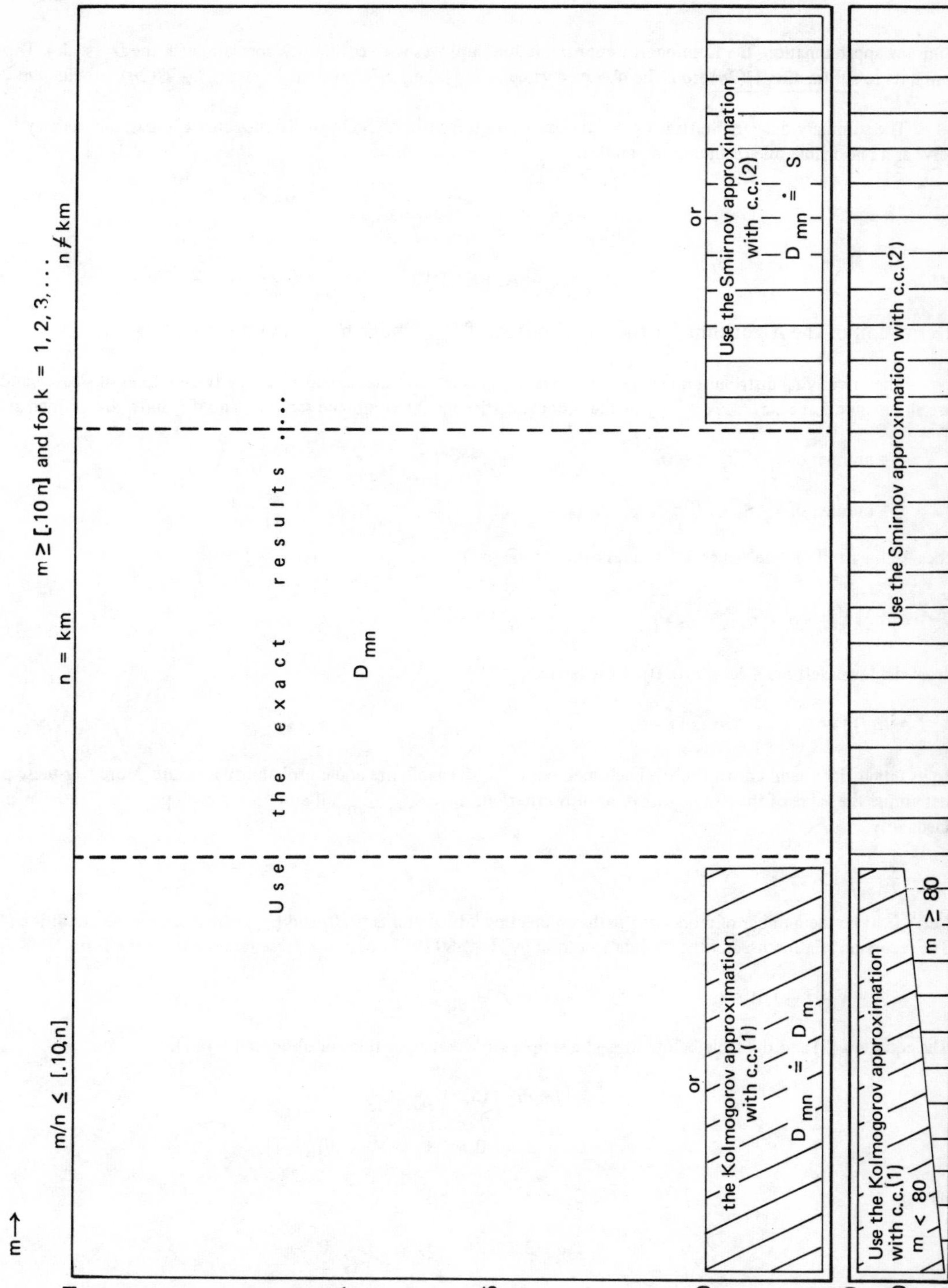

Diagram Illustrating Sample Size Restrictions for Large Sample Approximations

[1] with continuity correction $(-\frac{1}{2n}, m > 1)$

[2] with continuity correction $[1/(2\sqrt{n})]$

Smirnov approximation, the Kolmogorov approximation improves with continuity correction (in the D_m scale). This amounts to subtracting $1/(2n)$ from the observed value of D_{mn}, i.e., the corrected $D_m = D_{mn} - 1/(2n)$, provided $m > 1$.

The summary diagram illustrating sample size restrictions for the Smirnov-Kolmogorov approximations may serve as a reasonable rule of thumb in practice.

APPENDIX

Derivation of the Algorithm for the Distribution of D_{mn} under $H_0 : F(x) = F'(x)$

The underlying distribution being continuous, we may assume that x_i and x'_j of the two samples of sizes m and n are all distinct. Let $z_1, z_2, \ldots, z_{m+n}$ be the order statistics for the combined samples, and if i_r and j_r are defined as

$$i_r = \text{number of } x_i \leq z_r,\ i_0 = 0,$$

$$j_r = \text{number of } x'_j \leq z_r,\ j_0 = 0,$$

then each pair of sample values determines a monotone path

$$P = \{(i_r, j_r) \mid r = 0, 1, 2, \ldots, m+n\}$$

from the lower left-hand corner (0, 0) of the lattice

$$L = \{(i, j) \mid i = 0, 1, \ldots, m \text{ and } j = 0, 1, \ldots, n\}$$

to its upper right-hand corner (m, n). Each such path, $\binom{m+n}{n}$ in all, has equal probability measure. Moreover these paths determine the value of the Kolmogorov-Smirnov criterion, since $D_{mn} \leq \frac{c}{mn}$, if and only if each point in P satisfies the inequality

$$\left| \frac{i}{m} - \frac{j}{n} \right| \leq \frac{c}{mn}. \tag{1}$$

Let $N(i, j)$ be the number of monotone paths connecting lattice points (0, 0) and (i, j) which satisfy the condition (1). The recursion relation used in the "inside" method by Hodges (1957) is, in the present context, of the form

$$N(i, j) = N(i, j-1) + N(i-1, j). \tag{2}$$

The equation (2) and the table below suggest a simple recursion algorithm for obtaining $N(i, j)$.

$$(n, m) = (5, 3), \frac{c}{mn} = \frac{8}{15}$$

0	0	0	8	19	30
0	2	5	8	11	11
1	2	3	3	3	0
1	1	1	0	0	0

To be specific, let $C(i, j)$ be the indicator function for the condition (1), namely

$$C(i,j)=\begin{cases}1 \text{ if } \left|\frac{i}{m}-\frac{j}{n}\right|\leq\frac{c}{mn}\\[2ex] 0 \text{ otherwise}.\end{cases}$$

Then $N(i, j)$ may be obtained from the recursion relation

$$N(i,j)=C(i,j)\left[N(i,j-1)+N(i-1,j)\right], \tag{3}$$

subject to the initial condition

$$N(i,j)=C(i,j), \text{ when } i\cdot j=0.$$

The value of the cumulative distribution function is, then,

$$P\left(D_{mn}\leq\frac{c}{mn}\right)=\binom{m+n}{m}^{-1}N(m,n). \tag{4}$$

For the tables, we use an alternative algorithm that avoids a large value of $\binom{m+n}{m}$ and $N(m, n)$ (which can easily exceed floating-point exponent range).

Let $U(i,j)=\binom{i+n}{i}^{-1}N(i,j)$; then the recursion relation (3) becomes

$$U(i,j)=\frac{i}{i+n}C(i,j)\left[U(i,j-1)+U(i-1,j)\right], \tag{5}$$

subject to the initial condition

$$U(i,j)=\binom{i+n}{i}^{-1}C(i,j), \text{ when } i\cdot j=0$$

and the equation (4) is now replaced by the equation (6)

$$P\left(D_{mn}\leq\frac{c}{mn}\right)=U(m,n). \tag{6}$$

Storage is conserved by recording only the most recent column of U.

A FORTRAN Function Subroutine

A FORTRAN (IV) function subroutine for computing $P(D_{mn}\leq\frac{c}{mn})$ using the above algorithm is listed below for the possible use by the reader for the ranges of (m, n) and c which the tables do not cover.

Input: M = the first sample size (smaller of the two)

N = the second sample size

$D=\frac{c}{mn}$, the Kolmogorov-Smirnov criterion

U = work area, dimensioned to at least $N+1$

Output: *AKSCDF* (M, N, D, U) the functional value at D for M and N, i.e.,

the values of $P(D_{mn}\leq\frac{c}{mn})$.

An approximate machine time (with IBM 360/91) is given below.

Sample size N, $(M \leq N)$	10	15	20	25	50	100
Time in seconds	.00031	.00042	.00073	.00104	.00374	.01612

Program Listing.

```
      FUNCTION AKSCDF (M, N, D, U)
      DIMENSION U (N)
C     THE VARIABLE U IS DIMENSION TO AT LEAST N + 1 IN MAIN
      K = FLOAT (M * N) * D +.5
      U (1) = 1.
      DO 1  J = 1, N
      U (J + 1) = 1.
      IF (M * J . GT . K)   U (J + 1) = 0.
 1    CONTINUE
      DO 2   I = 1, M
      W = FLOAT (I) / FLOAT (I + N)
      U (1) = W * U (1)
      IF (N * I . GT . K)   U (1) = 0.
      DO 2   J = 1, N
      U (J + 1) = U (J) + U (J + 1) * W
      IF [IABS (N * I – M * J) . GT . K]    U (J + 1) = 0
 2    CONTINUE
      AKSCDF = U (N + 1)
      RETURN
      END
```

REFERENCES

1. Barton, D. E. and Mallows, C. L. (1965). "Some aspects of the random sequence," *Annals of Mathematical Statistics,* 36: 236–260.

2. Darling, D. A. (1957). "The Kolmogorov-Smirnov, Cramér-von Mises tests," *Annals of Mathematical Statistics,* 28: 223–238.

3. Feller, W. (1948). "On the Kolmogorov-Smirnov limit theorems for empirical distributions," *Annals of Mathematical Statistics,* 19: 177–189.

4. Hodges, J. L., Jr. (1957). "The significance probability of the Smirnov two sample test," *Arkiv för Matematik,* 3: No. 43, 469–486.

5. Kim, P. J. (1969). "On the exact and approximate sampling distribution of the two sample Kolmogorov-Smirnov criterion D_{mn}, $m \leq n$," *Journal of the American Statistical Association,* 64: 1625–37.

6. Marliss, G. S. and Zayachkowski, W. (1962). "The Smirnov two-sample statistic D_{mn}," Assumption University of Windsor, Ontario, Canada.

7. Massey, F. J., Jr. (1950). "A note on the power of a non-parametric test," *Annals of Mathematical Statistics,* 21: 440–443.

8. Massey, F. J., Jr. (1951). "The distribution of the maximum deviation between two sample cumulative step functions," *Annals of Mathematical Statistics,* 22: 125–128.

9. Miller, L. H. (1956). "Table of percentage points of Kolmogorov statistics," *Journal of the American Statistical Association,* 51: 111–121.

10. Smirnov, N. V. (1939). "Estimating the deviation between the empirical distribution functions of two independent samples," *Bulletin de l'Université de Moscou,* 2: nos. 2, 3.

11. Smirnov, N. V. (1948). "Table for estimating the goodness of fit of empirical distributions," *Annals of Mathematical Statistics,* 19: 279–281.

12. Steck, G. P. (1969). "The Smirnov two sample tests as rank tests," *Annals of Mathematical Statistics,* 40: 1449–1466.

TABLE I. UPPER TAIL AREAS

n	m	c	$P(D_{mn} \geq \frac{c}{mn})$
1	1	1	1.000000
2	1	2	.666667
2	2	4	.333333
3	1	3	.500000
3	2	6	.200000
3	3	9	.100000
4	1	4	.400000
4	2	8	.133333
4	3	9	.228571
		12	.057143
4	4	12	.228571
		16	.028571
5	1	5	.333333
5	2	8	.285714
		10	.095238
5	3	12	.142857
		15	.035714
5	4	15	.142857
		16	.079365
		20	.015873
5	5	15	.357143
		20	.079365
5	5	25	.007937
6	1	6	.285714
6	2	10	.214286
		12	.071429
6	3	12	.333333
		15	.095238
		18	.023810
6	4	16	.180952
		18	.095238
		20	.047619
		24	.009524
6	5	20	.108225
		24	.047619
		25	.025974
		30	.004329
6	6	24	.142857
		30	.025974
		36	.002165
7	1	7	.250000
7	2	12	.166667
		14	.055556
7	3	15	.166667
		18	.066667
		21	.016667
7	4	20	.121212
		21	.066667
		24	.030303
		28	.006061
7	5	23	.116162
		25	.065657
		28	.030303
		30	.015152
		35	.002525
7	6	24	.146853
		28	.090909
		29	.067599
		30	.038462
		35	.015152
		36	.008159
		42	.001166
7	7	28	.212121
		35	.053030
		42	.008159
		49	.000583
8	1	8	.222222
8	2	14	.133333
		16	.044444
8	3	18	.121212
		21	.048485
		24	.012121
8	4	20	.222222
		24	.084848
		28	.020202
		32	.004040
8	5	25	.125874
		27	.079254
		30	.041958
		32	.020202
		35	.009324
		40	.001554
8	6	28	.139194
		30	.092574
		32	.060606
		34	.042624
		36	.022644
		40	.009324
		42	.004662
		48	.000666
8	7	33	.118104
		34	.087024
		35	.055944
		40	.032634
		41	.024242
		42	.013054
		48	.004662
		49	.002486
		56	.000311
8	8	32	.282673
		40	.087024
		48	.018648
		56	.002486
		64	.000155
9	1	9	.200000
9	2	16	.109091
		18	.036364
9	3	18	.236364
		21	.090909
		24	.036364
		27	.009091
9	4	24	.114685
		27	.061538
		28	.041958
		32	.013986
		36	.002797

n	m	c	$P(D_{mn} \geq \frac{c}{mn})$
9	5	27	.118881
		30	.085914
		31	.055944
		35	.027972
		36	.013986
		40	.005994
		45	.000999
9	6	30	.175824
		33	.094705
		36	.061139
		39	.027972
		42	.013986
		45	.005994
		48	.002797
		54	.000400
9	7	35	.126748
		36	.097902
		38	.078671
		40	.055070
		42	.034091
		45	.020979
		47	.014860
		49	.007517
		54	.002797
		56	.001399
		63	.000175
9	8	39	.109420
		40	.078568
		45	.055944
		46	.046894
		47	.033566
		48	.020239
		54	.011189
		55	.008309
		56	.004278
		63	.001399
		64	.000740

n	m	c	$P(D_{mn} \geq \frac{c}{mn})$
9	9	45	.125874
		54	.033566
		63	.006294
		72	.000740
10	1	10	.181818
10	2	16	.181818
		18	.090909
		20	.030303
10	3	21	.139860
		24	.069930
		27	.027972
		30	.006993
10	4	26	.125874
		28	.083916
		30	.045954
		32	.029970
		36	.009990
		40	.001998
10	5	30	.165834
		35	.060606
		40	.019314
		45	.003996
		50	.000666
10	6	34	.125125
		36	.092158
		38	.066434
		40	.041958
		42	.031469
		44	.018981
		48	.008991
		50	.003996
		54	.001748
		60	.000250
10	7	39	.116516

n	m	c	$P(D_{mn} \geq \frac{c}{mn})$
10	7	40	.086898
		42	.071575
		43	.054093
		46	.036096
		49	.021699
		50	.013986
		53	.009461
		56	.004525
		60	.001748
		63	.000823
10	8	42	.120252
		44	.095114
		46	.070433
		48	.049865
		50	.037296
		52	.030257
		54	.020659
		56	.012021
		60	.006993
		62	.004982
		64	.002422
		70	.000823
10	9	45	.105577
		50	.083916
		51	.074628
		52	.060339
		53	.044556
		54	.030267
		60	.020979
		61	.017645
		62	.012341
		63	.007036
		70	.003702
		71	.002750
		72	.001364
		80	.000411
10	10	50	.167821
		60	.052448

n	m	c	$P(D_{mn} \geq \frac{c}{mn})$
10	10	70	.012341
		80	.002057
		90	.000217
11	1	11	.166667
11	2	18	.153846
		20	.076923
		22	.025641
11	3	24	.109890
		27	.054945
		30	.021978
		33	.005495
11	4	28	.143590
		29	.098168
		32	.063004
		33	.035165
		36	.021978
		40	.007326
		44	.001465
11	5	34	.105769
		35	.073718
		39	.043956
		40	.029304
		44	.013736
		45	.009615
		50	.002747
		55	.000458
11	6	37	.125566
		38	.091629
		42	.064480
		43	.047835
		44	.029735
		48	.021332
		49	.013251
		54	.005979
		55	.002747

n	m	c	$P(D_{mn} \geq \frac{c}{mn})$
11	6	60	.001131
		66	.000162
11	7	42	.105329
		44	.082768
		45	.070890
		48	.048894
		49	.035696
		52	.024384
		55	.014329
		56	.011061
		59	.006222
		63	.002828
		66	.001131
		70	.000503
11	8	47	.101215
		48	.081501
		50	.065333
		53	.046810
		55	.032918
		56	.027546
		58	.020137
		61	.013151
		64	.007436
		66	.004525
		69	.003096
		72	.001429
		77	.000503
11	9	50	.113372
		52	.089259
		54	.069612
		55	.057692
		57	.050107
		59	.039105
		61	.027852
		63	.018683
		66	.013575
		68	.011098

n	m	c	$P(D_{mn} \geq \frac{c}{mn})$
11	9	70	.007430
		72	.004096
		77	.002262
		79	.001619
		81	.000762
11	10	56	.106097
		57	.091501
		58	.074661
		59	.057820
		60	.043225
		66	.033937
		67	.030325
		68	.024212
		69	.017266
		70	.011153
		77	.007541
		78	.006362
		79	.004366
		80	.002370
		88	.001191
		89	.000885
11	11	55	.211476
		66	.074661
		77	.020739
		88	.004366
		99	.000655
12	1	12	.153846
12	2	20	.131868
		22	.065934
		24	.021978
12	3	24	.189011
		27	.087912
		30	.043956
		33	.017582
		36	.004396

n	m	c	$P(D_{mn} \geq \frac{c}{mn})$
12	4	32	.112088
		36	.048352
		40	.016484
		44	.005495
		48	.001099
12	5	35	.131222
		36	.096315
		38	.079509
		40	.053652
		43	.032644
		45	.021008
		48	.010019
		50	.006787
		55	.001939
		60	.000323
12	6	42	.114630
		48	.046326
		54	.014867
		60	.004094
		66	.000754
12	7	44	.126459
		46	.097841
		48	.074502
		49	.064777
		51	.050250
		53	.034373
		56	.024053
		58	.016909
		60	.009764
		63	.007264
		65	.004207
		70	.001826
		72	.000754
12	8	48	.149559
		52	.090672
		56	.055664
		60	.031992
		64	.018211

n	m	c	$P(D_{mn} \geq \frac{c}{mn})$
12	8	68	.008621
		72	.004763
		76	.001985
		80	.000873
12	9	54	.113833
		57	.078495
		60	.060633
		63	.041438
		66	.026040
		69	.017957
		72	.011969
		75	.007192
		78	.004620
		81	.002477
		84	.001429
		87	.000987
12	10	58	.111279
		60	.092703
		62	.073567
		64	.061746
		66	.048886
		68	.036979
		70	.027691
		72	.022624
		74	.019810
		76	.015307
		78	.010544
		80	.006733
		84	.004763
		86	.003919
		88	.002583
		90	.001358
		96	.000714
12	11	63	.108496
		64	.090598
		65	.073464
		66	.058926
		72	.049774

n	m	c	$P(D_{mn} \geq \frac{c}{mn})$
12	11	73	.046005
		74	.039409
		75	.031436
		76	.023463
		77	.016867
		84	.013098
		85	.011752
		86	.009291
		87	.006427
		88	.003966
		96	.002620
		97	.002216
		98	.001497
		99	.000778
12	12	60	.255775
		72	.099547
		84	.031436
		96	.007859
		108	.001497
		120	.000204
13	1	13	.142857
13	2	22	.114286
		24	.057143
		26	.019048
13	3	27	.125000
		30	.071429
		33	.035714
		36	.014286
		39	.003571
13	4	32	.132773
		35	.089076
		36	.066387
		39	.037815
		40	.029412
		44	.012605
		48	.004202
13	4	52	.000840
13	5	39	.100840
		40	.086835
		42	.060924
		45	.039916
		47	.024743
		50	.015406
		52	.007470
		55	.004902
		60	.001401
		65	.000233
13	6	42	.115067
		46	.086245
		47	.070913
		48	.050862
		52	.034056
		53	.029412
		54	.019460
		59	.010615
		60	.006929
		65	.002875
		66	.002064
		72	.000516
13	7	49	.111275
		50	.094118
		51	.072859
		52	.054076
		56	.045820
		57	.036378
		58	.024768
		63	.016641
		64	.011997
		65	.006837
		70	.004902
		71	.002915
		77	.001213
		78	.000516
13	8	52	.109661
		54	.098865
		56	.078225
		57	.065566
		59	.053182
		62	.039029
		64	.028807
		65	.022409
		67	.019195
		70	.012384
		72	.008669
		75	.005799
		78	.003145
		80	.002408
		83	.001307
		88	.000550
13	9	56	.115283
		59	.097974
		60	.080777
		63	.064453
		64	.055551
		65	.042282
		68	.037433
		69	.028427
		72	.021961
		73	.017760
		77	.011897
		78	.007917
		81	.006530
		82	.004785
		86	.002955
		90	.001548
		91	.000929
13	10	61	.114452
		64	.094261
		65	.077435
		67	.071024
		68	.058603
		70	.049205
13	10	71	.042622
		74	.032804
		77	.024343
		78	.018343
		80	.016165
		81	.013277
		84	.009936
		87	.006632
		90	.004210
		91	.003096
		94	.002484
		97	.001575
		100	.000806
13	11	66	.110559
		67	.099834
		69	.087585
		71	.073804
		73	.060231
		75	.048201
		77	.039014
		78	.034056
		80	.031005
		82	.025901
		84	.020068
		86	.014639
		88	.010555
		91	.008514
		93	.007504
		95	.005754
		97	.003851
		99	.002351
		104	.001622
		106	.001341
		108	.000873
13	12	70	.108339
		71	.091330
		72	.077053
		78	.068111
		79	.064267

n	m	c	$P(D_{mn} \geq \frac{c}{mn})$
13	12	80	.057407
		81	.048833
		82	.039712
		83	.031137
		84	.024278
		91	.020433
		92	.018969
		93	.016169
		94	.012649
		95	.009129
		96	.006330
		104	.004865
		105	.004381
		106	.003436
		107	.002314
		108	.001369
		117	.000885
13	13	78	.126488
		91	.044272
		104	.012649
		117	.002875
		130	.000500
14	1	14	.133333
14	2	24	.100000
		26	.050000
		28	.016667
14	3	30	.102941
		33	.058824
		36	.029412
		39	.011765
		42	.002941
14	4	36	.105882
		38	.071895
		40	.052288
		42	.030065
		44	.022876
		48	.009804
		52	.003268
		56	.000654
14	5	41	.108359
		42	.078947
		45	.066563
		46	.047472
		50	.030272
		51	.019092
		55	.011524
		56	.005676
		60	.003612
		65	.001032
		70	.000172
14	6	46	.113106
		48	.086275
		50	.066099
		52	.053096
		54	.037307
		56	.025542
		58	.021672
		60	.013932
		64	.007740
		66	.004902
		70	.002064
		72	.001445
		78	.000361
14	7	49	.176170
		56	.082697
		63	.033127
		70	.011782
		77	.003388
		84	.000826
14	8	56	.115108
		58	.090984
		60	.072671
		62	.056409
		64	.046402
		66	.038127
		68	.027995
		70	.020114
		72	.017563
		74	.013447
		76	.008631
		80	.005817
		82	.003990
		84	.002133
		88	.001582
		90	.000882
14	9	62	.100568
		63	.082065
		66	.070591
		67	.058256
		70	.045510
		71	.041804
		72	.032005
		75	.026163
		76	.021361
		80	.014929
		81	.010766
		84	.008076
		85	.007061
		89	.004320
		90	.003035
		94	.001938
		98	.000996
14	10	66	.107118
		68	.091314
		70	.075788
		72	.062325
		74	.049762
		76	.042383
		78	.033858
		80	.026479
		82	.022501
		84	.016444
		86	.014770
		88	.010787
		90	.008324
		92	.006604
		96	.004284
		98	.002714
		100	.002270
		102	.001615
		106	.000987
14	11	71	.106020
		73	.090080
		74	.080205
		76	.068377
		77	.057522
		79	.051231
		82	.040965
		84	.032495
		85	.030093
		87	.024163
		88	.020249
		90	.017410
		93	.013123
		96	.009392
		98	.006819
		99	.006063
		101	.004913
		104	.003657
		107	.002374
		110	.001440
		112	.001032
		115	.000835
14	12	76	.101529
		78	.087018
		80	.073337
		82	.061464
		84	.052494
		86	.044467
		88	.038947
		90	.032351
		92	.025718

n	m	c	$P(D_{mn} \geq \frac{c}{mn})$
14	12	94	.019914
		96	.015692
		98	.013622
		100	.012482
		102	.010394
		104	.007907
		106	.005581
		108	.003885
		112	.003096
		114	.002744
		116	.002092
		118	.001365
		120	.000799
14	13	77	.111137
		78	.097233
		84	.088543
		85	.084680
		86	.077703
		87	.068811
		88	.059030
		89	.049249
		90	.040357
		91	.033380
		98	.029515
		99	.027969
		100	.024932
		101	.020939
		102	.016625
		103	.012633
		104	.009595
		112	.008050
		113	.007500
		114	.006370
		115	.004900
		116	.003430
		117	.002299
		126	.001750
		127	.001580
		128	.001231
		129	.000810
14	14	84	.154935
		98	.059030
		112	.018782
		126	.004900
		140	.001021
		154	.000163
15	1	15	.125000
15	2	24	.147059
		26	.088235
		28	.044118
		30	.014706
15	3	30	.161765
		33	.085784
		36	.049020
		39	.024510
		42	.009804
		45	.002451
15	4	37	.116615
		40	.085655
		41	.058824
		44	.041796
		45	.024252
		48	.018060
		52	.007740
		56	.002580
		60	.000516
15	5	45	.109520
		50	.051858
		55	.023349
		60	.008772
		65	.002709
		70	.000774
15	6	48	.133569
		51	.087461
15	6	54	.065900
		57	.040432
		60	.027901
		63	.016254
		66	.010172
		69	.005750
		72	.003538
		75	.001511
		78	.001032
		84	.000258
15	7	55	.100760
		56	.079370
		60	.062588
		61	.057803
		62	.046721
		63	.034056
		68	.024416
		69	.020018
		70	.013568
		75	.008514
		76	.007482
		77	.004773
		83	.002392
		84	.001548
		90	.000575
15	8	59	.103919
		60	.085645
		64	.077742
		65	.067830
		66	.054394
		67	.041545
		72	.033224
		73	.027827
		74	.020485
		75	.014366
		80	.012286
		81	.009602
		82	.006147
		88	.003993
		89	.002802
15	8	90	.001481
		96	.001065
		97	.000608
15	9	66	.100631
		69	.073291
		72	.058973
		75	.041797
		78	.029728
		81	.022308
		84	.014848
		87	.010382
		90	.007284
		93	.004812
		96	.002928
		99	.001985
		102	.001300
		105	.000658
15	10	70	.118113
		75	.077405
		80	.049835
		85	.029582
		90	.018129
		95	.010033
		100	.005515
		105	.002833
		110	.001467
		115	.000634
15	11	75	.104064
		76	.098650
		77	.083936
		79	.076230
		80	.068256
		83	.056353
		84	.047924
		87	.039846
		88	.032215
		90	.028443
		91	.026549
		94	.020483

n	m	c	$P(D_{mn} \geq \frac{c}{mn})$
15	11	95	.017509
		98	.013579
		99	.010453
		102	.008770
		105	.006178
		106	.005614
		109	.003952
		110	.003051
		113	.002378
		117	.001501
		120	.000909
15	12	81	.101031
		84	.078345
		87	.060985
		90	.050718
		93	.039805
		96	.029669
		99	.021871
		102	.017034
		105	.013076
		108	.009551
		111	.006831
		114	.005060
		117	.003504
		120	.002457
		123	.001761
		126	.001305
		129	.000827
15	13	85	.101642
		87	.088168
		89	.076599
		90	.067904
		91	.064615
		92	.059934
		94	.054137
		96	.047017
		98	.039532
		100	.032454
		102	.026466
15	13	104	.022194
		105	.020124
		107	.018882
		109	.016528
		111	.013567
		113	.010533
		115	.007909
		117	.006078
		120	.005250
		122	.004836
		124	.004018
		126	.003009
		128	.002061
		130	.001388
		135	.001094
		137	.000974
15	14	91	.106832
		92	.099836
		93	.090814
		94	.080696
		95	.070232
		96	.060114
		97	.051092
		98	.044096
		105	.040248
		106	.038648
		107	.035450
		108	.031129
		109	.026248
		110	.021368
		111	.017047
		112	.013848
		120	.012249
		121	.011650
		122	.010371
		123	.008617
		124	.006695
		125	.004940
		126	.003662
		135	.003062
		136	.002862
15	14	137	.002424
		138	.001837
		139	.001251
		140	.000812
15	15	90	.184416
		105	.075464
		120	.026248
		135	.007656
		150	.001837
		165	.000353
16	1	16	.117647
16	2	26	.130719
		28	.078431
		30	.039216
		32	.013072
16	3	33	.115583
		36	.072239
		39	.041280
		42	.020640
		45	.008256
		48	.002064
16	4	40	.127967
		44	.070175
		48	.033849
		52	.014448
		56	.006192
		60	.002064
		64	.000413
16	5	45	.116762
		48	.088162
		49	.079807
		50	.060150
		54	.040985
		55	.030173
16	5	59	.018281
		60	.013563
		64	.006782
		65	.005504
		70	.002064
		75	.000590
16	6	52	.105344
		54	.086044
		56	.068594
		58	.051171
		60	.042057
		62	.031255
		64	.021230
		66	.018576
		68	.012384
		72	.007559
		74	.004342
		78	.002600
		80	.001126
		84	.000751
16	7	57	.107343
		59	.094030
		61	.076555
		63	.059815
		64	.048141
		66	.043980
		68	.034802
		70	.025013
		73	.018307
		75	.014717
		77	.009781
		80	.006265
		82	.005433
		84	.003377
		89	.001721
		91	.001085
		96	.000408
16	8	64	.125566

n	m	c	$P(D_{mn} \geq \frac{c}{mn})$
16	8	72	.057930
		80	.024238
		88	.008764
		96	.002798
		104	.000732
16	9	67	.118566
		69	.099934
		71	.084271
		72	.074785
		74	.065993
		76	.054336
		78	.043235
		80	.035247
		81	.032736
		83	.027979
		85	.021537
		87	.015884
		90	.012888
		92	.010510
		94	.007366
		96	.005040
		99	.004426
		101	.003343
		103	.002027
		108	.001328
		110	.000890
16	10	74	.101674
		76	.087550
		78	.076149
		80	.063403
		82	.053448
		84	.043967
		86	.035878
		88	.031290
		90	.024363
		92	.021038
		94	.016819
		96	.012681
		98	.011677

n	m	c	$P(D_{mn} \geq \frac{c}{mn})$
16	10	100	.008514
		102	.006954
		104	.005591
		108	.003736
		110	.002576
		112	.001915
		114	.001676
		118	.000971
16	11	79	.100372
		80	.086497
		83	.081812
		84	.071829
		85	.060318
		88	.053647
		89	.049019
		90	.040362
		94	.033566
		95	.028180
		96	.022359
		99	.020811
		100	.018566
		101	.014239
		105	.011898
		106	.009310
		110	.006974
		111	.005977
		112	.004158
		116	.003723
		117	.002639
		121	.001978
		122	.001579
		127	.000971
16	12	84	.113179
		88	.083697
		92	.063299
		96	.046914
		100	.033627
		104	.024272
		108	.016648

n	m	c	$P(D_{mn} \geq \frac{c}{mn})$
16	12	112	.011516
		116	.008105
		120	.005247
		124	.003307
		128	.002255
		132	.001406
		136	.000833
16	13	89	.101421
		91	.089213
		92	.083620
		95	.072115
		96	.062074
		98	.059346
		99	.051267
		101	.046566
		102	.040983
		104	.035345
		105	.032691
		108	.026973
		111	.021848
		112	.017793
		114	.016808
		115	.014146
		117	.012328
		118	.011172
		121	.008968
		124	.006822
		127	.005079
		128	.003957
		130	.003644
		131	.003186
		134	.002590
		137	.001890
		140	.001271
		143	.000862
16	14	94	.104575
		96	.093381
		98	.081638
		100	.071153
		102	.063690

n	m	c	$P(D_{mn} \geq \frac{c}{mn})$
16	14	104	.055625
		106	.047653
		108	.040364
		110	.034315
		112	.030051
		114	.026679
		116	.024120
		118	.020787
		120	.017183
		122	.013766
		124	.010938
		126	.009019
		128	.008166
		130	.007700
		132	.006740
		134	.005478
		136	.004164
		138	.003037
		140	.002280
		144	.001960
		146	.001813
		148	.001505
		150	.001111
		152	.000741
16	15	100	.104127
		101	.093326
		102	.082525
		103	.072281
		104	.063254
		105	.056303
		112	.052497
		113	.050866
		114	.047564
		115	.043024
		116	.037753
		117	.032243
		118	.026972
		119	.022432
		120	.019130
		128	.017499
		129	.016861

n	m	c	$P(D_{mn} \geq \frac{c}{mn})$
16	15	130	.015468
		131	.013488
		132	.011199
		133	.008910
		134	.006931
		135	.005538
		144	.004900
		145	.004674
		146	.004158
		147	.003423
		148	.002608
		149	.001872
		150	.001356
		160	.001131
		161	.001060
		162	.000896
16	16	96	.214535
		112	.093326
		128	.034998
		144	.011199
		160	.003015
		176	.000670
17	1	17	.111111
17	2	28	.116959
		30	.070175
		32	.035088
		34	.011696
17	3	34	.119298
		36	.098246
		39	.061404
		42	.035088
		45	.017544
		48	.007018
		51	.001754
17	4	43	.106934

n	m	c	$P(D_{mn} \geq \frac{c}{mn})$
17	4	44	.084879
		47	.058145
		48	.046115
		51	.027736
		52	.023392
		56	.011696
		60	.005013
		64	.001671
		68	.000334
17	5	48	.119769
		50	.093947
		51	.071846
		53	.064327
		55	.047619
		58	.032809
		60	.023696
		63	.014506
		65	.010557
		68	.005316
		70	.004253
		75	.001595
		80	.000456
17	6	55	.106214
		56	.084222
		60	.067382
		61	.054484
		62	.040318
		66	.032472
		67	.024488
		68	.016405
		72	.014146
		73	.009569
		78	.005706
		79	.003328
		84	.001942
		85	.000852
17	7	60	.119236
		61	.098901
		63	.087390

n	m	c	$P(D_{mn} \geq \frac{c}{mn})$
17	7	64	.073747
		67	.059011
		68	.045836
		70	.042282
		71	.033944
		74	.026316
		77	.018688
		78	.013938
		81	.010991
		84	.007177
		85	.004686
		88	.004010
		91	.002433
		95	.001260
		98	.000774
17	8	64	.113647
		68	.097407
		69	.092665
		70	.081348
		71	.067368
		72	.053688
		77	.043853
		78	.039137
		79	.031510
		80	.023540
		85	.017981
		86	.016822
		87	.013405
		88	.009275
		94	.006363
		95	.005231
		96	.003390
		102	.001997
		103	.001781
		104	.001108
		111	.000512
17	9	73	.105147
		74	.091148
		75	.076389

n	m	c	$P(D_{mn} \geq \frac{c}{mn})$
17	9	76	.063331
		81	.055122
		82	.049374
		83	.040905
		84	.032264
		85	.025735
		90	.023607
		91	.020496
		92	.015863
		93	.011526
		99	.009118
		100	.007563
		101	.005320
		102	.003558
		108	.003071
		109	.002363
		110	.001431
		117	.000907
17	10	76	.113709
		79	.097400
		80	.083744
		82	.075967
		83	.066636
		85	.056332
		86	.053460
		89	.043632
		90	.036184
		92	.031952
		93	.028261
		96	.022502
		99	.017310
		100	.014046
		102	.011907
		103	.011112
		106	.008228
		109	.005891
		110	.004850
		113	.003856
		116	.002582
		119	.001736

n	m	c	$P(D_{mn} \geq \frac{c}{mn})$
17	10	120	.001555
		123	.001137
		126	.000656
17	11	82	.104893
		85	.092067
		86	.088373
		87	.077516
		88	.065956
		91	.060001
		92	.054846
		93	.045953
		97	.038549
		98	.033442
		99	.026994
		102	.023931
		103	.022752
		104	.018423
		108	.014564
		109	.012826
		110	.009697
		114	.008238
		115	.006553
		119	.004753
		120	.004433
		121	.003154
		125	.002518
		126	.002046
		131	.001310
		132	.000892
17	12	88	.105677
		90	.092664
		91	.085599
		93	.076298
		95	.065505
		96	.057963
		98	.053624
		100	.045506
		102	.038549
		103	.036843

n	m	c	$P(D_{mn} \geq \frac{c}{mn})$
17	12	105	.031766
		107	.025995
		108	.023274
		110	.021148
		112	.016922
		115	.013981
		117	.011475
		119	.008894
		120	.008365
		122	.007401
		124	.005496
		127	.004621
		129	.003518
		132	.002604
		134	.002204
		136	.001483
		139	.001343
		141	.000920
17	13	93	.101324
		96	.091327
		97	.081024
		100	.071188
		101	.064195
		102	.055268
		104	.053001
		105	.049718
		106	.042187
		109	.038370
		110	.032663
		113	.028231
		114	.024598
		117	.020261
		118	.018722
		119	.014999
		122	.014217
		123	.011436
		126	.010036
		127	.008341
		130	.006702
		131	.006037
		135	.004508

n	m	c	$P(D_{mn} \geq \frac{c}{mn})$
17	13	136	.003338
		139	.003102
		140	.002400
		143	.001965
		144	.001703
		148	.001212
		152	.000802
17	14	98	.101250
		100	.095889
		102	.084647
		103	.081858
		105	.072085
		106	.067196
		108	.059338
		109	.053322
		111	.047890
		112	.041541
		114	.038959
		117	.032906
		119	.027660
		120	.026601
		122	.022513
		123	.020498
		125	.017835
		126	.015276
		128	.014126
		131	.011459
		134	.009057
		136	.007190
		137	.006830
		139	.005658
		140	.004937
		142	.004465
		145	.003551
		148	.002646
		151	.001914
		153	.001457
		154	.001349
		156	.001173
		159	.000954

n	m	c	$P(D_{mn} \geq \frac{c}{mn})$
17	15	104	.100108
		105	.094079
		106	.089695
		108	.082021
		110	.073581
		112	.065005
		114	.056808
		116	.049457
		118	.043428
		119	.039208
		120	.037836
		121	.035809
		123	.033096
		125	.029478
		127	.025428
		129	.021374
		131	.017699
		133	.014739
		135	.012766
		136	.011899
		138	.011389
		140	.010310
		142	.008830
		144	.007190
		146	.005631
		148	.004365
		150	.003543
		153	.003204
		155	.003033
		157	.002656
		159	.002140
		161	.001596
		163	.001133
		165	.000832
17	16	108	.106813
		109	.095899
		110	.085668
		111	.076716
		112	.069855
		119	.066107
		120	.064459

n	m	c	$P(D_{mn} \geq \frac{c}{mn})$
17	16	121	.061098
		122	.056420
		123	.050893
		124	.044953
		125	.039013
		126	.033486
		127	.028808
		128	.025447
		136	.023799
		137	.023131
		138	.021651
		139	.019499
		140	.016921
		141	.014200
		142	.011623
		143	.009470
		144	.007990
		153	.007323
		154	.007075
		155	.006494
		156	.005630
		157	.004611
		158	.003591
		159	.002727
		160	.002146
		170	.001898
		171	.001816
		172	.001615
		173	.001319
		174	.000987
17	17	119	.112377
		136	.044953
		153	.015561
		170	.004611
		187	.001153
		204	.000238
18	1	18	.105263

n	m	c	$P(D_{mn} \geq \frac{c}{mn})$
18	2	30	.105263
		32	.063158
		34	.031579
		36	.010526
18	3	36	.144361
		39	.084211
		42	.052632
		45	.030075
		48	.015038
		51	.006015
		54	.001504
18	4	44	.120301
		46	.090226
		48	.070540
		50	.048667
		52	.038004
		54	.022967
		56	.019139
		60	.009569
		64	.004101
		68	.001367
		72	.000273
18	5	50	.123748
		52	.098725
		54	.076496
		55	.069958
		57	.052424
		60	.038159
		62	.026568
		65	.018842
		67	.011650
		70	.008321
		72	.004220
		75	.003328
		80	.001248
		85	.000357
18	6	60	.101013

n	m	c	$P(D_{mn} \geq \frac{c}{mn})$
18	6	66	.053449
		72	.025409
		78	.010922
		84	.004369
		90	.001471
		96	.000416
18	7	63	.110828
		65	.094862
		66	.085126
		69	.068359
		70	.056335
		72	.046087
		73	.043112
		76	.032540
		77	.026362
		80	.020171
		83	.014179
		84	.011808
		87	.008325
		90	.005351
		91	.004810
		94	.003004
		98	.001781
		101	.000936
18	8	70	.104309
		72	.087822
		74	.072318
		76	.062524
		78	.051003
		80	.040479
		82	.033670
		84	.029657
		86	.023461
		88	.017391
		90	.013543
		92	.012570
		94	.009828
		96	.006709
		100	.004693

n	m	c	$P(D_{mn} \geq \frac{c}{mn})$
18	8	102	.003794
		104	.002413
		108	.001449
		110	.001279
		112	.000777
18	9	72	.172445
		81	.087945
		90	.041262
		99	.017302
		108	.006562
		117	.002169
		126	.000631
18	10	80	.116652
		82	.098786
		84	.086467
		86	.073765
		88	.062572
		90	.054754
		92	.047211
		94	.039806
		96	.032285
		98	.026300
		100	.023259
		102	.020511
		104	.016439
		106	.012523
		108	.009943
		110	.009297
		112	.007918
		114	.005895
		116	.004154
		120	.003343
		122	.002704
		124	.001817
		126	.001193
		130	.001054
		132	.000785
18	11	86	.108007
		88	.097457

n	m	c	$P(D_{mn} \geq \frac{c}{mn})$
18	11	89	.090636
		90	.078330
		92	.075106
		93	.066098
		96	.057057
		97	.048364
		99	.044009
		100	.040278
		103	.033273
		104	.028167
		107	.024007
		108	.019242
		110	.018265
		111	.016358
		114	.013011
		115	.010376
		118	.008987
		121	.006708
		122	.005802
		125	.004520
		126	.003301
		129	.003049
		132	.002130
		133	.001734
		136	.001380
		140	.000885
18	12	90	.135719
		96	.094507
		102	.062627
		108	.041714
		114	.026256
		120	.016286
		126	.009709
		132	.005713
		138	.003127
		144	.001735
		150	.000889
18	13	97	.110082
		99	.097862

n	m	c	$P(D_{mn} \geq \frac{c}{mn})$
18	13	100	.089267
		102	.081130
		104	.070992
		105	.066971
		107	.059170
		108	.051540
		110	.049686
		112	.042780
		113	.038005
		115	.034959
		117	.029316
		118	.027314
		120	.023869
		123	.019770
		125	.016514
		126	.014096
		128	.013483
		130	.010775
		131	.009848
		133	.008779
		136	.006836
		138	.005625
		141	.004519
		143	.003432
		144	.003035
		146	.002857
		149	.002059
		151	.001742
		154	.001292
		156	.000947
18	14	102	.110113
		104	.099766
		106	.089996
		108	.080252
		110	.071243
		112	.062511
		114	.055332
		116	.048003
		118	.043155
		120	.037618

n	m	c	$P(D_{mn} \geq \frac{c}{mn})$
18	14	122	.032663
		124	.029143
		126	.024587
		128	.022239
		130	.018435
		132	.016846
		134	.014047
		136	.012091
		138	.010401
		140	.008433
		142	.007774
		144	.006075
		146	.005797
		148	.004548
		150	.004007
		152	.003276
		154	.002613
		156	.002344
		160	.001714
		162	.001233
		164	.001154
		166	.000872
18	15	108	.109357
		111	.095611
		114	.081119
		117	.067231
		120	.055182
		123	.045956
		126	.039765
		129	.033368
		132	.027118
		135	.021636
		138	.017403
		141	.014420
		144	.011850
		147	.009502
		150	.007468
		153	.005882
		156	.004701
		159	.003632
		162	.002814

n	m	c	$P(D_{mn} \geq \frac{c}{mn})$
18	15	165	.002191
		168	.001729
		171	.001365
		174	.000997
18	16	114	.101697
		116	.093030
		118	.084068
		120	.075263
		122	.067009
		124	.059693
		126	.053734
		128	.048169
		130	.043353
		132	.039523
		134	.035133
		136	.030585
		138	.026230
		140	.022390
		142	.019351
		144	.017348
		146	.015931
		148	.014753
		150	.013091
		152	.011171
		154	.009227
		156	.007477
		158	.006106
		160	.005235
		162	.004882
		164	.004690
		166	.004252
		168	.003623
		170	.002910
		172	.002232
		174	.001690
		176	.001351
		180	.001220
		182	.001159
		184	.001016
		186	.000813

n	m	c	$P(D_{mn} \geq \frac{c}{mn})$
18	17	117	.100158
		118	.091337
		119	.084593
		126	.080914
		127	.079262
		128	.075873
		129	.071118
		130	.065432
		131	.059214
		132	.052822
		133	.046604
		134	.040918
		135	.036164
		136	.032774
		144	.031121
		145	.030433
		146	.028888
		147	.026605
		148	.023806
		149	.020748
		150	.017689
		151	.014890
		152	.012607
		153	.011062
		162	.010374
		163	.010109
		164	.009473
		165	.008502
		166	.007306
		167	.006032
		168	.004836
		169	.003864
		170	.003229
		180	.002964
		181	.002871
		182	.002637
		183	.002275
		184	.001840
		185	.001404
		186	.001043
		187	.000809
18	18	126	.132394
		144	.056018
		162	.020748
		180	.006669
		198	.001840
		216	.000429
19	1	19	.100000
19	2	30	.142857
		32	.095238
		34	.057143
		36	.028571
		38	.009524
19	3	39	.109091
		42	.072727
		45	.045455
		48	.025974
		51	.012987
		54	.005195
		57	.001299
19	4	48	.101637
		49	.076793
		52	.059176
		53	.041107
		56	.031621
		57	.019198
		60	.015810
		64	.007905
		68	.003388
		72	.001129
		76	.000226
19	5	55	.101967
		56	.082157
		57	.062959
		60	.057030
		61	.043149
19	5	65	.030915
		66	.021739
		70	.015152
		71	.009458
		75	.006635
		76	.003388
		80	.002635
		85	.000988
19	6	60	.101073
		64	.081988
		65	.070807
		66	.055776
		70	.042891
		71	.037741
		72	.028391
		76	.020124
		77	.018656
		78	.013518
		83	.008538
		84	.006166
		89	.003388
		90	.002541
		95	.001129
		96	.000949
19	7	67	.104789
		69	.087990
		70	.079477
		72	.067625
		74	.054147
		76	.043913
		77	.041338
		79	.033804
		81	.025375
		84	.020176
		86	.015652
		88	.010909
		91	.008933
		93	.006388
		95	.004047
19	7	98	.003600
		100	.002280
		105	.001323
		107	.000705
19	8	72	.110024
		74	.097136
		76	.081731
		77	.078074
		79	.065074
		80	.056764
		82	.048636
		85	.039123
		87	.030977
		88	.027608
		90	.022762
		93	.017708
		95	.013049
		96	.012240
		98	.009521
		101	.007308
		104	.004930
		106	.003512
		109	.002793
		112	.001746
		114	.001068
		117	.000932
19	9	79	.106682
		80	.091687
		81	.077825
		86	.068094
		87	.063280
		88	.055158
		89	.045798
		90	.037093
		95	.031323
		96	.030097
		97	.026284
		98	.021112
		99	.016095
		105	.012869

n	m	c	$P(D_{mn} \geq \frac{c}{mn})$
19	9	106	.011549
		107	.009098
		108	.006481
		114	.004796
		115	.004532
		116	.003584
		117	.002372
		124	.001557
		125	.001287
		126	.000805
19	10	84	.102152
		85	.089346
		90	.081391
		91	.075523
		92	.066556
		93	.056648
		94	.047482
		95	.040842
		100	.038720
		101	.035369
		102	.030032
		103	.024251
		104	.019430
		110	.016893
		111	.015092
		112	.012182
		113	.009208
		114	.007159
		120	.006628
		121	.005724
		122	.004289
		123	.002980
		130	.002344
		131	.001926
		132	.001299
		133	.000835
19	11	91	.104353
		92	.093542
		94	.084670
19	11	95	.073532
		97	.070796
		99	.060967
		100	.056459
		102	.049143
		103	.042607
		105	.039008
		108	.032339
		110	.026886
		111	.024449
		113	.021132
		114	.017499
		116	.016704
		119	.013140
		121	.010569
		122	.009333
		124	.008240
		127	.006398
		130	.004725
		132	.003751
		133	.003168
		135	.002973
		138	.002133
		141	.001465
		143	.001198
		146	.000946
19	12	97	.100258
		99	.090192
		101	.079129
		102	.070974
		104	.066155
		106	.057750
		108	.049863
		109	.046471
		111	.041628
		113	.035164
		114	.030504
		116	.029383
		118	.025153
		120	.020748
		121	.019000
19	12	123	.017313
		125	.014031
		128	.011588
		130	.009886
		132	.007695
		133	.006854
		135	.006536
		137	.005134
		140	.003972
		142	.003479
		144	.002527
		147	.002152
		149	.001674
		152	.001178
		154	.001101
		156	.000753
19	13	101	.102963
		104	.096525
		105	.091885
		106	.082504
		107	.072631
		111	.065749
		112	.059915
		113	.052166
		114	.045779
		117	.044207
		118	.041771
		119	.036293
		120	.030846
		124	.028324
		125	.025147
		126	.020922
		130	.018158
		131	.016956
		132	.014073
		133	.011557
		137	.011059
		138	.009475
		139	.007483
		143	.006640
		144	.006087
19	13	145	.004738
		150	.003818
		151	.003102
		152	.002314
		156	.002176
		157	.001940
		158	.001392
		163	.001155
		164	.000866
19	14	107	.103315
		110	.095214
		111	.086863
		112	.077324
		114	.073519
		115	.071580
		116	.063826
		119	.056609
		120	.053341
		121	.046579
		124	.042041
		125	.038206
		126	.032726
		129	.030817
		130	.026999
		133	.023039
		134	.022359
		135	.018940
		138	.016605
		139	.015370
		140	.012605
		143	.011709
		144	.010228
		148	.008278
		149	.006836
		152	.005737
		153	.005524
		154	.004319
		157	.003928
		158	.003524
		162	.002674
		163	.002195

n	m	c	$P(D_{mn} \geq \frac{c}{mn})$
19	14	167	.001728
		168	.001289
		171	.001131
		172	.001073
		176	.000751
19	15	112	.109270
		114	.099985
		115	.097581
		116	.088050
		118	.081699
		119	.077499
		120	.069002
		122	.066002
		123	.060853
		126	.053742
		127	.048324
		130	.042959
		131	.037754
		133	.034346
		134	.033431
		135	.028776
		137	.027331
		138	.025589
		141	.021761
		142	.019541
		145	.016750
		146	.014402
		149	.012726
		150	.010539
		152	.009888
		153	.009573
		156	.007761
		157	.007125
		160	.005824
		161	.005000
		164	.004250
		165	.003399
		168	.003126
		171	.002388
		172	.002290

n	m	c	$P(D_{mn} \geq \frac{c}{mn})$
19	15	175	.001755
		176	.001552
		179	.001249
		180	.000991
19	16	119	.105420
		120	.096345
		122	.089817
		123	.082536
		125	.075371
		126	.070343
		128	.063087
		129	.060696
		132	.053783
		133	.047578
		135	.046401
		136	.041171
		138	.038852
		139	.034843
		141	.031752
		142	.029168
		144	.025707
		145	.024599
		148	.021123
		151	.017934
		152	.015261
		154	.014823
		155	.012833
		157	.011903
		158	.010687
		160	.009406
		161	.008927
		164	.007493
		167	.006100
		170	.004906
		171	.004024
		173	.003876
		174	.003355
		176	.003024
		177	.002832
		180	.002372
		183	.001870

n	m	c	$P(D_{mn} \geq \frac{c}{mn})$
19	16	186	.001415
		189	.001071
		190	.000871
19	17	124	.104467
		126	.095277
		128	.086422
		130	.078217
		132	.070997
		133	.065142
		134	.063710
		135	.059631
		136	.056732
		137	.054777
		139	.050792
		141	.046150
		143	.041227
		145	.036348
		147	.031803
		149	.027865
		151	.024785
		152	.022772
		153	.022200
		154	.021326
		156	.020068
		158	.018257
		160	.016103
		162	.013831
		164	.011653
		166	.009759
		168	.008309
		170	.007403
		171	.007039
		173	.006829
		175	.006337
		177	.005607
		179	.004739
		181	.003852
		183	.003058
		185	.002451
		187	.002082
		190	.001942

n	m	c	$P(D_{mn} \geq \frac{c}{mn})$
19	17	192	.001871
		194	.001699
		196	.001442
		198	.001144
		200	.000860
19	18	126	.100357
		133	.096755
		134	.095108
		135	.091715
		136	.086928
		137	.081157
		138	.074771
		139	.068089
		140	.061408
		141	.055021
		142	.049250
		143	.044463
		144	.041069
		152	.039420
		153	.038717
		154	.037126
		155	.034746
		156	.031782
		157	.028466
		158	.025033
		159	.021717
		160	.018753
		161	.016373
		162	.014782
		171	.014079
		172	.013800
		173	.013120
		174	.012059
		175	.010715
		176	.009224
		177	.007733
		178	.006389
		179	.005328
		180	.004648
		190	.004369

n	m	c	$P(D_{mn} \geq \frac{c}{mn})$
19	18	191	.004267
		192	.004004
		193	.003583
		194	.003052
		195	.002482
		196	.001951
		197	.001531
		198	.001267
		209	.001165
		210	.001131
		211	.001039
		212	.000893
19	19	133	.153173
		152	.068089
		171	.026749
		190	.009224
		209	.002767
		228	.000714
20	1	19	.190476
		20	.095238
20	2	32	.129870
		34	.086580
		36	.051948
		38	.025974
		40	.008658
20	3	40	.110672
		42	.094862
		45	.063241
		48	.039526
		51	.022586
		54	.011293
		57	.004517
		60	.001129
20	4	48	.138340
		52	.086580
		56	.050066
20	4	60	.026539
		64	.013175
		68	.006588
		72	.002823
		76	.000941
20	5	55	.144099
		60	.084811
		65	.046941
		70	.025296
		75	.012309
		80	.005345
		85	.002108
		90	.000791
20	6	64	.100526
		66	.081892
		68	.067211
		70	.057308
		72	.044694
		74	.034783
		76	.030265
		78	.022473
		80	.016114
		82	.014837
		84	.010581
		88	.006750
		90	.004795
		94	.002658
		96	.001963
		100	.000877
20	7	71	.101301
		72	.085173
		73	.070680
		77	.063050
		78	.054271
		79	.043379
		80	.034650
		84	.032371
		85	.026790
20	7	86	.020026
		91	.015635
		92	.012283
		93	.008500
		98	.006844
		99	.004959
		100	.003101
		105	.002730
		106	.001752
		112	.000995
20	8	76	.115554
		80	.086906
		84	.061489
		88	.044055
		92	.030373
		96	.021137
		100	.013535
		104	.009244
		108	.005504
		112	.003676
		116	.002083
		120	.001282
		124	.000689
20	9	82	.105183
		84	.095223
		86	.083001
		88	.070689
		90	.060067
		91	.053356
		93	.049127
		95	.042240
		97	.034666
		99	.028063
		100	.024084
		102	.023023
		104	.019835
		106	.015700
		108	.011926
		111	.009702
20	9	113	.008611
		115	.006672
		117	.004717
		120	.003555
		122	.003340
		124	.002596
		126	.001696
		131	.001134
		133	.000924
20	10	90	.121809
		100	.062289
		110	.029047
		120	.012447
		130	.004797
		140	.001670
		150	.000509
20	11	94	.109147
		96	.096397
		98	.085492
		99	.077962
		100	.072693
		101	.070346
		103	.062420
		105	.053812
		107	.045885
		109	.039838
		110	.036826
		112	.033843
		114	.029157
		116	.024108
		118	.019784
		120	.017083
		121	.016426
		123	.014831
		125	.012265
		127	.009600
		129	.007590
		132	.006711
		134	.005909
		136	.004621

n	m	c	$P(D_{mn} \geq \frac{c}{mn})$
20	11	138	.003383
		140	.002631
		143	.002475
		145	.002099
		147	.001515
		149	.001025
		154	.000822
20	12	100	.114793
		104	.090629
		108	.072402
		112	.055532
		116	.042923
		120	.033012
		124	.024886
		128	.018267
		132	.013652
		136	.010038
		140	.007024
		144	.005180
		148	.003593
		152	.002421
		156	.001739
		160	.001147
		164	.000754
20	13	107	.104357
		108	.099258
		109	.089859
		110	.079780
		114	.072263
		115	.066924
		116	.059135
		117	.051907
		120	.048763
		121	.047535
		122	.042589
		123	.036588
		127	.032286
		128	.030353
		129	.026187
20	13	130	.022087
		134	.020464
		135	.018388
		136	.015218
		140	.012940
		141	.012562
		142	.010694
		143	.008595
		147	.007808
		148	.007181
		149	.005726
		154	.004617
		155	.003949
		156	.002992
		160	.002635
		161	.002533
		162	.001971
		167	.001481
		168	.001310
		169	.000929
20	14	112	.106338
		114	.094732
		116	.085996
		118	.076799
		120	.069802
		122	.062673
		124	.055417
		126	.049128
		128	.044085
		130	.038931
		132	.033647
		134	.030501
		136	.027502
		138	.023438
		140	.020403
		142	.018703
		144	.015936
		146	.013318
		148	.012323
		150	.010804
		152	.008789
20	14	154	.007632
		156	.007101
		158	.005760
		160	.004675
		162	.004506
		164	.003800
		166	.002935
		168	.002625
		170	.002394
		172	.001816
		176	.001461
		178	.001163
		180	.000852
20	15	120	.101918
		125	.078972
		130	.061089
		135	.046067
		140	.034688
		145	.026115
		150	.018992
		155	.013635
		160	.009931
		165	.007002
		170	.004850
		175	.003356
		180	.002268
		185	.001522
		190	.001018
		195	.000644
20	16	124	.108525
		128	.088882
		132	.072414
		136	.059773
		140	.048889
		144	.039312
		148	.031365
		152	.025115
		156	.019684
		160	.015421
20	16	164	.012102
		168	.009494
		172	.007191
		176	.005371
		180	.004099
		184	.003164
		188	.002339
		192	.001685
		196	.001222
		200	.000914
20	17	129	.105964
		130	.098631
		132	.091572
		133	.084008
		135	.079127
		136	.071719
		138	.069396
		140	.062473
		141	.061267
		143	.055105
		144	.052701
		146	.047528
		147	.044272
		149	.040315
		150	.036586
		152	.034040
		153	.030174
		155	.029083
		158	.025371
		160	.022056
		161	.021590
		163	.018855
		164	.017848
		166	.015831
		167	.014407
		169	.013182
		170	.011529
		172	.011048
		175	.009359
		178	.007805
		180	.006519

n	m	c	$P(D_{mn} \geq \frac{c}{mn})$
20	17	181	.006353
		183	.005423
		184	.005044
		186	.004503
		187	.003954
		189	.003755
		192	.003123
		195	.002502
		198	.001972
		200	.001588
		201	.001534
		203	.001316
		204	.001189
		206	.001113
		209	.000929
20	18	134	.107245
		136	.098454
		138	.090365
		140	.083279
		142	.076096
		144	.069156
		146	.063153
		148	.058329
		150	.053129
		152	.047854
		154	.042768
		156	.038114
		158	.034127
		160	.031035
		162	.028429
		164	.026238
		166	.024306
		168	.021961
		170	.019416
		172	.016872
		174	.014518
		176	.012518
		178	.011010
		180	.010080
		182	.009483

n	m	c	$P(D_{mn} \geq \frac{c}{mn})$
20	18	184	.008944
		186	.008126
		188	.007119
		190	.006040
		192	.005004
		194	.004115
		196	.003453
		198	.003060
		200	.002913
		202	.002833
		204	.002635
		206	.002326
		208	.001949
		210	.001561
		212	.001216
		214	.000957
20	19	143	.103674
		144	.097872
		145	.091398
		146	.084545
		147	.077570
		148	.070717
		149	.064244
		150	.058441
		151	.053654
		152	.050273
		160	.048635
		161	.047922
		162	.046299
		163	.043851
		164	.040766
		165	.037259
		166	.033542
		167	.029825
		168	.026317
		169	.023233
		170	.020785
		171	.019161
		180	.018448
		181	.018157
		182	.017442

n	m	c	$P(D_{mn} \geq \frac{c}{mn})$
20	19	183	.016308
		184	.014844
		185	.013171
		186	.011427
		187	.009754
		188	.008289
		189	.007156
		190	.006440
		200	.006149
		201	.006039
		202	.005750
		203	.005278
		204	.004661
		205	.003967
		206	.003273
		207	.002657
		208	.002184
		209	.001896
		220	.001785
		221	.001747
		222	.001641
		223	.001465
		224	.001238
		225	.000993
20	20	140	.174533
		160	.081058
		180	.033542
		200	.012299
		220	.003967
		240	.001116
		260	.000270
21	1	20	.181818
		21	.090909
21	2	34	.118577
		36	.079051
		38	.047431
		40	.023715
		42	.007905

n	m	c	$P(D_{mn} \geq \frac{c}{mn})$
21	3	42	.132411
		45	.083004
		48	.055336
		51	.034585
		54	.019763
		57	.009881
		60	.003953
		63	.000988
21	4	51	.119368
		52	.098814
		55	.074308
		56	.061028
		59	.042688
		60	.035573
		63	.022451
		64	.019921
		68	.011067
		72	.005534
		76	.002372
		80	.000791
21	5	59	.110672
		60	.090605
		63	.071146
		64	.066798
		65	.052964
		69	.038978
		70	.030465
		74	.020888
		75	.016327
		79	.010094
		80	.008148
		84	.004348
		85	.003831
		90	.001703
		95	.000638
21	6	66	.114456
		69	.082828
		72	.067018
		75	.046816

n	m	c	$P(D_{mn} \geq \frac{c}{mn})$
21	6	78	.036188
		81	.024506
		84	.017979
		87	.011919
		90	.008371
		93	.005392
		96	.003770
		99	.002108
		102	.001534
		105	.000689
21	7	70	.158390
		77	.092958
		84	.050546
		91	.025626
		98	.012253
		105	.005306
		112	.002095
		119	.000758
21	8	80	.104119
		81	.093363
		83	.082036
		84	.069411
		86	.066613
		88	.055957
		89	.048934
		91	.042909
		94	.034589
		96	.028214
		97	.023843
		99	.021450
		102	.016380
		104	.013048
		105	.010465
		107	.009900
		110	.007069
		112	.005631
		115	.004195
		118	.002776
		120	.002315

n	m	c	$P(D_{mn} \geq \frac{c}{mn})$
21	8	123	.001573
		126	.000954
21	9	87	.104544
		90	.083419
		93	.065254
		96	.055142
		99	.043288
		102	.032708
		105	.026562
		108	.020608
		111	.015143
		114	.011823
		117	.008961
		120	.006500
		123	.004957
		126	.003483
		129	.002492
		132	.001905
		135	.001230
		138	.000837
21	10	90	.105625
		95	.096151
		96	.091360
		97	.083114
		98	.073201
		99	.063035
		100	.054050
		105	.048239
		106	.046984
		107	.042974
		108	.037234
		109	.030984
		110	.025459
		116	.022065
		117	.020628
		118	.017783
		119	.014320
		120	.011171
		126	.009293
		127	.008996

n	m	c	$P(D_{mn} \geq \frac{c}{mn})$
21	10	128	.007847
		129	.006148
		130	.004499
		137	.003519
		138	.003179
		139	.002465
		140	.001683
		147	.001206
		148	.001149
		149	.000906
21	11	100	.105180
		101	.096070
		102	.085629
		103	.075170
		104	.065893
		105	.059305
		110	.057219
		111	.053749
		112	.048028
		113	.041400
		114	.035005
		115	.029950
		121	.027361
		122	.025404
		123	.022049
		124	.018209
		125	.014769
		126	.012545
		132	.011989
		133	.010942
		134	.009118
		135	.007112
		136	.00553[illegible]
		143	.004819
		144	.004292
		145	.003382
		146	.002459
		147	.001874
		154	.001749
		155	.001502
		156	.001092

n	m	c	$P(D_{mn} \geq \frac{c}{mn})$
21	11	157	.000729
21	12	105	.104608
		108	.090899
		111	.076649
		114	.062598
		117	.051564
		120	.044582
		123	.036438
		126	.029004
		129	.023742
		132	.019833
		135	.015730
		138	.012328
		141	.009918
		144	.007998
		147	.006228
		150	.004856
		153	.003693
		156	.002893
		159	.002254
		162	.001710
		165	.001217
		168	.000949
21	13	111	.103053
		113	.092605
		114	.088159
		116	.079397
		117	.071494
		119	.067855
		121	.059955
		122	.055409
		124	.050192
		126	.043880
		127	.042828
		129	.037496
		130	.033415
		132	.031446
		134	.026829
		135	.025212

n	m	c	$P(D_{mn} \geq \frac{c}{mn})$
21	13	137	.022272
		140	.018959
		142	.015987
		143	.014302
		145	.013292
		147	.010892
		148	.010580
		150	.009043
		153	.007580
		155	.006078
		156	.005577
		158	.005089
		161	.004002
		163	.003260
		166	.002745
		168	.002066
		169	.001986
		171	.001765
		174	.001350
		176	.001024
		179	.000894
21	14	119	.104783
		126	.074115
		133	.050449
		140	.033823
		147	.022043
		154	.014142
		161	.008698
		168	.005318
		175	.003113
		182	.001789
		189	.000988
21	15	123	.105955
		126	.090359
		129	.079998
		132	.067953
		135	.058188
		138	.049912
		141	.041592
21	15	144	.036499
		147	.030195
		150	.025673
		153	.021427
		156	.017643
		159	.014991
		162	.012163
		165	.010155
		168	.008278
		171	.006757
		174	.005563
		177	.004490
		180	.003576
		183	.002885
		186	.002289
		189	.001840
		192	.001482
		195	.001121
		198	.000917
21	16	129	.104211
		130	.095876
		131	.087498
		134	.083739
		135	.077517
		136	.070063
		139	.065480
		140	.061552
		141	.055222
		144	.050404
		145	.048746
		146	.043734
		147	.039102
		150	.038280
		151	.034726
		152	.030557
		155	.028995
		156	.026904
		157	.023388
		160	.021395
		161	.020593
21	16	162	.017870
		166	.015750
		167	.013890
		168	.011888
		171	.011603
		172	.010569
		173	.008852
		176	.008274
		177	.007909
		178	.006585
		182	.005828
		183	.004943
		187	.004146
		188	.003674
		189	.002952
		192	.002863
		193	.002707
		194	.002138
		198	.001949
		199	.001570
		203	.001322
		204	.001126
		208	.000873
21	17	134	.101417
		136	.095756
		137	.093080
		138	.085197
		141	.079022
		142	.072527
		145	.066315
		146	.061427
		147	.055512
		149	.054486
		150	.051379
		151	.046001
		153	.043985
		154	.042666
		155	.038031
		158	.035355
		159	.031644
		162	.028652
21	17	163	.025980
		166	.022962
		167	.021372
		168	.018563
		170	.018180
		171	.017565
		172	.015141
		175	.014330
		176	.012418
		179	.011298
		180	.009965
		183	.008700
		184	.007947
		187	.006693
		188	.006423
		189	.005311
		192	.005180
		193	.004299
		196	.004006
		197	.003403
		200	.002989
		201	.002661
		204	.002200
		205	.002089
		209	.001651
		210	.001294
		213	.001253
		214	.001008
		217	.000914
21	18	141	.100987
		144	.088531
		147	.078860
		150	.070766
		153	.062214
		156	.053740
		159	.045918
		162	.039251
		165	.034052
		168	.030197
		171	.026321

n	m	c	$P(D_{mn} \geq \frac{c}{mn})$
21	18	174	.022481
		177	.018909
		180	.015846
		183	.013419
		186	.011536
		189	.009862
		192	.008327
		195	.006939
		198	.005754
		201	.004797
		204	.004014
		207	.003292
		210	.002700
		213	.002222
		216	.001840
		219	.001533
		222	.001265
		225	.000998
21	19	146	.103367
		147	.096437
		148	.094987
		149	.089379
		150	.086409
		151	.082503
		152	.078348
		153	.076475
		155	.071524
		157	.066127
		159	.060561
		161	.055068
		163	.049865
		165	.045159
		167	.041160
		168	.038076
		169	.037465
		170	.035465
		171	.034094
		172	.033231
		174	.031202
		176	.028703
		178	.025935
21	19	180	.023089
		182	.020343
		184	.017858
		186	.015782
		188	.014235
		189	.013288
		190	.013049
		191	.012674
		193	.012095
		195	.011199
		197	.010072
		199	.008821
		201	.007563
		203	.006403
		205	.005436
		207	.004732
		209	.004318
		210	.004166
		212	.004078
		214	.003855
		216	.003499
		218	.003048
		220	.002557
		222	.002085
		224	.001685
		226	.001395
		228	.001229
		231	.001172
		233	.001142
		235	.001064
		237	.000938
21	20	153	.101962
		154	.094817
		155	.087671
		156	.080734
		157	.074232
		158	.068437
		159	.063675
		160	.060320
		168	.058698
		169	.057979
21	20	170	.056335
		171	.053840
		172	.050671
		173	.047025
		174	.043098
		175	.039079
		176	.035152
		177	.031506
		178	.028337
		179	.025842
		180	.024198
		189	.023479
		190	.023180
		191	.022435
		192	.021244
		193	.019681
		194	.017859
		195	.015905
		196	.013951
		197	.012130
		198	.010567
		199	.009375
		200	.008631
		210	.008331
		211	.008214
		212	.007904
		213	.007386
		214	.006695
		215	.005890
		216	.005045
		217	.004240
		218	.003549
		219	.003031
		220	.002720
		231	.002604
		232	.002561
		233	.002442
		234	.002239
		235	.001967
		236	.001657
		237	.001346
		238	.001074
21	20	239	.000871
21	21	147	.196312
		168	.094817
		189	.041089
		210	.015905
		231	.005467
		252	.001657
		273	.000439
22	1	21	.173913
		22	.086957
22	2	36	.108696
		38	.072464
		40	.043478
		42	.021739
		44	.007246
22	3	45	.104348
		48	.073043
		51	.048696
		54	.030435
		57	.017391
		60	.008696
		63	.003478
		66	.000870
22	4	54	.103679
		56	.084950
		58	.064214
		60	.052174
		62	.036656
		64	.030234
		66	.019130
		68	.016856
		72	.009365
		76	.004682
		80	.002007
		84	.000669

n	m	c	$P(D_{mn} \geq \frac{c}{mn})$
22	5	61	.104348
		63	.093720
		65	.076006
		66	.060151
		68	.056163
		70	.044023
		73	.032627
		75	.025195
		78	.017391
		80	.013427
		83	.008349
		85	.006664
		88	.003567
		90	.003122
		95	.001387
		100	.000520
22	6	68	.114148
		70	.095355
		72	.081977
		74	.068806
		76	.055349
		78	.047757
		80	.038573
		82	.029580
		84	.026161
		86	.020019
		88	.014525
		90	.013569
		92	.009662
		96	.006689
		98	.004348
		102	.002994
		104	.001688
		108	.001210
		110	.000547
22	7	76	.105153
		77	.089070
		81	.076198
		82	.071011
		83	.060931

n	m	c	$P(D_{mn} \geq \frac{c}{mn})$
22	7	84	.049675
		88	.040913
		89	.039380
		90	.033550
		91	.026204
		96	.020490
		97	.017664
		98	.013251
		103	.009703
		104	.008646
		105	.006243
		110	.004157
		111	.003921
		112	.002773
		118	.001625
		119	.001167
		125	.000584
22	8	82	.112291
		84	.097639
		86	.084006
		88	.074165
		90	.063628
		92	.053467
		94	.044468
		96	.039281
		98	.034054
		100	.027444
		102	.022113
		104	.020070
		106	.016838
		108	.012834
		110	.010072
		112	.009593
		114	.007694
		116	.005469
		120	.004285
		122	.003232
		124	.002122
		128	.001745
		130	.001201
		132	.000719

n	m	c	$P(D_{mn} \geq \frac{c}{mn})$
22	9	90	.105016
		91	.096345
		92	.083747
		95	.076119
		96	.066274
		99	.057369
		100	.051798
		101	.043480
		104	.040316
		105	.033913
		108	.029013
		109	.025586
		110	.020583
		113	.019806
		114	.015913
		117	.013693
		118	.011685
		122	.009007
		123	.006819
		126	.006073
		127	.004961
		131	.003726
		132	.002606
		135	.002461
		136	.001882
		140	.001416
		144	.000904
22	10	96	.105974
		98	.093607
		100	.083188
		102	.072357
		104	.065136
		106	.056758
		108	.048558
		110	.041762
		112	.036655
		114	.033196
		116	.028413
		118	.023433
		120	.019289
		122	.016956

n	m	c	$P(D_{mn} \geq \frac{c}{mn})$
22	10	124	.015732
		126	.013390
		128	.010659
		130	.008314
		132	.007023
		134	.006773
		136	.005837
		138	.004510
		140	.003288
		144	.002615
		146	.002340
		148	.001787
		150	.001211
		154	.000883
22	11	99	.158309
		110	.086611
		121	.043798
		132	.020582
		143	.008864
		154	.003507
		165	.001253
		176	.000404
22	12	108	.111324
		110	.098805
		112	.088401
		114	.079222
		116	.070183
		118	.062204
		120	.056235
		122	.050169
		124	.045102
		126	.039314
		128	.033759
		130	.029249
		132	.026508
		134	.024106
		136	.021150
		138	.017783
		140	.014717

n	m	c	$P(D_{mn} \geq \frac{c}{mn})$
22	12	142	.012547
		144	.011630
		146	.010702
		148	.009091
		150	.007302
		152	.005815
		154	.004975
		156	.004809
		158	.004341
		160	.003527
		162	.002671
		164	.002057
		168	.001823
		170	.001601
		172	.001224
		174	.000864
22	13	116	.100050
		117	.090183
		119	.085941
		120	.082164
		121	.073625
		124	.067324
		125	.060843
		128	.054079
		129	.050309
		130	.044075
		132	.041685
		133	.040804
		134	.035777
		137	.032047
		138	.028751
		141	.024726
		142	.023403
		143	.019842
		146	.018559
		147	.015971
		150	.013904
		151	.012561
		154	.010362
		155	.010111
		156	.008295

n	m	c	$P(D_{mn} \geq \frac{c}{mn})$
22	13	159	.007643
		160	.006513
		163	.005447
		164	.005052
		168	.003960
		169	.003149
		172	.002836
		173	.002442
		176	.001934
		177	.001872
		181	.001386
		182	.001079
		185	.000938
22	14	122	.101082
		124	.091630
		126	.083686
		128	.075378
		130	.068050
		132	.060861
		134	.054956
		136	.050287
		138	.044469
		140	.039597
		142	.035966
		144	.031952
		146	.027685
		148	.024990
		150	.022994
		152	.019817
		154	.017065
		156	.015774
		158	.013929
		160	.011647
		162	.010220
		164	.009607
		166	.008099
		168	.006676
		170	.006223
		172	.005547
		174	.004460
		176	.003760

n	m	c	$P(D_{mn} \geq \frac{c}{mn})$
22	14	178	.003658
		180	.003057
		182	.002390
		184	.002182
		186	.002005
		188	.001554
		192	.001237
		194	.001047
		196	.000768
22	15	129	.100023
		130	.091727
		131	.083323
		132	.076910
		135	.075373
		136	.072901
		137	.067027
		138	.060212
		139	.054413
		143	.051869
		144	.048454
		145	.043328
		146	.038385
		150	.035466
		151	.034170
		152	.030737
		153	.026796
		154	.023949
		158	.023425
		159	.021562
		160	.018683
		161	.016175
		165	.015240
		166	.014594
		167	.012746
		168	.010729
		173	.009624
		174	.008677
		175	.007220
		176	.006154
		180	.005993

n	m	c	$P(D_{mn} \geq \frac{c}{mn})$
22	15	181	.005689
		182	.004780
		183	.003885
		188	.003584
		189	.003137
		190	.002482
		195	.002125
		196	.001990
		197	.001585
		198	.001253
		203	.001209
		204	.001016
		205	.000760
22	16	134	.101322
		136	.093861
		138	.085786
		140	.077892
		142	.071086
		144	.064682
		146	.058924
		148	.052987
		150	.048419
		152	.044563
		154	.039674
		156	.036334
		158	.032684
		160	.028944
		162	.026672
		164	.023447
		166	.020908
		168	.019339
		170	.016692
		172	.015300
		174	.013667
		176	.011686
		178	.010974
		180	.009463
		182	.008166
		184	.007717
		186	.006455
		188	.005779

n	m	c	$P(D_{mn} \geq \frac{c}{mn})$
22	16	190	.005204
		192	.004261
		194	.004042
		196	.003451
		198	.002847
		200	.002778
		202	.002257
		204	.001954
		206	.001812
		208	.001415
		210	.001322
		212	.001140
		216	.000886
22	17	140	.101197
		142	.093098
		143	.088355
		145	.082407
		147	.075221
		148	.070133
		150	.066368
		152	.060275
		153	.055245
		154	.053652
		155	.052794
		157	.047964
		159	.043263
		160	.041604
		162	.038172
		164	.034004
		165	.031837
		167	.029814
		169	.026331
		170	.023954
		172	.023177
		174	.020496
		176	.018155
		177	.017845
		179	.016019
		181	.013896
		182	.013250

n	m	c	$P(D_{mn} \geq \frac{c}{mn})$
22	17	184	.012237
		186	.010459
		187	.009584
		189	.009226
		191	.007877
		194	.006912
		196	.006020
		198	.005091
		199	.004988
		201	.004516
		203	.003717
		204	.003491
		206	.003336
		208	.002727
		211	.002416
		213	.002019
		216	.001682
		218	.001479
		220	.001170
		221	.001139
		223	.001075
		225	.000832
22	18	146	.100158
		148	.092548
		150	.086065
		152	.079791
		154	.073484
		156	.067703
		158	.061788
		160	.056689
		162	.051354
		164	.047256
		166	.042681
		168	.039469
		170	.035818
		172	.032494
		174	.029871
		176	.026665
		178	.024699
		180	.021788
		182	.020310

n	m	c	$P(D_{mn} \geq \frac{c}{mn})$
22	18	184	.017835
		186	.016599
		188	.014666
		190	.013229
		192	.011890
		194	.010408
		196	.009655
		198	.008263
		200	.007850
		202	.006656
		204	.006324
		206	.005398
		208	.004914
		210	.004289
		212	.003724
		214	.003387
		216	.002818
		218	.002706
		220	.002199
		222	.002152
		224	.001755
		226	.001641
		228	.001375
		230	.001208
		232	.001069
		234	.000877
22	19	150	.106405
		152	.098601
		153	.096449
		154	.089081
		156	.087822
		157	.081084
		159	.078540
		160	.072598
		162	.069088
		163	.064128
		165	.060004
		166	.056226
		168	.051795
		169	.049371
		171	.044895

n	m	c	$P(D_{mn} \geq \frac{c}{mn})$
22	19	172	.043857
		175	.039559
		176	.035629
		178	.035116
		179	.031703
		181	.030568
		182	.027802
		184	.026146
		185	.024132
		187	.022127
		188	.020915
		190	.018741
		191	.018268
		194	.016094
		197	.014068
		198	.012308
		200	.012113
		201	.010711
		203	.010248
		204	.009261
		206	.008557
		207	.007993
		209	.007126
		210	.006922
		213	.005987
		216	.005077
		219	.004271
		220	.003630
		222	.003561
		223	.003119
		225	.002946
		226	.002704
		228	.002435
		229	.002353
		232	.002021
		235	.001673
		238	.001350
		241	.001088
		242	.000908
22	20	158	.100461
		160	.092457

n	m	c	$P(D_{mn} \geq \frac{c}{mn})$
22	20	162	.085600
		164	.080066
		166	.074294
		168	.068505
		170	.062891
		172	.057634
		174	.052919
		176	.048934
		178	.045248
		180	.041858
		182	.038900
		184	.036278
		186	.033331
		188	.030239
		190	.027169
		192	.024274
		194	.021698
		196	.019569
		198	.017996
		200	.016788
		202	.015799
		204	.014838
		206	.013605
		208	.012206
		210	.010749
		212	.009342
		214	.008082
		216	.007053
		218	.006314
		220	.005885
		222	.005634
		224	.005389
		226	.004989
		228	.004469
		230	.003881
		232	.003285
		234	.002739
		236	.002291
		238	.001974
		240	.001797
		242	.001736
		244	.001703
22	20	246	.001613
		248	.001463
		250	.001268
		252	.001053
		254	.000846
22	21	162	.105606
		163	.098380
		164	.091421
		165	.084935
		166	.079176
		167	.074456
		168	.071137
		176	.069533
		177	.068812
		178	.067157
		179	.064633
		180	.061407
		181	.057665
		182	.053588
		183	.049347
		184	.045106
		185	.041029
		186	.037287
		187	.034060
		188	.031537
		189	.029881
		198	.029160
		199	.028853
		200	.028087
		201	.026849
		202	.025208
		203	.023266
		204	.021141
		205	.018954
		206	.016829
		207	.014887
		208	.013246
		209	.012008
		210	.011241
		220	.010935
22	21	221	.010812
		222	.010483
		223	.009927
		224	.009170
		225	.008268
		226	.007290
		227	.006312
		228	.005409
		229	.004653
		230	.004097
		231	.003768
		242	.003645
		243	.003599
		244	.003468
		245	.003241
		246	.002927
		247	.002555
		248	.002162
		249	.001790
		250	.001476
		251	.001249
		252	.001118
		264	.001072
		265	.001056
		266	.001008
		267	.000924
22	22	176	.109262
		198	.049347
		220	.020047
		242	.007290
		264	.002358
		286	.000674
23	1	22	.166667
		23	.083333
23	2	38	.100000
		40	.066667
		42	.040000
		44	.020000
23	2	46	.006667
23	3	46	.104615
		48	.092308
		51	.064615
		54	.043077
		57	.026923
		60	.015385
		63	.007692
		66	.003077
		69	.000769
23	4	56	.113390
		57	.090598
		60	.073504
		61	.055840
		64	.044900
		65	.031681
		68	.025869
		69	.016410
		72	.014359
		76	.007977
		80	.003989
		84	.001709
		88	.000570
23	5	64	.107143
		65	.097619
		67	.079915
		69	.064245
		70	.060684
		72	.047558
		75	.036874
		77	.027513
		80	.021001
		82	.014591
		85	.011131
		87	.006960
		90	.005495
		92	.002951
		95	.002564

n	m	c	$P(D_{mn} \geq \frac{c}{mn})$
23	5	100	.001140
		105	.000427
23	6	72	.110690
		73	.096223
		74	.080077
		78	.068035
		79	.057589
		80	.046095
		84	.039316
		85	.032032
		86	.024391
		90	.021359
		91	.016488
		92	.011840
		96	.010997
		97	.007899
		102	.005393
		103	.003537
		108	.002400
		109	.001364
		114	.000964
23	7	78	.109803
		80	.099860
		82	.086309
		84	.072858
		85	.062989
		87	.058289
		89	.049476
		91	.040128
		92	.033410
		94	.032035
		96	.026980
		98	.020915
		101	.016534
		103	.014103
		105	.010477
		108	.007757
		110	.006853
		112	.004888
		115	.003289
23	7	117	.003089
		119	.002153
		124	.001273
		126	.000902
23	8	88	.100869
		89	.091934
		90	.080075
		91	.068428
		92	.059705
		96	.057480
		97	.051474
		98	.043307
		99	.035694
		104	.031146
		105	.027268
		106	.021986
		107	.017514
		112	.015739
		113	.013342
		114	.010159
		115	.007858
		120	.007446
		121	.006037
		122	.004276
		128	.003297
		129	.002515
		130	.001641
		136	.001330
		137	.000926
23	9	93	.108044
		94	.096267
		97	.084446
		98	.078023
		99	.067519
		102	.060894
		103	.053519
		106	.045491
		107	.042845
		108	.036041
23	9	111	.031834
		112	.027807
		115	.022668
		116	.022022
		117	.017999
		120	.015438
		121	.013625
		125	.010549
		126	.008421
		129	.006935
		130	.006330
		134	.004626
		135	.003649
		138	.002832
		139	.002715
		143	.001853
		144	.001485
		148	.001064
		152	.000673
23	10	100	.102819
		101	.095492
		104	.084355
		105	.074242
		107	.070460
		108	.062407
		110	.056158
		111	.051521
		114	.044443
		115	.037818
		117	.036852
		118	.031671
		120	.028711
		121	.025896
		124	.021905
		127	.017923
		128	.014794
		130	.013757
		131	.012124
		134	.010191
		137	.008027
		138	.006266
23	10	140	.006057
		141	.005156
		144	.004391
		147	.003347
		150	.002435
		151	.001966
		154	.001743
		157	.001311
		160	.000883
23	11	107	.104178
		108	.093461
		109	.083014
		110	.074006
		115	.068253
		116	.066993
		117	.062908
		118	.056898
		119	.049995
		120	.043142
		121	.037356
		127	.033888
		128	.032383
		129	.029301
		130	.025300
		131	.021150
		132	.017657
		138	.015659
		139	.015340
		140	.014053
		141	.012004
		142	.009703
		143	.007729
		150	.006632
		151	.006239
		152	.005342
		153	.004190
		154	.003155
		161	.002584
		162	.002516
		163	.002197

n	m	c	$P(D_{mn} \geq \frac{c}{mn})$
23	11	164	.001688
		165	.001189
		173	.000909
23	12	112	.106778
		113	.096235
		114	.087067
		115	.080619
		120	.078587
		121	.075076
		122	.069174
		123	.062097
		124	.054795
		125	.048116
		126	.042980
		132	.040389
		133	.038337
		134	.034704
		135	.030296
		136	.025871
		137	.022153
		138	.019831
		144	.019262
		145	.018115
		146	.016011
		147	.013472
		148	.011060
		149	.009294
		156	.008529
		157	.007919
		158	.006779
		159	.005442
		160	.004284
		161	.003609
		168	.003472
		169	.003164
		170	.002591
		171	.001957
		172	.001484
		180	.001297
		181	.001151
		182	.000887
23	13	119	.108699
		120	.099540
		122	.092265
		123	.085546
		125	.077403
		126	.074118
		129	.066158
		130	.059168
		132	.056380
		133	.050973
		135	.046462
		136	.043264
		138	.038131
		139	.037376
		142	.032464
		143	.028410
		145	.026851
		146	.024140
		148	.021506
		149	.020399
		152	.017400
		155	.014642
		156	.012582
		158	.011751
		159	.010660
		161	.009217
		162	.009010
		165	.007395
		168	.006006
		169	.005118
		171	.004697
		172	.004380
		175	.003644
		178	.002851
		181	.002237
		182	.001930
		184	.001729
		185	.001679
		188	.001331
		191	.000981
23	14	126	.103454
23	14	127	.099715
		128	.090945
		131	.083474
		132	.077944
		133	.070235
		136	.065917
		137	.059973
		138	.053888
		140	.052885
		141	.050776
		142	.045192
		145	.041261
		146	.037970
		147	.033312
		150	.031754
		151	.028180
		154	.024903
		155	.023767
		156	.020514
		159	.018859
		160	.017018
		161	.014516
		164	.014211
		165	.012225
		168	.010741
		169	.010159
		170	.008442
		173	.007940
		174	.006976
		178	.005805
		179	.004791
		182	.004254
		183	.003971
		184	.003163
		187	.003081
		188	.002611
		192	.002150
		193	.001679
		196	.001541
		197	.001411
		201	.001079
		202	.000868
23	15	133	.100109
		134	.091475
		135	.084139
		138	.081047
		139	.079826
		140	.074731
		141	.067950
		142	.061336
		146	.056927
		147	.054936
		148	.050354
		149	.044956
		150	.040448
		154	.038771
		155	.036520
		156	.032658
		157	.028690
		161	.026186
		162	.025780
		163	.023596
		164	.020564
		165	.017997
		169	.017131
		170	.016417
		171	.014509
		172	.012339
		177	.011004
		178	.010176
		179	.008663
		180	.007314
		184	.006889
		185	.006767
		186	.005976
		187	.004904
		192	.004239
		193	.004015
		194	.003359
		195	.002709
		200	.002512
		201	.002251
		202	.001780
		207	.001473

n	m	c	$P(D_{mn} \geq \frac{c}{mn})$
23	14	208	.001440
		209	.001202
		210	.000918
23	16	139	.100916
		141	.093175
		143	.085345
		144	.079061
		145	.076729
		146	.074048
		148	.068536
		150	.062157
		152	.056559
		153	.053401
		155	.050183
		157	.045368
		159	.040637
		160	.037447
		161	.036222
		162	.035655
		164	.032418
		166	.028654
		168	.025707
		169	.024665
		171	.022902
		173	.020155
		175	.017630
		176	.016353
		178	.015740
		180	.013976
		182	.011979
		184	.010683
		185	.010498
		187	.009593
		189	.008164
		191	.007000
		192	.006640
		194	.006349
		196	.005462
		198	.004524
		201	.004074
		203	.003640

n	m	c	$P(D_{mn} \geq \frac{c}{mn})$
23	16	205	.002970
		207	.002521
		208	.002467
		210	.002336
		212	.001928
		214	.001546
		217	.001436
		219	.001243
		221	.000964
23	17	145	.104133
		146	.096864
		147	.089230
		150	.084696
		151	.080774
		152	.074198
		153	.068122
		156	.066203
		157	.062069
		158	.056362
		161	.052017
		162	.051309
		163	.047280
		164	.042582
		167	.040036
		168	.038688
		169	.034980
		170	.031385
		173	.030412
		174	.028681
		175	.025459
		179	.023017
		180	.021157
		181	.018541
		184	.017201
		185	.016952
		186	.015152
		187	.013206
		190	.012737
		191	.012228
		192	.010628

n	m	c	$P(D_{mn} \geq \frac{c}{mn})$
23	17	196	.009358
		197	.008683
		198	.007378
		202	.006718
		203	.005990
		204	.005031
		207	.004817
		208	.004737
		209	.004051
		213	.003443
		214	.003271
		215	.002697
		219	.002394
		220	.002163
		221	.001738
		225	.001646
		226	.001401
		230	.001135
		231	.001111
		232	.000893
23	18	150	.106266
		152	.098457
		153	.092812
		155	.087484
		157	.080566
		158	.076987
		160	.071634
		161	.065772
		162	.064877
		163	.063360
		165	.058235
		166	.053591
		168	.051838
		170	.047136
		171	.043837
		173	.041513
		175	.037382
		176	.035438
		178	.032837
		180	.029406

n	m	c	$P(D_{mn} \geq \frac{c}{mn})$
23	18	181	.028660
		183	.026028
		184	.023397
		186	.023062
		188	.020580
		189	.018795
		191	.018085
		193	.015888
		194	.014899
		196	.013916
		198	.012104
		199	.011755
		201	.010637
		204	.009275
		206	.008151
		207	.007258
		209	.007142
		211	.006114
		212	.005644
		214	.005384
		216	.004528
		217	.004374
		219	.004003
		222	.003364
		224	.002942
		227	.002532
		229	.002119
		230	.001911
		232	.001874
		234	.001518
		235	.001454
		237	.001368
		240	.001098
		242	.000975
23	19	158	.100247
		159	.093666
		161	.087606
		162	.086534
		163	.080228
		165	.075654
		166	.073508

n	m	c	$P(D_{mn} \geq \frac{c}{mn})$
23	19	167	.067642
		169	.064726
		170	.061801
		171	.056536
		173	.055294
		174	.051910
		177	.047409
		178	.043843
		181	.040258
		182	.036729
		184	.034157
		185	.033732
		186	.030410
		188	.028882
		189	.027953
		190	.024982
		192	.024391
		193	.023057
		196	.020555
		197	.018970
		200	.017029
		201	.015346
		204	.014009
		205	.012364
		207	.011615
		208	.011458
		209	.009964
		211	.009697
		212	.009330
		215	.008075
		216	.007527
		219	.006568
		220	.005907
		223	.005266
		224	.004570
		227	.004227
		228	.003569
		230	.003455
		231	.003401
		234	.002838
		235	.002706
		238	.002276
23	19	239	.002074
		242	.001792
		243	.001549
		246	.001404
		247	.001155
		250	.001109
		253	.000887
23	20	162	.105985
		164	.098738
		165	.096173
		167	.089557
		168	.085993
		170	.080163
		171	.075936
		173	.071071
		174	.066475
		176	.062770
		177	.058047
		179	.055669
		180	.051028
		182	.050009
		184	.045630
		185	.045104
		187	.041140
		188	.039964
		190	.036543
		191	.034808
		193	.032046
		194	.029913
		196	.027907
		197	.025549
		199	.024345
		200	.021924
		202	.021455
		205	.019117
		207	.016988
		208	.016782
		210	.014961
		211	.014465
		213	.013028
23	20	214	.012261
		216	.011257
		217	.010291
		219	.009721
		220	.008648
		222	.008442
		225	.007356
		228	.006342
		230	.005469
		231	.005394
		233	.004713
		234	.004521
		236	.004057
		237	.003750
		239	.003498
		240	.003107
		242	.003022
		245	.002592
		248	.002170
		251	.001798
		253	.001506
		254	.001480
		256	.001284
		257	.001216
		259	.001113
		260	.001002
		262	.000969
23	21	169	.100427
		171	.094806
		173	.088895
		175	.082896
		177	.076984
		179	.071318
		181	.066056
		183	.061361
		184	.057409
		185	.056778
		186	.053745
		187	.052310
		188	.050351
		189	.048190
23	21	190	.047346
		192	.044629
		194	.041541
		196	.038252
		198	.034919
		200	.031686
		202	.028686
		204	.026047
		206	.023885
		207	.022297
		208	.022038
		209	.021075
		210	.020435
		211	.020057
		213	.019040
		215	.017719
		217	.016191
		219	.014562
		221	.012935
		223	.011407
		225	.010067
		227	.008989
		229	.008222
		230	.007782
		231	.007681
		232	.007521
		234	.007256
		236	.006819
		238	.006236
		240	.005559
		242	.004846
		244	.004157
		246	.003546
		248	.003057
		250	.002717
		252	.002530
		253	.002466
		255	.002429
		257	.002329
		259	.002157
		261	.001925

n	m	c	$P(D_{mn} \geq \frac{c}{mn})$
23	21	263	.001659
		265	.001388
		267	.001140
		269	.000940
23	22	172	.102731
		173	.096292
		174	.090591
		175	.085926
		176	.082649
		184	.081068
		185	.080347
		186	.078688
		187	.076149
		188	.072889
		189	.069084
		190	.064903
		191	.060504
		192	.056033
		193	.051634
		194	.047454
		195	.043648
		196	.040388
		197	.037850
		198	.036190
		207	.035468
		208	.035156
		209	.034373
		210	.033100
		211	.031397
		212	.029359
		213	.027096
		214	.024720
		215	.022344
		216	.020081
		217	.018044
		218	.016341
		219	.015068
		220	.014284
		230	.013972
		231	.013845
		232	.013500

n	m	c	$P(D_{mn} \geq \frac{c}{mn})$
23	22	233	.012911
		234	.012099
		235	.011113
		236	.010018
		237	.008886
		238	.007790
		239	.006804
		240	.005992
		241	.005404
		242	.005059
		253	.004931
		254	.004882
		255	.004741
		256	.004491
		257	.004139
		258	.003710
		259	.003241
		260	.002771
		261	.002342
		262	.001990
		263	.001740
		264	.001599
		276	.001550
		277	.001532
		278	.001479
		279	.001382
		280	.001244
		281	.001078
		282	.000902
23	23	184	.124295
		207	.058269
		230	.024720
		253	.009452
		276	.003241
		299	.000990
24	1	23	.160000
		24	.080000

n	m	c	$P(D_{mn} \geq \frac{c}{mn})$
24	2	38	.129231
		40	.092308
		42	.061538
		44	.036923
		46	.018462
		48	.006154
24	3	48	.123761
		51	.082051
		54	.057436
		57	.038291
		60	.023932
		63	.013675
		66	.006838
		69	.002735
		72	.000684
24	4	56	.145641
		60	.099145
		64	.063980
		68	.038877
		72	.022271
		76	.012308
		80	.006838
		84	.003419
		88	.001465
		92	.000488
24	5	66	.110934
		67	.092072
		70	.083230
		71	.068578
		72	.054684
		75	.051400
		76	.040537
		80	.031106
		81	.023359
		85	.017633
		86	.012328
		90	.009296
		91	.005844
		95	.004564

n	m	c	$P(D_{mn} \geq \frac{c}{mn})$
24	5	96	.002459
		100	.002122
		105	.000943
24	6	72	.148632
		78	.093224
		84	.056894
		90	.032622
		96	.017579
		102	.008987
		108	.004385
		114	.001940
		120	.000775
24	7	82	.103276
		84	.094346
		85	.083362
		88	.071398
		89	.060116
		91	.055924
		92	.048216
		95	.040502
		96	.032707
		98	.031488
		99	.026270
		102	.021878
		105	.016847
		106	.013455
		109	.011356
		112	.008363
		113	.006257
		116	.005481
		119	.003865
		120	.002626
		123	.002455
		126	.001688
		130	.001007
		133	.000704
24	8	88	.138756
		96	.082529

n	m	c	$P(D_{mn} \geq \frac{c}{mn})$
24	8	104	.046474
		112	.024923
		120	.012461
		128	.005836
		136	.002562
		144	.001025
		152	.000375
24	9	96	.110798
		99	.095531
		102	.077899
		105	.062748
		108	.053951
		111	.042742
		114	.034068
		117	.028426
		120	.021912
		123	.017319
		126	.014011
		129	.010607
		132	.008211
		135	.006474
		138	.004883
		141	.003561
		144	.002769
		147	.002077
		150	.001410
		153	.001114
		156	.000807
24	10	104	.101237
		106	.091855
		108	.082434
		110	.073056
		112	.064417
		114	.056049
		116	.050698
		118	.044375
		120	.038472
		122	.034282
		124	.028896

n	m	c	$P(D_{mn} \geq \frac{c}{mn})$
24	10	126	.026385
		128	.022373
		130	.019073
		132	.017048
		134	.013853
		136	.012981
		138	.010588
		140	.009001
		142	.007833
		144	.006111
		146	.005936
		148	.004601
		150	.003978
		152	.003337
		156	.002512
		158	.001825
		160	.001645
		162	.001312
		166	.000973
24	11	110	.104874
		111	.098561
		113	.091224
		115	.082475
		117	.073440
		119	.065035
		120	.058240
		121	.057108
		122	.053156
		124	.049544
		126	.044374
		128	.038636
		130	.033190
		132	.028861
		133	.026483
		135	.025171
		137	.022554
		139	.019269
		141	.016005
		143	.013410
		144	.012040
		146	.011765

n	m	c	$P(D_{mn} \geq \frac{c}{mn})$
24	11	148	.010685
		150	.009021
		152	.007227
		154	.005769
		157	.005017
		159	.004690
		161	.003969
		163	.003078
		165	.002317
		168	.001926
		170	.001870
		172	.001616
		174	.001224
		176	.000859
24	12	120	.113518
		132	.061219
		144	.030939
		156	.014514
		168	.006328
		180	.002537
		192	.000934
24	13	123	.107467
		125	.097859
		127	.088789
		129	.080938
		130	.075122
		131	.071941
		132	.069071
		134	.063785
		136	.057545
		138	.051197
		140	.045422
		142	.040860
		143	.038129
		144	.036286
		145	.035634
		147	.032411
		149	.028544
		151	.024682

n	m	c	$P(D_{mn} \geq \frac{c}{mn})$
24	13	153	.021383
		155	.019127
		156	.018193
		158	.017171
		160	.015312
		162	.013081
		164	.010938
		166	.009284
		168	.008387
		169	.008214
		171	.007672
		173	.006663
		175	.005473
		177	.004406
		179	.003700
		182	.003443
		184	.003170
		186	.002658
		188	.002077
		190	.001610
		192	.001369
		195	.001329
		197	.001199
		199	.000958
24	14	130	.105785
		132	.095701
		134	.087470
		136	.080255
		138	.074030
		140	.067007
		142	.060886
		144	.054625
		146	.049758
		148	.044789
		150	.040234
		152	.037002
		154	.032623
		156	.029937
		158	.026238
		160	.023889

n	m	c	$P(D_{mn} \geq \frac{c}{mn})$
24	14	162	.021264
		164	.018582
		166	.017238
		168	.014736
		170	.013742
		172	.011793
		174	.010499
		176	.009324
		178	.007859
		180	.007456
		182	.006159
		184	.005780
		186	.004881
		188	.004212
		190	.003791
		192	.003055
		194	.002989
		196	.002388
		198	.002205
		200	.001854
		202	.001532
		204	.001424
		208	.001087
		210	.000843
24	15	138	.101446
		141	.088856
		144	.076913
		147	.068853
		150	.059051
		153	.051522
		156	.044712
		159	.037858
		162	.033581
		165	.028408
		168	.024235
		171	.020892
		174	.017402
		177	.015038
		180	.012608
		183	.010548
		186	.008933

n	m	c	$P(D_{mn} \geq \frac{c}{mn})$
24	15	189	.007368
		192	.006205
		195	.005152
		198	.004266
		201	.003478
		204	.002857
		207	.002359
		210	.001905
		213	.001580
		216	.001237
		219	.001008
		222	.000829
24	16	144	.113086
		152	.081252
		160	.057552
		168	.039865
		176	.027198
		184	.018016
		192	.011808
		200	.007514
		208	.004697
		216	.002858
		224	.001712
		232	.000988
24	17	149	.104345
		151	.097103
		152	.094298
		153	.086955
		155	.084762
		156	.078660
		158	.073491
		159	.070122
		162	.064118
		163	.058750
		165	.055725
		166	.052245
		168	.047703
		169	.047095
		170	.042569

n	m	c	$P(D_{mn} \geq \frac{c}{mn})$
24	17	172	.041414
		173	.038097
		175	.035036
		176	.033898
		179	.030298
		180	.027327
		182	.025658
		183	.024225
		186	.021597
		187	.019094
		189	.018513
		190	.017008
		192	.015320
		193	.015114
		196	.013152
		197	.011733
		199	.010867
		200	.010452
		203	.009056
		204	.007832
		206	.007554
		207	.007015
		210	.006151
		213	.005189
		214	.004623
		216	.004202
		217	.004137
		220	.003460
		221	.002943
		223	.002817
		224	.002682
		227	.002273
		230	.001855
		231	.001678
		234	.001486
		237	.001190
		238	.001007
		240	.000953
24	18	156	.110901
		162	.088257
		168	.069975

n	m	c	$P(D_{mn} \geq \frac{c}{mn})$
24	18	174	.055279
		180	.042802
		186	.032837
		192	.025243
		198	.019057
		204	.014227
		210	.010564
		216	.007735
		222	.005625
		228	.004059
		234	.002863
		240	.002014
		246	.001413
		252	.000966
24	19	161	.103957
		164	.098537
		165	.092972
		166	.086315
		168	.082871
		169	.081955
		170	.076518
		171	.070869
		173	.069406
		174	.067594
		175	.062468
		178	.057983
		179	.055548
		180	.050887
		183	.047696
		184	.044922
		185	.040853
		188	.038968
		189	.036095
		190	.032729
		192	.032005
		193	.031652
		194	.028861
		197	.026284
		198	.025524
		199	.022954

n	m	c	$P(D_{mn} \geq \frac{c}{mn})$
24	19	202	.021206
		203	.020131
		204	.017891
		207	.016923
		208	.015668
		209	.013840
		212	.013499
		213	.012195
		216	.010832
		217	.010705
		218	.009461
		221	.008571
		222	.008279
		223	.007178
		226	.006711
		227	.006284
		228	.005386
		231	.005233
		232	.004729
		236	.004068
		237	.003551
		240	.003131
		241	.003089
		242	.002613
		245	.002402
		246	.002300
		247	.001905
		250	.001840
		251	.001688
		255	.001397
		256	.001219
		260	.001037
		261	.000860
24	20	168	.108431
		172	.094908
		176	.082043
		180	.070425
		184	.060528
		188	.052355
		192	.045055
		196	.038440
24	20	200	.032582
		204	.027586
		208	.023364
		212	.019569
		216	.016371
		220	.013695
		224	.011463
		228	.009558
		232	.007826
		236	.006360
		240	.005208
		244	.004302
		248	.003543
		252	.002852
		256	.002264
		260	.001799
		264	.001453
		268	.001178
		272	.000934
24	21	174	.107716
		177	.097637
		180	.087632
		183	.078156
		186	.069628
		189	.062402
		192	.056678
		195	.051727
		198	.046549
		201	.041339
		204	.036357
		207	.031859
		210	.028040
		213	.024973
		216	.022518
		219	.020102
		222	.017713
		225	.015432
		228	.013365
		231	.011596
		234	.010144
24	21	237	.008934
		240	.007836
		243	.006829
		246	.005906
		249	.005083
		252	.004377
		255	.003783
		258	.003263
		261	.002776
		264	.002361
		267	.002014
		270	.001727
		273	.001488
		276	.001283
		279	.001092
		282	.000903
24	22	180	.104221
		182	.098081
		184	.091961
		186	.086000
		188	.080332
		190	.075098
		192	.070445
		194	.065900
		196	.061451
		198	.057316
		200	.053696
		202	.050497
		204	.047053
		206	.043511
		208	.040005
		210	.036657
		212	.033588
		214	.030909
		216	.028728
		218	.026866
		220	.025239
		222	.023799
		224	.022403
		226	.020766
		228	.018989
24	22	230	.017170
		232	.015406
		234	.013782
		236	.012379
		238	.011262
		240	.010475
		242	.009920
		244	.009476
		246	.009003
		248	.008365
		250	.007607
		252	.006787
		254	.005964
		256	.005194
		258	.004527
		260	.004003
		262	.003644
		264	.003448
		266	.003342
		268	.003231
		270	.003038
		272	.002773
		274	.002458
		276	.002123
		278	.001798
		280	.001513
		282	.001289
		284	.001139
		286	.001059
		288	.001033
		290	.001020
		292	.000979
24	23	182	.102613
		183	.098013
		184	.094784
		192	.093226
		193	.092508
		194	.090851
		195	.088309
		196	.085032
		197	.081190

n	m	c	$P(D_{mn} \geq \frac{c}{mn})$
24	23	198	.076943
		199	.072437
		200	.067803
		201	.063170
		202	.058663
		203	.054417
		204	.050575
		205	.047298
		206	.044755
		207	.043097
		216	.042377
		217	.042062
		218	.041266
		219	.039965
		220	.038215
		221	.036103
		222	.033731
		223	.031203
		224	.028624
		225	.026096
		226	.023724
		227	.021612
		228	.019862
		229	.018561
		230	.017765
		240	.017450
		241	.017318
		242	.016960
		243	.016344
		244	.015485
		245	.014428
		246	.013232
		247	.011965
		248	.010699
		249	.009503
		250	.008446
		251	.007587
		252	.006970
		253	.006613
		264	.006481
		265	.006430

n	m	c	$P(D_{mn} \geq \frac{c}{mn})$
24	23	266	.006279
		267	.006009
		268	.005623
		269	.005143
		270	.004602
		271	.004040
		272	.003499
		273	.003019
		274	.002633
		275	.002362
		276	.002212
		288	.002160
		289	.002141
		290	.002082
		291	.001973
		292	.001815
		293	.001619
		294	.001401
		295	.001184
		296	.000988
24	24	192	.139823
		216	.067803
		240	.029914
		264	.011965
		288	.004321
		312	.001401
		336	.000406
25	1	24	.153846
		25	.076923
25	2	40	.119658
		42	.085470
		44	.056980
		46	.034188
		48	.017094
		50	.005698
25	3	51	.100733
		54	.073260

n	m	c	$P(D_{mn} \geq \frac{c}{mn})$
25	3	57	.051282
		60	.034188
		63	.021368
		66	.012210
		69	.006105
		72	.002442
		75	.000611
25	4	60	.109553
		63	.087154
		64	.073681
		67	.055998
		68	.047409
		71	.033851
		72	.029304
		75	.019283
		76	.017683
		80	.010610
		84	.005894
		88	.002947
		92	.001263
		96	.000421
25	5	70	.112388
		75	.071408
		80	.043816
		85	.026413
		90	.014905
		95	.007817
		100	.003817
		105	.001768
		110	.000786
25	6	77	.109469
		78	.093225
		82	.079043
		83	.070297
		84	.058317
		88	.047911
		89	.043022
		90	.034669
		94	.027264

n	m	c	$P(D_{mn} \geq \frac{c}{mn})$
25	6	95	.025091
		96	.019658
		100	.014576
		101	.013976
		102	.010743
		107	.007399
		108	.005688
		113	.003594
		114	.002820
		119	.001581
		120	.001309
		125	.000627
25	7	84	.112202
		86	.099887
		87	.092008
		90	.078644
		91	.068363
		93	.059493
		94	.055816
		97	.046192
		98	.039591
		100	.033405
		101	.032341
		104	.025781
		105	.022038
		108	.017878
		111	.013686
		112	.011903
		115	.009218
		118	.006733
		119	.006076
		122	.004419
		125	.003082
		126	.002938
		129	.001968
		133	.001336
		136	.000803
25	8	94	.100682
		95	.088662

n	m	c	$P(D_{mn} \geq \frac{c}{mn})$
25	8	96	.076842
		100	.068053
		101	.066520
		102	.060490
		103	.052232
		104	.043882
		109	.037883
		110	.034858
		111	.029649
		112	.024065
		117	.020115
		118	.018944
		119	.015973
		120	.012456
		125	.009953
		126	.009683
		127	.008199
		128	.006134
		134	.004616
		135	.003999
		136	.002885
		142	.002009
		143	.001815
		144	.001273
		150	.000797
25	9	100	.101883
		101	.099562
		103	.088723
		105	.077951
		107	.069038
		108	.064019
		110	.058462
		112	.050879
		114	.043503
		116	.038031
		117	.036001
		119	.032373
		121	.027324
		123	.022633
		125	.019696

n	m	c	$P(D_{mn} \geq \frac{c}{mn})$
25	9	126	.019209
		128	.016924
		130	.013745
		132	.011014
		135	.009713
		137	.008332
		139	.006452
		141	.005028
		144	.004599
		146	.003801
		148	.002769
		150	.002123
		153	.002041
		155	.001604
		157	.001085
		162	.000845
25	10	105	.127534
		110	.098255
		115	.074015
		120	.055659
		125	.040282
		130	.029641
		135	.020690
		140	.014833
		145	.010063
		150	.006911
		155	.004550
		160	.003018
		165	.001904
		170	.001235
		175	.000729
25	11	115	.102078
		117	.092152
		118	.085744
		120	.077885
		121	.070341
		123	.065827
		125	.058222
		126	.057228
		128	.050444

n	m	c	$P(D_{mn} \geq \frac{c}{mn})$
25	11	129	.047319
		131	.042115
		132	.037707
		134	.034906
		137	.030134
		139	.025796
		140	.024671
		142	.021400
		143	.019191
		145	.017519
		148	.014817
		150	.012240
		151	.012007
		153	.010063
		154	.009160
		156	.008202
		159	.006846
		162	.005441
		164	.004357
		165	.004086
		167	.003562
		170	.002979
		173	.002286
		175	.001723
		176	.001676
		178	.001405
		181	.001201
		184	.000898
25	12	119	.105353
		120	.096470
		125	.090833
		126	.089585
		127	.085500
		128	.079401
		129	.072200
		130	.064659
		131	.057510
		132	.051620
		138	.048139
		139	.046601

n	m	c	$P(D_{mn} \geq \frac{c}{mn})$
25	12	140	.043385
		141	.039066
		142	.034291
		143	.029705
		144	.026003
		150	.023937
		151	.023603
		152	.022224
		153	.019937
		154	.017173
		155	.014429
		156	.012228
		163	.011050
		164	.010617
		165	.009583
		166	.008141
		167	.006622
		168	.005388
		175	.004746
		176	.004669
		177	.004290
		178	.003626
		179	.002856
		180	.002208
		188	.001875
		189	.001774
		190	.001513
		191	.001161
		192	.000845
25	13	130	.102282
		131	.098782
		132	.092831
		133	.085555
		134	.077784
		135	.070184
		136	.063420
		137	.058296
		143	.055732
		144	.053631
		145	.049836

n	m	c	$P(D_{mn} \geq \frac{c}{mn})$
25	13	146	.045078
		147	.040016
		148	.035250
		149	.031380
		150	.029012
		156	.028438
		157	.027224
		158	.024927
		159	.022008
		160	.018966
		161	.016279
		162	.014391
		169	.013593
		170	.012920
		171	.011603
		172	.009935
		173	.008265
		174	.006933
		175	.006192
		182	.006045
		183	.005689
		184	.004978
		185	.004095
		186	.003271
		187	.002709
		195	.002498
		196	.002318
		197	.001959
		198	.001532
		199	.001174
		200	.000984
25	14	135	.102321
		136	.093839
		138	.090679
		140	.082729
		141	.079886
		143	.072996
		144	.068298
		146	.063000
		147	.057471
		149	.054347

n	m	c	$P(D_{mn} \geq \frac{c}{mn})$
25	14	150	.048745
		152	.048008
		154	.042925
		155	.041282
		157	.037180
		158	.034316
		160	.031604
		161	.028174
		163	.027075
		166	.023651
		168	.020698
		169	.019786
		171	.017647
		172	.015992
		174	.014878
		175	.012890
		177	.012680
		180	.010756
		182	.009218
		183	.008734
		185	.007787
		186	.006885
		188	.006554
		191	.005483
		194	.004498
		196	.003795
		197	.003550
		199	.003215
		200	.002753
		202	.002700
		205	.002166
		208	.001714
		210	.001443
		211	.001325
		213	.001239
		216	.001018
		219	.000776
25	15	140	.116581
		145	.095263
		150	.077431
		155	.062203

n	m	c	$P(D_{mn} \geq \frac{c}{mn})$
25	15	160	.049269
		165	.039166
		170	.030815
		175	.023740
		180	.018529
		185	.014119
		190	.010624
		195	.008089
		200	.005997
		205	.004415
		210	.003245
		215	.002353
		220	.001677
		225	.001198
		230	.000850
25	16	147	.108013
		149	.099938
		150	.092684
		151	.091537
		152	.086887
		154	.082105
		156	.075744
		158	.069369
		159	.064406
		160	.062528
		161	.060744
		163	.056419
		165	.051250
		167	.046585
		168	.043817
		170	.041682
		172	.037992
		174	.034039
		175	.031031
		176	.030645
		177	.029683
		179	.027594
		181	.024618
		183	.021843
		184	.020283

n	m	c	$P(D_{mn} \geq \frac{c}{mn})$
25	16	186	.019598
		188	.017739
		190	.015505
		192	.013802
		193	.013306
		195	.012501
		197	.010980
		199	.009457
		200	.008627
		202	.008509
		204	.007723
		206	.006584
		208	.005686
		209	.005441
		211	.005220
		213	.004544
		215	.003779
		218	.003364
		220	.003100
		222	.002583
		224	.002145
		225	.002030
		227	.001997
		229	.001745
		231	.001398
		234	.001203
		236	.001140
		238	.000936
25	17	155	.104044
		156	.096800
		157	.089608
		158	.083810
		162	.081319
		163	.077852
		164	.072389
		165	.066445
		166	.061271
		170	.058336
		171	.057008
		172	.053339
		173	.048687

n	m	c	$P(D_{mn} \geq \frac{c}{mn})$
25	17	174	.044267
		175	.041297
		179	.040767
		180	.038769
		181	.035413
		182	.031821
		183	.029060
		187	.028084
		188	.027390
		189	.025254
		190	.022513
		191	.020112
		196	.018909
		197	.017821
		198	.015911
		199	.013958
		200	.012726
		204	.012552
		205	.012204
		206	.011042
		207	.009570
		208	.008442
		213	.008100
		214	.007540
		215	.006538
		216	.005597
		221	.005164
		222	.004997
		223	.004409
		224	.003697
		225	.003253
		230	.003201
		231	.002930
		232	.002450
		233	.002058
		238	.001950
		239	.001875
		240	.001599
		241	.001293
		247	.001156
		248	.001034
		249	.000826

n	m	c	$P(D_{mn} \geq \frac{c}{mn})$
25	18	160	.105602
		162	.099020
		163	.096980
		164	.090091
		166	.086509
		167	.081702
		170	.075749
		171	.070149
		173	.066383
		174	.063562
		175	.058378
		177	.057728
		178	.053481
		180	.049806
		181	.048727
		182	.044395
		184	.043158
		185	.040290
		188	.036896
		189	.033475
		191	.031888
		192	.030321
		195	.027345
		196	.024851
		198	.023139
		199	.022593
		200	.020126
		202	.019897
		203	.018296
		206	.016627
		207	.014714
		209	.014244
		210	.013422
		213	.011918
		214	.010564
		216	.009936
		217	.009672
		220	.008411
		221	.007577
		224	.006893
		225	.005920
		227	.005846

n	m	c	$P(D_{mn} \geq \frac{c}{mn})$
25	18	228	.005439
		231	.004783
		232	.004107
		234	.003945
		235	.003823
		238	.003255
		239	.002850
		242	.002630
		245	.002185
		246	.001997
		249	.001758
		250	.001450
		252	.001427
		253	.001374
		256	.001152
		257	.000972
25	19	167	.103728
		168	.096817
		171	.092042
		172	.090394
		173	.085090
		174	.078963
		175	.074179
		178	.073371
		179	.069616
		180	.064380
		181	.059810
		184	.058232
		185	.056023
		186	.051760
		187	.047558
		190	.045469
		191	.044622
		192	.041399
		193	.037687
		197	.035351
		198	.033187
		199	.030060
		200	.027693
		203	.027391
		204	.026202

n	m	c	$P(D_{mn} \geq \frac{c}{mn})$
25	19	205	.023718
		206	.021475
		209	.020834
		210	.020418
		211	.018603
		212	.016596
		216	.015707
		217	.014541
		218	.012852
		222	.011836
		223	.011233
		224	.009908
		225	.008878
		228	.008774
		229	.008578
		230	.007634
		231	.006679
		235	.006442
		236	.005856
		237	.005039
		241	.004699
		242	.004410
		243	.003769
		247	.003377
		248	.003290
		249	.002838
		250	.002447
		254	.002413
		255	.002140
		256	.001792
		260	.001712
		261	.001583
		262	.001304
		266	.001188
		267	.001151
		268	.000954
25	20	175	.104611
		180	.088291
		185	.074128
		190	.062190
		195	.051598

n	m	c	$P(D_{mn} \geq \frac{c}{mn})$
25	20	200	.042712
		205	.035296
		210	.029080
		215	.023596
		220	.019016
		225	.015388
		230	.012451
		235	.009913
		240	.007816
		245	.006155
		250	.004862
		255	.003792
		260	.002934
		265	.002262
		270	.001735
		275	.001315
		280	.000996
25	21	181	.101642
		182	.097446
		183	.091256
		185	.088126
		186	.085444
		187	.079774
		189	.076083
		190	.074938
		191	.069915
		194	.065929
		195	.061680
		198	.057615
		199	.054242
		200	.050266
		202	.049803
		203	.047386
		204	.043631
		206	.042601
		207	.041165
		208	.037743
		210	.036230
		211	.035675
		212	.032686
		215	.030836

n	m	c	$P(D_{mn} \geq \frac{c}{mn})$
25	21	216	.028362
		219	.026326
		220	.024426
		223	.022339
		224	.021040
		225	.019019
		227	.018839
		228	.018114
		229	.016256
		231	.015823
		232	.015566
		233	.013948
		236	.013280
		237	.011960
		240	.011121
		241	.010133
		244	.009200
		245	.008548
		248	.007598
		249	.007254
		250	.006356
		252	.006290
		253	.006177
		254	.005386
		257	.005218
		258	.004575
		261	.004306
		262	.003832
		265	.003488
		266	.003185
		269	.002804
		270	.002650
		273	.002270
		274	.002223
		275	.001878
		278	.001856
		279	.001572
		282	.001512
		283	.001303
		286	.001205
		287	.001074

n	m	c	$P(D_{mn} \geq \frac{c}{mn})$
25	21	290	.000948
25	22	187	.100019
		189	.095179
		190	.090567
		192	.085496
		193	.081986
		195	.076866
		196	.074613
		198	.069601
		199	.068636
		200	.063869
		202	.063318
		203	.058915
		205	.057669
		206	.053731
		208	.051866
		209	.048495
		211	.046159
		212	.043455
		214	.040808
		215	.038853
		217	.036046
		218	.034876
		220	.032044
		221	.031589
		224	.028854
		225	.026319
		227	.026094
		228	.023847
		230	.023294
		231	.021408
		233	.020534
		234	.019066
		236	.017933
		237	.016918
		239	.015611
		240	.015039
		242	.013648
		243	.013443
		246	.012055
		249	.010747

n	m	c	$P(D_{mn} \geq \frac{c}{mn})$
25	22	250	.009583
		252	.009496
		253	.008529
		255	.008301
		256	.007565
		258	.007189
		259	.006697
		261	.006197
		262	.005934
		264	.005353
		265	.005265
		268	.004651
		271	.004051
		274	.003509
		275	.003057
		277	.003026
		278	.002686
		280	.002599
		281	.002378
		283	.002229
		284	.002116
		286	.001916
		287	.001880
		290	.001648
		293	.001408
		296	.001183
		299	.000993
25	23	193	.101495
		195	.095536
		197	.089903
		199	.084720
		200	.080125
		201	.079485
		202	.075630
		203	.074164
		204	.071212
		205	.068982
		206	.067077
		207	.064236
		208	.063416
		210	.060134

n	m	c	$P(D_{mn} \geq \frac{c}{mn})$
25	23	212	.056570
		214	.052864
		216	.049140
		218	.045513
		220	.042090
		222	.038976
		224	.036277
		225	.034089
		226	.033815
		227	.032220
		228	.031538
		229	.030576
		230	.029476
		231	.029100
		233	.027641
		235	.025912
		237	.024010
		239	.022028
		241	.020058
		243	.018184
		245	.016485
		247	.015033
		249	.013887
		250	.013085
		251	.012974
		252	.012518
		253	.012222
		254	.012057
		256	.011555
		258	.010867
		260	.010037
		262	.009120
		264	.008173
		266	.007254
		268	.006414
		270	.005701
		272	.005148
		274	.004773
		275	.004569
		276	.004527
		277	.004459

n	m	c	$P(D_{mn} \geq \frac{c}{mn})$
25	23	279	.004339
		281	.004127
		283	.003830
		285	.003468
		287	.003071
		289	.002671
		291	.002298
		293	.001979
		295	.001734
		297	.001572
		299	.001487
		300	.001460
		302	.001445
		304	.001399
		306	.001317
		308	.001200
		310	.001058
		312	.000906
25	24	203	.101035
		204	.097756
		205	.093898
		206	.089614
		207	.085039
		208	.080297
		209	.075500
		210	.070757
		211	.066183
		212	.061899
		213	.058041
		214	.054762
		215	.052224
		216	.050572
		225	.049856
		226	.049538
		227	.048732
		228	.047412
		229	.045626
		230	.043457
		231	.041001
		232	.038353
		233	.035611

n	m	c	$P(D_{mn} \geq \frac{c}{mn})$
25	24	234	.032869
		235	.030222
		236	.027766
		237	.025597
		238	.023810
		239	.022490
		240	.021685
		250	.021367
		251	.021232
		252	.020864
		253	.020225
		254	.019326
		255	.018209
		256	.016928
		257	.015546
		258	.014130
		259	.012749
		260	.011467
		261	.010350
		262	.009452
		263	.008813
		264	.008444
		275	.008309
		276	.008255
		277	.008097
		278	.007809
		279	.007393
		280	.006866
		281	.006261
		282	.005614
		283	.004968
		284	.004362
		285	.003836
		286	.003420
		287	.003132
		288	.002974
		300	.002919
		301	.002899
		302	.002835
		303	.002715
		304	.002538
		305	.002312

n	m	c	$P(D_{mn} \geq \frac{c}{mn})$
25	24	306	.002055
		307	.001787
		308	.001529
		309	.001304
		310	.001126
		311	.001006
		312	.000943
25	25	200	.155760
		225	.077898
		250	.035611
		275	.014838
		300	.005614
		325	.001921
		350	.000591

TABLE II. CRITICAL VALUES OF THE TWO-SAMPLE KOLMOGOROV-SMIRNOV CRITERION D_{mn}

N	M	.1	.05	.025	.01	.005	.001
1	1	-					
2	1	-					
	2	-					
3	1	-					
	2	-					
	3	9	-				
4	1	-					
	2	-					
	3	12	-				
	4	(16)	16	-			
5	1	-					
	2	10	-				
	3	(15)	15	-			
	4	16	(20)	20	-		
	5	20	(25)	(25)	25	-	
6	1	-					
	2	12	-				
	3	15	(18)	18	-		
	4	18	20	(24)	24	-	
	5	(24)	24	(30)	(30)	30	-
	6	(30)	30	(36)	(36)	36	-
7	1	-					
	2	14	-				
	3	18	(21)	21	-		
	4	21	24	(28)	28	-	
	5	25	28	30	(35)	35	-
	6	28	30	35	36	42	-
	7	35	(42)	(42)	42	(49)	49
8	1	-					
	2	(16)	16	-			
	3	(21)	21	24	-		
	4	24	(28)	28	(32)	32	-
	5	27	30	32	35	40	-
8	6	30	34	36	40	42	48
	7	34	40	41	(48)	48	56
	8	40	(48)	48	(56)	56	64
9	1	-					
	2	(18)	18	-			
	3	21	24	(27)	27	-	
	4	27	28	32	(36)	36	-
	5	30	35	36	40	(45)	45
	6	33	39	42	45	48	54
	7	36	42	45	49	54	63
	8	40	46	48	55	56	64
	9	(54)	54	(63)	63	(72)	72
10	1	-					
	2	18	20	-			
	3	24	27	(30)	30	-	
	4	28	30	(36)	36	40	-
	5	35	(40)	40	(45)	45	50
	6	36	40	44	48	50	60
	7	40	46	49	53	56	63
	8	44	48	54	60	62	70
	9	50	53	60	63	70	80
	10	60	(70)	70	(80)	80	90
11	1	-					
	2	20	22	-			
	3	27	(30)	30	33	-	
	4	29	33	36	40	44	-
	5	35	39	44	45	50	55
	6	38	43	48	54	55	66
	7	44	48	52	59	63	70
	8	48	53	58	64	66	77
	9	52	59	63	70	72	81
	10	57	60	68	77	79	89
	11	66	(77)	77	(88)	88	99
12	1	-					
	2	22	(24)	24	-		
	3	27	30	33	(36)	36	-
	4	(36)	36	40	44	48	-
	5	36	43	45	50	55	60
12	6	(48)	48	54	(60)	60	66
	7	46	53	56	60	65	72
	8	52	60	64	68	72	80
	9	57	63	69	75	78	87
	10	60	66	72	80	84	96
	11	64	72	76	86	88	99
	12	72	84	(96)	96	108	120
13	1	-					
	2	24	(26)	26	-		
	3	30	33	36	(39)	39	-
	4	35	39	44	(48)	48	52
	5	40	45	47	52	55	65
	6	46	52	54	60	65	72
	7	50	56	58	65	70	78
	8	54	62	65	72	78	88
	9	59	65	72	78	82	91
	10	64	70	77	84	90	100
	11	67	75	84	91	97	108
	12	71	81	84	95	104	117
	13	(91)	91	104	(117)	117	130
14	1	-					
	2	24	26	28	-		
	3	33	36	39	(42)	42	-
	4	38	42	44	48	52	56
	5	42	46	51	56	60	70
	6	48	54	58	64	66	78
	7	56	63	70	(77)	77	84
	8	58	64	70	76	82	90
	9	63	70	76	84	89	98
	10	68	74	82	90	96	106
	11	73	82	87	96	101	115
	12	78	86	94	104	108	120
	13	78	89	100	104	115	129
	14	98	(112)	112	(126)	126	154
15	1	-					
	2	26	28	30	-		
	3	33	36	39	42	45	-
	4	40	44	45	52	56	60
	5	50	(55)	55	60	65	70

N	M	.1	.05	.025	.01	.005	.001
15	6	51	57	63	69	72	84
	7	56	62	68	75	77	90
	8	60	67	74	81	88	97
	9	69	75	81	90	93	105
	10	75	80	90	100	105	115
	11	76	84	94	102	109	120
	12	84	93	99	108	117	129
	13	87	96	104	115	122	137
	14	92	98	110	123	125	140
	15	105	120	(135)	135	150	165
16	1	-					
	2	28	30	32	-		
	3	36	39	42	45	48	-
	4	44	48	52	56	60	64
	5	48	54	59	64	70	75
	6	54	60	64	72	74	84
	7	59	64	73	77	84	96
	8	72	(80)	80	88	96	104
	9	69	78	85	94	99	110
	10	76	84	90	100	108	118
	11	80	89	96	106	112	127
	12	88	96	104	116	124	136
	13	91	101	111	121	128	143
	14	96	106	116	126	136	152
	15	101	114	119	133	144	162
	16	112	128	144	(160)	160	176
17	1	-					
	2	30	32	34	-		
	3	36	42	45	48	51	-
	4	44	48	52	60	64	68
	5	50	55	60	68	70	80
	6	56	62	67	73	79	85
	7	61	68	77	84	85	98
	8	68	77	80	88	96	111
	9	74	82	90	99	102	117
	10	79	89	96	106	110	126
	11	85	93	102	110	119	132
	12	90	100	108	119	127	141
	13	96	105	114	127	135	152
	14	100	111	122	134	140	159

N	M	.1	.05	.025	.01	.005	.001
17	15	105	116	129	142	148	165
	16	109	124	136	143	157	174
	17	(136)	136	153	(170)	170	204
18	1	-					
	2	32	34	36	-		
	3	39	45	48	51	54	-
	4	46	50	54	60	64	72
	5	52	60	65	70	72	85
	6	66	72	78	(84)	84	96
	7	65	72	80	87	91	101
	8	72	80	86.	94	100	112
	9	81	90	99	108	117	126
	10	82	92	100	108	116	132
	11	88	97	107	118	125	140
	12	96	108	120	126	138	150
	13	99	110	120	131	141	156
	14	104	116	126	140	148	166
	15	111	123	135	147	156	174
	16	116	128	140	154	162	186
	17	118	133	148	164	168	187
	18	144	(162)	162	180	198	216
19	1	19	-				
	2	32	36	(38)	38	-	
	3	42	45	51	54	57	-
	4	49	53	57	64	68	76
	5	56	61	66	71	76	85
	6	64	70	76	83	89	96
	7	69	76	84	91	95	107
	8	74	82	90	98	104	117
	9	80	89	98	107	114	126
	10	85	94	103	113	122	133
	11	92	102	111	122	130	146
	12	99	108	120	130	140	156
	13	104	114	126	138	145	164
	14	110	121	133	148	154	176
	15	114	127	141	152	161	180
	16	120	133	145	160	170	190
	17	126	141	151	166	179	200
	18	133	142	159	176	180	212

N	M	.1	.05	.025	.01	.005	.001
19	19	152	171	(190)	190	209	228
20	1	20	-				
	2	34	38	(40)	40	-	
	3	42	48	51	(57)	57	-
	4	52	60	64	68	72	76
	5	60	65	75	80	85	90
	6	66	72	78	88	90	100
	7	72	79	86	93	99	112
	8	80	88	96	104	112	124
	9	84	93	100	111	117	133
	10	100	110	120	(130)	130	150
	11	96	107	116	127	136	154
	12	104	116	124	140	148	164
	13	108	120	130	143	154	169
	14	114	126	138	152	160	180
	15	125	135	150	160	170	195
	16	128	140	156	168	180	200
	17	130	146	160	175	186	209
	18	136	152	166	182	194	214
	19	144	160	169	187	204	225
	20	160	180	200	(220)	220	260
21	1	21	-				
	2	36	38	40	42	-	
	3	45	51	54	57	60	63
	4	52	59	63	72	76	80
	5	60	69	74	80	84	95
	6	69	75	81	90	96	105
	7	77	91	98	105	112	119
	8	81	89	97	107	115	126
	9	90	99	108	117	123	138
	10	95	105	116	126	130	149
	11	101	112	123	134	143	157
	12	108	120	129	141	150	168
	13	113	126	137	150	161	179
	14	126	140	147	161	175	189
	15	126	138	153	168	177	198
	16	130	145	157	173	183	208
	17	136	151	166	180	193	217
	18	144	159	174	189	201	225
	19	147	163	180	199	207	237

N	M	.1	.05	.025	.01	.005	.001
21	20	154	173	180	199	217	239
	21	168	189	210	231	252	273
22	1	22	-				
	2	38	40	42	44	-	
	3	48	51	57	60	63	66
	4	56	62	66	72	76	84
	5	63	70	78	83	88	100
	6	70	78	86	92	98	110
	7	77	84	96	103	110	125
	8	84	94	102	112	120	132
	9	91	101	110	122	127	144
	10	98	108	118	130	138	154
	11	110	121	132	143	154	176
	12	110	124	134	148	154	174
	13	117	130	141	156	168	185
	14	124	138	148	164	174	196
	15	130	144	154	173	182	205
	16	136	150	164	180	192	216
	17	142	157	170	187	199	225
	18	148	164	178	196	208	234
	19	152	169	185	204	219	242
	20	160	176	192	212	226	254
	21	163	183	203	223	229	267
	22	(198)	198	220	242	264	286
23	1	23	-				
	2	38	42	44	46	-	
	3	48	54	60	63	66	69
	4	57	64	69	76	80	88
	5	65	72	80	87	92	105
	6	73	80	86	97	103	114
	7	80	89	98	108	112	126
	8	89	98	106	115	122	137
	9	94	106	115	126	134	152
	10	101	114	124	137	144	160
	11	108	119	131	142	153	173
	12	113	125	137	149	160	182
	13	120	135	146	161	171	191
	14	127	142	154	170	179	202
	15	134	149	163	179	187	210
	16	141	157	169	187	198	221
	17	146	163	179	196	207	232
	18	152	170	184	204	216	242
	19	159	177	190	209	224	253
	20	164	184	199	219	233	262
	21	171	189	206	227	242	269
	22	173	194	214	237	253	282
	23	207	(230)	230	253	276	299
24	1	24	-				
	2	40	44	46	48	-	
	3	51	57	60	66	69	72
	4	60	68	72	80	84	92
	5	67	76	81	90	95	105
	6	78	90	96	102	108	120
	7	84	92	102	112	119	133
	8	96	104	112	128	136	152
	9	99	111	120	132	138	156
	10	106	118	128	140	148	166
	11	111	124	137	150	159	176
	12	132	144	156	168	180	192
	13	125	140	151	166	177	199
	14	132	146	160	176	186	210
	15	141	156	168	186	198	222
	16	152	168	184	200	208	232
	17	151	168	183	203	214	240
	18	162	180	198	216	228	252
	19	164	183	199	218	232	261
	20	172	192	208	228	244	272
	21	177	198	213	237	252	282
	22	182	204	222	242	258	292
	23	183	205	226	249	270	296
	24	216	240	264	(288)	288	336
25	1	25	-				
	2	42	46	48	50	-	
	3	54	60	63	69	72	75
	4	63	68	75	84	88	96
	5	75	80	90	95	100	110
	6	78	88	96	107	113	125
	7	86	97	105	115	122	136
	8	95	104	112	125	134	150
	9	101	114	123	135	144	162
	10	110	125	135	150	155	175
	11	117	129	140	154	164	184
	12	120	138	150	165	175	192
	13	131	145	158	172	184	200
	14	136	150	166	182	194	219
	15	145	160	175	195	205	230
	16	149	167	181	199	213	238
	17	156	173	190	207	222	249
	18	162	180	196	216	231	257
	19	168	187	205	224	241	268
	20	180	200	215	235	250	280
	21	182	202	220	244	258	290
	22	189	209	228	250	268	299
	23	195	216	237	262	274	312
	24	204	225	238	262	283	312
	25	225	250	275	300	325	350

N	M	.1	.05	.025	.01	.005	.001
26	1	26	-				
	2	44	48	50	52	-	
	3	54	60	66	72	75	78
	4	64	70	78	84	92	100
	5	73	80	89	99	104	115
	6	82	90	98	108	114	130
	7	90	100	109	119	126	142
	8	98	108	118	128	136	152
	9	103	118	128	138	147	164
	10	112	126	136	150	160	178
	11	120	132	145	160	171	190
	12	128	142	154	168	178	202
	13	143	156	169	(195)	195	221
	14	142	155	172	188	200	224
	15	148	163	178	199	208	234
	16	154	172	188	206	218	244
	17	160	178	194	212	228	255
	18	168	186	204	224	238	266
	19	174	195	212	233	247	278
	20	182	202	220	242	258	290
	21	187	208	228	249	265	300
	22	194	216	236	260	276	308
	23	200	223	243	267	284	319
	24	206	230	252	276	296	328
	25	215	240	263	289	297	344
	26	234	260	286	312	338	364
27	1	27	-				
	2	46	48	52	(54)	54	-
	3	57	63	69	72	75	81
	4	68	73	80	88	92	100
	5	76	85	93	100	105	120
	6	87	96	102	111	120	132
	7	93	101	113	121	128	147
	8	101	112	122	135	144	160
	9	117	126	135	153	162	180
	10	116	130	140	156	166	186
	11	123	135	150	162	176	194
	12	132	147	159	177	186	207
	13	139	154	167	181	194	220
	14	145	160	174	189	202	228
	15	153	171	186	204	216	243

N	M	.1	.05	.025	.01	.005	.001
27	16	159	176	192	213	224	254
	17	167	185	202	221	236	263
	18	180	198	216	234	252	279
	19	180	201	216	239	256	285
	20	186	206	226	250	265	297
	21	195	216	237	258	276	309
	22	199	222	242	268	285	320
	23	206	229	249	275	291	329
	24	213	237	258	285	303	342
	25	219	243	265	292	313	346
	26	226	252	277	304	312	359
	27	243	270	297	324	351	405
28	1	28	-				
	2	46	50	54	(56)	56	-
	3	57	66	69	75	78	84
	4	72	80	84	92	96	104
	5	79	87	95	102	107	125
	6	88	98	106	116	122	134
	7	98	112	119	133	140	154
	8	108	116	128	140	148	164
	9	113	125	135	150	159	178
	10	120	134	146	158	168	190
	11	129	142	154	170	181	202
	12	136	152	168	180	192	216
	13	142	159	174	191	202	228
	14	168	182	196	210	224	252
	15	158	175	190	209	222	248
	16	168	184	200	220	236	264
	17	172	190	207	228	244	273
	18	178	198	216	238	252	284
	19	185	204	223	248	260	296
	20	196	216	236	260	276	308
	21	203	224	245	273	287	322
	22	206	230	250	274	292	328
	23	211	236	257	284	302	340
	24	220	244	268	296	312	352
	25	224	250	273	300	320	360
	26	232	258	282	308	328	374
	27	237	264	290	318	343	375
	28	280	308	336	(364)	364	420

N	M	.1	.05	.025	.01	.005	.001
29	1	29	-				
	2	48	52	56	(58)	58	-
	3	58	66	72	78	81	87
	4	71	79	87	96	100	108
	5	81	87	96	106	111	125
	6	91	98	109	116	127	139
	7	98	111	119	132	140	154
	8	107	120	129	142	152	171
	9	116	129	140	153	162	185
	10	123	135	151	163	173	193
	11	132	145	159	174	187	209
	12	139	155	168	187	196	220
	13	145	164	177	195	206	232
	14	152	168	189	207	221	247
	15	158	182	198	215	229	258
	16	169	187	207	224	239	268
	17	176	197	214	236	251	280
	18	183	203	223	244	259	294
	19	189	209	228	254	271	303
	20	197	219	239	264	279	315
	21	204	227	248	273	290	327
	22	210	236	257	282	297	338
	23	218	241	264	292	310	345
	24	225	249	273	301	319	359
	25	231	259	281	310	330	371
	26	239	267	290	319	338	380
	27	246	273	297	329	347	393
	28	248	276	303	332	358	409
	29	290	319	348	377	406	435
30	1	30	-				
	2	50	54	56	(60)	60	-
	3	63	69	75	81	84	90
	4	74	82	88	96	104	112
	5	85	95	105	110	120	130
	6	96	108	114	126	132	150
	7	103	113	124	136	143	161
	8	112	124	134	148	156	172
	9	120	135	147	159	168	189
	10	140	150	160	180	190	210
	11	136	150	165	180	193	215
	12	150	162	180	192	204	234

N	M	.1	.05	.025	.01	.005	.001
30	13	150	167	183	201	214	240
	14	160	176	192	210	224	250
	15	180	195	210	240	255	270
	16	174	194	212	232	248	278
	17	180	202	220	242	255	289
	18	192	216	234	258	270	306
	19	195	217	236	262	279	311
	20	210	230	250	280	300	330
	21	210	234	255	282	300	336
	22	218	242	264	290	308	346
	23	223	250	273	298	319	358
	24	234	264	282	312	330	372
	25	240	270	290	320	340	385
	26	246	274	298	328	348	390
	27	255	282	309	339	360	405
	28	258	288	316	348	370	412
	29	259	288	316	346	372	428
	30	300	330	360	390	420	450
31	1	31	-				
	2	52	56	58	(62)	62	-
	3	63	72	78	84	87	93
	4	76	84	89	100	104	116
	5	85	95	104	114	119	135
	6	96	107	114	126	137	149
	7	105	117	127	140	147	165
	8	114	124	138	152	161	178
	9	123	136	149	163	172	194
	10	130	147	159	176	186	209
	11	140	154	169	184	195	222
	12	147	164	180	197	209	233
	13	155	173	190	208	221	247
	14	164	181	196	217	232	260
	15	173	191	207	224	239	269
	16	180	198	214	240	248	278
	17	186	207	225	248	262	298
	18	194	215	234	257	274	310
	19	201	223	244	268	284	320
	20	208	232	254	278	296	332
	21	215	237	263	287	307	340
	22	223	248	271	297	316	356
	23	229	256	279	306	328	367

N	M	.1	.05	.025	.01	.005	.001
31	24	238	264	288	317	336	377
	25	245	271	296	327	346	390
	26	252	279	304	335	356	402
	27	258	287	314	345	366	415
	28	266	296	322	355	377	427
	29	272	303	332	363	388	431
	30	284	316	344	378	388	444
	31	310	341	372	403	434	496
32	1	32	-				
	2	52	58	60	(64)	64	-
	3	64	75	81	87	90	96
	4	80	88	96	104	108	120
	5	88	96	108	118	123	140
	6	98	110	120	130	138	154
	7	108	119	132	143	153	171
	8	120	136	144	160	168	192
	9	126	142	153	170	179	198
	10	136	150	164	180	190	214
	11	145	158	170	190	201	223
	12	152	172	184	204	216	244
	13	159	178	192	211	224	254
	14	168	186	204	224	238	266
	15	176	194	213	234	249	281
	16	192	224	240	256	272	304
	17	190	212	231	254	269	301
	18	200	222	242	266	282	316
	19	207	228	251	276	291	329
	20	216	240	260	288	304	344
	21	219	248	270	293	313	354
	22	230	254	278	306	326	366
	23	236	263	287	315	336	378
	24	248	280	304	328	352	392
	25	250	279	304	334	355	398
	26	258	286	312	344	366	412
	27	265	296	323	355	377	424
	28	276	304	332	364	388	436
	29	278	310	339	373	396	446
	30	286	318	348	380	406	462
	31	296	329	360	394	422	461
	32	320	352	384	416	448	512

N	M	.1	.05	.025	.01	.005	.001
33	1	33	-				
	2	54	60	62	(66)	66	-
	3	69	75	81	87	93	99
	4	80	88	95	104	112	124
	5	92	102	110	120	127	145
	6	102	114	123	135	141	159
	7	111	123	135	149	156	175
	8	120	135	144	160	168	190
	9	132	147	159	174	183	204
	10	138	154	168	184	197	220
	11	154	176	187	209	220	242
	12	156	174	189	207	222	249
	13	165	182	199	219	233	260
	14	172	191	209	231	246	274
	15	180	201	219	240	255	288
	16	188	206	224	252	268	301
	17	194	214	240	260	277	311
	18	204	228	249	273	291	324
	19	211	235	257	282	301	338
	20	220	243	264	290	310	349
	21	228	252	276	303	324	363
	22	242	275	297	319	341	385
	23	241	269	294	322	344	387
	24	252	279	303	333	354	399
	25	255	286	312	344	362	411
	26	264	294	322	353	376	421
	27	273	303	330	363	387	435
	28	279	311	339	372	396	446
	29	286	317	346	383	406	457
	30	294	327	357	393	417	468
	31	300	333	363	400	424	482
	32	309	342	374	409	439	499
	33	330	(396)	396	(462)	462	528
34	1	34	-				
	2	56	60	64	66	68	-
	3	68	78	84	90	96	102
	4	82	90	98	108	116	124
	5	92	102	115	121	130	145
	6	104	116	126	136	146	162
	7	114	127	136	149	161	182
	8	124	138	150	166	176	196

N	M	.1	.05	.025	.01	.005	.001
34	9	134	148	161	177	186	211
	10	142	158	172	190	200	226
	11	151	165	184	198	216	241
	12	160	178	194	212	224	254
	13	168	187	204	225	238	267
	14	178	196	214	236	250	280
	15	186	206	224	246	262	295
	16	194	216	234	256	272	304
	17	221	238	255	289	306	340
	18	210	232	254	280	298	334
	19	217	241	263	289	306	346
	20	226	250	272	300	318	358
	21	232	258	282	310	331	371
	22	240	268	292	320	340	382
	23	247	276	302	328	350	395
	24	256	284	310	340	364	408
	25	264	292	319	350	373	421
	26	272	302	328	362	384	432
	27	276	309	337	370	395	444
	28	286	318	346	382	406	458
	29	293	326	356	391	415	468
	30	300	334	364	402	428	480
	31	306	340	372	411	438	491
	32	316	350	382	422	446	502
	33	321	355	388	424	455	519
	34	374	408	442	476	510	544
35	1	35	-				
	2	58	62	66	68	70	-
	3	70	81	87	93	96	102
	4	85	93	101	112	120	128
	5	100	110	120	130	135	150
	6	105	120	128	140	150	168
	7	126	133	147	161	168	189
	8	127	141	154	168	178	200
	9	137	153	165	182	191	217
	10	150	165	180	195	210	235
	11	155	172	187	207	218	245
	12	162	183	197	219	231	256
	13	173	193	208	229	245	273
	14	189	203	224	245	259	294
	15	195	215	235	255	270	305
	16	197	219	240	264	280	314
	17	203	231	251	271	288	323
	18	210	239	258	280	306	344
	19	221	247	269	296	315	354
	20	235	260	280	310	330	370
	21	245	273	294	322	343	385
	22	245	273	297	328	349	392
	23	252	283	308	341	356	402
	24	261	290	318	348	370	416
	25	270	305	330	365	385	435
	26	276	309	336	370	390	441
	27	284	315	346	381	405	454
	28	294	329	357	392	420	469
	29	300	333	364	399	423	479
	30	310	345	375	415	440	495
	31	315	350	382	420	448	503
	32	322	359	391	430	458	517
	33	328	368	401	442	471	522
	34	332	368	402	440	471	537
	35	385	420	455	490	525	595
36	1	36	-				
	2	60	64	68	70	72	-
	3	75	81	90	96	99	105
	4	88	100	104	116	120	132
	5	99	110	119	130	139	155
	6	114	126	138	150	156	174
	7	119	133	146	160	168	189
	8	132	148	160	176	184	208
	9	144	162	180	189	207	225
	10	150	166	182	200	212	238
	11	159	176	194	211	225	252
	12	180	192	216	228	252	276
	13	177	197	214	234	250	282
	14	186	206	224	246	264	294
	15	195	216	237	261	276	309
	16	204	228	248	272	288	324
	17	211	234	254	281	300	336
	18	234	252	288	306	324	360
	19	227	252	276	303	320	363
	20	236	264	288	316	336	376
	21	246	273	297	327	348	390
	22	252	280	306	336	358	402
	23	260	288	314	346	368	414
	24	276	312	336	372	384	432
	25	274	306	334	367	392	439
	26	284	316	344	378	402	452
	27	297	333	360	396	423	468
	28	300	336	364	400	428	480
	29	307	341	372	409	432	489
	30	318	354	384	426	450	504
	31	322	358	390	429	457	514
	32	332	368	400	440	468	528
	33	339	375	411	450	480	540
	34	344	384	418	460	490	554
	35	345	383	418	476	490	555
	36	396	432	468	504	540	612
37	1	37	-				
	2	60	66	70	72	74	-
	3	74	84	90	99	102	108
	4	91	99	107	116	124	136
	5	101	111	123	133	143	160
	6	113	125	136	149	161	179
	7	124	138	148	164	175	194
	8	134	148	161	177	190	211
	9	143	160	176	189	204	225
	10	153	170	185	206	216	243
	11	163	182	197	216	230	259
	12	173	190	210	227	240	274
	13	181	201	220	242	255	288
	14	189	211	230	253	268	300
	15	198	220	242	265	280	317
	16	206	231	252	278	294	331
	17	215	238	262	288	305	345
	18	226	248	268	301	320	357
	19	233	256	277	310	329	367
	20	240	267	292	321	343	383
	21	249	277	303	333	355	398
	22	256	285	313	343	365	410
	23	265	294	322	354	377	423
	24	273	304	332	365	389	437
	25	281	315	342	378	402	452

N	M	.1	.05	.025	.01	.005	.001
37	26	290	321	351	387	410	462
	27	296	331	362	399	422	474
	28	304	339	369	406	433	488
	29	312	348	381	418	444	500
	30	322	357	389	428	456	514
	31	329	366	399	437	467	524
	32	336	374	407	448	479	538
	33	344	382	417	460	489	550
	34	351	391	427	470	499	561
	35	359	400	435	478	509	577
	36	359	414	451	492	527	575
	37	407	444	481	518	555	629
38	1	38	-				
	2	62	68	72	74	76	-
	3	76	87	93	102	105	111
	4	92	102	110	120	128	140
	5	104	114	125	137	145	160
	6	116	128	140	152	162	180
	7	127	141	152	168	176	200
	8	138	152	166	182	194	216
	9	147	163	178	196	207	234
	10	158	176	190	208	222	246
	11	167	184	204	222	233	264
	12	176	196	214	236	250	280
	13	186	208	225	249	263	290
	14	196	216	236	260	276	308
	15	202	225	246	270	285	322
	16	212	236	258	282	300	336
	17	221	247	266	295	312	350
	18	230	256	278	304	324	368
	19	247	285	304	342	361	399
	20	246	274	298	330	350	392
	21	255	283	309	340	361	406
	22	264	292	320	352	374	418
	23	271	302	330	363	386	432
	24	280	312	340	374	398	446
	25	286	321	348	384	409	460
	26	296	330	360	396	420	474
	27	302	338	367	405	431	485
	28	312	348	378	416	444	500
	29	319	356	387	426	454	510
38	30	328	364	398	438	464	524
	31	334	372	406	450	475	536
	32	344	382	416	460	488	550
	33	351	390	427	470	499	561
	34	360	400	436	480	510	574
	35	366	407	445	490	519	586
	36	374	416	454	500	530	598
	37	387	429	467	509	544	617
	38	418	456	494	(570)	570	646
39	1	38	39	-			
	2	64	70	72	76	78	-
	3	81	87	96	102	108	114
	4	93	104	112	124	132	144
	5	106	117	130	140	146	165
	6	120	132	144	156	165	183
	7	129	146	156	171	181	203
	8	139	155	170	186	195	218
	9	153	168	183	201	213	237
	10	162	180	193	213	224	252
	11	170	190	207	229	241	273
	12	180	201	219	240	255	285
	13	195	221	247	260	273	312
	14	199	220	241	266	281	317
	15	210	234	255	279	294	330
	16	217	240	264	288	308	345
	17	227	250	272	300	320	361
	18	234	261	285	315	333	375
	19	242	266	297	321	340	380
	20	249	283	306	330	351	404
	21	261	291	318	348	369	417
	22	268	299	326	358	382	428
	23	276	309	336	371	394	442
	24	288	318	348	381	405	456
	25	294	325	357	393	418	468
	26	312	351	377	416	442	494
	27	312	345	378	414	441	498
	28	317	355	384	423	451	507
	29	326	364	395	434	462	521
	30	336	372	408	447	477	537
	31	341	379	416	456	487	548
39	32	350	389	425	468	496	560
	33	360	399	435	480	510	573
	34	366	407	445	490	519	586
	35	373	416	454	500	532	601
	36	384	426	465	513	543	612
	37	390	433	472	523	553	619
	38	401	443	483	525	561	637
	39	429	468	546	585	624	702
40	1	39	40	-			
	2	66	70	74	78	80	-
	3	80	90	99	105	111	117
	4	100	108	116	128	132	148
	5	110	125	135	145	155	170
	6	122	134	146	160	170	188
	7	132	147	160	177	186	210
	8	152	168	176	200	208	232
	9	154	172	186	204	217	240
	10	170	190	210	230	240	270
	11	174	195	213	232	247	276
	12	188	208	224	248	264	292
	13	193	217	233	258	272	310
	14	204	226	246	272	288	324
	15	215	240	260	285	305	340
	16	232	256	272	304	320	360
	17	231	255	280	306	328	368
	18	240	266	292	320	340	382
	19	247	277	300	332	353	395
	20	280	300	320	360	380	420
	21	266	294	323	354	375	426
	22	274	306	334	366	390	436
	23	283	314	342	377	400	451
	24	296	328	360	392	416	472
	25	305	335	365	405	430	480
	26	308	344	374	412	438	492
	27	316	354	383	423	450	504
	28	328	364	396	436	464	520
	29	333	371	404	445	471	532
	30	350	390	420	460	490	550
	31	349	389	424	465	496	558
	32	360	400	440	480	512	576
	33	366	407	444	487	519	585

N	M	.1	.05	.025	.01	.005	.001
40	34	374	416	454	500	532	598
	35	385	430	465	515	545	615
	36	392	436	476	524	556	624
	37	398	440	483	532	563	636
	38	406	452	492	544	580	652
	39	414	457	498	542	579	656
	40	440	520	560	600	640	720
41	1	40	41	-			
	2	66	72	76	80	82	-
	3	82	93	99	108	114	120
	4	99	108	119	128	136	148
	5	113	123	135	149	155	175
	6	123	138	150	163	174	193
	7	135	149	163	177	190	211
	8	148	164	176	197	207	232
	9	157	175	192	210	224	251
	10	168	188	205	226	238	267
	11	179	198	218	237	253	284
	12	189	210	228	251	266	300
	13	198	220	241	265	282	312
	14	204	229	254	273	296	326
	15	216	241	264	290	308	345
	16	227	252	275	302	320	359
	17	236	264	285	315	333	374
	18	245	273	297	327	348	391
	19	254	282	308	337	359	406
	20	258	293	316	353	375	417
	21	273	301	325	363	384	428
	22	279	311	339	374	396	446
	23	290	321	350	385	410	460
	24	297	331	360	396	420	477
	25	306	340	372	408	433	488
	26	315	349	382	419	448	501
	27	321	360	390	430	457	512
	28	331	368	402	443	471	529
	29	340	377	412	453	482	544
	30	348	387	424	465	495	555
	31	356	396	430	478	508	571
	32	364	405	444	487	518	583
	33	374	415	450	498	530	597
	34	380	424	461	509	542	610

N	M	.1	.05	.025	.01	.005	.001
41	35	389	431	472	519	553	623
	36	397	440	481	531	563	635
	37	405	450	491	541	574	650
	38	412	460	502	552	587	662
	39	421	468	514	563	600	678
	40	428	472	513	581	599	676
	41	492	533	574	615	656	738
42	1	41	42	-			
	2	68	74	78	82	84	-
	3	84	96	102	111	114	123
	4	100	110	120	132	140	152
	5	115	126	138	150	158	180
	6	132	144	156	174	180	198
	7	147	161	175	189	196	224
	8	150	168	182	200	210	236
	9	162	180	195	216	228	255
	10	172	192	208	228	242	270
	11	182	204	220	244	257	290
	12	198	216	240	258	276	306
	13	203	225	245	268	286	322
	14	224	252	266	294	308	350
	15	222	249	270	297	315	354
	16	232	258	280	308	328	368
	17	241	267	292	318	342	384
	18	252	282	306	336	360	402
	19	258	288	314	345	367	413
	20	268	298	326	356	378	430
	21	294	336	357	378	420	462
	22	286	318	346	382	406	454
	23	294	328	357	393	416	469
	24	306	342	372	408	432	486
	25	313	347	380	415	441	499
	26	322	358	390	428	456	514
	27	330	366	402	441	468	528
	28	350	392	420	462	490	546
	29	346	385	420	462	492	551
	30	360	396	432	480	510	570
	31	364	403	441	484	515	581
	32	372	414	452	498	528	594
	33	381	423	462	510	540	609

N	M	.1	.05	.025	.01	.005	.001
42	34	388	432	472	518	552	620
	35	399	448	490	532	567	637
	36	408	456	498	546	582	654
	37	413	460	502	550	586	660
	38	422	468	512	562	600	674
	39	432	480	522	576	612	687
	40	438	488	532	582	620	700
	41	441	487	530	600	640	722
	42	504	546	588	630	672	756
43	1	42	43	-			
	2	70	76	80	84	86	-
	3	86	96	105	114	117	126
	4	104	113	121	136	144	156
	5	117	129	140	152	162	180
	6	129	144	156	172	180	203
	7	140	158	172	188	201	223
	8	154	170	186	205	215	242
	9	166	184	200	220	231	258
	10	175	195	214	234	248	280
	11	188	204	224	246	265	290
	12	197	219	238	262	277	312
	13	206	230	249	275	292	330
	14	218	238	263	290	306	347
	15	226	252	276	301	321	360
	16	237	261	287	314	335	377
	17	245	271	297	329	347	390
	18	254	283	308	340	361	407
	19	264	294	321	351	375	421
	20	273	304	331	365	387	436
	21	285	311	344	375	397	455
	22	292	319	356	384	407	467
	23	300	334	364	401	426	481
	24	309	343	376	413	439	492
	25	317	353	385	425	450	507
	26	325	361	395	436	463	523
	27	336	373	407	446	477	536
	28	344	384	417	459	487	552
	29	353	395	427	470	499	559
	30	360	402	437	483	514	578
	31	371	411	450	495	526	593

N	M	.1	.05	.025	.01	.005	.001
43	32	380	422	457	507	540	606
	33	386	429	470	516	550	617
	34	395	439	481	529	561	633
	35	403	450	489	539	574	645
	36	413	457	500	550	586	658
	37	421	466	509	563	599	674
	38	429	477	520	573	611	687
	39	438	487	530	584	620	698
	40	445	496	540	594	634	714
	41	454	505	550	602	642	722
	42	455	520	568	617	659	743
	43	516	559	602	(688)	688	774
44	1	43	44	-			
	2	72	78	82	86	88	-
	3	88	99	108	117	120	129
	4	108	120	128	140	148	160
	5	120	131	145	156	166	185
	6	132	146	160	174	186	208
	7	145	161	175	192	203	229
	8	160	176	192	212	224	248
	9	166	184	202	220	236	263
	10	180	200	218	238	254	286
	11	198	220	242	264	275	308
	12	204	224	244	268	284	320
	13	211	234	256	281	299	336
	14	222	246	268	294	314	350
	15	230	258	277	305	322	364
	16	244	268	296	324	344	384
	17	250	279	304	334	355	396
	18	260	290	316	348	368	412
	19	269	300	326	358	382	427
	20	280	312	340	372	396	448
	21	287	320	349	385	410	458
	22	308	352	374	418	440	484
	23	305	340	370	407	432	491
	24	316	352	384	424	448	504
	25	324	361	393	431	461	518
	26	334	370	404	446	474	534
	27	342	381	415	457	486	549
	28	352	392	428	472	500	564
	29	359	401	433	477	506	576

N	M	.1	.05	.025	.01	.005	.001
44	30	368	410	448	492	522	590
	31	377	420	457	505	536	604
	32	388	432	472	516	552	620
	33	407	451	484	539	572	638
	34	404	448	490	538	572	644
	35	410	455	498	549	584	656
	36	424	468	512	564	600	672
	37	429	476	518	571	608	686
	38	438	486	530	584	620	698
	39	445	495	540	594	633	711
	40	456	508	552	608	648	728
	41	462	514	559	615	656	738
	42	470	522	570	630	668	748
	43	469	538	585	635	677	764
	44	528	572	616	704	748	836
45	1	44	45	-			
	2	74	80	84	88	90	-
	3	90	102	111	117	123	132
	4	107	119	128	140	148	164
	5	125	140	150	165	170	190
	6	135	153	165	180	189	210
	7	148	165	179	196	207	231
	8	160	177	193	214	225	251
	9	180	198	216	234	252	279
	10	185	205	225	245	260	290
	11	195	217	237	259	274	307
	12	207	228	249	273	291	324
	13	216	237	261	287	302	340
	14	224	251	273	301	318	358
	15	255	270	300	330	345	390
	16	246	272	297	328	348	393
	17	255	285	309	341	360	405
	18	270	306	333	360	387	432
	19	276	305	333	366	390	437
	20	285	320	350	380	405	455
	21	294	327	357	393	417	468
	22	302	339	368	408	433	481
	23	310	350	377	419	443	492
	24	324	357	390	429	456	513
	25	335	370	405	445	470	530

N	M	.1	.05	.025	.01	.005	.001
45	26	338	377	411	453	481	543
	27	360	396	432	468	504	567
	28	357	397	434	476	508	571
	29	366	407	445	488	520	585
	30	390	435	465	510	540	615
	31	384	427	464	513	545	613
	32	393	435	477	524	556	627
	33	402	447	489	537	570	642
	34	411	457	495	548	582	652
	35	420	470	510	565	600	675
	36	432	486	522	576	612	693
	37	436	485	530	583	620	699
	38	445	494	540	594	632	711
	39	453	504	552	606	645	726
	40	465	515	565	620	660	740
	41	470	523	570	628	669	752
	42	480	534	582	642	681	768
	43	486	540	589	654	695	783
	44	502	554	602	652	696	784
	45	540	585	675	720	765	855
46	1	45	46				
	2	74	80	86	90	92	-
	3	92	102	111	120	126	135
	4	110	122	130	144	152	168
	5	124	138	150	164	174	195
	6	138	154	166	182	194	212
	7	153	169	182	202	213	238
	8	164	182	198	218	228	258
	9	176	196	213	233	249	278
	10	188	208	226	248	264	298
	11	197	219	240	264	280	315
	12	210	232	254	280	294	330
	13	220	245	266	292	311	349
	14	230	256	280	306	326	364
	15	238	267	294	315	342	384
	16	250	278	304	332	356	396
	17	259	288	316	346	368	413
	18	270	300	328	360	384	430
	19	280	311	338	373	395	446
	20	290	322	352	386	412	460
	21	299	332	363	397	422	476

N	M	.1	.05	.025	.01	.005	.001
46	22	308	342	374	410	436	494
	23	345	368	414	437	460	529
	24	328	364	398	438	466	520
	25	336	374	408	449	475	537
	26	346	384	420	462	492	552
	27	355	394	431	474	504	566
	28	364	404	442	486	518	582
	29	372	414	453	499	529	596
	30	382	426	464	510	544	610
	31	392	437	472	519	550	625
	32	400	444	486	534	568	640
	33	407	453	497	544	578	654
	34	418	464	506	558	594	666
	35	425	474	517	570	605	682
	36	436	484	528	582	618	696
	37	441	494	539	594	632	707
	38	452	504	550	604	644	724
	39	460	512	559	615	654	736
	40	470	522	570	628	668	752
	41	478	531	581	638	679	766
	42	486	542	592	650	692	778
	43	495	550	601	662	704	791
	44	504	560	612	674	718	808
	45	517	570	618	672	715	806
	46	552	644	690	736	782	874
47	1	46	47	-			
	2	76	82	88	92	94	-
	3	94	105	114	123	129	138
	4	112	124	133	148	156	172
	5	126	140	153	168	178	195
	6	140	157	169	187	198	217
	7	154	172	186	205	214	240
	8	168	186	202	219	234	258
	9	179	198	216	235	252	282
	10	190	212	230	253	269	303
	11	202	224	244	269	285	318
	12	210	234	256	281	302	338
	13	225	249	272	298	316	356
	14	233	261	284	312	331	373
	15	245	273	296	326	345	388

N	M	.1	.05	.025	.01	.005	.001
47	16	257	280	309	340	358	404
	17	265	295	322	355	374	423
	18	275	306	333	367	391	438
	19	286	317	345	382	403	451
	20	293	326	358	393	419	470
	21	305	338	369	406	432	485
	22	313	349	381	418	447	500
	23	319	361	388	432	456	521
	24	328	369	398	441	467	534
	25	342	381	415	456	486	549
	26	352	390	426	468	499	561
	27	361	401	436	482	510	576
	28	370	410	447	493	527	588
	29	379	422	459	506	538	606
	30	388	431	470	517	551	622
	31	398	444	479	526	567	633
	32	406	452	494	543	576	650
	33	415	462	505	554	591	665
	34	423	472	516	568	603	680
	35	431	484	524	580	616	696
	36	442	492	536	590	629	709
	37	451	502	547	603	641	724
	38	459	511	558	616	654	737
	39	467	521	568	624	663	748
	40	477	531	578	638	678	765
	41	485	541	590	649	690	778
	42	494	550	600	661	703	792
	43	503	559	611	672	715	805
	44	512	569	623	685	727	820
	45	521	578	633	696	739	833
	46	532	585	635	716	762	857
	47	564	658	705	752	799	893
48	1	47	48	-			
	2	78	84	88	92	94	-
	3	96	108	117	126	132	138
	4	116	128	140	152	160	172
	5	129	144	157	170	180	200
	6	150	162	180	192	204	228
	7	156	175	190	210	219	246
	8	176	192	208	232	248	272
	9	183	204	222	243	258	288

N	M	.1	.05	.025	.01	.005	.001
48	10	194	216	236	258	274	306
	11	205	229	249	274	292	325
	12	228	252	276	300	312	348
	13	227	254	277	303	323	363
	14	240	266	290	318	338	380
	15	252	279	303	333	354	399
	16	272	304	336	368	384	432
	17	270	299	327	360	381	429
	18	282	318	342	378	402	450
	19	291	322	351	388	413	464
	20	304	336	368	404	428	480
	21	312	345	378	414	441	495
	22	320	356	388	426	454	510
	23	329	365	400	440	469	523
	24	360	408	432	480	504	552
	25	349	387	422	466	493	560
	26	358	398	434	478	508	572
	27	369	411	447	492	522	588
	28	380	420	460	504	536	604
	29	384	429	468	516	546	614
	30	402	444	486	534	564	636
	31	403	448	490	540	574	647
	32	432	480	512	560	608	672
	33	423	471	513	567	603	678
	34	432	480	524	576	614	692
	35	440	490	535	588	624	705
	36	456	516	552	612	648	732
	37	459	509	557	612	652	734
	38	468	520	568	624	664	748
	39	477	531	579	639	678	765
	40	488	544	592	656	696	784
	41	492	551	600	661	702	791
	42	510	564	612	678	720	810
	43	511	570	621	682	729	820
	44	524	580	636	696	744	836
	45	531	591	645	708	753	849
	46	538	598	654	716	762	858
	47	547	601	652	735	782	879
	48	576	672	720	768	816	912
49	1	48	49	-			

N	M	.1	.05	.025	.01	.005	.001
49	2	80	86	90	94	96	-
	3	98	111	120	129	135	141
	4	116	128	139	152	160	176
	5	131	146	160	175	181	205
	6	147	162	178	192	204	228
	7	168	182	196	217	231	252
	8	174	192	208	231	245	272
	9	186	207	225	245	262	293
	10	196	221	241	263	281	312
	11	210	233	254	277	294	331
	12	223	247	264	296	311	348
	13	232	259	281	311	328	370
	14	252	273	301	329	350	392
	15	254	283	307	340	360	404
	16	266	297	318	350	376	416
	17	274	305	333	367	388	437
	18	285	318	346	382	405	454
	19	294	328	358	395	418	471
	20	306	339	370	403	435	486
	21	322	357	385	427	455	511
	22	325	363	395	434	461	520
	23	335	372	407	447	476	537
	24	348	380	422	454	494	548
	25	356	389	432	478	506	560
	26	363	405	441	484	518	581
	27	374	414	453	499	529	597
	28	385	434	469	518	546	616
	29	392	436	478	525	558	627
	30	401	446	487	536	570	641
	31	410	458	499	549	584	659
	32	421	468	509	562	598	670
	33	432	480	519	570	614	683
	34	437	488	533	586	624	701
	35	455	504	546	602	644	721
	36	457	509	555	610	650	731
	37	464	520	563	622	660	745
	38	476	529	575	634	673	760
	39	484	536	585	649	689	777
	40	493	548	599	660	701	790
	41	503	560	610	669	715	805
	42	518	574	623	686	735	826
	43	519	579	630	696	740	832

N	M	.1	.05	.025	.01	.005	.001
49	44	528	588	640	707	752	848
	45	538	599	653	718	763	861
	46	545	609	664	730	776	878
	47	554	618	675	739	786	884
	48	561	617	694	753	802	900
	49	637	686	735	833	882	980
50	1	49	50	-			
	2	80	88	92	96	98	-
	3	100	111	120	132	135	144
	4	118	130	142	156	164	180
	5	140	150	165	180	190	210
	6	150	166	180	198	210	232
	7	165	181	196	217	229	258
	8	178	196	214	234	248	278
	9	189	211	229	252	269	297
	10	210	230	250	280	290	330
	11	214	236	258	284	302	339
	12	226	250	274	300	320	358
	13	236	264	286	316	333	374
	14	248	276	300	330	350	392
	15	260	290	315	350	370	415
	16	270	300	328	360	380	428
	17	279	311	342	376	396	446
	18	290	324	352	388	412	462
	19	300	334	365	400	427	479
	20	320	350	390	420	450	500
	21	320	357	390	429	454	514
	22	332	368	402	442	470	528
	23	341	379	413	454	483	544
	24	352	390	424	468	496	562
	25	375	425	450	500	525	600
	26	370	412	450	492	528	590
	27	380	422	461	507	538	607
	28	390	434	472	520	554	622
	29	399	443	484	533	567	638
	30	420	460	500	550	590	660
	31	418	464	507	557	594	668
	32	428	476	518	572	608	684
	33	438	488	526	586	622	693
	34	446	496	542	596	634	712
	35	460	510	555	610	650	730

N	M	.1	.05	.025	.01	.005	.001
50	36	464	518	564	620	660	742
	37	475	526	575	634	673	758
	38	484	539	586	646	686	772
	39	492	548	598	658	699	788
	40	510	570	620	680	720	810
	41	509	567	618	681	726	817
	42	520	578	630	694	738	830
	43	527	588	641	706	749	846
	44	538	598	652	718	764	860
	45	550	610	665	735	780	880
	46	556	618	674	742	790	890
	47	564	628	684	752	803	903
	48	572	638	694	768	818	922
	49	576	654	713	772	821	922
	50	650	700	750	850	900	1000
51	1	50	51	-			
	2	82	90	94	98	100	-
	3	102	114	123	132	138	147
	4	121	133	144	160	168	184
	5	138	153	165	180	190	215
	6	153	171	183	201	213	237
	7	166	183	201	220	234	264
	8	180	199	218	239	253	282
	9	195	216	234	258	273	303
	10	207	228	249	270	289	327
	11	218	242	262	290	306	345
	12	231	255	279	306	324	363
	13	239	266	290	317	341	380
	14	252	280	305	336	356	402
	15	264	294	321	351	375	420
	16	274	304	333	365	388	436
	17	306	340	357	391	425	476
	18	297	330	360	396	420	471
	19	304	340	372	408	434	487
	20	316	352	385	423	449	505
	21	327	363	396	438	465	522
	22	335	375	407	450	479	537
	23	346	385	422	463	492	552
	24	357	399	432	477	507	570
	25	367	411	444	492	520	591
	26	375	422	453	502	530	604

N	M	.1	.05	.025	.01	.005	.001
51	27	387	429	471	516	549	618
	28	396	440	480	528	561	633
	29	405	449	492	542	576	645
	30	417	462	504	555	591	666
	31	425	473	515	568	604	679
	32	433	483	527	579	617	694
	33	444	495	540	594	633	711
	34	476	527	561	612	663	748
	35	463	515	560	617	657	738
	36	474	525	573	633	672	756
	37	481	535	585	642	684	770
	38	490	543	593	655	695	783
	39	501	558	609	669	711	801
	40	509	566	618	680	723	814
	41	518	577	629	691	734	826
	42	528	588	642	705	750	846
	43	538	597	651	718	762	860
	44	545	608	662	729	777	873
	45	555	618	675	744	789	888
	46	564	626	683	754	800	902
	47	573	637	695	765	814	916
	48	585	648	708	780	828	933
	49	591	656	714	793	844	948
	50	591	675	731	791	841	945
	51	663	714	765	867	918	1020
52	1	51	52	-			
	2	84	92	96	100	102	-
	3	104	117	126	135	141	150
	4	124	140	148	160	172	188
	5	140	155	168	183	193	215
	6	154	172	188	206	218	242
	7	169	187	204	224	238	266
	8	184	204	224	244	260	288
	9	197	218	238	260	276	310
	10	210	232	252	278	296	330
	11	221	246	268	295	312	350
	12	236	260	284	312	332	372
	13	260	286	312	338	351	403
	14	256	286	312	342	364	406
	15	266	296	325	355	378	423
	16	280	312	340	372	396	444
	17	287	320	352	387	407	458
	18	300	334	366	402	426	478
	19	311	345	377	415	442	496
	20	324	360	392	432	460	516
	21	331	370	403	444	467	528
	22	342	380	416	458	486	548
	23	353	391	428	470	500	563
	24	364	404	444	488	516	580
	25	372	415	453	495	530	594
	26	416	442	494	546	572	624
	27	392	436	477	525	558	629
	28	404	448	492	540	572	644
	29	410	456	501	551	583	658
	30	422	470	512	564	600	674
	31	431	481	524	576	614	690
	32	444	492	536	592	628	708
	33	451	502	546	602	640	721
	34	460	512	558	616	654	738
	35	466	518	568	623	670	755
	36	480	536	584	644	684	768
	37	488	545	592	651	694	783
	38	498	554	606	666	708	798
	39	520	572	624	689	728	819
	40	520	576	628	692	736	828
	41	526	586	639	704	748	843
	42	536	596	650	716	762	858
	43	546	607	660	728	773	874
	44	556	620	676	744	788	888
	45	563	626	683	751	801	900
	46	572	636	696	764	814	918
	47	581	647	705	778	826	930
	48	592	660	720	792	840	948
	49	599	666	727	801	853	960
	50	608	676	738	814	866	974
	51	607	692	749	812	864	1000
	52	676	728	832	884	936	1040
53	1	52	53	-			
	2	86	92	98	102	104	-
	3	105	117	129	138	144	153
	4	127	139	151	164	176	192
	5	142	157	172	187	197	220
	6	157	175	188	206	222	246
	7	174	191	209	230	241	269
	8	187	208	225	248	262	294
	9	200	220	243	264	281	316
	10	212	235	258	284	298	337
	11	226	250	272	301	316	356
	12	237	264	287	316	335	376
	13	247	279	299	334	350	398
	14	261	290	315	348	368	415
	15	270	303	329	360	387	434
	16	284	316	344	376	402	450
	17	296	328	356	392	419	470
	18	308	335	368	404	435	489
	19	316	351	382	420	449	503
	20	327	363	396	436	463	522
	21	335	375	408	450	480	535
	22	347	387	422	462	495	554
	23	358	399	436	477	510	572
	24	368	408	446	492	523	588
	25	379	420	460	505	536	607
	26	385	433	466	515	544	619
	27	394	442	488	526	571	631
	28	409	454	496	546	580	653
	29	418	466	508	558	595	669
	30	428	476	520	573	608	686
	31	437	488	530	586	622	701
	32	449	496	545	600	634	718
	33	458	508	554	612	651	732
	34	468	520	568	623	664	748
	35	480	534	575	640	678	765
	36	487	540	592	648	694	781
	37	496	552	602	663	703	795
	38	505	563	612	677	718	809
	39	515	573	624	688	734	826
	40	525	581	634	699	742	843
	41	533	594	648	713	760	854
	42	543	605	659	728	772	869
	43	552	615	671	739	786	883
	44	561	624	683	753	799	897
	45	573	635	695	764	814	917

N	M	.1	.05	.025	.01	.005	.001
53	46	581	646	706	771	826	929
	47	591	656	716	781	839	945
	48	599	666	728	800	852	959
	49	609	676	740	813	864	974
	50	617	685	750	825	878	988
	51	627	696	761	837	890	1000
	52	644	709	767	860	914	1023
	53	689	742	848	901	954	1060
54	1	53	54	-			
	2	88	94	100	104	106	-
	3	108	120	129	141	147	156
	4	128	142	152	168	176	192
	5	145	161	175	191	201	225
	6	162	180	198	216	228	252
	7	175	195	212	233	247	275
	8	190	212	230	252	266	298
	9	207	234	252	279	297	324
	10	216	240	262	288	304	342
	11	226	255	278	301	322	364
	12	246	270	294	324	348	384
	13	254	282	308	336	360	401
	14	266	296	322	354	374	424
	15	279	309	336	369	393	441
	16	288	320	350	384	410	458
	17	300	333	364	401	424	476
	18	324	360	396	432	450	504
	19	321	357	389	429	456	512
	20	332	370	404	444	472	530
	21	345	384	417	459	489	549
	22	354	394	428	472	502	566
	23	365	405	442	483	517	581
	24	378	420	456	504	534	600
	25	385	427	465	514	544	614
	26	394	440	480	528	560	634
	27	432	486	513	567	594	675
	28	416	462	502	554	588	662
	29	425	472	516	567	604	679
	30	438	486	534	588	624	702
	31	445	493	539	593	631	709
	32	456	506	552	608	646	730
	33	465	519	567	621	663	744

N	M	.1	.05	.025	.01	.005	.001
54	34	476	528	576	634	674	760
	35	484	538	587	646	689	775
	36	504	558	612	684	720	810
	37	503	561	611	673	716	806
	38	514	572	624	686	730	822
	39	525	582	636	699	744	837
	40	534	592	646	712	756	852
	41	543	602	658	724	772	869
	42	558	618	672	744	786	888
	43	559	622	683	750	795	900
	44	572	636	694	762	812	914
	45	585	657	711	783	828	936
	46	590	656	716	788	838	944
	47	600	666	727	802	851	959
	48	612	678	744	816	870	978
	49	617	686	749	823	877	989
	50	628	698	762	838	892	1004
	51	636	708	774	852	906	1020
	52	646	718	786	860	914	1028
	53	661	725	785	880	935	1046
	54	702	810	864	918	972	1134
55	1	54	55	-			
	2	88	96	102	106	108	-
	3	110	123	132	144	150	159
	4	129	144	153	172	180	196
	5	150	165	180	195	205	230
	6	162	180	197	216	228	257
	7	178	198	213	238	247	275
	8	194	212	234	257	273	304
	9	206	231	251	275	294	324
	10	225	250	270	295	315	350
	11	242	264	286	319	341	374
	12	245	274	298	327	348	389
	13	258	286	312	345	364	410
	14	270	299	326	356	382	425
	15	285	315	345	380	400	450
	16	294	326	356	390	415	468
	17	304	338	368	405	431	485
	18	317	353	378	424	447	502
	19	326	363	396	434	464	521
	20	340	380	415	455	480	540

N	M	.1	.05	.025	.01	.005	.001
55	21	348	387	423	465	494	555
	22	363	407	440	484	517	583
	23	370	411	448	494	525	590
	24	380	423	461	507	540	609
	25	395	440	475	525	560	625
	26	401	446	487	536	569	640
	27	413	454	503	554	586	648
	28	425	463	513	567	597	678
	29	431	480	525	576	612	691
	30	445	495	540	595	630	710
	31	452	503	548	603	642	724
	32	461	512	561	617	656	736
	33	484	539	583	638	682	759
	34	481	536	585	643	684	770
	35	495	550	600	660	700	790
	36	501	557	607	672	712	803
	37	509	565	619	686	727	819
	38	521	578	633	696	741	834
	39	530	590	645	708	755	849
	40	545	605	660	725	770	870
	41	552	611	667	735	778	883
	42	560	623	680	748	796	897
	43	570	633	692	761	811	912
	44	583	649	715	781	836	935
	45	590	660	720	790	840	945
	46	598	666	724	797	850	959
	47	608	676	736	812	861	971
	48	616	685	747	822	876	987
	49	627	696	761	836	891	1002
	50	640	710	775	855	905	1020
	51	645	717	783	862	917	1034
	52	653	727	794	875	930	1048
	53	664	738	808	884	939	1069
	54	677	742	804	900	956	1069
	55	715	825	880	990	1045	1155
56	1	55	56	-			
	2	90	98	104	108	110	-
	3	111	126	135	147	153	162
	4	136	148	160	172	184	200
	5	150	165	180	199	209	235
	6	166	184	200	218	232	258

N	M	.1	.05	.025	.01	.005	.001
56	7	189	210	224	245	259	287
	8	200	224	240	264	280	312
	9	210	234	253	279	297	333
	10	224	250	270	296	316	352
	11	238	263	285	317	336	374
	12	252	280	304	336	356	396
	13	262	292	318	348	370	416
	14	280	322	350	378	406	448
	15	286	317	347	383	403	455
	16	304	336	368	400	424	480
	17	310	344	374	412	439	492
	18	322	356	390	428	456	512
	19	331	368	404	443	465	532
	20	344	384	420	460	488	548
	21	357	399	434	476	504	567
	22	366	406	444	488	516	582
	23	376	417	455	501	532	598
	24	392	432	472	520	552	624
	25	396	441	479	528	564	633
	26	408	454	494	544	578	650
	27	418	466	504	557	591	665
	28	448	504	532	588	644	700
	29	438	488	531	585	625	704
	30	448	500	544	600	638	716
	31	458	508	557	613	650	731
	32	472	528	576	632	672	752
	33	479	533	581	639	680	764
	34	490	544	594	654	696	782
	35	504	560	609	672	714	805
	36	512	568	620	684	724	816
	37	516	573	626	696	736	829
	38	530	588	642	706	752	848
	39	538	600	654	720	766	863
	40	552	616	672	736	784	880
	41	558	621	678	745	793	894
	42	574	644	700	770	826	924
	43	576	642	700	772	821	924
	44	588	656	716	788	836	944
	45	598	665	725	794	848	952
	46	606	676	736	812	864	972
	47	616	686	749	825	875	986
	48	632	704	768	840	896	1008
56	49	644	714	777	854	910	1022
	50	646	718	782	862	918	1032
	51	654	727	793	874	930	1048
	52	664	740	808	888	944	1064
	53	671	749	817	898	958	1077
	54	684	760	828	912	974	1096
	55	693	759	850	920	977	1092
	56	728	840	896	1008	1064	1176
57	1	56	57	-			
	2	92	100	104	110	112	-
	3	114	126	138	147	153	165
	4	135	148	160	176	188	204
	5	151	168	183	200	213	235
	6	171	189	204	225	237	261
	7	186	203	223	245	259	293
	8	199	222	240	264	280	312
	9	216	237	261	285	303	336
	10	228	253	275	302	320	359
	11	241	269	293	319	339	383
	12	255	282	309	339	360	402
	13	266	295	323	354	377	422
	14	278	307	335	374	391	445
	15	291	324	354	387	414	462
	16	303	335	366	405	430	480
	17	315	350	382	419	445	502
	18	327	363	396	435	462	519
	19	361	399	418	475	494	551
	20	349	387	423	464	493	555
	21	360	402	438	480	510	573
	22	371	411	449	494	525	591
	23	382	421	463	509	543	611
	24	393	438	477	525	558	627
	25	402	448	488	537	570	641
	26	413	460	501	553	585	658
	27	426	474	516	567	603	678
	28	437	485	525	580	612	694
	29	445	499	535	591	624	706
	30	456	507	555	609	648	729
	31	465	516	563	620	661	744
	32	476	529	576	636	675	763
57	33	486	543	591	651	693	777
	34	496	552	600	664	702	792
	35	506	563	614	677	720	809
	36	519	576	627	690	735	828
	37	526	585	639	703	747	841
	38	551	608	665	741	779	874
	39	549	609	663	732	777	876
	40	557	619	676	741	791	889
	41	566	630	687	756	804	908
	42	579	642	699	771	819	924
	43	583	651	709	781	835	937
	44	595	662	722	796	846	954
	45	606	675	735	810	861	969
	46	614	684	746	823	873	985
	47	625	694	759	836	888	999
	48	636	708	771	849	903	1017
	49	642	715	780	861	917	1032
	50	653	726	794	874	930	1046
	51	666	738	807	888	945	1065
	52	672	749	817	900	957	1077
	53	681	759	828	911	969	1091
	54	693	771	840	927	984	1110
	55	700	781	850	940	1001	1123
	56	709	778	871	940	998	1151
	57	798	855	912	1026	1083	1197
58	1	57	58	-			
	2	94	102	106	112	114	-
	3	116	129	141	150	156	168
	4	138	150	162	180	188	208
	5	155	172	187	202	215	240
	6	172	190	208	226	240	266
	7	189	208	225	248	262	294
	8	204	226	246	270	286	320
	9	217	243	262	289	306	339
	10	232	256	280	306	326	364
	11	244	272	296	326	346	387
	12	258	286	312	342	364	410
	13	270	301	327	360	380	430
	14	284	316	344	376	402	448
	15	295	329	357	394	417	471

N	M	.1	.05	.025	.01	.005	.001
58	16	308	342	372	410	436	490
	17	319	355	387	425	452	509
	18	332	368	402	442	468	526
	19	340	378	414	454	488	547
	20	354	394	430	472	502	566
	21	365	404	441	487	517	582
	22	376	418	456	502	534	600
	23	387	432	469	516	550	619
	24	398	442	484	532	566	634
	25	410	454	495	545	579	653
	26	420	466	510	560	596	670
	27	430	478	522	574	611	689
	28	442	490	536	590	628	708
	29	464	522	580	638	667	754
	30	462	514	560	618	654	738
	31	473	525	572	630	671	754
	32	484	536	586	644	686	772
	33	493	549	599	658	700	790
	34	504	560	612	672	714	806
	35	515	568	623	683	728	821
	36	524	582	636	700	744	838
	37	532	594	648	714	758	853
	38	544	606	660	728	772	868
	39	554	614	671	744	787	885
	40	564	628	684	754	802	902
	41	573	638	696	767	815	920
	42	584	650	710	780	830	934
	43	594	661	721	794	842	953
	44	604	672	734	806	860	966
	45	614	681	744	821	873	982
	46	624	694	758	834	888	998
	47	634	704	769	845	900	1015
	48	644	716	782	860	916	1032
	49	653	726	792	872	928	1046
	50	664	738	806	886	942	1062
	51	672	748	816	897	955	1077
	52	682	758	828	912	970	1092
	53	692	769	841	924	983	1107
	54	702	780	852	938	998	1122
	55	711	790	862	949	1010	1140
	56	720	802	872	966	1026	1150
	57	725	821	890	961	1023	1177

N	M	.1	.05	.025	.01	.005	.001
58	58	812	870	928	1044	1102	1218
59	1	57	59	-			
	2	94	102	108	114	116	-
	3	115	132	141	153	159	171
	4	140	153	165	184	192	212
	5	157	172	190	206	216	245
	6	175	193	211	229	246	271
	7	190	211	229	252	267	295
	8	207	228	250	274	290	322
	9	220	246	268	295	314	350
	10	234	262	283	312	332	371
	11	249	275	303	330	351	395
	12	259	291	316	350	372	412
	13	273	305	331	364	389	435
	14	288	319	349	383	406	458
	15	302	334	364	396	424	471
	16	312	346	378	416	442	496
	17	325	360	393	429	460	513
	18	336	373	407	448	476	535
	19	347	387	423	465	492	551
	20	362	402	430	483	507	567
	21	369	411	449	495	525	590
	22	381	424	462	509	541	608
	23	393	437	475	524	557	626
	24	405	448	490	538	573	645
	25	415	462	503	555	589	664
	26	426	472	518	570	604	681
	27	436	485	529	583	620	696
	28	447	496	543	596	635	714
	29	456	511	548	605	656	723
	30	465	521	574	617	668	755
	31	479	533	581	640	681	767
	32	489	543	594	655	697	783
	33	501	556	607	667	711	798
	34	510	568	620	680	724	815
	35	520	579	633	696	740	834
	36	530	589	644	710	754	849
	37	542	602	655	722	767	867
	38	551	614	669	737	783	883
	39	560	621	679	753	796	894
	40	572	637	694	762	814	915

N	M	.1	.05	.025	.01	.005	.001
59	41	583	647	706	778	826	930
	42	593	660	719	789	839	947
	43	602	669	729	804	856	964
	44	612	683	743	817	873	979
	45	622	691	755	831	884	996
	46	633	701	766	844	897	1009
	47	642	713	781	856	913	1030
	48	652	725	792	870	925	1043
	49	662	734	802	881	939	1057
	50	671	747	816	898	953	1075
	51	681	757	826	909	968	1090
	52	691	769	840	922	981	1108
	53	700	779	852	936	995	1123
	54	710	790	863	949	1010	1138
	55	720	801	874	962	1024	1154
	56	730	812	884	975	1037	1170
	57	738	821	896	988	1051	1179
	58	741	841	909	1013	1076	1201
	59	826	885	1003	1062	1121	1298
60	1	58	60	-			
	2	96	104	110	116	118	-
	3	120	132	144	156	162	174
	4	144	160	172	184	196	216
	5	165	180	195	215	225	250
	6	180	198	216	240	252	276
	7	195	216	233	258	272	304
	8	212	236	256	280	296	332
	9	225	249	273	300	318	354
	10	250	270	300	320	340	380
	11	253	278	305	336	354	398
	12	276	300	336	360	384	432
	13	278	311	338	371	394	441
	14	292	324	354	388	414	464
	15	315	345	375	420	450	495
	16	320	356	388	424	452	504
	17	329	365	398	440	466	525
	18	348	384	420	456	486	546
	19	353	393	428	470	501	562
	20	380	420	460	500	540	600
	21	378	420	456	504	534	600
	22	388	430	470	516	550	618

N	M	.1	.05	.025	.01	.005	.001
60	23	399	444	483	533	565	637
	24	420	468	504	552	588	660
	25	425	470	515	565	600	675
	26	432	480	524	578	614	690
	27	444	495	540	591	630	708
	28	456	508	552	608	648	728
	29	463	515	564	622	657	742
	30	510	570	600	660	690	780
	31	485	540	588	648	692	774
	32	500	556	604	664	708	796
	33	507	564	618	678	720	813
	34	518	576	628	692	736	828
	35	530	590	645	710	755	850
	36	552	612	660	732	780	876
	37	548	609	666	733	779	879
	38	560	622	678	748	796	894
	39	570	636	693	762	810	912
	40	600	660	720	800	840	940
	41	589	656	717	788	838	943
	42	606	672	732	804	858	966
	43	608	677	742	813	865	976
	44	624	692	756	832	884	996
	45	645	720	780	855	915	1020
	46	640	712	778	856	910	1026
	47	651	724	790	870	925	1042
	48	672	744	816	888	948	1068
	49	669	745	815	897	952	1072
	50	690	770	830	920	980	1100
	51	693	768	840	924	984	1107
	52	704	780	852	940	996	1124
	53	709	790	861	949	1009	1136
	54	726	804	876	966	1026	1158
	55	735	815	890	980	1040	1170
	56	744	824	900	992	1052	1188
	57	750	834	912	1002	1065	1203
	58	758	844	922	1012	1076	1220
	59	758	860	928	1036	1099	1225
	60	840	900	1020	1080	1140	1320
61	1	59	61	-			
	2	98	106	112	118	120	-
	3	120	135	147	159	165	174
61	4	144	159	171	188	200	216
	5	163	180	195	214	225	250
	6	180	202	216	238	252	281
	7	198	219	237	261	275	305
	8	212	236	257	281	297	334
	9	229	254	276	303	321	357
	10	240	269	290	320	340	380
	11	257	285	311	341	362	405
	12	271	299	324	359	382	430
	13	284	315	342	377	401	449
	14	296	330	359	395	419	471
	15	310	342	373	414	434	493
	16	322	358	390	431	456	511
	17	332	371	405	446	471	532
	18	346	386	419	461	488	549
	19	357	398	433	477	508	570
	20	372	413	452	495	519	594
	21	381	423	462	510	541	606
	22	394	438	477	524	558	626
	23	403	448	488	539	571	646
	24	416	463	503	555	590	666
	25	427	474	518	570	606	682
	26	439	485	532	586	622	700
	27	450	499	545	599	638	715
	28	460	512	558	614	653	736
	29	471	522	571	628	667	751
	30	476	533	588	648	683	772
	31	486	543	599	660	695	785
	32	502	560	610	671	714	803
	33	514	571	623	687	730	821
	34	525	582	636	698	744	839
	35	536	595	650	713	758	853
	36	545	606	663	728	775	872
	37	555	618	676	742	789	888
	38	567	629	689	758	804	904
	39	576	641	701	770	819	925
	40	588	652	712	786	834	939
	41	600	664	725	803	848	952
	42	608	676	738	812	865	972
	43	619	687	750	826	880	989
	44	629	699	763	839	893	1005
	45	638	711	775	853	908	1023
61	46	650	724	788	865	924	1035
	47	660	732	801	881	937	1055
	48	669	744	812	893	952	1071
	49	679	754	825	910	962	1084
	50	689	765	836	920	980	1103
	51	699	779	850	933	994	1117
	52	709	790	862	948	1009	1135
	53	718	800	873	961	1024	1152
	54	729	810	886	975	1037	1166
	55	738	822	896	987	1049	1183
	56	749	833	909	1001	1066	1199
	57	759	843	922	1014	1079	1215
	58	768	854	933	1026	1093	1229
	59	778	864	947	1037	1102	1252
	60	776	878	947	1058	1121	1249
	61	854	915	1037	1098	1220	1342
62	1	60	62	-			
	2	100	108	114	118	122	124
	3	121	138	150	162	168	177
	4	146	162	174	192	204	220
	5	165	181	200	218	228	255
	6	184	204	222	242	256	286
	7	199	220	240	266	281	310
	8	216	240	260	286	302	338
	9	229	255	281	308	326	362
	10	246	274	298	328	348	388
	11	261	288	316	346	367	412
	12	274	306	332	366	386	432
	13	288	320	350	382	406	455
	14	302	334	364	402	426	478
	15	313	350	382	418	444	496
	16	326	362	396	436	460	520
	17	338	376	410	450	479	540
	18	352	390	426	468	498	558
	19	363	403	441	484	513	579
	20	376	418	454	500	532	596
	21	386	428	469	512	551	615
	22	400	444	484	532	566	636
	23	410	456	496	548	582	656
	24	422	470	512	562	598	672
	25	431	480	527	579	616	691

N	M	.1	.05	.025	.01	.005	.001
62	26	444	494	540	594	630	710
	27	455	506	552	608	648	729
	28	466	520	566	622	662	744
	29	478	529	578	636	677	764
	30	488	544	592	650	696	778
	31	527	589	620	682	744	837
	32	510	568	620	684	724	818
	33	521	579	631	696	740	833
	34	532	592	646	710	756	850
	35	541	603	658	724	770	868
	36	554	616	672	740	786	884
	37	564	626	686	751	799	899
	38	574	640	698	768	816	920
	39	585	650	710	780	832	935
	40	596	662	722	796	846	952
	41	607	672	733	812	857	963
	42	616	686	748	822	876	984
	43	626	696	760	837	891	1001
	44	638	708	774	850	904	1018
	45	646	719	786	865	921	1035
	46	658	732	798	880	934	1052
	47	668	742	811	892	950	1068
	48	678	754	824	906	964	1086
	49	688	765	835	920	978	1101
	50	698	776	848	934	994	1120
	51	708	788	859	947	1007	1135
	52	720	800	872	960	1020	1152
	53	729	810	882	971	1033	1165
	54	740	822	898	986	1050	1184
	55	748	831	907	1001	1064	1196
	56	760	844	922	1014	1078	1214
	57	768	855	933	1026	1094	1233
	58	778	866	946	1040	1108	1248
	59	789	876	958	1053	1121	1262
	60	798	888	970	1064	1138	1282
	61	817	896	967	1079	1143	1277
	62	868	992	1054	1178	1240	1364
63	1	61	63	-			
	2	102	110	116	120	124	126
	3	126	138	150	162	171	180
	4	148	164	177	196	204	224
63	5	167	185	202	220	232	260
	6	186	207	225	246	261	291
	7	210	231	252	273	287	322
	8	219	243	265	290	306	345
	9	243	270	288	315	333	378
	10	250	278	301	331	351	394
	11	264	293	320	352	374	418
	12	279	312	339	372	393	441
	13	292	324	355	387	411	461
	14	308	343	378	413	434	490
	15	318	354	387	426	450	507
	16	328	370	403	438	468	530
	17	344	382	418	459	486	547
	18	360	405	441	486	513	576
	19	369	409	446	490	522	585
	20	380	424	460	507	539	608
	21	420	462	504	546	567	651
	22	404	449	490	539	572	644
	23	416	462	503	554	589	663
	24	429	477	519	573	609	684
	25	436	486	534	585	623	699
	26	450	500	546	600	639	721
	27	468	522	567	621	666	747
	28	476	532	581	637	679	763
	29	483	538	588	647	686	773
	30	495	552	603	663	705	792
	31	510	555	612	674	710	802
	32	520	580	623	686	741	817
	33	528	588	642	708	750	846
	34	540	599	653	719	765	862
	35	553	616	672	742	784	882
	36	567	630	684	756	801	900
	37	571	635	693	762	810	916
	38	579	649	704	777	828	930
	39	594	660	720	792	843	948
	40	604	671	731	805	856	965
	41	614	683	745	819	872	982
	42	651	714	777	861	903	1008
	43	635	706	770	849	902	1015
	44	645	716	784	860	916	1034
	45	666	738	801	882	936	1053
63	46	667	742	808	890	948	1065
	47	673	749	823	904	964	1079
	48	687	765	834	918	978	1101
	49	700	777	854	938	994	1120
	50	706	786	858	944	1005	1133
	51	720	798	873	960	1020	1149
	52	728	809	882	973	1034	1167
	53	738	821	896	985	1048	1181
	54	756	837	918	1008	1071	1206
	55	759	843	918	1012	1076	1215
	56	770	861	938	1029	1099	1239
	57	780	867	945	1041	1107	1248
	58	788	878	955	1053	1120	1264
	59	798	888	969	1068	1134	1278
	60	810	900	984	1080	1152	1296
	61	818	909	993	1095	1168	1308
	62	835	914	1016	1100	1165	1339
	63	882	1008	1071	1197	1260	1386
64	1	62	64				
	2	102	112	118	122	126	128
	3	126	141	153	165	174	183
	4	152	168	180	196	208	228
	5	171	187	205	225	236	265
	6	188	210	228	248	264	294
	7	207	229	249	273	287	322
	8	232	256	272	304	320	352
	9	239	266	287	315	333	376
	10	254	282	308	338	358	398
	11	268	297	325	356	378	424
	12	284	316	344	376	400	448
	13	293	329	356	394	419	470
	14	310	344	376	412	438	490
	15	322	359	390	430	457	512
	16	352	384	416	464	496	544
	17	348	383	423	463	493	555
	18	362	402	438	482	512	574
	19	373	416	454	499	530	594
	20	388	432	472	516	548	616
	21	395	438	480	525	563	629
	22	410	456	498	546	582	652
	23	421	469	511	562	598	672

N	M	.1	.05	.025	.01	.005	.001
64	24	440	488	536	584	624	696
	25	446	494	540	594	632	710
	26	458	508	554	610	648	728
	27	468	520	567	623	663	750
	28	480	536	584	644	684	768
	29	491	545	596	655	696	783
	30	502	558	610	670	712	802
	31	512	571	623	687	728	821
	32	544	608	672	736	768	864
	33	534	594	649	715	758	853
	34	546	608	662	730	776	874
	35	557	620	674	743	790	890
	36	572	632	692	760	808	912
	37	578	644	702	773	822	925
	38	590	656	716	788	838	944
	39	599	669	728	801	853	960
	40	616	688	744	824	872	984
	41	622	691	754	831	883	995
	42	632	704	768	846	896	1014
	43	638	716	781	863	912	1022
	44	656	728	796	876	932	1048
	45	663	738	805	887	944	1061
	46	674	750	818	902	958	1080
	47	685	762	831	913	972	1096
	48	704	784	864	944	1008	1120
	49	704	783	856	941	1001	1129
	50	716	796	870	956	1018	1148
	51	724	810	878	967	1030	1159
	52	740	820	896	988	1048	1180
	53	747	831	906	999	1063	1195
	54	758	842	920	1012	1078	1212
	55	766	855	930	1023	1088	1231
	56	784	872	952	1048	1112	1248
	57	787	874	957	1051	1121	1263
	58	798	888	968	1066	1134	1280
	59	806	898	981	1080	1149	1293
	60	820	912	996	1096	1164	1312
	61	827	922	1005	1106	1178	1327
	62	838	934	1016	1124	1194	1338
	63	853	932	1040	1121	1189	1364
	64	896	1024	1088	1216	1280	1408

N	M	.1	.05	.025	.01	.005	.001
65	1	63	65	-			
	2	104	112	120	124	128	130
	3	127	144	156	168	174	186
	4	152	168	183	200	212	232
	5	175	195	210	230	240	270
	6	189	212	230	253	270	295
	7	210	232	252	276	294	325
	8	224	252	272	301	318	358
	9	242	268	293	322	340	381
	10	260	290	315	345	365	410
	11	270	301	331	363	386	431
	12	287	318	347	381	405	453
	13	312	351	377	416	429	481
	14	314	348	380	417	445	500
	15	330	365	400	440	465	525
	16	342	379	413	455	479	542
	17	353	394	428	469	500	561
	18	366	409	445	488	518	583
	19	379	421	459	505	536	606
	20	395	440	480	525	560	625
	21	404	450	490	538	574	643
	22	411	465	508	556	583	665
	23	427	474	519	569	607	683
	24	439	488	533	585	624	701
	25	455	505	550	605	645	725
	26	468	520	572	624	663	754
	27	474	528	576	634	674	756
	28	487	538	590	648	687	777
	29	497	553	604	663	706	793
	30	510	570	620	680	725	815
	31	519	577	631	693	738	831
	32	532	594	636	701	757	854
	33	541	604	663	730	770	868
	34	553	615	670	738	785	884
	35	570	630	690	755	805	905
	36	574	639	697	768	818	920
	37	587	653	710	783	831	934
	38	598	664	724	796	849	953
	39	624	689	754	819	871	988
	40	620	690	755	830	880	995
	41	630	700	763	840	895	1010
	42	640	710	775	855	910	1023

N	M	.1	.05	.025	.01	.005	.001
65	43	654	724	789	873	921	1046
	44	662	736	804	882	940	1056
	45	675	750	820	900	960	1080
	46	683	759	828	913	970	1093
	47	693	771	841	926	985	1110
	48	703	782	854	940	1001	1128
	49	712	792	870	954	1017	1147
	50	730	810	885	970	1035	1165
	51	736	816	893	982	1043	1177
	52	754	845	910	1014	1066	1209
	53	757	840	917	1010	1075	1212
	54	765	851	928	1023	1088	1228
	55	780	865	945	1040	1110	1245
	56	786	875	955	1052	1119	1260
	57	796	885	966	1063	1135	1275
	58	807	899	979	1080	1147	1294
	59	818	908	991	1091	1162	1309
	60	830	925	1010	1110	1180	1330
	61	838	933	1017	1120	1190	1344
	62	847	944	1029	1135	1206	1358
	63	858	956	1040	1149	1220	1369
	64	870	952	1061	1144	1248	1389
	65	910	1040	1105	1235	1300	1495
66	1	64	66	-			
	2	106	114	120	126	130	132
	3	132	147	159	171	177	189
	4	154	170	186	204	216	236
	5	175	194	210	230	244	270
	6	198	222	240	258	276	306
	7	214	236	256	281	295	330
	8	230	254	278	304	322	360
	9	246	273	297	327	345	387
	10	262	290	316	346	366	412
	11	286	319	341	374	396	440
	12	294	330	354	390	414	462
	13	306	337	370	403	428	481
	14	318	354	386	424	452	506
	15	333	372	405	444	471	528
	16	346	384	418	460	490	550
	17	358	399	435	477	507	571
	18	378	420	456	498	528	594

N	M	.1	.05	.025	.01	.005	.001
66	19	[illegible]	424	467	509	544	611
	20	396	440	480	530	562	632
	21	411	456	498	546	582	654
	22	440	484	528	572	616	682
	23	433	481	525	578	614	690
	24	450	498	546	600	636	714
	25	458	508	556	609	649	731
	26	470	522	568	626	666	750
	27	483	537	585	642	684	768
	28	492	548	598	658	698	788
	29	503	560	610	672	716	804
	30	522	576	630	690	738	828
	31	526	585	639	701	750	843
	32	538	598	652	718	762	856
	33	594	627	693	759	825	924
	34	560	624	678	746	798	900
	35	572	636	693	763	811	912
	36	588	654	714	780	834	936
	37	593	660	720	792	842	949
	38	604	672	734	808	860	968
	39	618	687	750	825	876	987
	40	626	698	760	838	890	1002
	41	637	709	775	851	907	1021
	42	654	726	792	870	924	1044
	43	659	733	801	880	936	1056
	44	682	770	836	924	968	1100
	45	681	759	828	912	969	1089
	46	692	770	840	924	982	1108
	47	702	779	852	938	995	1125
	48	714	798	870	954	1014	1146
	49	723	804	877	968	1026	1156
	50	734	816	892	980	1044	1176
	51	747	828	906	996	1059	1194
	52	756	840	916	1008	1074	1210
	53	763	852	931	1024	1090	1224
	54	780	870	948	1038	1104	1248
	55	792	880	968	1056	1122	1265
	56	798	886	968	1064	1132	1276
	57	807	900	981	1080	1149	1293
	58	818	910	992	1092	1162	1308
	59	826	921	1003	1106	1176	1325
	60	840	936	1020	1122	1194	1344

N	M	.1	.05	.025	.01	.005	.001
66	61	847	943	1029	1132	1207	1359
	62	858	956	1042	1146	1220	1376
	63	870	969	1053	1161	1236	1392
	64	878	978	1066	1174	1246	1414
	65	887	999	1081	1201	1273	1414
	66	990	1056	1122	1254	1320	1518
67	1	65	67	-			
	2	108	116	122	128	132	134
	3	131	147	159	174	180	192
	4	157	173	188	204	216	240
	5	178	196	215	233	248	275
	6	197	220	238	262	276	306
	7	216	238	259	286	301	335
	8	232	258	280	309	327	365
	9	249	276	299	330	348	392
	10	265	293	320	350	372	419
	11	281	308	339	372	395	440
	12	294	327	358	390	416	468
	13	309	343	375	412	438	490
	14	323	359	392	432	455	513
	15	335	372	408	446	475	535
	16	350	389	424	468	496	555
	17	363	400	443	482	515	581
	18	377	418	455	503	534	598
	19	388	434	473	520	550	617
	20	400	446	486	536	570	640
	21	414	460	502	552	586	661
	22	430	476	520	568	609	679
	23	439	487	531	587	621	702
	24	452	501	545	602	640	720
	25	464	515	553	616	657	740
	26	475	527	577	633	674	760
	27	489	543	588	651	692	775
	28	499	555	603	664	708	797
	29	510	568	619	682	724	816
	30	521	579	634	696	742	833
	31	534	592	647	711	757	850
	32	544	605	660	727	775	871
	33	553	617	679	747	785	885
	34	562	627	690	759	798	899
	35	578	643	702	773	821	925

N	M	.1	.05	.025	.01	.005	.001
67	36	590	656	714	786	837	941
	37	602	669	729	803	854	960
	38	613	680	743	817	870	978
	39	623	693	756	830	885	997
	40	635	704	770	849	902	1012
	41	645	717	783	861	917	1031
	42	657	728	794	877	933	1050
	43	667	740	809	891	946	1068
	44	677	755	824	908	961	1086
	45	686	770	839	914	977	1107
	46	700	777	848	934	994	1120
	47	709	789	862	949	1010	1137
	48	722	801	876	964	1023	1155
	49	733	814	888	978	1040	1171
	50	743	827	898	994	1059	1192
	51	752	838	914	1007	1071	1206
	52	763	850	927	1020	1084	1224
	53	774	861	940	1036	1099	1240
	54	784	874	954	1049	1116	1256
	55	795	886	965	1063	1130	1273
	56	803	898	979	1079	1146	1292
	57	817	909	991	1091	1159	1306
	58	827	919	1003	1105	1175	1324
	59	836	931	1017	1119	1191	1340
	60	848	943	1029	1133	1205	1357
	61	857	954	1041	1145	1218	1375
	62	868	966	1054	1162	1235	1390
	63	878	976	1066	1173	1248	1407
	64	889	988	1078	1187	1265	1422
	65	898	1000	1091	1199	1273	1447
	66	904	1021	1101	1225	1297	1440
	67	1005	1072	1206	1273	1340	1541
68	1	66	68				
	2	108	118	124	130	132	136
	3	133	150	162	177	183	195
	4	160	180	192	208	220	240
	5	180	199	217	237	250	280
	6	200	222	242	264	280	310
	7	219	242	263	289	305	340
	8	236	264	288	316	332	372
	9	252	279	305	336	355	396

N	M	.1	.05	.025	.01	.005	.001
68	10	268	298	324	356	376	424
	11	284	315	342	375	401	449
	12	300	332	364	400	424	476
	13	313	348	381	416	442	495
	14	328	364	398	436	464	522
	15	342	380	412	455	480	540
	16	356	396	432	476	504	568
	17	391	425	459	510	544	595
	18	382	424	462	508	540	608
	19	395	437	479	525	558	627
	20	408	456	496	544	580	652
	21	418	465	507	559	595	669
	22	432	480	524	578	612	692
	23	445	492	538	587	629	714
	24	460	512	556	612	652	732
	25	469	521	570	625	664	748
	26	482	536	584	642	684	768
	27	494	549	599	658	699	783
	28	508	564	616	676	720	808
	29	518	576	626	690	733	826
	30	528	588	642	706	750	846
	31	540	599	654	721	766	864
	32	552	616	672	740	784	884
	33	561	625	680	748	801	904
	34	612	680	714	782	850	952
	35	586	652	711	784	831	936
	36	600	668	728	800	852	956
	37	608	677	738	813	863	973
	38	620	690	752	828	880	992
	39	629	699	765	842	893	1008
	40	644	716	780	860	916	1028
	41	651	728	795	869	924	1044
	42	664	738	806	888	944	1064
	43	675	751	819	901	959	1080
	44	688	764	836	920	976	1100
	45	693	778	847	935	986	1118
	46	708	788	860	944	1006	1136
	47	719	798	871	960	1022	1151
	48	732	812	888	976	1040	1172
	49	741	823	899	989	1052	1187
	50	752	836	912	1004	1068	1204
68	51	782	867	935	1037	1105	1241
	52	776	860	940	1036	1100	1240
	53	783	872	950	1048	1115	1253
	54	794	884	964	1060	1128	1274
	55	805	894	977	1074	1142	1289
	56	816	908	992	1092	1160	1308
	57	825	920	1003	1104	1174	1323
	58	838	930	1016	1118	1190	1340
	59	847	942	1028	1132	1205	1355
	60	860	956	1044	1148	1220	1376
	61	869	965	1054	1159	1235	1392
	62	878	978	1066	1174	1250	1408
	63	889	987	1079	1187	1264	1424
	64	900	1004	1092	1204	1280	1444
	65	910	1010	1105	1214	1291	1458
	66	920	1022	1118	1226	1306	1476
	67	922	1041	1121	1243	1320	1506
	68	1020	1088	1224	1292	1428	1564
69	1	67	69	-			
	2	110	120	126	132	134	138
	3	138	153	165	177	186	198
	4	163	179	195	212	224	244
	5	182	201	221	241	251	285
	6	204	225	246	267	285	315
	7	220	246	267	294	309	344
	8	240	265	289	318	336	374
	9	258	285	309	339	360	402
	10	272	302	331	362	382	431
	11	286	319	348	381	406	456
	12	303	339	366	402	429	480
	13	318	353	383	422	448	504
	14	329	368	398	439	467	523
	15	348	384	420	462	489	549
	16	359	398	435	477	509	572
	17	371	417	454	493	525	593
	18	387	432	471	516	549	615
	19	399	444	486	532	567	638
	20	412	459	500	550	586	657
	21	426	474	516	567	603	678
	22	437	486	530	583	621	696
	23	460	529	575	621	667	736
69	24	465	516	564	618	657	741
	25	474	529	575	634	674	756
	26	488	541	592	652	689	775
	27	501	558	606	669	711	798
	28	512	570	620	682	725	818
	29	522	580	634	699	743	836
	30	537	597	651	717	762	855
	31	547	608	663	730	775	873
	32	557	620	678	747	794	891
	33	573	636	693	762	810	912
	34	574	640	704	774	832	937
	35	597	651	715	786	848	953
	36	606	672	735	810	858	969
	37	615	685	748	822	876	985
	38	627	698	759	838	890	1003
	39	639	711	777	855	909	1023
	40	649	722	788	867	922	1040
	41	661	735	801	883	939	1058
	42	672	750	816	900	957	1077
	43	684	761	830	911	970	1091
	44	694	771	843	927	984	1110
	45	708	786	858	942	1005	1131
	46	736	828	897	989	1035	1173
	47	728	808	883	971	1034	1164
	48	741	822	837	987	1050	1182
	49	749	833	908	1001	1065	1201
	50	759	846	922	1017	1080	1217
	51	774	858	936	1032	1098	1236
	52	784	872	945	1046	1114	1253
	53	793	882	963	1059	1127	1269
	54	804	894	978	1074	1143	1287
	55	811	904	987	1084	1153	1303
	56	824	917	1001	1102	1171	1321
	57	837	930	1014	1119	1188	1338
	58	846	941	1028	1130	1203	1355
	59	855	952	1040	1146	1217	1374
	60	870	966	1053	1161	1233	1392
	61	879	976	1066	1172	1248	1407
	62	887	989	1079	1187	1263	1424
	63	900	1002	1092	1203	1281	1443
	64	909	1012	1104	1215	1294	1457
	65	920	1022	1116	1229	1309	1473

N	M	.1	.05	.025	.01	.005	.001
69	66	933	1035	1131	1245	1323	1491
	67	941	1045	1144	1254	1343	1505
	68	939	1061	1142	1270	1343	1536
	69	1035	1104	1242	1380	1449	1587
70	1	68	70	-			
	2	112	122	128	134	136	140
	3	137	153	168	180	189	201
	4	164	182	194	216	228	248
	5	190	210	225	245	260	285
	6	206	228	248	272	288	320
	7	231	252	273	301	322	357
	8	242	268	292	322	340	380
	9	260	287	314	344	364	409
	10	280	310	340	370	400	440
	11	292	325	352	387	413	461
	12	308	342	372	408	434	486
	13	321	357	390	429	455	508
	14	350	392	420	462	490	546
	15	355	395	430	470	500	560
	16	364	406	442	486	516	580
	17	379	420	456	504	536	604
	18	392	436	476	520	554	622
	19	404	449	490	538	575	645
	20	430	470	510	570	600	670
	21	434	483	525	581	616	693
	22	444	494	538	592	630	708
	23	455	503	549	613	642	730
	24	468	522	568	624	664	750
	25	485	540	585	645	685	770
	26	494	550	598	658	700	788
	27	505	560	613	673	716	807
	28	532	588	644	700	742	840
	29	530	590	642	707	752	846
	30	550	610	670	730	780	870
	31	554	616	672	736	786	883
	32	566	628	686	754	804	902
	33	576	641	699	770	819	922
	34	588	654	714	788	834	940
	35	630	700	770	840	875	980
	36	612	680	742	818	868	980
	37	622	692	756	832	886	997

N	M	.1	.05	.025	.01	.005	.001
70	38	634	706	770	848	902	1016
	39	645	719	784	862	916	1033
	40	670	740	810	890	940	1060
	41	668	744	812	894	950	1070
	42	686	770	840	924	980	1106
	43	692	768	838	923	982	1106
	44	702	782	852	940	1000	1124
	45	715	795	870	955	1015	1145
	46	726	806	880	970	1028	1160
	47	737	813	897	977	1044	1181
	48	748	830	906	998	1062	1196
	49	763	847	924	1015	1085	1218
	50	780	860	940	1040	1100	1240
	51	780	865	946	1042	1107	1247
	52	792	880	960	1058	1124	1268
	53	801	891	972	1072	1142	1284
	54	814	904	986	1086	1156	1302
	55	825	920	1005	1105	1175	1320
	56	840	938	1022	1120	1204	1344
	57	846	940	1026	1128	1201	1354
	58	856	952	1038	1144	1216	1372
	59	867	964	1050	1158	1230	1388
	60	890	980	1070	1180	1260	1410
	61	887	988	1077	1187	1260	1424
	62	898	1000	1090	1200	1278	1440
	63	917	1015	1106	1218	1295	1463
	64	920	1024	1116	1230	1308	1474
	65	935	1040	1135	1245	1325	1495
	66	942	1046	1142	1258	1338	1508
	67	952	1056	1156	1272	1353	1526
	68	962	1070	1168	1288	1372	1534
	69	958	1080	1164	1292	1367	1563
	70	1050	1190	1260	1400	1470	1610
71	1	69	71	-			
	2	114	124	130	136	138	142
	3	139	156	171	183	192	204
	4	165	184	197	213	232	252
	5	188	208	229	249	260	290
	6	207	230	252	272	289	324
	7	228	252	273	301	320	355
	8	244	273	297	323	345	385

N	M	.1	.05	.025	.01	.005	.001
71	9	263	291	317	346	370	415
	10	279	309	338	370	396	440
	11	294	327	355	392	415	464
	12	312	342	376	412	436	494
	13	326	361	393	432	458	516
	14	341	376	415	455	483	543
	15	354	394	429	473	503	564
	16	368	410	447	493	523	587
	17	384	425	463	510	541	608
	18	399	439	477	528	562	633
	19	410	454	496	544	581	653
	20	423	470	514	565	599	676
	21	437	434	529	582	618	697
	22	449	500	544	600	637	715
	23	461	515	561	615	657	734
	24	470	531	580	632	677	753
	25	487	541	591	650	691	775
	26	500	555	605	668	708	798
	27	512	568	621	683	727	819
	28	524	582	636	699	746	836
	29	536	596	650	715	762	857
	30	548	608	665	732	779	877
	31	560	623	679	748	796	894
	32	571	635	694	763	812	914
	33	583	649	709	780	829	933
	34	595	662	722	796	847	952
	35	612	681	730	802	864	971
	36	622	693	742	834	877	985
	37	629	701	764	841	894	1008
	38	642	714	779	858	912	1026
	39	654	726	793	872	931	1048
	40	665	740	807	887	945	1064
	41	677	752	820	905	961	1083
	42	687	766	835	919	977	1100
	43	700	777	847	933	994	1117
	44	710	790	862	949	1010	1136
	45	723	802	877	963	1026	1156
	46	733	814	889	977	1042	1173
	47	744	831	906	999	1053	1192
	48	754	840	916	1008	1074	1212
	49	766	852	930	1023	1089	1228
	50	778	865	944	1040	1106	1245

N	M	.1	.05	.025	.01	.005	.001
71	51	788	877	957	1053	1121	1263
	52	800	888	969	1068	1136	1280
	53	809	898	984	1087	1147	1301
	54	822	914	998	1096	1167	1316
	55	832	925	1010	1112	1183	1334
	56	843	937	1024	1127	1198	1351
	57	851	948	1034	1144	1216	1371
	58	865	960	1049	1157	1230	1385
	59	878	975	1061	1167	1247	1402
	60	887	986	1076	1183	1260	1418
	61	898	1000	1091	1200	1275	1438
	62	909	1009	1104	1212	1291	1452
	63	919	1022	1116	1227	1306	1471
	64	929	1034	1128	1241	1320	1489
	65	939	1045	1140	1257	1335	1506
	66	952	1058	1155	1271	1353	1525
	67	961	1068	1166	1284	1367	1541
	68	972	1081	1181	1300	1383	1558
	69	983	1092	1194	1318	1399	1568
	70	1004	1099	1219	1314	1394	1589
	71	1065	1207	1278	1420	1491	1704
72	1	70	72	-			
	2	116	124	132	138	140	144
	3	144	159	171	186	192	207
	4	172	188	204	220	232	256
	5	190	210	230	250	263	295
	6	216	240	258	282	300	330
	7	231	257	278	304	322	360
	8	256	280	304	336	352	392
	9	270	306	333	360	378	423
	10	284	314	342	376	398	448
	11	300	333	362	396	423	473
	12	324	360	396	432	456	504
	13	330	367	400	440	467	524
	14	346	384	418	460	488	546
	15	360	402	438	480	510	573
	16	384	424	456	504	536	600
	17	388	431	469	516	550	614
	18	414	468	504	558	576	648
	19	415	461	503	553	587	659
	20	432	480	520	572	608	684
72	21	444	492	537	591	627	705
	22	456	506	552	608	644	726
	23	468	520	568	625	663	746
	24	504	552	600	672	696	792
	25	493	548	598	658	700	787
	26	506	562	614	674	718	806
	27	522	585	639	702	747	837
	28	532	592	644	712	756	848
	29	542	601	658	727	772	862
	30	558	624	678	744	792	888
	31	565	632	687	758	802	904
	32	584	648	712	776	832	928
	33	591	657	717	789	840	945
	34	602	670	732	804	856	964
	35	613	682	746	820	871	987
	36	648	720	792	864	936	1044
	37	637	709	774	850	908	1021
	38	650	722	788	868	924	1038
	39	663	738	804	885	942	1059
	40	680	752	824	904	960	1080
	41	684	758	829	911	972	1095
	42	702	780	846	936	996	1116
	43	709	785	857	943	1002	1130
	44	720	800	876	964	1024	1152
	45	738	819	891	981	1044	1179
	46	742	824	900	992	1054	1188
	47	752	837	913	1006	1070	1205
	48	792	864	960	1032	1104	1248
	49	776	861	941	1035	1100	1241
	50	786	874	956	1050	1118	1260
	51	798	888	969	1068	1134	1278
	52	812	900	984	1084	1152	1296
	53	820	912	995	1094	1165	1313
	54	846	936	1026	1134	1206	1350
	55	841	935	1022	1125	1197	1349
	56	856	952	1040	1144	1216	1376
	57	864	963	1050	1155	1230	1386
	58	876	974	1062	1170	1244	1402
	59	887	986	1073	1183	1259	1419
	60	912	1008	1104	1212	1284	1452
	61	907	1008	1101	1212	1290	1455
	62	918	1022	1116	1228	1306	1472
72	63	936	1044	1134	1251	1332	1494
	64	944	1048	1144	1264	1344	1512
	65	950	1057	1155	1271	1351	1524
	66	966	1074	1170	1290	1374	1542
	67	972	1080	1180	1299	1382	1557
	68	984	1096	1196	1316	1400	1576
	69	996	1107	1206	1329	1413	1593
	70	1004	1118	1218	1344	1426	1616
	71	1025	1118	1243	1338	1457	1616
	72	1080	1224	1296	1440	1512	1728
73	1	71	73	-			
	2	116	126	134	140	142	146
	3	143	162	174	189	195	207
	4	171	188	203	224	236	260
	5	194	214	232	255	267	300
	6	214	238	258	282	299	335
	7	235	259	281	309	329	365
	8	253	279	304	335	352	398
	9	269	297	324	359	378	423
	10	287	318	345	380	404	451
	11	302	338	365	401	427	478
	12	320	356	384	420	453	503
	13	334	371	405	446	472	530
	14	349	388	424	467	495	556
	15	363	404	441	486	516	578
	16	378	421	460	503	535	602
	17	392	436	476	524	555	626
	18	408	448	496	538	574	646
	19	421	467	509	560	595	670
	20	435	482	524	577	615	689
	21	446	497	540	593	634	709
	22	459	510	559	612	654	735
	23	473	525	574	632	673	755
	24	489	543	592	645	690	768
	25	499	554	605	666	710	795
	26	511	569	621	681	727	816
	27	525	583	636	699	744	837
	28	536	597	652	716	761	856
	29	548	608	664	735	780	880
	30	561	625	681	749	797	898

N	M	.1	.05	.025	.01	.005	.001
73	31	574	637	696	767	814	917
	32	584	651	711	780	830	935
	33	598	664	725	798	849	955
	34	610	678	738	815	867	975
	35	621	691	754	830	882	993
	36	634	706	775	850	893	1004
	37	644	717	787	863	907	1041
	38	657	730	797	876	933	1051
	39	669	743	812	894	949	1070
	40	680	756	827	908	966	1087
	41	692	769	838	924	983	1107
	42	704	784	854	941	999	1126
	43	716	794	867	956	1017	1145
	44	726	810	884	973	1032	1164
	45	739	822	897	987	1049	1179
	46	750	834	910	1002	1065	1199
	47	760	846	922	1016	1081	1219
	48	773	859	938	1034	1095	1236
	49	787	869	946	1042	1112	1257
	50	795	883	964	1062	1129	1274
	51	806	895	979	1077	1147	1289
	52	818	911	994	1090	1162	1308
	53	828	920	1006	1106	1178	1325
	54	840	933	1020	1123	1193	1345
	55	851	944	1034	1141	1214	1363
	56	862	960	1046	1152	1225	1381
	57	873	971	1060	1165	1241	1399
	58	884	983	1075	1182	1256	1416
	59	895	996	1086	1196	1273	1433
	60	907	1007	1100	1211	1289	1452
	61	919	1019	1115	1225	1301	1469
	62	929	1032	1128	1240	1320	1487
	63	940	1044	1140	1255	1336	1505
	64	948	1055	1155	1267	1349	1522
	65	962	1068	1166	1286	1368	1539
	66	972	1080	1180	1299	1382	1556
	67	983	1093	1192	1313	1398	1574
	68	993	1105	1207	1328	1412	1592
	69	1004	1117	1219	1342	1427	1609
	70	1014	1130	1231	1356	1443	1624
	71	1026	1141	1241	1372	1455	1650
	72	1044	1139	1265	1400	1483	1643

N	M	.1	.05	.025	.01	.005	.001
73	73	1095	1241	1314	1460	1533	1752
74	1	72	74	-			
	2	118	128	136	142	144	148
	3	145	162	177	189	198	210
	4	174	190	206	228	240	264
	5	196	216	236	256	271	305
	6	216	240	260	284	304	336
	7	237	262	286	314	332	370
	8	256	284	310	338	360	400
	9	273	304	331	361	385	430
	10	290	322	352	386	410	458
	11	308	341	370	408	433	485
	12	324	358	392	430	456	512
	13	338	376	410	450	479	537
	14	354	394	428	472	500	562
	15	367	410	443	487	517	587
	16	384	426	464	510	542	608
	17	396	441	481	530	561	632
	18	412	458	498	550	584	658
	19	425	472	515	567	603	677
	20	440	488	532	586	624	700
	21	451	502	546	605	640	722
	22	466	518	566	622	660	744
	23	479	533	581	639	680	765
	24	492	548	596	658	700	784
	25	507	560	610	679	711	807
	26	518	576	628	692	736	828
	27	530	590	644	706	752	847
	28	544	604	660	726	772	868
	29	556	617	674	740	788	888
	30	568	632	690	760	806	908
	31	581	645	705	775	824	929
	32	592	660	720	792	842	948
	33	605	672	731	805	860	968
	34	618	686	748	824	876	986
	35	629	699	762	840	892	1005
	36	642	712	776	854	912	1026
	37	703	777	814	888	962	1073
	38	664	740	806	890	942	1060
	39	677	751	821	903	961	1082
	40	688	766	836	920	978	1102

N	M	.1	.05	.025	.01	.005	.001
74	41	701	777	850	934	992	1121
	42	712	792	864	950	1012	1140
	43	723	804	878	968	1028	1159
	44	736	818	892	982	1046	1178
	45	747	829	906	996	1062	1195
	46	758	844	920	1014	1078	1212
	47	770	857	933	1027	1093	1233
	48	782	870	948	1044	1110	1252
	49	796	877	966	1052	1122	1268
	50	804	894	976	1074	1144	1290
	51	814	906	989	1090	1159	1307
	52	828	920	1004	1106	1174	1322
	53	839	929	1014	1118	1191	1340
	54	850	944	1032	1134	1208	1360
	55	860	957	1045	1148	1221	1380
	56	872	970	1058	1166	1240	1396
	57	882	982	1070	1178	1256	1414
	58	894	994	1086	1194	1270	1432
	59	907	1009	1100	1207	1282	1445
	60	916	1020	1112	1224	1304	1468
	61	927	1031	1126	1239	1318	1484
	62	938	1044	1140	1254	1336	1502
	63	949	1056	1152	1268	1350	1520
	64	960	1068	1166	1284	1366	1540
	65	971	1080	1180	1299	1382	1557
	66	982	1092	1194	1312	1396	1574
	67	993	1104	1205	1328	1412	1589
	68	1004	1119	1220	1342	1428	1608
	69	1014	1128	1231	1355	1443	1626
	70	1026	1142	1246	1372	1453	1644
	71	1035	1153	1257	1384	1473	1661
	72	1048	1166	1270	1398	1484	1680
	73	1062	1161	1286	1426	1507	1712
	74	1110	1258	1332	1480	1554	1776
75	1	73	75	-			
	2	120	130	136	144	146	150
	3	147	165	180	192	201	213
	4	176	193	209	225	244	264
	5	200	220	240	265	275	305
	6	219	246	267	291	309	339
	7	240	265	289	317	336	373

N	M	.1	.05	.025	.01	.005	.001
75	8	258	287	312	343	362	408
	9	279	309	336	369	390	435
	10	295	330	360	395	415	465
	11	310	345	376	413	437	490
	12	327	363	396	435	462	519
	13	343	380	416	455	484	545
	14	358	399	434	476	507	569
	15	390	435	465	510	540	600
	16	388	431	469	517	549	617
	17	403	447	488	537	570	641
	18	417	465	507	558	591	663
	19	428	480	520	575	612	688
	20	450	495	540	595	635	710
	21	459	510	558	612	651	732
	22	473	523	572	629	668	752
	23	483	539	589	647	688	774
	24	498	555	606	666	708	795
	25	525	600	650	700	750	850
	26	524	582	636	698	744	836
	27	537	597	654	717	762	858
	28	549	612	668	735	781	876
	29	562	625	683	750	799	898
	30	585	645	705	780	825	930
	31	588	652	712	784	833	939
	32	597	668	725	799	851	958
	33	612	681	744	819	870	978
	34	624	693	757	834	884	998
	35	640	710	775	855	905	1020
	36	648	723	786	867	921	1038
	37	657	731	802	879	945	1061
	38	666	741	813	891	960	1076
	39	687	762	831	915	972	1095
	40	700	775	850	935	990	1115
	41	707	788	858	946	1007	1131
	42	720	801	876	963	1023	1152
	43	733	812	887	974	1037	1167
	44	742	825	902	993	1055	1191
	45	765	855	930	1020	1080	1215
	46	768	853	929	1023	1090	1228
	47	777	867	944	1038	1104	1245
	48	792	879	960	1056	1125	1266
	49	801	892	971	1069	1138	1282
75	50	825	925	1000	1100	1175	1325
	51	825	918	1002	1101	1173	1320
	52	835	928	1014	1117	1187	1337
	53	846	942	1027	1131	1204	1354
	54	861	957	1044	1149	1221	1377
	55	875	970	1060	1165	1240	1395
	56	882	981	1073	1173	1248	1413
	57	894	993	1083	1194	1269	1431
	58	903	1005	1097	1208	1285	1447
	59	915	1017	1111	1221	1300	1466
	60	945	1050	1140	1245	1335	1500
	61	937	1042	1138	1251	1331	1500
	62	948	1053	1151	1268	1348	1519
	63	960	1068	1167	1284	1365	1539
	64	970	1080	1179	1296	1380	1554
	65	985	1095	1195	1315	1400	1575
	66	993	1104	1206	1329	1413	1593
	67	1004	1115	1218	1340	1425	1607
	68	1016	1128	1232	1356	1443	1624
	69	1026	1143	1248	1371	1461	1644
	70	1040	1155	1260	1390	1475	1665
	71	1045	1165	1270	1400	1490	1676
	72	1059	1179	1284	1416	1506	1698
	73	1069	1189	1297	1425	1519	1710
	74	1081	1214	1308	1450	1531	1745
	75	1125	1275	1425	1500	1650	1800
76	1	74	76	-			
	2	122	132	138	144	148	152
	3	149	168	180	195	204	216
	4	180	200	212	232	244	268
	5	200	223	244	264	279	310
	6	222	248	268	292	312	344
	7	241	268	290	322	338	379
	8	264	292	320	348	372	412
	9	280	312	339	371	393	438
	10	298	330	360	396	420	470
	11	313	347	378	420	443	497
	12	332	372	404	444	468	528
	13	348	385	422	463	489	550
	14	362	404	440	482	514	576
	15	375	418	460	506	536	599
76	16	396	440	480	524	556	628
	17	405	453	495	540	574	648
	18	422	468	512	562	598	670
	19	456	494	551	589	627	703
	20	452	504	548	600	640	720
	21	463	515	562	618	659	739
	22	478	530	580	636	678	762
	23	491	545	595	655	696	781
	24	508	564	612	676	716	808
	25	517	571	622	692	739	821
	26	530	590	644	708	750	846
	27	543	603	657	723	771	869
	28	560	620	676	744	792	892
	29	568	632	690	759	807	908
	30	582	646	706	776	826	928
	31	595	660	720	792	843	950
	32	608	676	740	812	864	972
	33	618	688	749	825	879	990
	34	632	702	766	842	898	1010
	35	644	716	780	857	914	1029
	36	656	732	796	876	932	1052
	37	666	743	809	893	946	1063
	38	722	798	874	950	988	1140
	39	691	769	841	925	982	1110
	40	704	784	856	944	1000	1128
	41	716	794	869	957	1016	1145
	42	728	810	884	972	1034	1164
	43	739	822	898	988	1050	1185
	44	752	836	916	1004	1068	1204
	45	762	848	925	1018	1084	1222
	46	776	862	940	1036	1102	1240
	47	787	874	954	1050	1117	1258
	48	800	892	972	1068	1136	1280
	49	810	899	984	1082	1149	1296
	50	822	914	998	1098	1166	1314
	51	830	927	1007	1109	1183	1334
	52	848	940	1028	1132	1204	1356
	53	855	951	1040	1143	1216	1370
	54	868	966	1054	1160	1234	1388
	55	878	978	1067	1173	1249	1408
	56	892	992	1084	1192	1268	1428
	57	912	1026	1121	1216	1292	1463

N	M	.1	.05	.025	.01	.005	.001
76	58	914	1016	1108	1220	1298	1462
	59	924	1027	1120	1236	1313	1481
	60	940	1044	1140	1252	1332	1500
	61	949	1056	1150	1267	1347	1516
	62	958	1066	1164	1280	1362	1534
	63	969	1077	1178	1294	1378	1554
	64	984	1092	1192	1312	1396	1572
	65	992	1102	1201	1327	1407	1589
	66	1004	1116	1218	1340	1426	1606
	67	1013	1128	1231	1353	1442	1625
	68	1028	1144	1248	1372	1460	1644
	69	1035	1151	1258	1385	1474	1661
	70	1048	1164	1272	1400	1490	1678
	71	1058	1175	1285	1413	1504	1696
	72	1072	1192	1300	1432	1524	1716
	73	1079	1200	1310	1442	1535	1731
	74	1092	1212	1326	1454	1556	1742
	75	1099	1236	1329	1473	1556	1773
	76	1216	1292	1444	1596	1672	1824
77	1	75	77	-			
	2	122	134	140	146	150	154
	3	151	168	183	198	207	219
	4	180	199	215	236	248	272
	5	203	225	245	268	283	315
	6	224	248	271	296	313	349
	7	252	280	301	329	350	385
	8	265	294	320	352	374	416
	9	283	315	342	376	399	449
	10	302	335	365	400	426	476
	11	330	363	396	429	462	517
	12	335	372	407	446	472	532
	13	353	392	422	470	497	560
	14	371	413	448	497	525	588
	15	382	425	462	508	539	609
	16	398	441	482	528	563	633
	17	413	458	500	550	585	654
	18	426	474	517	569	603	679
	19	437	490	530	586	624	701
	20	455	506	552	607	646	724
	21	476	525	574	630	672	756

N	M	.1	.05	.025	.01	.005	.001
77	22	495	550	594	649	693	781
	23	496	551	603	663	704	794
	24	511	566	618	681	724	814
	25	522	582	636	695	742	838
	26	533	600	655	712	762	862
	27	549	611	666	733	780	877
	28	567	630	686	756	805	903
	29	575	641	699	767	816	921
	30	586	654	713	785	834	940
	31	598	670	731	798	856	963
	32	613	682	742	818	870	979
	33	638	704	770	847	902	1012
	34	637	710	774	853	907	1019
	35	658	728	798	875	931	1043
	36	662	737	804	885	939	1060
	37	674	751	819	902	959	1078
	38	680	756	828	908	977	1094
	39	692	783	840	942	990	1109
	40	710	790	863	949	1009	1138
	41	722	804	876	967	1028	1157
	42	742	826	896	987	1050	1183
	43	748	832	906	998	1063	1196
	44	770	858	935	1023	1089	1221
	45	770	858	935	1031	1096	1232
	46	780	870	948	1042	1114	1253
	47	794	884	965	1061	1128	1273
	48	805	898	978	1075	1148	1293
	49	826	917	1001	1099	1169	1316
	50	830	923	1007	1109	1180	1329
	51	837	935	1016	1118	1193	1345
	52	854	949	1035	1139	1214	1368
	53	865	962	1049	1155	1227	1385
	54	876	975	1062	1169	1245	1401
	55	902	1001	1089	1199	1276	1430
	56	903	1008	1099	1204	1281	1449
	57	911	1013	1107	1218	1295	1459
	58	920	1029	1115	1228	1306	1477
	59	935	1040	1135	1248	1326	1497
	60	946	1050	1145	1263	1342	1513
	61	956	1064	1161	1278	1360	1532
	62	967	1076	1175	1295	1376	1549
	63	987	1092	1197	1316	1400	1575

N	M	.1	.05	.025	.01	.005	.001
77	64	992	1099	1201	1326	1406	1585
	65	1001	1114	1215	1339	1424	1606
	66	1023	1133	1243	1364	1452	1628
	67	1024	1138	1242	1369	1456	1640
	68	1036	1150	1258	1384	1472	1660
	69	1047	1163	1271	1400	1490	1677
	70	1064	1183	1288	1421	1512	1701
	71	1069	1188	1298	1429	1519	1714
	72	1080	1201	1312	1444	1536	1732
	73	1090	1213	1324	1458	1551	1748
	74	1101	1225	1339	1471	1567	1767
	75	1111	1236	1353	1488	1588	1775
	76	1117	1257	1351	1497	1581	1800
	77	1232	1309	1463	1617	1694	1925
78	1	76	78	-			
	2	124	134	142	148	152	156
	3	153	171	186	201	210	222
	4	182	202	218	240	252	276
	5	207	227	247	272	287	320
	6	234	258	276	306	324	360
	7	250	277	300	329	349	390
	8	268	298	324	356	378	420
	9	288	321	348	381	405	453
	10	306	338	368	406	430	482
	11	324	360	392	428	451	506
	12	342	384	414	456	486	540
	13	364	403	442	481	507	572
	14	372	412	450	496	524	588
	15	387	432	471	516	549	615
	16	402	446	488	536	570	638
	17	417	464	505	556	590	663
	18	438	486	528	582	618	690
	19	446	497	541	596	634	712
	20	462	512	558	614	652	732
	21	477	528	576	633	675	756
	22	488	544	592	652	692	778
	23	502	558	609	669	711	802
	24	522	576	630	690	738	828
	25	529	588	642	705	750	844
	26	572	624	676	754	806	884

N	M	.1	.05	.025	.01	.005	.001
78	27	558	618	675	744	789	888
	28	570	632	690	760	806	908
	29	581	646	705	777	826	927
	30	600	666	726	798	846	954
	31	604	676	738	812	864	972
	32	620	690	752	828	882	992
	33	633	705	768	846	900	1014
	34	646	718	784	862	916	1032
	35	657	731	797	878	933	1053
	36	672	750	816	900	954	1074
	37	681	759	828	911	968	1091
	38	694	772	844	928	988	1114
	39	741	819	897	975	1053	1170
	40	720	800	872	958	1024	1148
	41	731	813	887	978	1037	1170
	42	750	834	906	996	1062	1194
	43	755	840	916	1009	1074	1208
	44	768	854	932	1026	1090	1228
	45	780	867	948	1044	1110	1248
	46	792	880	962	1058	1124	1266
	47	804	896	976	1072	1137	1289
	48	822	912	996	1092	1164	1308
	49	827	921	1003	1106	1175	1323
	50	840	934	1018	1120	1192	1344
	51	852	948	1035	1137	1209	1362
	52	884	988	1066	1170	1248	1404
	53	874	972	1060	1166	1242	1401
	54	888	990	1080	1188	1266	1422
	55	897	998	1089	1199	1274	1437
	56	910	1012	1104	1216	1292	1456
	57	921	1026	1119	1230	1311	1476
	58	932	1036	1132	1246	1324	1494
	59	944	1049	1144	1259	1340	1510
	60	960	1068	1164	1278	1362	1536
	61	967	1074	1172	1291	1374	1548
	62	978	1088	1188	1308	1390	1568
	63	990	1101	1203	1323	1407	1587
	64	1002	1114	1216	1338	1422	1604
	65	1027	1131	1235	1365	1456	1638
	66	1026	1140	1248	1374	1458	1644
	67	1034	1153	1256	1380	1469	1657
	68	1046	1164	1270	1398	1488	1676
78	69	1059	1179	1284	1413	1506	1695
	70	1070	1188	1298	1428	1520	1714
	71	1080	1202	1310	1444	1536	1729
	72	1098	1218	1326	1464	1554	1752
	73	1102	1225	1336	1473	1565	1765
	74	1114	1238	1352	1488	1584	1784
	75	1125	1251	1368	1503	1599	1803
	76	1134	1262	1380	1522	1616	1826
	77	1136	1278	1374	1520	1610	1828
	78	1248	1326	1482	1638	1716	1950
79	1	76	78	79	–		
	2	126	136	144	150	154	158
	3	155	174	189	204	210	225
	4	184	204	220	237	256	280
	5	207	231	251	275	290	325
	6	231	257	280	304	323	359
	7	252	280	302	334	353	395
	8	273	300	329	361	379	425
	9	290	323	351	384	409	456
	10	306	343	373	411	433	483
	11	326	362	394	434	460	515
	12	342	382	414	456	485	545
	13	360	400	438	479	506	571
	14	375	417	455	501	532	598
	15	390	435	474	521	553	622
	16	406	453	489	545	579	645
	17	422	468	512	563	597	671
	18	437	486	530	581	620	693
	19	451	502	548	602	638	717
	20	466	513	566	624	663	744
	21	480	534	582	640	681	766
	22	494	548	599	658	701	788
	23	507	565	616	679	721	810
	24	521	579	632	696	740	832
	25	535	594	648	714	760	852
	26	543	613	667	725	775	878
	27	562	624	681	749	799	901
	28	575	640	697	768	816	918
	29	587	653	713	786	836	940
	30	600	667	729	802	854	960
79	31	614	682	744	820	872	982
	32	627	697	761	837	891	1002
	33	638	710	775	854	907	1020
	34	651	727	789	868	924	1045
	35	664	740	804	889	943	1065
	36	675	752	823	904	961	1083
	37	689	767	836	920	979	1104
	38	701	780	852	937	998	1122
	39	720	800	872	960	1007	1152
	40	731	812	888	973	1022	1169
	41	737	820	895	986	1050	1182
	42	752	836	912	1002	1068	1201
	43	764	850	924	1020	1082	1223
	44	777	860	939	1034	1103	1243
	45	786	875	954	1054	1121	1258
	46	800	887	971	1069	1136	1280
	47	811	902	985	1086	1154	1298
	48	825	916	1000	1100	1172	1319
	49	837	928	1014	1116	1188	1338
	50	847	942	1030	1133	1204	1357
	51	860	957	1042	1148	1222	1376
	52	871	970	1056	1162	1236	1394
	53	884	986	1071	1177	1255	1414
	54	893	995	1084	1195	1271	1432
	55	906	1008	1099	1211	1288	1451
	56	918	1020	1114	1227	1305	1471
	57	929	1034	1128	1243	1322	1490
	58	941	1047	1142	1258	1337	1506
	59	955	1057	1155	1273	1352	1528
	60	964	1073	1171	1288	1372	1544
	61	976	1086	1184	1304	1388	1564
	62	986	1098	1199	1318	1403	1583
	63	1001	1112	1210	1334	1416	1602
	64	1011	1124	1226	1350	1437	1619
	65	1021	1137	1241	1365	1453	1638
	66	1035	1152	1257	1379	1470	1655
	67	1045	1162	1268	1395	1487	1674
	68	1057	1175	1281	1412	1504	1692
	69	1069	1189	1297	1427	1516	1712
	70	1080	1200	1309	1442	1536	1730
	71	1091	1213	1323	1457	1551	1749
	72	1102	1224	1339	1472	1567	1764

N	M	.1	.05	.025	.01	.005	.001
79	73	1112	1238	1349	1487	1584	1784
	74	1124	1251	1365	1503	1598	1800
	75	1136	1262	1377	1517	1614	1819
	76	1147	1273	1393	1533	1630	1835
	77	1158	1286	1405	1551	1644	1863
	78	1155	1298	1433	1545	1678	1856
	79	1264	1422	1501	1659	1738	1975
80	1	77	79	80	-		
	2	128	138	146	152	156	160
	3	157	177	189	204	213	228
	4	188	208	224	244	260	284
	5	215	235	255	280	295	325
	6	234	260	282	308	328	364
	7	256	284	306	337	357	399
	8	280	312	336	368	392	440
	9	292	326	354	389	415	461
	10	320	360	390	420	450	500
	11	330	367	400	439	467	522
	12	348	388	424	464	492	552
	13	363	404	440	484	516	579
	14	380	424	462	506	538	604
	15	400	445	485	530	565	630
	16	432	464	512	560	592	672
	17	428	474	518	569	605	680
	18	442	492	536	590	626	704
	19	456	507	554	608	646	728
	20	480	540	580	640	680	760
	21	486	539	588	649	688	775
	22	500	556	606	666	710	798
	23	512	569	624	683	728	820
	24	536	592	648	712	752	848
	25	545	605	660	725	770	865
	26	554	616	672	740	790	884
	27	572	631	686	762	799	901
	28	584	648	708	780	828	932
	29	594	659	722	795	841	949
	30	620	680	750	820	870	980
	31	620	689	753	828	880	992
	32	640	720	784	864	912	1024
	33	646	719	785	861	917	1034
	34	660	734	800	880	936	1054
80	35	675	750	820	900	960	1080
	36	688	764	832	916	976	1096
	37	695	774	847	929	990	1113
	38	710	790	862	948	1008	1136
	39	722	802	875	962	1027	1154
	40	760	840	920	1000	1080	1200
	41	746	830	904	999	1059	1193
	42	760	844	922	1014	1078	1214
	43	772	858	936	1031	1096	1234
	44	788	872	952	1048	1116	1256
	45	800	890	970	1065	1135	1275
	46	808	898	980	1080	1148	1294
	47	821	911	995	1095	1165	1311
	48	848	944	1024	1120	1200	1344
	49	844	940	1025	1127	1199	1351
	50	860	960	1050	1150	1230	1380
	51	869	965	1053	1161	1234	1388
	52	884	980	1072	1180	1252	1412
	53	892	995	1080	1187	1265	1425
	54	904	1006	1098	1208	1286	1448
	55	920	1020	1115	1225	1305	1470
	56	936	1040	1136	1248	1328	1488
	57	941	1045	1138	1252	1332	1503
	58	952	1058	1156	1272	1352	1522
	59	963	1069	1169	1286	1368	1541
	60	1000	1100	1200	1320	1400	1580
	61	986	1096	1196	1317	1401	1580
	62	998	1110	1212	1334	1420	1598
	63	1008	1123	1225	1347	1436	1616
	64	1040	1152	1248	1376	1472	1648
	65	1035	1150	1255	1385	1470	1655
	66	1044	1162	1268	1396	1484	1674
	67	1056	1175	1283	1409	1501	1689
	68	1068	1188	1296	1428	1520	1712
	69	1078	1200	1309	1440	1533	1728
	70	1100	1220	1330	1460	1560	1750
	71	1100	1223	1337	1471	1567	1764
	72	1120	1240	1360	1496	1592	1792
	73	1124	1250	1364	1501	1598	1800
	74	1136	1264	1378	1518	1614	1820
	75	1150	1280	1395	1535	1635	1840
80	76	1160	1288	1408	1548	1648	1856
	77	1169	1300	1418	1563	1661	1873
	78	1180	1312	1432	1578	1674	1894
	79	1176	1319	1460	1573	1705	1886
	80	1280	1440	1520	1680	1760	2000
81	1	78	80	81	-		
	2	130	140	148	154	158	162
	3	159	177	192	207	216	231
	4	188	208	227	248	260	284
	5	214	235	259	280	299	330
	6	237	264	285	312	330	369
	7	258	287	310	342	363	402
	8	270	310	336	363	391	438
	9	306	342	369	405	423	477
	10	317	350	380	419	447	499
	11	336	372	405	446	472	528
	12	354	390	426	468	498	558
	13	369	408	446	489	520	583
	14	385	427	466	511	544	611
	15	402	447	486	534	567	636
	16	414	461	507	556	589	656
	17	431	480	525	576	612	687
	18	450	504	549	603	639	720
	19	463	513	560	616	655	736
	20	475	532	576	635	676	757
	21	492	546	597	657	698	783
	22	505	562	614	673	717	806
	23	517	575	630	690	735	827
	24	534	594	648	714	759	852
	25	547	609	664	728	777	872
	26	560	622	680	748	796	894
	27	594	675	729	783	837	945
	28	587	653	712	784	835	938
	29	600	669	728	803	854	960
	30	615	684	747	822	873	984
	31	628	698	761	836	891	1004
	32	640	713	777	856	908	1021
	33	654	729	795	873	930	1047
	34	666	741	809	890	947	1066
	35	679	756	824	907	966	1085
	36	702	774	846	927	990	1116

N	M	.1	.05	.025	.01	.005	.001
81	37	704	783	854	940	999	1126
	38	716	797	870	958	1019	1148
	39	729	813	885	975	1038	1170
	40	744	826	904	989	1062	1189
	41	754	837	916	1002	1077	1204
	42	768	855	933	1026	1092	1227
	43	779	866	946	1041	1109	1247
	44	791	880	960	1059	1126	1267
	45	810	900	981	1080	1152	1296
	46	816	907	991	1090	1158	1306
	47	829	921	1005	1107	1179	1326
	48	843	936	1023	1125	1197	1347
	49	852	948	1035	1139	1210	1367
	50	864	963	1050	1156	1230	1385
	51	879	978	1065	1173	1248	1404
	52	889	987	1079	1188	1264	1421
	53	900	1000	1093	1204	1280	1442
	54	945	1053	1134	1242	1323	1485
	55	925	1029	1122	1235	1313	1481
	56	937	1043	1137	1250	1331	1499
	57	951	1056	1152	1269	1350	1521
	58	959	1070	1165	1282	1363	1536
	59	971	1081	1178	1300	1382	1558
	60	987	1095	1197	1317	1401	1578
	61	998	1107	1208	1330	1413	1594
	62	1008	1119	1223	1347	1433	1614
	63	1026	1143	1242	1368	1458	1638
	64	1031	1146	1252	1377	1466	1652
	65	1045	1160	1262	1390	1484	1666
	66	1056	1173	1281	1410	1500	1689
	67	1066	1185	1294	1425	1517	1707
	68	1078	1198	1308	1439	1531	1727
	69	1089	1212	1323	1458	1548	1746
	70	1100	1224	1336	1469	1564	1763
	71	1111	1234	1351	1485	1578	1780
	72	1134	1260	1368	1512	1602	1809
	73	1133	1262	1376	1514	1611	1820
	74	1146	1274	1391	1531	1630	1836
	75	1158	1290	1407	1548	1647	1857
	76	1169	1300	1417	1563	1660	1872
	77	1181	1312	1433	1577	1678	1891
	78	1194	1329	1446	1593	1695	1911

N	M	.1	.05	.025	.01	.005	.001
81	79	1202	1337	1457	1607	1703	1927
	80	1226	1339	1483	1638	1730	1963
	81	1296	1458	1539	1701	1782	2025
82	1	79	81	82			
	2	130	142	150	156	160	164
	3	161	180	195	210	219	234
	4	192	210	228	252	264	288
	5	216	240	260	283	300	335
	6	240	266	290	316	336	372
	7	261	291	314	345	366	408
	8	282	312	340	374	394	444
	9	302	333	365	401	423	474
	10	320	356	388	426	450	504
	11	338	376	409	448	475	535
	12	356	394	430	474	500	560
	13	371	414	451	495	526	591
	14	390	432	472	518	552	620
	15	406	451	490	540	573	641
	16	422	468	510	560	598	668
	17	437	485	530	582	618	695
	18	452	502	548	602	640	720
	19	466	517	566	623	661	743
	20	482	536	586	642	684	770
	21	497	551	602	663	705	791
	22	512	568	620	682	724	814
	23	524	583	636	700	744	836
	24	540	600	654	720	764	860
	25	552	613	670	738	784	883
	26	568	630	688	758	806	906
	27	583	642	699	775	828	930
	28	594	662	720	794	842	946
	29	607	675	737	810	862	969
	30	622	690	754	830	882	994
	31	634	704	770	847	899	1013
	32	648	720	786	864	920	1034
	33	663	733	799	881	933	1059
	34	674	750	818	900	956	1078
	35	686	765	835	918	976	1095
	36	700	778	848	934	994	1120
	37	712	791	864	950	1011	1139

N	M	.1	.05	.025	.01	.005	.001
82	38	726	806	880	968	1030	1158
	39	736	819	893	984	1047	1180
	40	750	834	908	1004	1064	1202
	41	820	902	943	1066	1107	1271
	42	774	862	942	1036	1102	1242
	43	788	876	956	1050	1118	1261
	44	800	890	972	1068	1136	1280
	45	812	902	986	1086	1153	1300
	46	826	918	1002	1102	1172	1320
	47	838	932	1016	1120	1191	1342
	48	850	946	1030	1136	1208	1358
	49	861	958	1042	1153	1222	1382
	50	874	972	1060	1168	1242	1400
	51	885	984	1076	1185	1260	1418
	52	898	1000	1090	1200	1276	1438
	53	910	1011	1103	1214	1291	1455
	54	922	1026	1118	1230	1310	1474
	55	938	1033	1136	1248	1329	1495
	56	946	1052	1148	1264	1344	1516
	57	958	1065	1163	1280	1362	1534
	58	970	1080	1178	1296	1378	1552
	59	982	1091	1191	1311	1396	1573
	60	994	1106	1206	1328	1412	1592
	61	1005	1118	1220	1345	1431	1612
	62	1018	1132	1236	1360	1446	1630
	63	1030	1145	1247	1373	1461	1649
	64	1042	1158	1264	1390	1480	1668
	65	1052	1170	1277	1406	1495	1687
	66	1064	1184	1292	1422	1514	1706
	67	1077	1197	1307	1436	1531	1725
	68	1088	1210	1322	1454	1546	1744
	69	1099	1223	1334	1469	1561	1761
	70	1112	1236	1348	1486	1580	1780
	71	1123	1249	1363	1500	1594	1799
	72	1134	1262	1378	1516	1614	1818
	73	1145	1273	1391	1529	1628	1837
	74	1158	1288	1406	1546	1646	1856
	75	1169	1300	1418	1561	1661	1873
	76	1180	1312	1434	1578	1678	1892
	77	1192	1325	1447	1592	1694	1909
	78	1204	1338	1460	1608	1712	1928
	79	1214	1351	1474	1623	1725	1948

N	M	.1	.05	.025	.01	.005	.001
82	80	1226	1364	1486	1634	1742	1958
	81	1249	1361	1506	1665	1756	1993
	82	1312	1476	1558	1722	1886	2050
83	1	80	82	83	-		
	2	132	144	152	158	162	166
	3	163	183	198	213	222	237
	4	193	213	229	249	268	292
	5	219	242	262	287	302	340
	6	242	270	290	319	337	373
	7	262	294	317	350	371	413
	8	285	316	344	378	399	448
	9	305	338	369	405	430	480
	10	322	360	391	428	457	510
	11	341	379	415	454	483	539
	12	361	399	435	474	507	567
	13	376	419	456	502	533	598
	14	395	439	478	522	552	621
	15	410	456	495	546	581	652
	16	426	474	516	568	603	679
	17	442	490	535	583	624	703
	18	457	507	554	608	648	727
	19	473	524	573	630	670	753
	20	487	540	591	650	690	776
	21	498	555	612	673	716	802
	22	515	575	626	688	731	824
	23	529	591	643	708	753	846
	24	544	606	660	728	774	870
	25	560	621	678	745	794	894
	26	573	637	694	765	813	913
	27	587	653	713	782	830	935
	28	599	661	733	796	850	960
	29	613	683	745	820	870	981
	30	627	697	762	838	891	1002
	31	639	711	775	855	908	1022
	32	655	727	794	873	930	1046
	33	669	740	806	888	951	1068
	34	679	757	826	909	968	1086
	35	693	772	842	925	985	1109
	36	707	785	857	943	1003	1130
	37	718	800	873	958	1021	1150

N	M	.1	.05	.025	.01	.005	.001
83	38	732	814	889	977	1041	1171
	39	745	828	904	994	1057	1192
	40	756	842	919	1011	1077	1211
	41	767	851	932	1019	1095	1223
	42	778	863	944	1036	1108	1240
	43	795	884	966	1062	1130	1273
	44	808	898	980	1078	1147	1293
	45	821	912	995	1096	1165	1314
	46	834	926	1010	1111	1184	1332
	47	845	940	1027	1128	1201	1353
	48	858	954	1041	1146	1217	1374
	49	870	968	1055	1162	1236	1393
	50	879	977	1072	1175	1255	1408
	51	894	994	1086	1196	1272	1433
	52	908	1006	1098	1211	1286	1450
	53	919	1022	1115	1227	1303	1470
	54	931	1036	1130	1244	1323	1492
	55	947	1043	1145	1258	1339	1506
	56	956	1063	1160	1277	1360	1530
	57	967	1076	1175	1290	1373	1549
	58	980	1090	1190	1307	1391	1566
	59	991	1103	1203	1325	1410	1587
	60	1004	1115	1217	1341	1427	1607
	61	1015	1129	1232	1358	1442	1625
	62	1027	1146	1250	1376	1461	1647
	63	1040	1155	1260	1388	1478	1664
	64	1050	1170	1275	1403	1493	1684
	65	1063	1182	1290	1420	1511	1703
	66	1076	1195	1306	1438	1527	1723
	67	1086	1208	1319	1451	1544	1740
	68	1099	1221	1333	1468	1560	1757
	69	1111	1236	1348	1485	1581	1777
	70	1122	1247	1361	1498	1594	1798
	71	1133	1261	1376	1512	1608	1818
	72	1144	1272	1388	1531	1626	1835
	73	1157	1287	1403	1546	1644	1853
	74	1168	1299	1417	1562	1663	1870
	75	1180	1312	1432	1577	1676	1890
	76	1192	1325	1444	1593	1692	1909
	77	1202	1337	1461	1605	1710	1925
	78	1214	1350	1475	1622	1727	1945
	79	1224	1364	1488	1638	1743	1963

N	M	.1	.05	.025	.01	.005	.001
83	80	1237	1377	1499	1653	1760	1982
	81	1249	1390	1514	1664	1780	1992
	82	1269	1383	1529	1689	1781	2022
	83	1328	1494	1660	1743	1909	2075
84	1	81	83	84			
	2	134	144	152	160	164	168
	3	165	183	201	216	225	240
	4	196	216	236	256	272	296
	5	221	242	266	291	306	336
	6	252	276	300	324	348	384
	7	273	301	329	357	378	420
	8	292	324	352	384	408	452
	9	309	342	372	411	435	486
	10	328	364	396	434	462	518
	11	346	383	418	460	488	544
	12	372	420	456	492	528	588
	13	380	424	462	508	536	605
	14	406	448	490	546	574	644
	15	417	462	504	552	588	660
	16	432	480	524	576	612	688
	17	447	497	536	598	633	704
	18	468	516	564	619	660	738
	19	478	530	577	638	676	760
	20	496	548	600	660	700	788
	21	525	588	630	693	735	819
	22	522	580	634	696	740	832
	23	537	595	648	716	761	854
	24	564	624	684	744	792	888
	25	565	628	687	755	803	901
	26	580	644	702	772	822	924
	27	594	660	720	792	843	948
	28	644	700	756	840	896	1008
	29	621	689	753	828	880	992
	30	636	708	774	852	906	1020
	31	648	720	786	865	919	1034
	32	664	736	804	884	940	1060
	33	675	750	819	903	960	1080
	34	688	764	834	918	978	1100
	35	707	784	854	938	1001	1127
	36	720	804	876	960	1020	1152

N	M	.1	.05	.025	.01	.005	.001
84	37	727	808	882	970	1031	1162
	38	740	824	898	988	1052	1184
	39	753	837	915	1005	1071	1206
	40	768	852	932	1024	1088	1228
	41	777	866	946	1040	1105	1247
	42	840	924	1008	1092	1176	1302
	43	804	894	976	1070	1144	1283
	44	820	908	992	1092	1160	1308
	45	831	924	1008	1107	1179	1329
	46	842	936	1022	1124	1196	1348
	47	854	949	1037	1141	1213	1366
	48	876	972	1056	1164	1236	1392
	49	882	980	1071	1176	1253	1414
	50	892	992	1082	1190	1268	1426
	51	906	1005	1098	1209	1284	1449
	52	920	1020	1112	1224	1304	1468
	53	927	1032	1126	1240	1319	1484
	54	942	1050	1146	1260	1338	1512
	55	952	1059	1155	1272	1353	1524
	56	980	1092	1204	1316	1400	1568
	57	978	1086	1188	1305	1389	1566
	58	990	1100	1200	1322	1406	1584
	59	1002	1113	1214	1338	1424	1602
	60	1020	1140	1236	1368	1452	1632
	61	1026	1141	1242	1370	1455	1639
	62	1038	1154	1260	1386	1474	1662
	63	1071	1176	1302	1428	1512	1701
	64	1064	1184	1292	1420	1512	1700
	65	1072	1193	1302	1434	1526	1717
	66	1092	1212	1320	1452	1548	1740
	67	1098	1218	1333	1467	1555	1755
	68	1112	1236	1348	1484	1580	1780
	69	1122	1248	1362	1497	1593	1797
	70	1148	1274	1386	1526	1624	1820
	71	1142	1271	1387	1528	1625	1832
	72	1164	1296	1416	1560	1656	1860
	73	1167	1298	1416	1560	1660	1870
	74	1180	1312	1432	1576	1676	1888
	75	1191	1326	1446	1593	1695	1908
	76	1204	1340	1464	1608	1712	1928
	77	1218	1358	1477	1631	1729	1953
	78	1230	1368	1494	1644	1746	1968

N	M	.1	.05	.025	.01	.005	.001
84	79	1237	1376	1501	1653	1760	1983
	80	1252	1392	1520	1672	1776	2004
	81	1260	1404	1530	1686	1794	2022
	82	1272	1414	1544	1696	1814	2040
	83	1289	1441	1551	1714	1807	2050
	84	1344	1512	1680	1848	1932	2184
85	1	82	84	85	-		
	2	136	146	154	162	166	170
	3	167	186	201	219	228	243
	4	199	219	236	260	272	300
	5	225	250	270	295	315	345
	6	246	275	299	328	347	384
	7	271	300	326	357	378	425
	8	291	323	352	385	409	459
	9	312	344	378	414	438	491
	10	335	370	405	445	470	525
	11	348	388	422	465	493	551
	12	369	407	443	490	516	583
	13	386	428	467	513	545	611
	14	402	446	487	532	571	641
	15	420	470	510	560	595	670
	16	435	483	528	578	615	690
	17	459	510	561	612	646	731
	18	467	519	567	622	662	742
	19	480	536	584	641	685	765
	20	500	555	605	665	710	795
	21	515	565	623	686	729	816
	22	527	586	640	704	748	841
	23	542	603	657	725	769	865
	24	557	620	677	741	790	887
	25	575	640	695	765	815	915
	26	585	650	710	779	830	934
	27	599	666	725	800	850	959
	28	610	686	746	811	864	975
	29	627	696	760	836	890	1002
	30	645	715	780	860	915	1025
	31	654	726	794	872	927	1043
	32	667	741	807	890	945	1069
	33	682	758	825	909	966	1089
	34	714	782	850	935	1003	1122
	35	710	790	860	950	1010	1135

N	M	.1	.05	.025	.01	.005	.001
85	36	720	802	874	961	1023	1154
	37	734	816	890	979	1042	1175
	38	748	830	906	998	1061	1195
	39	760	845	922	1015	1080	1217
	40	775	865	940	1035	1100	1240
	41	786	874	954	1050	1117	1256
	42	791	878	960	1074	1126	1286
	43	803	900	975	1088	1141	1302
	44	823	916	999	1099	1171	1320
	45	840	935	1020	1120	1195	1345
	46	849	944	1031	1135	1206	1360
	47	861	958	1046	1153	1227	1379
	48	874	971	1061	1168	1243	1399
	49	888	986	1076	1185	1262	1421
	50	905	1005	1095	1205	1280	1445
	51	935	1037	1122	1241	1309	1479
	52	924	1028	1122	1236	1315	1478
	53	938	1044	1137	1249	1332	1502
	54	948	1055	1153	1269	1348	1519
	55	965	1075	1170	1290	1370	1545
	56	973	1082	1183	1301	1384	1558
	57	983	1095	1202	1320	1405	1578
	58	998	1110	1212	1334	1420	1599
	59	1010	1123	1226	1351	1436	1618
	60	1025	1140	1245	1370	1455	1640
	61	1036	1150	1256	1381	1472	1657
	62	1047	1164	1271	1400	1488	1676
	63	1059	1177	1286	1415	1506	1696
	64	1073	1187	1295	1435	1523	1715
	65	1085	1210	1320	1450	1545	1740
	66	1093	1216	1328	1460	1554	1752
	67	1107	1231	1343	1480	1575	1772
	68	1139	1258	1377	1513	1598	1802
	69	1130	1257	1373	1511	1607	1812
	70	1145	1275	1390	1530	1630	1835
	71	1156	1284	1399	1541	1640	1851
	72	1167	1298	1416	1560	1659	1869
	73	1176	1309	1429	1575	1673	1890
	74	1190	1325	1444	1591	1691	1906
	75	1205	1340	1460	1610	1710	1930
	76	1214	1349	1474	1622	1725	1944
	77	1224	1362	1488	1637	1743	1962

N	M	.1	.05	.025	.01	.005	.001
85	78	1237	1374	1500	1654	1756	1982
	79	1249	1389	1516	1668	1776	2002
	80	1265	1405	1535	1690	1795	2025
	81	1270	1414	1542	1699	1808	2038
	82	1281	1426	1558	1715	1824	2055
	83	1295	1440	1573	1733	1845	2084
	84	1309	1466	1574	1739	1836	2079
	85	1360	1530	1700	1870	1955	2210
86	1	83	85	86	-		
	2	136	148	156	164	168	172
	3	168	189	204	219	231	243
	4	200	222	238	258	276	304
	5	225	250	274	299	314	350
	6	250	278	302	330	352	388
	7	273	302	329	362	383	427
	8	296	326	356	390	412	460
	9	315	350	381	418	444	495
	10	334	372	406	444	472	528
	11	353	393	428	470	499	558
	12	372	414	450	494	526	588
	13	389	433	472	519	550	618
	14	406	452	492	542	574	644
	15	424	470	512	564	598	673
	16	440	490	534	586	624	700
	17	455	506	555	608	644	728
	18	472	526	572	630	670	752
	19	488	541	589	650	692	777
	20	504	560	610	672	712	802
	21	518	576	627	688	732	828
	22	534	592	648	712	758	850
	23	548	610	664	732	778	871
	24	562	626	682	750	798	898
	25	577	642	699	770	818	921
	26	592	658	718	788	840	944
	27	605	674	735	809	859	966
	28	620	688	750	826	880	992
	29	627	705	766	848	904	1019
	30	648	720	786	864	920	1034
	31	660	735	801	882	939	1056
	32	676	750	818	902	958	1078
	33	688	765	836	917	976	1101
86	34	702	780	852	938	996	1120
	35	715	795	868	954	1015	1142
	36	728	810	884	972	1036	1166
	37	743	822	901	991	1053	1188
	38	756	840	916	1008	1072	1206
	39	769	855	933	1027	1090	1230
	40	782	868	948	1044	1108	1250
	41	794	883	962	1060	1126	1269
	42	806	896	980	1076	1148	1286
	43	860	946	1032	1118	1204	1333
	44	832	924	1008	1112	1180	1336
	45	844	940	1026	1128	1200	1350
	46	858	954	1042	1146	1220	1374
	47	871	967	1055	1163	1235	1391
	48	884	982	1072	1180	1256	1414
	49	893	998	1087	1197	1271	1431
	50	908	1010	1104	1214	1290	1454
	51	922	1025	1117	1229	1309	1474
	52	934	1038	1134	1246	1328	1494
	53	945	1051	1149	1262	1342	1514
	54	958	1066	1164	1282	1362	1534
	55	971	1079	1178	1297	1380	1555
	56	984	1094	1194	1314	1398	1574
	57	991	1104	1212	1330	1416	1590
	58	1008	1122	1222	1348	1434	1614
	59	1021	1134	1238	1361	1451	1634
	60	1032	1148	1254	1380	1468	1652
	61	1045	1163	1267	1396	1485	1673
	62	1058	1176	1282	1412	1502	1692
	63	1069	1189	1298	1426	1517	1711
	64	1082	1202	1312	1444	1538	1734
	65	1093	1214	1325	1460	1555	1750
	66	1106	1230	1342	1476	1572	1772
	67	1116	1241	1355	1490	1585	1786
	68	1130	1256	1370	1510	1606	1810
	69	1143	1268	1386	1525	1626	1832
	70	1154	1284	1400	1542	1640	1848
	71	1165	1296	1414	1557	1655	1866
	72	1178	1310	1428	1572	1672	1886
	73	1189	1321	1443	1587	1690	1905
	74	1202	1336	1458	1604	1708	1924
	75	1214	1348	1472	1622	1724	1944
86	76	1226	1362	1436	1638	1742	1962
	77	1237	1375	1501	1652	1756	1982
	78	1248	1388	1516	1663	1776	2000
	79	1260	1401	1530	1683	1791	2018
	80	1272	1414	1544	1700	1808	2038
	81	1283	1426	1557	1714	1825	2057
	82	1294	1440	1572	1732	1842	2076
	83	1305	1452	1587	1747	1858	2094
	84	1318	1466	1602	1766	1874	2118
	85	1328	1489	1597	1763	1910	2109
	86	1462	1548	1720	1892	1978	2236
87	1	84	86	87	-		
	2	138	150	158	166	170	174
	3	171	189	207	222	231	246
	4	201	224	241	261	280	308
	5	230	251	275	300	318	355
	6	255	282	306	336	354	393
	7	277	306	334	365	386	428
	8	298	330	361	394	418	466
	9	321	354	387	423	450	501
	10	338	375	410	449	479	536
	11	356	398	432	475	500	564
	12	378	420	456	501	531	597
	13	394	437	476	524	557	624
	14	411	458	498	546	581	654
	15	429	477	519	570	606	681
	16	445	495	539	592	630	708
	17	460	513	558	613	655	731
	18	477	531	579	636	678	759
	19	493	548	597	658	699	786
	20	509	565	616	678	719	810
	21	525	582	636	699	744	834
	22	537	600	649	714	758	862
	23	553	615	671	740	786	883
	24	570	633	690	759	807	909
	25	581	645	705	779	828	929
	26	596	665	725	796	848	952
	27	612	681	744	819	870	978
	28	626	695	758	835	888	999
	29	667	725	812	870	928	1044
	30	654	729	795	873	930	1047

N	M	.1	.05	.025	.01	.005	.001
87	31	668	742	810	891	948	1066
	32	681	758	827	909	968	1088
	33	696	774	846	930	987	1113
	34	709	788	860	946	1007	1133
	35	723	799	879	967	1023	1158
	36	738	819	894	984	1047	1179
	37	748	833	908	1002	1065	1200
	38	763	846	925	1019	1083	1220
	39	777	864	942	1038	1104	1242
	40	789	877	957	1051	1119	1262
	41	802	890	972	1069	1138	1283
	42	816	906	990	1089	1158	1305
	43	834	926	1011	1105	1184	1322
	44	846	938	1024	1118	1200	1338
	45	855	951	1035	1140	1212	1365
	46	865	963	1051	1155	1231	1385
	47	880	977	1068	1174	1249	1407
	48	894	993	1083	1191	1269	1428
	49	904	1005	1097	1207	1283	1446
	50	917	1021	1113	1225	1302	1469
	51	930	1035	1131	1245	1323	1491
	52	945	1050	1141	1260	1342	1505
	53	954	1062	1159	1276	1357	1528
	54	969	1077	1176	1293	1377	1551
	55	979	1089	1189	1309	1392	1569
	56	992	1104	1204	1326	1410	1589
	57	1008	1119	1221	1344	1428	1611
	58	1044	1160	1247	1392	1479	1653
	59	1029	1145	1250	1375	1464	1650
	60	1044	1161	1266	1392	1482	1671
	61	1052	1172	1278	1407	1496	1686
	62	1064	1186	1296	1423	1512	1708
	63	1080	1200	1311	1443	1536	1728
	64	1090	1214	1324	1457	1552	1749
	65	1100	1227	1337	1469	1572	1769
	66	1116	1242	1356	1491	1587	1788
	67	1128	1255	1369	1505	1604	1805
	68	1139	1268	1383	1522	1620	1824
	69	1152	1284	1401	1539	1638	1845
	70	1163	1295	1414	1556	1656	1866
	71	1175	1307	1427	1572	1671	1883
	72	1191	1323	1443	1590	1689	1905
87	73	1199	1333	1456	1602	1706	1922
	74	1212	1348	1470	1618	1722	1940
	75	1224	1362	1488	1638	1740	1962
	76	1235	1375	1497	1649	1756	1982
	77	1248	1388	1515	1666	1773	1999
	78	1260	1401	1530	1686	1791	2019
	79	1270	1413	1544	1697	1807	2036
	80	1283	1426	1558	1714	1824	2055
	81	1296	1443	1572	1731	1842	2076
	82	1307	1453	1585	1745	1857	2096
	83	1319	1466	1600	1762	1873	2112
	84	1332	1479	1617	1779	1893	2133
	85	1340	1491	1630	1796	1905	2152
	86	1348	1511	1623	1789	1939	2190
	87	1479	1566	1740	1914	2001	2262
88	1	85	87	88	-		
	2	140	152	160	168	170	176
	3	173	192	210	225	234	249
	4	208	228	244	268	284	308
	5	232	257	280	305	322	360
	6	256	284	308	338	360	398
	7	280	310	336	370	391	437
	8	304	344	368	408	432	480
	9	323	357	391	427	454	508
	10	342	380	414	454	482	540
	11	374	407	451	484	517	583
	12	384	424	464	508	540	604
	13	398	442	481	528	562	633
	14	416	462	504	554	588	658
	15	432	480	524	576	612	685
	16	456	504	552	608	640	720
	17	466	518	565	622	659	741
	18	482	536	586	642	682	766
	19	498	553	605	664	707	792
	20	516	572	624	688	732	820
	21	528	588	641	704	749	842
	22	572	616	682	748	792	880
	23	560	622	677	746	792	891
	24	584	648	704	776	824	920
	25	586	657	711	785	835	937
88	26	604	672	732	806	856	962
	27	618	687	748	823	877	985
	28	636	704	768	848	900	1012
	29	651	717	779	862	919	1036
	30	662	734	802	884	940	1058
	31	675	749	818	899	958	1079
	32	696	768	840	928	984	1104
	33	715	792	858	946	1001	1133
	34	716	796	870	956	1018	1144
	35	729	814	886	975	1031	1168
	36	744	828	904	996	1056	1192
	37	757	841	918	1008	1073	1209
	38	770	856	934	1028	1094	1230
	39	784	871	951	1043	1110	1254
	40	800	888	976	1072	1136	1280
	41	808	900	982	1082	1148	1294
	42	822	914	998	1098	1168	1316
	43	836	927	1012	1116	1184	1342
	44	880	968	1056	1188	1232	1408
	45	961	957	1044	1153	1223	1380
	46	874	972	1062	1168	1242	1400
	47	888	985	1077	1184	1260	1419
	48	904	1008	1096	1208	1288	1448
	49	914	1014	1110	1219	1298	1463
	50	926	1030	1124	1238	1316	1482
	51	939	1042	1139	1254	1333	1503
	52	952	1060	1156	1272	1356	1524
	53	963	1069	1172	1283	1368	1543
	54	978	1086	1186	1306	1388	1564
	55	1001	1111	1210	1331	1419	1595
	56	1008	1120	1224	1344	1432	1608
	57	1014	1127	1231	1355	1439	1624
	58	1028	1142	1248	1372	1460	1642
	59	1042	1159	1255	1394	1482	1663
	60	1052	1172	1280	1408	1496	1684
	61	1062	1183	1291	1420	1513	1704
	62	1076	1198	1306	1438	1530	1724
	63	1089	1213	1318	1454	1544	1744
	64	1104	1232	1344	1480	1568	1768
	65	1114	1238	1351	1487	1582	1783
	66	1144	1276	1386	1518	1628	1826
	67	1138	1264	1331	1520	1616	1821

N	M	.1	.05	.025	.01	.005	.001
88	68	1152	1280	1400	1540	1636	1844
	69	1162	1292	1410	1553	1652	1861
	70	1174	1306	1426	1568	1670	1880
	71	1186	1320	1439	1584	1686	1899
	72	1208	1336	1464	1608	1712	1928
	73	1210	1347	1468	1616	1720	1938
	74	1224	1360	1484	1634	1738	1960
	75	1235	1372	1499	1649	1756	1978
	76	1248	1388	1516	1668	1776	2000
	77	1265	1408	1540	1694	1804	2024
	78	1272	1414	1544	1700	1808	2036
	79	1283	1426	1558	1714	1823	2057
	80	1304	1448	1576	1736	1848	2080
	81	1307	1453	1584	1747	1858	2094
	82	1318	1466	1600	1762	1874	2112
	83	1330	1479	1614	1777	1892	2132
	84	1344	1496	1632	1796	1912	2152
	85	1353	1504	1643	1810	1925	2169
	86	1364	1518	1658	1826	1936	2186
	87	1368	1532	1688	1820	1966	2224
	88	1496	1672	1760	1936	2024	2288
89	1	86	88	89	-		
	2	142	154	162	170	172	178
	3	174	195	210	223	237	252
	4	207	228	247	272	288	312
	5	232	257	281	306	325	356
	6	259	288	313	342	361	402
	7	282	314	340	373	396	443
	8	304	336	368	405	429	478
	9	326	362	391	433	460	514
	10	344	383	415	461	484	543
	11	363	405	440	484	516	579
	12	384	426	465	510	542	609
	13	402	448	487	537	567	639
	14	420	466	508	559	595	667
	15	438	485	529	579	619	694
	16	454	504	551	606	643	720
	17	470	523	571	627	665	749
	18	489	544	586	652	689	777
	19	504	560	608	671	714	801
	20	519	577	630	692	737	826

N	M	.1	.05	.025	.01	.005	.001
89	21	533	595	648	711	758	852
	22	547	611	670	737	784	876
	23	565	628	685	752	801	901
	24	580	645	703	773	823	924
	25	594	661	721	793	843	949
	26	608	678	740	812	864	974
	27	623	693	756	833	886	995
	28	638	709	773	853	905	1021
	29	653	725	792	873	925	1039
	30	669	736	815	884	942	1061
	31	680	757	826	909	967	1087
	32	695	773	844	928	987	1111
	33	708	789	860	946	1005	1136
	34	723	804	876	965	1027	1155
	35	737	818	893	982	1047	1176
	36	751	833	912	1003	1064	1200
	37	762	849	925	1020	1086	1220
	38	770	861	945	1036	1101	1240
	39	789	878	960	1056	1123	1262
	40	804	893	975	1074	1142	1284
	41	817	908	990	1092	1160	1305
	42	830	923	1008	1108	1180	1328
	43	843	936	1023	1126	1198	1347
	44	860	953	1041	1136	1218	1358
	45	871	964	1053	1150	1232	1402
	46	883	981	1070	1179	1254	1412
	47	896	996	1086	1196	1272	1435
	48	908	1011	1101	1211	1292	1456
	49	921	1023	1119	1231	1308	1475
	50	933	1038	1133	1249	1327	1494
	51	948	1052	1151	1267	1345	1513
	52	961	1067	1163	1282	1363	1535
	53	974	1082	1180	1300	1383	1559
	54	986	1097	1197	1315	1402	1578
	55	998	1111	1213	1334	1420	1598
	56	1011	1123	1227	1350	1436	1617
	57	1024	1138	1242	1367	1455	1639
	58	1036	1152	1258	1384	1473	1658
	59	1050	1168	1265	1404	1493	1675
	60	1061	1179	1288	1417	1508	1700
	61	1074	1194	1302	1435	1525	1718
	62	1086	1209	1318	1450	1543	1740

N	M	.1	.05	.025	.01	.005	.001
89	63	1098	1222	1333	1467	1560	1760
	64	1111	1236	1350	1484	1579	1780
	65	1122	1250	1363	1501	1597	1799
	66	1136	1262	1377	1518	1614	1819
	67	1148	1280	1394	1530	1636	1839
	68	1160	1290	1407	1548	1650	1859
	69	1172	1305	1423	1566	1668	1879
	70	1186	1318	1436	1582	1683	1899
	71	1199	1328	1451	1595	1700	1914
	72	1210	1345	1468	1615	1719	1937
	73	1221	1358	1484	1631	1736	1955
	74	1236	1372	1494	1651	1755	1978
	75	1245	1385	1512	1665	1770	1995
	76	1258	1399	1526	1680	1790	2016
	77	1270	1413	1543	1697	1804	2036
	78	1284	1423	1555	1711	1822	2055
	79	1292	1439	1569	1727	1841	2072
	80	1305	1454	1587	1747	1856	2094
	81	1317	1464	1600	1761	1874	2110
	82	1330	1479	1615	1779	1890	2130
	83	1343	1492	1629	1794	1908	2151
	84	1353	1506	1644	1808	1926	2169
	85	1366	1517	1653	1824	1941	2188
	86	1378	1530	1671	1841	1957	2209
	87	1388	1544	1684	1856	1970	2220
	88	1388	1554	1715	1890	1993	2254
	89	1512	1691	1780	1958	2136	2314
90	1	87	89	90	-		
	2	144	156	164	172	174	180
	3	177	198	213	231	240	255
	4	210	230	250	270	288	316
	5	240	265	285	315	330	365
	6	264	294	318	348	372	408
	7	287	317	344	379	400	443
	8	308	342	372	408	432	484
	9	333	369	405	441	468	522
	10	360	400	430	470	500	560
	11	370	410	447	491	518	583
	12	390	438	474	522	552	618
	13	407	448	494	538	575	641
	14	424	472	514	564	600	674

N	M	.1	.05	.025	.01	.005	.001
90	15	450	510	555	600	645	720
	16	460	510	556	612	650	730
	17	476	528	577	633	674	757
	18	504	558	612	666	720	792
	19	509	566	615	677	719	809
	20	530	590	640	710	750	840
	21	540	603	657	720	768	861
	22	556	618	674	740	786	888
	23	572	635	691	760	810	911
	24	588	654	714	786	834	942
	25	605	670	730	805	855	960
	26	616	686	748	822	874	984
	27	639	711	774	846	900	1017
	28	646	718	782	862	916	1030
	29	658	733	800	880	935	1054
	30	690	780	840	930	990	1110
	31	688	764	835	919	976	1099
	32	702	780	852	938	996	1122
	33	717	798	870	957	1017	1146
	34	730	812	886	974	1038	1166
	35	745	830	905	995	1060	1190
	36	774	864	936	1026	1098	1224
	37	771	856	935	1030	1094	1231
	38	786	872	952	1048	1116	1254
	39	801	888	969	1068	1134	1278
	40	820	910	990	1090	1160	1310
	41	824	915	1002	1102	1169	1317
	42	840	936	1020	1122	1194	1344
	43	851	946	1032	1137	1210	1360
	44	866	962	1050	1156	1228	1384
	45	900	1035	1125	1215	1305	1440
	46	890	992	1082	1188	1268	1422
	47	905	1005	1097	1208	1283	1447
	48	924	1026	1116	1230	1308	1470
	49	929	1035	1127	1240	1321	1486
	50	950	1060	1150	1270	1350	1520
	51	957	1065	1161	1278	1359	1533
	52	970	1078	1176	1294	1378	1550
	53	982	1090	1190	1312	1396	1571
	54	1008	1116	1224	1350	1422	1602
	55	1010	1125	1225	1350	1435	1615

N	M	.1	.05	.025	.01	.005	.001
90	56	1020	1136	1238	1362	1450	1632
	57	1035	1149	1254	1380	1470	1656
	58	1046	1164	1270	1398	1486	1674
	59	1058	1176	1284	1412	1502	1694
	60	1110	1230	1320	1470	1560	1740
	61	1084	1206	1315	1447	1540	1733
	62	1096	1220	1330	1464	1558	1754
	63	1116	1242	1350	1485	1584	1782
	64	1122	1246	1362	1498	1594	1796
	65	1135	1265	1380	1520	1615	1820
	66	1152	1278	1398	1536	1632	1842
	67	1158	1287	1405	1548	1645	1858
	68	1172	1302	1422	1564	1666	1876
	69	1185	1317	1437	1581	1683	1896
	70	1200	1340	1460	1610	1710	1920
	71	1207	1343	1466	1613	1717	1935
	72	1242	1368	1494	1638	1746	1962
	73	1233	1370	1497	1648	1751	1974
	74	1246	1384	1510	1664	1770	1994
	75	1275	1410	1545	1695	1800	2025
	76	1270	1412	1540	1696	1804	2034
	77	1280	1422	1552	1714	1819	2050
	78	1296	1440	1572	1734	1842	2076
	79	1305	1451	1584	1746	1855	2091
	80	1330	1470	1610	1770	1880	2120
	81	1341	1485	1620	1782	1899	2142
	82	1342	1492	1630	1794	1908	2150
	83	1354	1506	1644	1810	1922	2170
	84	1368	1524	1662	1830	1944	2190
	85	1380	1535	1675	1845	1960	2210
	86	1390	1546	1688	1856	1976	2228
	87	1404	1560	1701	1872	1995	2247
	88	1412	1572	1712	1886	2012	2256
	89	1409	1576	1739	1917	2019	2284
	90	1530	1710	1800	1980	2160	2430
91	1	88	90	91	-		
	2	144	156	166	172	176	182
	3	179	198	216	234	243	258
	4	212	233	252	273	292	320
	5	239	265	289	314	334	370
	6	264	294	318	348	371	413

N	M	.1	.05	.025	.01	.005	.001
91	7	294	322	350	385	406	455
	8	311	346	375	413	437	488
	9	332	368	403	441	468	522
	10	354	390	428	469	499	559
	11	373	414	451	495	527	591
	12	392	437	474	521	553	620
	13	416	468	507	559	585	663
	14	434	483	525	574	609	686
	15	446	494	538	596	629	705
	16	463	516	560	618	656	736
	17	480	535	583	640	680	765
	18	498	553	605	662	700	790
	19	513	571	624	684	727	818
	20	529	588	643	707	749	845
	21	553	609	665	735	777	875
	22	561	624	680	748	796	894
	23	579	636	698	767	814	924
	24	592	656	717	789	838	943
	25	608	675	735	809	860	970
	26	637	702	767	845	897	1001
	27	636	707	773	849	903	1016
	28	658	728	798	875	931	1043
	29	665	740	806	888	943	1063
	30	680	749	829	900	957	1078
	31	693	772	842	925	985	1107
	32	708	787	860	946	1007	1133
	33	722	805	877	964	1028	1154
	34	735	819	893	983	1047	1177
	35	756	840	917	1008	1071	1204
	36	765	850	928	1020	1087	1221
	37	778	866	944	1039	1106	1244
	38	793	879	961	1059	1126	1265
	39	819	910	988	1092	1157	1300
	40	818	910	994	1094	1163	1309
	41	832	926	1010	1110	1182	1332
	42	854	945	1029	1134	1204	1358
	43	859	955	1043	1149	1222	1375
	44	871	972	1060	1166	1239	1396
	45	885	979	1069	1192	1251	1424
	46	896	991	1082	1208	1266	1441
	47	914	1015	1108	1218	1296	1460
	48	926	1028	1123	1237	1315	1482

N	M	.1	.05	.025	.01	.005	.001
91	49	945	1050	1148	1260	1337	1505
	50	950	1058	1154	1272	1353	1522
	51	964	1073	1171	1288	1371	1543
	52	938	1105	1196	1313	1404	1573
	53	990	1102	1202	1323	1407	1587
	54	1003	1116	1218	1341	1428	1607
	55	1016	1131	1233	1357	1444	1628
	56	1036	1148	1253	1379	1470	1652
	57	1042	1158	1262	1394	1478	1669
	58	1055	1173	1280	1411	1497	1687
	59	1068	1187	1297	1427	1518	1710
	60	1080	1203	1309	1445	1538	1728
	61	1087	1209	1325	1452	1560	1750
	62	1106	1229	1342	1477	1570	1769
	63	1127	1246	1365	1498	1596	1799
	64	1131	1258	1372	1509	1608	1810
	65	1157	1287	1404	1534	1638	1846
	66	1157	1285	1402	1546	1644	1853
	67	1169	1300	1418	1560	1660	1871
	68	1177	1310	1437	1579	1673	1895
	69	1193	1328	1448	1593	1696	1910
	70	1211	1344	1470	1617	1722	1939
	71	1218	1354	1479	1627	1732	1951
	72	1231	1368	1493	1644	1750	1970
	73	1246	1379	1506	1664	1762	1994
	74	1255	1397	1523	1677	1784	2012
	75	1267	1409	1539	1694	1802	2031
	76	1282	1422	1555	1709	1817	2046
	77	1295	1442	1575	1729	1841	2079
	78	1313	1456	1599	1755	1872	2106
	79	1318	1464	1598	1759	1872	2109
	80	1329	1478	1613	1777	1890	2130
	81	1341	1492	1626	1789	1907	2150
	82	1353	1506	1643	1808	1925	2165
	83	1365	1519	1658	1824	1940	2188
	84	1386	1540	1680	1848	1967	2212
	85	1388	1545	1686	1855	1976	2225
	86	1400	1557	1702	1874	1991	2244
	87	1414	1573	1715	1889	2010	2265
	88	1425	1586	1729	1903	2026	2284
	89	1437	1598	1741	1915	2050	2310
	90	1432	1598	1763	1943	2046	2314

N	M	.1	.05	.025	.01	.005	.001
91	91	1547	1729	1820	2002	2184	2457
92	1	89	91	92	-		
	2	146	158	168	174	178	184
	3	178	201	219	234	246	261
	4	216	236	256	280	296	324
	5	241	266	290	318	335	375
	6	268	296	322	354	376	418
	7	292	322	350	385	411	455
	8	316	352	380	420	444	496
	9	336	374	406	446	473	529
	10	358	396	432	474	502	564
	11	377	418	457	501	534	594
	12	400	444	480	528	560	628
	13	414	463	504	554	584	660
	14	434	482	524	578	614	686
	15	450	502	547	599	639	716
	16	472	524	568	628	664	748
	17	486	540	588	646	689	774
	18	502	558	608	670	712	802
	19	519	576	629	692	736	826
	20	536	596	652	716	760	856
	21	551	613	668	734	781	878
	22	568	630	688	756	804	902
	23	598	667	736	805	851	943
	24	600	668	728	800	852	956
	25	614	681	741	816	870	975
	26	628	698	762	838	892	1002
	27	642	715	780	858	912	1026
	28	660	.32	800	880	936	1052
	29	671	748	816	898	952	1073
	30	686	764	832	916	976	1098
	31	697	783	849	939	999	1124
	32	720	796	872	956	1020	1144
	33	729	812	885	975	1036	1168
	34	744	828	904	994	1058	1190
	35	758	843	920	1012	1076	1214
	36	776	860	940	1032	1100	1236
	37	785	875	952	1045	1116	1261
	38	800	890	970	1068	1136	1280
	39	813	904	986	1086	1155	1302
	40	828	920	1004	1108	1176	1324

N	M	.1	.05	.025	.01	.005	.001
92	41	838	936	1019	1122	1193	1346
	42	854	950	1036	1142	1214	1368
	43	868	965	1053	1158	1232	1386
	44	884	980	1072	1180	1252	1412
	45	893	995	1087	1193	1273	1426
	46	966	1058	1150	1242	1334	1518
	47	920	1023	1117	1228	1307	1479
	48	936	1040	1136	1248	1328	1496
	49	947	1052	1150	1266	1346	1517
	50	960	1068	1166	1282	1364	1538
	51	975	1080	1181	1302	1384	1558
	52	988	1100	1200	1320	1404	1580
	53	1000	1112	1214	1335	1419	1600
	54	1014	1126	1230	1354	1440	1622
	55	1023	1141	1242	1370	1460	1644
	56	1040	1156	1264	1388	1480	1664
	57	1051	1169	1276	1405	1493	1682
	58	1064	1184	1292	1422	1512	1704
	59	1077	1198	1308	1440	1532	1726
	60	1092	1216	1324	1460	1552	1748
	61	1097	1231	1334	1479	1573	1762
	62	1116	1240	1354	1488	1584	1788
	63	1127	1254	1369	1508	1602	1806
	64	1144	1272	1388	1528	1624	1828
	65	1152	1282	1400	1541	1639	1845
	66	1166	1298	1416	1558	1658	1870
	67	1178	1311	1429	1572	1675	1888
	68	1192	1328	1448	1592	1696	1912
	69	1219	1357	1472	1633	1725	1955
	70	1216	1354	1476	1626	1730	1950
	71	1229	1368	1491	1640	1747	1970
	72	1244	1384	1508	1660	1768	1992
	73	1253	1394	1522	1675	1782	2008
	74	1266	1408	1536	1694	1800	2030
	75	1279	1422	1552	1707	1816	2049
	76	1292	1440	1568	1728	1836	2072
	77	1303	1448	1582	1741	1852	2089
	78	1316	1464	1598	1758	1872	2108
	79	1326	1479	1612	1772	1888	2126
	80	1344	1492	1628	1792	1908	2148
	81	1353	1504	1641	1808	1922	2167
	82	1366	1518	1658	1824	1940	2186

N	M	.1	.05	.025	.01	.005	.001
92	83	1377	1531	1671	1840	1956	2205
	84	1392	1548	1688	1860	1976	2228
	85	1400	1556	1702	1873	1993	2243
	86	1414	1572	1716	1888	2010	2266
	87	1425	1584	1730	1904	2027	2284
	88	1440	1600	1748	1924	2044	2304
	89	1450	1613	1759	1935	2063	2324
	90	1462	1626	1774	1950	2084	2352
	91	1488	1621	1787	1969	2074	2344
	92	1564	1748	1932	2116	2208	2484
93	1	90	92	93			
	2	148	160	168	176	180	186
	3	183	204	219	237	249	264
	4	216	239	259	284	300	328
	5	244	269	295	322	337	372
	6	273	300	327	357	378	420
	7	295	327	357	390	413	458
	8	318	352	384	422	446	499
	9	342	378	411	453	480	534
	10	360	400	435	478	508	570
	11	381	421	459	506	536	601
	12	402	447	486	534	567	633
	13	419	465	506	557	594	663
	14	437	487	529	581	618	696
	15	456	507	552	609	645	726
	16	473	526	575	630	671	752
	17	489	546	595	651	692	777
	18	510	564	615	678	720	807
	19	525	582	637	698	741	831
	20	540	601	655	721	763	862
	21	558	621	675	744	789	888
	22	572	637	694	764	813	913
	23	589	657	709	779	827	938
	24	606	672	732	807	858	963
	25	619	688	750	825	878	988
	26	634	705	771	846	901	1012
	27	651	723	789	867	921	1038
	28	665	740	806	883	943	1059
	29	678	754	823	907	962	1084
	30	696	771	843	927	984	1110

N	M	.1	.05	.025	.01	.005	.001
93	31	744	806	899	961	1023	1147
	32	722	802	876	966	1027	1155
	33	738	819	894	984	1047	1179
	34	751	834	911	1003	1067	1199
	35	764	848	927	1020	1088	1220
	36	780	867	948	1041	1107	1248
	37	791	882	959	1054	1125	1271
	38	806	898	979	1078	1146	1291
	39	822	915	999	1098	1167	1314
	40	833	927	1014	1113	1182	1337
	41	848	943	1029	1133	1205	1357
	42	864	960	1047	1152	1227	1380
	43	876	972	1063	1170	1243	1400
	44	889	987	1079	1187	1261	1422
	45	903	1005	1098	1206	1284	1446
	46	910	1008	1100	1226	1311	1461
	47	921	1041	1136	1240	1328	1477
	48	945	1050	1143	1260	1341	1509
	49	954	1062	1159	1276	1358	1531
	50	969	1077	1176	1292	1377	1549
	51	984	1095	1194	1314	1398	1575
	52	995	1107	1207	1329	1414	1593
	53	1006	1122	1226	1347	1430	1614
	54	1023	1137	1242	1365	1452	1638
	55	1035	1151	1255	1382	1470	1656
	56	1050	1165	1274	1403	1485	1672
	57	1062	1182	1290	1419	1509	1701
	58	1074	1192	1305	1435	1527	1716
	59	1086	1208	1319	1453	1546	1741
	60	1101	1224	1335	1470	1563	1761
	61	1112	1237	1350	1484	1580	1782
	62	1147	1271	1395	1519	1643	1829
	63	1140	1266	1383	1521	1620	1824
	64	1150	1278	1395	1538	1633	1842
	65	1165	1295	1410	1556	1651	1864
	66	1179	1311	1428	1572	1674	1884
	67	1189	1322	1443	1588	1690	1904
	68	1201	1337	1458	1605	1706	1924
	69	1215	1350	1476	1623	1728	1947
	70	1225	1364	1484	1641	1739	1967
	71	1239	1379	1504	1656	1762	1986
	72	1254	1395	1521	1674	1782	2007

N	M	.1	.05	.025	.01	.005	.001
93	73	1265	1406	1533	1690	1797	2023
	74	1278	1420	1549	1707	1817	2045
	75	1290	1437	1566	1725	1833	2067
	76	1300	1449	1578	1738	1848	2084
	77	1314	1462	1595	1757	1869	2105
	78	1329	1476	1611	1773	1887	2127
	79	1338	1490	1625	1789	1904	2146
	80	1351	1504	1640	1806	1920	2168
	81	1365	1518	1656	1824	1941	2187
	82	1376	1530	1671	1839	1956	2203
	83	1385	1545	1686	1855	1975	2224
	84	1401	1560	1701	1872	1992	2247
	85	1413	1571	1714	1888	2008	2263
	86	1426	1585	1729	1905	2028	2284
	87	1440	1599	1746	1923	2046	2304
	88	1450	1611	1759	1939	2061	2324
	89	1462	1625	1774	1953	2078	2341
	90	1476	1641	1791	1971	2097	2361
	91	1487	1653	1804	1987	2114	2387
	92	1511	1646	1811	1995	2107	2376
	93	1581	1767	1953	2139	2232	2511
94	1	91	93	94			
	2	150	162	170	178	182	188
	3	183	204	222	240	249	267
	4	218	242	260	282	300	332
	5	246	272	296	325	341	376
	6	274	304	328	358	380	422
	7	298	330	358	393	417	463
	8	322	356	389	426	452	504
	9	343	380	415	455	483	537
	10	364	404	440	484	514	576
	11	384	428	465	511	542	609
	12	404	450	490	536	570	640
	13	424	470	512	563	598	673
	14	442	492	536	588	626	700
	15	460	510	557	613	651	731
	16	478	532	580	636	678	760
	17	493	551	601	661	701	789
	18	512	570	622	684	726	814
	19	525	591	637	707	747	841
	20	546	606	662	728	772	870

N	M	.1	.05	.025	.01	.005	.001
94	21	561	624	678	749	793	896
	22	578	642	702	770	820	922
	23	593	658	722	793	841	950
	24	610	678	738	812	862	972
	25	624	693	756	834	886	998
	26	640	712	778	854	908	1022
	27	654	731	791	871	926	1044
	28	670	746	814	894	950	1070
	29	685	763	830	916	974	1095
	30	700	778	850	934	994	1118
	31	709	795	862	953	1015	1140
	32	730	812	886	976	1038	1168
	33	743	826	901	994	1056	1188
	34	758	844	920	1012	1076	1212
	35	773	858	937	1031	1095	1234
	36	786	874	954	1050	1118	1258
	37	801	890	972	1069	1137	1282
	38	816	906	988	1088	1156	1304
	39	829	920	1004	1105	1176	1324
	40	842	938	1022	1126	1196	1348
	41	857	952	1040	1144	1216	1369
	42	870	968	1056	1162	1236	1392
	43	883	982	1073	1179	1254	1414
	44	898	998	1088	1198	1276	1436
	45	910	1012	1103	1217	1293	1457
	46	924	1028	1122	1232	1312	1484
	47	987	1081	1175	1316	1363	1551
	48	950	1056	1152	1272	1348	1522
	49	963	1071	1169	1286	1370	1543
	50	978	1088	1186	1306	1390	1566
	51	990	1102	1202	1323	1408	1586
	52	1004	1118	1220	1342	1428	1608
	53	1018	1131	1235	1360	1447	1630
	54	1032	1146	1252	1376	1466	1650
	55	1044	1161	1267	1394	1484	1672
	56	1058	1176	1282	1412	1502	1692
	57	1070	1190	1300	1431	1521	1713
	58	1084	1204	1314	1448	1540	1734
	59	1096	1219	1330	1464	1557	1756
	60	1110	1234	1346	1482	1576	1776
	61	1121	1247	1363	1500	1594	1796
	62	1136	1262	1378	1518	1616	1816

N	M	.1	.05	.025	.01	.005	.001
94	63	1147	1274	1396	1528	1624	1838
	64	1162	1292	1410	1552	1650	1860
	65	1174	1305	1422	1566	1667	1877
	66	1186	1320	1440	1584	1686	1902
	67	1200	1334	1456	1603	1700	1923
	68	1212	1348	1472	1620	1722	1942
	69	1225	1362	1485	1635	1741	1960
	70	1238	1376	1502	1654	1758	1984
	71	1250	1390	1516	1672	1775	2004
	72	1264	1404	1532	1688	1796	2022
	73	1277	1416	1549	1705	1812	2042
	74	1288	1432	1564	1722	1830	2064
	75	1303	1442	1574	1737	1850	2078
	76	1314	1460	1594	1754	1866	2104
	77	1324	1475	1606	1768	1885	2125
	78	1338	1488	1624	1788	1902	2144
	79	1351	1500	1639	1804	1920	2162
	80	1364	1516	1654	1822	1938	2184
	81	1376	1529	1670	1837	1955	2203
	82	1388	1544	1684	1854	1974	2222
	83	1400	1557	1699	1871	1990	2244
	84	1414	1572	1714	1888	2008	2262
	85	1425	1583	1730	1904	2026	2281
	86	1438	1598	1746	1920	2044	2304
	87	1450	1611	1760	1936	2060	2321
	88	1462	1626	1774	1954	2080	2344
	89	1473	1640	1789	1970	2095	2362
	90	1486	1654	1804	1986	2112	2382
	91	1497	1666	1819	2001	2130	2401
	92	1512	1680	1836	2024	2146	2422
	93	1532	1710	1835	2021	2184	2463
	94	1598	1786	1974	2162	2256	2538
95	1	92	94	95			
	2	150	164	172	180	184	190
	3	184	207	225	243	252	270
	4	221	244	261	285	304	332
	5	250	280	305	330	350	385
	6	276	307	332	362	385	427
	7	301	335	363	398	423	472
	8	323	361	393	427	457	506
	9	346	385	418	459	486	548

N	M	.1	.05	.025	.01	.005	.001
95	10	370	410	450	490	520	585
	11	387	431	470	515	548	616
	12	410	450	495	543	578	649
	13	427	474	517	569	604	677
	14	446	496	539	594	631	709
	15	470	520	565	620	660	740
	16	484	534	582	644	679	770
	17	500	556	606	665	708	796
	18	516	575	626	688	733	823
	19	551	608	665	722	779	874
	20	555	615	670	740	785	880
	21	566	629	689	755	806	902
	22	583	648	707	779	825	928
	23	598	665	727	799	850	956
	24	612	682	749	822	872	987
	25	635	705	770	845	900	1010
	26	646	716	783	862	914	1033
	27	660	737	805	885	941	1052
	28	676	753	820	904	960	1078
	29	692	768	840	924	982	1106
	30	710	790	960	945	1005	1130
	31	722	801	877	964	1026	1156
	32	741	815	838	975	1038	1167
	33	750	833	911	1002	1064	1200
	34	765	851	928	1022	1085	1221
	35	785	870	950	1045	1110	1250
	36	792	883	962	1060	1127	1272
	37	807	897	980	1079	1145	1292
	38	836	931	1007	1121	1178	1330
	39	837	928	1013	1117	1186	1336
	40	855	950	1035	1140	1210	1365
	41	863	960	1048	1153	1227	1380
	42	877	975	1066	1172	1246	1404
	43	891	992	1081	1191	1268	1425
	44	905	1005	1099	1210	1285	1449
	45	920	1025	1120	1230	1310	1475
	46	933	1037	1130	1244	1325	1491
	47	936	1058	1154	1257	1347	1500
	48	966	1071	1167	1272	1362	1546
	49	972	1081	1179	1300	1381	1557
	50	990	1100	1200	1320	1405	1580

N	M	.1	.05	.025	.01	.005	.001
95	51	1000	1112	1213	1336	1420	1600
	52	1014	1127	1229	1352	1439	1621
	53	1028	1143	1247	1371	1459	1642
	54	1039	1155	1262	1388	1478	1666
	55	1055	1175	1280	1410	1500	1690
	56	1066	1187	1294	1424	1516	1708
	57	1102	1216	1330	1463	1558	1748
	58	1091	1215	1327	1459	1553	1749
	59	1106	1230	1341	1477	1572	1770
	60	1120	1250	1360	1500	1595	1795
	61	1131	1259	1374	1510	1606	1812
	62	1144	1272	1389	1528	1627	1833
	63	1156	1283	1406	1539	1652	1850
	64	1169	1302	1421	1564	1663	1876
	65	1185	1320	1440	1585	1685	1900
	66	1196	1330	1451	1598	1701	1917
	67	1210	1344	1467	1615	1719	1936
	68	1220	1356	1486	1635	1734	1953
	69	1233	1371	1498	1647	1756	1975
	70	1250	1390	1520	1670	1775	2000
	71	1264	1406	1530	1678	1791	2024
	72	1272	1416	1547	1701	1809	2038
	73	1284	1429	1560	1718	1829	2058
	74	1297	1443	1578	1737	1845	2078
	75	1315	1460	1595	1755	1865	2105
	76	1349	1482	1615	1786	1900	2147
	77	1337	1486	1622	1785	1899	2142
	78	1349	1500	1637	1802	1916	2160
	79	1364	1513	1653	1815	1937	2178
	80	1380	1530	1670	1840	1955	2205
	81	1386	1543	1683	1853	1973	2222
	82	1398	1556	1699	1868	1989	2242
	83	1410	1571	1715	1883	2007	2261
	84	1424	1584	1727	1902	2023	2282
	85	1440	1600	1745	1925	2045	2305
	86	1449	1610	1759	1936	2058	2320
	87	1461	1626	1774	1952	2078	2340
	88	1474	1639	1780	1970	2095	2362
	89	1487	1653	1803	1986	2111	2379
	90	1500	1670	1820	2005	2135	2405
	91	1510	1680	1833	2017	2148	2421
	92	1522	1693	1850	2037	2165	2441

N	M	.1	.05	.025	.01	.005	.001
95	93	1536	1707	1867	2055	2178	2455
	94	1554	1735	1860	2048	2214	2497
	95	1615	1805	1995	2185	2280	2565
96	1	93	95	96	-		
	2	152	166	174	182	186	192
	3	189	210	228	246	255	273
	4	224	248	268	292	308	336
	5	253	278	304	330	349	390
	6	282	312	342	372	390	438
	7	305	338	366	401	424	473
	8	336	368	400	440	464	520
	9	351	390	426	465	492	552
	10	372	414	450	494	524	588
	11	392	435	474	521	553	622
	12	420	468	504	564	588	660
	13	432	480	523	575	611	684
	14	452	500	546	600	638	718
	15	471	522	570	627	666	747
	16	496	560	608	656	704	784
	17	506	559	612	671	716	801
	18	528	582	636	702	744	834
	19	542	601	657	718	759	854
	20	560	620	676	744	792	888
	21	573	639	696	765	813	915
	22	590	656	714	786	836	938
	23	605	672	732	807	858	964
	24	648	720	768	840	912	1008
	25	637	708	773	849	904	1016
	26	652	726	792	872	926	1040
	27	669	744	810	891	948	1068
	28	684	760	832	912	972	1092
	29	698	776	847	931	992	1117
	30	720	798	870	954	1014	1146
	31	727	808	884	972	1034	1161
	32	768	864	928	1024	1088	1216
	33	759	843	921	1011	1077	1212
	34	772	858	936	1032	1098	1234
	35	785	872	956	1052	1115	1255
	36	816	900	984	1080	1152	1296
	37	816	906	988	1090	1158	1305
	38	830	922	1008	1108	1180	1324

N	M	.1	.05	.025	.01	.005	.001
96	39	846	939	1026	1128	1200	1350
	40	864	960	1048	1152	1224	1376
	41	871	968	1060	1160	1239	1392
	42	888	990	1080	1188	1260	1422
	43	899	1000	1091	1202	1278	1440
	44	916	1016	1112	1224	1300	1464
	45	927	1032	1125	1239	1317	1485
	46	942	1046	1142	1256	1338	1506
	47	955	1060	1156	1278	1352	1527
	48	1008	1104	1200	1344	1440	1584
	49	980	1090	1191	1310	1398	1565
	50	994	1106	1208	1328	1414	1592
	51	1011	1122	1224	1347	1434	1617
	52	1024	1140	1244	1368	1456	1640
	53	1035	1151	1256	1383	1471	1658
	54	1056	1170	1278	1404	1494	1686
	55	1063	1178	1286	1421	1506	1698
	56	1080	1200	1312	1440	1536	1728
	57	1089	1212	1323	1455	1548	1746
	58	1102	1226	1338	1472	1566	1766
	59	1113	1238	1354	1488	1584	1784
	60	1140	1260	1380	1512	1608	1812
	61	1140	1269	1384	1523	1620	1828
	62	1156	1284	1402	1542	1642	1848
	63	1170	1299	1419	1560	1659	1869
	64	1216	1344	1472	1600	1696	1920
	65	1194	1327	1448	1594	1696	1911
	66	1212	1344	1470	1614	1722	1938
	67	1220	1355	1480	1631	1734	1955
	68	1236	1372	1500	1648	1756	1976
	69	1248	1386	1512	1665	1770	1995
	70	1258	1400	1528	1680	1788	2014
	71	1271	1414	1543	1698	1806	2036
	72	1296	1440	1584	1728	1848	2088
	73	1297	1442	1574	1731	1843	2077
	74	1310	1456	1592	1750	1862	2098
	75	1323	1473	1605	1767	1881	2118
	76	1336	1488	1624	1784	1900	2140
	77	1351	1495	1631	1799	1914	2162
	78	1368	1518	1656	1824	1938	2184
	79	1373	1527	1666	1833	1951	2199
	80	1408	1552	1696	1872	1984	2240

N	M	.1	.05	.025	.01	.005	.001
96	81	1401	1557	1698	1869	1989	2241
	82	1412	1570	1712	1886	2004	2260
	83	1425	1582	1727	1901	2023	2281
	84	1452	1608	1752	1932	2052	2316
	85	1449	1611	1757	1935	2059	2321
	86	1462	1626	1774	1952	2078	2342
	87	1476	1641	1791	1971	2097	2361
	88	1496	1656	1808	1992	2120	2384
	89	1498	1666	1818	2003	2129	2400
	90	1518	1686	1836	2022	2154	2424
	91	1522	1692	1848	2035	2166	2440
	92	1536	1708	1864	2052	2184	2464
	93	1548	1722	1881	2070	2202	2481
	94	1560	1734	1894	2086	2212	2492
	95	1574	1759	1886	2077	2242	2528
	96	1632	1824	2016	2208	2304	2592
97	1	94	96	97	-		
	2	154	166	176	184	188	194
	3	189	213	231	249	258	276
	4	224	248	268	296	312	340
	5	255	281	308	333	353	395
	6	281	312	340	372	395	437
	7	308	339	371	407	429	477
	8	332	367	399	439	464	520
	9	353	393	427	470	499	555
	10	375	419	453	499	529	593
	11	396	440	479	525	558	624
	12	417	464	503	551	587	659
	13	437	484	528	580	614	691
	14	454	507	550	606	646	719
	15	474	527	573	632	670	752
	16	492	543	592	654	699	782
	17	509	565	618	679	722	812
	18	527	585	639	702	747	837
	19	544	605	659	727	771	868
	20	562	624	679	749	798	895
	21	578	642	701	772	819	920
	22	594	660	721	792	843	946
	23	610	679	740	814	865	974
	24	622	693	761	835	885	1002
	25	642	714	780	858	911	1025

N	M	.1	.05	.025	.01	.005	.001
97	26	658	732	798	878	935	1051
	27	673	748	818	899	957	1077
	28	689	766	837	921	977	1102
	29	704	782	856	937	1002	1127
	30	720	800	873	960	1020	1151
	31	735	817	891	980	1042	1174
	32	753	827	914	1006	1055	1204
	33	764	849	926	1018	1084	1218
	34	779	866	945	1040	1105	1245
	35	794	882	962	1058	1127	1268
	36	808	898	980	1079	1148	1292
	37	822	914	997	1098	1168	1316
	38	835	931	1015	1117	1188	1339
	39	849	945	1036	1137	1213	1356
	40	865	962	1049	1156	1230	1385
	41	879	978	1066	1174	1250	1405
	42	893	993	1084	1192	1269	1428
	43	909	1008	1103	1212	1288	1450
	44	921	1026	1116	1230	1309	1476
	45	936	1040	1135	1249	1329	1495
	46	948	1055	1152	1266	1347	1516
	47	962	1069	1166	1286	1367	1541
	48	982	1086	1184	1291	1381	1568
	49	994	1099	1197	1332	1396	1584
	50	1002	1116	1218	1340	1425	1606
	51	1018	1131	1233	1358	1443	1629
	52	1031	1146	1250	1375	1465	1649
	53	1045	1161	1267	1392	1481	1673
	54	1055	1174	1281	1410	1505	1690
	55	1070	1191	1299	1429	1522	1715
	56	1085	1205	1316	1449	1542	1736
	57	1098	1222	1332	1467	1560	1756
	58	1113	1233	1348	1483	1579	1775
	59	1124	1251	1365	1502	1597	1799
	60	1138	1264	1380	1518	1615	1822
	61	1150	1280	1398	1538	1635	1843
	62	1164	1294	1413	1556	1654	1864
	63	1177	1308	1429	1573	1674	1885
	64	1190	1324	1446	1587	1691	1904
	65	1209	1342	1465	1606	1705	1929
	66	1217	1353	1477	1625	1729	1947
	67	1230	1367	1493	1642	1746	1969

N	M	.1	.05	.025	.01	.005	.001
97	68	1241	1383	1509	1657	1764	1987
	69	1255	1398	1525	1677	1786	2010
	70	1269	1410	1539	1695	1803	2030
	71	1282	1424	1555	1712	1823	2051
	72	1295	1439	1570	1730	1841	2073
	73	1304	1450	1589	1741	1859	2099
	74	1320	1467	1603	1763	1877	2114
	75	1333	1481	1617	1780	1895	2134
	76	1345	1496	1634	1799	1913	2154
	77	1358	1511	1643	1815	1932	2177
	78	1371	1524	1664	1833	1949	2198
	79	1385	1539	1680	1849	1967	2217
	80	1397	1555	1696	1868	1985	2239
	81	1412	1565	1769	1885	2001	2258
	82	1422	1580	1727	1899	2020	2279
	83	1434	1598	1740	1918	2042	2295
	84	1447	1608	1756	1932	2059	2318
	85	1461	1622	1767	1949	2072	2337
	86	1474	1630	1789	1967	2092	2360
	87	1485	1652	1802	1983	2112	2378
	88	1499	1666	1818	2002	2131	2399
	89	1510	1679	1833	2018	2148	2422
	90	1522	1694	1847	2034	2165	2441
	91	1535	1708	1861	2049	2182	2460
	92	1548	1722	1877	2067	2200	2479
	93	1559	1733	1894	2085	2218	2501
	94	1571	1746	1909	2102	2234	2520
	95	1584	1760	1923	2117	2247	2530
	96	1595	1782	1957	2154	2270	2559
	97	1746	1843	2037	2231	2425	2716
98	1	95	97	98			
	2	156	168	178	186	190	196
	3	190	213	231	249	261	279
	4	228	250	270	294	316	344
	5	257	284	310	337	357	392
	6	284	316	342	374	396	442
	7	315	350	378	413	441	490
	8	334	370	404	442	470	524
	9	355	397	433	471	504	560
	10	380	422	458	504	534	596

N	M	.1	.05	.025	.01	.005	.001
98	11	400	444	486	531	564	630
	12	420	468	510	560	594	668
	13	441	490	533	585	623	700
	14	476	518	574	630	658	742
	15	478	531	581	637	675	762
	16	496	552	602	662	702	792
	17	514	571	624	686	729	819
	18	534	592	646	710	754	848
	19	549	610	666	733	779	875
	20	568	630	688	756	804	906
	21	588	651	714	784	833	938
	22	600	668	728	800	852	956
	23	615	684	747	822	874	983
	24	634	704	768	846	898	1008
	25	649	722	787	866	920	1037
	26	664	740	806	888	942	1060
	27	680	756	825	910	965	1086
	28	714	784	854	938	994	1120
	29	712	790	862	949	1010	1137
	30	726	808	882	970	1032	1160
	31	741	824	899	989	1051	1185
	32	758	842	916	1010	1074	1206
	33	771	863	936	1032	1098	1232
	34	786	874	954	1050	1116	1258
	35	805	896	980	1078	1141	1288
	36	816	908	990	1090	1158	1304
	37	830	924	1007	1109	1179	1329
	38	844	940	1024	1128	1200	1350
	39	855	952	1045	1146	1222	1377
	40	874	972	1060	1166	1240	1396
	41	886	986	1075	1186	1261	1418
	42	910	1008	1106	1218	1288	1456
	43	916	1019	1112	1222	1301	1466
	44	930	1034	1128	1242	1320	1486
	45	943	1048	1145	1260	1341	1510
	46	958	1064	1162	1280	1360	1532
	47	970	1080	1178	1296	1378	1554
	48	986	1094	1196	1314	1402	1572
	49	1029	1176	1274	1372	1470	1666
	50	1012	1126	1230	1350	1438	1624
	51	1026	1140	1245	1371	1458	1642
	52	1040	1156	1262	1388	1478	1664
98	53	1053	1172	1278	1406	1496	1686
	54	1068	1186	1296	1426	1516	1708
	55	1080	1201	1311	1442	1536	1728
	56	1106	1232	1344	1470	1568	1764
	57	1107	1232	1343	1479	1574	1772
	58	1122	1246	1360	1498	1594	1794
	59	1132	1264	1380	1519	1615	1816
	60	1148	1276	1392	1534	1632	1838
	61	1162	1290	1408	1552	1650	1859
	62	1174	1306	1426	1568	1668	1880
	63	1190	1323	1449	1589	1694	1904
	64	1200	1336	1456	1604	1706	1922
	65	1218	1351	1478	1617	1717	1941
	66	1228	1364	1488	1640	1744	1968
	67	1240	1379	1504	1654	1762	1984
	68	1254	1394	1520	1674	1782	2008
	69	1266	1408	1535	1691	1800	2028
	70	1288	1428	1568	1722	1834	2058
	71	1292	1436	1568	1726	1836	2068
	72	1306	1452	1584	1744	1854	2090
	73	1319	1467	1599	1761	1876	2108
	74	1332	1480	1616	1778	1892	2134
	75	1343	1493	1630	1796	1908	2150
	76	1358	1510	1646	1814	1928	2174
	77	1372	1526	1666	1834	1953	2198
	78	1384	1538	1678	1846	1964	2214
	79	1396	1551	1693	1864	1984	2234
	80	1408	1566	1710	1382	2002	2256
	81	1421	1580	1725	1899	2019	2275
	82	1434	1594	1740	1916	2038	2296
	83	1446	1610	1755	1934	2056	2319
	84	1470	1638	1778	1960	2086	2352
	85	1473	1637	1787	1965	2090	2357
	86	1486	1652	1802	1984	2112	2380
	87	1497	1663	1815	1999	2128	2400
	88	1510	1680	1832	2018	2146	2420
	89	1521	1691	1848	2035	2162	2437
	90	1536	1708	1864	2052	2184	2460
	91	1554	1729	1883	2072	2205	2485
	92	1560	1736	1894	2086	2218	2500
	93	1571	1750	1907	2102	2235	2519
	94	1586	1762	1924	2118	2254	2540
98	95	1597	1775	1938	2134	2271	2559
	96	1610	1790	1952	2148	2294	2588
	97	1616	1805	1985	2184	2298	2590
	98	1764	1960	2058	2254	2450	2744
99	1	95	98	99	-		
	2	158	170	180	183	192	198
	3	195	216	234	252	264	279
	4	229	253	273	297	316	348
	5	260	286	311	341	360	396
	6	288	321	348	378	402	447
	7	314	348	378	413	439	488
	8	338	375	407	447	474	530
	9	369	405	441	486	513	576
	10	383	424	463	510	541	602
	11	418	462	495	550	583	649
	12	426	474	516	567	600	672
	13	444	494	538	592	628	704
	14	461	515	559	615	656	739
	15	483	537	585	645	684	768
	16	501	557	608	667	711	795
	17	519	577	630	692	735	828
	18	549	603	657	720	765	864
	19	554	617	672	739	786	884
	20	569	631	689	765	807	907
	21	591	657	717	786	837	939
	22	616	682	748	814	869	979
	23	621	691	756	829	882	993
	24	639	711	777	852	906	1020
	25	657	731	788	878	931	1037
	26	670	746	813	894	951	1071
	27	693	774	837	927	981	1107
	28	703	779	850	936	998	1119
	29	717	798	869	956	1020	1146
	30	735	816	891	978	1041	1173
	31	747	832	908	996	1062	1194
	32	762	848	925	1018	1082	1221
	33	825	891	990	1056	1122	1287
	34	794	882	963	1059	1125	1267
	35	808	899	978	1077	1146	1292
	36	828	918	1008	1107	1179	1323

N	M	.1	.05	.025	.01	.005	.001
99	37	836	932	1018	1119	1191	1341
	38	850	948	1032	1138	1208	1360
	39	867	966	1053	1158	1233	1386
	40	880	980	1068	1176	1252	1410
	41	894	995	1085	1196	1271	1430
	42	912	1014	1104	1215	1293	1455
	43	923	1027	1121	1233	1311	1477
	44	946	1056	1144	1265	1342	1507
	45	963	1062	1161	1278	1359	1530
	46	965	1074	1170	1290	1371	1545
	47	979	1088	1188	1307	1390	1567
	48	993	1107	1206	1326	1413	1590
	49	1009	1114	1214	1352	1417	1606
	50	1020	1127	1228	1366	1460	1623
	51	1035	1152	1257	1383	1470	1656
	52	1048	1167	1272	1400	1491	1678
	53	1062	1180	1289	1418	1510	1699
	54	1080	1206	1314	1440	1539	1728
	55	1100	1221	1331	1463	1562	1749
	56	1103	1226	1338	1472	1566	1767
	57	1119	1242	1356	1494	1587	1788
	58	1129	1257	1371	1508	1606	1808
	59	1145	1270	1386	1525	1624	1834
	60	1158	1287	1404	1548	1647	1854
	61	1170	1300	1421	1561	1664	1872
	62	1184	1318	1434	1580	1680	1892
	63	1206	1341	1458	1602	1710	1926
	64	1209	1344	1470	1616	1721	1937
	65	1222	1361	1484	1635	1738	1958
	66	1254	1419	1518	1683	1782	2013
	67	1249	1390	1517	1668	1776	2002
	68	1264	1403	1532	1687	1796	2021
	69	1278	1422	1551	1707	1815	2046
	70	1290	1435	1564	1721	1832	2065
	71	1303	1449	1580	1742	1851	2086
	72	1323	1467	1602	1764	1881	2115
	73	1328	1478	1613	1775	1889	2128
	74	1344	1494	1623	1792	1912	2142
	75	1356	1509	1644	1812	1926	2172
	76	1368	1520	1661	1827	1943	2191
	77	1386	1540	1683	1859	1969	2222
	78	1395	1551	1692	1863	1983	2235
99	79	1410	1567	1709	1874	2003	2248
	80	1419	1579	1721	1896	2017	2273
	81	1440	1602	1746	1917	2043	2304
	82	1446	1606	1753	1931	2054	2314
	83	1458	1621	1770	1947	2073	2335
	84	1473	1638	1788	1965	2091	2358
	85	1482	1650	1803	1978	2105	2373
	86	1496	1664	1817	1998	2127	2398
	87	1512	1680	1833	2019	2148	2418
	88	1529	1705	1859	2046	2178	2442
	89	1532	1707	1864	2052	2182	2459
	90	1557	1728	1881	2079	2205	2484
	91	1561	1735	1893	2084	2218	2499
	92	1573	1748	1908	2099	2234	2517
	93	1587	1764	1926	2118	2256	2541
	94	1597	1775	1938	2135	2270	2560
	95	1610	1789	1954	2151	2289	2580
	96	1623	1806	1968	2169	2307	2601
	97	1633	1817	1980	2178	2331	2630
	98	1637	1827	2011	2211	2326	2621
	99	1782	1980	2079	2277	2475	2772
100	1	96	99	100	-		
	2	158	172	182	190	194	200
	3	194	219	237	255	267	282
	4	232	256	280	304	320	352
	5	265	295	320	345	365	405
	6	290	322	350	382	404	450
	7	316	350	379	418	444	493
	8	344	380	412	452	480	536
	9	364	404	440	484	512	573
	10	390	440	480	520	550	620
	11	406	451	493	539	572	646
	12	432	480	520	572	608	680
	13	448	498	544	597	635	714
	14	468	520	568	624	664	742
	15	490	545	595	650	695	775
	16	508	564	616	676	720	808
	17	524	583	637	698	743	837
	18	542	604	658	724	770	862
	19	560	622	679	748	793	892
	20	600	660	720	780	840	940
100	21	593	661	719	792	843	945
	22	612	680	742	816	866	976
	23	627	697	759	839	889	1003
	24	648	720	784	860	916	1032
	25	675	750	825	900	950	1075
	26	678	752	822	904	960	1080
	27	692	771	841	925	984	1106
	28	712	792	860	948	1008	1132
	29	724	805	878	966	1030	1159
	30	750	830	910	1000	1060	1190
	31	754	840	916	1007	1071	1207
	32	772	860	936	1032	1096	1232
	33	783	877	949	1047	1114	1249
	34	800	890	972	1070	1138	1280
	35	820	910	990	1090	1160	1305
	36	832	924	1008	1112	1180	1328
	37	845	939	1024	1128	1200	1350
	38	860	956	1044	1148	1220	1376
	39	874	971	1060	1166	1242	1398
	40	900	1000	1100	1200	1280	1440
	41	902	1004	1094	1206	1281	1444
	42	918	1020	1114	1226	1302	1468
	43	933	1037	1127	1240	1324	1493
	44	948	1052	1148	1264	1344	1516
	45	965	1070	1170	1285	1365	1540
	46	974	1084	1182	1300	1384	1560
	47	987	1098	1198	1320	1403	1580
	48	1004	1116	1216	1340	1424	1604
	49	1015	1130	1234	1353	1442	1630
	50	1100	1200	1300	1400	1500	1700
	51	1042	1160	1266	1394	1478	1668
	52	1060	1176	1284	1416	1504	1696
	53	1070	1192	1299	1429	1522	1715
	54	1086	1206	1318	1450	1542	1736
	55	1100	1225	1335	1470	1565	1760
	56	1116	1240	1352	1483	1584	1784
	57	1123	1250	1365	1505	1596	1805
	58	1140	1268	1382	1522	1620	1824
	59	1153	1282	1399	1540	1638	1846
	60	1180	1320	1440	1580	1680	1880
	61	1180	1312	1433	1577	1677	1889
	62	1194	1328	1448	1594	1696	1912

N	M	.1	.05	.025	.01	.005	.001
100	63	1207	1342	1464	1611	1714	1933
	64	1224	1360	1484	1632	1736	1956
	65	1235	1375	1500	1650	1755	1980
	66	1248	1388	1514	1664	1770	1996
	67	1257	1394	1524	1686	1789	2021
	68	1276	1420	1548	1704	1812	2040
	69	1286	1430	1561	1717	1829	2060
	70	1310	1450	1590	1740	1860	2090
	71	1314	1460	1593	1754	1866	2103
	72	1328	1476	1612	1776	1888	2128
	73	1340	1489	1624	1789	1904	2147
	74	1354	1504	1642	1808	1924	2168
	75	1400	1550	1675	1850	1975	2200
	76	1380	1536	1676	1844	1960	2212
	77	1391	1546	1689	1859	1976	2229
	78	1406	1562	1706	1878	1996	2250
	79	1418	1577	1720	1894	2015	2272
	80	1440	1600	1760	1920	2060	2300
	81	1444	1604	1752	1928	2052	2313
	82	1458	1620	1768	1946	2070	2332
	83	1470	1634	1784	1963	2088	2351
	84	1484	1652	1800	1984	2108	2376
	85	1500	1665	1820	2000	2130	2400
	86	1510	1678	1832	2016	2146	2418
	87	1521	1691	1846	2030	2162	2436
	88	1536	1708	1864	2052	2184	2460
	89	1544	1719	1875	2064	2197	2475
	90	1570	1740	1900	2090	2220	2510
	91	1571	1749	1907	2098	2234	2517
	92	1588	1764	1924	2120	2256	2540
	93	1597	1776	1939	2134	2269	2561
	94	1610	1790	1954	2152	2290	2580
	95	1625	1810	1975	2170	2310	2605
	96	1640	1820	1988	2188	2328	2624
	97	1648	1833	1998	2200	2342	2639
	98	1660	1846	2012	2214	2366	2666
	99	1659	1850	2037	2238	2355	2654
	100	1800	2000	2100	2300	2500	2800

CRITICAL VALUES AND PROBABILITY LEVELS FOR THE WILCOXON RANK SUM TEST AND THE WILCOXON SIGNED RANK TEST

Frank Wilcoxon*
S. K. Katti†
Roberta A. Wilcox‡

***Deceased, November 18, 1965.**
†Department of Statistics, University of Missouri, Columbia.
‡CIBA Pharmaceutical Company, Division of CIBA Corporation.

INTRODUCTION

Several individuals and organizations have assisted in the production of these tables. Probabilities were computed using an IBM 709 computer at the Computing Center of the Florida State University.

Programming was done by Dr. S. K. Katti of the Department of Statistics, Florida State University, starting from generating functions furnished by Dr. F. Wilcoxon.

The assistance of the National Science Foundation in the establishment of the Florida State University Computing Center made this work possible.

The tables were prepared and distributed under the supervision of Roberta A. Wilcox, formerly of the Department of Statistical Design and Analysis, Lederle Laboratories Division, American Cyanamid Company and currently of the Clinical Research Department, CIBA Pharmaceutical Company, Division of CIBA Corporation.

Thanks are due to Mrs. C. DeTora, Mrs. R. Haring, and Mrs. M. Rackowski for typing and proofreading.

TABLES FOR THE RANK SUM TEST

Let $x_1, x_2, \ldots, x_M$ be a random sample of size M from a population π_1 and $y_1, y_2, \ldots, y_N$, a random sample of size N from a population π_2. Without loss of generality, we may assume that $M \leq N$. The minimum possible sum of the ranks of the x's when the x's and the y's are arranged in increasing order of magnitude is obviously $M(M+1)/2$. Hence, the sum of the ranks of the x's in any experiment can be denoted by

$$M(M+1)/2 + U$$

where U is an integer greater than or equal to zero. It has been shown that the number of ways of obtaining a rank sum with a given value for U is the coefficient of t^U in the expansion of the following function – called generating function – in powers of t:

$$g(t) = \frac{(1-t^{M+1})(1-t^{M+2})\ldots(1-t^{M+N})}{(1-t)(1-t^2)\ldots(1-t^N)}$$

or

$$g(t) = \prod_{i=1}^{N} \frac{(1-t^{M+i})}{(1-t^i)}. \tag{1}$$

See Mann and Whitney (1947) and MacMahon (1916).

The total number of ways of obtaining any rank sum is $\binom{M+N}{N}$. Hence if $\pi_1 \equiv \pi_2$, the probability of a rank sum is given by the ratio of the number of ways of obtaining the rank sum and this total number. The aim of the table is to give one-sided critical regions corresponding to type I error rates 0.005, 0.01, 0.025 and 0.05 and for $3 \leq M \leq N \leq 50$. This was achieved in the following two stages:

(i) Developing a method of collecting the coefficient of t^U in the expansion of formula (1), and

(ii) Using the coefficients in (i) to compute successive cumulative probabilities and to obtain critical regions.

rief discussions are given below on these stages.

ollecting Coefficients

Let us denote the number of ways, i.e., frequency, of obtaining a given U by $f(U)$. From the definition of formula 1), we have

$$g(t) = \prod_{i=1}^{N} \frac{(1 - t^{M+i})}{(1 - t^{i})} = f(0) + tf(1) + \ldots + t^{U} f(U) + \ldots . \tag{2}$$

ence,

$$\log g(t) = \sum_{i=1}^{N} \log (1 - t^{M+i}) - \sum_{i=1}^{N} \log (1 - t^{i})$$

r

$$\frac{g'(t)}{g(t)} = \sum_{i=1}^{N} \frac{-(M+i)\, t^{M+i-1}}{1 - t^{M+i}} + \sum_{i=1}^{N} \frac{it^{i-1}}{1 - t^{i}} . \tag{3}$$

n expanding the terms on the right hand side and rearranging the terms, we get

$$(1) + 2tf(2) + \ldots + Ut^{U-1} f(U) + \ldots = [f(0) + tf(1) + \ldots + t^{U} f(U) + \ldots]$$

$$\{\sum_{i=1}^{N} \sum_{j=1}^{\infty} -(M+i)\, t^{(M+i)j-1} + \sum_{i=1}^{N} \sum_{j=1}^{\infty} it^{ij-1}\}. \tag{4}$$

Let z_i denote the coefficient of t^i in the braces. Then, on comparing the coefficient of t^{U-1} on the two sides of equa- tion (4), we get

$$f(U) = \frac{1}{U} \sum_{i=0}^{U-1} f(i)\, z_{U-i-1} \quad (U = 1, 2, 3, \ldots) . \tag{5}$$

It is to be noted that $f(0) = 1$. Hence, to compute $f(U)$ using (5), we need to compute the z's.

For this, we first obtain an upper bound for the number of z's needed. A convenient upper bound is $k = (M+50)^2/2 - M(M+1)/2$. First set $z_i = 0, i = 1, \ldots, k$. For a given i and j, compute the indices $k_1 = (M+i)j - 1$ and $k_2 = ij - 1$, add $-(M+i)$ to z_{k_1} if $k_1 \leq k$ and add i to z_{k_2} if $k_2 \leq k$. Start with $i = 1$ and do this for successive $j = 1, 2, \ldots$ etc. until both k_1 and k_2 exceed k. When that j is reached, change i to $i + 1$ and repeat for $j = 1, 2, \ldots$. Repeat the process until (and including) $i = n$ and j such that both k_1 and k_2 exceed k.

It is to be noted that if the z's have been computed for some (M, N), then the z's corresponding to $(M, N + 1)$ can be obtained by adding $-(M+i)$ to z_{k_1} and $+i$ to z_{k_2} for $i = M + 1$ and $j = 0, 1, 2, \ldots$ etc., until both k_1 and k_2 exceed k.

Description of the Tables

Tables were to be constructed for one-tail type I error rates 0.005, 0.01, 0.025 and 0.05. Both left-tail and right-tail critical regions were to be given so that if one uses both the tails, one gets two-tail tests with critical regions corresponding to error rates 0.01, 0.02, 0.05 and 0.10. It is clear that if the left-tail is from 0 to r (both ends included), the right-tail would be from $M(M+N+1)-r$ to $M(M+N+1)$ (both ends included). The distributions being discrete, critical regions cannot be obtained to correspond to the exact specified type I error rates. Hence two numbers $r-1$ and r were obtained such that P (rank sum $\leq r-1) \leq$ specified type I error rate $\leq P$ (rank sum $\leq r$) and the exact probabilities were recorded so that the user knows how far off he is from the required type I error rate. For example, an entry in the table like

$$M=10,\ N=30: \quad \begin{array}{l} 151, 259\ .0476 \\ 152, 258\ .0508 \end{array}$$

means that if we take the critical region as less than or equal to 152 or as greater than or equal to 258 the type I error r would be 0.0508 and if we take it as less than or equal to 151 or as greater than or equal to 259, it would be 0.0476. If we use a two-tail test for which the critical region consists of values less than or equal to 152 and greater than or equal to 258, type I error rate would be 2 x 0.0508, i.e., 0.1016 and if we use one for which the critical region consists of values less than or equal to 151 and greater than or equal to 259, the type I error rate would be 2 x 0.0476, i.e., 0.0952 The exact one-tail probability sought here is 0.05 and the exact two-tail probability sought is 0.10.

The critical rank totals are tabulated for sample sizes $M=N=3$ up to $M=N=50$, including all combinations of M and N within these limits.

Example

The data from two samples, x_i and y_j, with 15 x observations and 12 y observations are assigned ranks 1 to 27. The sum of the x ranks is 286 and that of the y ranks is 92. The experimenter wishes to know whether such a result is unlikely if the two samples came from the same population; he is willing to accept a 1 per cent chance of wrongly decidin that the samples came from different populations. Referring to the table under the section $P = .01$ (two-sided), for $M =$ 12 and $N = 15$, the following numbers are found:

115,221 .0044
116,220 .0051

The interpretation of these entries is as follows: The probability that the rank total for the smaller size sample, y in this case, is equal to or less than 115 or equal to or greater than 221 is 2(.0044) = .0088. The corresponding probability for rank totals of 116 or 220 is 2(.0051) = .0102. Since the rank total for the y sample is 92, one may reject, at the 1 per cent probability level, the hypothesis that the two samples came from the same population. To be conservative in rejecting the null hypothesis, the upper pair of totals 115 and 221 are used; if, however, the experimenter is willing to overstep slightly the 1 per cent significance level, the lower totals of 116 and 220 are used. A third possibility is to use the total nearest to the desired value of 1 per cent.

The methodology for the Wilcoxon rank sum test is given by Wilcoxon (1945). This method is described in such well known textbooks as those by Dixon and Massey (1951) and Siegel (1956) and in a monograph by Wilcoxon and Wilcox (1964).

TABLES FOR THE SIGNED RANK TEST

Let $x_1, \ldots, x_N$ be N independently distributed random variables. Assume that the probability distribution of each

$_i$ is symmetrical around the same point μ. In the signed rank test, we test the hypothesis that $\mu = \mu_0$ where μ_0 is a pecified number. The criterion W for the test is computed as follows: First compute $y_i = x_i - \mu_0$ and rank the y's by heir absolute values. Define $c_i = 0$ if the y with rank i is negative and $c_i = 1$ otherwise. Then

$$W = \sum_{i=1}^{N} ic_i,$$

.e., W is the sum of the positive y's. The range of W clearly is $0, 1, \ldots, N(N+1)/2$. As an example of computing W, if he observed x's are $-5.2, -3.0, +0.8, +2.3, +4.8$ and $+5.9$ and the aim is to test $\mu = 1.2$, then we first compute the y's s $-6.4, -4.2, -0.4, +1.1, +3.6$ and $+4.7$. Their ranks, as ranked by their absolute values, are 6, 4, 1, 2, 3 and 5 respec- ively. W then is $2 + 3 + 5$ which is 10.

An interesting applied case arises when paired experiments are made and the x_i is the difference in the ith paired ob- ervations. A null hypothesis in such case is that both observations in the pair are identically distributed. Under this null ypothesis, the x_i will have probability distributions which are symmetrical around zero. W then is the sum of the ranks of positive x's when all the x's are ranked by their absolute values.

Distribution of W

From the definition of W, each sample gives rise to a vector $c = (c_1, \ldots, c_N)$. Since each c_i is either zero or one, there are 2^N possible vectors. Under the null hypothesis, all the vectors are equally likely. Hence, $P_N(w)$, the probability that in a sample of size N, W equals w, is given by $f_N(w)/2^N$ where $f_N(w)$ is the number of vectors c which give w.

To compute $f_N(w)$, we first define the generating function:

$$g_N(t) = \sum_{w=0}^{N(N+1)/2} f_N(w)\, t^w.$$

It is well known that

$$g_{N+1}(t) = g_N(t)(1 + t^{N+1})$$

or, more explicitly,

$$\sum_{w=0}^{(N+1)(N+2)/2} f_{N+1}(w)t^w = \sum_{w=0}^{N(N+1)/2} f_N(w)t^w + \sum_{w=0}^{N(N+1)/2} f_N(w)t^{w+N+1}.$$

Given $f_N(w)$ for all w for some N, a scheme for computing $f_{N+1}(w)$ for all w is first to set $f_{N+1}(w) = 0$ for all w, i.e. $w = 0, 1, \ldots, (N+1)(N+2)/2$ and then add $f_N(w)$ to $f_{N+1}(w)$ and $f_{N+1}(w + N + 1)$ for $w = 0, 1, \ldots, N(N+1)/2$. For $N = 2$, we can list c and w and note that $f_2(0) = f_2(1) = f_2(2) = f_2(3) = 1$. Hence, $f_N(w)$ for any N can be computed successively starting from $N = 2$.

Since the distribution of W is symmetrical, only half of the distribution is tabulated. If one wishes to set the critical points for a right-tail test on W, which is not available in these tables, one computes the sum of the negative y's and makes a left-tail test. If one is making a two-tail test, one computes the smaller of the sums of the ranks of the positive and negative y's and uses a left-tail test with half the required error rate.

Use of the Table

Exact probability levels (P) are given to four decimal places for all possible rank totals (T) which yield a different probability level at the fourth decimal place from 0.0001 up to and including 0.5000. The rank totals (T) are tabulated for all sample sizes from $N = 5$ up to $N = 50$. An asterisk is placed to the left of the smallest rank total for which the probability level is equal to or less than 0.0500.

Probability levels between 0.5000 and 1.0000 are not tabulated because the distribution is symmetrical. To conserve space, the first rank total for which the probability level is nonzero is the first given. Thereafter the rank total is tabulated only if the probability level changes from the previous one tabulated. For example for $N = 42$, 154 is the smallest rank total which gives a nonzero probability level in the rourth decimal place. The smallest rank total which has a probability level of 0.0002 is 173. Any number above 154 and below 173 will have a probability level of 0.0001.

Example

The sums of the ranked differences from two paired samples of eight observations each are +29 and −7. The experimenter wishes to know whether such a result is unlikely if the two samples came from the same population; he is willing to accept a 5 per cent chance of wrongly deciding that the samples came from different populations. Referring to Table II for $N = 8$, one finds the following:

$$T = 7, P = .0742$$

Thus, the probability that the smallest rank total in absolute value is equal to or less than 7 is 2(.0742) = .1484. Therefore, there is not sufficient evidence at the 5 per cent probability level to reject the hypothesis that the two samples came from the same population.

The methodology for the signed rank test is given in Wilcoxon (1945). This method is described extensively in such well known text books as Dixon and Massey (1951) and Siegel (1956) and in the monograph by Wilcoxon and Wilcox (1964).

REFERENCES

1. Dixon, W. J. and Massey, F. J. (1951). *Introduction to Statistical Analysis,* New York: McGraw-Hill.

2. MacMahon, Percy A. (1916). *Combinatory Analysis,* Vol. II, London: Cambridge University Press.

3. Mann, H. B. and Whitney, D. R. (1947). "On a test of whether one of two random variables is stochastically larger than the other," *Ann. Math. Statist.* 18: 50–60.

4. Siegel, Sidney (1956). *Nonparametric Statistics for the Behavioral Sciences,* New York: McGraw-Hill.

5. Wilcoxon, Frank (1945). "Individual comparisons by ranking methods," *Biometrics,* 1:80–83.

6. Wilcoxon, Frank, and Wilcox, R. A. (1964). *Some Rapid Approximate Statistical Procedures,* Pearl River, New York: Lederle Laboratories, Division of the American Cyanamid Company.

TABLE I. CRITICAL VALUES AND PROBABILITY LEVELS FOR THE WILCOXON RANK SUM TEST

P = .05 one-sided

P = .10 two-sided

N	M = 3	M = 4	M = 5	M = 6	M = 7	M = 8
3	5,16 .0000 6,15 .0500					
4	6,18 .0286 7,17 .0571	11,25 .0286 12,24 .0571				
5	7,20 .0357 8,19 .0714	12,28 .0317 13,27 .0556	19,36 .0476 20,35 .0754			
6	8,22 .0476 9,21 .0833	13,31 .0333 14,30 .0571	20,40 .0411 21,39 .0628	28,50 .0465 29,49 .0660		
7	8,25 .0333 9,24 .0583	14,34 .0364 15,33 .0545	21,44 .0366 22,43 .0530	29,55 .0367 30,54 .0507	39,66 .0487 40,65 .0641	
8	9,27 .0424 10,26 .0667	15,37 .0364 16,36 .0545	23,47 .0466 24,46 .0637	31,59 .0406 32,58 .0539	41,71 .0469 42,70 .0603	51,85 .0415 52,84 .0524
9	9,30 .0318 10,29 .0500	16,40 .0378 17,39 .0531	24,51 .0415 25,50 .0559	33,63 .0440 34,62 .0567	43,76 .0454 44,75 .0571	54,90 .0464 55,89 .0570
10	10,32 .0385 11,31 .0559	17,43 .0380 18,42 .0529	26,54 .0496 27,53 .0646	35,67 .0467 36,66 .0589	45,81 .0439 46,80 .0544	56,96 .0416 57,95 .0506
11	11,34 .0440 12,33 .0632	18,46 .0388 19,45 .0520	27,58 .0449 28,57 .0575	37,71 .0491 38,70 .0608	47,86 .0427 48,85 .0521	59,101 .0454 60,100 .0543
12	11,37 .0352 12,36 .0505	19,49 .0390 20,48 .0516	28,62 .0409 29,61 .0519	38,76 .0415 39,75 .0512	49,91 .0416 50,90 .0501	62,106 .0489 63,105 .0576
13	12,39 .0411 13,38 .0554	20,52 .0395 21,51 .0508	30,65 .0473 31,64 .0586	40,80 .0437 41,79 .0530	52,95 .0484 53,94 .0573	64,112 .0445 65,111 .0521
14	13,41 .0456 14,40 .0603	21,55 .0395 22,54 .0507	31,69 .0435 32,68 .0534	42,84 .0457 43,83 .0547	54,100 .0469 55,99 .0550	67,117 .0475 68,116 .0550
15	13,44 .0380 14,43 .0502	22,58 .0400 23,57 .0501	33,72 .0491 34,71 .0593	44,88 .0474 45,87 .0561	56,105 .0455 57,104 .0531	69,123 .0437 70,122 .0503
16	14,46 .0423 15,45 .0547	24,60 .0497 25,59 .0609	34,76 .0455 35,75 .0546	46,92 .0490 47,91 .0574	58,110 .0443 59,109 .0513	72,128 .0463 73,127 .0528
17	15,48 .0465 16,47 .0588	25,63 .0493 26,62 .0600	35,80 .0425 36,79 .0506	47,97 .0433 48,96 .0505	61,114 .0497 62,113 .0569	75,133 .0487 76,132 .0552
18	15,51 .0398 16,50 .0504	26,66 .0491 27,65 .0589	37,83 .0472 38,82 .0555	49,101 .0448 50,100 .0518	63,119 .0484 64,118 .0550	77,139 .0452 78,138 .0510
19	16,53 .0435 17,52 .0539	27,69 .0487 28,68 .0582	38,87 .0442 39,86 .0517	51,105 .0462 52,104 .0530	65,124 .0471 66,123 .0533	80,144 .0475 81,143 .0532
20	17,55 .0469 18,54 .0576	28,72 .0485 29,71 .0573	40,90 .0485 41,89 .0562	53,109 .0475 54,108 .0541	67,129 .0460 68,128 .0518	83,149 .0495 84,148 .0552

P = .05 one-sided

P = .10 two-sided

N	M = 3	M = 4	M = 5	M = 6	M = 7	M = 8
21	17,58 .0410 18,57 .0504	29,75 .0481 30,74 .0567	41,94 .0457 42,93 .0527	55,113 .0487 56,112 .0551	69,134 .0449 70,133 .0504	85,155 .0464 86,154 .0515
22	18,60 .0443 19,59 .0535	30,78 .0480 31,77 .0560	43,97 .0496 44,96 .0567	57,117 .0498 58,116 .0561	72,138 .0492 73,137 .0548	88,160 .0483 89,159 .0533
23	19,62 .0473 20,61 .0565	31,81 .0477 32,80 .0554	44,101 .0469 45,100 .0535	58,122 .0452 59,121 .0508	74,143 .0481 75,142 .0534	90,166 .0454 91,165 .0501
24	19,65 .0421 20,64 .0503	32,84 .0475 33,83 .0548	45,105 .0445 46,104 .0505	60,126 .0463 61,125 .0518	76,148 .0470 77,147 .0521	93,171 .0472 94,170 .0518
25	20,67 .0449 21,66 .0531	33,87 .0473 34,86 .0544	47,108 .0480 48,107 .0541	62,130 .0473 63,129 .0527	78,153 .0461 79,152 .0508	96,176 .0488 97,175 .0534
26	21,69 .0476 22,68 .0558	34,90 .0471 35,89 .0539	48,112 .0456 49,111 .0514	64,134 .0483 65,133 .0536	81,157 .0497 82,156 .0546	98,182 .0462 99,181 .0504
27	21,72 .0429 22,71 .0502	35,93 .0469 36,92 .0535	50,115 .0489 51,114 .0547	66,138 .0492 67,137 .0544	83,162 .0487 84,161 .0533	101,187 .0478 102,186 .0520
28	22,74 .0454 23,73 .0527	36,96 .0468 37,95 .0531	51,119 .0466 52,118 .0521	67,143 .0454 68,142 .0501	85,167 .0477 86,166 .0522	104,192 .0493 105,191 .0534
29	23,76 .0478 24,75 .0552	37,99 .0466 38,98 .0527	53,122 .0497 54,121 .0552	69,147 .0463 70,146 .0509	87,172 .0469 88,171 .0511	106,198 .0468 107,197 .0507
30	23,79 .0434 24,78 .0502	38,102 .0465 39,101 .0523	54,126 .0475 55,125 .0527	71,151 .0471 72,150 .0517	89,177 .0460 90,176 .0501	109,203 .0482 110,202 .0521
31	24,81 .0458 25,80 .0525	39,105 .0464 40,104 .0520	55,130 .0456 56,129 .0504	73,155 .0479 74,154 .0524	92,181 .0491 93,180 .0533	112,208 .0496 113,207 .0534
32	25,83 .0480 26,82 .0547	40,108 .0462 41,107 .0517	57,133 .0483 58,132 .0532	75,159 .0487 76,158 .0531	94,186 .0483 95,185 .0522	114,214 .0473 115,213 .0509
33	25,86 .0440 26,85 .0501	41,111 .0461 42,110 .0514	58,137 .0464 59,136 .0510	77,163 .0495 78,162 .0538	96,191 .0474 97,190 .0512	117,219 .0486 118,218 .0522
34	26,88 .0461 27,87 .0523	42,114 .0460 43,113 .0511	60,140 .0490 61,139 .0537	78,168 .0462 79,167 .0502	98,196 .0467 99,195 .0503	120,224 .0499 121,223 .0534
35	27,90 .0481 28,89 .0543	43,117 .0459 44,116 .0509	61,144 .0472 62,143 .0516	80,172 .0470 81,171 .0509	101,200 .0495 102,199 .0532	122,230 .0477 123,229 .0510
36	27,93 .0444 28,92 .0501	44,120 .0458 45,119 .0506	63,147 .0497 64,146 .0542	82,176 .0477 83,175 .0515	103,205 .0487 104,204 .0522	125,235 .0489 126,234 .0522
37	28,95 .0464 29,94 .0520	45,123 .0457 46,122 .0504	64,151 .0479 65,150 .0521	84,180 .0484 85,179 .0522	105,210 .0479 106,209 .0513	127,241 .0470 128,240 .0501
38	29,97 .0482 30,96 .0539	46,126 .0456 47,125 .0502	65,155 .0462 66,154 .0503	86,184 .0490 87,183 .0528	107,215 .0472 108,214 .0505	130,246 .0481 131,245 .0512

P = .05 one-sided

P = .10 two-sided

N	M = 3	M = 4	M = 5	M = 6	M = 7	M = 8
39	29,100 .0448	48,128 .0500	67,158 .0485	88,188 .0496	110,219 .0497	133,251 .0492
	30,99 .0501	49,127 .0547	68,157 .0526	89,187 .0533	111,218 .0531	134,250 .0522
40	30,102 .0466	49,131 .0498	68,162 .0469	89,193 .0468	112,224 .0490	135,257 .0473
	31,101 .0519	50,130 .0544	69,161 .0508	90,192 .0503	113,223 .0522	136,256 .0502
41	31,104 .0483	50,134 .0496	70,165 .0491	91,197 .0474	114,229 .0483	138,262 .0484
	32,103 .0536	51,133 .0541	71,164 .0530	92,196 .0508	115,228 .0514	139,261 .0513
42	31,107 .0451	51,137 .0494	71,169 .0475	93,201 .0480	116,234 .0476	141,267 .0494
	32,106 .0500	52,136 .0538	72,168 .0513	94,200 .0514	117,233 .0506	142,266 .0523
43	32,109 .0468	52,140 .0492	73,172 .0496	95,205 .0486	119,238 .0499	143,273 .0477
	33,108 .0517	53,139 .0535	74,171 .0534	96,204 .0519	120,237 .0530	144,272 .0504
44	33,111 .0484	53,143 .0490	74,176 .0481	97,209 .0492	121,243 .0492	146,278 .0486
	34,110 .0533	54,142 .0535	75,175 .0517	98,208 .0525	122,242 .0522	147,277 .0513
45	33,114 .0454	54,146 .0489	75,180 .0467	99,213 .0497	123,248 .0486	149,283 .0496
	34,113 .0500	55,145 .0530	76,179 .0501	100,212 .0530	124,247 .0514	150,282 .0523
46	34,116 .0469	55,149 .0487	77,183 .0486	100,218 .0472	125,253 .0479	151,289 .0479
	35,115 .0516	56,148 .0527	78,182 .0521	101,217 .0503	126,252 .0507	152,288 .0505
47	35,118 .0485	56,152 .0486	78,187 .0473	102,222 .0478	127,258 .0474	154,294 .0488
	36,117 .0531	57,151 .0525	79,186 .0506	103,221 .0508	128,257 .0501	155,293 .0514
48	36,120 .0500	57,155 .0484	80,190 .0491	104,226 .0483	130,262 .0494	157,299 .0497
	37,119 .0546	58,154 .0523	81,189 .0525	105,225 .0513	131,261 .0522	158,298 .0523
49	36,123 .0471	58,158 .0483	81,194 .0478	106,230 .0488	132,267 .0488	159,305 .0482
	37,122 .0514	59,157 .0520	82,193 .0510	107,229 .0518	133,266 .0515	160,304 .0506
50	37,125 .0485	59,161 .0482	83,197 .0496	108,234 .0493	134,272 .0482	162,210 .0490
	38,124 .0529	60,160 .0518	84,196 .0529	109,233 .0522	135,271 .0508	163,309 .0514

P = .05 one-sided

P = .10 two-sided

N	M = 9	M = 10	M = 11	M = 12	M = 13	M = 14
9	66,105 .0470					
	67,104 .0567					
10	69,111 .0474	82,128 .0446				
	70,110 .0564	83,127 .0526				
11	72,117 .0476	86,134 .0493	100,153 .0440			
	73,116 .0560	87,133 .0572	101,152 .0507			

P = .05 one-sided

P = .10 two-sided

N	M = 9	M = 10	M = 11	M = 12	M = 13	M = 14
12	75,123 .0477	89,141 .0465	104,160 .0454	120,180 .0444		
	76,122 .0555	90,140 .0536	105,159 .0519	121,179 .0503		
13	78,129 .0478	92,148 .0441	108,167 .0467	125,187 .0488	142,209 .0454	
	79,128 .0551	93,147 .0505	109,166 .0528	126,186 .0548	143,208 .0507	
14	81,135 .0478	96,154 .0478	112,174 .0477	129,195 .0475	147,217 .0472	166,240 .0469
	82,134 .0547	97,153 .0542	113,173 .0536	130,194 .0530	148,216 .0524	167,239 .0518
15	84,141 .0478	99,161 .0455	116,181 .0486	133,203 .0463	152,225 .0489	171,249 .0466
	85,140 .0542	100,160 .0513	117,180 .0543	134,202 .0514	153,224 .0539	172,248 .0512
16	87,147 .0477	103,167 .0487	120,188 .0494	138,210 .0500	156,234 .0458	176,258 .0463
	88,146 .0538	104,166 .0545	121,187 .0549	139,209 .0551	157,233 .0503	177,257 .0506
17	90,153 .0476	106,174 .0465	123,196 .0453	142,218 .0486	161,242 .0472	182,266 .0500
	91,152 .0534	107,173 .0517	124,195 .0501	143,217 .0534	162,241 .0516	183,265 .0543
18	93,159 .0475	110,180 .0493	127,203 .0461	146,226 .0474	166,250 .0485	187,275 .0495
	94,158 .0531	111,179 .0546	128,202 .0508	147,225 .0519	167,249 .0528	188,274 .0536
19	96,165 .0474	113,187 .0472	131,210 .0468	150,234 .0463	171,258 .0497	192,284 .0489
	97,164 .0527	114,186 .0521	132,209 .0513	151,233 .0505	172,257 .0539	193,283 .0528
20	99,171 .0473	117,193 .0498	135,217 .0474	155,241 .0493	175,267 .0470	197,293 .0484
	100,170 .0524	118,192 .0546	136,216 .0518	156,240 .0535	176,266 .0508	198,292 .0521
21	102,177 .0472	120,200 .0478	139,224 .0480	159,249 .0481	180,275 .0481	202,302 .0480
	103,176 .0520	121,199 .0523	140,223 .0522	160,248 .0521	181,274 .0518	203,301 .0515
22	105,183 .0471	123,207 .0459	143,231 .0486	163,257 .0471	185,283 .0491	207,311 .0475
	106,182 .0517	124,206 .0501	144,230 .0526	164,256 .0508	186,282 .0528	208,310 .0509
23	108,189 .0470	127,213 .0482	147,238 .0490	168,264 .0496	189,292 .0467	212,320 .0471
	109,188 .0514	128,212 .0524	148,237 .0530	169,263 .0534	190,291 .0500	213,319 .0503
24	111,195 .0469	130,220 .0465	151,245 .0495	172,272 .0486	194,300 .0476	218,328 .0498
	112,194 .0517	131,219 .0504	152,244 .0533	173,271 .0521	195,299 .0509	219,327 .0530
25	114,201 .0468	134,226 .0486	155,252 .0499	176,280 .0475	199,308 .0485	223,337 .0492
	115,200 .0509	135,225 .0525	156,251 .0536	177,279 .0509	200,307 .0518	224,336 .0524
26	117,207 .0467	137,233 .0469	158,260 .0468	181,287 .0498	204,316 .0493	228,346 .0488
	118,206 .0507	138,232 .0506	159,259 .0503	182,286 .0532	205,315 .0525	229,345 .0518
27	120,213 .0466	141,239 .0488	162,267 .0473	185,295 .0488	208,325 .0471	233,355 .0483
	121,212 .0504	142,238 .0525	163,266 .0506	186,294 .0521	209,324 .0501	234,354 .0512
28	123,219 .0465	144,246 .0473	166,274 .0477	189,303 .0479	213,333 .0479	238,364 .0479
	124,218 .0502	145,245 .0507	167,273 .0510	190,302 .0510	214,332 .0509	239,363 .0506

P = .05 one-sided

P = .10 two-sided

N	M = 9		M = 10		M = 11		M = 12		M = 13		M = 14	
29	127,224	.0500	148,252	.0491	170,281	.0481	194,310	.0500	218,341	.0487	243,373	.0474
	128,223	.0537	149,251	.0525	171,280	.0513	195,309	.0531	219,340	.0516	244,372	.0501
30	130,230	.0498	151,259	.0476	174,288	.0484	198,318	.0490	223,349	.0494	249,381	.0496
	131,229	.0534	152,258	.0508	175,287	.0515	199,317	.0520	224,348	.0522	250,380	.0523
31	133,236	.0496	155,265	.0493	178,295	.0488	202,326	.0481	227,358	.0474	254,390	.0492
	134,235	.0531	156,264	.0525	179,294	.0518	203,325	.0510	228,357	.0501	255,389	.0518
32	136,242	.0494	158,272	.0478	182,302	.0491	206,334	.0473	232,366	.0481	259,399	.0487
	137,241	.0528	159,271	.0509	183,301	.0520	207,333	.0500	233,365	.0507	260,398	.0512
33	139,248	.0492	162,278	.0494	186,309	.0494	211,341	.049[illegible]	237,374	.0488	264,408	.0483
	140,247	.0525	163,277	.0525	187,308	.0523	212,340	.0519	238,373	.0513	265,407	.0507
34	142,254	.0490	165,285	.0480	190,316	.0496	215,349	.0483	242,382	.0494	269,417	.0479
	143,253	.0523	166,284	.0510	191,315	.0525	216,348	.0509	243,381	.0519	270,416	.0503
35	145,260	.0489	169,291	.0496	194,323	.0499	219,357	.0475	247,390	.0500	275,425	.0498
	146,259	.0520	170,290	.0525	195,322	.0527	220,356	.0500	248,389	.0525	276,424	.0522
36	148,266	.0487	172,298	.0482	197,331	.0476	224,364	.0492	251,399	.0482	280,434	.0494
	149,265	.0518	173,297	.0510	198,330	.0502	225,363	.0517	252,398	.0505	281,433	.0517
37	151,272	.0486	176,304	.0497	201,338	.0478	228,372	.0484	256,407	.0488	285,443	.0489
	152,271	.0515	177,303	.0525	202,337	.0504	229,371	.0509	257,406	.0511	286,442	.0512
38	154,278	.0484	179,311	.0484	205,345	.0481	232,380	.0477	261,415	.0493	290,452	.0486
	155,277	.0513	180,310	.0511	206,344	.0506	233,379	.0500	262,414	.0516	291,451	.0507
39	157,284	.0483	183,317	.0498	209,352	.0484	237,387	.0492	266,423	.0499	295,461	.0482
	158,283	.0511	184,316	.0525	210,351	.0508	238,386	.0516	267,422	.0521	296,460	.0503
40	160,290	.0482	186,324	.0485	213,359	.0486	241,395	.0485	270,432	.0482	301,469	.0499
	161,289	.0509	187,323	.0511	214,358	.0511	242,394	.0508	271,431	.0504	302,468	.0520
41	163,296	.0480	190,330	.0499	217,366	.0489	245,403	.0478	275,440	.0487	306,478	.0495
	164,295	.0507	191,329	.0524	218,365	.0512	246,402	.0500	276,439	.0509	307,477	.0515
42	166,302	.0479	193,337	.0487	221,373	.0491	250,410	.0493	280,448	.0492	311,487	.0491
	167,301	.0505	194,336	.0511	222,372	.0514	251,409	.0515	281,447	.0513	312,486	.0511
43	169,308	.0478	197,343	.0499	225,380	.0493	254,418	.0486	285,456	.0497	316,496	.0487
	170,307	.0503	198,342	.0524	226,379	.0516	255,417	.0507	286,455	.0518	317,495	.0506
44	172,314	.0477	200,350	.0488	229,387	.0495	259,425	.0500	289,465	.0482	321,505	.0484
	173,313	.0502	201,349	.0512	230,386	.0518	260,424	.0521	290,464	.0502	322,504	.0502
45	176,319	.0500	204,356	.0500	233,394	.0497	263,433	.0493	294,473	.0487	327,513	.0499
	177,318	.0525	205,355	.0524	234,393	.0519	264,432	.0513	295,472	.0506	328,512	.0518

P = .05 one-sided

P = .10 two-sided

N	M = 9	M = 10	M = 11	M = 12	M = 13	M = 14
46	179,325 .0498	207,363 .0489	237,401 .0499	267,441 .0486	299,481 .0491	332,522 .0495
	180,324 .0523	208,362 .0512	238,400 .0521	268,440 .0506	300,480 .0511	333,521 .0513
47	182,331 .0497	210,370 .0479	240,409 .0480	272,448 .0499	304,489 .0496	337,531 .0491
	183,330 .0521	211,369 .0501	241,408 .0501	273,447 .0519	305,488 .0515	338,530 .0509
48	185,337 .0495	214,376 .0490	244,416 .0482	276,456 .0493	308,498 .0482	342,540 .0488
	186,336 .0519	215,375 .0512	245,415 .0503	277,455 .0512	309,497 .0500	343,539 .0506
49	188,343 .0494	217,383 .0480	248,423 .0485	280,464 .0486	313,506 .0486	347,549 .0485
	189,342 .0517	218,382 .0501	249,422 .0505	281,463 .0505	314,505 .0504	348,548 .0502
50	191,349 .0492	221,389 .0491	252,430 .0486	285,471 .0499	318,514 .0491	353,557 .0498
	192,348 .0515	222,388 .0512	253,429 .0506	286,470 .0518	319,513 .0508	354,556 .0516

P = .05 one-sided

P = .10 two-sided

N	M = 15	M = 16	M = 17	M = 18	M = 19	M = 20
15	192,273 .0488					
	193,272 .0532					
16	197,283 .0466	219,309 .0469				
	198,282 .0507	220,308 .0508				
17	203,292 .0485	225,319 .0471	249,346 .0493			
	204,291 .0525	226,318 .0509	250,345 .0530			
18	208,302 .0465	231,329 .0473	255,357 .0479	280,386 .0485		
	209,301 .0502	232,328 .0509	256,356 .0514	281,385 .0519		
19	214,311 .0482	237,339 .0474	262,367 .0499	287,397 .0490	313,428 .0482	
	215,310 .0518	238,338 .0508	263,366 .0533	288,396 .0523	314,427 .0513	
20	220,320 .0497	243,349 .0475	268,378 .0485	294,408 .0495	320,440 .0474	348,472 .0482
	221,319 .0533	244,348 .0508	269,377 .0518	295,407 .0526	321,439 .0503	349,471 .0511
21	225,330 .0478	249,359 .0475	274,389 .0473	301,419 .0499	328,451 .0494	356,484 .0490
	226,329 .0511	250,358 .0507	275,388 .0503	302,418 .0529	329,450 .0523	357,483 .0518
22	231,339 .0492	255,369 .0476	281,399 .0490	307,431 .0474	335,463 .0486	364,496 .0497
	232,338 .0525	256,368 .0506	282,398 .0520	308,430 .0502	336,462 .0513	365,495 .0524
23	236,349 .0474	261,379 .0476	287,410 .0477	314,442 .0478	342,475 .0478	371,509 .0478
	237,348 .0505	262,378 .0505	288,409 .0505	315,441 .0505	343,474 .0504	372,508 .0503

P = .05 one-sided

P = .10 two-sided

N	M = 15	M = 16	M = 17	M = 18	M = 19	M = 20
24	242,358 .0486	267,389 .0476	294,420 .0492	321,453 .0481	350,486 .0496	379,521 .0484
	243,357 .0517	268,388 .0504	295,419 .0521	322,452 .0508	351,485 .0522	380,520 .0509
25	248,367 .0499	273,399 .0476	300,431 .0480	328,464 .0484	357,498 .0488	387,533 .0490
	249,366 .0529	274,398 .0503	301,430 .0507	329,463 .0510	358,497 .0512	388,532 .0514
26	253,377 .0482	279,409 .0475	307,441 .0495	335,475 .0487	364,510 .0480	395,545 .0496
	254,376 .0510	280,408 .0502	308,440 .0521	336,474 .0512	365,509 .0504	396,544 .0519
27	259,386 .0493	285,419 .0475	313,452 .0483	342,486 .0490	372,521 .0495	402,558 .0478
	260,385 .0521	286,418 .0501	314,451 .0508	343,485 .0514	373,520 .0519	403,557 .0500
28	264,396 .0477	292,428 .0500	320,462 .0496	349,497 .0492	379,533 .0488	410,570 .0483
	265,395 .0503	293,427 .0526	321,461 .0521	350,496 .0516	380,532 .0510	411,569 .0505
29	270,405 .0487	298,438 .0499	326,473 .0485	356,508 .0494	386,545 .0480	418,582 .0488
	271,404 .0514	299,437 .0524	327,472 .0509	357,507 .0517	387,544 .0502	419,581 .0509
30	276,414 .0497	304,448 .0498	333,483 .0497	363,519 .0496	394,556 .0494	426,594 .0493
	277,413 .0523	305,447 .0522	334,482 .0521	364,518 .0519	395,555 .0516	427,593 .0513
31	281,424 .0482	310,458 .0496	339,494 .0486	370,530 .0498	401,568 .0487	434,606 .0497
	282,423 .0507	311,457 .0520	340,493 .0509	371,529 .0520	402,567 .0508	435,605 .0517
32	287,433 .0492	316,468 .0495	346,504 .0498	377,541 .0499	408,580 .0480	441,619 .0481
	288,432 .0516	317,467 .0518	347,503 .0520	378,540 .0521	409,579 .0500	442,618 .0501
33	292,443 .0478	322,478 .0494	352,515 .0487	383,553 .0480	416,591 .0493	449,631 .0485
	293,442 .0501	323,477 .0517	353,514 .0509	384,552 .0501	417,590 .0513	450,630 .0504
34	298,452 .0487	328,488 .0493	359,525 .0498	390,564 .0482	423,603 .0486	457,643 .0489
	299,451 .0509	329,487 .0515	360,524 .0519	391,563 .0502	424,602 .0505	458,642 .0508
35	304,461 .0495	334,498 .0492	365,536 .0488	397,575 .0484	431,614 .0498	465,655 .0493
	305,460 .0518	335,497 .0513	366,535 .0508	398,574 .0503	432,613 .0517	466,654 .0511
36	309,471 .0482	340,508 .0491	372,546 .0498	404,586 .0485	438,626 .0491	473,667 .0496
	310,470 .0503	341,507 .0512	373,545 .0518	405,585 .0504	439,625 .0510	474,666 .0514
37	315,480 .0490	346,518 .0490	378,557 .0489	411,597 .0487	445,638 .0485	481,679 .0499
	316,479 .0511	347,517 .0510	379,556 .0508	412,596 .0505	446,637 .0503	482,678 .0517
38	321,489 .0498	352,528 .0489	385,568 .0498	418,608 .0488	453,649 .0496	488,692 .0485
	322,488 .0519	353,527 .0508	386,566 .0518	419,607 .0506	454,648 .0514	489,691 .0502
39	326,499 .0485	358,538 .0488	391,578 .0489	425,619 .0489	460,661 .0489	496,704 .0489
	327,498 .0505	359,537 .0507	392,577 .0508	426,618 .0507	461,660 .0507	497,703 .0505
40	332,508 .0493	364,548 .0487	398,588 .0498	432,630 .0491	468,672 .0500	504,716 .0492
	333,507 .0513	365,547 .0505	399,587 .0517	433,629 .0508	469,671 .0517	505,715 .0508

P = .05 one-sided

P = .10 two-sided

N	M = 15	M = 16	M = 17	M = 18	M = 19	M = 20
41	337,518 .0481 338,517 .0500	370,558 .0486 371,557 .0504	404,599 .0489 405,598 .0507	439,641 .0492 440,640 .0509	475,684 .0493 476,683 .0510	512,728 .0495 513,727 .0511
42	343,527 .0488 344,526 .0507	376,568 .0485 377,567 .0503	411,609 .0498 412,608 .0516	446,652 .0493 447,651 .0510	482,696 .0487 483,695 .0503	520,740 .0497 521,739 .0513
43	349,536 .0495 350,535 .0514	382,578 .0484 383,577 .0501	417,620 .0489 418,619 .0506	453,663 .0494 454,662 .0510	490,707 .0497 491,706 .0513	528,752 .0500 529,751 .0516
44	354,546 .0484 355,545 .0502	388,588 .0483 389,587 .0500	424,630 .0498 425,629 .0515	460,674 .0495 461,673 .0511	497,719 .0491 498,718 .0507	535,765 .0487 536,764 .0502
45	360,555 .0490 361,554 .0508	395,597 .0499 396,596 .0516	430,641 .0489 431,640 .0506	467,685 .0495 468,684 .0511	504,731 .0485 505,730 .0501	543,777 .0490 544,776 .0505
46	366,564 .0497 367,563 .0515	401,607 .0497 402,606 .0515	437,651 .0497 438,650 .0514	474,696 .0496 475,695 .0512	512,742 .0495 513,741 .0510	551,789 .0493 552,788 .0507
47	371,574 .0486 372,573 .0503	407,617 .0496 408,616 .0513	443,662 .0489 444,661 .0505	481,707 .0497 482,706 .0512	519,754 .0489 520,753 .0504	559,801 .0495 560,800 .0509
48	377,583 .0492 378,582 .0509	413,627 .0495 414,626 .0511	450,672 .0497 451,671 .0513	488,718 .0498 489,717 .0513	527,765 .0498 528,764 .0513	567,813 .0497 568,812 .0512
49	383,592 .0498 384,591 .0515	419,637 .0494 420,636 .0510	456,683 .0489 457,682 .0504	495,729 .0498 496,728 .0513	534,777 .0492 535,776 .0507	575,825 .0500 576,824 .0514
50	388,602 .0488 389,601 .0504	425,647 .0493 426,646 .0509	463,693 .0496 464,692 .0512	502,740 .0499 503,739 .0514	541,789 .0487 542,788 .0501	582,838 .0488 583,837 .0502

P = .05 one-sided

P = .10 one-sided

N	M = 21	M = 22	M = 23	M = 24	M = 25	M = 26
21	385,518 .0486 386,517 .0512					
22	393,531 .0482 394,530 .0507	424,566 .0491 425,565 .0516				
23	401,544 .0478 402,543 .0502	432,580 .0477 433,579 .0500	465,616 .0499 466,615 .0522			
24	410,556 .0497 411,555 .0521	441,593 .0486 442,592 .0509	474,630 .0497 475,629 .0520	507,669 .0486 508,668 .0508		
25	418,569 .0492 419,568 .0515	450,606 .0494 451,605 .0516	483,644 .0495 484,643 .0517	517,683 .0496 518,682 .0517	552,723 .0497 553,722 .0518	

P = .05 one-sided

P = .10 two-sided

N	M = 21	M = 22	M = 23	M = 24	M = 25	M = 26
26	426,582 .0488	458,620 .0480	492,658 .0493	526,698 .0486	562,738 .0498	598,780 .0490
	427,581 .0510	459,619 .0502	493,657 .0514	527,697 .0506	563,737 .0517	599,779 .0509
27	434,595 .0483	467,633 .0488	501,672 .0491	536,712 .0495	572,753 .0498	608,796 .0482
	435,594 .0505	468,632 .0509	502,671 .0512	537,711 .0515	573,752 .0517	609,795 .0501
28	443,607 .0500	476,646 .0495	510,686 .0490	545,727 .0485	582,768 .0498	619,811 .0493
	444,606 .0521	477,645 .0515	511,685 .0509	546,726 .0504	583,767 .0517	620,810 .0511
29	451,620 .0495	484,660 .0482	519,700 .0488	555,741 .0493	592,783 .0498	629,827 .0485
	452,619 .0516	485,659 .0501	520,699 .0507	556,740 .0512	593,782 .0516	630,826 .0503
30	459,633 .0490	493,673 .0488	528,714 .0486	564,756 .0483	602,798 .0498	640,842 .0495
	460,632 .0511	494,672 .0508	529,713 .0504	565,755 .0501	603,797 .0516	641,841 .0512
31	467,646 .0486	502,686 .0494	537,728 .0484	574,770 .0491	612,813 .0498	650,858 .0487
	468,645 .0505	503,685 .0513	538,727 .0502	575,769 .0509	613,812 .0515	651,857 .0504
32	475,659 .0482	510,700 .0482	547,741 .0500	584,784 .0499	622,828 .0498	661,873 .0496
	476,658 .0501	511,699 .0500	548,740 .0518	585,783 .0516	623,827 .0515	662,872 .0513
33	484,671 .0496	519,713 .0488	556,755 .0497	593,799 .0489	632,843 .0497	671,889 .0489
	485,670 .0515	520,712 .0506	557,754 .0515	594,798 .0506	633,842 .0514	672,888 .0505
34	492,684 .0492	528,726 .0494	565,769 .0495	603,813 .0496	642,858 .0497	682,904 .0498
	493,683 .0510	529,725 .0511	566,768 .0512	604,812 .0513	643,857 .0513	683,903 .0513
35	500,697 .0487	537,739 .0499	574,783 .0493	612,828 .0487	652,873 .0497	692,920 .0490
	501,696 .0505	538,738 .0516	575,782 .0509	613,827 .0503	653,872 .0512	693,919 .0506
36	508,710 .0483	545,753 .0487	583,797 .0491	622,842 .0494	662,888 .0496	703,935 .0498
	509,709 .0500	546,752 .0504	584,796 .0507	623,841 .0509	663,887 .0512	704,934 .0513
37	517,722 .0496	554,766 .0492	592,811 .0489	631,857 .0485	672,903 .0496	713,951 .0491
	518,721 .0513	555,765 .0509	593,810 .0504	632,856 .0500	673,902 .0511	714,950 .0506
38	525,735 .0492	563,779 .0497	601,825 .0486	641,871 .0491	682,918 .0495	724,966 .0499
	526,734 .0508	564,778 .0513	602,824 .0502	642,870 .0506	683,917 .0510	725,965 .0513
39	533,748 .0487	571,793 .0486	611,838 .0499	651,885 .0497	692,933 .0495	734,982 .0492
	534,747 .0504	572,792 .0502	612,837 .0515	652,884 .0512	693,932 .0509	735,981 .0506
40	542,760 .0499	580,806 .0491	620,852 .0497	660,900 .0489	702,948 .0494	745,997 .0499
	543,759 .0515	581,805 .0506	621,851 .0512	661,899 .0503	703,947 .0508	746,996 .0513
41	550,773 .0495	589,819 .0495	629,866 .0495	670,914 .0494	712,963 .0493	755,1013 .0492
	551,772 .0511	590,818 .0510	630,865 .0510	671,913 .0509	713,962 .0507	756,1012 .0506

P = .05 one-sided

P = .10 two-sided

N	M = 21	M = 22	M = 23	M = 24	M = 25	M = 26
42	558,786 .0491 559,785 .0506	598,832 .0499 599,831 .0514	638,880 .0493 639,879 .0507	680,928 .0500 681,927 .0514	722,978 .0493 723,977 .0506	766,1028 .0499 767,1027 .0512
43	566,799 .0487 567,798 .0502	606,846 .0489 607,845 .0504	647,894 .0491 648,893 .0505	689,943 .0492 690,942 .0505	732,993 .0492 733,992 .0506	776,1044 .0493 777,1043 .0506
44	575,811 .0498 576,810 .0513	615,859 .0493 616,858 .0507	656,908 .0488 657,907 .0502	699,957 .0497 700,956 .0510	742,1008 .0492 743,1007 .0505	787,1059 .0499 788,1058 .0512
45	583,824 .0494 584,823 .0508	624,872 .0497 625,871 .0511	666,921 .0500 667,920 .0514	708,972 .0489 709,971 .0502	752,1023 .0491 753,1022 .0504	797,1075 .0493 798,1074 .0505
46	591,837 .0490 592,836 .0504	632,886 .0487 633,885 .0501	675,935 .0498 676,934 .0511	718,986 .0494 719,985 .0507	762,1038 .0490 763,1037 .0503	808,1090 .0499 809,1089 .0511
47	599,850 .0486 600,849 .0500	641,899 .0491 642,898 .0505	684,949 .0495 685,948 .0509	728,1000 .0499 729,999 .0512	772,1053 .0490 773,1052 .0502	818,1106 .0493 819,1105 .0505
48	608,862 .0496 609,861 .0510	650,912 .0495 651,911 .0508	693,963 .0493 694,962 .0506	737,1015 .0491 738,1014 .0504	782,1068 .0489 783,1067 .0501	829,1121 .0498 830,1120 .0510
49	616,875 .0493 617,874 .0506	659,925 .0499 660,924 .0512	702,977 .0491 703,976 .0504	747,1029 .0496 748,1028 .0508	792,1083 .0488 793,1082 .0500	839,1137 .0493 840,1136 .0504
50	624,888 .0489 625,887 .0502	667,939 .0489 668,938 .0502	711,991 .0489 712,990 .0501	756,1044 .0489 757,1043 .0501	803,1097 .0499 804,1096 .0511	850,1152 .0498 851,1151 .0510

P = .05 one-sided

P = .10 two-sided

N	M = 27	M = 28	M = 29	M = 30	M = 31	M = 32
27	646,839 .0485 647,838 .0503					
28	657,855 .0487 658,854 .0505	697,899 .0499 698,898 .0517				
29	668,871 .0490 669,870 .0507	708,916 .0494 709,915 .0511	749,962 .0498 750,961 .0514			
30	679,887 .0492 680,886 .0508	719,933 .0489 720,932 .0505	760,980 .0485 761,979 .0501	803,1027 .0498 804,1026 .0513		
31	690,903 .0494 691,902 .0510	731,949 .0499 732,948 .0515	772,997 .0489 773,996 .0504	815,1045 .0494 816,1044 .0509	859,1094 .0499 860,1093 .0514	
32	701,919 .0495 702,918 .0511	742,966 .0494 743,965 .0509	784,1014 .0492 785,1013 .0507	827,1063 .0491 828,1062 .0505	871,1113 .0489 872,1112 .0503	916,1164 .0487 917,1163 .0501

P = .05 one-sided

P = .10 two-sided

N	M = 27		M = 28		M = 29		M = 30		M = 31		M = 32	
33	712,935	.0497	753,983	.0489	796,1031	.0495	839,1081	.0487	884,1131	.0493	930,1182	.0499
	713,934	.0512	754,982	.0504	797,1030	.0510	840,1080	.0502	885,1130	.0507	931,1181	.0513
34	723,951	.0498	765,999	.0498	808,1048	.0498	852,1098	.0498	897,1149	.0498	943,1201	.0497
	724,950	.0513	766,998	.0513	809,1047	.0513	853,1097	.0512	898,1148	.0511	944,1200	.0511
35	734,967	.0499	776,1016	.0493	819,1066	.0487	864,1116	.0494	909,1168	.0488	956,1220	.0495
	735,966	.0514	777,1015	.0507	820,1065	.0501	865,1115	.0508	910,1167	.0501	957,1219	.0508
36	744,984	.0486	787,1033	.0488	831,1083	.0489	876,1134	.0491	922,1186	.0492	969,1239	.0493
	745,983	.0500	788,1032	.0502	832,1082	.0503	877,1133	.0504	923,1185	.0505	970,1238	.0506
37	755,1000	.0487	799,1049	.0497	843,1100	.0492	888,1152	.0488	935,1204	.0496	982,1258	.0491
	756,999	.0501	800,1048	.0510	844,1099	.0505	889,1151	.0501	936,1203	.0509	983,1257	.0504
38	766,1016	.0488	810,1066	.0491	855,1117	.0494	901,1169	.0497	948,1222	.0499	995,1277	.0489
	767,1015	.0502	811,1065	.0505	856,1116	.0508	902,1168	.0510	949,1221	.0512	996,1276	.0501
39	777,1032	.0489	822,1082	.0500	867,1134	.0497	913,1187	.0493	960,1241	.0490	1009,1295	.0499
	778,1031	.0503	823,1081	.0513	868,1133	.0510	914,1186	.0506	961,1240	.0503	1010,1294	.0511
40	788,1048	.0490	833,1099	.0495	879,1151	.0499	925,1205	.0490	973,1259	.0494	1022,1314	.0497
	789,1047	.0504	834,1098	.0508	880,1150	.0511	926,1204	.0502	974,1258	.0506	1023,1313	.0509
41	799,1064	.0491	844,1116	.0490	890,1169	.0488	938,1222	.0499	986,1277	.0497	1035,1333	.0495
	800,1063	.0504	845,1115	.0503	891,1168	.0501	939,1221	.0511	987,1276	.0509	1036,1332	.0506
42	810,1080	.0492	856,1132	.0497	902,1186	.0490	950,1240	.0495	999,1295	.0500	1048,1352	.0493
	811,1079	.0505	857,1131	.0510	903,1185	.0503	951,1239	.0507	1000,1294	.0512	1049,1351	.0504
43	821,1096	.0493	867,1149	.0493	914,1203	.0492	962,1258	.0492	1011,1314	.0491	1061,1371	.0491
	822,1095	.0505	868,1148	.0505	915,1202	.0504	963,1257	.0504	1012,1313	.0503	1062,1370	.0502
44	832,1112	.0493	879,1165	.0500	926,1220	.0494	975,1275	.0500	1024,1332	.0494	1075,1389	.0499
	833,1111	.0506	880,1164	.0512	927,1219	.0506	976,1274	.0512	1025,1331	.0505	1076,1388	.0510
45	843,1128	.0494	890,1182	.0495	938,1237	.0496	987,1293	.0497	1037,1350	.0497	1088,1408	.0497
	844,1127	.0506	891,1181	.0507	939,1236	.0508	988,1292	.0508	1038,1349	.0508	1089,1407	.0508
46	854,1144	.0495	901,1199	.0490	950,1254	.0498	999,1311	.0493	1050,1368	.0500	1101,1427	.0495
	855,1143	.0507	902,1198	.0502	951,1253	.0509	1000,1310	.0504	1051,1367	.0510	1102,1426	.0506
47	865,1160	.0495	913,1215	.0497	962,1271	.0499	1011,1329	.0490	1062,1387	.0491	1114,1446	.0493
	866,1159	.0507	914,1214	.0509	963,1270	.0510	1012,1328	.0501	1063,1386	.0502	1115,1445	.0503
48	876,1176	.0496	924,1232	.0493	973,1289	.0490	1024,1346	.0497	1075,1405	.0494	1127,1465	.0491
	877,1175	.0507	925,1231	.0504	974,1288	.0501	1025,1345	.0508	1076,1404	.0504	1128,1464	.0501
49	887,1192	.0496	936,1248	.0499	985,1306	.0491	1036,1364	.0494	1088,1423	.0496	1141,1483	.0499
	888,1191	.0507	937,1247	.0510	986,1305	.0502	1037,1363	.0505	1089,1422	.0507	1142,1482	.0509
50	898,1208	.0496	947,1265	.0495	997,1323	.0493	1048,1382	.0491	1101,1441	.0499	1154,1502	.0496
	899,1207	.0508	948,1264	.0506	998,1322	.0503	1049,1381	.0501	1102,1440	.0509	1155,1501	.0506

P = .05 one-sided

P = .10 two-sided

N	M = 33	M = 34	M = 35	M = 36	M = 37
33	976,1235 .0491 977,1234 .0505				
34	990,1254 .0497 991,1253 .0510	1038,1308 .0496 1039,1307 .0509			
35	1003,1274 .0489 1004,1273 .0502	1052,1328 .0495 1053,1327 .0508	1101,1384 .0489 1102,1383 .0502		
36	1017,1293 .0494 1018,1292 .0507	1066,1348 .0495 1067,1347 .0507	1116,1404 .0495 1117,1403 .0507	1167,1461 .0496 1168,1460 .0508	
37	1031,1312 .0499 1032,1311 .0511	1080,1368 .0494 1081,1367 .0506	1130,1425 .0489 1131,1424 .0501	1182,1482 .0496 1183,1481 .0508	1234,1541 .0492 1235,1540 .0503
38	1044,1332 .0491 1045,1331 .0503	1094,1388 .0493 1095,1387 .0505	1145,1445 .0495 1146,1444 .0506	1197,1503 .0496 1198,1502 .0508	1250,1562 .0498 1251,1561 .0509
39	1058,1351 .0496 1059,1350 .0508	1108,1408 .0492 1109,1407 .0504	1159,1466 .0489 1160,1465 .0500	1212,1524 .0497 1213,1523 .0508	1265,1584 .0493 1266,1583 .0504
40	1072,1370 .0500 1073,1369 .0512	1122,1428 .0491 1123,1427 .0503	1174,1486 .0494 1175,1485 .0505	1227,1545 .0497 1228,1544 .0508	1281,1605 .0499 1282,1604 .0510
41	1085,1390 .0493 1086,1389 .0504	1136,1448 .0491 1137,1447 .0502	1189,1506 .0499 1190,1505 .0510	1242,1566 .0497 1243,1565 .0508	1296,1627 .0495 1297,1626 .0505
42	1099,1409 .0497 1100,1408 .0508	1150,1468 .0490 1151,1467 .0500	1203,1527 .0493 1204,1526 .0504	1257,1587 .0497 1258,1586 .0507	1311,1649 .0490 1312,1648 .0500
43	1112,1429 .0490 1113,1428 .0501	1165,1487 .0499 1166,1486 .0510	1218,1547 .0498 1219,1546 .0509	1272,1608 .0497 1273,1607 .0507	1327,1670 .0496 1328,1669 .0506
44	1126,1448 .0494 1127,1447 .0504	1179,1507 .0498 1180,1506 .0509	1232,1568 .0492 1233,1567 .0503	1287,1629 .0497 1288,1628 .0507	1342,1692 .0491 1343,1691 .0501
45	1140,1467 .0497 1141,1466 .0508	1193,1527 .0497 1194,1526 .0507	1247,1588 .0497 1248,1587 .0507	1302,1650 .0497 1303,1649 .0507	1358,1713 .0496 1359,1712 .0506
46	1153,1487 .0490 1154,1486 .0501	1207,1547 .0496 1208,1546 .0506	1261,1609 .0491 1262,1608 .0501	1317,1671 .0496 1318,1670 .0506	1373,1735 .0492 1374,1734 .0501
47	1167,1506 .0494 1168,1505 .0504	1221,1567 .0495 1222,1566 .0505	1276,1629 .0496 1277,1628 .0505	1332,1692 .0496 1333,1591 .0506	1389,1756 .0497 1390,1755 .0506
48	1181,1525 .0497 1182,1524 .0507	1235,1587 .0494 1236,1586 .0504	1291,1649 .0500 1292,1648 .0509	1347,1713 .0496 1348,1712 .0505	1404,1778 .0492 1405,1777 .0502
49	1194,1545 .0491 1195,1544 .0501	1249,1607 .0493 1250,1606 .0502	1305,1670 .0494 1306,1669 .0504	1362,1734 .0496 1363,1733 .0505	1420,1799 .0497 1421,1798 .0506
50	1208,1564 .0494 1209,1563 .0504	1263,1627 .0492 1264,1626 .0501	1320,1690 .0498 1321,1689 .0508	1377,1755 .0495 1378,1754 .0505	1435,1821 .0493 1436,1820 .0502

P = .05 one-sided

P = .10 two-sided

N	M = 38	M = 39	M = 40	M = 41	M = 42
38	1304,1622 .0499 1305,1621 .0510				
39	1319,1645 .0490 1320,1644 .0500	1375,1706 .0497 1376,1705 .0507			
40	1335,1667 .0491 1336,1666 .0501	1391,1729 .0493 1392,1728 .0503	1448,1792 .0495 1449,1791 .0505		
41	1351,1689 .0492 1352,1688 .0502	1408,1751 .0500 1409,1750 .0510	1465,1815 .0497 1466,1814 .0507	1523,1880 .0495 1524,1879 .0504	
42	1367,1711 .0493 1368,1710 .0503	1424,1774 .0496 1425,1773 .0506	1482,1838 .0499 1483,1837 .0509	1540,1904 .0492 1541,1903 .0502	1600,1970 .0495 1601,1969 .0504
43	1383,1733 .0494 1384,1732 .0504	1440,1797 .0493 1441,1796 .0502	1498,1862 .0491 1499,1861 .0501	1558,1927 .0499 1559,1926 .0509	1618,1994 .0498 1619,1993 .0507
44	1399,1755 .0495 1400,1754 .0505	1457,1819 .0499 1458,1818 .0508	1515,1885 .0493 1516,1884 .0502	1575,1951 .0497 1576,1950 .0506	1636,2018 .0500 1637,2017 .0509
45	1415,1777 .0496 1416,1776 .0505	1473,1842 .0495 1474,1841 .0505	1532,1908 .0495 1533,1907 .0504	1592,1975 .0494 1593,1974 .0503	1653,2043 .0493 1654,2042 .0502
46	1431,1799 .0497 1432,1798 .0506	1489,1865 .0492 1490,1864 .0501	1549,1931 .0496 1550,1930 .0505	1609,1999 .0492 1610,1998 .0500	1671,2067 .0496 1672,2066 .0504
47	1447,1821 .0497 1448,1820 .0506	1506,1887 .0497 1507,1886 .0507	1566,1954 .0498 1567,1953 .0507	1627,2022 .0498 1628,2021 .0506	1689,2091 .0498 1690,2090 .0506
48	1463,1843 .0498 1464,1842 .0507	1522,1910 .0494 1523,1909 .0503	1583,1977 .0499 1584,1976 .0508	1644,2046 .0495 1645,2045 .0504	1707,2115 .0500 1708,3114 .0508
49	1479,1865 .0498 1480,1864 .0507	1539,1932 .0499 1540,1931 .0508	1599,2001 .0492 1600,2000 .0500	1661,2070 .0493 1662,2069 .0501	1724,2140 .0494 1725,2139 .0502
50	1495,1887 .0499 1496,1886 .0508	1555,1955 .0496 1556,1954 .0504	1616,2024 .0493 1617,2023 .0501	1679,2093 .0498 1680,2092 .0507	1742,2164 .0495 1743,2163 .0504

P = .05 one-sided

P = .10 two-sided

N	M = 43	M = 44	M = 45	M = 46	M = 47
43	1679,2062 .0496 1680,2061 .0505				
44	1697,2087 .0494 1698,2086 .0503	1760,2156 .0497 1761,2155 .0506			
45	1715,2112 .0493 1716,2111 .0501	1778,2182 .0492 1779,2181 .0500	1843,2252 .0499 1844,2251 .0508		

P = .05 one-sided

P = .10 two-sided

N	M = 43	M = 44	M = 45	M = 46	M = 47
46	1734,2136 .0500 1735,2135 .0508	1797,2207 .0495 1798,2206 .0503	1862,2278 .0498 1863,2277 .0507	1927,2351 .0494 1928,2350 .0502	
47	1752,2161 .0498 1753,2160 .0506	1816,2232 .0498 1817,2231 .0506	1881,2304 .0497 1882,2303 .0506	1947,2377 .0497 1948,2376 .0505	2014,2451 .0497 2015,2450 .0505
48	1770,2186 .0496 1771,2185 .0504	1834,2258 .0492 1835,2257 .0500	1900,2330 .0496 1901,2329 .0504	1966,2404 .0493 1967,2403 .0501	2034,2478 .0497 2035,2477 .0504
49	1788,2211 .0494 1789,2210 .0502	1853,2283 .0495 1854,2282 .0503	1919,2356 .0495 1920,2355 .0503	1986,2430 .0496 1987,2429 .0504	2054,2505 .0496 2055,2504 .0504
50	1806,2236 .0493 1807,2235 .0501	1872,2308 .0497 1873,2307 .0505	1938,2382 .0494 1939,2381 .0502	2006,2456 .0499 2007,2455 .0507	2074,2532 .0496 2075,2531 .0504

P = .05 one-sided

P = .10 two-sided

N	M = 48	M = 49	M = 50
48	2102,2554 .0493 2103,2553 .0501		
49	2123,2581 .0497 2124,2580 .0504	2193,2658 .0497 2194,2657 .0504	
50	2143,2609 .0493 2144,2608 .0500	2214,2686 .0497 2215,2685 .0505	2285,2765 .0494 2286,2764 .0501

P = .025 one-sided

P = .05 two-sided

N	M = 3	M = 4	M = 5	M = 6	M = 7	M = 8
3	5,16 .0000 6,15 .0500					
4	5,19 .0000 6,18 .0286	10,26 .0143 11,25 .0286				
5	6,21 .0179 7,20 .0357	11,29 .0159 12,28 .0317	17,38 .0159 18,37 .0278			
6	7,23 .0238 8,22 .0476	12,32 .0190 13,31 .0333	18,42 .0152 19,41 .0260	26,52 .0206 27,51 .0325		
7	7,26 .0167 8,25 .0333	13,35 .0212 14,34 .0364	20,45 .0240 21,44 .0366	27,57 .0175 28,56 .0256	36,69 .0189 37,68 .0265	
8	8,28 .0242 9,27 .0424	14,38 .0242 15,37 .0364	21,49 .0225 22,48 .0326	29,61 .0213 30,60 .0296	38,74 .0200 39,73 .0270	49,87 .0249 50,86 .0325
9	8,31 .0182 9,30 .0318	14,42 .0168 15,41 .0252	22,53 .0210 23,52 .0300	31,65 .0248 32,64 .0332	40,79 .0209 41,78 .0274	51,93 .0232 52,92 .0296
10	9,33 .0245 10,32 .0385	15,45 .0180 16,44 .0270	23,57 .0200 24,56 .0276	32,70 .0210 33,69 .0280	42,84 .0215 43,83 .0277	53,99 .0217 54,98 .0273
11	9,36 .0192 10,35 .0302	16,48 .0198 17,47 .0278	24,61 .0190 25,60 .0259	34,74 .0238 35,73 .0308	44,89 .0221 45,88 .0278	55,105 .0204 56,104 .0253
12	10,38 .0242 11,37 .0352	17,51 .0209 18,50 .0291	26,64 .0242 27,63 .0318	35,79 .0207 36,78 .0264	46,94 .0225 47,93 .0278	58,110 .0237 59,109 .0287
13	10,41 .0196 11,40 .0286	18,54 .0223 19,53 .0298	27,68 .0230 28,67 .0296	37,83 .0231 38,82 .0289	48,99 .0228 49,98 .0278	60,116 .0223 61,115 .0267
14	11,43 .0235 12,42 .0338	19,57 .0232 20,56 .0307	28,72 .0218 29,71 .0279	38,88 .0204 39,87 .0253	50,104 .0230 51,103 .0278	62,122 .0211 63,121 .0251
15	11,46 .0196 12,45 .0282	20,60 .0243 21,59 .0312	29,76 .0209 30,75 .0263	40,92 .0224 41,91 .0274	52,109 .0233 53,108 .0278	65,127 .0237 66,126 .0278
16	12,48 .0237 13,47 .0320	21,63 .0250 22,62 .0320	30,80 .0201 31,79 .0250	42,96 .0244 43,95 .0294	54,114 .0234 55,113 .0277	67,133 .0224 68,132 .0262
17	12,51 .0202 13,50 .0272	21,67 .0202 22,66 .0259	32,83 .0238 33,82 .0292	43,101 .0219 44,100 .0262	56,119 .0236 57,118 .0277	70,138 .0247 71,137 .0286
18	13,53 .0233 14,52 .0308	22,70 .0212 23,69 .0265	33,87 .0229 34,86 .0277	45,105 .0236 46,104 .0279	58,124 .0237 59,123 .0276	72,144 .0235 73,143 .0270
19	13,56 .0201 14,55 .0266	23,73 .0219 24,72 .0272	34,91 .0220 35,90 .0264	46,110 .0214 47,109 .0252	60,129 .0238 61,128 .0275	74,150 .0224 75,149 .0256

P = .025 one-sided

P = .05 two-sided

N	M = 3	M = 4	M = 5	M = 6	M = 7	M = 8
20	14,58 .0232	24,76 .0227	35,95 .0212	48,114 .0229	62,134 .0239	77,155 .0244
	15,57 .0299	25,75 .0278	36,94 .0253	49,113 .0268	63,133 .0274	78,154 .0276
21	14,61 .0203	25,79 .0233	37,98 .0243	50,118 .0244	64,139 .0240	79,161 .0233
	15,60 .0262	26,78 .0284	38,97 .0287	51,117 .0283	65,138 .0274	80,160 .0263
22	15,63 .0230	26,82 .0240	38,102 .0234	51,123 .0224	66,144 .0240	81,167 .0223
	16,62 .0291	27,81 .0288	39,101 .0275	52,122 .0258	67,143 .0273	82,166 .0251
23	15,66 .0204	27,85 .0246	39,106 .0226	53,127 .0237	68,149 .0241	84,172 .0240
	16,65 .0258	28,84 .0293	40,105 .0264	54,126 .0272	69,148 .0272	85,171 .0269
24	16,68 .0229	27,89 .0211	40,110 .0219	54,132 .0219	70,154 .0241	86,178 .0231
	17,67 .0284	28,88 .0252	41,109 .0255	55,131 .0250	71,153 .0272	87,177 .0257
25	16,71 .0205	28,92 .0217	42,113 .0246	56,136 .0231	72,159 .0242	89,183 .0247
	17,70 .0253	29,91 .0256	43,112 .0284	57,135 .0263	73,158 .0271	90,182 .0274
26	17,73 .0227	29,95 .0222	43,117 .0238	58,140 .0243	74,164 .0242	91,189 .0237
	18,72 .0279	30,94 .0262	44,116 .0273	59,139 .0275	75,163 .0270	92,188 .0263
27	17,76 .0204	30,98 .0228	44,121 .0231	59,145 .0226	76,169 .0242	93,195 .0229
	18,75 .0251	31,97 .0266	45,120 .0264	60,144 .0255	77,168 .0270	94,194 .0252
28	18,78 .0227	31,101 .0233	45,125 .0224	61,149 .0237	78,174 .0243	96,200 .0243
	19,77 .0274	32,100 .0271	46,124 .0256	62,148 .0266	79,173 .0269	97,199 .0267
29	19,80 .0248	32,104 .0238	47,128 .0248	63,153 .0248	80,179 .0243	98,206 .0235
	20,79 .0296	33,103 .0274	48,127 .0281	64,152 .0277	81,178 .0268	99,205 .0257
30	19,83 .0225	33,107 .0242	48,132 .0241	64,158 .0232	82,184 .0243	101,211 .0248
	20,82 .0269	34,106 .0279	49,131 .0272	65,157 .0259	83,183 .0268	102,210 .0271
31	20,85 .0246	34,110 .0247	49,136 .0234	66,162 .0242	84,189 .0243	103,217 .0240
	21,84 .0291	35,109 .0282	50,135 .0264	67,161 .0269	85,188 .0267	104,216 .0262
32	20,88 .0225	34,114 .0219	50,140 .0228	67,167 .0228	86,194 .0243	105,223 .0232
	21,87 .0266	35,113 .0251	51,139 .0256	68,166 .0252	87,193 .0267	106,222 .0253
33	21,90 .0244	35,117 .0224	52,143 .0249	69,171 .0237	88,199 .0243	108,228 .0245
	22,89 .0286	36,116 .0255	53,142 .0278	70,170 .0262	89,198 .0266	109,227 .0265
34	21,93 .0224	36,120 .0228	53,147 .0243	71,175 .0247	90,204 .0244	110,234 .0237
	22,92 .0263	37,119 .0258	54,146 .0270	72,174 .0271	91,203 .0266	111,233 .0257
35	22,95 .0242	37,123 .0232	54,151 .0237	72,180 .0233	92,209 .0244	113,239 .0249
	23,94 .0281	38,122 .0262	55,150 .0263	73,179 .0256	93,208 .0265	114,238 .0269
36	22,98 .0223	38,126 .0236	55,155 .0231	74,184 .0241	94,214 .0244	115,245 .0242
	23,97 .0259	39,125 .0266	56,154 .0256	75,183 .0264	95,213 .0265	116,244 .0260

P = .025 one-sided

P = .05 two-sided

N	M = 3	M = 4	M = 5	M = 6	M = 7	M = 8
37	23,100 .0240	39,129 .0240	56,159 .0226	76,188 .0250	96,219 .0244	117,251 .0235
	24,99 .0277	40,128 .0269	57,158 .0250	77,187 .0273	97,218 .0264	118,250 .0253
38	23,103 .0222	40,132 .0243	58,162 .0244	77,193 .0237	98,224 .0244	120,256 .0245
	24,102 .0257	41,131 .0272	59,161 .0269	78,192 .0258	99,223 .0264	121,255 .0264
39	24,105 .0239	41,135 .0247	59,166 .0239	79,197 .0245	100,229 .0244	122,262 .0239
	25,104 .0274	42,134 .0275	60,165 .0263	80,196 .0267	101,228 .0263	123,261 .0256
40	24,108 .0222	41,139 .0224	60,170 .0234	80,202 .0233	102,234 .0244	125,267 .0249
	25,107 .0254	42,138 .0250	61,169 .0256	81,201 .0253	103,233 .0263	126,266 .0267
41	25,110 .0237	42,142 .0228	61,174 .0229	82,206 .0241	104,239 .0244	127,273 .0243
	26,109 .0270	43,141 .0253	62,173 .0251	83,205 .0261	105,238 .0262	128,272 .0259
42	25,113 .0221	43,145 .0231	63,177 .0245	84,210 .0248	106,244 .0244	129,279 .0236
	26,112 .0252	44,144 .0256	64,176 .0268	85,209 .0268	107,243 .0262	130,278 .0252
43	26,115 .0236	44,148 .0235	64,181 .0240	85,215 .0236	108,249 .0244	132,284 .0246
	27,114 .0267	45,147 .0259	65,180 .0262	86,214 .0256	109,248 .0262	133,283 .0262
44	26,118 .0221	45,151 .0238	65,185 .0235	87,219 .0244	110,254 .0244	134,290 .0240
	27,117 .0250	46,150 .0262	66,184 .0257	88,218 .0263	111,253 .0261	135,289 .0255
45	27,120 .0235	46,154 .0241	66,189 .0231	88,224 .0233	112,259 .0244	137,295 .0249
	28,119 .0265	47,153 .0265	67,188 .0251	89,223 .0251	113,258 .0261	138,294 .0265
46	28,122 .0249	47,157 .0244	68,192 .0246	90,228 .0240	114,264 .0244	139,301 .0243
	29,121 .0279	48,156 .0268	69,191 .0267	91,227 .0258	115,263 .0261	140,300 .0258
47	28,125 .0234	48,160 .0247	69,196 .0242	92,232 .0247	116,269 .0244	141,307 .0238
	29,124 .0262	49,159 .0270	70,195 .0262	93,231 .0265	117,268 .0260	142,306 .0252
48	29,127 .0247	49,163 .0249	70,200 .0237	93,237 .0236	118,274 .0244	144,312 .0246
	30,126 .0276	50,162 .0273	71,199 .0257	94,236 .0253	119,273 .0260	145,311 .0261
49	29,130 .0233	49,167 .0231	71,204 .0233	95,241 .0243	120,279 .0244	146,318 .0241
	30,129 .0260	50,166 .0252	72,203 .0252	96,240 .0260	121,278 .0260	147,317 .0255
50	30,132 .0245	50,170 .0234	73,207 .0247	97,245 .0249	122,284 .0244	149,323 .0249
	31,131 .0273	51,169 .0255	74,206 .0266	98,244 .0266	123,283 .0259	150,322 .0263

P = .025 one-sided

P = .05 two-sided

N	M = 9	M = 10	M = 11	M = 12	M = 13	M = 14
9	62,109 .0200 63,108 .0252					
10	65,115 .0217 66,114 .0267	78,132 .0216 79,131 .0262				
11	68,121 .0232 69,120 .0281	81,139 .0215 82,138 .0257	96,157 .0237 97,156 .0278			
12	71,127 .0245 72,126 .0292	84,146 .0213 85,145 .0252	99,165 .0219 100,164 .0256	115,185 .0225 116,184 .0259		
13	73,134 .0217 74,133 .0257	88,152 .0247 89,151 .0287	103,172 .0237 104,171 .0274	119,193 .0229 120,192 .0262	136,215 .0221 137,214 .0251	
14	76,140 .0228 77,139 .0267	91,159 .0242 92,158 .0280	106,180 .0221 107,179 .0254	123,201 .0232 124,200 .0263	141,223 .0241 142,222 .0271	160,246 .0249 161,245 .0278
15	79,146 .0238 80,145 .0276	94,166 .0238 95,165 .0273	110,187 .0236 111,186 .0269	127,209 .0234 128,208 .0264	145,232 .0232 146,231 .0260	164,256 .0229 165,255 .0255
16	82,152 .0247 83,151 .0284	97,173 .0234 98,172 .0266	113,195 .0221 114,194 .0250	131,217 .0236 132,216 .0265	150,240 .0250 151,239 .0278	169,265 .0236 170,264 .0261
17	84,159 .0223 85,158 .0255	100,180 .0230 101,179 .0260	117,202 .0235 118,201 .0263	135,225 .0238 136,224 .0265	154,249 .0240 155,248 .0266	174,274 .0242 175,273 .0267
18	87,165 .0231 88,164 .0263	103,187 .0226 104,186 .0255	121,209 .0247 122,208 .0276	139,233 .0239 140,232 .0265	158,258 .0232 159,257 .0256	179,283 .0247 180,282 .0271
19	90,171 .0239 91,170 .0269	107,193 .0250 108,192 .0279	124,217 .0233 125,216 .0259	143,241 .0240 144,240 .0265	163,266 .0247 164,265 .0271	183,293 .0230 184,292 .0252
20	93,177 .0245 94,176 .0276	110,200 .0245 111,199 .0273	128,224 .0244 129,223 .0269	147,249 .0241 148,248 .0265	167,275 .0238 168,274 .0261	188,302 .0235 189,301 .0256
21	95,184 .0225 96,183 .0252	113,207 .0241 114,206 .0267	131,232 .0230 132,231 .0254	151,257 .0242 152,256 .0265	171,284 .0231 172,283 .0252	193,311 .0239 194,310 .0260
22	98,190 .0231 99,189 .0258	116,214 .0237 117,213 .0262	135,239 .0240 136,238 .0264	155,265 .0242 156,264 .0265	176,292 .0243 177,291 .0264	198,320 .0243 199,319 .0263
23	101,196 .0237 102,195 .0263	119,221 .0233 120,220 .0257	139,246 .0250 140,245 .0273	159,273 .0243 160,272 .0264	180,301 .0236 181,300 .0255	203,329 .0247 204,328 .0267
24	104,202 .0243 105,201 .0269	122,228 .0230 123,227 .0252	142,254 .0237 143,253 .0259	163,281 .0243 164,280 .0264	185,309 .0247 186,308 .0267	207,339 .0233 208,338 .0250
25	107,208 .0249 108,207 .0274	126,234 .0248 127,233 .0271	146,261 .0246 147,260 .0268	167,289 .0243 168,288 .0263	189,318 .0240 190,317 .0258	212,348 .0236 213,347 .0253

P = .025 one-sided

P = .05 two-sided

N	M = 9	M = 10	M = 11	M = 12	M = 13	M = 14
26	109,215 .0231	129,241 .0244	149,269 .0235	171,297 .0243	193,327 .0233	217,357 .0239
	110,214 .0254	130,240 .0266	150,268 .0254	172,296 .0263	194,326 .0251	218,356 .0256
27	112,221 .0236	132,248 .0240	153,276 .0243	175,305 .0243	198,335 .0243	222,366 .0242
	113,220 .0258	133,247 .0261	154,275 .0262	176,304 .0262	199,334 .0261	223,365 .0259
28	115,227 .0241	135,255 .0237	156,284 .0232	179,313 .0243	202,344 .0236	227,375 .0245
	116,226 .0263	136,254 .0257	157,283 .0250	180,312 .0262	203,343 .0253	228,374 .0261
29	118,233 .0245	138,262 .0234	160,291 .0239	183,321 .0243	207,352 .0246	232,384 .0248
	119,232 .0267	139,261 .0253	161,290 .0258	184,320 .0261	208,351 .0263	233,383 .0264
30	121,239 .0250	142,268 .0249	164,298 .0247	187,329 .0243	211,361 .0240	236,394 .0235
	122,238 .0271	143,267 .0269	165,297 .0265	188,328 .0261	212,360 .0255	237,393 .0250
31	123,246 .0234	145,275 .0245	167,306 .0237	191,337 .0243	216,369 .0249	241,403 .0238
	124,245 .0254	146,274 .0264	168,305 .0254	192,336 .0260	217,368 .0264	242,402 .0252
32	126,252 .0239	148,282 .0242	171,313 .0243	195,345 .0243	220,378 .0242	246,412 .0240
	127,251 .0258	149,281 .0260	172,312 .0261	196,344 .0259	221,377 .0257	247,411 .0255
33	129,258 .0243	151,289 .0239	174,321 .0234	199,353 .0243	224,387 .0236	251,421 .0242
	130,257 .0262	152,288 .0256	175,320 .0250	200,352 .0259	225,386 .0251	252,420 .0257
34	132,264 .0247	154,296 .0236	178,328 .0240	203,361 .0243	229,395 .0244	256,430 .0245
	133,263 .0265	155,295 .0253	179,327 .0256	204,360 .0258	230,394 .0259	257,429 .0258
35	134,271 .0233	158,302 .0249	182,335 .0247	207,369 .0243	233,404 .0239	261,439 .0247
	135,270 .0250	159,301 .0266	183,334 .0263	208,368 .0258	234,403 .0252	262,438 .0260
36	137,277 .0237	161,309 .0246	185,343 .0238	211,377 .0243	238,412 .0246	266,448 .0249
	138,276 .0254	162,308 .0263	186,342 .0253	212,376 .0257	239,411 .0260	267,447 .0262
37	140,283 .0240	164,316 .0243	189,350 .0244	215,385 .0243	242,421 .0241	270,458 .0238
	141,282 .0257	165,315 .0259	190,349 .0259	216,384 .0257	243,420 .0254	271,457 .0251
38	143,289 .0244	167,323 .0240	193,357 .0249	219,393 .0242	247,429 .0248	275,467 .0240
	144,288 .0261	168,322 .0256	194,356 .0264	220,392 .0256	248,428 .0261	276,466 .0252
39	146,295 .0247	170,330 .0237	196,365 .0241	223,401 .0242	251,438 .0243	280,476 .0242
	147,294 .0264	171,329 .0252	197,364 .0255	224,400 .0256	252,437 .0255	281,475 .0254
40	148,302 .0235	174,336 .0249	200,372 .0246	227,409 .0242	256,446 .0250	285,485 .0244
	149,301 .0250	175,335 .0264	201,371 .0260	228,408 .0255	257,445 .0262	286,484 .0256
41	151,308 .0238	177,343 .0246	203,380 .0238	231,417 .0242	260,455 .0244	290,494 .0245
	152,307 .0254	178,342 .0261	204,379 .0252	232,416 .0255	261,454 .0257	291,493 .0257
42	154,314 .0241	180,350 .0243	207,387 .0243	235,425 .0242	264,464 .0239	295,503 .0247
	155,313 .0257	181,349 .0258	208,386 .0257	236,424 .0254	265,463 .0251	296,502 .0257

P = .025 one-sided

P = .05 two-sided

N	M = 9	M = 10	M = 11	M = 12	M = 13	M = 14
43	157,320 .0244 158,319 .0259	183,357 .0241 184,356 .0255	211,394 .0248 212,393 .0262	239,433 .0242 240,432 .0254	269,472 .0246 270,471 .0258	300,512 .0249 301,511 .0260
44	160,326 .0247 161,325 .0262	186,364 .0238 187,363 .0252	214,402 .0241 215,401 .0253	243,441 .0241 244,440 .0253	273,481 .0241 274,480 .0252	304,522 .0239 305,521 .0250
45	162,333 .0236 163,332 .0250	190,370 .0249 191,369 .0262	218,409 .0246 219,408 .0258	247,449 .0241 248,448 .0253	278,489 .0247 279,488 .0259	309,531 .0241 310,530 .0251
46	165,339 .0239 166,338 .0253	193,377 .0246 194,376 .0259	221,417 .0238 222,416 .0250	251,457 .0241 252,456 .0253	282,498 .0242 283,497 .0253	314,540 .0242 315,539 .0253
47	168,345 .0242 169,344 .0256	196,384 .0244 197,383 .0256	225,424 .0243 226,423 .0255	255,465 .0241 256,464 .0252	287,506 .0248 288,505 .0259	319,549 .0244 320,548 .0254
48	171,351 .0245 172,350 .0258	199,391 .0241 200,390 .0254	229,431 .0248 230,430 .0260	259,473 .0241 260,472 .0252	291,515 .0244 292,514 .0254	324,558 .0245 325,557 .0255
49	174,357 .0248 175,356 .0261	202,398 .0239 203,397 .0251	232,439 .0241 233,438 .0252	263,481 .0240 264,480 .0251	296,523 .0249 297,522 .0260	329,567 .0247 330,566 .0257
50	176,364 .0237 177,363 .0250	206,404 .0248 207,403 .0261	236,446 .0245 237,445 .0256	267,489 .0240 268,488 .0251	300,532 .0245 301,531 .0255	334,576 .0248 335,575 .0258

P = .025 one-sided

P = .05 two-sided

N	M = 15	M = 16	M = 17	M = 18	M = 19	M = 20
15	184,281 .0227 185,280 .0251					
16	190,290 .0247 191,289 .0272	211,317 .0234 212,316 .0257				
17	195,300 .0243 196,299 .0266	217,327 .0243 218,326 .0266	240,355 .0243 241,354 .0265			
18	200,310 .0239 201,309 .0261	222,338 .0231 223,337 .0252	246,366 .0243 247,365 .0264	270,396 .0235 271,395 .0254		
19	205,320 .0235 206,319 .0256	228,348 .0239 229,347 .0259	252,377 .0243 253,376 .0262	277,407 .0246 278,406 .0265	303,438 .0248 304,437 .0266	
20	210,330 .0232 211,329 .0251	234,358 .0247 235,357 .0267	258,388 .0242 259,387 .0261	283,419 .0238 284,418 .0256	309,451 .0234 310,450 .0250	337,483 .0245 338,482 .0262

P = .025 one-sided

P = .05 two-sided

N	M = 15	M = 16	M = 17	M = 18	M = 19	M = 20
21	216,339 .0247	239,369 .0235	264,399 .0242	290,430 .0247	316,463 .0236	344,496 .0241
	217,338 .0267	240,368 .0254	265,398 .0260	291,429 .0265	317,462 .0252	345,495 .0256
22	221,349 .0243	245,379 .0242	270,410 .0241	296,442 .0240	323,475 .0238	351,509 .0236
	222,348 .0262	246,378 .0260	271,409 .0258	297,441 .0256	324,474 .0254	352,508 .0251
23	226,359 .0239	251,389 .0248	276,421 .0240	303,453 .0248	330,487 .0240	359,521 .0246
	227,358 .0257	252,388 .0266	277,420 .0257	304,452 .0264	331,486 .0255	360,520 .0261
24	231,369 .0235	256,400 .0238	282,432 .0239	309,465 .0240	337,499 .0241	366,534 .0242
	232,368 .0253	257,399 .0254	283,431 .0255	310,464 .0256	338,498 .0256	367,533 .0256
25	237,378 .0248	262,410 .0243	288,443 .0238	316,476 .0248	344,511 .0243	373,547 .0237
	238,377 .0266	263,409 .0260	289,442 .0254	317,475 .0263	345,510 .0257	374,546 .0251
26	242,388 .0244	268,420 .0249	294,454 .0237	322,488 .0241	351,523 .0244	381,559 .0246
	243,387 .0261	269,419 .0265	295,453 .0252	323,487 .0255	352,522 .0258	382,558 .0259
27	247,398 .0241	273,431 .0239	300,465 .0237	329,499 .0248	358,535 .0245	388,572 .0241
	248,397 .0257	274,430 .0254	301,464 .0251	330,498 .0262	359,534 .0258	389,571 .0254
28	252,408 .0237	279,441 .0244	307,475 .0249	335,511 .0241	365,547 .0245	396,584 .0249
	253,407 .0252	280,440 .0259	308,474 .0264	336,510 .0254	366,546 .0259	397,583 .0262
29	258,417 .0248	285,451 .0248	313,486 .0248	342,522 .0247	372,559 .0246	403,597 .0245
	259,416 .0264	286,450 .0263	314,485 .0262	343,521 .0261	373,558 .0259	404,596 .0257
30	263,427 .0245	290,462 .0239	319,497 .0247	348,534 .0241	379,571 .0247	410,610 .0241
	264,426 .0259	291,461 .0253	320,496 .0260	349,533 .0253	380,570 .0259	411,609 .0252
31	268,437 .0241	296,472 .0244	325,508 .0245	355,545 .0247	386,583 .0247	418,622 .0248
	269,436 .0255	297,471 .0257	326,507 .0258	356,544 .0259	387,582 .0260	419,621 .0260
32	273,447 .0238	302,482 .0248	331,519 .0244	361,557 .0240	393,595 .0248	425,635 .0244
	274,446 .0251	303,481 .0261	332,518 .0257	362,556 .0252	394,594 .0260	426,634 .0255
33	279,456 .0248	307,493 .0239	337,530 .0243	368,568 .0246	400,607 .0248	432,648 .0239
	280,455 .0261	308,492 .0252	338,529 .0255	369,567 .0258	401,606 .0260	433,647 .0250
34	284,466 .0244	313,503 .0243	343,541 .0242	374,580 .0240	407,619 .0249	440,660 .0246
	285,465 .0257	314,502 .0256	344,540 .0254	375,579 .0251	408,618 .0260	441,659 .0257
35	289,476 .0241	319,513 .0247	349,552 .0240	381,591 .0245	414,631 .0249	447,673 .0242
	290,475 .0254	320,512 .0259	350,551 .0252	382,590 .0256	415,630 .0260	448,672 .0252
36	294,486 .0238	324,524 .0239	355,563 .0239	387,603 .0239	421,643 .0249	455,685 .0248
	295,485 .0250	325,523 .0251	356,562 .0251	388,602 .0250	422,642 .0260	456,684 .0259
37	300,495 .0247	330,534 .0243	362,573 .0249	394,614 .0244	428,655 .0250	462,698 .0244
	301,494 .0259	331,533 .0254	363,572 .0261	395,613 .0255	429,654 .0260	463,697 .0254

P = .025 one-sided

P = .05 two-sided

N	M = 15	M = 16	M = 17	M = 18	M = 19	M = 20
38	305,505 .0243 306,504 .0255	336,544 .0246 337,543 .0258	368,584 .0248 369,583 .0259	401,625 .0249 402,624 .0260	435,667 .0250 436,666 .0260	469,711 .0240 470,710 .0250
39	310,515 .0240 311,514 .0252	342,554 .0249 343,553 .0261	374,595 .0247 375,594 .0257	407,637 .0243 408,636 .0254	442,679 .0250 443,678 .0260	477,723 .0246 478,722 .0256
40	316,524 .0249 317,523 .0260	347,565 .0242 348,564 .0253	380,606 .0245 381,605 .0256	414,648 .0248 415,647 .0258	448,692 .0240 449,691 .0250	484,736 .0242 485,735 .0252
41	321,534 .0246 322,533 .0257	353,575 .0245 354,574 .0256	386,617 .0244 387,616 .0254	420,660 .0243 421,659 .0252	455,704 .0241 456,703 .0250	492,748 .0248 493,747 .0257
42	326,544 .0243 327,543 .0253	359,585 .0248 360,584 .0259	392,628 .0243 393,627 .0253	427,671 .0247 428,670 .0257	462,716 .0241 463,715 .0250	499,761 .0244 500,760 .0253
43	331,554 .0240 332,553 .0250	364,596 .0241 365,595 .0251	398,639 .0242 399,638 .0251	433,683 .0242 434,682 .0251	469,728 .0241 470,727 .0250	507,773 .0249 508,772 .0258
44	337,563 .0247 338,562 .0258	370,606 .0244 371,605 .0254	404,650 .0240 405,649 .0250	440,694 .0246 441,693 .0255	476,740 .0241 477,739 .0250	514,786 .0246 515,785 .0254
45	342,573 .0244 343,572 .0255	376,616 .0247 377,615 .0257	411,660 .0249 412,659 .0258	447,705 .0250 448,704 .0259	483,752 .0242 484,751 .0250	521,799 .0242 522,798 .0250
46	347,583 .0242 348,582 .0252	382,626 .0250 383,625 .0260	417,671 .0247 418,670 .0257	453,717 .0245 454,716 .0254	490,764 .0242 491,763 .0250	529,811 .0247 530,810 .0255
47	353,592 .0249 354,591 .0259	387,637 .0243 388,636 .0252	423,682 .0246 424,681 .0255	460,728 .0249 461,727 .0257	497,776 .0242 498,775 .0250	536,824 .0243 537,823 .0252
48	358,602 .0246 359,601 .0256	393,647 .0246 394,646 .0255	429,693 .0245 430,692 .0254	466,740 .0244 467,739 .0252	504,788 .0242 505,787 .0250	544,836 .0248 545,835 .0256
49	363,612 .0243 364,611 .0253	399,657 .0248 400,656 .0258	435,704 .0244 436,703 .0253	473,751 .0247 474,750 .0256	511,800 .0242 512,799 .0250	551,849 .0245 552,848 .0253
50	369,621 .0250 370,620 .0259	404,668 .0242 405,667 .0251	441,715 .0243 442,714 .0251	479,763 .0243 480,762 .0251	518,812 .0242 519,811 .0250	559,861 .0249 560,860 .0257

P = .025 one-sided

P = .05 two-sided

N	M = 21	M = 22	M = 23	M = 24	M = 25	M = 26
21	373,530 .0245 374,529 .0260					
22	381,543 .0249 382,542 .0264	411,579 .0247 412,578 .0261				

P = .025 one-sided

P = .05 two-sided

N	M = 21	M = 22	M = 23	M = 24	M = 25	M = 26
23	388,557 .0238 389,556 .0252	419,593 .0244 420,592 .0258	451,630 .0249 452,629 .0263			
24	396,570 .0242 397,569 .0256	427,607 .0242 428,606 .0255	459,645 .0242 460,644 .0254	492,684 .0241 493,683 .0254		
25	404,583 .0245 405,582 .0258	435,621 .0240 436,620 .0252	468,659 .0246 469,658 .0259	501,699 .0241 502,698 .0253	536,739 .0247 537,738 .0259	
26	412,596 .0248 413,595 .0261	444,634 .0250 445,633 .0262	476,674 .0239 477,673 .0251	510,714 .0240 511,713 .0252	545,755 .0242 546,754 .0253	581,797 .0243 582,796 .0254
27	419,610 .0238 420,609 .0251	452,648 .0247 453,647 .0259	485,688 .0243 486,687 .0255	519,729 .0240 520,728 .0251	555,770 .0247 556,769 .0258	591,813 .0244 592,812 .0254
28	427,623 .0241 428,622 .0253	460,662 .0244 461,661 .0256	494,702 .0247 495,701 .0259	528,744 .0239 529,743 .0250	564,786 .0242 565,785 .0252	601,829 .0244 602,828 .0255
29	435,636 .0244 436,635 .0255	468,676 .0242 469,675 .0253	502,717 .0240 503,716 .0251	538,758 .0249 539,757 .0260	574,801 .0247 575,800 .0257	611,845 .0245 612,844 .0255
30	443,649 .0246 444,648 .0257	476,690 .0240 477,689 .0251	511,731 .0244 512,730 .0255	547,773 .0248 548,772 .0259	583,817 .0242 584,816 .0252	621,861 .0245 622,860 .0255
31	451,662 .0248 452,661 .0259	485,703 .0248 486,702 .0259	520,745 .0248 521,744 .0258	556,788 .0247 557,787 .0257	593,832 .0246 594,831 .0256	631,877 .0246 632,876 .0255
32	458,676 .0239 459,675 .0250	493,717 .0245 494,716 .0256	528,760 .0241 529,759 .0251	565,803 .0246 566,802 .0256	602,848 .0241 603,847 .0251	641,893 .0246 642,892 .0255
33	466,689 .0241 467,688 .0252	501,731 .0243 502,730 .0253	537,774 .0244 538,773 .0254	574,818 .0245 575,817 .0255	612,863 .0246 613,862 .0255	651,909 .0246 652,908 .0255
34	474,702 .0243 475,701 .0254	509,745 .0241 510,744 .0250	546,788 .0247 547,787 .0257	583,833 .0244 584,832 .0253	622,878 .0250 623,877 .0259	661,925 .0246 662,924 .0255
35	482,715 .0245 483,714 .0255	518,758 .0248 519,757 .0258	554,803 .0241 555,802 .0250	592,848 .0243 593,847 .0252	631,894 .0245 632,893 .0254	671,941 .0247 672,940 .0255
36	490,728 .0247 491,727 .0257	526,772 .0245 527,771 .0255	563,817 .0244 564,816 .0253	601,863 .0242 602,862 .0251	641,909 .0249 642,908 .0258	681,957 .0247 682,956 .0255
37	498,741 .0248 499,740 .0258	534,786 .0243 535,785 .0252	572,831 .0246 573,830 .0256	611,877 .0250 612,876 .0259	650,925 .0244 651.924 .0252	691,973 .0247 692,972 .0255
38	506,754 .0250 507,753 .0260	543,799 .0250 544,798 .0259	581,845 .0249 582,844 .0258	620,892 .0248 621,891 .0257	660,940 .0248 661,939 .0256	701,989 .0247 702,988 .0255

P = .025 one-sided

P = .05 two-sided

N	M = 21	M = 22	M = 23	M = 24	M = 25	M = 26
39	513,768 .0242 514,767 .0251	551,813 .0247 552,812 .0256	589,860 .0243 590,859 .0252	629,907 .0247 630,906 .0256	669,956 .0243 670,955 .0251	711,1005 .0247 712,1104 .0255
40	521,781 .0244 522,780 .0253	559,827 .0245 560,826 .0254	598,874 .0246 599,873 .0254	638,922 .0246 639,921 .0254	679,971 .0246 680,970 .0254	721,1021 .0246 722,1020 .0254
41	529,794 .0245 530,793 .0254	567,841 .0243 568,840 .0251	607,888 .0248 608,887 .0256	647,937 .0245 648,936 .0253	689,986 .0250 690,985 .0258	731,1037 .0246 732,1036 .0254
42	537,807 .0247 538,806 .0255	576,854 .0249 577,853 .0257	615,903 .0242 616,902 .0250	656,952 .0244 657,951 .0252	698,1002 .0245 699,1001 .0253	741,1053 .0246 742,1052 .0254
43	545,820 .0248 546,819 .0256	584,868 .0246 585,867 .0255	624,917 .0245 625,916 .0253	665,967 .0243 666,966 .0250	708,1017 .0248 709,1016 .0256	751,1069 .0246 752,1068 .0253
44	553,833 .0249 554,832 .0258	592,882 .0244 593,881 .0252	633,931 .0247 634,930 .0255	675,981 .0249 676,980 .0257	717,1033 .0244 718,1032 .0251	761,1085 .0246 762,1084 .0253
45	560,847 .0242 561,846 .0250	601,895 .0250 602,894 .0258	642,945 .0249 643,944 .0257	684,996 .0248 685,995 .0256	727,1048 .0247 728,1047 .0254	771,1101 .0246 772,1100 .0253
46	568,860 .0243 569,859 .0251	609,909 .0247 610,908 .0255	650,960 .0244 651,959 .0251	693,1011 .0247 694,1010 .0254	737,1063 .0250 738,1062 .0257	781,1117 .0245 782,1116 .0252
47	576,873 .0245 577,872 .0252	617,923 .0245 618,922 .0253	659,974 .0246 660,973 .0253	702,1026 .0246 703,1025 .0253	746,1079 .0246 747,1078 .0253	791,1133 .0245 792,1132 .0252
48	584,886 .0246 585,885 .0253	625,937 .0243 626,936 .0251	668,988 .0248 669,987 .0255	711,1041 .0245 712,1040 .0252	756,1094 .0248 757,1093 .0255	801,1149 .0245 802,1148 .0252
49	592,899 .0247 593,898 .0254	634,950 .0248 635,949 .0256	677,1002 .0250 678,1001 .0257	720,1056 .0243 721,1055 .0250	765,1110 .0244 766,1109 .0251	811,1165 .0245 812,1164 .0251
50	600,912 .0248 601,911 .0255	642,964 .0246 643,963 .0253	685,1017 .0244 686,1016 .0251	730,1070 .0249 731,1069 .0256	775,1125 .0247 776,1124 .0254	821,1181 .0245 822,1180 .0251

P = .025 one-sided

P = .05 two-sided

N	M = 27	M = 28	M = 29	M = 30	M = 31	M = 32
27	628,857 .0240 629,856 .0250					
28	639,873 .0246 640,872 .0257	678,918 .0248 679,917 .0258				

P = .025 one-sided

P = .05 two-sided

N	M = 27	M = 28	M = 29	M = 30	M = 31	M = 32
29	649,890 .0243 650,889 .0252	688,936 .0241 689,935 .0250	729,982 .0248 730,981 .0257			
30	660,906 .0249 661,905 .0258	699,953 .0242 700,952 .0252	740,1000 .0245 741,999 .0254	782,1048 .0248 783,1047 .0257		
31	670,923 .0245 671,922 .0254	710,970 .0244 711,969 .0253	751,1018 .0243 752,1017 .0252	793,1067 .0242 794,1066 .0251	837,1116 .0249 838,1115 .0258	
32	680,940 .0241 681,939 .0250	721,987 .0246 722,986 .0254	763,1035 .0249 764,1034 .0258	805,1085 .0245 806,1084 .0253	849,1135 .0248 850,1134 .0256	893,1187 .0243 894,1186 .0251
33	691,956 .0247 692,955 .0256	732,1004 .0247 733,1003 .0256	774,1053 .0247 775,1052 .0255	817,1103 .0247 818,1102 .0255	861,1154 .0247 862,1153 .0255	906,1206 .0247 907,1205 .0255
34	701,973 .0243 702,972 .0252	743,1021 .0248 744,1020 .0257	785,1071 .0245 786,1070 .0253	829,1121 .0249 830,1120 .0257	873,1173 .0246 874,1172 .0253	919,1225 .0250 920,1224 .0257
35	712,989 .0248 713,988 .0257	754,1038 .0249 755,1037 .0258	796,1089 .0242 797,1088 .0250	840,1140 .0243 841,1139 .0251	885,1192 .0244 886,1191 .0252	931,1245 .0245 932,1244 .0253
36	722,1006 .0244 723,1005 .0253	764,1056 .0242 765,1055 .0250	808,1106 .0248 809,1105 .0256	852,1158 .0245 853,1157 .0253	897,1211 .0243 898,1210 .0250	944,1264 .0248 945,1263 .0255
37	733,1022 .0249 734,1021 .0257	775,1073 .0243 776,1072 .0251	819,1124 .0245 820,1123 .0253	864,1176 .0247 865,1175 .0255	910,1229 .0249 911,1228 .0256	956,1284 .0243 957,1283 .0251
38	743,1039 .0245 744,1038 .0253	786,1090 .0244 787,1089 .0252	830,1142 .0243 831,1141 .0251	876,1194 .0249 877,1193 .0257	922,1248 .0248 923,1247 .0255	969,1303 .0246 970,1302 .0253
39	754,1055 .0250 755,1054 .0258	797,1107 .0245 798,1106 .0253	842,1159 .0248 843,1158 .0256	887,1213 .0244 888,1212 .0251	934,1267 .0246 935,1266 .0253	982,1322 .0249 983,1321 .0255
40	764,1072 .0246 765,1071 .0254	808,1124 .0246 809,1123 .0254	853,1177 .0246 854,1176 .0253	899,1231 .0245 900,1230 .0252	946,1286 .0245 947,1285 .0252	994,1342 .0244 995,1341 .0251
41	774,1089 .0243 775,1088 .0250	819,1141 .0247 820,1140 .0254	864,1195 .0244 865,1194 .0251	911,1249 .0247 912,1248 .0254	958,1305 .0243 959,1304 .0250	1007,1361 .0247 1008.1360 .0253
42	785,1105 .0247 786,1104 .0254	830,1158 .0248 831,1157 .0255	876,1212 .0248 877,1211 .0255	923,1267 .0248 924,1266 .0255	971,1323 .0249 972,1322 .0255	1020,1380 .0249 1021,1379 .0255
43	795,1122 .0244 796,1121 .0251	841,1175 .0248 842,1174 .0255	887,1230 .0246 888,1229 .0253	935,1285 .0250 936,1284 .0257	983,1342 .0247 984,1341 .0254	1032,1400 .0245 1033,1399 .0251
44	806,1138 .0248 807,1137 .0255	852,1192 .0249 853,1191 .0256	898,1248 .0244 899,1247 .0250	946,1304 .0245 947,1303 .0251	995,1361 .0246 996,1360 .0252	1045,1419 .0247 1046,1418 .0253
45	816,1155 .0244 817,1154 .0251	863,1209 .0250 864,1208 .0256	910,1265 .0248 911,1264 .0255	958,1322 .0246 959,1321 .0253	1007,1380 .0244 1008,1379 .0251	1058,1438 .0249 1059,1437 .0255

P = .025 one-sided

P = .05 two-sided

N	M = 27	M = 28	M = 29	M = 30	M = 31	M = 32
46	827,1171 .0248	873,1227 .0244	921,1283 .0246	970,1340 .0248	1020,1398 .0249	1070,1458 .0245
	828,1170 .0255	874,1226 .0250	922,1282 .0252	971,1339 .0254	1021,1397 .0255	1071,1457 .0251
47	837,1188 .0245	884,1244 .0244	933,1300 .0250	982,1358 .0249	1032,1417 .0248	1083,1477 .0247
	838,1187 .0251	885,1243 .0251	934,1299 .0256	983,1357 .0255	1033,1416 .0254	1084,1476 .0253
48	848,1204 .0248	895,1261 .0245	944,1318 .0247	994,1376 .0250	1044,1436 .0246	1096,1496 .0248
	849,1203 .0255	896,1260 .0251	945,1317 .0254	995,1375 .0256	1045,1435 .0252	1097,1495 .0254
49	858,1221 .0245	906,1278 .0245	955,1336 .0245	1005,1395 .0245	1056,1455 .0245	1108,1516 .0245
	859,1220 .0252	907,1277 .0252	956,1335 .0251	1006,1394 .0251	1057,1454 .0251	1109,1515 .0250
50	869,1237 .0248	917,1295 .0246	967,1353 .0249	1017,1413 .0246	1069,1473 .0249	1121,1535 .0246
	870,1236 .0255	918,1294 .0252	968,1352 .0255	1018,1412 .0252	1070,1472 .0255	1122,1534 .0252

P = .025 one-sided

P = .05 two-sided

N	M = 33	M = 34	M = 35	M = 36	M = 37
33	952,1259 .0246				
	953,1258 .0254				
34	965,1279 .0246	1013,1333 .0250			
	966,1278 .0254	1014,1332 .0257			
35	978,1299 .0246	1026,1354 .0246	1075,1410 .0247		
	979,1298 .0253	1027,1353 .0254	1076,1409 .0254		
36	991,1319 .0245	1040,1374 .0250	1089,1431 .0247	1139,1489 .0245	
	992,1318 .0253	1041,1373 .0257	1090,1430 .0254	1140,1488 .0251	
37	1004,1339 .0245	1053,1395 .0246	1103,1452 .0248	1154,1510 .0249	1206,1569 .0250
	1005,1338 .0252	1054,1394 .0253	1104,1451 .0254	1155,1509 .0255	1207,1568 .0256
38	1017,1359 .0245	1067,1415 .0250	1117,1473 .0248	1168,1532 .0246	1220,1592 .0245
	1018,1358 .0251	1068,1414 .0256	1118,1472 .0255	1169,1531 .0253	1221,1591 .0251
39	1030,1379 .0244	1080,1436 .0246	1131,1494 .0248	1182,1554 .0244	1235,1614 .0246
	1031,1378 .0251	1081,1435 .0253	1132,1493 .0255	1183,1553 .0250	1236,1613 .0252
40	1043,1399 .0244	1094,1456 .0249	1145,1515 .0248	1197,1575 .0247	1250,1636 .0246
	1044,1398 .0250	1095,1455 .0256	1146,1514 .0255	1198,1574 .0254	1251,1635 .0253

P = .025 one-sided

P = .05 two-sided

N	M = 33	M = 34	M = 35	M = 36	M = 37
41	1057,1418 .0249	1107,1477 .0246	1159,1536 .0248	1211,1597 .0245	1265,1658 .0247
	1058,1417 .0256	1108,1476 .0252	1160,1535 .0255	1212,1596 .0251	1266,1657 .0253
42	1070,1438 .0249	1121,1497 .0249	1173,1557 .0249	1226,1618 .0248	1280,1680 .0248
	1071,1437 .0255	1122,1496 .0255	1174,1556 .0255	1227,1617 .0254	1281,1679 .0254
43	1083,1458 .0248	1134,1518 .0245	1187,1578 .0249	1240,1640 .0246	1295,1702 .0249
	1084,1457 .0254	1135,1517 .0251	1188,1577 .0255	1241,1639 .0252	1296,1701 .0255
44	1096,1478 .0247	1148,1538 .0248	1201,1599 .0249	1255,1661 .0249	1310,1724 .0249
	1097,1477 .0254	1149,1537 .0254	1202,1598 .0255	1256,1660 .0255	1311,1723 .0255
45	1109,1498 .0247	1161,1559 .0245	1215,1620 .0249	1269,1683 .0247	1325,1746 .0250
	1110,1497 .0253	1162,1558 .0251	1216,1619 .0254	1270,1682,.0252	1326,1745 .0256
46	1122,1518 .0246	1175,1579 .0247	1229,1641 .0249	1284,1704 .0250	1339,1769 .0245
	1123,1517 .0252	1176,1578 .0253	1230,1640 .0254	1285,1703 .0255	1340,1768 .0251
47	1135,1538 .0245	1189,1599 .0250	1243,1662 .0248	1298,1726 .0247	1354,1791 .0246
	1136,1537 .0251	1190,1598 .0256	1244,1661 .0254	1299,1725 .0253	1355,1790 .0251
48	1148,1558 .0245	1202,1620 .0247	1257,1683 .0248	1313,1747 .0250	1369,1813 .0246
	1149,1557 .0250	1203,1619 .0252	1258,1682 .0254	1314,1746 .0255	1370,1812 .0251
49	1162,1577 .0250	1216,1640 .0249	1271,1704 .0248	1327,1769 .0248	1384,1835 .0247
	1163,1576 .0255	1217,1639 .0255	1272,1703 .0254	1328,1768 .0253	1385,1834 .0252
50	1175,1597 .0249	1229,1661 .0246	1285,1725 .0248	1341,1791 .0245	1399,1857 .0247
	1176,1596 .0254	1230,1660 .0251	1286,1724 .0253	1342,1790 .0250	1400,1856 .0252

P = .025 one-sided

P = .05 two-sided

N	M = 38	M = 39	M = 40	M = 41	M = 42
38	1274,1652 .0249				
	1275,1651 .0255				
39	1289,1675 .0247	1344,1737 .0249			
	1290,1674 .0253	1345,1736 .0255			
40	1304,1698 .0246	1359,1761 .0245	1416,1824 .0249		
	1305,1697 .0251	1360,1760 .0250	1417,1823 .0255		
41	1320,1720 .0250	1375,1784 .0246	1432,1848 .0248	1489,1914 .0245	
	1321,1719 .0255	1376,1783 .0252	1433,1847 .0254	1490,1913 .0250	

P = .025 one-sided
P = .05 two-sided

I	M = 38	M = 39	M = 40	M = 41	M = 42
42	1335,1743 .0248 1336,1742 .0253	1391,1807 .0247 1392,1806 .0253	1448,1872 .0247 1449,1871 .0252	1506,1938 .0247 1507,1937 .0252	1565,2005 .0246 1566,2004 .0251
43	1350,1766 .0246 1351,1765 .0252	1407,1830 .0249 1408,1829 .0254	1464,1896 .0246 1465,1895 .0251	1523,1962 .0248 1524,1961 .0254	1582,2030 .0246 1583,2029 .0251
44	1366,1788 .0250 1367,1787 .0255	1423,1853 .0250 1424,1852 .0255	1480,1920 .0245 1481,1919 .0250	1539,1987 .0245 1540,1986 .0250	1599,2055 .0245 1600,2054 .0250
45	1381,1811 .0248 1382,1810 .0253	1438,1877 .0246 1439,1876 .0251	1497,1943 .0249 1498,1942 .0254	1556,2011 .0247 1557,2010 .0252	1617,2079 .0250 1618,2078 .0255
46	1396,1834 .0246 1397,1833 .0251	1454,1900 .0247 1455,1899 .0252	1513,1967 .0248 1514,1966 .0253	1573,2035 .0248 1574,2034 .0253	1634,2104 .0249 1635,2103 .0254
47	1412,1856 .0249 1413,1855 .0255	1470,1923 .0248 1471,1922 .0253	1529,1991 .0246 1530,1990 .0251	1590,2059 .0250 1591,2058 .0255	1651,2129 .0248 1642,2128 .0253
48	1427,1879 .0248 1428,1878 .0253	1486,1946 .0249 1487,1945 .0254	1545,2015 .0245 1546,2014 .0250	1606,2084 .0246 1607,2083 .0251	1668,2154 .0248 1669,2153 .0252
49	1442,1902 .0246 1443,1901 .0251	1502,1969 .0250 1503,1968 .0255	1562,2038 ,0249 1563,2037 .0254	1623,2108 .0248 1624,2107 .0253	1685,2179 .0247 1686,2178 .0252
50	1458,1924 .0249 1459,1923 .0254	1517,1993 .0246 1518,1992 .0251	1578,2062 .0248 1579,2061 .0253	1640,2132 .0249 1641,2131 .0254	1702,2204 .0246 1703,2203 .0251

P = .025 one-sided
P = .05 two-sided

N	M = 43	M = 44	M = 45	M = 46	M = 47
43	1643,2098 .0248 1644,2097 .0253				
44	1660,2124 .0245 1661,2123 .0250	1722,2194 .0245 1723,2193 .0250			
45	1678,2149 .0247 1679,2148 .0252	1741,2219 .0250 1742,2218 .0255	1804,2291 .0248 1805,2290 .0253		
46	1696,2174 .0249 1697,2173 .0254	1759,2245 .0250 1760,2244 .0255	1822,2318 .0246 1823,2317 .0250	1887,2391 .0246 1888,2390 .0251	
47	1713,2200 .0247 1714,2199 .0251	1777,2271 .0250 1778,2270 .0254	1841,2344 .0248 1842,2343 .0253	1906,2418 .0246 1907,2417 .0251	1973,2492 .0249 1974,2491 .0254

P = .025 one-sided

P = .05 two-sided

N	M = 43	M = 44	M = 45	M = 46	M = 47
48	1731,2225 .0249	1795,2297 .0250	1859,2371 .0246	1925,2445 .0247	1992,2520 .0248
	1732,2224 .0253	1796,2296 .0254	1860,2370 .0250	1926,2444 .0251	1993,2519 .0252
49	1748,2251 .0246	1813,2323 .0249	1878,2397 .0248	1944,2472 .0247	2011,2548 .0246
	1749,2250 .0250	1814,2322 .0254	1879,2396 .0253	1945,2471 .0251	2012,2547 .0250
50	1766,2276 .0248	1831,2349 .0249	1896,2424 .0246	1963,2499 .0247	2031,2575 .0249
	1767,2275 .0252	1832,2348 .0254	1897,2423 .0250	1964,2498 .0252	2032,2574 .0253

P = .025 one-sided

P = .05 two-sided

N	M = 48	M = 49	M = 50
48	2060,2596 .0248		
	2061,2595 .0253		
49	2080,2624 .0249	2149,2702 .0248	
	2081,2623 .0253	2150,2701 .0252	
50	2100,2652 .0250	2169,2731 .0247	2240,2810 .0248
	2101,2651 .0254	2170,2730 .0251	2241,2809 .0252

P = .01 one-sided

P = .02 two-sided

N	M = 3	M = 4	M = 5	M = 6	M = 7	M = 8
3	5,16 .0000 6,15 .0500					
4	5,19 .0000 6,18 .0286	9,27 .0000 10,26 .0143				
5	5,22 .0000 6,21 .0179	10,30 .0079 11,29 .0159	16,39 .0079 17,38 .0159			
6	5,25 .0000 6,24 .0119	11,33 .0095 12,32 .0190	17,43 .0087 18,42 .0152	24,54 .0076 25,53 .0130		
7	6,27 .0083 7,26 .0167	11,37 .0061 12,36 .0121	18,47 .0088 19,46 .0152	25,59 .0070 26,58 .0111	34,71 .0087 35,70 .0131	
8	6,30 .0061 7,29 .0121	12,40 .0081 13,39 .0141	19,51 .0093 20,50 .0148	27,63 .0100 28,62 .0147	35,77 .0070 36,76 .0103	45,91 .0074 46,90 .0103
9	7,32 .0091 8,31 .0182	13,43 .0098 14,42 .0168	20,55 .0095 21,54 .0145	28,68 .0088 29,67 .0128	37,82 .0082 38,81 .0115	47,97 .0076 48,96 .0103
10	7,35 .0070 8,34 .0140	13,47 .0070 14,46 .0120	21,59 .0097 22,58 .0140	29,73 .0080 30,72 .0112	39,87 .0093 40,86 .0125	49,103 .0078 50,102 .0103
11	7,38 .0055 8,37 .0110	14,50 .0088 15,49 .0132	22,63 .0096 23,62 .0137	30,78 .0073 31,77 .0101	40,93 .0077 41,92 .0102	51,109 .0079 52,108 .0102
12	8,40 .0088 9,39 .0154	15,53 .0099 16,52 .0148	23,67 .0097 24,66 .0134	32,82 .0091 33,81 .0122	42,98 .0085 43,97 .0111	53,115 .0079 54,114 .0101
13	8,43 .0071 9,42 .0125	15,57 .0076 16,56 .0113	24,71 .0097 25,70 .0132	33,87 .0084 34,86 .0110	44,103 .0093 45,102 .0118	56,120 .0099 57,119 .0123
14	8,46 .0059 9,45 .0103	16,60 .0088 17,59 .0124	25,75 .0097 26,74 .0129	34,92 .0077 35,91 .0100	45,109 .0079 46,108 .0100	58,126 .0098 59,125 .0120
15	9,48 .0086 10,47 .0135	17,63 .0098 18,62 .0137	26,79 .0097 27,78 .0127	36,96 .0092 37,95 .0117	47,114 .0086 48,113 .0106	60,132 .0097 61,131 .0117
16	9,51 .0072 10,50 .0114	17,67 .0078 18,66 .0109	27,83 .0097 28,82 .0125	37,101 .0085 38,100 .0107	49,119 .0092 50,118 .0112	62,138 .0096 63,137 .0115
17	10,53 .0096 11,52 .0140	18,70 .0089 19,69 .0119	28,87 .0096 29,86 .0123	39,105 .0099 40,104 .0122	51,124 .0097 52,123 .0118	64,144 .0095 65,143 .0113
18	10,56 .0083 11,55 .0120	19,73 .0097 20,72 .0129	29,91 .0096 30,90 .0121	40,110 .0091 41,109 .0112	52,130 .0085 53,129 .0103	66,150 .0094 67,149 .0110

P = .01 one-sided

P = .02 two-sided

N	M = 3	M = 4	M = 5	M = 6	M = 7	M = 8
19	10,59 .0071	19,77 .0080	30,95 .0096	41,115 .0085	54,135 .0090	68,156 .0093
	11,58 .0104	20,76 .0106	31,94 .0120	42,114 .0104	55,134 .0108	69,155 .0108
20	11,61 .0090	20,80 .0088	31,99 .0096	43,119 .0097	56,140 .0095	70,162 .0092
	12,60 .0130	21,79 .0114	32,98 .0118	44,118 .0116	57,139 .0112	71,161 .0107
21	11,64 .0079	21,83 .0096	32,103 .0095	44,124 .0091	58,145 .0099	72,168 .0091
	12,63 .0114	22,82 .0123	33,102 .0117	45,123 .0109	59,144 .0116	73,167 .0105
22	11,67 .0070	21,87 .0081	33,107 .0095	45,129 .0085	59,151 .0089	74,174 .0090
	12,66 .0100	22,86 .0104	34,106 .0116	46,128 .0102	60,150 .0104	75,173 .0103
23	12,69 .0088	22,90 .0088	34,111 .0095	47,133 .0095	61,156 .0093	76,180 .0089
	13,68 .0119	23,89 .0111	35,110 .0114	48,132 .0112	62,155 .0108	77,179 .0102
24	12,72 .0079	23,93 .0095	35,115 .0095	48,138 .0090	63,161 .0097	78,186 .0088
	13,71 .0106	24,92 .0118	36,114 .0113	49,137 .0106	64,160 .0111	79,185 .0100
25	13,74 .0095	23,97 .0082	36,119 .0094	50,142 .0100	64,167 .0088	81,191 .0099
	14,73 .0125	24,96 .0101	37,118 .0112	51,141 .0116	65,166 .0100	82,190 .0112
26	13,77 .0085	24,100 .0088	37,123 .0094	51,147 .0094	66,172 .0091	83,197 .0098
	14,76 .0112	25,99 .0108	38,122 .0111	52,146 .0109	67,171 .0104	84,196 .0110
27	13,80 .0076	25,103 .0094	38,127 .0094	52,152 .0089	68,177 .0095	85,203 .0097
	14,79 .0101	26,102 .0114	39,126 .0110	53,151 .0103	69,176 .0107	86,202 .0108
28	14,82 .0091	26,106 .0100	39,131 .0094	54,156 .0098	70,182 .0098	87,209 .0096
	15,81 .0118	27,105 .0120	40,130 .0110	55,155 .0112	71,181 .0111	88,208 .0107
29	14,85 .0083	26,110 .0088	40,135 .0094	55,161 .0093	71,188 .0090	89,215 .0095
	15,84 .0107	27,109 .0105	41,134 .0109	56,160 .0106	72,187 .0101	90,214 .0105
30	15,87 .0097	27,113 .0093	41,139 .0093	56,166 .0089	73,193 .0093	91,221 .0094
	16,86 .0123	28,112 .0111	42,138 .0108	57,165 .0101	74,192 .0104	92,220 .0104
31	15,90 .0089	28,116 .0098	42,143 .0093	58,170 .0096	75,198 .0096	93,227 .0093
	16,89 .0112	29,115 .0116	43,142 .0107	59,169 .0109	76,197 .0107	94,226 .0103
32	15,93 .0081	28,120 .0087	43,147 .0093	59,175 .0092	77,203 .0098	95,233 .0092
	16,92 .0102	29,119 .0103	44,146 .0107	60,174 .0104	78,202 .0110	96,232 .0101
33	16,95 .0094	29,123 .0092	44,151 .0093	61,179 .0099	78,209 .0091	97,239 .0091
	17,94 .0116	30,122 .0109	45,150 .0106	62,178 .0112	79,208 .0101	98,238 .0100
34	16,98 .0086	30,126 .0097	45,155 .0093	62,184 .0095	80,214 .0094	100,244 .0099
	17,97 .0107	31,125 .0113	46,154 .0105	63,183 .0107	81,213 .0104	101,243 .0109
35	17,100 .0098	30,130 .0087	46,159 .0092	63,189 .0091	82,219 .0096	102,250 .0098
	18,99 .0121	31,129 .0102	47,158 .0105	64,188 .0102	83,218 .0106	103,249 .0108
36	17,103 .0091	31,133 .0092	47,163 .0092	65,193 .0098	84,224 .0099	104,256 .0097
	18,102 .0112	32,132 .0106	48,162 .0104	66,192 .0109	85,223 .0109	105,255 .0106

P = .01 one-sided

P = .02 two-sided

N	M = 3	M = 4	M = 5	M = 6	M = 7	M = 8
37	17,106 .0084 18,105 .0103	32,136 .0096 33,135 .0111	48,167 .0092 49,166 .0104	66,198 .0094 67,197 .0104	85,230 .0092 86,229 .0101	106,262 .0096 107,261 .0105
38	18,108 .0096 19,107 .0115	32,140 .0087 33,139 .0100	49,171 .0092 50,170 .0103	67,203 .0090 68,202 .0100	87,235 .0094 88,234 .0104	108,268 .0095 109,267 .0104
39	18,111 .0089 19,110 .0107	33,143 .0091 34,142 .0105	50,175 .0092 51,174 .0103	69,207 .0096 70,206 .0107	89,240 .0097 90,239 .0106	110,274 .0095 111,273 .0103
40	19,113 .0100 20,112 .0119	34,146 .0095 35,145 .0109	51,179 .0091 52,178 .0102	70,212 .0093 71,211 .0103	91,245 .0099 92,244 .0108	112,280 .0094 113,279 .0102
41	19,116 .0093 20,115 .0111	35,149 .0099 36,148 .0113	52,183 .0091 53,182 .0102	72,216 .0099 73,215 .0109	92,251 .0093 93,250 .0101	114,286 .0093 115,285 .0101
42	19,119 .0087 20,118 .0104	35,153 .0091 36,152 .0103	53,187 .0091 54,186 .0102	73,221 .0095 74,220 .0105	94,256 .0095 95,255 .0103	117,291 .0100 118,290 .0108
43	20,121 .0097 21,120 .0115	36,156 .0094 37,155 .0107	54,191 .0091 55,190 .0101	74,226 .0092 75,225 .0101	96,261 .0097 97,260 .0106	119,297 .0099 120,296 .0107
44	20,124 .0091 21,123 .0107	37,159 .0098 38,158 .0111	55,195 .0091 56,194 .0101	76,230 .0097 77,229 .0107	98,266 .0099 99,265 .0108	121,303 .0098 122,302 .0106
45	20,127 .0085 21,126 .0101	37,163 .0090 38,162 .0102	56,199 .0091 57,198 .0101	77,235 .0094 78,234 .0103	99,272 .0093 100,271 .0101	123,309 .0097 124,308 .0104
46	21,129 .0094 22,128 .0111	38,166 .0094 39,165 .0105	57,203 .0091 58,202 .0100	79,239 .0100 80,238 .0109	101,277 .0095 102,276 .0103	125,315 .0096 126,314 .0104
47	21,132 .0089 22,131 .0104	39,169 .0097 40,168 .0109	59,206 .0100 60,205 .0110	80,244 .0096 81,243 .0105	103,282 .0097 104,281 .0105	127,321 .0096 128,320 .0103
48	22,134 .0098 23,133 .0114	39,173 .0090 40,172 .0101	60,210 .0100 61,209 .0109	81,249 .0093 82,248 .0101	105,287 .0099 106,286 .0107	129,327 .0095 130,326 .0102
49	22,137 .0092 23,136 .0107	40,176 .0093 41,175 .0104	61,214 .0099 62,213 .0109	83,253 .0098 84,252 .0107	106,293 .0094 107,292 .0101	131,333 .0094 132,332 .0101
50	22,140 .0087 23,139 .0101	41,179 .0096 42,178 .0107	62,218 .0099 63,217 .0108	84,258 .0095 85,257 .0103	108,298 .0096 109,297 .0103	134,338 .0100 135,337 .0107

P = .01 one-sided

P = .02 two-sided

N	M = 9	M = 10	M = 11	M = 12	M = 13	M = 14
9	59,112 .0094 60,111 .0122					
10	61,119 .0086 62,118 .0110	74,136 .0093 75,135 .0116				
11	63,126 .0079 64,125 .0100	77,143 .0098 78,142 .0121	91,162 .0096 92,161 .0117			
12	66,132 .0092 67,131 .0114	79,151 .0084 80,150 .0103	94,170 .0094 95,169 .0113	109,191 .0086 110,190 .0102		
13	68,139 .0085 69,138 .0104	82,158 .0089 83,157 .0107	97,178 .0092 98,177 .0109	113,199 .0094 114,198 .0110	130,221 .0095 131,220 .0111	
14	71,145 .0096 72,144 .0115	85,165 .0093 86,164 .0110	100,186 .0090 101,185 .0105	116,208 .0087 117,207 .0101	134,230 .0097 135,229 .0111	152,254 .0093 153,253 .0106
15	73,152 .0089 74,151 .0106	88,172 .0096 89,171 .0113	103,194 .0088 104,193 .0102	120,216 .0093 121,215 .0107	138,239 .0097 139,238 .0111	156,264 .0089 157,263 .0101
16	76,158 .0098 77,157 .0116	91,179 .0099 92,178 .0115	107,201 .0099 108,200 .0114	124,224 .0099 125,223 .0113	142,248 .0098 143,247 .0111	161,273 .0097 162,272 .0109
17	78,165 .0091 79,164 .0107	93,187 .0088 94,186 .0102	110,209 .0096 111,208 .0110	127,233 .0092 128,232 .0104	146,257 .0098 147,256 .0111	165,283 .0093 166,282 .0104
18	81,171 .0100 82,170 .0116	96,194 .0090 97,193 .0104	113,217 .0094 114,216 .0107	131,241 .0097 132,240 .0109	150,266 .0099 151,265 .0110	170,292 .0100 171,291 .0111
19	83,178 .0093 84,177 .0108	99,201 .0093 100,200 .0106	116,225 .0092 117,224 .0104	134,250 .0090 135,249 .0101	154,275 .0099 155,274 .0110	174,302 .0096 175,301 .0106
20	85,185 .0088 86,184 .0101	102,208 .0095 103,207 .0108	119,233 .0089 120,232 .0101	138,258 .0094 139,257 .0106	158,284 .0099 159,283 .0109	178,312 .0092 179,311 .0102
21	88,191 .0095 89,190 .0108	105,215 .0097 106,214 .0109	123,240 .0098 124,239 .0110	142,266 .0099 143,265 .0110	162,293 .0099 163,292 .0109	183,321 .0098 184,320 .0108
22	90,198 .0089 91,197 .0101	108,222 .0099 109,221 .0111	126,248 .0096 127,247 .0107	145,275 .0092 146,274 .0103	166,302 .0098 167,301 .0108	187,331 .0094 188,330 .0104
23	93,204 .0096 94,203 .0108	110,230 .0089 111,229 .0100	129,256 .0093 130,255 .0104	149,283 .0096 150,282 .0106	170,311 .0098 171,310 .0108	192,340 .0100 193,339 .0109
24	95,211 .0090 96,210 .0102	113,237 .0091 114,236 .0102	132,264 .0091 133,263 .0101	153,291 .0100 154,290 .0110	174,320 .0098 175,319 .0107	196,350 .0096 197,349 .0105
25	98,217 .0096 99,216 .0108	116,244 .0093 117,243 .0103	136,271 .0099 137,270 .0109	156,300 .0094 157,299 .0103	178,329 .0098 179,328 .0107	200,360 .0093 201,359 .0101

P = .01 one-sided

P = .02 two-sided

N	M = 9	M = 10	M = 11	M = 12	M = 13	M = 14
26	100,224 .0091	119,251 .0094	139,279 .0096	160,308 .0097	182,338 .0098	205,369 .0097
	101,223 .0102	120,250 .0105	140,278 .0106	161,307 .0106	183,337 .0106	206,368 .0105
27	103,230 .0097	122,258 .0096	142,287 .0094	163,317 .0092	186,347 .0097	209,379 .0094
	104,229 .0108	123,257 .0106	143,286 .0103	164,316 .0100	187,346 .0106	210,378 .0102
28	105,237 .0092	125,265 .0097	145,295 .0092	167,325 .0095	190,356 .0097	214,388 .0098
	106,236 .0102	126,264 .0107	146,294 .0101	168,324 .0103	191,355 .0105	215,387 .0106
29	108,243 .0097	128,272 .0098	149,302 .0099	171,333 .0098	194,365 .0097	218,398 .0095
	109,242 .0107	129,271 .0108	150,301 .0107	172,332 .0106	195,364 .0104	219,397 .0102
30	110,250 .0093	131,279 .0100	152,310 .0096	174,342 .0093	198,374 .0096	223,407 .0099
	111,249 .0102	132,278 .0109	153,309 .0105	175,341 .0101	199,373 .0104	224,406 .0106
31	113,256 .0098	133,287 .0092	155,318 .0094	178,350 .0096	202,383 .0096	227,417 .0096
	114,255 .0107	134,286 .0101	156,317 .0102	179,349 .0103	203,382 .0103	228,416 .0103
32	115,263 .0093	136,294 .0093	158,326 .0093	182,358 .0098	206,392 .0096	232,426 .0100
	116,262 .0102	137,293 .0102	159,325 .0100	183,357 .0106	207,391 .0103	233,425 .0107
33	118,269 .0098	139,301 .0095	162,333 .0098	185,367 .0094	210,401 .0095	236,436 .0097
	119,268 .0107	140,300 .0103	163,332 .0106	186,366 .0101	211,400 .0102	237.435 .0103
34	120,276 .0094	142,308 .0096	165,341 .0096	189,375 .0096	214,410 .0095	240,446 .0094
	121,275 .0102	143,307 .0104	166,340 .0104	190,374 .0103	215,409 .0102	241,445 .0100
35	123,282 .0098	145,315 .0097	168,349 .0094	193,383 .0098	218,419 .0095	245,455 .0097
	124,281 .0107	146,314 .0105	169,348 .0102	194,382 .0105	219,418 .0101	246,454 .0104
36	125,289 .0094	148,322 .0098	172,356 .0100	196,392 .0094	222,428 .0095	249,465 .0095
	126,288 .0102	149,321 .0105	173,355 .0107	197,391 .0101	223,427 .0101	250,464 .0101
37	128,295 .0098	151,329 .0099	175,364 .0098	200,400 .0096	226,437 .0094	254,474 .0098
	129,294 .0106	152,328 .0106	176,363 .0105	201,399 .0103	227,436 .0100	255,473 .0104
38	130,302 .0094	154,336 .0099	178,372 .0096	204,408 .0098	230,446 .0094	258,484 .0095
	131,301 .0102	155,335 .0107	179,371 .0103	205,407 .0105	231,445 .0100	259,483 .0101
39	133,308 .0098	156,344 .0093	181,380 .0094	207,417 .0094	235,454 .0100	263,493 .0098
	134,307 .0106	157,343 .0100	182,379 .0101	208,416 .0101	236,453 .0106	264,492 .0104
40	135,315 .0095	159,351 .0094	185,387 .0099	211,425 .0096	239,463 .0099	267,503 .0096
	136,314 .0102	160,350 .0101	186,386 .0106	212,424 .0103	240,462 .0105	268,502 .0101
41	138,321 .0098	162,358 .0095	188,395 .0097	215,433 .0098	243,472 .0099	272,512 .0099
	139,320 .0106	163,357 .0102	189,394 .0104	216,432 .0105	244,471 .0105	273,511 .0104
42	140,328 .0095	165,365 .0096	191,403 .0096	218,442 .0095	247,481 .0098	276,522 .0096
	141,327 .0102	166,364 .0103	192,402 .0102	219,441 .0100	248,480 .0104	277,521 .0101

P = .01 one-sided

P = .02 two-sided

N	M = 9	M = 10	M = 11	M = 12	M = 13	M = 14
43	143,334 .0098 144,333 .0106	168,372 .0097 169,371 .0103	194,411 .0094 195,410 .0100	222,450 .0097 223,449 .0102	251,490 .0098 252,489 .0104	281,531 .0099 282,530 .0104
44	145,341 .0095 146,340 .0102	171,379 .0097 172,378 .0104	198,418 .0098 199,417 .0104	226,458 .0098 227,457 .0104	255,499 .0098 256,498 .0103	285,541 .0096 286,540 .0102
45	148,347 .0098 149,346 .0105	174,386 .0098 175,385 .0105	201,426 .0097 202,425 .0103	229,467 .0095 230,466 .0100	259,508 .0097 260,507 .0103	290,550 .0099 291,549 .0104
46	150,354 .0095 151,353 .0102	177,393 .0099 178,392 .0105	204,434 .0095 205,433 .0101	233,475 .0097 234,474 .0102	263,517 .0097 264,516 .0102	294,560 .0097 295,559 .0102
47	153,360 .0099 154,359 .0105	180,400 .0100 181,399 .0106	208,441 .0099 209,440 .0105	237,483 .0098 238,482 .0104	267,526 .0097 268,525 .0102	299,569 .0099 300,568 .0104
48	155,367 .0095 156,366 .0102	182,408 .0094 183,407 .0100	211,449 .0098 212,448 .0103	240,492 .0095 241,491 .0100	271,535 .0096 272,534 .0101	303,579 .0097 304,578 .0102
49	158,373 .0099 159,372 .0105	185,415 .0095 186,414 .0101	214,457 .0096 215,456 .0102	244,500 .0097 245,499 .0102	275,544 .0096 276,543 .0101	308,588 .0099 309,587 .0104
50	160,380 .0096 161,379 .0102	188,422 .0096 189,421 .0101	217,465 .0095 218,464 .0100	248,508 .0098 249,507 .0103	279,553 .0096 280,552 .0100	312,598 .0097 313,597 .0102

P = .01 one-sided

P = .02 two-sided

N	M = 15	M = 16	M = 17	M = 18	M = 19	M = 20
15	176,289 .0093 177,288 .0105					
16	181,299 .0096 182,298 .0107	202,326 .0095 203,325 .0105				
17	186,309 .0099 187,308 .0110	207,337 .0093 208,336 .0104	230,365 .0098 231,364 .0108			
18	190,320 .0091 191,319 .0101	212,348 .0092 213,347 .0102	235,377 .0093 236,376 .0102	259,407 .0094 260,406 .0102		
19	195,330 .0093 196,329 .0103	218,358 .0100 219,357 .0110	241,388 .0097 242,387 .0106	265,419 .0094 266,418 .0102	291,450 .0099 292,449 .0108	
20	200,340 .0095 201,339 .0105	223,369 .0098 224,368 .0107	246,400 .0092 247,399 .0100	271,431 .0094 272,430 .0102	297,463 .0096 298,462 .0104	324,496 .0098 325,495 .0105

P = .01 one-sided

P = .02 two-sided

N	M = 15	M = 16	M = 17	M = 18	M = 19	M = 20
21	205,350 .0097	228,380 .0096	252,411 .0095	277,443 .0094	303,476 .0093	331,509 .0099
	206,349 .0107	229,379 .0105	253,410 .0104	278,442 .0102	304,475 .0101	332,508 .0107
22	210,360 .0099	233,391 .0095	258,422 .0099	283,455 .0094	310,488 .0098	337,523 .0093
	211,359 .0108	234,390 .0103	259,421 .0107	284,454 .0102	311,487 .0105	338,522 .0100
23	214,371 .0092	238,402 .0093	263,434 .0094	289,467 .0094	316,501 .0095	344,536 .0095
	215,370 .0101	239,401 .0101	264,433 .0102	290,466 .0102	317,500 .0102	345,535 .0101
24	219,381 .0094	244,412 .0099	269,445 .0097	295,479 .0094	323,513 .0098	351,549 .0096
	220,380 .0102	245,411 .0107	270,444 .0104	296,478 .0101	324,512 .0105	352,548 .0102
25	224,391 .0095	249,423 .0098	275,456 .0100	301,491 .0094	329,526 .0096	358,562 .0097
	225,390 .0103	250,422 .0105	276,455 .0107	302,490 .0101	330,525 .0102	359,561 .0103
26	229,401 .0097	254,434 .0096	280,468 .0095	307,503 .0094	336,538 .0099	365,575 .0098
	230,400 .0104	255,433 .0103	281,467 .0102	308,502 .0101	337,537 .0106	366,574 .0104
27	234,411 .0098	259,445 .0094	286,479 .0098	313,515 .0094	342,551 .0096	372,588 .0098
	235,410 .0106	260,444 .0101	287,478 .0104	314,514 .0100	343,550 .0102	373,587 .0104
28	239,421 .0099	265,455 .0100	292,490 .0100	320,526 .0100	349,563 .0099	379,601 .0099
	240,420 .0107	266,454 .0107	293,489 .0107	321,525 .0106	350,562 .0106	380,600 .0105
29	243,432 .0093	270,466 .0098	297,502 .0096	326,538 .0099	355,576 .0097	386,614 .0100
	244,431 .0100	271,465 .0105	298,501 .0102	327,537 .0106	356,575 .0103	387,613 .0106
30	248,442 .0095	275,477 .0096	303,513 .0098	332,550 .0099	362,588 .0100	392,628 .0095
	249,441 .0101	276,476 .0103	304,512 .0104	333,549 .0105	363,587 .0106	393,627 .0100
31	253,452 .0096	280,488 .0095	309,524 .0100	338,562 .0098	368,601 .0097	399,641 .0095
	254,451 .0102	281,487 .0101	310,523 .0106	339,561 .0104	369,600 .0103	400,640 .0101
32	258,462 .0097	286,498 .0099	314,536 .0096	344,574 .0098	375,613 .0100	406,654 .0096
	259,461 .0103	287,497 .0106	315,535 .0102	345,573 .0104	376,612 .0105	407,653 .0101
33	263,472 .0097	291,509 .0098	320,547 .0098	350,586 .0098	381,626 .0097	413,667 .0097
	264,471 .0104	292,508 .0104	321,546 .0104	351,585 .0103	382,625 .0102	414,666 .0102
34	268,482 .0098	296,520 .0096	326,558 .0100	356,598 .0097	388,638 .0100	420,680 .0097
	269,481 .0105	297,519 .0102	327,557 .0105	357,597 .0102	389,637 .0105	421,679 .0102
35	273,492 .0099	301,531 .0095	331,570 .0096	362,610 .0097	394,651 .0097	427,693 .0097
	274,491 .0105	302,530 .0100	332,569 .0101	363,609 .0102	395,650 .0102	428,692 .0102
36	278,502 .0100	307,541 .0099	337,581 .0098	368,622 .0096	401,663 .0100	434,706 .0098
	279,501 .0106	308,540 .0105	338,580 .0103	369,621 .0101	402,662 .0105	435,705 .0103
37	282,513 .0095	312,552 .0097	343,592 .0099	374,634 .0096	407,676 .0097	441,719 .0098
	283,512 .0101	313,551 .0103	344,591 .0105	375,633 .0101	408,675 .0102	442,718 .0103

P = .01 one-sided

P = .02 two-sided

N	M = 15	M = 16	M = 17	M = 18	M = 19	M = 20
38	287,523 .0096	317,563 .0096	348,604 .0096	380,646 .0095	414,688 .0099	448,732 .0098
	288,522 .0101	318,562 .0101	349,603 .0101	381,645 .0100	415,687 .0104	449,731 .0103
39	292,533 .0097	323,573 .0100	354,615 .0097	387,657 .0100	420,701 .0097	455,745 .0099
	293,532 .0102	324,572 .0105	355,614 .0102	388,656 .0105	421,700 .0102	456,744 .0103
40	297,543 .0097	328,584 .0098	360,626 .0099	393,669 .0099	427,713 .0099	462,758 .0099
	298,542 .0103	329,583 .0103	361,625 .0104	394,668 .0104	428,712 .0104	463,757 .0103
41	302,553 .0098	333,595 .0097	365,638 .0096	399,681 .0099	433,726 .0097	469,771 .0099
	303,552 .0103	334,594 .0102	366,637 .0100	400,680 .0103	434,725 .0101	470,770 .0104
42	307,563 .0099	338,606 .0096	371,649 .0097	405,693 .0098	440,738 .0099	476,784 .0100
	308,562 .0104	339,605 .0100	372,648 .0102	406,692 .0103	441,737 .0103	477,783 .0104
43	312,573 .0099	344,616 .0099	377,660 .0098	411,705 .0098	446,751 .0097	483,797 .0100
	313,572 .0104	345,615 .0104	378,659 .0103	412,704 .0102	447,750 .0101	484,796 .0104
44	317,583 .0100	349,627 .0098	383,671 .0100	417,717 .0097	453,763 .0099	490,810 .0100
	318,582 .0105	350,626 .0102	384,670 .0104	418,716 .0102	454,762 .0103	491,809 .0104
45	321,594 .0096	354,638 .0096	388,683 .0097	423,729 .0097	459,776 .0097	496,824 .0096
	322,593 .0100	355,637 .0101	389,682 .0101	424,728 .0101	460,775 .0101	497,823 .0100
46	326,604 .0096	360,648 .0099	394,694 .0098	429,741 .0096	466,788 .0098	503,837 .0096
	327,603 .0101	361,647 .0104	395,693 .0102	430,740 .0100	467,787 .0102	504,836 .0100
47	331,614 .0097	365,659 .0098	400,705 .0099	436,752 .0100	472,801 .0096	510,850 .0097
	332,613 .0101	366,658 .0103	401,704 .0104	437,751 .0104	473,800 .0100	511,849 .0100
48	336,624 .0097	370,670 .0097	405,717 .0096	442,764 .0099	479,813 .0098	517,863 .0097
	337,623 .0102	371,669 .0101	406,716 .0100	443,763 .0103	480,812 .0102	518,862 .0101
49	341,634 .0098	376,680 .0100	411,728 .0098	448,776 .0099	486,825 .0100	524,876 .0097
	342,633 .0102	377,679 .0104	412,727 .0102	449,775 .0103	487,824 .0104	525,875 .0101
50	346,644 .0098	381,691 .0099	417,739 .0099	454,788 .0098	492,838 .0098	531,889 .0097
	347,643 .0103	382,690 .0103	418,738 .0103	455,787 .0102	493,837 .0102	532,888 .0101

P = .01 one-sided

P = .02 two-sided

N	M = 21	M = 22	M = 23	M = 24	M = 25	M = 26
21	359,544 .0098					
	360,543 .0105					

P = .01 one-sided

P = .02 two-sided

N	M = 21	M = 22	M = 23	M = 24	M = 25	M = 26
22	366,558 .0096 367,557 .0103	396,594 .0098 397,593 .0105				
23	373,572 .0095 374,571 .0101	403,609 .0094 404,608 .0101	434,647 .0094 435,646 .0100			
24	381,585 .0099 382,584 .0106	411,623 .0097 412,622 .0103	443,661 .0100 444,660 .0106	475,701 .0097 476,700 .0103		
25	388,599 .0098 389,598 .0104	419,637 .0098 420,636 .0104	451,676 .0099 452,675 .0105	484,716 .0100 485,715 .0105	517,758 .0095 518,757 .0100	
26	395,613 .0096 396,612 .0102	426,652 .0095 427,651 .0100	459,691 .0099 460,690 .0104	492,732 .0097 493,731 .0102	526,774 .0095 527,773 .0100	562,816 .0099 563,815 .0104
27	402,627 .0095 403,626 .0100	434,666 .0096 435,665 .0102	467,706 .0098 468,705 .0103	501,747 .0099 502,746 .0105	535,790 .0096 536,789 .0101	571,833 .0097 572,832 .0102
28	410,640 .0099 411,639 .0104	442,680 .0098 443,679 .0103	475,721 .0097 476,720 .0103	509,763 .0097 510,762 .0102	544,806 .0096 545,805 .0101	581,849 .0100 582,848 .0105
29	417,654 .0097 418,653 .0102	450,694 .0099 451,693 .0105	483,736 .0097 484,735 .0102	518,778 .0099 519,777 .0104	553,822 .0096 554,821 .0101	590,866 .0098 591,865 .0103
30	424,668 .0095 425,667 .0101	457,709 .0096 458,708 .0101	491,751 .0096 492,750 .0101	526,794 .0096 527,793 .0101	562,838 .0096 563,837 .0101	599,883 .0096 600,882 .0101
31	432,681 .0099 433,680 .0104	465,723 .0097 466,722 .0102	499,766 .0095 500,765 .0100	535,809 .0098 536,808 .0103	571,854 .0096 572,853 .0101	609,899 .0099 610,898 .0103
32	439,695 .0097 440,694 .0102	473,737 .0098 474,736 .0103	508,780 .0099 509,779 .0104	543,825 .0096 544,824 .0100	580,870 .0096 581,869 .0101	618,916 .0097 619,915 .0101
33	446,709 .0096 447,708 .0101	481,751 .0100 482,750 .0105	516,795 .0099 517,794 .0103	552,840 .0098 553,839 .0102	589,886 .0097 590,885 .0101	628,932 .0100 629,931 .0104
34	454,722 .0099 455,721 .0104	488,766 .0096 489,765 .0101	524,810 .0098 525,809 .0102	561,855 .0099 562,854 .0104	598,902 .0097 599,901 .0101	637,949 .0098 638,948 .0102
35	461,736 .0097 462,735 .0102	496,780 .0097 497,779 .0102	532,825 .0097 533,824 .0102	569,871 .0097 570,870 .0101	607,918 .0097 608,917 .0101	646,966 .0096 647,965 .0100
36	468,750 .0096 469,749 .0101	504,794 .0099 505,793 .0103	540,840 .0097 541,839 .0101	578,886 .0099 579,885 .0103	616,934 .0097 617,933 .0101	656,982 .0098 657,981 .0102
37	476,763 .0099 477,762 .0103	512,808 .0100 513,807 .0104	549,854 .0100 550,853 .0104	586,902 .0096 587,901 .0100	625,950 .0096 626,949 .0100	665,999 .0097 666,998 .0100
38	483,777 .0097 484,776 .0102	519,823 .0096 520,822 .0101	557,869 .0099 558,868 .0103	595,917 .0098 596,916 .0102	634,966 .0096 635,965 .0100	675,1015 .0099 676,1014 .0102

P = .01 one-sided

P = .02 two-sided

N	M = 21	M = 22	M = 23	M = 24	M = 25	M = 26
39	490,791 .0096 491,790 .0100	527,837 .0097 528,836 .0101	565,884 .0098 566,883 .0102	604,932 .0099 605,931 .0103	643,982 .0096 644,981 .0100	684,1032 .0097 685,1031 .0101
40	498,804 .0099 499,803 .0103	535,851 .0098 536,850 .0102	573,899 .0098 574,898 .0102	612,948 .0097 613,947 .0101	653,997 .0100 654,996 .0104	694,1048 .0099 695,1047 .0103
41	505,818 .0097 506,817 .0101	543,865 .0099 544,864 .0103	581,914 .0097 582,913 .0101	621,963 .0098 622,962 .0102	662,1013 .0100 663,1012 .0103	703,1065 .0097 704,1064 .0101
42	513,831 .0100 514,830 .0104	551,879 .0100 552,878 .0104	590,928 .0100 591,927 .0104	630,978 .0100 631,977 .0103	671,1029 .0100 672,1028 .0103	713,1081 .0099 714,1080 .0103
43	520,845 .0098 521,844 .0102	558,894 .0097 559,893 .0101	598,943 .0099 599,942 .0103	638,994 .0098 639,993 .0101	680,1045 .0099 681,1044 .0103	722,1098 .0098 723,1097 .0101
44	527,859 .0097 528,858 .0101	566,908 .0098 567,907 .0102	606,958 .0098 607,957 .0102	647,1009 .0099 648,1008 .0102	689,1061 .0099 690,1060 .0103	732,1114 .0099 733,1113 .0103
45	535,872 .0099 536,871 .0103	574,922 .0099 575,921 .0102	614,973 .0098 615,972 .0101	655,1025 .0097 656,1024 .0100	698,1077 .0099 699,1076 .0102	741,1131 .0098 742,1130 .0101
46	542,886 .0098 543,885 .0102	582,936 .0099 583,935 .0103	622,988 .0097 623,987 .0100	664,1040 .0098 665,1039 .0101	707,1093 .0099 708,1092 .0102	751,1147 .0099 752,1146 .0103
47	549,900 .0097 550,899 .0100	589,951 .0097 590,950 .0100	631,1002 .0100 632,1001 .0103	673,1055 .0099 674,1054 .0102	716,1109 .0098 717,1108 .0102	760,1164 .0098 761,1163 .0101
48	557,913 .0099 558,912 .0103	597,965 .0097 598,964 .0101	639,1017 .0099 640,1016 .0102	681,1071 .0097 682,1070 .0100	725,1125 .0098 726,1124 .0101	770,1180 .0099 771,1179 .0103
49	564,927 .0098 565,926 .0101	605,979 .0098 606,978 .0101	647,1032 .0098 648,1031 .0101	690,1086 .0098 691,1085 .0101	734,1141 .0098 735,1140 .0101	779,1197 .0098 780,1196 .0101
50	572,940 .0100 573,939 .0103	613,993 .0099 614,992 .0102	655,1047 .0097 656,1046 .0101	699,1101 .0099 700,1100 .0102	743,1157 .0098 744,1156 .0101	789,1213 .0099 790,1212 .0102

P = .01 one-sided

P = .02 two-sided

N	M = 27	M = 28	M = 29	M = 30	M = 31
27	608,877 .0098 609,876 .0103				
28	618,894 .0099 619,893 .0104	656,940 .0098 657,939 .0102			

P = .01 one-sided

P = .02 two-sided

N	M = 27	M = 28	M = 29	M = 30	M = 31
29	628,911 .0100 629,910 .0104	666,958 .0097 667,957 .0101	706,1005 .0098 707,1004 .0103		
30	637,929 .0096 638,928 .0100	676,976 .0096 677,975 .0100	717,1023 .0100 718,1022 .0104	758,1072 .0100 759,1071 .0104	
31	647,946 .0097 648,945 .0101	687,993 .0099 688,992 .0103	727,1042 .0097 728,1041 .0101	769,1091 .0099 770,1090 .0103	811,1142 .0097 812,1141 .0101
32	657,963 .0098 658,962 .0102	697,1011 .0098 698,1010 .0102	738,1060 .0098 739,1059 .0102	780,1110 .0099 781,1109 .0103	823,1161 .0099 824,1160 .0103
33	667,980 .0098 668,979 .0102	707,1029 .0097 708,1028 .0101	749,1078 .0100 750,1077 .0104	791,1129 .0098 792,1128 .0102	834,1181 .0097 835,1180 .0101
34	677,997 .0099 678,996 .0103	718,1046 .0100 719,1045 .0104	759,1097 .0097 760,1096 .0101	802,1148 .0098 803,1147 .0102	846,1200 .0099 847,1199 .0102
35	687,1014 .0099 688,1013 .0103	728,1064 .0099 729,1063 .0103	770,1115 .0098 771,1114 .0102	813,1167 .0097 814,1166 .0101	857,1220 .0097 858,1219 .0100
36	697,1031 .0100 698,1030 .0104	738,1082 .0098 739,1081 .0101	781,1133, 0099 782,1132 .0103	824,1186 .0097 825,1185 .0100	869,1239 .0098 870,1238 .0102
37	706,1049 .0097 707,1048 .0100	748,1100 .0097 749,1099 .0100	791,1152 .0097 792,1151 .0100	836,1204 .0100 837,1203 .0103	881,1258 .0100 882,1257 .0103
38	716,1066 .0097 717,1065 .0101	759,1117 .0099 760,1116 .0103	802,1170 .0098 803,1169 .0101	847,1223 .0099 848,1222 .0103	892,1278 .0098 893,1277 .0101
39	726,1083 .0098 727,1082 .0101	769,1135 .0098 770,1134 .0102	813,1188 .0098 814,1187 .0102	858,1242 .0099 859,1241 .0102	904,1297 .0099 905,1296 .0102
40	736,1100 .0098 737,1099 .0102	779,1153 .0097 780,1152 .0100	824,1206 .0099 825,1205 .0103	869,1261 .0098 870,1260 .0101	915,1317 .0097 916,1316 .0100
41	746,1117 .0098 747,1116 .0102	790,1170 .0099 791,1169 .0103	834,1225 .0097 835,1224 .0100	880,1280 .0098 881,1279 .0101	927,1336 .0098 928,1335 .0102
42	756,1134 .0099 757,1133 .0102	800,1188 .0098 801,1187 .0102	845,1243 .0098 846,1242 .0101	891,1299 .0097 892,1298 .0100	939,1355 .0100 940,1354 .0103
43	766,1151 .0099 767,1150 .0102	810,1206 .0097 811,1205 .0100	856,1261 .0099 857,1260 .0102	903,1317 .0100 904,1316 .0103	950,1375 .0098 951,1374 .0101
44	776,1168 .0099 777,1167 .0103	821,1223 .0099 822,1222 .0103	867,1279 .0099 868,1278 .0102	914,1336 .0099 915,1335 .0102	962,1394 .0099 963,1393 .0102
45	786,1185 .0100 787,1184 .0103	831,1241 .0098 832,1240 .0101	878,1297 .0100 879,1296 .0103	925,1355 .0098 926,1354 .0101	974,1413 .0100 975,1412 .0103

P = .01 one-sided

P = .02 two-sided

N	M = 27	M = 28	M = 29	M = 30	M = 31
46	796,1202 .0100 797,1201 .0103	841,1259 .0097 842,1258 .0100	888,1316 .0098 889,1315 .0101	936,1374 .0098 937,1373 .0101	985,1433 .0098 986,1432 .0101
47	805,1220 .0097 806,1219 .0100	852,1276 .0099 853,1275 .0102	899,1334 .0098 900,1333 .0101	947,1393 .0097 948,1392 .0100	997,1452 .0099 998,1451 .0102
48	815,1237 .0097 816,1236 .0100	862,1294 .0098 863,1293 .0101	910,1352 .0099 911,1351 .0102	959,1411 .0100 960,1410 .0102	1008,1472 .0097 1009,1471 .0100
49	825,1254 .0098 826,1253 .0101	872,1312 .0097 873,1311 .0100	921,1370 .0100 922,1369 .0102	970,1430 .0099 971,1429 .0102	1020,1491 .0098 1021,1490 .0101
50	835,1271 .0098 836,1270 .0101	883,1329 .0099 884,1328 .0102	931,1389 .0097 932,1388 .0100	981,1449 .0098 982,1448 .0101	1032,1510 .0099 1033,1509 .0102

P = .01 one-sided

P = .02 two-sided

N	M = 32	M = 33	M = 34	M = 35	M = 36
32	867,1213 .0099 868,1212 .0103				
33	879,1233 .0099 880,1232 .0103	924,1287 .0098 925,1286 .0101			
34	891,1253 .0099 892,1252 .0103	937,1307 .0100 938,1306 .0103	983,1363 .0097 984,1362 .0100		
35	903,1273 .0099 904,1272 .0103	949,1328 .0098 950,1327 .0102	996,1384 .0098 997,1383 .0101	1044,1441 .0097 1045,1440 .0100	
36	915,1293 .0099 916,1292 .0103	961,1349 .0097 962,1348 .0100	1009,1405 .0098 1010,1404 .0101	1058,1462 .0099 1059,1461 .0102	1107,1521 .0097 1108,1520 .0100
37	927,1313 .0099 928,1312 .0103	974,1369 .0099 975,1368 .0102	1022,1426 .0099 1023,1425 .0102	1071,1484 .0098 1072,1483 .0101	1121,1543 .0098 1122,1542 .0101
38	939,1333 .0099 940,1332 .0103	986,1390 .0098 987,1389 .0101	1035,1447 .0099 1036,1446 .0102	1084,1506 .0097 1085,1505 .0100	1135,1565 .0099 1136,1564 .0102
39	951,1353 .0099 952,1352 .0102	999,1410 .0099 1000,1409 .0102	1048,1468 .0099 1049,1467 .0102	1098,1527 .0099 1099,1526 .0102	1149,1587 .0099 1150,1586 .0102
40	963,1373 .0099 964,1372 .0102	1011,1431 .0098 1012,1430 .0101	1061,1489 .0100 1062,1488 .0103	1111,1549 .0098 1112,1548 .0101	1162,1610 .0097 1163,1609 .0100

P = .01 one-sided

P = .02 two-sided

N	M = 32	M = 33	M = 34	M = 35	M = 36
41	975,1393 .0099 976,1392 .0102	1024,1451 .0100 1025,1450 .0103	1074,1510 .0100 1075,1509 .0103	1124,1571 .0098 1125,1570 .0100	1176,1632 .0098 1177,1631 .0101
42	987,1413 .0099 988,1412 .0102	1036,1472 .0098 1037,1471 .0101	1086,1532 .0097 1087,1531 .0100	1138,1592 .0099 1139,1591 .0102	1190,1654 .0099 1191,1653 .0101
43	999,1433 .0099 1000,1432 .0102	1049,1492 .0100 1050,1491 .0103	1099,1553 .0098 1100,1552 .0100	1151,1614 .0098 1152,1613 .0101	1204,1676 .0099 1205,1675 .0102
44	1011,1453 .0099 1012,1452 .0101	1061,1513 .0098 1062,1512 .0101	1112,1574 .0098 1113,1573 .0101	1164,1636 .0098 1165,1635 .0100	1218,1698 .0100 1219,1697 .0102
45	1023,1473 .0098 1024,1472 .0101	1074,1533 .0100 1075,1532 .0103	1125,1595 .0098 1126,1594 .0101	1178,1657 .0099 1179,1656 .0102	1231,1721 .0098 1232,1720 .0100
46	1035,1493 .0098 1036,1492 .0101	1086,1554 .0098 1087,1553 .0101	1138,1616 .0098 1139,1615 .0101	1191,1679 .0098 1192,1678 .0101	1245,1743 .0098 1246,1742 .0101
47	1047,1513 .0098 1048,1512 .0101	1099,1574 .0100 1100,1573 .0102	1151,1637 .0099 1152,1636 .0101	1205,1700 .0100 1206,1699 .0103	1259,1765 .0099 1260,1764 .0101
48	1059,1533 .0098 1060,1532 .0101	1111,1595 .0098 1112,1594 .0101	1164,1658 .0099 1165,1657 .0101	1218,1722 .0099 1219,1721 .0102	1273,1787 .0099 1274,1786 .0102
49	1071,1553 .0098 1072,1552 .0100	1124,1615 .0100 1125,1614 .0102	1177,1679 .0099 1178,1678 .0101	1231,1744 .0098 1232,1743 .0101	1287,1809 .0100 1288,1808 .0102
50	1083,1573 .0098 1084,1572 .0100	1136,1636 .0098 1137,1635 .0101	1190,1700 .0099 1191,1699 .0102	1245,1765 .0100 1246,1764 .0102	1300,1832 .0098 1301,1831 .0100

P = .01 one-sided

P = .02 two-sided

N	M = 37	M = 38	M = 39	M = 40	M = 41
37	1172,1603 .0097 1173,1602 .0100				
38	1187,1625 .0100 1188,1624 .0103	1239,1687 .0098 1240,1686 .0101			
39	1201,1648 .0099 1202,1647 .0102	1254,1710 .0099 1255,1709 .0102	1308,1773 .0099 1309,1772 .0102		
40	1215,1671 .0099 1216,1670 .0102	1268,1734 .0098 1269,1733 .0100	1323,1797 .0099 1324,1796 .0102	1378,1862 .0098 1379,1861 .0100	

P = .01 one-sided

P = .02 two-sided

N	M = 37	M = 38	M = 39	M = 40	M = 41
41	1229,1694 .0098 1230,1693 .0101	1283,1757 .0099 1284,1756 .0101	1338,1821 .0099 1339,1820 .0102	1394,1886 .0099 1395,1885 .0102	1451,1952 .0099 1452,1951 .010[illegible]
42	1243,1717 .0098 1244,1716 .0100	1298,1780 .0100 1299,1779 .0102	1353,1845 .0099 1354,1844 .0101	1409,1911 .0098 1410,1910 .0100	1467,1977 .0099 1468,1976 .0102
43	1258,1739 .0100 1259,1738 .0103	1312,1804 .0098 1313,1803 .0101	1368,1869 .0099 1369,1868, .0101	1425,1935 .0099 1426,1934 .0102	1483,2002 .0100 1484,2001 .0102
44	1272,1762 .0099 1273,1761 .0102	1327,1827 .0099 1328,1826 .0101	1383,1893 .0098 1384,1892 .0101	1440,1960 .0098 1441,1959 .0100	1499,2027 .0100 1500,2026 .0102
45	1286,1785 .0099 1287,1784 .0102	1342,1850 .0100 1343,1849 .0102	1398,1917 .0098 1399,1916 .0101	1456,1984 .0099 1457,1983 .0101	1515,2052 .0100 1516,2051 .0102
46	1300,1808 .0098 1301,1807 .0101	1356,1874 .0098 1357,1873 .0101	1413,1941 .0098 1414,1940 .0100	1471,2009 .0098 1472,2008 .0100	1531,2077 .0100 1532,2076 .0102
47	1314,1831 .0098 1315,1830 .0100	1371,1897 .0099 1372,1896 .0101	1428,1965 .0098 1429,1964 .0100	1487,2033 .0099 1488,2032 .0101	1547,2102 .0100 1548,2101 .0102
48	1329,1853 .0099 1330,1852 .0102	1386,1920 .0100 1387,1919 .0102	1444,1988 .0100 1445,1987 .0102	1503,2057 .0100 1504,2056 .0102	1563,2127 .0100 1564,2126 .0102
49	1343,1876 .0099 1344,1875 .0101	1400,1944 .0098 1401,1943 .0100	1459,2012 .0099 1460,2011 .0102	1518,2082 .0099 1519,2081 .0101	1579,2152 .0100 1580,2151 .0102
50	1357,1899 .0098 1358,1898 .0101	1415,1967 .0099 1416,1966 .0101	1474,2036 .0099 1475,2035 .0101	1534,2106 .0100 1535,2105 .0102	1595,2177 .0100 1596,2176 .0102

N	M = 42	M = 43	M = 44	M = 45	M = 46
42	1525,2045 .0099 1526,2044 .0101				
43	1541,2071 .0098 1542,2070 .0100	1601,2140 .0098 1602,2139 .0100			
44	1558,2096 .0099 1559,2095 .0101	1618,2166 .0099 1619,2165 .0101	1679,2237 .0098 1680,2236 .0100		
45	1574,2122 .0098 1575,2121 .0101	1635,2192 .0099 1636,2191 .0101	1697,2263 .0100 1698,2262 .0102	1759,2336 .0098 1760,2335 .0100	
46	1591,2147 .0100 1592,2146 .0102	1652,2218 .0099 1653,2217 .0102	1714,2290 .0099 1715,2289 .0101	1777,2363 .0099 1778,2362 .0101	1841,2437 .0098 1842,2436 .0100

P = .01 one-sided

P = .02 two-sided

N	M = 42	M = 43	M = 44	M = 45	M = 46
47	1607,2173 .0099	1669,2244 .0100	1731,2317 .0098	1795,2390 .0099	1859,2465 .0098
	1608,2172 .0101	1670,2243 .0102	1732,2316 .0101	1796,2389 .0101	1860,2464 .0100
48	1624,2198 .0100	1686,2270 .0100	1749,2343 .0100	1813,2417 .0100	1878,2492 .0100
	1625,2197 .0102	1687,2269 .0102	1750,2342 .0102	1814,2416 .0102	1879,2491 .0102
49	1640,2224 .0099	1702,2297 .0098	1766,2370 .0099	1830,2445 .0098	1896,2520 .0099
	1641,2223 .0101	1703,2296 .0100	1767,2369 .0101	1831,2444 .0100	1897,2519 .0101
50	1656,2250 .0098	1719,2323 .0098	1783,2397 .0099	1848,2472 .0099	1914,2548 .0099
	1657,2249 .0100	1720,2322 .0100	1784,2396 .0101	1849,2471 .0101	1915,2547 .0101

P = .01 one-sided

P = .02 two-sided

N	M = 47	M = 48	M = 49	M = 50
47	1925,2540 .0099			
	1926,2539 .0101			
48	1944,2568 .0100	2011,2645 .0100		
	1945,2567 .0102	2012,2644 .0102		
49	1962,2597 .0098	2030,2674 .0099	2098,2753 .0098	
	1963,2596 .0100	2031,2673 .0101	2099,2752 .0100	
50	1981,2625 .0099	2049,2703 .0099	2118,2782 .0099	2188,2862 .0099
	1982,2624 .0101	2050,2702 .0101	2119,2781 .0101	2189,2861 .0101

P = .005 one-sided

P = .01 two-sided

N	M = 3	M = 4	M = 5	M = 6	M = 7	M = 8
3	5,16 .0000 6,15 .0500					
4	5,19 .0000 6,18 .0286	9,27 .0000 10,26 .0143				
5	5,22 .0000 6,21 .0179	9,31 .0000 10,30 .0079	15,40 .0040 16,39 .0079			
6	5,25 .0000 6,24 .0119	10,34 .0048 11,33 .0095	16,44 .0043 17,43 .0087	23,55 .0043 24,54 .0076		
7	5,28 .0000 6,27 .0083	10,38 .0030 11,37 .0061	16,49 .0025 17,48 .0051	24,60 .0041 25,59 .0070	32,73 .0035 33,72 .0055	
8	5,31 .0000 6,30 .0061	11,41 .0040 12,40 .0081	17,53 .0031 18,52 .0054	25,65 .0040 26,64 .0063	34,78 .0047 35,77 .0070	43,93 .0035 44,92 .0052
9	6,33 .0045 7,32 .0091	11,45 .0028 12,44 .0056	18,57 .0035 19,56 .0060	26,70 .0038 27,69 .0060	35,84 .0039 36,83 .0058	45,99 .0039 46,98 .0056
10	6,36 .0035 7,35 .0070	12,48 .0040 13,47 .0070	19,61 .0040 20,60 .0063	27,75 .0037 28,74 .0055	37,89 .0048 38,88 .0068	47,105 .0043 48,104 .0058
11	6,39 .0027 7,38 .0055	12,52 .0029 13,51 .0051	20,65 .0043 21,64 .0066	28,80 .0036 29,79 .0052	38,95 .0041 39,94 .0057	49,111 .0046 50,110 .0060
12	7,41 .0044 8,40 .0088	13,55 .0038 14,54 .0066	21,69 .0047 22,68 .0068	30,84 .0048 31,83 .0067	40,100 .0049 41,99 .0065	51,117 .0048 52,116 .0062
13	7,44 .0036 8,43 .0071	13,59 .0029 14,58 .0050	22,73 .0049 23,72 .0070	31,89 .0046 32,88 .0062	41,106 .0042 42,105 .0056	53,123 .0050 54,122 .0063
14	7,47 .0029 8,46 .0059	14,62 .0039 15,61 .0059	22,78 .0036 23,77 .0052	32,94 .0044 33,93 .0059	43,111 .0048 44,110 .0062	54,130 .0041 55,129 .0051
15	8,49 .0049 9,48 .0086	15,65 .0046 16,64 .0070	23,82 .0039 24,81 .0054	33,99 .0042 34,98 .0055	44,117 .0043 45,116 .0054	56,136 .0042 57,135 .0053
16	8,52 .0041 9,51 .0072	15,69 .0037 16,68 .0056	24,86 .0041 25,85 .0056	34,104 .0040 35,103 .0052	46,122 .0048 47,121 .0060	58,142 .0044 59,141 .0054
17	8,55 .0035 9,5[illegible] .00[illegible]	16,72 .0045 17,71 .0063	25,90 .0043 26,89 .0057	36,108 .0049 37,107 .0063	47,128 .0043 48,127 .0053	60,148 .0045 61,147 .0055
18	8,58 .0030 9,57 .0053	16,76 .0037 17,75 .0052	26,94 .0045 27,93 .0059	37,113 .0047 38,112 .0059	49,133 .0047 50,132 .0058	62,154 .0046 63,153 .0056

P = .005 one-sided

P = .01 two-sided

N	M = 3	M = 4	M = 5	M = 6	M = 7	M = 8
19	9,60 .0045 10,59 .0071	17,79 .0043 18,78 .0060	27,98 .0046 28,97 .0060	38,118 .0045 39,117 .0056	50,139 .0042 51,138 .0052	64,160 .0047 65,159 .0056
20	9,63 .0040 10,62 .0062	18,82 .0050 19,81 .0067	28,102 .0048 29,101 .0061	39,123 .0043 40,122 .0053	52,144 .0047 53,143 .0056	66,166 .0048 67,165 .0057
21	9,66 .0035 10,65 .0054	18,86 .0042 19,85 .0056	29,106 .0049 30,105 .0062	40,128 .0042 41,127 .0051	53,150 .0042 54,149 .0051	68,172 .0049 69,171 .0057
22	10,68 .0048 11,67 .0070	19,89 .0047 20,88 .0063	29,111 .0040 30,110 .0051	42,132 .0049 43,131 .0059	55,155 .0046 56,154 .0055	70,178 .0050 71,177 .0058
23	10,71 .0042 11,70 .0062	19,93 .0040 20,92 .0054	30,115 .0042 31,114 .0052	43,137 .0047 44,136 .0057	57,160 .0050 58,159 .0058	71,185 .0043 72,184 .0050
24	10,74 .0038 11,73 .0055	20,96 .0046 21,95 .0059	31,119 .0043 32,118 .0053	44,142 .0045 45,141 .0054	58,166 .0045 59,165 .0053	73,191 .0044 74,190 .0051
25	11,76 .0049 12,75 .0070	20,100 .0040 21,99 .0051	32,123 .0044 33,122 .0054	45,147 .0044 46,146 .0052	60,171 .0049 61,170 .0057	75,197 .0045 76,196 .0051
26	11,79 .0044 12,78 .0063	21,103 .0044 22,102 .0057	33,127 .0045 34,126 .0055	46,152 .0042 47,151 .0050	61,177 .0045 62,176 .0052	77,203 .0045 78,202 .0052
27	11,82 .0039 12,81 .0057	22,106 .0049 23,105 .0062	34,131 .0046 35,130 .0056	48,156 .0048 49,155 .0057	63,182 .0048 64,181 .0055	79,209 .0046 80,208 .0052
28	11,85 .0036 12,84 .0051	22,110 .0043 23,109 .0054	35,135 .0047 36,134 .0057	49,161 .0047 50,160 .0055	64,188 .0044 65,187 .0051	81,215 .0047 82,214 .0053
29	12,87 .0046 13,86 .0063	23,113 .0047 24,112 .0052	36,139 .0048 37,138 .0058	50,166 .0045 51,165 .0053	66,193 .0047 67,192 .0054	83,221 .0047 84,220 .0053
30	12,90 .0042 13,89 .0057	23,117 .0042 24,116 .0052	37,143 .0049 38,142 .0058	51,171 .0044 52,170 .0051	68,198 .0050 69,197 .0057	85,227 .0048 86,226 .0054
31	12,93 .0038 13,92 .0052	24,120 .0046 25,119 .0056	37,148 .0042 38,147 .0050	53,175 .0049 54,174 .0057	69,204 .0046 70,203 .0053	87,233 .0048 88,232 .0054
32	13,95 .0047 14,94 .0063	24,124 .0041 25,123 .0050	38,152 .0043 39,151 .0051	54,180 .0048 55,179 .0055	71,209 .0049 72,208 .0055	89,239 .0049 90,238 .0054
33	13,98 .0043 14,97 .0057	25,127 .0045 26,126 .0054	39,156 .0044 40,155 .0052	55,185 .0046 56,184 .0053	72,215 .0046 73,214 .0052	91,245 .0049 92,244 .0055
34	13,101 .0040 14,100 .0053	26,130 .0049 27,129 .0058	40,160 .0045 41,159 .0053	56,190 .0045 57,189 .0051	74,220 .0043 75,219 .0054	93,251 .0049 94,250 .0055

P = .005 one-sided

P = .01 two-sided

N	M = 3	M = 4	M = 5	M = 6	M = 7	M = 8
35	14,103 .0049	26,134 .0044	41,164 .0046	58,194 .0050	75,226 .0045	95,257 .0050
	15,102 .0063	27,133 .0052	42,163 .0053	59,193 .0057	76,225 .0051	96,256 .0055
36	14,106 .0045	27,137 .0047	42,168 .0047	59,199 .0048	77,231 .0048	96,264 .0045
	15,105 .0058	28,136 .0056	43,167 .0054	60,198 .0055	78,230 .0053	97,263 .0050
37	14,109 .0041	27,141 .0043	43,172 .0048	60,204 .0047	79,236 .0050	98,270 .0046
	15,108 .0054	28,140 .0051	44,171 .0055	61,203 .0053	80,235 .0055	99,269 .0050
38	15,111 .0050	28,144 .0046	44,176 .0048	61,209 .0046	80,242 .0047	100,276 .0046
	16,110 .0063	29,143 .0054	45,175 .0055	62,208 .0052	81,241 .0052	101,275 .0051
39	15,114 .0046	29,147 .0049	45,180 .0049	62,214 .0045	82,247 .0049	102,282 .0046
	16,113 .0058	30,146 .0058	46,179 .0056	63,213 .0050	83,246 .0054	103,281 .0051
40	15,117 .0043	29,151 .0045	46,184 .0050	64,218 .0049	83,253 .0046	104,288 .0047
	16,116 .0054	30,150 .0053	47,183 .0057	65,217 .0055	84,252 .0051	105,287 .0051
41	15,120 .0040	30,154 .0048	46,189 .0044	65,223 .0048	85,258 .0048	106,294 .0047
	16,119 .0051	31,153 .0056	47,188 .0050	66,222 .0053	86,257 .0053	107,293 .0052
42	16,122 .0047	30,158 .0044	47,193 .0045	66,228 .0047	86,264 .0046	108,300 .0048
	17,121 .0058	31,157 .0051	48,192 .0051	67,227 .0052	87,263 .0050	109,299 .0052
43	16,125 .0044	31,161 .0047	48,197 .0046	67,233 .0046	88,269 .0048	110,306 .0048
	17,124 .0055	32,160 .0055	49,196 .0052	68,232 .0051	89,268 .0052	111,305 .0052
44	16,128 .0041	31,165 .0043	49,201 .0046	69,237 .0050	90,274 .0050	112,312 .0048
	17,127 .0051	32,164 .0050	50,200 .0052	70,236 .0055	91,273 .0054	113,311 .0052
45	17,130 .0048	32,168 .0046	50,205 .0047	70,242 .0048	91,280 .0047	114,318 .0048
	18,129 .0059	33,167 .0053	51,204 .0053	71,241 .0053	92,279 .0052	115,317 .0053
46	17,133 .0045	33,171 .0049	51,209 .0048	71,247 .0047	93,285 .0049	116,324 .0049
	18,132 .0055	34,170 .0056	52,208 .0053	72,246 .0052	94,284 .0053	117,323 .0053
47	17,136 .0042	33,175 .0045	52,213 .0048	72,252 .0046	94,291 .0047	118,330 .0049
	18,135 .0052	34,174 .0052	53,212 .0054	73,251 .0051	95,290 .0051	119,329 .0053
48	18,138 .0049	34,178 .0048	53,217 .0049	74,256 .0050	96,296 .0048	120,336 .0049
	19,137 .0059	35,177 .0055	54,216 .0054	75,255 .0055	97,295 .0053	121,335 .0053
49	18,141 .0046	34,182 .0044	54,221 .0049	75,261 .0049	97,302 .0046	122,342 .0049
	19,140 .0056	35,181 .0050	55,220 .0055	76,260 .0054	98,301 .0050	123,341 .0053
50	18,144 .0044	35,185 .0047	55,225 .0050	76,266 .0048	99,307 .0048	124,348 .0050
	19,143 .0053	36,184 .0053	56,224 .0055	77,265 .0052	100,306 .0052	125,347 .0053

P = .005 one-sided

P = .01 two sided

N	M = 9	M = 10	M = 11	M = 12	M = 13	M = 14
9	56,115 .0039 57,114 .0053					
10	58,122 .0038 59,121 .0051	71,139 .0045 72,138 .0057				
11	61,128 .0048 62,127 .0062	73,147 .0040 74,146 .0050	87,166 .0042 88,165 .0052			
12	63,135 .0046 64,134 .0059	76,154 .0045 77,153 .0056	90,174 .0043 91,173 .0053	105,195 .0041 106,194 .0050		
13	65,142 .0045 66,141 .0056	79,161 .0049 80,160 .0060	93,182 .0044 94,181 .0054	109,203 .0048 110,202 .0057	125,226 .0043 126,225 .0051	
14	67,149 .0043 68,148 .0053	81,169 .0044 82,168 .0054	96,190 .0045 97,189 .0054	112,212 .0046 113,211 .0054	129,235 .0046 130,234 .0053	147,259 .0046 148,258 .0053
15	69,156 .0041 70,155 .0050	84,176 .0048 85,175 .0058	99,198 .0046 100,197 .0054	115,221 .0044 116,220 .0051	133,244 .0048 134,243 .0056	151,269 .0046 152,268 .0052
16	72,162 .0048 73,161 .0058	86,184 .0043 87,183 .0052	102,206 .0046 103,205 .0054	119,229 .0049 120,228 .0056	136,254 .0044 137,253 .0050	155,279 .0045 156,278 .0052
17	74,169 .0046 75,168 .0055	89,191 .0047 90,190 .0055	105,214 .0047 106,213 .0054	122,238 .0046 123,237 .0053	140,263 .0046 141,262 .0052	159,289 .0045 160,288 .0051
18	76,176 .0044 77,175 .0053	92,198 .0050 93,197 .0058	108,222 .0047 109,221 .0054	125,247 .0044 126,246 .0051	144,272 .0048 145,271 .0054	163,299 .0045 164,298 .0050
19	78,183 .0043 79,182 .0050	94,206 .0045 95,205 .0053	111,230 .0047 112,229 .0054	129,255 .0048 130,254 .0055	148,281 .0049 149,280 .0056	168,308 .0050 169,307 .0056
20	81,189 .0048 82,188 .0056	97,213 .0048 98,212 .0055	114,238 .0047 115,237 .0054	132,264 .0046 133,263 .0052	151,291 .0045 152,290 .0051	172,318 .0049 173,317 .0055
21	83,196 .0047 84,195 .0054	99,221 .0044 100,220 .0051	117,246 .0047 118,245 .0054	136,272 .0050 137,271 .0056	155,300 .0047 156,299 .0052	176,328 .0048 177,327 .0054
22	85,203 .0045 86,202 .0052	102,228 .0047 103,227 .0053	120,254 .0047 121,253 .0054	139,281 .0048 140,280 .0054	159,309 .0048 160,308 .0053	180,338 .0048 181,337 .0053
23	88,209 .0050 89,208 .0057	105,235 .0049 106,234 .0055	123,262 .0047 124,261 .0053	142,290 .0046 143,289 .0051	163,318 .0049 164,317 .0054	184,348 .0047 185,347 .0052
24	90,216 .0048 91,215 .0055	107,243 .0045 108,242 .0051	126,270 .0047 127,269 .0053	146,298 .0049 147,297 .0055	166,328 .0045 167,327 .0050	188,358 .0046 189,357 .0051

P = .005 one-sided

P = .01 two-sided

N	M = 9	M = 10	M = 11	M = 12	M = 13	M = 14
25	92,223 .0046	110,250 .0047	129,278 .0047	149,307 .0047	170,337 .0047	192,368 .0046
	93,222 .0053	111,249 .0053	130,277 .0053	150,306 .0052	171,336 .0051	193,367 .0050
26	94,230 .0045	113,257 .0049	132,286 .0047	152,316 .0045	174,346 .0048	197,377 .0049
	95,229 .0051	114,256 .0055	133,285 .0053	153,315 .0050	175,345 .0052	198,376 .0054
27	97,236 .0049	115,265 .0046	135,294 .0047	156,324 .0048	178,355 .0049	201,387 .0049
	98,235 .0055	116,264 .0051	136,293 .0052	157,323 .0053	179,354 .0053	202,386 .0053
28	99,243 .0048	118,272 .0048	138,302 .0047	159,333 .0046	182,364 .0050	205,397 .0048
	100,242 .0053	119,271 .0053	139,301 .0052	160,332 .0051	183,363 .0054	206,396 .0052
29	101,250 .0046	121,279 .0050	141,310 .0047	163,341 .0049	185,374 .0046	209,407 .0047
	102,249 .0052	122,278 .0055	142,309 .0052	164,340 .0054	186,373 .0050	210,406 .0051
30	103,257 .0045	123,287 .0046	144,318 .0047	166,350 .0047	189,383 .0047	213,417 .0047
	104,256 .0050	124,286 .0051	145,317 .0052	167,349 .0052	190,382 .0051	214,416 .0051
31	106,263 .0049	126,294 .0048	147,326 .0047	170,358 .0050	193,392 .0048	218,426 .0050
	107,262 .0054	127,293 .0053	148,325 .0052	171,357 .0054	194,391 .0052	219,425 .0054
32	108,270 .0047	129,301 .0050	150,334 .0047	173,367 .0048	197,401 .0049	222,436 .0049
	109,269 .0052	130,300 .0055	151,333 .0051	174,366 .0052	198,400 .0053	223,435 .0053
33	110,277 .0046	131,309 .0047	153,342 .0047	176,376 .0047	201,410 .0050	226,446 .0048
	111,276 .0051	132,308 .0051	154,341 .0051	177,375 .0050	202,409 .0053	227,445 .0052
34	113,283 .0049	134,316 .0048	156,350 .0047	180,384 .0049	204,420 .0047	230,456 .0048
	114,282 .0054	135,315 .0053	157,349 .0051	181,383 .0053	205,419 .0050	231,455 .0051
35	115,290 .0048	137,323 .0050	159,358 .0047	183,393 .0047	208,429 .0047	234,466 .0047
	116,289 .0053	138,322 .0054	160,357 .0051	184,392 .0051	209,428 .0051	235,465 .0050
36	117,297 .0047	139,331 .0047	162,366 .0047	187,401 .0049	212,438 .0048	239,475 .0050
	118,296 .0051	140,330 .0051	163,365 .0050	188,400 .0053	213,437 .0052	240,474 .0053
37	120,303 .0050	142,338 .0048	165,374 .0046	190,410 .0048	216,447 .0049	243,485 .0049
	121,302 .0054	143,337 .0052	166,373 .0050	191,409 .0051	217,446 .0052	244,484 .0052
38	122,310 .0048	145,345 .0050	168,382 .0046	194,418 .0050	220,456 .0049	247,495 .0048
	123,309 .0053	146,344 .0054	169,381 .0050	195,417 .0054	221,455 .0053	248,494 .0052
39	124,317 .0047	147,353 .0047	172,389 .0050	197,427 .0048	224,465 .0050	251,505 .0048
	125,316 .0051	148,352 .0051	173,288 .0054	198,426 .0052	225,464 .0053	252,504 .0051
40	126,324 .0046	150,360 .0048	175,397 .0050	200,436 .0047	227,475 .0047	255,515 .0047
	127,323 .0050	151,359 .0052	176,396 .0053	201,435 .0050	228,474 .0050	256,514 .0050
41	129,330 .0049	153,367 .0050	178,405 .0049	204,444 .0049	231,484 .0048	260,524 .0050
	130,329 .0053	154,366 .0053	179,404 .0053	205,443 .0052	232,483 .0051	261,523 .0053
42	131,337 .0048	155,375 .0047	181,413 .0049	207,453 .0048	235,493 .0048	264,534 .0049
	132,336 .0052	156,374 .0051	182,412 .0053	208,452 .0051	236,492 .0052	265,533 .0052

P = .005 one-sided

P = .01 two-sided

N	M = 9	M = 10	M = 11	M = 12	M = 13	M = 14
43	133,344 .0047 134,343 .0051	158,382 .0048 159,381 .0052	184,421 .0049 185,420 .0053	211,461 .0049 212,460 .0053	239,502 .0049 240,501 .0052	268,544 .0048 269,543 .0051
44	136,350 .0049 137,349 .0053	161,389 .0050 162,388 .0053	187,429 .0049 188,428 .0052	214,470 .0048 215,469 .0051	243,511 .0050 244,510 .0053	272,554 .0048 273,553 .0051
45	138,357 .0048 139,356 .0052	163,397 .0047 164,396 .0051	190,437 .0049 191,436 .0052	218,478 .0050 219,477 .0053	246,521 .0047 247,520 .0050	277,563 .0050 278,562 .0053
46	140,364 .0047 141,363 .0051	166,404 .0048 167,403 .0052	193,445 .0049 194,444 .0052	221,487 .0048 222,486 .0051	250,530 .0048 251,529 .0050	281,573 .0049 282,572 .0052
47	143,370 .0050 144,369 .0053	169,411 .0049 170,410 .0053	196,453 .0049 197,452 .0052	224,496 .0047 225,495 .0050	254,539 .0048 255,538 .0051	285,583 .0049 286,582 .0051
48	145,377 .0049 146,376 .0052	171,419 .0047 172,418 .0051	199,461 .0048 200,460 .0051	228,504 .0049 229,503 .0052	258,548 .0049 259,547 .0051	289,593 .0048 290,592 .0051
49	147,384 .0048 148,383 .0051	174,426 .0048 175,425 .0052	202,469 .0048 203,468 .0051	231,513 .0048 232,512 .0050	262,557 .0049 263,556 .0052	293,603 .0048 294,602 .0050
50	149,391 .0047 150,390 .0050	177,433 .0049 178,432 .0053	205,477 .0048 206,476 .0051	235,521 .0049 236,520 .0052	266,566 .0050 267,565 .0052	298,612 .0050 299,611 .0052

P = .005 one-sided

P = .01 two-sided

N	M = 15	M = 16	M = 17	M = 18	M = 19	M = 20
15	171,294 .0049 172,293 .0056					
16	175,305 .0047 176,304 .0053	196,332 .0048 197,331 .0054				
17	180,315 .0050 181,314 .0056	201,343 .0049 202,342 .0054	223,372 .0047 224,371 .0053			
18	184,326 .0047 185,325 .0053	206,354 .0049 207,353 .0055	228,384 .0046 229,383 .0051	252,414 .0048 253,413 .0053		
19	189,336 .0050 190,335 .0056	210,366 .0045 211,365 .0050	234,395 .0050 235,394 .0055	258,426 .0050 259,425 .0055	283,458 .0050 284,457 .0054	
20	193,347 .0047 194,346 .0053	215,377 .0046 216,376 .0051	239,407 .0049 240,406 .0053	263,439 .0047 264,438 .0051	289,471 .0049 290,470 .0054	315,505 .0047 316,504 .0052

P = .005 one-sided

P = .01 two-sided

N	M = 15	M = 16	M = 17	M = 18	M = 19	M = 20
21	198,357 .0050 199,356 .0055	220,388 .0046 221,387 .0051	244,419 .0047 245,418 .0052	269,451 .0048 270,450 .0053	295,484 .0049 296,483 .0053	322,518 .0050 323,517 .0054
22	202,368 .0047 203,367 .0052	225,399 .0047 226,398 .0051	249,431 .0046 250,430 .0050	275,463 .0049 276,462 .0054	301,497 .0049 302,496 .0053	328,532 .0048 329,531 .0052
23	207,378 .0049 208,377 .0054	230,410 .0047 231,409 .0052	255,442 .0049 256,441 .0053	280,476 .0047 281,475 .0051	307,510 .0048 308,509 .0052	335,545 .0050 336,544 .0053
24	211,389 .0047 212,388 .0051	235,421 .0047 236,420 .0052	260,454 .0048 261,453 .0052	286,488 .0048 287,487 .0052	313,523 .0048 314,522 .0052	341,559 .0048 342,558 .0051
25	216,399 .0049 217,398 .0053	240,432 .0048 241,431 .0052	265,466 .0047 266,465 .0050	292,500 .0049 293,499 .0053	319,536 .0047 320,535 .0051	348,572 .0049 349,571 .0053
26	220,410 .0047 221,409 .0051	245,443 .0048 246,442 .0052	271,477 .0049 272,476 .0053	298,512 .0050 299,511 .0054	325,549 .0047 326,548 .0050	354,586 .0048 355,585 .0051
27	225,420 .0049 226,419 .0053	250,454 .0048 251,453 .0052	276,489 .0048 277,488 .0051	303,525 .0047 304,524 .0051	332,561 .0050 333,560 .0053	361,599 .0049 362,598 .0052
28	229,431 .0046 230,430 .0050	255,465 .0048 256,464 .0052	281,501 .0047 282,500 .0050	309,537 .0048 310,536 .0052	338,574 .0049 339,573 .0053	367,613 .0047 368,612 .0051
29	234,441 .0048 235,440 .0052	260,476 .0049 261,475 .0052	287,512 .0049 288,511 .0052	315,549 .0049 316,548 .0052	344,587 .0049 345,586 .0052	374,626 .0049 375,625 .0052
30	239,451 .0050 240,450 .0053	265,487 .0049 266,486 .0052	292,524 .0048 293,523 .0051	321,561 .0050 322,560 .0053	350,600 .0048 351,599 .0052	380,640 .0047 381,639 .0050
31	243,462 .0048 244,461 .0051	270,498 .0049 271,497 .0052	298,535 .0050 299,534 .0053	326,574 .0047 327,573 .0050	356,613 .0048 357,612 .0051	387,653 .0048 388,652 .0051
32	248,472 .0049 249,471 .0053	275,509 .0049 276,508 .0052	303,547 .0048 304,546 .0052	332,586 .0048 333,585 .0051	362,626 .0047 363,625 .0050	394,666 .0050 395,665 .0053
33	252,483 .0047 253,482 .0050	280,520 .0049 281,519 .0052	308,559 .0047 309,558 .0050	338,598 .0049 339,597 .0052	369,638 .0050 370,637 .0053	400,680 .0048 401,679 .0051
34	257,493 .0048 258,492 .0052	285,531 .0049 286,530 .0052	314,570 .0049 315,569 .0052	344,610 .0049 345,609 .0052	375,651 .0049 376,650 .0052	407,693 .0049 408,692 .0052
35	262,503 .0050 263,502 .0053	290,542 .0049 291,541 .0052	319,582 .0048 320,581 .0051	350,622 .0050 351,621 .0053	381,664 .0049 382,663 .0051	413,707 .0048 414,706 .0050
36	266,514 .0048 267,513 .0051	295,553 .0049 296,552 .0052	325,593 .0050 326,592 .0053	355,635 .0048 356,634 .0050	387,677 .0048 388,676 .0051	420,720 .0049 421,719 .0051
37	271,524 .0049 272,523 .0052	300,564 .0049 301,563 .0052	330,605 .0049 331,604 .0052	361,647 .0048 362,646 .0051	393,690 .0048 394,689 .0050	427,733 .0050 428,732 .0052

P = .005 one-sided

P = .01 two-sided

N	M = 15	M = 16	M = 17	M = 18	M = 19	M = 20
38	275,535 .0047 276,534 .0050	305,575 .0049 306,574 .0052	335,617 .0048 336,616 .0050	367,659 .0049 368,658 .0052	400,702 .0050 401,701 .0052	433,747 .0048 434,746 .0051
39	280,545 .0049 281,544 .0052	310,586 .0049 311,585 .0052	341,628 .0049 342,627 .0052	373,671 .0049 374,670 .0052	406,715 .0049 407,714 .0052	440,760 .0049 441,759 .0051
40	285,555 .0050 286,554 .0053	315,597 .0049 316,596 .0052	346,640 .0048 347,639 .0051	379,683 .0050 380,682 .0052	412,728 .0049 413,727 .0051	447,773 .0050 448,772 .0052
41	289,566 .0048 290,565 .0051	320,608 .0049 321,607 .0052	352,651 .0050 353,650 .0052	384,696 .0048 385,695 .0050	418,741 .0048 419,740 .0051	453,787 .0048 454,786 .0051
42	294,576 .0049 295,575 .0052	325,619 .0049 326,618 .0052	357,663 .0049 358,662 .0051	390,708 .0048 391,707 .0051	424,754 .0048 425,753 .0050	460,800 .0049 461,799 .0052
43	298,587 .0047 299,586 .0050	330,630 .0049 331,629 .0052	362,675 .0048 363,674 .0050	396,720 .0049 397,719 .0051	431,766 .0050 432,765 .0052	466,814 .0048 467,813 .0050
44	303,597 .0049 304,596 .0051	335,641 .0049 336,640 .0052	368,686 .0049 369,685 .0052	402,732 .0049 403,731 .0052	437,779 .0049 438,778 .0051	473,827 .0049 474,826 .0051
45	308,607 .0050 309,606 .0052	340,652 .0049 341,651 .0051	373,698 .0048 374,697 .0051	408,744 .0050 409,743 .0052	443,792 .0049 444,791 .0051	480,840 .0050 481,839 .0052
46	312,618 .0048 313,617 .0051	345,663 .0049 346,662 .0051	379,709 .0050 380,708 .0052	413,757 .0048 414,756 .0050	449,805 .0048 450,804 .0050	486,854 .0048 487,853 .0050
47	317,628 .0049 318,627 .0052	350,674 .0049 351,673 .0051	384,721 .0049 385,720 .0051	419,769 .0048 420,768 .0050	456,817 .0050 457,816 .0052	493,867 .0049 494,866 .0051
48	322,638 .0050 323,637 .0052	355,685 .0049 356,684 .0051	390,732 .0050 391,731 .0052	425,781 .0049 426,780 .0051	462,830 .0049 463,829 .0051	500,880 .0050 501,879 .0052
49	326,649 .0048 327,648 .0051	360,696 .0049 361,695 .0051	395,744 .0049 396,743 .0051	431,793 .0049 432,792 .0051	468,843 .0049 469,842 .0051	506,894 .0048 507,893 .0050
50	331,659 .0049 332,658 .0052	365,707 .0049 366,706 .0051	400,756 .0048 401,755 .0050	437,805 .0049 438,804 .0051	474,856 .0048 475,855 .0050	513,907 .0049 514,906 .0051

P = .005 one-sided

P = .01 two-sided

N	M = 21	M = 22	M = 23	M = 24	M = 25	M = 26
21	349,554 .0046 350,553 .0050					
22	356,568 .0047 357,567 .0050	386,604 .0049 387,603 .0053				

P = .005 one-sided

P = .01 two-sided

N	M = 21	M = 22	M = 23	M = 24	M = 25	M = 26
23	363,582 .0047	393,619 .0048	424,657 .0049			
	364,581 .0051	394,618 .0052	425,656 .0053			
24	370,596 .0048	400,634 .0047	431,673 .0047	464,712 .0050		
	371,595 .0051	401,633 .0051	432,672 .0050	465,711 .0053		
25	377,610 .0048	408,648 .0050	439,688 .0048	472,728 .0049	505,770 .0048	
	378,609 .0051	409,647 .0053	440,687 .0051	473,727 .0053	506,769 .0051	
26	384,624 .0048	415,663 .0048	447,703 .0049	480,744 .0049	514,786 .0049	549,829 .0049
	385,623 .0051	416,662 .0052	448,702 .0052	481,743 .0052	515,785 .0052	550,828 .0052
27	391,638 .0048	422,678 .0047	455,718 .0049	488,760 .0049	522,803 .0048	558,846 .0049
	392,637 .0051	423,677 .0050	456,717 .0053	489,759 .0051	523,802 .0050	559,845 .0052
28	398,652 .0048	430,692 .0049	462,734 .0047	496,776 .0048	531,819 .0049	567,863 .0049
	399,651 .0052	431,691 .0052	463,733 .0050	497,775 .0051	532,818 .0051	568,862 .0052
29	405,666 .0049	437,707 .0048	470,749 .0048	504,792 .0048	540,835 .0050	576,880 .0049
	406,665 .0052	438,706 .0051	471,748 .0051	505,791 .0050	541,834 .0052	577,879 .0052
30	412,680 .0049	444,722 .0047	478,764 .0049	513,807 .0050	548,852 .0048	585,897 .0049
	413,679 .0052	445,721 .0050	479,763 .0051	514,806 .0052	549,851 .0051	586,896 .0052
31	419,694 .0049	452,736 .0049	486,779 .0049	521,823 .0049	557,868 .0049	594,914 .0049
	420,693 .0052	453,735 .0052	487,778 .0052	522,822 .0052	558,867 .0052	595,913 .0052
32	426,708 .0049	459,751 .0048	494,794 .0050	529,839 .0049	565,885 .0048	603,931 .0049
	427,707 .0052	460,750 .0051	495,793 .0052	530,838 .0051	566,884 .0050	604,930 .0052
33	433,722 .0049	467,765 .0050	501,810 .0048	537,855 .0048	574,901 .0049	612,948 .0049
	434,721 .0052	468,764 .0052	502,809 .0050	538,854 .0051	575,900 .0051	613,947 .0051
34	440,736 .0049	474,780 .0048	509,825 .0048	545,871 .0048	583,917 .0049	621,965 .0049
	441,735 .0051	475,779 .0051	510,824 .0051	546,870 .0050	584,916 .0052	622,964 .0051
35	447,750 .0049	482,794 .0050	517,840 .0049	554,886 .0049	591,934 .0048	630,982 .0049
	448,749 .0051	483,793 .0052	518,839 .0051	555,885 .0052	592,933 .0050	631,981 .0051
36	454,764 .0049	489,809 .0049	525,855 .0049	562,902 .0049	600,950 .0049	639,999 .0049
	455,763 .0051	490,808 .0051	526,854 .0051	563,901 .0051	601,949 .0051	640,998 .0051
37	461,778 .0049	496,824 .0048	533,870 .0049	570,918 .0048	609,966 .0050	648,1016 .0049
	462,777 .0051	497,823 .0050	534,869 .0052	571,917 .0051	610,965 .0052	649,1015 .0051
38	468,792 .0049	504,838 .0049	541,885 .0050	578,934 .0048	617,983 .0048	657,1033 .0048
	469,791 .0051	505,837 .0052	542,884 .0052	579,933 .0050	618,982 .0050	658,1032 .0050
39	475,806 .0049	511,853 .0048	548,901 .0048	587,949 .0049	626,999 .0049	666,1050 .0048
	476,805 .0051	512,852 .0051	549,900 .0050	588,948 .0052	627,998 .0051	667,1049 .0050

P = .005 one-sided

P = .01 two-sided

N	M = 21	M = 22	M = 23	M = 24	M = 25	M = 26
40	482,820 .0049	519,867 .0050	556,916 .0048	595,965 .0049	635,1015 .0050	675,1067 .0048
	483,819 .0051	520,866 .0052	557,915 .0050	596,964 .0051	636,1014 .0052	676,1066 .0050
41	489,834 .0049	526,882 .0049	564,931 .0049	603,981 .0048	643,1032 .0048	685,1083 .0050
	490,833 .0051	527,881 .0051	565,930 .0051	604,980 .0050	644,1031 .0050	686,1082 .0052
42	496,848 .0049	534,896 .0050	572,946 .0049	612,996 .0050	652,1048 .0049	694,1100 .0050
	497,847 .0051	535,895 .0052	573,945 .0051	613,995 .0052	653,1047 .0051	695,1099 .0052
43	503,862 .0048	541,911 .0049	580,961 .0049	620,1012 .0049	661,1064 .0049	703,1117 .0049
	504,861 .0051	542,910 .0051	581,960 .0051	621,1011 .0051	662,1063 .0051	704,1116 .0051
44	510,876 .0048	549,925 .0050	588,976 .0049	628,1028 .0049	670,1080 .0050	712,1134 .0049
	511,875 .0051	550,924 .0052	589,975 .0051	629,1027 .0051	671,1079 .0052	713,1133 .0051
45	517,890 .0048	556,940 .0049	596,991 .0050	636,1044 .0048	678,1097 .0049	721,1151 .0049
	518,889 .0050	557,939 .0051	597,990 .0052	637,1043 .0050	679,1096 .0051	722,1150 .0051
46	524,904 .0048	563,955 .0048	604,1006 .0050	645,1059 .0050	687,1113 .0049	730,1168 .0049
	525,903 .0050	564,954 .0050	605,1005 .0052	646,1058 .0051	688,1112 .0051	731,1167 .0051
47	531,918 .0048	571,969 .0049	611,1022 .0048	653,1075 .0049	696,1129 .0050	739,1185 .0049
	532,917 .0050	572,968 .0051	612,1021 .0050	654,1074 .0051	697,1128 .0052	740,1184 .0050
48	538,932 .0048	578,984 .0048	619,1037 .0048	661,1091 .0049	704,1146 .0049	748,1202 .0048
	539,931 .0050	579,983 .0050	620,1036 .0050	662,1090 .0050	705,1145 .0050	749,1201 .0050
49	546,945 .0050	586,998 .0049	627,1052 .0049	670,1106 .0050	713,1162 .0049	758,1218 .0050
	547,944 .0052	587,997 .0051	628,1051 .0051	671,1105 .0052	714,1161 .0051	759,1217 .0052
50	553,959 .0050	593,1013 .0049	635,1067 .0049	678,1122 .0049	722,1178 .0049	767,1235 .0050
	554,958 .0052	594,1012 .0050	636,1066 .0051	679,1121 .0051	723,1177 .0051	768,1234 .0051

P = .005 one-sided

P = .01 two-sided

N	M = 27	M = 28	M = 29	M = 30	M = 31
27	594,891 .0048				
	595,890 .0051				
28	604,908 .0050	641,955 .0048			
	605,907 .0052	642,954 .0050			
29	613,926 .0049	651,973 .0048	690,1021 .0048		
	614,925 .0051	652,972 .0051	691,1020 .0050		
30	622,944 .0048	661,991 .0049	701,1039 .0049	741,1089 .0048	
	623,943 .0050	662,990 .0051	702,1038 .0052	742,1088 .0050	

P = .005 one-sided

P = .01 two-sided

N	M = 27	M = 28	M = 29	M = 30	M = 31
31	632,961 .0049	671,1009 .0049	711,1058 .0049	752,1108 .0049	794,1159 .0049
	633,960 .0052	672,1008 .0051	712,1057 .0051	753,1107 .0051	795,1158 .0051
32	641,979 .0048	681,1027 .0049	721,1077 .0048	763,1127 .0049	805,1179 .0048
	642,978 .0050	682,1026 .0052	722,1076 .0050	764,1126 .0052	806,1178 .0050
33	651,996 .0049	691,1045 .0050	732,1095 .0050	773,1147 .0048	816,1199 .0048
	652,995 .0052	692,1044 .0052	733,1094 .0052	774,1146 .0050	817,1198 .0050
34	660,1014 .0048	701,1063 .0050	742,1114 .0049	784,1166 .0049	828,1218 .0050
	661,1013 .0051	702,1062 .0052	743,1113 .0051	785,1165 .0051	829,1217 .0052
35	670,1031 .0049	710,1082 .0048	752,1133 .0049	795,1185 .0049	839,1238 .0050
	671,1030 .0052	711,1081 .0050	753,1132 .0051	796,1184 .0051	840,1237 .0052
36	679,1049 .0048	720,1100 .0048	763,1151 .0050	806,1204 .0050	850,1258 .0049
	680,1048 .0051	721,1099 .0050	764,1150 .0052	807,1203 .0052	851,1257 .0051
37	689,1066 .0050	730,1118 .0048	773,1170 .0049	816,1224 .0048	861,1278 .0049
	690,1065 .0052	731,1117 .0050	774,1169 .0051	817,1223 .0050	862,1277 .0051
38	698,1084 .0049	740,1136 .0049	783,1189 .0049	827,1243 .0049	872,1298 .0049
	699,1083 .0051	741,1135 .0051	784,1188 .0051	828,1242 .0051	873,1297 .0050
39	708,1101 .0050	750,1154 .0049	794,1207 .0050	838,1262 .0049	883,1318 .0048
	709,1100 .0052	751,1153 .0051	795,1206 .0052	839,1261 .0051	884,1317 .0050
40	717,1119 .0049	760,1172 .0049	804,1226 .0049	849,1281 .0050	895,1337 .0050
	718,1118 .0050	761,1171 .0051	805,1225 .0051	850,1280 .0051	896,1336 .0052
41	727,1136 .0049	770,1190 .0049	814,1245 .0049	860,1300 .0050	906,1357 .0049
	728,1135 .0051	771,1189 .0051	815,1244 .0050	861,1299 .0052	907,1356 .0051
42	736,1154 .0049	780,1208 .0049	825,1263 .0050	870,1320 .0049	917,1377 .0049
	737,1153 .0050	781,1207 .0051	826,1262 .0052	871,1319 .0050	918,1376 .0051
43	746,1171 .0049	790,1226 .0049	835,1282 .0049	881,1339 .0049	928,1397 .0049
	747,1170 .0051	791,1225 .0051	836,1281 .0051	882,1338 .0051	929,1396 .0050
44	755,1189 .0048	800,1244 .0049	845,1301 .0049	892,1358 .0049	939,1417 .0048
	756,1188 .0050	801,1243 .0051	846,1300 .0050	893,1357 .0051	940,1416 .0050
45	765,1206 .0049	810,1262 .0049	856,1319 .0050	903,1377 .0050	951,1436 .0050
	766,1205 .0051	811,1261 .0051	857,1318 .0051	904,1376 .0051	952,1435 .0051
46	774,1224 .0048	820,1280 .0050	866,1338 .0049	914,1396 .0050	962,1456 .0049
	775,1223 .0050	821,1279 .0051	867,1337 .0051	915,1395 .0052	963,1455 .0051
47	784,1241 .0049	830,1298 .0050	877,1356 .0050	924,1416 .0049	973,1476 .0049
	785,1240 .0051	831,1297 .0051	878,1355 .0052	925,1415 .0050	974,1475 .0051

P = .005 one-sided

P = .01 two-sided

N	M = 27	M = 28	M = 29	M = 30	M = 31
48	794,1258 .0050 795,1257 .0052	840,1316 .0050 841,1315 .0051	887,1375 .0049 888,1374 .0051	935,1435 .0049 936,1434 .0051	984,1496 .0049 985,1495 .0050
49	803,1276 .0049 804,1275 .0051	850,1334 .0050 851,1333 .0051	897,1394 .0049 898,1393 .0050	946,1454 .0049 947,1453 .0051	996,1515 .0050 997,1514 .0051
50	813,1293 .0050 814,1292 .0051	860,1352 .0050 861,1351 .0051	908,1412 .0050 909,1411 .0051	957,1473 .0050 958,1472 .0051	1007,1535 .0049 1008,1534 .0051

P = .005 one-sided

P = .01 two-sided

N	M = 32	M = 33	M = 34	M = 35	M = 36
32	849,1231 .0049 850,1230 .0051				
33	860,1252 .0048 861,1251 .0050	905,1306 .0048 906,1305 .0050			
34	872,1272 .0049 873,1271 .0051	917,1327 .0048 918,1326 .0050	964,1382 .0050 965,1381 .0051		
35	884,1292 .0050 885,1291 .0052	929,1348 .0049 930,1347 .0050	976,1404 .0049 977,1403 .0051	1024,1461 .0049 1025,1460 .0051	
36	895,1313 .0049 896,1312 .0051	941,1369 .0049 942,1368 .0050	989,1425 .0050 990,1424 .0052	1037,1483 .0049 1038,1482 .0051	1086,1542 .0049 1087,1541 .0051
37	907,1333 .0050 908,1332 .0052	953,1390 .0049 954,1389 .0050	1001,1447 .0049 1002,1446 .0051	1050,1505 .0050 1051,1504 .0051	1099,1565 .0049 1100,1564 .0050
38	918,1354 .0049 919,1353 .0050	965,1411 .0049 966,1410 .0050	1013,1469 .0048 1014,1468 .0050	1063,1527 .0050 1064,1526 .0052	1113,1587 .0050 1114,1586 .0051
39	930,1374 .0049 931,1373 .0051	977,1432 .0049 978,1431 .0050	1026,1490 .0049 1027,1489 .0051	1075,1550 .0049 1076,1549 .0050	1126,1610 .0049 1127,1609 .0051
40	941,1395 .0048 942,1394 .0050	989,1453 .0049 990,1452 .0050	1038,1512 .0049 1039,1511 .0050	1088,1572 .0049 1089,1571 .0050	1139,1633 .0049 1140,1632 .0050
41	953,1415 .0049 954,1414 .0051	1001,1474 .0048 1002,1473 .0050	1051,1533 .0050 1052,1532 .0051	1101,1594 .0049 1102,1593 .0051	1153,1655 .0050 1154,1654 .0051

P = .005 one-sided

P = .01 two-sided

N	M = 32	M = 33	M = 34	M = 35	M = 36
42	965,1435 .0050 966,1434 .0051	1013,1495 .0048 1014,1494 .0050	1063,1555 .0049 1064,1554 .0050	1114,1616 .0049 1115,1615 .0051	1166,1678 .0049 1167,1677 .0051
43	976,1456 .0049 977,1455 .0050	1026,1515 .0050 1027,1514 .0052	1076,1576 .0050 1077,1575 .0051	1127,1638 .0049 1128,1637 .0051	1179,1701 .0049 1180,1700 .0050
44	988,1476 .0049 989,1475 .0051	1038,1536 .0050 1039,1535 .0051	1088,1598 .0049 1089,1597 .0050	1140,1660 .0049 1141,1659 .0051	1193,1723 .0050 1194,1722 .0051
45	1000,1496 .0050 1001,1495 .0051	1050,1557 .0050 1051,1556 .0051	1101,1619 .0050 1102,1618 .0051	1153,1682 .0050 1154,1681 .0051	1206,1746 .0049 1207,1745 .0051
46	1011,1517 .0049 1012,1516 .0050	1062,1578 .0050 1063,1577 .0051	1113,1641 .0049 1114,1640 .0050	1166,1704 .0050 1167,1703 .0051	1219,1769 .0049 1220,1768 .0050
47	1023,1537 .0049 1024,1536 .0051	1074,1599 .0049 1075,1598 .0051	1126,1662 .0050 1127,1661 .0051	1179,1726 .0050 1180,1725 .0051	1233,1791 .0050 1234,1790 .0051
48	1035,1557 .0050 1036,1556 .0051	1086,1620 .0049 1087,1619 .0051	1138,1684 .0049 1139,1683 .0050	1192,1748 .0050 1193,1747 .0051	1246,1814 .0049 1247,1813 .0051
49	1046,1578 .0049 1047,1577 .0050	1098,1641 .0049 1099,1640 .0051	1151,1705 .0050 1152,1704 .0051	1205,1770 .0050 1206,1769 .0051	1259,1837 .0049 1260,1836 .0050
50	1058,1598 .0049 1059,1597 .0051	1110,1662 .0049 1111,1661 .0051	1163,1727 .0049 1164,1726 .0050	1218,1792 .0050 1219,1791 .0051	1273,1859 .0050 1274,1858 .0051

P = .005 one-sided

P = .01 two-sided

N	M = 37	M = 38	M = 39	M = 40	M = 41
37	1150,1625 .0049 1151,1624 .0051				
38	1164,1648 .0050 1165,1647 .0051	1216,1710 .0049 1217,1709 .0051			
39	1177,1672 .0049 1178,1671 .0050	1230,1734 .0049 1231,1733 .0051	1284,1797 .0050 1285,1796 .0051		
40	1191,1695 .0049 1192,1694 .0050	1244,1758 .0049 1245,1757 .0050	1298,1822 .0049 1299,1821 .0050	1353,1887 .0049 1354,1886 .0050	
41	1205,1718 .0049 1206,1717 .0051	1258,1782 .0049 1259,1781 .0050	1313,1846 .0050 1314,1845 .0051	1368,1912 .0049 1369,1911 .0050	1425,1978 .0050 1426,1977 .0051

P = .005 one-sided

P = .01 two-sided

N	M = 37	M = 38	M = 39	M = 40	M = 41
42	1219,1741 .0050 1220,1740 .0051	1273,1805 .0050 1274,1804 .0051	1327,1871 .0049 1328,1870 .0050	1383,1937 .0049 1384,1936 .0050	1440,2004 .0049 1441,2003 .0051
43	1232,1765 .0049 1233,1764 .0050	1287,1829 .0050 1288,1828 .0051	1342,1895 .0049 1343,1894 .0051	1398,1962 .0049 1399,1961 .0050	1456,2029 .0050 1457,2028 .0051
44	1246,1788 .0049 1247,1787 .0050	1301,1853 .0049 1302,1852 .0051	1357,1919 .0050 1358,1918 .0051	1413,1987 .0049 1414,1986 .0050	1471,2055 .0049 1472,2054 .0051
45	1260,1811 .0049 1261,1810 .0051	1315,1877 .0049 1316,1876 .0051	1371,1944 .0049 1372,1943 .0050	1428,2012 .0049 1429,2011 .0050	1487,2080 .0050 1488,2079 .0051
46	1274,1834 .0050 1275,1833 .0051	1329,1901 .0049 1330,1900 .0050	1386,1968 .0050 1387,1967 .0051	1443,2037 .0049 1444,2036 .0050	1502,2106 .0049 1503,2105 .0051
47	1288,1857 .0050 1289,1856 .0051	1344,1924 .0050 1345,1923 .0051	1401,1992 .0050 1402,1991 .0051	1459,2061 .0050 1460,2060 .0051	1518,2131 .0050 1519,2130 .0051
48	1301,1881 .0049 1302,1880 .0050	1358,1948 .0050 1359,1947 .0051	1415,2017 .0049 1416,2016 .0050	1474,2086 .0050 1475,2085 .0051	1533,2157 .0049 1534,2156 .0051
49	1315,1904 .0049 1316,1903 .0050	1372,1972 .0049 1373,1971 .0051	1430,2041 .0050 1431,2040 .0051	1489,2111 .0050 1490,2110 .0051	1549,2182 .0050 1550,2181 .0051
50	1329,1927 .0049 1330,1926 .0051	1386,1996 .0049 1387,1995 .0050	1444,2066 .0049 1445,2065 .0050	1504,2136 .0050 1505,2135 .0051	1564,2208 .0049 1565,2207 .0051

P = .005 one-sided

P = .01 two-sided

N	M = 42	M = 43	M = 44	M = 45	M = 46
42	1498,2072 .0049 1499,2071 .0051				
43	1514,2098 .0050 1515,2097 .0051	1573,2168 .0049 1574,2167 .0050			
44	1530,2124 .0050 1531,2123 .0051	1590,2194 .0050 1591,2193 .0051	1650,2266 .0049 1651,2265 .0050		
45	1546,2150 .0050 1547,2149 .0051	1606,2221 .0050 1607,2220 .0051	1667,2293 .0049 1668,2292 .0051	1729,2366 .0049 1730,2365 .0050	
46	1562,2176 .0050 1563,2175 .0051	1622,2248 .0049 1623,2247 .0050	1684,2320 .0050 1685,2319 .0051	1746,2394 .0049 1747,2393 .0050	1810,2468 .0049 1811,2467 .0050

P = .005 one-sided

P = .01 two-sided

N	M = 42	M = 43	M = 44	M = 45	M = 46
47	1578,2202 .0050 1579,2201 .0051	1639,2274 .0050 1640,2273 .0051	1701,2347 .0050 1702,2346 .0051	1764,2421 .0050 1765,2420 .0051	1828,2496 .0050 1829,2495 .0051
48	1593,2229 .0049 1594,2228 .0050	1655,2301 .0049 1656,2300 .0051	1717,2375 .0049 1718,2374 .0050	1781,2449 .0049 1782,2448 .0051	1845,2525 .0049 1846,2524 .0050
49	1609,2255 .0049 1610,2254 .0050	1671,2328 .0049 1672,2327 .0050	1734,2402 .0049 1735,2401 .0050	1798,2477 .0049 1799,2476 .0050	1863,2553 .0049 1864,2552 .0050
50	1625,2281 .0049 1626,2280 .0050	1688,2354 .0050 1689,2353 .0051	1751,2429 .0049 1752,2428 .0050	1815,2505 .0049 1816,2504 .0050	1881,2581 .0050 1882,2580 .0051

P = .005 one-sided

P = .01 two-sided

N	M = 47	M = 48	M = 49	M = 50
47	1893,2572 .0050 1894,2571 .0051			
48	1911,2601 .0049 1912,2600 .0051	1978,2678 .0050 1979,2677 .0051		
49	1929,2630 .0049 1930,2629 .0050	1996,2708 .0049 1997,2707 .0050	2064,2787 .0049 2065,2786 .0051	
50	1947,2659 .0049 1948,2658 .0050	2015,2737 .0050 2016,2736 .0051	2083,2817 .0049 2084,2816 .0051	2152,2898 .0049 2153,2897 .0050

TABLE II. PROBABILITY LEVELS FOR THE WILCOXON SIGNED RANK TEST

N = 5

T	P
0	.0313
1	.0625
2	.0938
3	.1563
4	.2188
5	.3125
6	.4063
7	.5000

N = 6

T	P
0	.0156
1	.0313
2	.0469
3	.0781
4	.1094
5	.1563
6	.2188
7	.2813
8	.3438
9	.4219
10	.5000

N = 7

T	P
0	.0078
1	.0156
2	.0234
*3	.0391
4	.0547
5	.0781
6	.1094
7	.1484
8	.1875
9	.2344
10	.2891
11	.3438
12	.4063
13	.4688
14	.5313

N = 8

T	P
0	.0039
1	.0078
2	.0117
3	.0195
4	.0273
*5	.0391
6	.0547
7	.0742
8	.0977
9	.1250
10	.1563
11	.1914
12	.2305
13	.2734
14	.3203
15	.3711
16	.4219
17	.4727
18	.5273

N = 9

T	P
0	.0020
1	.0039
2	.0059
3	.0098
4	.0137
5	.0195
6	.0273
7	.0371
*8	.0488
9	.0645
10	.0820
11	.1016
12	.1250
13	.1504
14	.1797
15	.2129
16	.2480
17	.2852
18	.3262
19	.3672
20	.4102
21	.4551
22	.5000

N = 10

T	P
0	.0010
1	.0020
2	.0029
3	.0049
4	.0068
5	.0098
6	.0137
7	.0186
8	.0244
9	.0322
*10	.0420
11	.0527
12	.0654
13	.0801
14	.0967
15	.1162
16	.1377
17	.1611
18	.1875
19	.2158
20	.2461
21	.2783
22	.3125
23	.3477
24	.3848
25	.4229
26	.4609
27	.5000

N = 11

T	P
0	.0005
1	.0010
2	.0015
3	.0024
4	.0034
5	.0049
6	.0068
7	.0093
8	.0122
9	.0161
10	.0210
11	.0269
12	.0337
*13	.0415
14	.0508
15	.0615
16	.0737
17	.0874
18	.1030
19	.1201
20	.1392
21	.1602
22	.1826
23	.2065
24	.2324
25	.2598
26	.2886
27	.3188
28	.3501
29	.3823
30	.4155
31	.4492
32	.4829
33	.5171

N = 12

T	P
0	.0002
1	.0005
2	.0007
3	.0012
4	.0017
5	.0024
6	.0034
7	.0046
8	.0061
9	.0081
10	.0105
11	.0134
12	.0171
13	.0212
14	.0261
15	.0320
16	.0386
*17	.0461
18	.0549
19	.0647
20	.0757
21	.0881
22	.1018
23	.1167
24	.1331
25	.1506
26	.1697
27	.1902
28	.2119
29	.2349
30	.2593
31	.2847
32	.3110
33	.3386
34	.3667
35	.3955
36	.4250
37	.4548
38	.4849
39	.5151

N = 13

T	P
0	.0001
1	.0002
2	.0004
3	.0006
4	.0009
5	.0012
6	.0017
7	.0023
8	.0031
9	.0040
10	.0052
11	.0067
12	.0085
13	.0107
14	.0133
15	.0164
16	.0199
17	.0239
18	.0287
19	.0341
20	.0402
*21	.0471
22	.0549
23	.0636
24	.0732
25	.0839
26	.0955
27	.1082
28	.1219
29	.1367
30	.1527
31	.1698
32	.1879
33	.2072
34	.2274
35	.2487
36	.2709
37	.2939
38	.3177
39	.3424
40	.3677
41	.3934
42	.4197
43	.4463
44	.4730
45	.5000

N = 14

T	P
0	.0001
2	.0002
3	.0003
4	.0004
5	.0006
6	.0009
7	.0012
8	.0015
9	.0020
10	.0026
11	.0034
12	.0043
13	.0054
14	.0067
15	.0083
16	.0101
17	.0123
18	.0148
19	.0176
20	.0209
21	.0247
22	.0290
23	.0338
24	.0392
*25	.0453
26	.0520
27	.0594
28	.0676
29	.0765
30	.0863
31	.0969
32	.1083
33	.1206
34	.1338
35	.1479
36	.1629
37	.1788
38	.1955
39	.2131
40	.2316
41	.2508
42	.2708
43	.2915
44	.3129
45	.3349
46	.3574
47	.3804
48	.4039
49	.4276
50	.4516
51	.4758
52	.5000

N = 15

T	P
1	.0001
3	.0002
5	.0003
6	.0004
7	.0006
8	.0008
9	.0010
10	.0013
11	.0017
12	.0021
13	.0027
14	.0034
15	.0042
16	.0051
17	.0062
18	.0075
19	.0090
20	.0108
21	.0128
22	.0151
23	.0177
24	.0206
25	.0240
26	.0277
27	.0319
28	.0365
29	.0416
*30	.0473
31	.0535
32	.0603
33	.0677
34	.0757
35	.0844
36	.0938
37	.1039
38	.1147
39	.1262
40	.1384
41	.1514
42	.1651
43	.1796
44	.1947
45	.2106
46	.2271
47	.2444
48	.2622
49	.2807
50	.2997
51	.3193
52	.3394
53	.3599
54	.3808
55	.4020
56	.4235
57	.4452
58	.4670
59	.4890
60	.5110

N = 16

T	P
3	.0001
5	.0002
7	.0003
8	.0004
9	.0005
10	.0007
11	.0008
12	.0011
13	.0013
14	.0017
15	.0021
16	.0026
17	.0031
18	.0038
19	.0046
20	.0055
21	.0065
22	.0078
23	.0091
24	.0107
25	.0125
26	.0145
27	.0168
28	.0193
29	.0222
30	.0253
31	.0288
32	.0327
33	.0370
34	.0416
*35	.0467
36	.0523
37	.0583
38	.0649
39	.0719
40	.0795
41	.0877
42	.0964
43	.1057
44	.1156
45	.1261
46	.1372
47	.1489
48	.1613
49	.1742
50	.1877
51	.2019
52	.2166
53	.2319
54	.2477
55	.2641
56	.2809
57	.2983
58	.3161
59	.3343
60	.3529
61	.3718
62	.3910
63	.4104
64	.4301
65	.4500
66	.4699
67	.4900
68	.5100

N = 17

T	P
4	.0001
8	.0002
9	.0003
11	.0004
12	.0005
13	.0007
14	.0008
15	.0010
16	.0013
17	.0016
18	.0019
19	.0023
20	.0028
21	.0033
22	.0040
23	.0047
24	.0055
25	.0064
26	.0075
27	.0087
28	.0101
29	.0116
30	.0133
31	.0153
32	.0174
33	.0198
34	.0224
35	.0253
36	.0284
37	.0319
38	.0357
39	.0398
40	.0443
*41	.0492
42	.0544
43	.0601
44	.0662
45	.0727
46	.0797
47	.0871
48	.0950
49	.1034
50	.1123
51	.1218
52	.1317
53	.1421
54	.1530
55	.1645
56	.1764
57	.1889
58	.2019
59	.2153
60	.2293
61	.2437
62	.2585
63	.2738
64	.2895
65	.3056
66	.3221
67	.3389
68	.3559
69	.3733
70	.3910
71	.4088
72	.4268
73	.4450
74	.463
75	.481
76	.500

N = 18

T	P
6	.000
10	.000
12	.000
14	.000
15	.000
16	.000
17	.000
18	.001
19	.001
20	.001
21	.001
22	.002
23	.002
24	.002
25	.003
26	.003
27	.004
28	.005
29	.006
30	.006
31	.008
32	.009
33	.010
34	.011
35	.013
36	.015
37	.017
38	.019
39	.021
40	.024
41	.026
42	.030
43	.033
44	.036
45	.040
46	.044
*47	.049
48	.054
49	.059
50	.064
51	.070
52	.077
53	.083
54	.090

N = 18

T	P
5	.0982
6	.1061
7	.1144
8	.1231
9	.1323
0	.1419
1	.1519
2	.1624
3	.1733
4	.1846
5	.1964
6	.2086
7	.2211
8	.2341
9	.2475
0	.2613
1	.2754
2	.2899
3	.3047
4	.3198
5	.3353
6	.3509
7	.3669
8	.3830
9	.3994
0	.4159
1	.4325
2	.4493
3	.4661
4	.4831
5	.5000

N = 19

T	P
9	.0001
13	.0002
15	.0003
17	.0004
18	.0005
19	.0006
20	.0007
21	.0008
22	.0010
23	.0012
24	.0014
25	.0017
26	.0020
27	.0023
28	.0027
29	.0031

N = 19

T	P
30	.0036
31	.0041
32	.0047
33	.0054
34	.0062
35	.0070
36	.0080
37	.0090
38	.0102
39	.0115
40	.0129
41	.0145
42	.0162
43	.0180
44	.0201
45	.0223
46	.0247
47	.0273
48	.0301
49	.0331
50	.0364
51	.0399
52	.0437
*53	.0478
54	.0521
55	.0567
56	.0616
57	.0668
58	.0723
59	.0782
60	.0844
61	.0909
62	.0978
63	.1051
64	.1127
65	.1206
66	.1290
67	.1377
68	.1467
69	.1562
70	.1660
71	.1762
72	.1868
73	.1977
74	.2090
75	.2207
76	.2327
77	.2450
78	.2576

N = 19

T	P
79	.2706
80	.2839
81	.2974
82	.3113
83	.3254
84	.3397
85	.3543
86	.3690
87	.3840
88	.3991
89	.4144
90	.4298
91	.4453
92	.4609
93	.4765
94	.4922
95	.5078

N = 20

T	P
11	.0001
16	.0002
19	.0003
20	.0004
22	.0005
23	.0006
24	.0007
25	.0008
26	.0010
27	.0012
28	.0014
29	.0016
30	.0018
31	.0021
32	.0024
33	.0028
34	.0032
35	.0036
36	.0042
37	.0047
38	.0053
39	.0060
40	.0068
41	.0077
42	.0086
43	.0096
44	.0107
45	.0120
46	.0133
47	.0148

N = 20

T	P
48	.0164
49	.0181
50	.0200
51	.0220
52	.0242
53	.0266
54	.0291
55	.0319
56	.0348
57	.0379
58	.0413
59	.0448
*60	.0487
61	.0527
62	.0570
63	.0615
64	.0664
65	.0715
66	.0768
67	.0825
68	.0884
69	.0947
70	.1012
71	.1081
72	.1153
73	.1227
74	.1305
75	.1387
76	.1471
77	.1559
78	.1650
79	.1744
80	.1841
81	.1942
82	.2045
83	.2152
84	.2262
85	.2375
86	.2490
87	.2608
88	.2729
89	.2853
90	.2979
91	.3108
92	.3238
93	.3371
94	.3506
95	.3643
96	.3781

N = 20

T	P
97	.3921
98	.4062
99	.4204
100	.4347
101	.4492
102	.4636
103	.4782
104	.4927
105	.5073

N = 21

T	P
14	.0001
20	.0002
22	.0003
24	.0004
26	.0005
27	.0006
28	.0007
29	.0008
30	.0009
31	.0011
32	.0012
33	.0014
34	.0016
35	.0019
36	.0021
37	.0024
38	.0028
39	.0031
40	.0036
41	.0040
42	.0045
43	.0051
44	.0057
45	.0063
46	.0071
47	.0079
48	.0088
49	.0097
50	.0108
51	.0119
52	.0132
53	.0145
54	.0160
55	.0175
56	.0192
57	.0210
58	.0230
59	.0251
60	.0273

N = 21

T	P
61	.0298
62	.0323
63	.0351
64	.0380
65	.0411
66	.0444
*67	.0479
68	.0516
69	.0555
70	.0597
71	.0640
72	.0686
73	.0735
74	.0786
75	.0839
76	.0895
77	.0953
78	.1015
79	.1078
80	.1145
81	.1214
82	.1286
83	.1361
84	.1439
85	.1519
86	.1602
87	.1688
88	.1777
89	.1869
90	.1963
91	.2060
92	.2160
93	.2262
94	.2367
95	.2474
96	.2584
97	.2696
98	.2810
99	.2927
100	.3046
101	.3166
102	.3289
103	.3414
104	.3540
105	.3667
106	.3796
107	.3927
108	.4058
109	.4191

N = 21	
T	P
110	.4324
111	.4459
112	.4593
113	.4729
114	.4864
115	.5000

N = 22	
18	.0001
23	.0002
26	.0003
29	.0004
30	.0005
32	.0006
33	.0007
34	.0008
35	.0010
36	.0011
37	.0013
38	.0014
39	.0016
40	.0018
41	.0021
42	.0023
43	.0026
44	.0030
45	.0033
46	.0037
47	.0042
48	.0046
49	.0052
50	.0057
51	.0064
52	.0070
53	.0078
54	.0086
55	.0095
56	.0104
57	.0115
58	.0126
59	.0138
60	.0151
61	.0164
62	.0179
63	.0195
64	.0212
65	.0231
66	.0250

N = 22	
T	P
67	.0271
68	.0293
69	.0317
70	.0342
71	.0369
72	.0397
73	.0427
74	.0459
*75	.0492
76	.0527
77	.0564
78	.0603
79	.0644
80	.0687
81	.0733
82	.0780
83	.0829
84	.0881
85	.0935
86	.0991
87	.1050
88	.1111
89	.1174
90	.1240
91	.1308
92	.1378
93	.1451
94	.1527
95	.1604
96	.1685
97	.1767
98	.1853
99	.1940
100	.2030
101	.2122
102	.2217
103	.2314
104	.2413
105	.2514
106	.2618
107	.2723
108	.2830
109	.2940
110	.3051
111	.3164
112	.3278
113	.3394
114	.3512
115	.3631

N = 22	
T	P
116	.3751
117	.3873
118	.3995
119	.4119
120	.4243
121	.4368
122	.4494
123	.4620
124	.4746
125	.4873
126	.5000

N = 23	
21	.0001
28	.0002
31	.0003
33	.0004
35	.0005
36	.0006
38	.0007
39	.0008
40	.0009
41	.0011
42	.0012
43	.0014
44	.0015
45	.0017
46	.0019
47	.0022
48	.0024
49	.0027
50	.0030
51	.0034
52	.0037
53	.0041
54	.0046
55	.0051
56	.0056
57	.0061
58	.0068
59	.0074
60	.0082
61	.0089
62	.0098
63	.0107
64	.0117
65	.0127
66	.0138
67	.0150

N = 23	
T	P
68	.0163
69	.0177
70	.0192
71	.0208
72	.0224
73	.0242
74	.0261
75	.0281
76	.0303
77	.0325
78	.0349
79	.0374
80	.0401
81	.0429
82	.0459
*83	.0490
84	.0523
85	.0557
86	.0593
87	.0631
88	.0671
89	.0712
90	.0755
91	.0801
92	.0848
93	.0897
94	.0948
95	.1001
96	.1056
97	.1113
98	.1172
99	.1234
100	.1297
101	.1363
102	.1431
103	.1501
104	.1573
105	.1647
106	.1723
107	.1802
108	.1883
109	.1965
110	.2050
111	.2137
112	.2226
113	.2317
114	.2410
115	.2505
116	.2601

N = 23	
T	P
117	.2700
118	.2800
119	.2902
120	.3005
121	.3110
122	.3217
123	.3325
124	.3434
125	.3545
126	.3657
127	.3770
128	.3884
129	.3999
130	.4115
131	.4231
132	.4348
133	.4466
134	.4584
135	.4703
136	.4822
137	.4941
138	.5060

N = 24	
25	.0001
32	.0002
36	.0003
38	.0004
40	.0005
42	.0006
43	.0007
44	.0008
45	.0009
46	.0010
47	.0011
48	.0013
49	.0014
50	.0016
51	.0018
52	.0020
53	.0022
54	.0024
55	.0027
56	.0029
57	.0033
58	.0036
59	.0040
60	.0044
61	.0048

N = 24	
T	P
62	.005
63	.005
64	.006
65	.006
66	.007
67	.008
68	.008
69	.009
70	.010
71	.011
72	.012
73	.013
74	.014
75	.015
76	.017
77	.018
78	.019
79	.021
80	.022
81	.024
82	.026
83	.028
84	.030
85	.032
86	.034
87	.036
88	.039
89	.042
90	.044
*91	.047
92	.050
93	.053
94	.057
95	.060
96	.064
97	.0678
98	.0717
99	.0758
100	.080
101	.0844
102	.089
103	.0938
104	.0987
105	.1038
106	.1091
107	.1146
108	.1203
109	.1261
110	.1322

N = 24

T	P
11	.1384
12	.1448
13	.1515
14	.1583
15	.1653
16	.1724
7	.1798
18	.1874
19	.1951
0	.2031
21	.2112
2	.2195
3	.2279
24	.2366
5	.2454
6	.2544
27	.2635
8	.2728
9	.2823
0	.2919
1	.3017
32	.3115
3	.3216
34	.3317
5	.3420
36	.3524
37	.3629
8	.3735
39	.3841
0	.3949
1	.4058
2	.4167
3	.4277
4	.4387
5	.4498
6	.4609
7	.4721
8	.4832
9	.4944
50	.5056

N = 25

T	P
29	.0001
37	.0002
41	.0003
43	.0004
45	.0005
47	.0006
48	.0007

N = 25

T	P
50	.0008
51	.0009
52	.0010
53	.0011
54	.0013
55	.0014
56	.0015
57	.0017
58	.0019
59	.0021
60	.0023
61	.0025
62	.0028
63	.0031
64	.0034
65	.0037
66	.0040
67	.0044
68	.0048
69	.0053
70	.0057
71	.0062
72	.0068
73	.0074
74	.0080
75	.0087
76	.0094
77	.0101
78	.0110
79	.0118
80	.0128
81	.0137
82	.0148
83	.0159
84	.0171
85	.0183
86	.0197
87	.0211
88	.0226
89	.0241
90	.0258
91	.0275
92	.0294
93	.0313
94	.0334
95	.0355
96	.0377
97	.0401
98	.0426

N = 25

T	P
99	.0452
*100	.0479
101	.0507
102	.0537
103	.0567
104	.0600
105	.0633
106	.0668
107	.0705
108	.0742
109	.0782
110	.0822
111	.0865
112	.0909
113	.0954
114	.1001
115	.1050
116	.1100
117	.1152
118	.1205
119	.1261
120	.1317
121	.1376
122	.1436
123	.1498
124	.1562
125	.1627
126	.1694
127	.1763
128	.1833
129	.1905
130	.1979
131	.2054
132	.2131
133	.2209
134	.2289
135	.2371
136	.2454
137	.2539
138	.2625
139	.2712
140	.2801
141	.2891
142	.2983
143	.3075
144	.3169
145	.3264
146	.3360
147	.3458

N = 25

T	P
148	.3556
149	.3655
150	.3755
151	.3856
152	.3957
153	.4060
154	.4163
155	.4266
156	.4370
157	.4474
158	.4579
159	.4684
160	.4789
161	.4895
162	.5000

N = 26

T	P
34	.0001
42	.0002
46	.0003
49	.0004
51	.0005
53	.0006
55	.0007
56	.0008
57	.0009
58	.0010
59	.0011
60	.0012
61	.0013
62	.0015
63	.0016
64	.0018
65	.0020
66	.0021
67	.0023
68	.0026
69	.0028
70	.0031
71	.0033
72	.0036
73	.0040
74	.0043
75	.0047
76	.0051
77	.0055
78	.0060
79	.0065
80	.0070

N = 26

T	P
81	.0076
82	.0082
83	.0088
84	.0095
85	.0102
86	.0110
87	.0118
88	.0127
89	.0136
90	.0146
91	.0156
92	.0167
93	.0179
94	.0191
95	.0204
96	.0217
97	.0232
98	.0247
99	.0263
100	.0279
101	.0297
102	.0315
103	.0334
104	.0355
105	.0376
106	.0398
107	.0421
108	.0445
109	.0470
*110	.0497
111	.0524
112	.0553
113	.0582
114	.0613
115	.0646
116	.0679
117	.0714
118	.0750
119	.0787
120	.0825
121	.0865
122	.0907
123	.0950
124	.0994
125	.1039
126	.1086
127	.1135
128	.1185
129	.1236

N = 26

T	P
130	.1289
131	.1344
132	.1399
133	.1457
134	.1516
135	.1576
136	.1638
137	.1702
138	.1767
139	.1833
140	.1901
141	.1970
142	.2041
143	.2114
144	.2187
145	.2262
146	.2339
147	.2417
148	.2496
149	.2577
150	.2658
151	.2741
152	.2826
153	.2911
154	.2998
155	.3085
156	.3174
157	.3264
158	.3355
159	.3447
160	.3539
161	.3633
162	.3727
163	.3822
164	.3918
165	.4014
166	.4111
167	.4208
168	.4306
169	.4405
170	.4503
171	.4602
172	.4702
173	.4801
174	.4900
175	.5000

N = 27

T	P
39	.0001
47	.0002
52	.0003
55	.0004
57	.0005
59	.0006
61	.0007
62	.0008
64	.0009
65	.0010
66	.0011
67	.0012
68	.0014
69	.0015
70	.0016
71	.0018
72	.0019
73	.0021
74	.0023
75	.0025
76	.0027
77	.0030
78	.0032
79	.0035
80	.0038
81	.0041
82	.0044
83	.0048
84	.0052
85	.0056
86	.0060
87	.0065
88	.0070
89	.0075
90	.0081
91	.0087
92	.0093
93	.0100
94	.0107
95	.0115
96	.0123
97	.0131
98	.0140
99	.0150
100	.0159
101	.0170
102	.0181
103	.0193
104	.0205

N = 27

T	P
105	.0218
106	.0231
107	.0246
108	.0260
109	.0276
110	.0292
111	.0309
112	.0327
113	.0346
114	.0366
115	.0386
116	.0407
117	.0430
118	.0453
*119	.0477
120	.0502
121	.0528
122	.0555
123	.0583
124	.0613
125	.0643
126	.0674
127	.0707
128	.0741
129	.0776
130	.0812
131	.0849
132	.0888
133	.0927
134	.0968
135	.1010
136	.1054
137	.1099
138	.1145
139	.1193
140	.1242
141	.1292
142	.1343
143	.1396
144	.1450
145	.1506
146	.1563
147	.1621
148	.1681
149	.1742
150	.1804
151	.1868
152	.1932
153	.1999

N = 27

T	P
154	.2066
155	.2135
156	.2205
157	.2277
158	.2349
159	.2423
160	.2498
161	.2574
162	.2652
163	.2730
164	.2810
165	.2890
166	.2972
167	.3055
168	.3138
169	.3223
170	.3308
171	.3395
172	.3482
173	.3570
174	.3659
175	.3748
176	.3838
177	.3929
178	.4020
179	.4112
180	.4204
181	.4297
182	.4390
183	.4483
184	.4577
185	.4670
186	.4764
187	.4859
188	.4953
189	.5047

N = 28

T	P
44	.0001
53	.0002
58	.0003
61	.0004
64	.0005
66	.0006
68	.0007
69	.0008
70	.0009
72	.0010
73	.0011

N = 28

T	P
74	.0012
75	.0013
76	.0015
77	.0016
78	.0017
79	.0019
80	.0020
81	.0022
82	.0024
83	.0026
84	.0028
85	.0030
86	.0033
87	.0035
88	.0038
89	.0041
90	.0044
91	.0048
92	.0051
93	.0055
94	.0059
95	.0064
96	.0068
97	.0073
98	.0078
99	.0084
100	.0089
101	.0096
102	.0102
103	.0109
104	.0116
105	.0124
106	.0132
107	.0140
108	.0149
109	.0159
110	.0168
111	.0179
112	.0190
113	.0201
114	.0213
115	.0226
116	.0239
117	.0252
118	.0267
119	.0282
120	.0298
121	.0314
122	.0331

N = 28

T	P
123	.0349
124	.0368
125	.0387
126	.0407
127	.0428
128	.0450
129	.0473
*130	.0496
131	.0521
132	.0546
133	.0573
134	.0600
135	.0628
136	.0657
137	.0688
138	.0719
139	.0751
140	.0785
141	.0819
142	.0855
143	.0891
144	.0929
145	.0968
146	.1008
147	.1049
148	.1091
149	.1135
150	.1180
151	.1225
152	.1273
153	.1321
154	.1370
155	.1421
156	.1473
157	.1526
158	.1580
159	.1636
160	.1693
161	.1751
162	.1810
163	.1870
164	.1932
165	.1995
166	.2059
167	.2124
168	.2190
169	.2257
170	.2326
171	.2395

N = 28

T	P
172	.246
173	.253
174	.261
175	.268
176	.275
177	.283
178	.291
179	.299
180	.306
181	.314
182	.322
183	.330
184	.339
185	.347
186	.355
187	.364
188	.372
189	.381
190	.389
191	.398
192	.407
193	.415
194	.424
195	.433
196	.442
197	.451
198	.459
199	.468
200	.477
201	.486
202	.495
203	.504

N = 29

T	P
50	.000
59	.000
65	.000
68	.000
71	.000
73	.000
75	.000
76	.000
78	.000
79	.001
80	.001
81	.001
82	.001
83	.001
84	.001

N = 29 T	P
5	.0016
6	.0018
7	.0019
8	.0021
9	.0022
0	.0024
1	.0026
2	.0028
3	.0030
4	.0032
5	.0035
6	.0037
7	.0040
8	.0043
9	.0046
0	.0049
1	.0053
2	.0057
3	.0061
4	.0065
5	.0069
6	.0074
7	.0079
8	.0084
9	.0089
0	.0095
1	.0101
2	.0108
3	.0115
4	.0122
5	.0129
6	.0137
7	.0145
8	.0154
9	.0163
20	.0173
21	.0183
2	.0193
23	.0204
24	.0216
25	.0228
26	.0240
27	.0253
28	.0267
29	.0281
30	.0296
31	.0311
32	.0328
33	.0344

N = 29 T	P
134	.0362
135	.0380
136	.0399
137	.0418
138	.0439
139	.0460
*140	.0482
141	.0504
142	.0528
143	.0552
144	.0577
145	.0603
146	.0630
147	.0658
148	.0687
149	.0716
150	.0747
151	.0778
152	.0811
153	.0844
154	.0879
155	.0914
156	.0951
157	.0988
158	.1027
159	.1066
160	.1107
161	.1149
162	.1191
163	.1235
164	.1280
165	.1326
166	.1373
167	.1421
168	.1471
169	.1521
170	.1572
171	.1625
172	.1679
173	.1733
174	.1789
175	.1846
176	.1904
177	.1963
178	.2023
179	.2085
180	.2147
181	.2210
182	.2274

N = 29 T	P
183	.2340
184	.2406
185	.2473
186	.2541
187	.2611
188	.2681
189	.2752
190	.2824
191	.2896
192	.2970
193	.3044
194	.3120
195	.3196
196	.3272
197	.3350
198	.3428
199	.3507
200	.3586
201	.3666
202	.3747
203	.3828
204	.3909
205	.3991
206	.4074
207	.4157
208	.4240
209	.4324
210	.4408
211	.4492
212	.4576
213	.4661
214	.4745
215	.4830
216	.4915
217	.5000

N = 30 T	P
55	.0001
66	.0002
71	.0003
75	.0004
78	.0005
80	.0006
82	.0007
84	.0008
85	.0009
87	.0010
88	.0011
89	.0012

N = 30 T	P
90	.0013
91	.0014
92	.0015
93	.0016
94	.0017
95	.0019
96	.0020
97	.0022
98	.0023
99	.0025
100	.0027
101	.0029
102	.0031
103	.0033
104	.0036
105	.0038
106	.0041
107	.0044
108	.0047
109	.0050
110	.0053
111	.0057
112	.0060
113	.0064
114	.0068
115	.0073
116	.0077
117	.0082
118	.0087
119	.0093
120	.0098
121	.0104
122	.0110
123	.0117
124	.0124
125	.0131
126	.0139
127	.0147
128	.0155
129	.0164
130	.0173
131	.0182
132	.0192
133	.0202
134	.0213
135	.0225
136	.0236
137	.0249
138	.0261

N = 30 T	P
139	.0275
140	.0288
141	.0303
142	.0318
143	.0333
144	.0349
145	.0366
146	.0384
147	.0402
148	.0420
149	.0440
150	.0460
*151	.0481
152	.0502
153	.0524
154	.0547
155	.0571
156	.0595
157	.0621
158	.0647
159	.0674
160	.0701
161	.0730
162	.0759
163	.0790
164	.0821
165	.0853
166	.0886
167	.0920
168	.0955
169	.0990
170	.1027
171	.1065
172	.1103
173	.1143
174	.1183
175	.1225
176	.1267
177	.1311
178	.1355
179	.1400
180	.1447
181	.1494
182	.1543
183	.1592
184	.1642
185	.1694
186	.1746
187	.1799

N = 30 T	P
188	.1854
189	.1909
190	.1965
191	.2022
192	.2081
193	.2140
194	.2200
195	.2261
196	.2323
197	.2386
198	.2449
199	.2514
200	.2579
201	.2646
202	.2713
203	.2781
204	.2849
205	.2919
206	.2989
207	.3060
208	.3132
209	.3204
210	.3277
211	.3351
212	.3425
213	.3500
214	.3576
215	.3652
216	.3728
217	.3805
218	.3883
219	.3961
220	.4039
221	.4118
222	.4197
223	.4276
224	.4356
225	.4436
226	.4516
227	.4596
228	.4677
229	.4758
230	.4838
231	.4919
232	.5000

N = 31	
T	P
62	.0001
73	.0002
79	.0003
82	.0004
85	.0005
88	.0006
90	.0007
92	.0008
93	.0009
95	.0010
96	.0011
97	.0012
98	.0013
100	.0015
101	.0016
102	.0017
103	.0018
104	.0019
105	.0021
106	.0022
107	.0024
108	.0025
109	.0027
110	.0029
111	.0031
112	.0033
113	.0035
114	.0038
115	.0040
116	.0043
117	.0046
118	.0049
119	.0052
120	.0055
121	.0059
122	.0062
123	.0066
124	.0070
125	.0074
126	.0079
127	.0083
128	.0088
129	.0093
130	.0099
131	.0105
132	.0110
133	.0117
134	.0123
135	.0130

N = 31	
T	P
136	.0137
137	.0145
138	.0152
139	.0161
140	.0169
141	.0178
142	.0187
143	.0197
144	.0207
145	.0217
146	.0228
147	.0239
148	.0251
149	.0263
150	.0276
151	.0289
152	.0303
153	.0317
154	.0332
155	.0347
156	.0363
157	.0380
158	.0397
159	.0414
160	.0433
161	.0451
162	.0471
*163	.0491
164	.0512
165	.0533
166	.0555
167	.0578
168	.0602
169	.0626
170	.0651
171	.0677
172	.0703
173	.0730
174	.0758
175	.0787
176	.0817
177	.0847
178	.0878
179	.0910
180	.0943
181	.0977
182	.1012
183	.1047
184	.1083

N = 31	
T	P
185	.1120
186	.1158
187	.1197
188	.1237
189	.1278
190	.1319
191	.1362
192	.1405
193	.1450
194	.1495
195	.1541
196	.1588
197	.1636
198	.1685
199	.1734
200	.1785
201	.1836
202	.1889
203	.1942
204	.1996
205	.2051
206	.2107
207	.2164
208	.2221
209	.2280
210	.2339
211	.2399
212	.2460
213	.2521
214	.2584
215	.2647
216	.2711
217	.2776
218	.2841
219	.2907
220	.2974
221	.3042
222	.3110
223	.3179
224	.3248
225	.3318
226	.3388
227	.3460
228	.3531
229	.3603
230	.3676
231	.3749
232	.3823
233	.3897

N = 31	
T	P
234	.3971
235	.4046
236	.4121
237	.4196
238	.4272
239	.4348
240	.4424
241	.4500
242	.4577
243	.4654
244	.4731
245	.4807
246	.4884
247	.4961
248	.5039

N = 32	
68	.0001
80	.0002
86	.0003
90	.0004
93	.0005
96	.0006
98	.0007
100	.0008
102	.0009
103	.0010
104	.0011
106	.0012
107	.0013
108	.0014
109	.0015
110	.0016
111	.0017
112	.0018
113	.0019
114	.0021
115	.0022
116	.0024
117	.0025
118	.0027
119	.0029
120	.0030
121	.0032
122	.0034
123	.0037
124	.0039
125	.0041
126	.0044

N = 32	
T	P
127	.0047
128	.0050
129	.0053
130	.0056
131	.0059
132	.0063
133	.0066
134	.0070
135	.0074
136	.0078
137	.0083
138	.0087
139	.0092
140	.0097
141	.0103
142	.0108
143	.0114
144	.0120
145	.0126
146	.0133
147	.0140
148	.0147
149	.0155
150	.0162
151	.0171
152	.0179
153	.0188
154	.0197
155	.0207
156	.0217
157	.0227
158	.0238
159	.0249
160	.0260
161	.0272
162	.0285
163	.0298
164	.0311
165	.0325
166	.0339
167	.0354
168	.0369
169	.0385
170	.0402
171	.0419
172	.0436
173	.0454
174	.0473
*175	.0492

N = 32	
T	P
176	.051
177	.053
178	.055
179	.057
180	.059
181	.062
182	.064
183	.066
184	.069
185	.071
186	.074
187	.077
188	.080
189	.082
190	.085
191	.088
192	.091
193	.095
194	.098
195	.101
196	.105
197	.108
198	.112
199	.115
200	.119
201	.123
202	.127
203	.130
204	.135
205	.139
206	.143
207	.147
208	.151
209	.156
210	.160
211	.165
212	.170
213	.174
214	.179
215	.184
216	.189
217	.194
218	.200
219	.205
220	.210
221	.216
222	.221
223	.227
224	.232

N = 32 T	P
5	.2384
6	.2442
7	.2500
8	.2560
9	.2620
0	.2680
1	.2742
2	.2804
3	.2867
4	.2930
5	.2994
6	.3058
7	.3124
8	.3189
9	.3256
0	.3323
1	.3390
2	.3458
3	.3526
4	.3595
5	.3664
6	.3734
7	.3804
8	.3875
9	.3946
0	.4017
1	.4088
2	.4160
3	.4232
4	.4305
5	.4377
6	.4450
7	.4523
8	.4596
9	.4669
0	.4743
1	.4816
2	.4890
3	.4963
4	.5037

N = 33 T	P
5	.0001
8	.0002
4	.0003
8	.0004
2	.0005
4	.0006
7	.0007

N = 33 T	P
109	.0008
111	.0009
112	.0010
114	.0011
115	.0012
117	.0014
118	.0015
119	.0016
120	.0017
121	.0018
122	.0019
123	.0020
124	.0021
125	.0023
126	.0024
127	.0026
128	.0027
129	.0029
130	.0031
131	.0033
132	.0035
133	.0037
134	.0039
135	.0041
136	.0044
137	.0047
138	.0049
139	.0052
140	.0055
141	.0058
142	.0061
143	.0065
144	.0068
145	.0072
146	.0076
147	.0080
148	.0085
149	.0089
150	.0094
151	.0099
152	.0104
153	.0109
154	.0115
155	.0121
156	.0127
157	.0133
158	.0140
159	.0146
160	.0154

N = 33 T	P
161	.0161
162	.0169
163	.0177
164	.0185
165	.0194
166	.0203
167	.0212
168	.0222
169	.0232
170	.0242
171	.0253
172	.0264
173	.0276
174	.0288
175	.0300
176	.0313
177	.0326
178	.0340
179	.0354
180	.0369
181	.0384
182	.0399
183	.0415
184	.0432
185	.0449
186	.0467
*187	.0485
188	.0503
189	.0523
190	.0543
191	.0563
192	.0584
193	.0605
194	.0628
195	.0650
196	.0674
197	.0698
198	.0722
199	.0748
200	.0774
201	.0800
202	.0827
203	.0855
204	.0884
205	.0913
206	.0943
207	.0974
208	.1005
209	.1038

N = 33 T	P
210	.1070
211	.1104
212	.1138
213	.1173
214	.1209
215	.1245
216	.1283
217	.1321
218	.1359
219	.1399
220	.1439
221	.1480
222	.1522
223	.1564
224	.1607
225	.1651
226	.1696
227	.1742
228	.1788
229	.1835
230	.1882
231	.1931
232	.1980
233	.2030
234	.2080
235	.2132
236	.2184
237	.2237
238	.2290
239	.2344
240	.2399
241	.2454
242	.2511
243	.2567
244	.2625
245	.2683
246	.2742
247	.2801
248	.2861
249	.2921
250	.2982
251	.3044
252	.3106
253	.3169
254	.3232
255	.3296
256	.3360
257	.3425
258	.3490

N = 33 T	P
259	.3555
260	.3621
261	.3688
262	.3755
263	.3822
264	.3889
265	.3957
266	.4025
267	.4094
268	.4162
269	.4231
270	.4300
271	.4370
272	.4439
273	.4509
274	.4579
275	.4649
276	.4719
277	.4789
278	.4859
279	.4930
280	.5000

N = 34 T	P
83	.0001
96	.0002
102	.0003
107	.0004
110	.0005
113	.0006
116	.0007
118	.0008
120	.0009
121	.0010
123	.0011
124	.0012
126	.0013
127	.0014
128	.0015
129	.0016
130	.0017
131	.0018
132	.0019
133	.0020
134	.0022
135	.0023
136	.0024
137	.0026
138	.0027

N = 34 T	P
139	.0029
140	.0031
141	.0033
142	.0034
143	.0036
144	.0038
145	.0041
146	.0043
147	.0045
148	.0048
149	.0051
150	.0053
151	.0056
152	.0059
153	.0062
154	.0066
155	.0069
156	.0073
157	.0077
158	.0080
159	.0085
160	.0089
161	.0093
162	.0098
163	.0103
164	.0108
165	.0113
166	.0119
167	.0124
168	.0130
169	.0137
170	.0143
171	.0150
172	.0157
173	.0164
174	.0171
175	.0179
176	.0187
177	.0195
178	.0204
179	.0213
180	.0222
181	.0232
182	.0242
183	.0252
184	.0263
185	.0274
186	.0285
187	.0297

N = 34

T	P
188	.0309
189	.0322
190	.0335
191	.0348
192	.0362
193	.0376
194	.0391
195	.0406
196	.0421
197	.0437
198	.0454
199	.0471
*200	.0488
201	.0506
202	.0525
203	.0544
204	.0563
205	.0583
206	.0604
207	.0625
208	.0647
209	.0669
210	.0692
211	.0715
212	.0739
213	.0764
214	.0789
215	.0815
216	.0841
217	.0868
218	.0896
219	.0924
220	.0953
221	.0982
222	.1013
223	.1043
224	.1075
225	.1107
226	.1140
227	.1173
228	.1208
229	.1243
230	.1278
231	.1314
232	.1351
233	.1389
234	.1427
235	.1466
236	.1506

N = 34

T	P
237	.1546
238	.1587
239	.1629
240	.1671
241	.1714
242	.1758
243	.1802
244	.1848
245	.1893
246	.1940
247	.1987
248	.2035
249	.2083
250	.2133
251	.2182
252	.2233
253	.2284
254	.2336
255	.2388
256	.2441
257	.2494
258	.2549
259	.2603
260	.2659
261	.2715
262	.2771
263	.2828
264	.2886
265	.2944
266	.3003
267	.3062
268	.3121
269	.3181
270	.3242
271	.3303
272	.3365
273	.3427
274	.3489
275	.3552
276	.3615
277	.3678
278	.3742
279	.3806
280	.3871
281	.3936
282	.4001
283	.4066
284	.4132
285	.4198

N = 34

T	P
286	.4264
287	.4330
288	.4396
289	.4463
290	.4530
291	.4597
292	.4664
293	.4731
294	.4798
295	.4865
296	.4933
297	.5000

N = 35

T	P
90	.0001
104	.0002
111	.0003
116	.0004
120	.0005
123	.0006
125	.0007
127	.0008
129	.0009
131	.0010
133	.0011
134	.0012
135	.0013
137	.0014
138	.0015
139	.0016
140	.0017
141	.0018
142	.0019
143	.0020
144	.0021
145	.0023
146	.0024
147	.0025
148	.0027
149	.0028
150	.0030
151	.0032
152	.0033
153	.0035
154	.0037
155	.0039
156	.0041
157	.0043
158	.0046

N = 35

T	P
159	.0048
160	.0051
161	.0053
162	.0056
163	.0059
164	.0062
165	.0065
166	.0068
167	.0072
168	.0075
169	.0079
170	.0083
171	.0087
172	.0091
173	.0096
174	.0100
175	.0105
176	.0110
177	.0115
178	.0120
179	.0126
180	.0132
181	.0137
182	.0144
183	.0150
184	.0157
185	.0164
186	.0171
187	.0178
188	.0186
189	.0194
190	.0202
191	.0210
192	.0219
193	.0228
194	.0238
195	.0247
196	.0257
197	.0268
198	.0278
199	.0289
200	.0301
201	.0312
202	.0325
203	.0337
204	.0350
205	.0363
206	.0377
207	.0391

N = 35

T	P
208	.0405
209	.0420
210	.0435
211	.0451
212	.0467
*213	.0484
214	.0501
215	.0518
216	.0536
217	.0555
218	.0574
219	.0593
220	.0613
221	.0634
222	.0655
223	.0676
224	.0698
225	.0721
226	.0744
227	.0768
228	.0792
229	.0817
230	.0842
231	.0868
232	.0894
233	.0921
234	.0948
235	.0977
236	.1006
237	.1035
238	.1065
239	.1096
240	.1127
241	.1159
242	.1191
243	.1225
244	.1258
245	.1293
246	.1327
247	.1363
248	.1399
249	.1436
250	.1474
251	.1512
252	.1550
253	.1590
254	.1630
255	.1670
256	.1712

N = 35

T	P
257	.175
258	.179
259	.183
260	.188
261	.192
262	.197
263	.201
264	.206
265	.211
266	.215
267	.220
268	.225
269	.230
270	.235
271	.240
272	.245
273	.250
274	.255
275	.261
276	.266
277	.271
278	.277
279	.282
280	.288
281	.293
282	.299
283	.305
284	.310
285	.316
286	.322
287	.328
288	.334
289	.339
290	.345
291	.351
292	.357
293	.364
294	.370
295	.376
296	.382
297	.388
298	.394
299	.401
300	.407
301	.413
302	.419
303	.426
304	.432
305	.438

N = 35	
T	P
06	.4453
07	.4517
08	.4581
09	.4646
10	.4710
11	.4774
12	.4839
13	.4903
14	.4968
15	.5032

N = 36	
98	.0001
13	.0002
20	.0003
25	.0004
29	.0005
32	.0006
35	.0007
37	.0008
39	.0009
41	.0010
43	.0011
44	.0012
46	.0013
47	.0014
48	.0015
49	.0016
50	.0017
52	.0018
53	.0020
54	.0021
55	.0022
56	.0023
57	.0024
58	.0026
59	.0027
60	.0028
61	.0030
62	.0032
63	.0033
64	.0035
65	.0037
66	.0039
67	.0041
68	.0043
69	.0045
70	.0047
71	.0050

N = 36	
T	P
172	.0052
173	.0055
174	.0058
175	.0060
176	.0063
177	.0066
178	.0070
179	.0073
180	.0076
181	.0080
182	.0084
183	.0088
184	.0092
185	.0096
186	.0100
187	.0105
188	.0110
189	.0114
190	.0119
191	.0125
192	.0130
193	.0136
194	.0142
195	.0148
196	.0154
197	.0160
198	.0167
199	.0174
200	.0181
201	.0189
202	.0196
203	.0204
204	.0213
205	.0221
206	.0230
207	.0239
208	.0248
209	.0258
210	.0268
211	.0278
212	.0289
213	.0299
214	.0311
215	.0322
216	.0334
217	.0346
218	.0359
219	.0372
220	.0385

N = 36	
T	P
221	.0399
222	.0413
223	.0427
224	.0442
225	.0457
226	.0473
*227	.0489
228	.0505
229	.0522
230	.0540
231	.0558
232	.0576
233	.0594
234	.0614
235	.0633
236	.0653
237	.0674
238	.0695
239	.0716
240	.0739
241	.0761
242	.0784
243	.0808
244	.0832
245	.0856
246	.0881
247	.0907
248	.0933
249	.0960
250	.0987
251	.1015
252	.1043
253	.1072
254	.1102
255	.1132
256	.1162
257	.1194
258	.1225
259	.1258
260	.1291
261	.1324
262	.1358
263	.1393
264	.1428
265	.1464
266	.1500
267	.1537
268	.1575
269	.1613

N = 36	
T	P
270	.1651
271	.1691
272	.1731
273	.1771
274	.1812
275	.1854
276	.1896
277	.1939
278	.1982
279	.2026
280	.2070
281	.2115
282	.2161
283	.2207
284	.2253
285	.2301
286	.2348
287	.2397
288	.2445
289	.2495
290	.2544
291	.2595
292	.2645
293	.2697
294	.2748
295	.2801
296	.2853
297	.2906
298	.2960
299	.3014
300	.3069
301	.3123
302	.3179
303	.3234
304	.3290
305	.3347
306	.3404
307	.3461
308	.3518
309	.3576
310	.3634
311	.3693
312	.3752
313	.3811
314	.3870
315	.3930
316	.3989
317	.4050
318	.4110

N = 36	
T	P
319	.4170
320	.4231
321	.4292
322	.4353
323	.4414
324	.4475
325	.4537
326	.4598
327	.4660
328	.4722
329	.4783
330	.4845
331	.4907
332	.4969
333	.5031

N = 37	
107	.0001
122	.0002
130	.0003
135	.0004
139	.0005
142	.0006
145	.0007
147	.0008
149	.0009
151	.0010
153	.0011
155	.0012
156	.0013
158	.0014
159	.0015
160	.0016
161	.0017
162	.0018
163	.0019
164	.0020
165	.0021
166	.0022
167	.0023
168	.0024
169	.0025
170	.0027
171	.0028
172	.0030
173	.0031
174	.0033
175	.0034
176	.0036

N = 37	
T	P
177	.0038
178	.0040
179	.0042
180	.0044
181	.0046
182	.0048
183	.0050
184	.0053
185	.0055
186	.0058
187	.0061
188	.0063
189	.0066
190	.0069
191	.0073
192	.0076
193	.0079
194	.0083
195	.0087
196	.0090
197	.0094
198	.0099
199	.0103
200	.0107
201	.0112
202	.0117
203	.0122
204	.0127
205	.0132
206	.0137
207	.0143
208	.0149
209	.0155
210	.0161
211	.0168
212	.0174
213	.0181
214	.0189
215	.0196
216	.0203
217	.0211
218	.0219
219	.0228
220	.0236
221	.0245
222	.0254
223	.0264
224	.0273
225	.0283

N = 37

T	P
226	.0294
227	.0304
228	.0315
229	.0326
230	.0338
231	.0350
232	.0362
233	.0374
234	.0387
235	.0400
236	.0414
237	.0428
238	.0442
239	.0456
240	.0471
*241	.0487
242	.0503
243	.0519
244	.0535
245	.0552
246	.0570
247	.0587
248	.0606
249	.0624
250	.0643
251	.0663
252	.0683
253	.0703
254	.0724
255	.0745
256	.0767
257	.0789
258	.0812
259	.0835
260	.0859
261	.0883
262	.0908
263	.0933
264	.0958
265	.0985
266	.1011
267	.1038
268	.1066
269	.1094
270	.1123
271	.1152
272	.1182
273	.1212
274	.1243

N = 37

T	P
275	.1274
276	.1306
277	.1339
278	.1371
279	.1405
280	.1439
281	.1473
282	.1509
283	.1544
284	.1580
285	.1617
286	.1654
287	.1692
288	.1730
289	.1769
290	.1808
291	.1848
292	.1889
293	.1930
294	.1971
295	.2013
296	.2056
297	.2099
298	.2142
299	.2186
300	.2231
301	.2276
302	.2321
303	.2367
304	.2414
305	.2461
306	.2508
307	.2556
308	.2605
309	.2653
310	.2703
311	.2752
312	.2803
313	.2853
314	.2904
315	.2956
316	.3008
317	.3060
318	.3112
319	.3165
320	.3219
321	.3273
322	.3327
323	.3381

N = 37

T	P
324	.3436
325	.3491
326	.3546
327	.3602
328	.3658
329	.3714
330	.3771
331	.3828
332	.3885
333	.3942
334	.4000
335	.4057
336	.4115
337	.4173
338	.4232
339	.4290
340	.4349
341	.4408
342	.4466
343	.4525
344	.4585
345	.4644
346	.4703
347	.4762
348	.4822
349	.4881
350	.4941
351	.5000

N = 38

T	P
116	.0001
131	.0002
139	.0003
145	.0004
149	.0005
152	.0006
155	.0007
158	.0008
160	.0009
162	.0010
164	.0011
166	.0012
167	.0013
169	.0014
170	.0015
171	.0016
173	.0017
174	.0018
175	.0019

N = 38

T	P
176	.0020
177	.0021
178	.0022
179	.0023
180	.0025
181	.0026
182	.0027
183	.0029
184	.0030
185	.0031
186	.0033
187	.0035
188	.0036
189	.0038
190	.0040
191	.0042
192	.0044
193	.0046
194	.0048
195	.0050
196	.0052
197	.0055
198	.0057
199	.0060
200	.0062
201	.0065
202	.0068
203	.0071
204	.0074
205	.0077
206	.0081
207	.0084
208	.0088
209	.0092
210	.0095
211	.0099
212	.0104
213	.0108
214	.0112
215	.0117
216	.0122
217	.0126
218	.0132
219	.0137
220	.0142
221	.0148
222	.0154
223	.0160
224	.0166

N = 38

T	P
225	.0172
226	.0179
227	.0185
228	.0192
229	.0199
230	.0207
231	.0214
232	.0222
233	.0230
234	.0239
235	.0247
236	.0256
237	.0265
238	.0275
239	.0284
240	.0294
241	.0304
242	.0315
243	.0325
244	.0336
245	.0348
246	.0359
247	.0371
248	.0383
249	.0396
250	.0409
251	.0422
252	.0436
253	.0450
254	.0464
255	.0478
*256	.0493
257	.0509
258	.0524
259	.0540
260	.0557
261	.0573
262	.0591
263	.0608
264	.0626
265	.0644
266	.0663
267	.0682
268	.0702
269	.0722
270	.0742
271	.0763
272	.0784
273	.0806

N = 38

T	P
274	.08
275	.08
276	.08
277	.08
278	.09
279	.09
280	.09
281	.09
282	.10
283	.10
284	.10
285	.11
286	.11
287	.11
288	.11
289	.12
290	.12
291	.12
292	.13
293	.13
294	.13
295	.14
296	.14
297	.14
298	.15
299	.15
300	.15
301	.16
302	.16
303	.16
304	.17
305	.17
306	.17
307	.18
308	.18
309	.1
310	.19
311	.19
312	.20
313	.20
314	.21
315	.21
316	.21
317	.22
318	.22
319	.23
320	.23
321	.24
322	.24

= 38

P
.2499
.2545
.2591
.2638
.2686
.2733
.2781
.2830
.2879
.2928
.2978
.3028
.3078
.3129
.3180
.3231
.3283
.3335
.3387
.3440
.3493
.3546
.3600
.3654
.3708
.3762
.3817
.3872
.3927
.3982
.4037
.4093
.4149
.4205
.4261
.4317
.4374
.4430
.4487
.4544
.4600
.4657
.4714
.4771
.4829
.4886
.4943
.5000

N = 39

T	P
125	.0001
141	.0002
150	.0003
155	.0004
160	.0005
163	.0006
166	.0007
169	.0008
172	.0009
173	.0010
175	.0011
177	.0012
179	.0013
180	.0014
182	.0015
183	.0016
184	.0017
185	.0018
186	.0019
188	.0020
189	.0021
190	.0023
191	.0024
192	.0025
193	.0026
194	.0027
195	.0028
196	.0030
197	.0031
198	.0033
199	.0034
200	.0036
201	.0037
202	.0039
203	.0041
204	.0043
205	.0045
206	.0047
207	.0049
208	.0051
209	.0053
210	.0056
211	.0058
212	.0060
213	.0063
214	.0066
215	.0069
216	.0071
217	.0074

N = 39

T	P
218	.0078
219	.0081
220	.0084
221	.0088
222	.0091
223	.0095
224	.0099
225	.0103
226	.0107
227	.0111
228	.0115
229	.0120
230	.0124
231	.0129
232	.0134
233	.0139
234	.0144
235	.0150
236	.0155
237	.0161
238	.0167
239	.0173
240	.0180
241	.0186
242	.0193
243	.0200
244	.0207
245	.0214
246	.0222
247	.0230
248	.0238
249	.0246
250	.0254
251	.0263
252	.0272
253	.0281
254	.0290
255	.0300
256	.0310
257	.0320
258	.0330
259	.0341
260	.0352
261	.0363
262	.0375
263	.0387
264	.0399
265	.0411
266	.0424

N = 39

T	P
267	.0437
268	.0451
269	.0464
270	.0478
*271	.0493
272	.0507
273	.0522
274	.0538
275	.0554
276	.0570
277	.0586
278	.0603
279	.0620
280	.0637
281	.0655
282	.0674
283	.0692
284	.0711
285	.0731
286	.0751
287	.0771
288	.0791
289	.0812
290	.0834
291	.0856
292	.0878
293	.0901
294	.0924
295	.0947
296	.0971
297	.0996
298	.1020
299	.1046
300	.1071
301	.1097
302	.1124
303	.1151
304	.1178
305	.1206
306	.1235
307	.1264
308	.1293
309	.1323
310	.1353
311	.1383
312	.1415
313	.1446
314	.1478
315	.1511

N = 39

T	P
316	.1544
317	.1577
318	.1611
319	.1645
320	.1680
321	.1715
322	.1751
323	.1787
324	.1824
325	.1861
326	.1898
327	.1936
328	.1975
329	.2013
330	.2053
331	.2093
332	.2133
333	.2173
334	.2214
335	.2256
336	.2298
337	.2340
338	.2383
339	.2426
340	.2470
341	.2514
342	.2558
343	.2603
344	.2648
345	.2694
346	.2739
347	.2786
348	.2833
349	.2880
350	.2927
351	.2975
352	.3023
353	.3071
354	.3120
355	.3169
356	.3219
357	.3268
358	.3318
359	.3369
360	.3419
361	.3470
362	.3521
363	.3573
364	.3624

N = 39

T	P
365	.3676
366	.3728
367	.3781
368	.3833
369	.3886
370	.3939
371	.3993
372	.4046
373	.4100
374	.4153
375	.4207
376	.4261
377	.4315
378	.4370
379	.4424
380	.4479
381	.4533
382	.4588
383	.4643
384	.4698
385	.4753
386	.4807
387	.4862
388	.4917
389	.4972
390	.5028

N = 40

134	.0001
152	.0002
160	.0003
166	.0004
171	.0005
174	.0006
178	.0007
180	.0008
183	.0009
185	.0010
187	.0011
189	.0012
190	.0013
192	.0014
193	.0015
195	.0016
196	.0017
197	.0018
199	.0019
200	.0020
201	.0021

N = 40

T	P
202	.0022
203	.0023
204	.0024
205	.0025
206	.0027
207	.0028
208	.0029
209	.0030
210	.0032
211	.0033
212	.0035
213	.0036
214	.0038
215	.0040
216	.0041
217	.0043
218	.0045
219	.0047
220	.0049
221	.0051
222	.0053
223	.0055
224	.0058
225	.0060
226	.0063
227	.0065
228	.0068
229	.0071
230	.0073
231	.0076
232	.0079
233	.0083
234	.0086
235	.0089
236	.0093
237	.0096
238	.0100
239	.0104
240	.0108
241	.0112
242	.0116
243	.0120
244	.0125
245	.0129
246	.0134
247	.0139
248	.0144
249	.0149
250	.0155
251	.0160
252	.0166
253	.0172
254	.0178
255	.0184
256	.0191
257	.0197
258	.0204
259	.0211
260	.0218
261	.0225
262	.0233
263	.0241
264	.0249
265	.0257
266	.0265
267	.0274
268	.0283
269	.0292
270	.0301
271	.0311
272	.0321
273	.0331
274	.0341
275	.0352
276	.0363
277	.0374
278	.0385
279	.0397
280	.0409
281	.0421
282	.0433
283	.0446
284	.0459
285	.0472
*286	.0486
287	.0500
288	.0514
289	.0529
290	.0544
291	.0559
292	.0575
293	.0591
294	.0607
295	.0624
296	.0641
297	.0658
298	.0675
299	.0693
300	.0712
301	.0730
302	.0750
303	.0769
304	.0789
305	.0809
306	.0830
307	.0850
308	.0872
309	.0893
310	.0916
311	.0938
312	.0961
313	.0984
314	.1008
315	.1032
316	.1057
317	.1081
318	.1107
319	.1133
320	.1159
321	.1185
322	.1212
323	.1240
324	.1267
325	.1296
326	.1324
327	.1353
328	.1383
329	.1413
330	.1443
331	.1474
332	.1505
333	.1537
334	.1569
335	.1602
336	.1634
337	.1668
338	.1702
339	.1736
340	.1770
341	.1806
342	.1841
343	.1877
344	.1913
345	.1950
346	.1987
347	.2025
348	.2063
349	.2101
350	.2140
351	.2179
352	.2219
353	.2259
354	.2299
355	.2340
356	.2381
357	.2423
358	.2465
359	.2507
360	.2550
361	.2593
362	.2636
363	.2680
364	.2724
365	.2769
366	.2813
367	.2859
368	.2904
369	.2950
370	.2996
371	.3043
372	.3089
373	.3136
374	.3184
375	.3232
376	.3280
377	.3328
378	.3376
379	.3425
380	.3474
381	.3523
382	.3573
383	.3623
384	.3673
385	.3723
386	.3773
387	.3824
388	.3875
389	.3926
390	.3977
391	.4029
392	.4080
393	.4132
394	.4184
395	.4236
396	.4288
397	.4340
398	.4393
399	.4445
400	.4498
401	.4550
402	.4603
403	.4656
404	.4709
405	.4762
406	.4814
407	.4867
408	.4920
409	.4973
410	.5027

N = 41

T	P
144	.0001
162	.0002
171	.0003
177	.0004
182	.0005
186	.0006
189	.0007
192	.0008
195	.0009
197	.0010
199	.0011
201	.0012
203	.0013
204	.0014
206	.0015
207	.0016
209	.0017
210	.0018
211	.0019
212	.0020
213	.0021
214	.0022
215	.0023
216	.0024
217	.0025
218	.0026
219	.0027
220	.0028
221	.0029
222	.0031
223	.0032
224	.0033
225	.0035
226	.0036

N = 4

T	
227	.0
228	.0
229	.0
230	.0
231	.0
232	.0
233	.0
234	.0
235	.0
236	.0
237	.0
238	.0
239	.0
240	.0
241	.
242	.
243	.
244	.
245	.
246	.
247	.
248	.
249	.
250	.
251	.
252	.
253	.
254	.
255	.
256	.
257	.
258	.
259	.
260	.
261	.
262	.
263	.
264	.
265	.
266	.
267	.
268	.
269	.
270	.
271	.
272	.
273	.
274	.
275	.

= 41
P
.0226
.0233
.0241
.0248
.0256
.0264
.0273
.0281
.0290
.0299
.0308
.0317
.0327
.0337
.0347
.0357
.0368
.0378
.0389
.0401
.0412
.0424
.0436
.0449
.0461
.0471
.0488
.0501
.0515
.0529
.0543
.0558
.0573
.0588
.0604
.0620
.0636
.0653
.0669
.0687
.0704
.0722
.0740
.0759
.0778
.0797
.0817
.0836
.0857

N = 41	
T	P
325	.0877
326	.0898
327	.0920
328	.0942
329	.0964
330	.0986
331	.1009
332	.1032
333	.1056
334	.1080
335	.1104
336	.1129
337	.1154
338	.1180
339	.1206
340	.1232
341	.1259
342	.1286
343	.1313
344	.1341
345	.1369
346	.1398
347	.1427
348	.1457
349	.1486
350	.1517
351	.1547
352	.1578
353	.1610
354	.1642
355	.1674
356	.1707
357	.1740
358	.1773
359	.1807
360	.1841
361	.1876
362	.1911
363	.1946
364	.1982
365	.2018
366	.2055
367	.2091
368	.2129
369	.2166
370	.2204
371	.2243
372	.2282
373	.2321

N = 41	
T	P
374	.2360
375	.2400
376	.2441
377	.2481
378	.2522
379	.2563
380	.2605
381	.2647
382	.2689
383	.2732
384	.2775
385	.2818
386	.2862
387	.2906
388	.2950
389	.2994
390	.3039
391	.3084
392	.3130
393	.3175
394	.3221
395	.3267
396	.3314
397	.3361
398	.3408
399	.3455
400	.3502
401	.3550
402	.3598
403	.3646
404	.3694
405	.3743
406	.3791
407	.3840
408	.3889
409	.3939
410	.3988
411	.4038
412	.4088
413	.4137
414	.4187
415	.4238
416	.4288
417	.4338
418	.4389
419	.4439
420	.4490
421	.4541
422	.4592

N = 41	
T	P
423	.4643
424	.4694
425	.4745
426	.4796
427	.4847
428	.4898
429	.4949
430	.5000

N = 42	
154	.0001
173	.0002
182	.0003
189	.0004
194	.0005
198	.0006
201	.0007
204	.0008
207	.0009
209	.0010
211	.0011
213	.0012
215	.0013
217	.0014
218	.0015
220	.0016
221	.0017
223	.0018
224	.0019
225	.0020
226	.0021
227	.0022
229	.0023
230	.0024
231	.0026
232	.0027
233	.0028
234	.0029
235	.0030
236	.0031
237	.0033
238	.0034
239	.0035
240	.0037
241	.0038
242	.0040
243	.0042
244	.0043
245	.0045

N = 42	
T	P
246	.0047
247	.0049
248	.0051
249	.0053
250	.0055
251	.0057
252	.0059
253	.0061
254	.0063
255	.0066
256	.0068
257	.0071
258	.0074
259	.0076
260	.0079
261	.0082
262	.0085
263	.0088
264	.0091
265	.0095
266	.0098
267	.0102
268	.0105
269	.0109
270	.0113
271	.0117
272	.0121
273	.0125
274	.0129
275	.0133
276	.0138
277	.0143
278	.0147
279	.0152
280	.0157
281	.0163
282	.0168
283	.0174
284	.0179
285	.0185
286	.0191
287	.0197
288	.0203
289	.0210
290	.0216
291	.0223
292	.0230
203	.0237
294	.0245

N = 42	
T	P
295	.0252
296	.0260
297	.0268
298	.0276
299	.0284
300	.0292
301	.0301
302	.0310
303	.0319
304	.0328
305	.0338
306	.0348
307	.0358
308	.0368
309	.0378
310	.0389
311	.0400
312	.0411
313	.0422
314	.0434
315	.0446
316	.0458
317	.0470
318	.0483
*319	.0496
320	.0509
321	.0523
322	.0536
323	.0550
324	.0565
325	.0579
326	.0594
327	.0609
328	.0625
329	.0640
330	.0657
331	.0673
332	.0690
333	.0706
334	.0724
335	.0741
336	.0759
337	.0777
338	.0796
339	.0815
340	.0834
341	.0854
342	.0874
343	.0894

N = 42

T	P
344	.0914
345	.0935
346	.0956
347	.0978
348	.1000
349	.1022
350	.1045
351	.1068
352	.1091
353	.1115
354	.1139
355	.1163
356	.1188
357	.1213
358	.1239
359	.1264
360	.1291
361	.1317
362	.1344
363	.1371
364	.1399
365	.1427
366	.1456
367	.1484
368	.1514
369	.1543
370	.1573
371	.1603
372	.1634
373	.1665
374	.1696
375	.1728
376	.1760
377	.1793
378	.1826
379	.1859
380	.1892
381	.1926
382	.1961
383	.1995
384	.2030
385	.2066
386	.2101
387	.2137
388	.2174
389	.2211
390	.2248
391	.2285
392	.2323

N = 42

T	P
393	.2361
394	.2400
395	.2439
396	.2478
397	.2517
398	.2557
399	.2597
400	.2638
401	.2678
402	.2719
403	.2761
404	.2802
405	.2845
406	.2887
407	.2929
408	.2972
409	.3015
410	.3058
411	.3102
412	.3146
413	.3190
414	.3235
415	.3279
416	.3324
417	.3369
418	.3415
419	.3460
420	.3506
421	.3552
422	.3598
423	.3645
424	.3691
425	.3738
426	.3785
427	.3832
428	.3880
429	.3927
430	.3975
431	.4023
432	.4071
433	.4119
434	.4167
435	.4215
436	.4264
437	.4312
438	.4361
439	.4410
440	.4459
441	.4508

N = 42

T	P
442	.4557
443	.4606
444	.4655
445	.4704
446	.4753
447	.4803
448	.4852
449	.4901
450	.4951
451	.5000

N = 43

T	P
165	.0001
184	.0002
194	.0003
201	.0004
206	.0005
210	.0006
214	.0007
217	.0008
219	.0009
222	.0010
224	.0011
226	.0012
228	.0013
230	.0014
232	.0015
233	.0016
235	.0017
236	.0018
237	.0019
239	.0020
240	.0021
241	.0022
242	.0023
243	.0024
244	.0025
245	.0026
246	.0027
247	.0028
248	.0029
249	.0030
250	.0032
251	.0033
252	.0034
253	.0036
254	.0037
255	.0038
256	.0040

N = 43

T	P
257	.0042
258	.0043
259	.0045
260	.0047
261	.0048
262	.0050
263	.0052
264	.0054
265	.0056
266	.0058
267	.0060
268	.0062
269	.0065
270	.0067
271	.0070
272	.0072
273	.0075
274	.0077
275	.0080
276	.0083
277	.0086
278	.0089
279	.0092
280	.0095
281	.0098
282	.0102
283	.0105
284	.0109
285	.0112
286	.0116
287	.0120
288	.0124
289	.0128
290	.0133
291	.0137
292	.0141
293	.0146
294	.0151
295	.0155
296	.0160
297	.0166
298	.0171
299	.0176
300	.0182
301	.0187
302	.0193
303	.0199
304	.0205
305	.0211

N = 43

T	P
306	.0218
307	.0224
308	.0231
309	.0238
310	.0245
311	.0252
312	.0260
313	.0267
314	.0275
315	.0283
316	.0291
317	.0300
318	.0308
319	.0317
320	.0326
321	.0335
322	.0344
323	.0354
324	.0364
325	.0374
326	.0384
327	.0394
328	.0405
329	.0416
330	.0427
331	.0438
332	.0450
333	.0461
334	.0473
335	.0486
*336	.0498
337	.0511
338	.0524
339	.0537
340	.0551
341	.0565
342	.0579
343	.0593
344	.0608
345	.0623
346	.0638
347	.0653
348	.0669
349	.0685
350	.0701
351	.0718
352	.0735
353	.0752
354	.0769

N = 4[illegible]

T	P
355	.0[illegible]
356	.0[illegible]
357	.0[illegible]
358	.0[illegible]
359	.0[illegible]
360	.0[illegible]
361	.0[illegible]
362	.0[illegible]
363	.0[illegible]
364	.0[illegible]
365	.0[illegible]
366	.1[illegible]
367	.1[illegible]
368	.1[illegible]
369	.1[illegible]
370	.1[illegible]
371	.1[illegible]
372	.1[illegible]
373	.1[illegible]
374	.1[illegible]
375	.1[illegible]
376	.1[illegible]
377	.1[illegible]
378	.1[illegible]
379	.1[illegible]
380	.1[illegible]
381	.1[illegible]
382	.1[illegible]
383	.1[illegible]
384	.1[illegible]
385	.1[illegible]
386	.1[illegible]
387	.1[illegible]
388	.1[illegible]
389	.1[illegible]
390	.1[illegible]
391	.1[illegible]
392	.1[illegible]
393	.1[illegible]
394	.1[illegible]
395	.1[illegible]
396	.1[illegible]
397	.1[illegible]
398	.1[illegible]
399	.1[illegible]
400	.1[illegible]
401	.1[illegible]
402	.1[illegible]
403	.2[illegible]

= 43	N = 43		N = 44		N = 44		N = 44		N = 44	
P	T	P	T	P	T	P	T	P	T	P
.2060	453	.4079	260	.0027	309	.0147	358	.0559	407	.1552
.2094	454	.4125	261	.0028	310	.0152	359	.0572	408	.1580
.2129	455	.4172	262	.0029	311	.0156	360	.0586	409	.1609
.2164	456	.4218	263	.0030	312	.0161	361	.0600	410	.1637
.2200	457	.4265	264	.0031	313	.0166	362	.0614	411	.1666
.2235	458	.4312	265	.0033	314	.0171	363	.0629	412	.1695
.2271	459	.4359	266	.0034	315	.0176	364	.0643	413	.1725
.2308	460	.4406	267	.0035	316	.0182	365	.0658	414	.1755
.2345	461	.4453	268	.0037	317	.0187	366	.0674	415	.1785
.2382	462	.4501	269	.0038	318	.0193	367	.0689	416	.1816
.2419	463	.4548	270	.0039	319	.0199	368	.0705	417	.1847
.2457	464	.4595	271	.0041	320	.0204	369	.0721	418	.1878
.2495	465	.4643	272	.0042	321	.0210	370	.0737	419	.1910
.2533	466	.4690	273	.0044	322	.0217	371	.0754	420	.1941
.2571	467	.4738	274	.0046	323	.0223	372	.0771	421	.1974
.2610	468	.4786	275	.0047	324	.0229	373	.0788	422	.2006
.2649	469	.4833	276	.0049	325	.0236	374	.0806	423	.2039
.2689	470	.4881	277	.0051	326	.0243	375	.0823	424	.2072
.2729	471	.4929	278	.0053	327	.0250	376	.0841	425	.2105
.2769	472	.4976	279	.0055	328	.0257	377	.0860	426	.2139
.2809	473	.5024	280	.0057	329	.0264	378	.0878	427	.2173
.2849			281	.0059	330	.0272	379	.0897	428	.2207
.2890	N = 44		282	.0061	331	.0279	380	.0917	429	.2242
.2931	176	.0001	283	.0063	332	.0287	381	.0936	430	.2277
.2973	196	.0002	284	.0065	333	.0294	382	.0956	431	.2312
.3014	206	.0003	285	.0067	334	.0303	383	.0976	432	.2348
.3056	213	.0004	286	.0070	335	.0311	384	.0996	433	.2383
.3098	219	.0005	287	.0072	336	.0320	385	.1017	434	.2420
.3141	223	.0006	288	.0075	337	.0328	386	.1038	435	.2456
.3183	227	.0007	289	.0077	338	.0337	387	.1060	436	.2493
.3226	230	.0008	290	.0080	339	.0346	388	.1081	437	.2530
.3269	233	.0009	291	.0082	340	.0356	389	.1103	438	.2567
.3313	235	.0010	292	.0085	341	.0365	390	.1125	439	.2604
.3356	237	.0011	293	.0088	342	.0375	391	.1148	440	.2642
.3400	240	.0012	294	.0091	343	.0385	392	.1171	441	.2680
.3444	241	.0013	295	.0094	344	.0395	393	.1194	442	.2719
.3488	243	.0014	296	.0097	345	.0405	394	.1218	443	.2757
.3532	245	.0015	297	.0101	346	.0415	395	.1241	444	.2796
.3577	247	.0016	298	.0104	347	.0426	396	.1265	445	.2835
.3622	248	.0017	299	.0107	348	.0437	397	.1290	446	.2875
.3667	250	.0018	300	.0111	349	.0448	398	.1315	447	.2914
.3712	251	.0019	301	.0114	350	.0460	399	.1340	448	.2954
.3757	252	.0020	302	.0118	351	.0471	400	.1365	449	.2994
.3803	254	.0021	303	.0122	352	.0483	401	.1391	450	.3034
.3848	255	.0022	304	.0126	*353	.0495	402	.1417	451	.3075
.3894	256	.0023	305	.0130	354	.0507	403	.1443	452	.3116
.3940	257	.0024	306	.0134	355	.0520	404	.1470	453	.3157
.3986	258	.0025	307	.0138	356	.0533	405	.1497	454	.3198
.4032	259	.0026	308	.0143	357	.0546	406	.1525	455	.3240

N = 44

T	P
456	.3281
457	.3323
458	.3366
459	.3408
460	.3450
461	.3493
462	.3536
463	.3579
464	.3622
465	.3666
466	.3709
467	.3753
468	.3797
469	.3841
470	.3886
471	.3930
472	.3974
473	.4019
474	.4064
475	.4109
476	.4154
477	.4199
478	.4244
479	.4289
480	.4335
481	.4380
482	.4426
483	.4472
484	.4517
485	.4563
486	.4609
487	.4655
488	.4701
489	.4747
490	.4793
491	.4839
492	.4885
493	.4931
494	.4977
495	.5023

N = 45

T	P
187	.0001
208	.0002
219	.0003
226	.0004
231	.0005
236	.0006
240	.0007

N = 45

T	P
243	.0008
246	.0009
249	.0010
251	.0011
253	.0012
255	.0013
257	.0014
259	.0015
261	.0016
262	.0017
264	.0018
265	.0019
266	.0020
268	.0021
269	.0022
270	.0023
271	.0024
272	.0025
273	.0026
274	.0027
275	.0028
276	.0029
277	.0030
278	.0031
279	.0032
280	.0033
281	.0034
282	.0036
283	.0037
284	.0038
285	.0040
286	.0041
287	.0043
288	.0044
289	.0046
290	.0047
291	.0049
292	.0051
293	.0052
294	.0054
295	.0056
296	.0058
297	.0060
298	.0062
299	.0064
300	.0067
301	.0069
302	.0071
303	.0073

N = 45

T	P
304	.0076
305	.0078
306	.0081
307	.0084
308	.0086
309	.0089
310	.0092
311	.0095
312	.0098
313	.0101
314	.0105
315	.0108
316	.0111
317	.0115
318	.0118
319	.0122
320	.0126
321	.0130
322	.0134
323	.0138
324	.0142
325	.0146
326	.0151
327	.0155
328	.0160
329	.0165
330	.0169
331	.0174
332	.0179
333	.0185
334	.0190
335	.0196
336	.0201
337	.0207
338	.0213
339	.0219
340	.0225
341	.0231
342	.0238
343	.0244
344	.0251
345	.0258
346	.0265
347	.0272
348	.0280
349	.0287
350	.0295
351	.0303
352	.0311

N = 45

T	P
353	.0319
354	.0327
355	.0336
356	.0344
357	.0353
358	.0362
359	.0372
360	.0381
361	.0391
362	.0401
363	.0411
364	.0421
365	.0431
366	.0442
367	.0453
368	.0464
369	.0475
370	.0487
*371	.0498
372	.0510
373	.0523
374	.0535
375	.0547
376	.0560
377	.0573
378	.0587
379	.0600
380	.0614
381	.0628
382	.0642
383	.0657
384	.0671
385	.0686
386	.0702
387	.0717
388	.0733
389	.0749
390	.0765
391	.0782
392	.0798
393	.0816
394	.0833
395	.0850
396	.0868
397	.0886
398	.0905
399	.0924
400	.0943
401	.0962

N = 45

T	P
402	.0981
403	.1001
404	.1021
405	.1042
406	.1062
407	.1083
408	.1104
409	.1126
410	.1148
411	.1170
412	.1192
413	.1215
414	.1238
415	.1261
416	.1285
417	.1309
418	.1333
419	.1358
420	.1382
421	.1408
422	.1433
423	.1459
424	.1485
425	.1511
426	.1538
427	.1564
428	.1592
429	.1619
430	.1647
431	.1675
432	.1704
433	.1732
434	.1761
435	.1791
436	.1820
437	.1850
438	.1880
439	.1911
440	.1942
441	.1973
442	.2004
443	.2036
444	.2068
445	.2100
446	.2133
447	.2166
448	.2199
449	.2232
450	.2266

N = 4

T	
451	.2
452	.2
453	.2
454	.2
455	.2
456	.2
457	.2
458	.2
459	.2
460	.2
461	.2
462	.2
463	.2
464	.2
465	.2
466	.2
467	.2
468	.2
469	.2
470	.2
471	.3
472	.3
473	.3
474	.3
475	.3
476	.3
477	.3
478	.3
479	.3
480	.3
481	.3
482	.3
483	.3
484	.3
485	.3
486	.3
487	.3
488	.3
489	.3
490	.3
491	.3
492	.3
493	.3
494	.3
495	.4
496	.4
497	.4
498	.4
499	.4

= 45
P
.4247
.4290
.4334
.4378
.4422
.4467
.4511
.4555
.4599
.4644
.4688
.4733
.4777
.4822
.4866
.4911
.4955
.5000

= 46
.0001
.0002
.0003
.0004
.0005
.0006
.0007
.0008
.0009
.0010
.0011
.0012
.0013
.0014
.0015
.0016
.0017
.0018
.0019
.0020
.0021
.0022
.0023
.0024
.0025
.0026
.0027
.0028
.0029

N = 46		N = 46		N = 46		N = 46		N = 46	
T	P	T	P	T	P	T	P	T	P
292	.0030	341	.0142	390	.0508	439	.1363	488	.2867
293	.0031	342	.0148	391	.0520	440	.1387	489	.2904
294	.0032	343	.0152	392	.0532	441	.1412	490	.2941
295	.0033	344	.0157	393	.0544	442	.1436	491	.2978
296	.0034	345	.0161	394	.0556	443	.1461	492	.3016
297	.0035	346	.0166	395	.0569	444	.1486	493	.3054
298	.0037	347	.0170	396	.0582	445	.1512	494	.3092
299	.0038	348	.0175	397	.0595	446	.1538	495	.3130
300	.0039	349	.0180	398	.0608	447	.1564	496	.3169
301	.0041	350	.0185	399	.0621	448	.1590	497	.3208
302	.0042	351	.0191	400	.0635	449	.1617	498	.3247
303	.0044	352	.0196	401	.0649	450	.1644	499	.3286
304	.0045	353	.0201	402	.0663	451	.1671	500	.3325
305	.0047	354	.0207	403	.0677	452	.1698	501	.3364
306	.0048	355	.0212	404	.0692	453	.1726	502	.3404
307	.0050	356	.0218	405	.0707	454	.1754	503	.3444
308	.0052	357	.0224	406	.0722	455	.1782	504	.3484
309	.0053	358	.0230	407	.0737	456	.1811	505	.3524
310	.0055	359	.0237	408	.0753	457	.1840	506	.3564
311	.0057	360	.0243	409	.0768	458	.1869	507	.3605
312	.0059	361	.0249	410	.0784	459	.1898	508	.3645
313	.0061	362	.0256	411	.0801	460	.1928	509	.3686
314	.0063	363	.0263	412	.0817	461	.1958	510	.3727
315	.0065	364	.0270	413	.0834	462	.1988	511	.3768
316	.0067	365	.0277	414	.0851	463	.2019	512	.3809
317	.0069	366	.0284	415	.0868	464	.2050	513	.3851
318	.0071	367	.0291	416	.0886	465	.2081	514	.3892
319	.0074	368	.0299	417	.0904	466	.2112	515	.3934
320	.0076	369	.0307	418	.0922	467	.2144	516	.3975
321	.0079	370	.0314	419	.0940	468	.2175	517	.4017
322	.0081	371	.0322	420	.0959	469	.2208	518	.4059
323	.0084	372	.0331	421	.0978	470	.2240	519	.4101
324	.0086	373	.0339	422	.0997	471	.2273	520	.4143
325	.0089	374	.0347	423	.1016	472	.2306	521	.4186
326	.0092	375	.0356	424	.1036	473	.2339	522	.4228
327	.0095	376	.0365	425	.1056	474	.2372	523	.4270
328	.0098	377	.0374	426	.1076	475	.2406	524	.4313
329	.0101	378	.0383	427	.1096	476	.2440	525	.4355
330	.0104	379	.0392	428	.1117	477	.2474	526	.4398
331	.0107	380	.0402	429	.1138	478	.2509	527	.4441
332	.0110	381	.0412	430	.1159	479	.2543	528	.4483
333	.0114	382	.0422	431	.1181	480	.2578	529	.4526
334	.0117	383	.0432	432	.1201	481	.2614	530	.4569
335	.0121	384	.0442	433	.1225	482	.2649	531	.4612
336	.0124	385	.0453	434	.1247	483	.2685	532	.4655
337	.0128	386	.0463	435	.1270	484	.2721	533	.4698
338	.0132	387	.0474	436	.1293	485	.2757	534	.4741
339	.0136	388	.0485	437	.1316	486	.2793	535	.4784
340	.0140	*389	.0497	438	.1339	487	.2830	536	.4827

N = 46	
T	P
537	.4871
538	.4914
539	.4957
540	.5000
N = 47	
211	.0001
234	.0002
245	.0003
253	.0004
258	.0005
263	.0006
267	.0007
270	.0008
274	.0009
277	.0010
279	.0011
282	.0012
284	.0013
286	.0014
288	.0015
290	.0016
291	.0017
293	.0018
295	.0019
296	.0020
297	.0021
299	.0022
300	.0023
301	.0024
302	.0025
304	.0026
305	.0027
306	.0028
307	.0029
308	.0030
309	.0031
310	.0032
311	.0034
312	.0035
313	.0036
314	.0037
315	.0038
316	.0040
317	.0041
318	.0043
319	.0044
320	.0045
321	.0047
322	.0048

N = 47	
T	P
323	.0050
324	.0052
325	.0053
326	.0055
327	.0057
328	.0059
329	.0061
330	.0063
331	.0065
332	.0067
333	.0069
334	.0071
335	.0073
336	.0075
337	.0078
338	.0080
339	.0083
340	.0085
341	.0088
342	.0090
343	.0093
344	.0096
345	.0099
346	.0102
347	.0105
348	.0108
349	.0111
350	.0114
351	.0118
352	.0121
353	.0125
354	.0128
355	.0132
356	.0136
357	.0140
358	.0144
359	.0148
360	.0152
361	.0156
362	.0160
363	.0165
364	.0169
365	.0174
366	.0179
367	.0184
368	.0189
369	.0194
370	.0199
371	.0204

N = 47	
T	P
372	.0210
373	.0215
374	.0221
375	.0227
376	.0233
377	.0239
378	.0245
379	.0251
380	.0258
381	.0264
382	.0271
383	.0278
384	.0285
385	.0292
386	.0299
387	.0307
388	.0314
389	.0322
390	.0330
391	.0338
392	.0346
393	.0355
394	.0363
395	.0372
396	.0381
397	.0390
398	.0399
399	.0408
400	.0418
401	.0428
402	.0438
403	.0448
404	.0458
405	.0468
406	.0479
*407	.0490
408	.0501
409	.0512
410	.0524
411	.0535
412	.0547
413	.0559
414	.0571
415	.0584
416	.0596
417	.0609
418	.0622
419	.0635
420	.0649

N = 47	
T	P
421	.0662
422	.0676
423	.0690
424	.0705
425	.0719
426	.0734
427	.0749
428	.0764
429	.0780
430	.0796
431	.0811
432	.0828
433	.0844
434	.0861
435	.0878
436	.0895
437	.0912
438	.0930
439	.0947
440	.0966
441	.0984
442	.1003
443	.1021
444	.1040
445	.1060
446	.1079
447	.1099
448	.1119
449	.1140
450	.1160
451	.1181
452	.1202
453	.1224
454	.1245
455	.1267
456	.1290
457	.1312
458	.1335
459	.1358
460	.1381
461	.1404
462	.1428
463	.1452
464	.1477
465	.1501
466	.1526
467	.1551
468	.1576
469	.1602

N = 47	
T	P
470	.1628
471	.1654
472	.1681
473	.1707
474	.1734
475	.1761
476	.1789
477	.1817
478	.1845
479	.1873
480	.1902
481	.1930
482	.1959
483	.1989
484	.2018
485	.2048
486	.2078
487	.2109
488	.2139
489	.2170
490	.2201
491	.2232
492	.2264
493	.2296
494	.2328
495	.2360
496	.2393
497	.2426
498	.2459
499	.2492
500	.2525
501	.2559
502	.2593
503	.2627
504	.2662
505	.2696
506	.2731
507	.2766
508	.2802
509	.2837
510	.2873
511	.2909
512	.2945
513	.2981
514	.3018
515	.3055
516	.3092
517	.3129
518	.3166

N = [illegible]	
T	[illegible]
519	[illegible]
520	[illegible]
521	[illegible]
522	[illegible]
523	[illegible]
524	[illegible]
525	[illegible]
526	[illegible]
527	[illegible]
528	[illegible]
529	[illegible]
530	[illegible]
531	[illegible]
532	[illegible]
533	[illegible]
534	[illegible]
535	[illegible]
536	[illegible]
537	[illegible]
538	[illegible]
539	[illegible]
540	[illegible]
541	[illegible]
542	[illegible]
543	[illegible]
544	[illegible]
545	[illegible]
546	[illegible]
547	[illegible]
548	[illegible]
549	[illegible]
550	[illegible]
551	[illegible]
552	[illegible]
553	[illegible]
554	[illegible]
555	[illegible]
556	[illegible]
557	[illegible]
558	[illegible]
559	[illegible]
560	[illegible]
561	[illegible]
562	[illegible]
563	[illegible]
564	[illegible]
N = [illegible]	
223	[illegible]

= 48	N = 48		N = 48		N = 48		N = 48		N = 48	
P	T	P	T	P	T	P	T	P	T	P
.0002	344	.0058	393	.0227	442	.0683	491	.1627	540	.3146
.0003	345	.0060	394	.0233	443	.0697	492	.1652	541	.3182
.0004	346	.0062	395	.0238	444	.0711	493	.1678	542	.3218
.0005	347	.0064	396	.0244	445	.0725	494	.1704	543	.3255
.0006	348	.0065	397	.0251	446	.0739	495	.1730	544	.3291
.0007	349	.0067	398	.0257	447	.0754	496	.1756	545	.3328
.0008	350	.0070	399	.0263	448	.0769	497	.1783	546	.3365
.0009	351	.0072	400	.0270	449	.0784	498	.1810	547	.3403
.0010	352	.0074	401	.0276	450	.0799	499	.1837	548	.3440
.0011	353	.0076	402	.0283	451	.0815	500	.1864	549	.3478
.0012	354	.0078	403	.0290	452	.0830	501	.1891	550	.3515
.0013	355	.0081	404	.0297	453	.0846	502	.1919	551	.3553
.0014	356	.0083	405	.0304	454	.0862	503	.1947	552	.3591
.0015	357	.0086	406	.0311	455	.0879	504	.1976	553	.3629
.0016	358	.0088	407	.0319	456	.0895	505	.2004	554	.3667
.0017	359	.0091	408	.0326	457	.0912	506	.2033	555	.3706
.0018	360	.0093	409	.0334	458	.0929	507	.2062	556	.3744
.0019	361	.0096	410	.0342	459	.0946	508	.2091	557	.3783
.0020	362	.0099	411	.0350	460	.0964	509	.2121	558	.3822
.0021	363	.0102	412	.0358	461	.0982	510	.2151	559	.3860
.0022	364	.0105	413	.0367	462	.1000	511	.2181	560	.3899
.0023	365	.0108	414	.0375	463	.1018	512	.2211	561	.3938
.0024	366	.0111	415	.0384	464	.1036	513	.2241	562	.3978
.0025	367	.0114	416	.0392	465	.1055	514	.2272	563	.4017
.0026	368	.0117	417	.0401	466	.1074	515	.2303	564	.4056
.0027	369	.0120	418	.0411	467	.1093	516	.2334	565	.4096
.0028	370	.0124	419	.0420	468	.1113	517	.2365	566	.4135
.0029	371	.0127	420	.0429	469	.1132	518	.2397	567	.4175
.0030	372	.0131	421	.0439	470	.1152	519	.2429	568	.4215
.0031	373	.0134	422	.0449	471	.1172	520	.2461	569	.4255
.0032	374	.0138	423	.0459	472	.1193	521	.2493	570	.4294
.0033	375	.0142	424	.0469	473	.1213	522	.2526	571	.4334
.0034	376	.0146	425	.0479	474	.1234	523	.2558	572	.4374
.0035	377	.0150	426	.0490	475	.1255	524	.2591	573	.4414
.0036	378	.0154	*427	.0500	476	.1277	525	.2624	574	.4454
.0037	379	.0158	428	.0511	477	.1298	526	.2658	575	.4495
.0038	380	.0162	429	.0522	478	.1320	527	.2691	576	.4535
.0040	381	.0167	430	.0533	479	.1342	528	.2725	577	.4575
.0041	382	.0171	431	.0545	480	.1365	529	.2759	578	.4615
.0042	383	.0176	432	.0556	481	.1387	530	.2793	579	.4656
.0044	384	.0180	433	.0568	482	.1410	531	.2828	580	.4696
.0045	385	.0185	434	.0580	483	.1433	532	.2862	581	.4737
.0047	386	.0190	435	.0592	484	.1456	533	.2897	582	.4777
.0048	387	.0195	436	.0605	485	.1480	534	.2932	583	.4818
.0050	388	.0200	437	.0617	486	.1504	535	.2967	584	.4858
.0051	389	.0205	438	.0630	487	.1528	536	.3002	585	.4899
.0053	390	.0210	439	.0643	488	.1552	537	.3038	586	.4939
.0054	391	.0216	440	.0656	489	.1577	538	.3074	587	.4980
.0056	392	.0221	441	.0669	490	.1602	539	.3109	588	.5020

N = 49		N = 49		N = 49		N = 49		N = 49		N = 4	
T	P	T	P	T	P	T	P	T	P	T	P
236	.0001	360	.0056	409	.0214	458	.0632	507	.1496	556	[illegible]
261	.0002	361	.0058	410	.0219	459	.0644	508	.1519	557	[illegible]
273	.0003	362	.0060	411	.0224	460	.0657	509	.1542	558	[illegible]
281	.0004	363	.0062	412	.0230	461	.0670	510	.1566	559	[illegible]
287	.0005	364	.0064	413	.0236	462	.0683	511	.1590	560	[illegible]
292	.0006	365	.0065	414	.0241	463	.0696	512	.1614	561	[illegible]
297	.0007	366	.0067	415	.0247	464	.0710	513	.1639	562	[illegible]
300	.0008	367	.0069	416	.0253	465	.0724	514	.1664	563	[illegible]
304	.0009	368	.0071	417	.0259	466	.0738	515	.1688	564	[illegible]
307	.0010	369	.0074	418	.0265	467	.0752	516	.1714	565	[illegible]
310	.0011	370	.0076	419	.0272	468	.0766	517	.1739	566	[illegible]
312	.0012	371	.0078	420	.0278	469	.0781	518	.1765	567	[illegible]
314	.0013	372	.0080	421	.0285	470	.0796	519	.1790	568	[illegible]
317	.0014	373	.0082	422	.0292	471	.0810	520	.1817	569	[illegible]
319	.0015	374	.0085	423	.0298	472	.0826	521	.1843	570	[illegible]
321	.0016	375	.0087	424	.0305	473	.0841	522	.1869	571	[illegible]
322	.0017	376	.0090	425	.0312	474	.0857	523	.1896	572	[illegible]
324	.0018	377	.0092	426	.0320	475	.0872	524	.1923	573	[illegible]
326	.0019	378	.0095	427	.0327	476	.0888	525	.1951	574	[illegible]
327	.0020	379	.0098	428	.0335	477	.0905	526	.1978	575	[illegible]
329	.0021	380	.0100	429	.0342	478	.0921	527	.2006	576	[illegible]
330	.0022	381	.0103	430	.0350	479	.0938	528	.2034	577	[illegible]
331	.0023	382	.0106	431	.0358	480	.0955	529	.2062	578	[illegible]
333	.0024	383	.0109	432	.0366	481	.0972	530	.2090	579	[illegible]
334	.0025	384	.0112	433	.0374	482	.0989	531	.2119	580	[illegible]
335	.0026	385	.0115	434	.0383	483	.1007	532	.2148	581	[illegible]
336	.0027	386	.0118	435	.0391	484	.1024	533	.2177	582	[illegible]
338	.0028	387	.0121	436	.0400	485	.1042	534	.2206	583	[illegible]
339	.0029	388	.0125	437	.0408	486	.1061	535	.2235	584	[illegible]
340	.0030	389	.0128	438	.0417	487	.1079	536	.2265	585	[illegible]
341	.0031	390	.0132	439	.0427	488	.1098	537	.2295	586	[illegible]
342	.0032	391	.0135	440	.0436	489	.1116	538	.2325	587	[illegible]
343	.0033	392	.0139	441	.0445	490	.1136	539	.2356	588	[illegible]
344	.0034	393	.0142	442	.0455	491	.1155	540	.2386	589	[illegible]
345	.0036	394	.0146	443	.0465	492	.1174	541	.2417	590	[illegible]
346	.0037	395	.0150	444	.0475	493	.1194	542	.2448	591	[illegible]
347	.0038	396	.0154	445	.0485	494	.1214	543	.2479	592	[illegible]
348	.0039	397	.0158	*446	.0495	495	.1235	544	.2511	593	[illegible]
349	.0040	398	.0162	447	.0505	496	.1255	545	.2542	594	[illegible]
350	.0042	399	.0166	448	.0516	497	.1276	546	.2574	595	[illegible]
351	.0043	400	.0171	449	.0527	498	.1297	547	.2606	596	[illegible]
352	.0044	401	.0175	450	.0538	499	.1318	548	.2638	597	[illegible]
353	.0046	402	.0180	451	.0549	500	.1339	549	.2671	598	[illegible]
354	.0047	403	.0184	452	.0560	501	.1361	550	.2703	599	[illegible]
355	.0049	404	.0189	453	.0572	502	.1383	551	.2736	600	[illegible]
356	.0050	405	.0194	454	.0583	503	.1405	552	.2769	601	[illegible]
357	.0052	406	.0198	455	.0595	504	.1427	553	.2802	602	[illegible]
358	.0053	407	.0203	456	.0607	505	.1450	554	.2836	603	[illegible]
359	.0055	408	.0209	457	.0619	506	.1473	555	.2869	604	[illegible]

= 49
P
.4725
.4764
.4803
.4843
.4882
.4921
.4961
.5000

= 50
.0001
.0002
.0003
.0004
.0005
.0006
.0007
.0008
.0009
.0010
.0011
.0012
.0013
.0014
.0015
.0016
.0017
.0018
.0019
.0020
.0021
.0022
.0023
.0024
.0025
.0026
.0027
.0028
.0029
.0030
.0031
.0032
.0033
.0034
.0035
.0036
.0037
.0038
.0039

N = 50		N = 50		N = 50		N = 50		N = 50		N = 50	
T	P	T	P	T	P	T	P	T	P	T	P
366	.0040	415	.0156	464	.0476	513	.1168	562	.2362	611	.4018
367	.0042	416	.0160	465	.0485	514	.1187	563	.2392	612	.4055
368	.0043	417	.0164	*466	.0495	515	.1206	564	.2422	613	.4092
369	.0044	418	.0168	467	.0506	516	.1226	565	.2452	614	.4129
370	.0046	419	.0173	468	.0516	517	.1245	566	.2482	615	.4166
371	.0047	420	.0177	469	.0526	518	.1265	567	.2512	616	.4204
372	.0048	421	.0181	470	.0537	519	.1286	568	.2543	617	.4241
373	.0050	422	.0186	471	.0548	520	.1306	569	.2574	618	.4279
374	.0051	423	.0190	472	.0559	521	.1327	570	.2605	619	.4316
375	.0053	424	.0195	473	.0570	522	.1348	571	.2636	620	.4354
376	.0054	425	.0200	474	.0581	523	.1369	572	.2668	621	.4392
377	.0056	426	.0205	475	.0592	524	.1390	573	.2700	622	.4429
378	.0058	427	.0210	476	.0604	525	.1411	574	.2731	623	.4467
379	.0059	428	.0215	477	.0616	526	.1433	575	.2763	624	.4505
380	.0061	429	.0220	478	.0628	527	.1455	576	.2796	625	.4543
381	.0063	430	.0225	479	.0640	528	.1477	577	.2828	626	.4581
382	.0065	431	.0231	480	.0652	529	.1500	578	.2860	627	.4619
383	.0067	432	.0236	481	.0665	530	.1522	579	.2893	628	.4657
384	.0068	433	.0242	482	.0677	531	.1545	580	.2926	629	.4695
385	.0070	434	.0247	483	.0690	532	.1568	581	.2959	630	.4733
386	.0072	435	.0253	484	.0703	533	.1592	582	.2992	631	.4771
387	.0074	436	.0259	485	.0717	534	.1615	583	.3026	632	.4809
388	.0077	437	.0265	486	.0730	535	.1639	584	.3059	633	.4847
389	.0079	438	.0271	487	.0744	536	.1663	585	.3093	634	.4886
390	.0081	439	.0277	488	.0758	537	.1687	586	.3127	635	.4924
391	.0083	440	.0284	489	.0772	538	.1711	587	.3161	636	.4962
392	.0086	441	.0290	490	.0786	539	.1736	588	.3195	637	.5000
393	.0088	442	.0297	491	.0800	540	.1761	589	.3229		
394	.0090	443	.0304	492	.0815	541	.1786	590	.3264		
395	.0093	444	.0310	493	.0829	542	.1811	591	.3298		
396	.0095	445	.0317	494	.0844	543	.1837	592	.3333		
397	.0098	446	.0324	495	.0860	544	.1862	593	.3368		
398	.0101	447	.0332	496	.0875	545	.1888	594	.3403		
399	.0103	448	.0339	497	.0891	546	.1915	595	.3438		
400	.0106	449	.0347	498	.0906	547	.1941	596	.3474		
401	.0109	450	.0354	499	.0922	548	.1967	597	.3509		
402	.0112	451	.0362	500	.0938	549	.1994	598	.3545		
403	.0115	452	.0370	501	.0955	550	.2021	599	.3580		
404	.0118	453	.0378	502	.0971	551	.2049	600	.3616		
405	.0121	454	.0386	503	.0988	552	.2076	601	.3652		
406	.0124	455	.0394	504	.1005	553	.2104	602	.3688		
407	.0128	456	.0403	505	.1022	554	.2131	603	.3724		
408	.0131	457	.0411	506	.1040	555	.2160	604	.3761		
409	.0134	458	.0420	507	.1057	556	.2188	605	.3797		
410	.0138	459	.0429	508	.1075	557	.2216	606	.3834		
411	.0141	460	.0438	509	.1093	558	.2245	607	.3870		
412	.0145	461	.0447	510	.1112	559	.2274	608	.3907		
413	.0149	462	.0457	511	.1130	560	.2303	609	.3944		
414	.0152	463	.0466	512	.1149	561	.2332	610	.3981		

THE NULL DISTRIBUTION OF THE FIRST THREE PRODUCT-MOMENT STATISTICS FOR EXPONENTIAL, GAMMA ($\lambda = 0.5$), AND NORMAL SCORES

P. A. W. Lewis*
A. S. Goodman*

*BM Research Division, Yorktown Heights, New York.

INTRODUCTION

Let $x_1, x_2, \ldots, x_n$ be n observations on a stochastic process taken at n successive times. The serially ordered obse vations are usually called a time-series. It is desired to test the hypothesis that the n observations are a random sample Thus the joint distribution of the parent population is assumed, under the null hypothesis, to be

$$F_{X_1, X_2, \ldots, X_n}(x_1, x_2, \ldots, x_n) = F_{X_1}(x_1) F_{X_2}(x_2) \ldots F_{X_n}(x_n),$$
$$= F(x_1) F(x_2) \ldots F(x_n),$$

where the common marginal distribution $F(x)$ is generally unspecified. Alternative hypotheses are stationary, dependent time series and time series with trends.

The tests considered here are based on the idea that if the series are random samples they could have occurred with equal probability in each of the $n!$ serial orderings obtained from all possible permutations of the x_i's. This test seems to have first been put forward by Young (1941).

The statistics or functions of the sample $x_1, \ldots, x_n$ used in these tests are the product-moment statistics for various lags j;

$$R_j(n) = x_1 x_{1+j} + x_2 x_{(2+j)} + \ldots + x_{(n-j)} x_n.$$

However, in order to obviate the necessity of computing the probability points of $R_j(n)$ for each sample $x_1, \ldots, x_n$, i is usual to substitute for the actual values $x_1, \ldots, x_n$ a monotonically increasing function of the rank, r_i, of the ith member of the series. Call this function $a(r_i, n)$. This function could be the ranks themselves, in which case the series

5. 73, 4. 62, 4. 13, 8. 40, 6. 00

becomes

3, 2, 1, 5, 4.

The actual function $a(r_i, n)$ used depends on (real or supposed) power considerations. For a discussion of this see Cox and Lewis (1966, p. 166). The functions $a(\, .\, ,\, .\,)$ are generally score functions, i. e., the expected values of the order statistics for random samples from a population with given marginal distribution $F(x)$.

The asymptotic normality of the score product-moment statistics has been established by various statisticians, start ing with the work of Wald and Wolfowitz (1943). See Jogdeo (1968) for a summary and extension of these results.

The scores used in the tables given here were the following:

(i) Normal Scores.

These are the expected values of the order statistics from a population with a unit normal distribution. They cannc be obtained in closed form, but were tabulated by Harter (1961) for series of length $n = 1$ to $n = 100$.

(ii) Exponential Scores.

These are the expected values of the order statistics from a population with an exponential distribution with mean

$$F(x) = 1 - e^{-x} \qquad (x \geq 0$$
$$= 0 \qquad (x < 0$$

e exponential distribution is a Gamma distribution with shape parameter $\lambda = 1$. The exponential scores are given (Cox l Lewis, 1966, p. 55) by

$$a(r_j, n) = s(j, n) = \frac{1}{n} + \ldots + \frac{1}{n-j+1} \qquad (j = 1, \ldots, n).$$

) *Half-gamma Scores.*

We will use this term to denote the expected values of the order statistics from a population with a Gamma distribu- n (Cox and Lewis, 1966, p. 136) with mean 1 and shape parameter $\lambda = 0.5$. (Chi-square with one degree of freedom.) have

$$F(x) = \begin{cases} \int_0^{x/2} v^{-1/2} e^{-v} dv / (\pi)^{1/2} & (x \geq 0), \\ 0 & (x < 0). \end{cases}$$

The half-gamma scores cannot be obtained in closed form but were tabulated by Harter (1964) for $n = 1$ to $n = 40$.

The motivation for using score product-moment statistics and a permutation hypothesis in testing for serial correla- n in time series as opposed, say, to the use of the ordinary serial correlation coefficient was given by Cox and Lewis 66, p. 166). In addition the high positive skewness of the distributions given here for the exponential and half- nma score product-moment statistics, as well as their rather slow convergence to the asymptotic normal distribution, gests that the use of "parametric" serial correlation coefficient tests with large sample approximations can lead to ious errors.

METHOD OF COMPUTATION

The computations were performed on IBM 360/67 and IBM 360/91 computers using an extensively tested pseudo- dom number generator (Lewis et al, 1969). For a series of length n the numbers 1 through n were permuted ng the pseudorandom numbers and a standard scheme described by Moses and Oakford (1963). The scores corres- nding to each number in the permuted series were then used to compute the score product-moment statistics of s 1, 2 and 3. To compute the α-probability points, where $\alpha = 0.001, 0.002, 0.005, 0.010, 0.020, 0.025, 0.050,$ 00, 0.900, 0.950, 0.975, 0.980, 0.990, 0.995, 0.998, 0.999, of the score product-moment statistics, the following eme was used.

Let M be the number of permutations (synthetic samples) generated to estimate the probability points. Consider example, $R_1^{(1)}(n)$, the normal score product-moment statistic of lag 1 for a series of length n and estimation of e $\alpha = 0.001$ probability point of the distribution of $R_1^{(1)}(n)$. It would be preferable to order the M values of $_1^{1)}(n)$ obtained and if M were 999,999, for example, use the $\alpha \cdot (M + 1) = 1000$th value in magnitude of the ob- ved values $R_1^{(1)}(n)$ to estimate the 0.001 probability point. Ordering is, however, extremely time consuming on a mputer and was therefore avoided. Instead the range from the 0.001 point to the 0.100 point and the range from e 0.900 point to the 0.999 point of the statistic in question were estimated in a pilot run. These ranges were then ided into equidistant intervals and an ogive obtained. Up to 1000,000 intervals were available for use with each tistic. The probability points were then interpolated from the ogive.

Now let the value at the α-probability point of the first derivative of the distribution $G(x)$ which is being esti- ted be denoted by $g(x_a)$. If M is very large the standard deviation of the estimate of the α-probability point is proximately

$$\left(\frac{a(1-a)}{M}\right)^{1/2} \frac{1}{g(x_a)}. \qquad (1)$$

This follows from binomial theory and elementary calculus considerations; see, for instance, Loynes (1966).

The values of M used in computing the tables and the number of divisions of the range of the statitic used were la enough to insure a small standard deviation and negligible bias in the estimates. The estimated probability points give in the tables are the raw data rounded to within the accuracy of the interpolation error. In theory the sampling error the estimates could be reduced by (locally) "smoothing" the data, which could then be rounded to reflect the accura of the estimation procedure. However, three points should be kept in mind in attempting this procedure, which is no always as simple as it seems:

(i) For the large values of M used in computing these tables the unsmoothed tabulated values are sufficiently accurate for the purpose of testing for serial correlation in time series. This is particularly true in view of the arbitrari ness involved in picking the level for the test.

(ii) The sampling standard deviation of the estimates is sufficiently small that a poor choice of the (local) smoot ing function may introduce sufficient bias into the smoothed estimates to give an increase in the overall mean-square error of the estimates. To be more specific, the second line of Table A gives an estimated value of $g(x_a)$ for the expor ential score product-moment statistic of lag 1 for $n = 20$ and several values of α. The approximate standard deviation the error, from (1), is given on the next line. In estimating these standard deviations we have been very conservative; true values are probably smaller by a factor of 1/2.

Table A. Estimated error standard deviations and smoothed probability point estimates for the exponential score product-moment statistic of lag one at $n = 20$.

α	0.900	0.950	0.975	0.980	0.990	0.995	0.998	0.999
$g(x_a)$	0.040	0.030	0.019	0.013	0.008	0.004	0.002	0.001
st. dev.	0.008	0.007	0.008	0.010	0.012	0.017	0.022	0.033
x_a	22.802	24.184	25.357	25.705	26.661	27.508	28.427	29.040
$\bar{x}_a$	22.786	24.164	25.340	25.682	26.635	27.464	28.398	28.997

The second last line of Table A gives the estimates x_a from Table 4b. The next line gives a smoothed estimate obtained by adding to x_a the corresponding x_a's for $n = 16, 17, 18, 19$ and $n = 21, 22, 23, 24$ and dividing by 9. The standard deviations of the $\bar{x}_a$'s are 1/3 the standard deviations of the x_a's.

Uniform weights were used in the smoothing procedure because graphical inspection of the $\bar{x}_a$'s revealed no devia from linearity. However, note that all the $\bar{x}_a$'s are less than the corresponding $\bar{x}_a$'s. This is true too for the estimates (shown) for $\alpha = .001$ to $\alpha = .100$. Thus the true x_a's are certainly not even locally linear in n, this being the condition under which uniform weights in the smoothing process are optimal. More elaborate smoothing schemes would theref be required if the tabular values are to be locally smoothed; see, for example, Daniels (1962) for a non-parametric method.

(iii) It appears difficult to find a 'good' regression function on n for the x_a's for the exponential score product-moment statistic. The results of the local smoothing suggest trying at least an n^2 term in the regression function; asymptotic expansions for the first two moments of the statistic (Cox and Lewis 1966, p. 167) suggest using powers $n^{1/2}$ in the regression. However, a regression analysis using the first five powers of $n^{1/2}$ did not give good predictions the probability points for $n = 750$. These probability points had been previously estimated with $M = 16{,}000{,}000$. Th "global" smoothing of the tabular values appears to be difficult also.

Note too in this connection that the probability points of the normal score product-moment statistics are "smoot on odd and even values of n. This occurs because when n is odd the middle normal score is zero. Thus the score prod moment statistic is the sum of only $n - 2 - j$ non-zero terms, unless the zero-valued score occurs at the beginning or of the series. In that case the statistic is the sum of $n - 1 - j$ non-zero terms.

PROBABILITY-POINT TABLES

General Considerations

The rate of convergence of the score product-moment statistics to a normal distribution depends quite critically on the skewness of distribution of the population from which the scores are obtained. For normal scores the normal approximation holds well for n greater than 40, as shown on the last lines of Tables 1a and 1b. Convergence is essentially complete by $n = 50$. For uniform scores (ranks), not tabulated here, the normal approximation holds for $n \geq 75$, but for exponential scores it holds completely only for $n \geq 10{,}000$. For half-gamma scores the convergence is even slower.

General expressions for moments of the score product-moment statistics were given by Wald and Wolfowitz (1943).

Specific Considerations

(i) Normal Scores.

The normal score product-moment statistics of lags 1, 2, and 3, for series of length n are denoted by $R_1^{(1)}(n)$, $R_2^{(1)}(n)$ and $R_3^{(1)}(n)$ respectively and their estimated probability points are given in Tables 1a and 1b, 2a and 2b, and 3a and 3b. In these computations $M = 1{,}000{,}000$ synthetic samples were generated. Beyond $n = 40$ a normal approximation to the distribution can be used.

To apply the normal approximation for series of length $n \geq 40$, the mean and standard deviation of the statistics can be taken to be

$$E\{R_1^{(1)}(n)\} = -\frac{S_2(n)}{n};$$

$$E\{R_2^{(1)}(n)\} = \frac{n-2}{n-1}\, E\{R_1^{(1)}(n)\};$$

$$E\{R_3^{(1)}(n)\} = \frac{n-3}{n-1}\, E\{R_1^{(1)}(n)\};$$

$$\text{st. dev. } \{R_1^{(1)}(n)\} = S_2(n)\frac{(n-2)}{n(n-1)^{(1/2)}};$$

$$\text{st. dev. } \{R_2^{(1)}(n)\} = \frac{(n-2)^{(1/2)}}{(n-1)^{(1/2)}} \text{ st. dev. } \{R_1^{(1)}(n)\};$$

$$\text{st. dev. } \{R_3^{(1)}(n)\} = \frac{(n-3)^{(1/2)}}{(n-1)^{(1/2)}} \text{ st. dev. } \{R_1^{(1)}(n)\};$$

where $S_2(n)$ is the sum of the squares of the normal scores. These approximations are very good, as can be shown both theoretically and empirically.

The next to last line of Table la and 1b shows the values of the estimated percentiles of $R_1^{(1)}(n)$, for $n = 40$, minus the estimated mean and divided by the estimated standard deviation. The approximation to the probability points of a normal distribution, given on the last line of the table, is good.

(ii) Exponential Scores.

The exponential-score product-moment statistics of lags 1, 2, and 3 for series of length n are denoted respectively by $R_1^{(2)}(n)$, $R_2^{(2)}(n)$ and $R_3^{(2)}(n)$ and their estimated probability points are given in Tables 4a and 4b, 5a and 5b, and 6a and 6b, respectively. In these computations $M = 1{,}000{,}000$ synthetic samples were used for $n = 11$ to $n = 150$, $M = 500{,}000$ synthetic samples for $n = 151$ to $n = 500$, and $M = 500{,}000$ for $n = 900$, $M = 200{,}000$ for $n = 2{,}500$, $M = 100{,}000$ for $n = 5{,}000$, and $M = 50{,}000$ for $n = 9{,}000$. From $n = 250$ to $n = 500$, every 10th value of n was used because of computation time limitations. Results of the computations for odd values of n between $n = 175$ and $n = 250$

have been omitted because of calculation-space limitations. Linear interpolation for intermediate values can be used because the probability point curves are relatively smooth.

The distributions have a high positive skewness and converge very slowly to a normal distribution ($\sim n = 10{,}000$). However, a normal approximation can be used for $n \geq 500$ with very small error. The exponential scores are very useful for testing serial dependence in highly skewed data because they can be computed exactly for all n. Moreover, the result of these computations indicates that using the ordinary serial correlation coefficient statistics and a normal approximation can be highly erroneous with skewed data when n is small, say less than 250.

Asymptotic expansions for the first two moments of the exponential-score product-moment statistics were given by Cox and Lewis (1966, p. 167).

(iii) Half-gamma Scores.

The half-gamma-score product-moment statistics of lags 1, 2, and 3 for series of length n are denoted respectively by $R_1^{(3)}(n)$, $R_2^{(3)}(n)$ and $R_3^{(3)}(n)$ and their estimated probability points are given in Tables 7a and 7b, 8a and 8b, and 9a and 9b, respectively. Estimated means and variances are also given. In these computations $M = 4{,}000{,}000$ synthetic samples were used for $n = 11$ to $n = 40$. The scores are not tabulated above $n = 40$ and approximations to the means and standard deviations of the product-moment statistics are not known.

The distributions have a high positive skewness and appear to converge more slowly to a normal distribution than do the distributions of the exponential-score product-moment statistics.

AN APPLICATION OF THE TABLES

Lewis (1964) has given a series of 255 times between failures of a computer. For this data the estimated serial correlation coefficient of lag 1 has been computed and found to have the value $\hat{\rho}_1 = +0.133$. Multiplication by $(255)^{(1/2)}$ yields the value +2.234, which is significant at the 2.5 per cent level (two-tailed) when tested as a variate with a unit normal distribution. However, in the data it can be seen that the two longest intervals occur contiguously, and they are twice as long as any other intervals between failures. This raises the possibility of recording errors; the use of scores is helpful in controlling their effect. Using exponential scores instead of actual values and computing the product-moment statistic $R_1^{(2)}(255)$ we get

$$R_1^{(2)}(255) = 285.45$$

$$\frac{R_1^{(2)}(255) - E\,[R_1^{(2)}(255)]}{\text{st. dev. } [R_1^{(2)}(255)]} = 2.08.$$

Tested as a unit normal variate the value 2.08 is just significant at the 3.6 per cent level (two-tailed). From Table 4b, however, the value 285.45 is significant at approximately the 6 per cent level. The differences in significance levels for this value of n are small but relatively important; by $n = 500$ the differences are negligible for testing purposes.

REFERENCES

1. Cox, D. R. and Lewis, P. A. W. (1966). *The Statistical Analysis of Series of Events,* London, Methuen; New York, Wiley.

2. Daniels, H. E. (1962). "The estimation of spectral densities," *J. Roy. Statist. Soc.* B, 24: 185–198.

3. Harter, H. L. (1961). "Expected values of normal order statistics," *Biometrika,* 48: 151–165.

4. Harter, H. L. (1964). "Expected values of exponential, Weibull and Gamma order statistics," Aerospace Research Laboratories Report, ARL 64–31.

5. Jogdeo, K. (1968). "Asymptotic normality in nonparametric methods," *Ann. Math. Statist,* 39: 905–922.

6. Lewis, P. A. W. (1964). "A branching Poisson process model for the analysis of computer failure patterns," *J. Roy. Statist. Soc.,* B, 26: 398–456.

7. Lewis, P. A. W., Goodman, A. S. and Miller, J. M. (1969). "A pseudo-random number generator for the System/360," *IBM Systems Journal,* 8: 2.

8. Loynes, R. M. (1966). "Some aspects of the estimation of quantities," *J. Roy. Statist. Soc.,* B, 28: 497–512.

9. Moses, L. E. and Oakford, R. F. (1963). *Tables of Random Permutations,* Stanford: Stanford University Press.

10. Wald, A. and Wolfowitz, J. (1943). "An exact test of randomness in the nonparametric case based on serial correlation," *Ann. Math. Statist.,* 14: 378–388.

11. Young, L. C. (1941). "On randomness in ordered sequences," *Ann. Math. Statist.,* 12: 293–300.

Table 1a. Estimated probability points, x_α, of the normal score product-moment statistic, $R_1^{(1)}(n)$, from a Monte Carlo experiment with $M = 1,000,000$ repetitions for series of length $n = 11$ to $n = 40$.

$n \backslash \alpha$	Mean	0.001	0.002	0.005	0.010	0.020	0.025	0.050	0.100
11	-0.810	-7.003	-6.714	-6.289	-5.906	-5.437	-5.268	-4.662	-3.893
12	-0.817	-7.522	-7.232	-6.760	-6.335	-5.817	-5.632	-4.962	-4.128
13	-0.833	-8.091	-7.746	-7.213	-6.746	-6.179	-5.971	-5.254	-4.353
14	-0.845	-8.653	-8.270	-7.681	-7.149	-6.527	-6.305	-5.526	-4.562
15	-0.853	-9.101	-8.693	-8.064	-7.506	-6.860	-6.623	-5.787	-4.771
16	-0.851	-9.587	-9.137	-8.457	-7.860	-7.164	-6.905	-6.025	-4.953
17	-0.872	-10.034	-9.582	-8.866	-8.216	-7.474	-7.206	-6.282	-5.154
18	-0.867	-10.518	-10.002	-9.229	-8.544	-7.749	-7.471	-6.502	-5 327
19	-0.880	-10.942	-10.406	-9.593	-8.868	-8.048	-7.758	-6.732	-5.505
20	-0.883	-11.355	-10.788	-9.944	-9.184	-8.326	-8.022	-6.952	-5.674
21	-0.890	-11.711	-11.133	-10.229	-9.447	-8.567	-8.258	-7.165	-5.839
22	-0.894	-12.113	-11.520	-10.593	-9.768	-8.830	-8.497	-7.370	-5.995
23	-0.896	-12.552	-11.898	-10.927	-10.078	-9.089	-8.753	-7.576	-6.156
24	-0.907	-12.889	-12.213	-11.213	-10.341	-9.345	-8.985	-7.770	-6.312
25	-0.911	-13.210	-12.507	-11.462	-10.566	-9.551	-9.187	-7.935	-6.448
26	-0.909	-13.623	-12.893	-11.807	-10.871	-9.795	-9.415	-8.126	-6.598
27	-0.913	-13.995	-13.223	-12.107	-11.106	-10.006	-9.626	-8.302	-6.728
28	-0.915	-14.279	-13.501	-12.352	-11.362	-10.236	-9.840	-8.491	-6.873
29	-0.912	-14.655	-13.834	-12.636	-11.616	-10.444	-10.034	-8.652	-6.998
30	-0.921	-14.969	-14.130	-12.906	-11.855	-10.670	-10.242	-8.825	-7.152
31	-0.913	-15.227	-14.397	-13.138	-12.059	-10.856	-10.425	-8.968	-7.247
32	-0.926	-15.577	-14.724	-13.048	-12.301	-11.082	-10.646	-9.172	-7.406
33	-0.917	-15.871	-14.997	-13.641	-12.535	-11.261	-10.818	-9.301	-7.510
34	-0.928	-16.138	-15.271	-13.934	-12.785	-11.491	-11.028	-9.469	-7.632
35	-0.924	-16.466	-15.565	-14.151	-12.969	-11.661	-11.200	-9.612	-7.748
36	-0.931	-16.773	-15.823	-14.415	-13.195	-11.847	-11.380	-9.768	-7.870
37	-0.930	-17.048	-16.023	-14.613	-13.428	-12.048	-11.573	-9.924	-7.974
38	-0.923	-17.343	-16.356	-14.877	-13.622	-12.227	-11.742	-10.071	-8.101
39	-0.929	-17.613	-16.616	-15.096	-13.873	-12.422	-11.921	-10.213	-8.209
40	-0.941	-17.929	-16.896	-15.346	-14.023	-12.583	-12.079	-10.355	-8.334
		(-2.969)	(-2.789)	(-2.518)	(-2.287)	(-2.035)	(-1.947)	(-1.645)	(-1.292)
		(-3.090)	(-2.878)	(-2.576)	(-2.326)	(-2.054)	(-1.960)	(-1.645)	(-1.282)

Table 1b. Estimated probability points, x_α, of the normal score product-moment statistic, $R_1^{(1)}(n)$, from a Monte Carlo experiment with $M = 1,000,000$ repetitions for series of length $n = 11$ to $n = 40$.

n \ α	0.900	0.950	0.975	0.980	0.990	0.995	0.998	0.999	σ
11	2.287	3.085	3.719	3.904	4.395	4.815	5.272	5.559	2.346
12	2.498	3.352	4.046	4.244	4.792	5.260	5.769	6.105	2.519
13	2.705	3.631	4.384	4.595	5.177	5.679	6.239	6.617	2.689
14	2.880	3.867	4.666	4.892	5.525	6.078	6.697	7.109	2.847
15	3.076	4.109	4.959	5.204	5.895	6.472	7.139	7.581	3.001
16	3.258	4.344	5.246	5.507	6.218	6.857	7.574	8.029	3.144
17	3.418	4.558	5.518	5.790	6.559	7.216	7.990	8.486	3.287
18	3.605	4.800	5.790	6.074	6.868	7.552	8.363	8.555	3.425
19	3.760	5.005	6.033	6.336	7.168	7.903	8.728	9.287	3.557
20	3.920	5.203	6.268	6.575	7.442	8.214	9.085	9.644	3.683
21	4.067	5.398	6.514	6.834	7.751	8.551	9.467	10.095	3.809
22	4.214	5.599	6.753	7.089	8.033	8.873	9.866	10.505	3.931
23	4.370	5.787	6.975	7.321	8.306	9.159	10.170	10.804	4.051
24	4.497	5.964	7.204	7.564	8.570	9.465	10.499	11.189	4.164
25	4.640	6.148	7.415	7.777	8.806	9.730	10.774	11.496	4.271
26	4.784	6.324	7.614	7.990	9.047	9.994	11.117	11.865	4.383
27	4.913	6.492	7.825	8.208	9.331	10.319	11.468	12.236	4.489
28	5.049	6.677	8.048	8.441	9.592	10.613	11.759	12.567	4.599
29	5.182	6.828	8.230	8.635	9.784	10.835	12.026	12.827	4.699
30	5.311	6.991	8.415	8.819	10.024	11.110	12.332	13.212	4.804
31	5.443	7.162	8.621	9.059	10.287	11.347	12.634	13.489	4.894
32	5.539	7.302	8.796	9.235	10.481	11.602	12.887	13.847	4.999
33	5.674	7.470	9.002	9.450	10.737	11.908	13.217	14.121	5.092
34	5.786	7.623	9.177	9.628	10.929	12.103	13.475	14.406	5.186
35	5.905	7.769	9.349	9.810	11.137	12.334	13.746	14.706	5.277
36	6.017	7.921	9.537	10.012	11.344	12.571	13.987	14.908	5.370
37	6.133	8.067	9.718	10.193	11.583	12.812	14.229	15.264	5.461
38	6.255	8.221	9.890	10.375	11.771	13.045	14.569	15.533	5.551
39	6.348	8.347	10.057	10.561	12.002	13.258	14.818	15.816	5.636
40	6.460	8.486	10.197	10.706	12.158	13.447	14.991	16.032	5.721
	(1.294)	(1.648)	(1.947)	(2.036)	(2.289)	(2.515)	(2.785)	(2.967)	
	(1.282)	(1.645)	(1.960)	(2.054)	(2.326)	(2.576)	(2.878)	(3.090)	

Table 2a. Estimated probability points, x_α, of the normal score product-moment statistic, $R_2^{(1)}(n)$, from a Monte Carlo experiment with $M = 1,000,000$ repetitions for series of length $n = 11$ to $n = 40$.

$n \backslash \alpha$	Mean	0.001	0.002	0.005	0.010	0.020	0.025	0.050	0.100
11	-0.726	-6.625	-6.370	-5.956	-5.580	-5.140	-4.972	-4.388	-3.658
12	-0.747	-7.233	-6.937	-6.461	-6.036	-5.530	-5.349	-4.712	-3.912
13	-0.766	-7.767	-7.424	-6.922	-6.455	-5.900	-5.705	-5.004	-4.140
14	-0.778	-8.302	-7.937	-7.361	-6.847	-6.262	-6.043	-5.293	-4.358
15	-0.789	-8.827	-8.408	-7.769	-7.226	-6.584	-6.359	-5.553	-4.569
16	-0.810	-9.281	-8.850	-8.192	-7.605	-6.916	-6.673	-5.822	-4.779
17	-0.814	-9.716	-9.266	-8.566	-7.952	-7.223	-6.959	-6.057	-4.969
18	-0.823	-10.201	-9.721	-8.964	-8.279	-7.509	-7.234	-6.292	-5.147
19	-0.828	-10.647	-10.113	-9.300	-8.608	-7.808	-7.519	-6.533	-5.321
20	-0.835	-11.048	-10.482	-9.655	-8.914	-8.074	-7.780	-6.745	-5.506
21	-0.849	-11.475	-10.883	-9.989	-9.227	-8.348	-8.036	-6.963	-5.673
22	-0.846	-11.857	-11.215	-10.326	-9.515	-8.606	-8.288	-7.173	-5.834
23	-0.856	-12.223	-11.591	-10.623	-9.801	-8.871	-8.544	-7.384	-6.004
24	-0.860	-12.608	-11.938	-10.953	-10.081	-9.097	-8.756	-7.568	-6.136
25	-0.870	-12.959	-12.295	-11.268	-10.358	-9.346	-8.993	-7.759	-6.297
26	-0.871	-13.353	-12.622	-11.574	-10.642	-9.580	-9.209	-7.946	-6.447
27	-0.878	-13.682	-12.949	-11.824	-10.887	-9.821	-9.444	-8.139	-6.588
28	-0.872	-14.070	-13.283	-12.123	-11.135	-10.035	-9.645	-8.298	-6.709
29	-0.885	-14.375	-13.577	-12.400	-11.407	-10.258	-9.867	-8.487	-6.855
30	-0.890	-14.731	-13.907	-12.665	-11.647	-10.466	-10.052	-8.655	-6.994
31	-0.891	-15.029	-14.179	-12.928	-11.869	-10.701	-10.277	-8.829	-7.128
32	-0.888	-15.264	-14.403	-13.165	-12.090	-10.875	-10.446	-8.984	-7.246
33	-0.891	-15.693	-14.803	-13.503	-12.363	-11.091	-10.657	-9.139	-7.378
34	-0.901	-15.901	-15.006	-13.714	-12.551	-11.287	-10.836	-9.316	-7.515
35	-0.909	-16.267	-15.296	-13.944	-12.786	-11.469	-11.021	-9.463	-7.624
36	-0.899	-16.617	-15.643	-14.193	-13.006	-11.676	-11.210	-9.602	-7.727
37	-0.910	-16.888	-15.898	-14.457	-13.242	-11.873	-11.399	-9.770	-7.863
38	-0.890	-17.113	-16.140	-14.681	-13.462	-12.050	-11.564	-9.906	-7.972
39	-0.918	-17.376	-16.369	-14.875	-13.637	-12.256	-11.764	-10.086	-8.114
40	-0.910	-17.635	-16.642	-15.105	-13.819	-12.388	-11.902	-10.201	-8.205

Table 2b. Estimated probability points, x_α, of the normal score product-moment statistic, $R_2^{(1)}(n)$, from a Monte Carlo experiment with $M = 1,000,000$ repetitions for series of length $n = 11$ to $n = 40$.

n \ α	0.900	0.950	0.975	0.980	0.990	0.995	0.998	0.999	σ
11	2.215	2.973	3.576	3.755	4.255	4.697	5.208	5.516	2.233
12	2.417	3.237	3.901	4.096	4.621	5.089	5.630	6.003	2.411
13	2.624	3.503	4.227	4.435	5.021	5.526	6.090	6.469	2.581
14	2.815	3.756	4.523	4.751	5.363	5.916	6.533	6.941	2.741
15	3.000	4.003	4.825	5.062	5.724	6.302	6.967	7.409	2.896
16	3.165	4.220	5.089	5.340	6.034	6.649	7.348	7.820	3.042
17	3.351	4.457	5.370	5.629	6.376	7.017	7.765	8.294	3.189
18	3.508	4.670	5.632	5.905	6.692	7.362	8.133	8.658	3.322
19	3.681	4.888	5.882	6.172	6.985	7.695	8.530	9.096	3.458
20	3.836	5.083	6.118	6.427	7.295	8.043	8.914	9.487	3.588
21	3.980	5.280	6.362	6.670	7.561	8.341	9.233	9.862	3.714
22	4.145	5.489	6.611	6.937	7.858	8.664	9.622	10.285	3.838
23	4.290	5.680	6.849	7.178	8.134	8.960	9.948	10.580	3.960
24	4.422	5.857	7.080	7.426	8.402	9.263	10.302	10.925	4.071
25	4.569	6.032	7.284	7.643	8.664	9.569	10.618	11.317	4.187
26	4.710	6.224	7.489	7.863	8.919	9.863	10.961	11.711	4.299
27	4.843	6.394	7.689	8.074	9.167	10.152	11.277	12.047	4.407
28	4.975	6.561	7.908	8.295	9.407	10.411	11.554	12.339	4.510
29	5.104	6.738	8.122	8.523	9.656	10.667	11.890	12.668	4.616
30	5.226	6.897	8.299	8.713	9.895	10.938	12.160	12.999	4.718
31	5.343	7.042	8.493	8.919	10.113	11.168	12.481	13.313	4.817
32	5.464	7.209	8.668	9.097	10.329	11.452	12.755	13.593	4.910
33	5.606	7.384	8.890	9.335	10.567	11.698	13.035	13.952	5.017
34	5.720	7.531	9.073	9.533	10.824	11.993	13.345	14.278	5.114
35	5.818	7.659	9.213	9.669	10.988	12.137	13.527	14.476	5.195
36	5.941	7.821	9.420	9.879	11.209	12.392	13.822	14.779	5.290
37	6.048	7.958	9.579	10.063	11.434	12.678	14.155	15.141	5.381
38	6.168	8.112	9.765	10.249	11.623	12.859	14.330	15.354	5.470
39	6.281	8.254	9.921	10.420	11.847	13.128	14.623	15.626	5.565
40	6.384	8.404	10.116	10.624	12.067	13.338	14.865	15.903	5.647

Table 3a. Estimated probability points, x_α, of the normal score product-moment statistic, $R_3^{(1)}(n)$, from a Monte Carlo experiment with $M = 1,000,000$ repetitions for series of length $n = 11$ to $n = 40$.

n \ α	Mean	0.001	0.002	0.005	0.010	0.020	0.025	0.050	0.100
11	-0.647	-6.123	-5.912	-5.566	-5.242	-4.829	-4.681	-4.125	-3.430
12	-0.672	-6.767	-6.519	-6.104	-5.722	-5.248	-5.070	-4.443	-3.666
13	-0.697	-7.374	-7.079	-6.596	-6.153	-5.623	-5.430	-4.759	-3.923
14	-0.711	-7.906	-7.574	-7.026	-6.539	-5.968	-5.763	-5.039	-4.149
15	-0.731	-8.437	-8.056	-7.472	-6.944	-6.319	-6.096	-5.328	-4.382
16	-0.745	-8.937	-8.529	-7.887	-7.303	-6.637	-6.400	-5.578	-4.575
17	-0.756	-9.424	-8.961	-8.269	-7.666	-6.958	-6.706	-5.842	-4.781
18	-0.771	-9.870	-9.408	-8.666	-8.003	-7.255	-6.995	-6.074	-4.972
19	-0.784	-10.335	-9.795	-9.021	-8.332	-7.547	-7.274	-6.321	-5.161
20	-0.800	-10.737	-10.172	-9.367	-8.649	-7.841	-7.555	-6.548	-5.334
21	-0.799	-11.199	-10.601	-9.723	-8.977	-8.120	-7.809	-6.755	-5.502
22	-0.803	-11.571	-10.938	-10.049	-9.259	-8.370	-8.051	-6.962	-5.666
23	-0.823	-11.953	-11.310	-10.381	-9.570	-8.646	-8.320	-7.190	-5.840
24	-0.825	-12.325	-11.661	-10.670	-9.818	-8.860	-8.516	-7.359	-5.975
25	-0.828	-12.657	-11.953	-10.971	-10.110	-9.131	-8.780	-7.579	-6.140
26	-0.835	-13.093	-12.412	-11.343	-10.404	-9.371	-9.012	-7.756	-6.282
27	-0.836	-13.412	-12.672	-11.605	-10.683	-9.630	-9.257	-7.970	-6.447
28	-0.851	-13.781	-13.032	-11.894	-10.920	-9.818	-9.433	-8.135	-6.587
29	-0.848	-14.152	-13.311	-12.175	-11.178	-10.049	-9.653	-8.305	-6.704
30	-0.863	-14.423	-13.637	-12.446	-11.430	-10.279	-9.881	-8.496	-6.858
31	-0.858	-14.764	-13.917	-12.674	-11.654	-10.474	-10.071	-8.663	-6.986
32	-0.869	-15.023	-14.190	-12.966	-11.895	-10.701	-10.273	-8.828	-7.125
33	-0.866	-15.428	-14.523	-13.230	-12.121	-10.891	-10.471	-8.987	-7.238
34	-0.874	-15.713	-14.815	-13.497	-12.376	-11.132	-10.678	-9.171	-7.373
35	-0.868	-15.979	-15.083	-13.753	-12.585	-11.288	-10.843	-9.299	-7.494
36	-0.876	-16.349	-15.363	-13.988	-12.782	-11.482	-11.027	-9.484	-7.627
37	-0.870	-16.568	-15.632	-14.263	-13.056	-11.701	-11.224	-9.620	-7.738
38	-0.882	-16.821	-15.867	-14.431	-13.215	-11.856	-11.380	-9.754	-7.830
39	-0.885	-17.244	-16.171	-14.721	-13.479	-12.073	-11.586	-9.939	-7.983
40	-0.883	-17.386	-16.437	-14.950	-13.676	-12.239	-11.739	-10.055	-8.063

Table 3b. Estimated probability points, x_α, of the normal score product-moment statistic, $R_3^{(1)}(n)$, from a Monte Carlo experiment with M = 1,000,000 repetitions for series of length n = 11 to n = 40.

n \ α	0.900	0.950	0.975	0.980	0.990	0.995	0.998	0.999	σ
11	2.151	2.899	3.520	3.689	4.153	4.540	4.952	5.195	2.124
12	2.337	3.144	3.828	4.022	4.554	4.999	5.454	5.742	2.299
13	2.536	3.389	4.111	4.318	4.899	5.395	5.935	6.279	2.471
14	2.731	3.646	4.400	4.622	5.246	5.770	6.380	6.752	2.633
15	2.923	3.887	4.688	4.918	5.578	6.146	6.798	7.239	2.794
16	3.088	4.112	4.949	5.190	5.879	6.478	7.172	7.653	2.938
17	3.267	4.332	5.211	5.473	6.205	6.837	7.583	8.087	3.084
18	3.429	4.552	5.476	5.751	6.509	7.179	7.937	8.477	3.222
19	3.587	4.755	5.735	6.015	6.813	7.511	8.288	8.841	3.359
20	3.738	4.953	5.979	6.269	7.107	7.839	8.685	9.273	3.488
21	3.908	5.157	6.222	6.535	7.407	8.184	9.084	9.707	3.618
22	4.060	5.361	6.457	6.779	7.703	8.533	9.443	10.066	3.743
23	4.191	5.559	6.692	7.020	7.968	8.792	9.771	10.449	3.865
24	4.337	5.746	6.926	7.275	8.243	9.114	10.108	10.823	3.974
25	4.484	5.924	7.125	7.478	8.487	9.381	10.440	11.123	4.096
26	4.630	6.121	7.350	7.717	8.764	9.679	10.774	11.534	4.208
27	4.768	6.285	7.557	7.936	8.990	9.958	11.070	11.829	4.323
28	4.888	6.450	7.760	8.145	9.237	10.223	11.341	12.085	4.425
29	5.020	6.626	7.971	8.373	9.512	10.523	11.651	12.461	4.530
30	5.140	6.777	8.160	8.555	9.717	10.744	11.954	12.755	4.634
31	5.272	6.958	8.388	8.804	9.989	11.065	12.329	13.152	4.738
32	5.392	7.098	8.544	8.953	10.177	11.252	12.475	13.373	4.835
33	5.517	7.263	8.745	9.188	10.423	11.533	12.812	13.684	4.932
34	5.631	7.411	8.931	9.368	10.638	11.783	13.104	13.992	5.029
35	5.745	7.570	9.104	9.567	10.855	12.027	13.402	14.394	5.124
36	5.864	7.728	9.303	9.769	11.076	12.235	13.612	14.559	5.219
37	6.004	7.881	9.474	9.947	11.301	12.463	13.915	14.833	5.312
38	6.079	8.003	9.633	10.114	11.491	12.704	14.149	15.132	5.388
39	6.203	8.153	9.809	10.304	11.721	12.975	14.414	15.411	5.490
40	6.318	8.293	9.987	10.468	11.881	13.135	14.625	15.670	5.570

Table 4a. Estimated lower probability points, x_α, of the exponential score product-moment statistic, $R_1^{(2)}(n)$, from a Monte Carlo experiment with $M = 1,000,000$ repetitions, for series of length $n = 11\ (1)\ 150$; $M = 500,000$ for $n = 151(1)\ 174,\ 176(2)\ 250,\ 260(10)\ 500,\ 900$; $M = 200,000$ for $n = 2,\ 500$; $M = 100,000$ for $n = 5,000$; and $M = 50,000$ for $n = 9,000$.

n \ α	0.001	0.002	0.005	0.010	0.020	0.025	0.050	0.100
11	4.389	4.556	4.826	5.086	5.403	5.519	5.953	6.516
12	4.986	5.160	5.468	5.746	6.099	6.228	6.694	7.309
13	5.566	5.775	6.113	6.426	6.807	6.948	7.455	8.113
14	6.180	6.398	6.771	7.110	7.519	7.673	8.217	8.919
15	6.773	7.036	7.439	7.808	8.250	8.411	8.989	9.732
16	7.436	7.704	8.137	8.527	8.995	9.165	9.780	10.562
17	8.053	8.355	8.818	9.232	9.734	9.917	10.566	11.387
18	8.710	9.019	9.521	9.962	10.488	10.673	11.360	12.224
19	9.344	9.681	10.205	10.677	11.231	11.431	12.150	13.056
20	10.024	10.389	10.948	11.441	12.021	12.231	12.969	13.909
21	10.722	11.088	11.662	12.173	12.787	13.005	13.779	14.754
22	11.386	11.782	12.382	12.926	13.556	13.787	14.592	15.611
23	12.052	12.465	13.110	13.677	14.329	14.564	15.407	16.450
24	12.788	13.197	13.859	14.428	15.099	15.347	16.214	17.312
25	13.453	13.917	14.596	15.202	15.914	16.168	17.062	18.176
26	14.208	14.662	15.358	15.979	16.692	16.960	17.895	19.046
27	14.896	15.362	16.106	16.747	17.498	17.766	18.722	19.912
28	15.582	16.099	16.846	17.519	18.304	18.574	19.562	20.783
29	16.324	16.822	17.612	18.296	19.097	19.375	20.381	21.641
30	17.018	17.572	18.378	19.092	19.909	20.203	21.231	22.517
31	17.798	18.338	19.143	19.883	20.734	21.028	22.092	23.410
32	18.515	19.066	19.934	20.661	21.525	21.838	22.929	24.273
33	19.211	19.801	20.696	21.454	22.346	22.660	23.774	25.160
34	19.959	20.558	21.462	22.249	23.166	23.496	24.632	26.036
35	20.705	21.310	22.245	23.066	23.999	24.326	25.488	26.929
36	21.493	22.117	23.050	23.858	24.813	25.156	26.346	27.814
37	22.336	22.877	23.844	24.678	25.644	25.983	27.210	28.695
38	23.006	23.645	24.625	25.488	26.480	26.837	28.080	29.596
39	23.782	24.430	25.423	26.330	27.342	27.696	28.954	30.488
40	24.492	25.173	26.183	27.072	28.113	28.492	29.789	31.367
41	25.279	25.992	27.030	27.941	28.996	29.365	30.672	32.280
42	25.988	26.723	27.808	28.755	29.821	30.209	31.544	33.162

Table 4a (continued)

n \ α	0.001	0.002	0.005	0.010	0.020	0.025	0.050	0.100
43	26.850	27.594	28.660	29.592	30.686	31.073	32.412	34.057
44	27.598	28.341	29.446	30.413	31.521	31.912	33.287	34.967
45	28.355	29.082	30.212	31.205	32.330	32.733	34.134	35.850
46	29.087	29.872	31.044	32.028	33.200	33.606	35.030	36.762
47	29.925	30.669	31.831	32.856	34.032	34.443	35.885	37.648
48	30.646	31.461	32.648	33.704	34.880	35.307	36.756	38.551
49	31.446	32.227	33.454	34.525	35.754	36.184	37.662	39.469
50	32.243	33.092	34.295	35.361	36.582	37.023	38.526	40.369
51	33.047	33.881	35.138	36.225	37.473	37.914	39.430	41.272
52	33.806	34.620	35.959	37.052	38.299	38.737	40.298	42.182
53	34.634	35.491	36.760	37.880	39.173	39.616	41.200	43.107
54	35.415	36.262	37.616	38.720	40.011	40.462	42.065	43.999
55	36.140	37.084	38.426	39.564	40.863	41.332	42.939	44.904
56	36.989	37.901	39.235	40.414	41.741	42.213	43.826	45.805
57	37.831	38.760	40.075	41.249	42.607	43.074	44.709	46.722
58	38.633	39.533	40.924	42.108	43.463	43.944	45.608	47.630
59	39.368	40.305	41.762	42.964	44.321	44.804	46.496	48.545
60	40.222	41.171	42.591	43.805	45.195	45.688	47.404	49.473
61	41.080	42.006	43.381	44.632	46.036	46.535	48.259	50.362
62	41.842	42.842	44.294	45.514	46.927	47.433	49.190	51.313
63	42.625	43.646	45.095	46.371	47.802	48.300	50.060	52.210
64	43.403	44.415	45.902	47.191	48.652	49.170	50.957	53.119
65	44.259	45.262	46.758	48.057	49.511	50.043	51.850	54.041
66	45.055	46.070	47.622	48.918	50.404	50.929	52.760	54.965
67	45.802	46.882	48.453	49.781	51.282	51.806	53.639	55.880
68	46.651	47.741	49.288	50.601	52.124	52.676	54.547	56.800
69	47.539	48.599	50.113	51.480	53.023	53.569	55.435	57.703
70	48.263	49.358	50.963	52.321	53.887	54.446	56.345	58.626
71	49.123	50.224	51.794	53.172	54.740	55.297	57.209	59.519
72	50.017	51.097	52.730	54.099	55.662	56.217	58.141	60.485
73	50.809	51.896	53.532	54.939	56.517	57.076	59.027	61.383
74	51.577	52.707	54.355	55.774	57.385	57.968	59.942	62.311
75	52.407	53.562	55.213	56.659	58.287	58.870	60.852	63.232

Table 4a (continued)

n \ α	0.001	0.002	0.005	0.010	0.020	0.025	0.050	0.100
76	53.321	54.436	56.074	57.553	59.189	59.771	61.747	64.181
77	54.108	55.206	56.927	58.397	60.049	60.620	62.626	65.070
78	54.787	55.993	57.741	59.227	60.911	61.485	63.519	65.987
79	55.652	56.776	58.565	60.086	61.789	62.398	64.475	66.946
80	56.526	57.745	59.489	60.992	62.683	63.285	65.377	67.874
81	57.421	58.576	60.325	61.868	63.557	64.163	66.246	68.781
82	58.121	59.375	61.124	62.678	64.405	65.020	67.141	69.698
83	58.927	60.188	62.008	63.555	65.327	65.945	68.067	70.633
84	59.873	61.044	62.868	64.447	66.211	66.836	68.988	71.571
85	60.668	61.924	63.788	65.328	67.127	67.746	69.897	72.498
86	61.551	62.753	64.628	66.224	68.027	68.653	70.837	73.441
87	62.322	63.616	65.443	67.059	68.872	69.505	71.711	74.348
88	63.224	64.473	66.339	67.960	69.767	70.412	72.628	75.284
89	64.079	65.322	67.205	68.846	70.677	71.324	73.526	76.201
90	64.807	66.174	68.072	69.719	71.560	72.204	74.464	77.166
91	65.689	67.063	68.994	70.619	72.467	73.130	75.378	78.068
92	66.578	67.899	69.779	71.453	73.351	74.005	76.269	78.990
93	67.468	68.744	70.688	72.345	74.216	74.879	77.174	79.928
94	68.193	69.533	71.529	73.215	75.160	75.803	78.101	80.875
95	69.033	70.327	72.344	74.069	75.994	76.684	79.016	81.801
96	69.981	71.279	73.272	74.981	76.911	77.585	79.924	82.720
97	70.674	72.121	74.117	75.857	77.806	78.469	80.816	83.636
98	71.604	72.957	75.038	76.751	78.706	79.387	81.722	84.579
99	72.447	73.881	75.906	77.615	79.571	80.251	82.636	85.518
100	73.354	74.687	76.732	78.504	80.469	81.166	83.558	86.444
101	74.153	75.512	77.591	79.371	81.378	82.078	84.480	87.390
102	74.873	76.368	78.441	80.241	82.249	82.944	85.389	88.317
103	75.828	77.252	79.384	81.129	83.153	83.883	86.320	89.250
104	76.763	78.169	80.232	82.000	84.062	84.788	87.243	90.214
105	77.538	78.997	81.067	82.883	84.933	85.660	88.147	91.120
106	78.403	79.774	81.927	83.821	85.864	86.587	89.082	92.070
107	79.266	80.748	82.912	84.699	86.770	87.500	89.990	93.003
108	80.032	81.509	83.692	85.543	87.647	88.401	90.927	93.935

Table 4a (continued)

n \ α	0.001	0.002	0.005	0.010	0.020	0.025	0.050	0.100
109	80.928	82.450	84.665	86.471	88.592	89.343	91.863	94.900
110	81.749	83.251	85.489	87.362	89.459	90.185	92.730	95.794
111	82.528	84.073	86.351	88.212	90.340	91.091	93.661	96.737
112	83.513	85.050	87.251	89.118	91.243	91.987	94.571	97.675
113	84.452	85.971	88.175	90.019	92.168	92.942	95.516	98.628
114	85.223	86.792	89.002	90.938	93.071	93.826	96.415	99.518
115	86.108	87.623	89.852	91.787	93.946	94.704	97.359	100.500
116	86.876	88.354	90.682	92.687	94.876	95.644	98.279	101.440
117	87.747	89.387	91.648	93.603	95.802	96.575	99.214	102.360
118	88.652	90.233	92.482	94.435	96.650	97.419	100.124	103.298
119	89.537	91.070	93.440	95.395	97.598	98.386	101.062	104.265
120	90.258	91.839	94.229	96.243	98.467	99.247	101.935	105.180
121	91.109	92.783	95.147	97.148	99.378	100.160	102.884	106.112
122	92.063	93.681	96.019	98.027	100.302	101.075	103.807	107.047
123	93.016	94.556	96.925	98.967	101.202	102.001	104.740	108.019
124	93.792	95.370	97.720	99.797	102.080	102.882	105.647	108.940
125	94.618	96.299	98.702	100.747	103.062	103.871	106.605	109.897
126	95.639	97.267	99.587	101.633	103.896	104.699	107.518	110.845
127	96.789	98.217	100.585	102.610	104.917	105.753	108.515	111.812
128	97.237	99.045	101.417	103.438	105.815	106.638	109.417	112.742
129	98.044	99.885	102.287	104.311	106.652	107.464	110.242	113.620
130	99.050	100.683	103.205	105.225	107.551	108.383	111.206	114.583
131	99.749	101.630	104.198	106.182	108.452	109.303	112.118	115.505
132	100.689	102.310	104.745	106.927	109.391	110.199	113.096	116.504
133	101.584	103.465	105.987	108.012	110.327	111.133	113.977	117.403
134	102.439	104.268	106.704	108.832	111.253	112.074	114.908	118.365
135	103.299	105.059	107.642	109.687	112.090	112.942	115.884	119.334
136	103.897	105.735	108.460	110.556	113.034	113.877	116.773	120.255
137	104.739	106.669	109.399	111.510	113.939	114.801	117.671	121.172
138	105.754	107.530	110.180	112.403	114.827	115.668	118.593	122.159
139	106.604	108.379	111.189	113.362	115.779	116.623	119.596	123.111
140	107.467	109.195	112.040	114.283	116.779	117.656	120.590	124.073
141	108.372	110.139	112.794	115.023	117.558	118.428	121.410	125.012

Table 4a (continued)

n \ α	0.001	0.002	0.005	0.010	0.020	0.025	0.050	0.100
142	109.330	111.188	113.710	115.937	118.509	119.369	122.340	125.920
143	110.097	112.000	114.640	116.874	119.339	120.260	123.255	125.846
144	110.705	112.776	115.568	117.779	120.291	121.142	124.169	127.822
145	112.219	113.959	116.584	118.715	121.247	122.125	125.112	128.745
146	112.844	114.649	117.333	119.643	122.062	122.937	125.998	129.606
147	113.627	115.344	118.127	120.500	123.031	123.941	126.942	130.593
148	114.734	116.580	119.140	121.350	123.934	124.844	127.946	131.602
149	115.359	117.328	119.920	122.321	124.799	125.742	128.850	132.524
150	116.624	118.342	120.900	123.117	125.751	126.670	129.745	133.489
151	117.227	119.185	121.867	124.152	126.766	127.652	130.716	134.402
152	118.399	120.047	122.759	125.042	127.671	128.585	131.649	135.324
153	118.769	120.598	123.637	125.921	128.527	129.435	132.542	136.339
154	119.949	121.840	124.525	126.829	129.491	130.398	133.520	137.265
155	120.919	122.752	125.402	127.745	130.358	131.258	134.475	138.247
156	121.802	123.654	126.404	128.712	131.335	132.241	135.398	139.199
157	122.607	124.380	127.274	129.596	132.149	133.098	136.323	140.067
158	123.370	125.194	128.102	130.467	133.098	134.023	137.234	141.066
159	124.247	126.224	129.074	131.410	133.932	134.852	138.157	142.032
160	124.879	126.823	129.741	132.230	134.997	135.930	139.071	142.938
161	126.259	128.258	130.860	133.319	135.988	136.927	140.084	143.927
162	126.905	128.815	131.537	133.974	136.827	137.726	140.958	144.790
163	127.570	129.618	132.551	134.997	137.779	138.715	141.946	145.839
164	128.379	130.453	133.386	135.858	138.585	139.573	142.839	146.748
165	129.679	131.498	134.284	136.707	139.427	140.383	143.708	147.635
166	130.514	132.443	135.338	137.757	140.447	141.392	144.679	148.650
167	131.582	133.475	136.255	138.711	141.429	142.399	145.669	149.591
168	131.884	134.072	136.959	139.558	142.306	143.275	146.612	150.485
169	132.840	134.973	138.030	140.423	143.241	144.232	147.499	151.459
170	133.907	135.982	138.708	141.210	144.015	145.045	148.398	152.383
171	134.817	136.909	139.744	142.340	145.095	146.094	149.445	153.376
172	135.972	138.043	140.880	143.257	146.073	147.060	150.399	154.412
173	136.569	138.769	141.785	144.178	147.032	148.037	151.304	155.298
174	137.292	139.414	142.495	144.972	147.794	148.814	152.215	156.257

Table 4a (continued)

n \ α	0.001	0.002	0.005	0.010	0.020	0.025	0.050	0.100
176	139.419	141.484	144.397	146.873	149.752	150.758	154.125	158.195
178	141.154	143.205	146.115	148.544	151.492	152.515	155.988	160.112
180	142.264	144.630	147.795	150.424	153.338	154.337	157.836	161.954
182	144.554	146.602	149.754	152.355	155.270	156.266	159.760	163.876
184	146.345	148.389	151.454	154.042	156.987	158.056	161.597	165.739
186	147.990	150.160	153.267	156.001	158.855	159.914	163.449	167.692
188	149.604	151.941	155.147	157.785	160.796	161.822	165.379	169.598
190	151.589	153.778	156.960	159.694	162.580	163.637	167.187	171.420
192	153.225	155.398	158.564	161.373	164.417	165.463	169.101	173.374
194	155.209	157.387	160.621	163.225	166.322	167.402	171.028	175.299
196	156.844	159.158	162.259	164.981	168.054	169.125	172.825	177.136
198	158.695	160.840	164.282	167.030	170.067	171.106	174.754	179.088
200	160.529	162.594	165.932	168.671	171.831	172.889	176.627	181.006
202	162.699	164.865	167.850	170.670	173.764	174.915	178.548	182.875
204	163.722	166.312	169.567	172.356	175.516	176.587	180.348	184.816
206	165.937	168.369	171.667	174.319	177.459	178.540	182.200	186.695
208	167.845	170.164	173.289	176.075	179.226	180.361	184.145	188.605
210	169.590	171.762	175.204	178.019	181.248	182.371	186.098	190.598
212	171.094	173.589	176.957	179.800	182.975	184.077	187.903	192.512
214	173.129	175.645	178.996	181.760	184.948	186.083	189.858	194.482
216	175.002	177.279	180.715	183.596	186.762	187.916	191.835	196.337
218	176.652	179.026	182.379	185.265	188.539	189.596	193.518	198.143
220	178.249	180.665	184.217	187.142	190.451	191.666	195.533	200.137
222	180.041	182.322	186.048	188.975	192.293	193.464	197.433	202.081
224	181.972	184.485	188.010	190.971	194.193	195.300	199.284	203.989
226	183.827	186.208	189.678	192.716	195.992	197.151	201.180	205.867
228	185.219	187.885	191.467	194.580	197.858	199.002	202.991	207.758
230	187.238	189.744	193.390	196.392	199.720	200.906	204.890	209.640
232	189.478	191.797	195.349	198.309	201.629	202.787	206.819	211.586
234	190.869	193.367	197.014	200.119	203.500	204.705	208.684	213.444
236	192.887	195.364	198.922	201.967	205.333	206.549	210.608	215.398
238	194.424	197.025	200.705	203.761	207.178	208.354	212.429	217.262
240	196.484	198.939	202.622	205.678	209.112	210.342	214.420	219.220

Table 4a (continued)

n \ α	0.001	0.002	0.005	0.010	0.020	0.025	0.050	0.100
242	198.204	200.530	204.220	207.347	210.852	212.089	216.221	221.131
244	200.269	202.698	206.194	209.260	212.833	214.027	218.191	223.065
246	201.798	204.309	207.982	211.227	214.682	215.905	220.046	224.992
248	203.643	206.194	209.880	213.022	216.526	217.743	221.931	226.849
250	205.456	208.109	211.760	214.859	218.393	219.612	223.837	228.794
260	214.399	217.059	220.857	224.209	227.787	229.003	233.313	238.451
270	223.249	225.883	230.097	233.487	237.174	238.412	242.786	247.961
280	232.929	235.488	239.389	242.752	246.508	247.856	252.320	257.596
290	241.422	244.279	248.288	251.788	255.728	257.119	261 857	267.266
300	250.557	253.410	257.734	261.239	265.173	266.544	271.293	276.811
310	260.269	262.980	267.310	270.846	274.796	276.183	288.911	286.517
320	269.207	272.254	276.393	280.106	284.132	285.575	290.406	296.147
330	277.794	281.034	285.586	289.417	293.542	294.970	299.898	305.780
340	287.019	290.398	294.913	298.772	302.945	304.440	309.488	315.430
350	296.630	299.702	304.315	308.174	312.317	313.774	318.906	325.110
360	305.438	308.870	313.475	317.508	321.863	323.349	328.541	334.603
370	314.850	318.132	322.812	326.833	331.216	332.773	338.142	344.305
380	323.754	327.308	332.158	336.164	340.793	342.329	347.611	353.997
390	332.785	336.223	341.209	345.495	350.140	351.780	357.342	363.705
400	342.052	345.631	350.695	354.949	359.650	361.228	366.801	373.295
410	351.593	355.281	360.280	364.472	369.167	370.859	376.474	383.057
420	360.728	364.214	369.354	373.854	378.613	380.279	385.979	392.662
430	369.656	373.737	379.108	383.311	388.124	389.831	395.619	402.334
440	379.596	383.221	388.356	392.719	397.628	399.352	405.176	412.014
450	388.708	392.426	397.784	402.260	207.134	408.850	414.818	421.744
460	398.120	401.796	406.947	411.561	416.780	418.553	424.430	431.424
470	406.469	410.654	416.248	420.878	426.140	427.935	433.980	441.138
480	416.411	420.199	425.865	430.562	435.810	437.599	443.616	450.797
490	425.748	429.653	435.325	440.046	445.192	447.040	453.229	460.541
500	435.245	439.306	444.946	449.618	454.759	456.633	462.962	470.257
900	810.745	816.248	824.580	831.540	838.818	841.406	850.053	860.227
2500	2349.054	2358.066	2372.543	2384.633	2397.633	2401.243	2416.273	2433.481
5000	4781.866	4797.142	4812.946	4832.514	4851.610	4857.410	4879.174	4904.208
9000	8692.095	8705.814	8737.746	8763.361	8789.208	8797.825	8825.937	8860.094

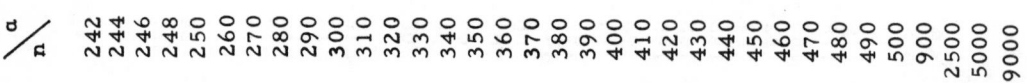

Table 4b. Estimated upper probability points, x_q, of the exponential score product-moment statistic, $R_1^{(2)}(n)$, from a Monte Carlo experiment with M = 1,000,000 repetitions, for series of length n = 11(1) 150; M = 500,000 for n = 151(1) 174, 176(2) 250, 260(10) 500, 900; M = 200,000 for n = 2, 500; M = 100,000 for n = 5,000; and M = 50,000 for n = 9,000.

n \ α	0.900	0.950	0.975	0.980	0.990	0.995	0.998	0.999
11	12.242	13.109	13.810	14.008	14.532	14.935	15.338	15.577
12	13.440	14.362	15.127	15.340	15.913	16.370	16.844	17.133
13	14.632	15.624	16.448	16.675	17.301	17.818	18.359	18.694
14	15.819	16.870	17.749	18.002	18.688	19.264	19.868	20.244
15	16.994	18.093	19.024	19.290	20.035	20.649	21.300	21.721
16	18.156	19.323	20.302	20.581	21.353	22.037	22.774	23.218
17	19.321	20.550	21.595	21.888	22.713	23.425	24.193	24.682
18	20.493	21.769	22.859	23.166	24.039	24.787	25.643	26.165
19	21.642	22.968	24.097	24.429	25.364	26.159	27.042	27.617
20	22.802	24.184	25.357	25.705	26.661	27.508	28.427	29.040
21	23.950	25.382	26.607	26.972	27.974	28.831	29.806	30.472
22	25.097	26.581	27.843	28.214	29.254	30.167	31.199	31.874
23	26.246	27.770	29.086	29.472	30.529	31.461	32.565	33.259
24	27.369	28.954	30.312	30.708	31.827	32.802	33.934	34.650
25	28.520	30.142	31.541	31.954	33.094	34.130	35.290	36.078
26	29.664	31.337	32.772	33.190	34.379	35.423	36.594	37.370
27	30.780	32.477	33.946	34.380	35.618	36.710	37.974	38.821
28	31.925	33.680	35.184	35.629	36.907	37.999	39.322	40.161
29	33.038	34.838	36.387	36.846	38.152	39.294	40.624	41.509
30	34.163	35.998	37.586	38.047	39.375	40.524	41.896	42.805
31	35.274	37.168	38.778	39.264	40.618	41.857	43.304	44.195
32	36.405	38.332	39.995	40.492	41.888	43.145	44.549	45.505
33	37.520	39.481	41.168	41.656	43.082	44.368	44.857	46.880
34	38.633	40.634	42.385	42.899	44.362	45.615	47.133	48.173
35	39.765	41.804	43.558	44.084	45.561	46.867	48.436	49.489
36	40.868	42.938	44.747	45.272	46.794	48.185	49.800	50.921
37	41.990	44.123	45.945	46.481	48.024	49.402	51.022	52.102
38	43.098	45.239	47.115	47.666	49.224	50.642	52.281	53.440
39	44.217	46.401	42.285	48.841	50.440	51.902	53.591	54.709
40	45.324	47.553	49.474	50.043	51.668	53.092	54.796	56.011
41	46.427	48.698	50.652	51.234	52.904	54.428	56.191	57.399
42	47.525	49.809	51.794	52.379	54.074	55.598	57.468	58.688

Table 4b (continued)

n \ α	0.900	0.950	0.975	0.980	0.990	0.995	0.998	0.999
43	48.624	50.960	52.981	53.580	55.312	56.887	58.726	59.945
44	49.746	52.125	54.163	54.774	56.516	58.099	59.955	61.300
45	50.846	53.222	55.308	55.916	57.683	59.281	61.178	62.483
46	51.937	54.370	56.456	57.075	58.861	60.499	62.404	63.672
47	53.018	55.466	57.634	57.998	60.123	61.760	63.702	65.038
48	54.112	56.607	58.774	59.423	61.297	62.981	64.971	66.277
49	55.229	57.749	59.942	60.601	62.487	64.186	66.151	67.555
50	56.310	58.868	61.059	61.713	63.621	65.359	67.454	68.858
51	57.409	59.984	62.224	62.916	64.844	66.569	68.618	70.042
52	58.501	61.123	63.402	64.088	66.045	67.830	69.918	71.328
53	59.592	62.218	64.520	65.218	67.228	69.088	71.245	72.521
54	60.714	63.363	65.690	66.386	68.397	70.260	72.403	73.769
55	61.792	64.481	66.812	67.500	69.548	71.414	73.529	75.001
56	62.860	65.582	67.967	68.679	70.732	72.606	74.860	76.343
57	63.968	66.726	69.138	69.849	71.938	73.831	76.115	77.678
58	65.064	67.847	70.288	71.024	73.122	74.974	77.206	78.786
59	66.166	68.994	71.439	72.167	74.253	76.209	78.541	80.135
60	67.216	70.073	72.555	73.322	75.445	77.404	79.714	81.310
61	68.323	71.210	73.718	74.470	76.644	78.614	80.972	82.607
62	69.390	72.285	74.800	75.565	77.350	79.768	82.117	83.689
63	70.476	73.422	75.978	76.738	78.946	80.942	83.293	84.910
64	71.534	74.468	77.030	77.786	80.036	82.046	84.473	86.206
65	72.651	75.606	78.215	78.978	81.232	83.240	85.727	87.486
66	73.721	76.749	79.369	80.169	82.454	84.506	87.086	88.771
67	74.808	77.832	80.497	81.289	83.601	85.680	88.210	90.015
68	75.909	78.973	81.624	82.441	84.753	86.890	89.361	91.203
69	76.968	80.055	82.746	83.544	85.904	88.057	90.593	92.386
70	78.080	81.172	83.910	84.736	87.101	89.261	91.798	93.578
71	79.142	82.263	85.006	85.840	88.207	90.382	92.937	94.822
72	80.222	83.384	86.126	86.943	89.347	91.528	94.127	96.017
73	81.268	84.466	87.236	88.070	90.512	92.715	95.297	97.118
74	82.392	85.622	88.442	89.272	91.715	93.939	96.577	98.434
75	83.449	86.693	89.552	90.402	92.851	95.061	97.834	99.649

Table 4b (continued)

n \ α	0.900	0.950	0.975	0.980	0.990	0.995	0.998	0.999
76	84.526	87.766	90.630	91.495	93.977	96.256	99.023	100.946
77	85.617	88.895	91.792	92.642	95.144	97.428	100.172	102.069
78	86.670	89.969	92.862	93.709	96.211	98.497	101.184	103.088
79	87.776	91.092	94.008	94.882	97.393	99.680	102.449	104.339
80	88.848	92.194	95.138	96.013	98.512	100.873	103.714	105.667
81	89.890	93.270	96.217	97.126	99.692	102.047	104.892	106.854
82	91.002	94.368	97.346	98.240	100.822	103.179	106.059	107.969
83	92.079	95.512	98.470	99.370	101.989	104.346	107.251	109.352
84	93.150	96.591	99.604	100.505	103.133	105.571	108.422	110.420
85	94.191	97.667	100.727	101.628	104.265	106.774	109.699	111.700
86	95.298	98.794	101.866	102.783	105.410	107.847	110.780	112.754
87	96.367	99.878	102.988	103.888	106.559	108.988	111.868	113.939
88	97.421	100.932	104.022	104.952	107.640	110.147	113.048	115.088
89	98.504	102.037	105.157	106.083	108.826	111.264	114.242	116.233
90	99.582	103.118	106.268	107.205	109.909	112.423	115.494	117.567
91	100.608	104.243	107.418	108.353	111.115	113.582	116.663	118.713
92	101.734	105.329	108.516	109.435	112.175	114.713	117.759	119.873
93	102.757	106.428	109.626	110.559	113.361	115.882	119.039	121.193
94	103.883	107.548	110.728	111.685	114.473	117.055	120.138	122.293
95	104.914	108.606	111.861	112.853	115.616	118.236	121.323	123.418
96	105.969	109.684	112.933	113.915	116.797	119.471	122.578	124.753
97	107.038	110.751	114.020	115.008	117.844	120.394	123.452	125.744
98	108.117	111.877	115.191	116.182	119.086	121.685	124.922	127.034
99	109.170	112.939	116.240	117.220	120.132	122.807	126.043	128.293
100	110.219	114.005	117.323	118.315	121.229	123.871	127.071	129.326
101	111.318	115.146	118.514	119.523	122.459	125.187	128.414	130.658
102	112.361	116.215	119.562	120.572	123.493	126.207	129.459	131.817
103	113.458	117.319	120.717	121.702	124.675	127.406	130.768	133.069
104	114.528	118.377	121.786	122.818	125.771	128.527	131.751	134.035
105	115.600	119.478	122.900	123.929	126.939	129.706	133.065	135.264
106	116.661	120.576	124.005	125.035	128.051	130.803	134.249	136.565
107	117.710	121.647	125.104	126.160	129.179	131.924	135.299	137.679
108	118.752	122.701	126.150	127.198	130.179	132.946	136.307	138.558

Table 4b (continued)

n \ α	0.900	0.950	0.975	0.980	0.990	0.995	0.998	0.999
109	119.828	123.766	127.252	128.314	131.333	134.184	137.627	139.933
110	120.902	124.895	128.402	129.466	132.530	135.308	138.728	141.092
111	121.932	125.947	129.433	130.495	133.533	136.372	139.784	142.108
112	123.038	127.076	130.616	131.660	134.753	137.580	140.952	143.544
113	124.093	128.164	131.725	132.771	135.827	138.684	142.182	144.583
114	125.136	129.186	132.748	133.812	136.922	139.814	143.172	145.724
115	126.197	130.253	133.845	134.882	138.028	140.906	144.374	146.948
116	127.269	131.384	134.980	136.068	139.207	142.157	145.667	148.058
117	128.332	132.468	136.099	137.203	140.418	143.324	146.795	149.356
118	129.393	133.561	137.172	138.238	141.426	144.328	147.934	150.434
119	130.450	134.604	138.267	139.364	142.503	145.376	149.009	151.484
120	131.513	135.690	139.330	140.424	143.685	146.657	150.332	152.855
121	132.586	136.780	140.483	141.618	144.841	147.885	151.530	153.937
122	133.651	137.864	141.567	142.671	145.904	148.900	152.561	155.089
123	134.655	138.876	142.591	143.715	146.981	149.941	153.672	156.157
124	135.772	140.020	143.757	144.875	148.128	151.131	154.814	157.367
125	136.810	141.093	144.867	146.000	149.249	152.266	155.907	158.468
126	137.792	142.082	145.802	146.925	150.287	153.254	157.109	159.303
127	138.891	143.192	146.887	148.046	151.315	154.425	158.138	160.628
128	139.980	144.270	148.130	149.320	152.800	155.944	159.719	162.500
129	141.021	145.452	149.190	150.306	153.862	156.945	160.432	162.788
130	142.065	146.405	150.217	151.372	154.737	157.695	161.628	164.408
131	143.082	147.383	151.283	152.515	155.810	158.819	162.764	165.419
132	144.226	148.623	152.482	153.658	157.079	160.167	163.945	166.743
133	145.293	149.634	153.538	154.705	158.057	161.125	164.925	167.718
134	146.378	150.835	154.715	155.872	159.322	162.430	166.137	168.702
135	147.437	151.880	155.757	156.905	160.349	163.456	167.129	169.639
136	148.495	152.915	156.788	157.953	161.400	164.604	168.224	170.705
137	149.503	153.983	157.920	159.082	162.527	165.827	169.750	172.517
138	150.526	154.983	158.934	160.067	163.528	166.690	170.454	173.060
139	151.651	156.182	160.122	161.277	164.825	168.160	172.107	174.640
140	152.744	157.263	161.202	162.358	165.917	169.205	173.164	175.903
141	153.784	158.333	162.365	163.549	167.002	170.289	174.010	176.428

Table 4b (continued)

n \ α	0.900	0.950	0.975	0.980	0.990	0.995	0.998	0.999
142	154.823	159.339	163.373	164.642	168.148	171.534	175.445	178.279
143	155.835	160.390	164.442	165.632	169.085	172.248	176.143	178.972
144	156.897	161.530	165.467	166.682	170.289	173.471	177.440	180.280
145	157.950	162.594	166.577	167.847	171.544	174.717	178.994	181.794
146	159.023	163.670	167.616	168.835	172.428	175.722	179.680	182.547
147	160.086	164.788	168.829	170.065	173.593	176.858	180.935	183.735
148	161.085	165.714	169.849	171.054	174.619	177.830	181.850	184.528
149	162.181	166.898	171.074	172.280	175.840	179.180	183.244	186.232
150	163.292	167.963	172.131	173.412	177.032	180.557	184.663	187.624
151	164.283	168.955	173.064	174.294	178.061	181.347	185.450	188.113
152	165.332	170.046	174.268	175.527	179.177	182.485	186.493	189.469
153	166.379	171.081	175.314	176.554	180.275	183.657	187.759	190.551
154	167.424	172.176	176.396	177.690	181.462	184.812	188.840	191.708
155	168.492	173.262	177.458	178.717	182.523	186.037	190.235	192.923
156	169.561	174.310	178.509	179.722	183.476	186.897	190.910	193.805
157	170.588	175.347	179.652	180.911	184.610	188.039	192.188	194.768
158	171.674	176.507	180.787	182.068	185.848	189.240	193.225	196.042
159	172.720	177.550	181.805	183.066	186.774	190.140	194.519	197.480
160	173.790	178.583	182.835	184.099	187.797	191.133	195.249	198.334
161	174.762	179.667	184.032	185.325	188.962	192.493	196.693	199.830
162	175.813	180.690	184.936	186.238	190.051	193.489	199.670	200.529
163	176.905	181.839	186.123	187.364	191.081	194.530	199.085	202.140
164	178.061	182.973	187.301	188.609	192.471	196.025	200.408	203.410
165	178.968	183.937	188.342	189.670	193.297	196.844	200.994	203.959
166	180.024	184.933	189.225	190.496	194.282	197.729	202.330	205.465
167	181.163	186.173	190.544	191.876	195.819	199.399	203.530	206.353
168	182.178	187.239	191.572	192.870	196.821	200.132	204.404	207.473
169	183.228	188.195	192.544	193.861	197.777	201.349	205.859	208.783
170	184.244	189.193	193.510	194.841	198.683	202.164	206.141	209.139
171	185.256	190.367	194.885	196.219	199.990	203.627	208.148	211.186
172	186.333	191.380	195.864	197.212	201.221	204.540	208.867	211.925
173	187.421	192.487	196.852	198.143	202.051	205.665	210.151	213.458
174	188.501	193.489	198.027	199.372	203.280	206.819	210.994	214.437

Table 4b (continued)

n \ α	0.900	0.950	0.975	0.980	0.990	0.995	0.998	0.999
176	190.655	195.653	200.196	201.507	205.492	209.152	213.522	216.578
178	192.694	197.870	202.456	203.828	207.753	211.365	215.930	218.702
180	194.759	199.898	204.400	205.869	209.892	213.725	218.304	221.362
182	196.918	202.148	206.742	208.107	212.010	215.723	220.424	223.592
184	198.927	204.154	208.716	210.104	214.225	217.887	222.294	225.534
186	201.078	206.226	210.831	212.233	216.349	220.088	224.810	227.858
188	203.174	208.427	213.084	214.482	218.555	222.220	226.696	229.955
190	205.272	210.604	215.297	216.647	220.605	224.444	228.924	232.328
192	207.310	212.627	217.285	218.612	222.717	226.687	231.262	234.629
194	209.371	214.646	219.270	220.703	224.837	228.819	233.535	237.035
196	211.598	216.872	221.616	223.029	227.078	230.829	235.544	238.937
198	213.632	219.066	223.878	225.243	229.431	233.309	238.311	241.702
200	215.736	221.055	225.838	227.259	231.497	235.287	239.970	243.340
202	217.755	223.262	228.195	229.597	233.769	237.649	242.389	246.028
204	219.866	225.399	230.257	231.735	236.087	240.044	244.773	248.510
206	221.938	227.367	232.258	233.705	238.022	242.043	246.613	249.878
208	224.008	229.546	234.420	235.919	240.313	244.457	249.440	252.880
210	226.224	231.786	236.686	238.154	242.522	246.624	251.475	254.847
212	228.350	233.932	238.898	240.447	244.688	248.645	253.763	257.310
214	230.380	235.914	240.908	242.392	246.629	250.687	255.347	258.899
216	232.480	238.106	243.121	244.670	248.979	253.062	257.753	261.474
218	234.468	240.138	245.234	246.761	251.110	255.160	260.167	263.704
220	236.645	242.381	247.347	248.815	253.329	257.455	262.238	265.728
222	238.662	244.365	249.398	250.925	255.350	259.405	264.347	267.604
224	240.818	246.592	251.752	253.292	257.758	261.960	267.084	270.681
226	242.899	248.664	253.693	255.260	259.747	263.833	268.849	272.444
228	244.999	250.803	255.891	257.462	261.890	266.093	271.044	274.688
230	247.090	252.979	258.184	259.724	264.291	268.427	273.508	277.132
232	249.174	254.941	260.133	261.690	266.102	270.315	275.440	278.854
234	251.183	257.072	262.293	263.878	268.451	272.662	277.893	281.497
236	253.335	259.280	264.522	266.094	270.691	275.002	280.088	283.714
238	255.429	261.330	266.516	268.089	272.617	276.789	282.187	286.254
240	257.510	263.420	268.700	270.264	274.950	279.275	284.439	288.165

Table 4b (continued)

n \ α	0.900	0.950	0.975	0.980	0.990	0.995	0.998	0.999
242	259.492	265.425	270.783	272.334	276.995	281.192	286.388	290.329
244	261.635	267.623	273.043	274.649	279.269	283.662	289.058	292.764
246	263.777	269.760	275.126	276.774	281.360	285.724	291.169	294.770
248	265.797	271.842	277.227	278.822	283.512	287.714	293.179	296.900
250	267.968	274.089	279.430	281.032	285.733	290.136	295.519	299.532
260	278.323	284.516	289.977	291.646	296.565	300.958	306.379	310.485
270	288.729	295.065	300.550	302.223	307.102	311.662	317.205	321.214
280	299.038	305.491	311.119	312.808	317.852	322.452	327.919	332.004
290	309.512	316.066	321.818	323.551	328.621	333.429	339.359	343.398
300	319.792	326.470	332.371	334.191	339.311	344.185	349.974	354.192
310	330.257	336.976	342.944	344.720	350.001	354.752	360.423	364.642
320	340.602	347.466	353.601	355.419	360.768	365.620	371.657	375.905
330	350.935	357.825	364.035	365.937	371.245	376.269	382.459	386.477
340	363.383	368.411	374.665	376.576	381.896	386.677	392.854	397.474
350	371.669	378.774	385.206	387.052	392.678	397.723	403.572	408.291
360	381.971	389.219	395.604	397.478	403.245	408.249	212.918	419.456
370	392.397	399.791	406.307	408.351	414.055	419.223	425.885	430.433
380	402.590	210.030	416.743	418.702	424.494	429.846	436.142	441.156
390	413.061	420.655	427.301	429.338	435.127	440.763	447.153	452.170
400	423.365	431.020	437.789	439.768	445.779	451.198	457.748	462.744
410	433.778	441.509	448.460	450.468	456.537	461.941	468.719	473.779
420	443.924	451.940	458.846	460.966	467.093	472.735	479.831	484.544
430	454.291	462.212	469.162	471.266	477.515	483.170	490.018	494.669
440	464.624	472.686	479.748	481.902	488.176	493.863	500.879	505.713
450	474.921	483.111	490.275	492.424	498.766	504.569	511.774	516.780
460	485.191	493.406	500.802	502.942	509.297	515.020	522.363	527.290
470	495.443	503.779	511.742	513.182	519.737	525.766	532.908	537.802
480	505.895	514.228	521.518	523.800	530.181	536.178	543.508	548.705
490	516.122	524.602	532.023	534.292	540.653	546.576	554.311	559.623
500	526.452	535.088	542.649	544.895	551.777	557.810	565.245	570.868
900	936.149	947.587	957.620	960.576	969.384	977.693	987.706	994.723
2500	2561.241	2579.889	2596.700	2601.609	2615.822	2629.187	2644.203	2655.961
5000	5081.945	5110.248	5132.718	5139.160	5158.670	5176.478	5198.950	5214.330
9000	9103.000	9139.120	9168.785	9178.238	9205.096	9215.654	9231.308	9273.687

Table 5a. Estimated lower probability points, x_α, of the exponential score product-moment statistic, $R_2^{(2)}(n)$, from a Monte Carlo experiment with M = 1,000,000 repetitions, for series of length n = 11 (1) 150; M = 500,000 for n = 151(1) 174, 176(2) 250, 260(10) 500, 900; M = 200,000 for n = 2, 500; M = 100,000 for n = 5,000; and M = 50,000 for n = 9,000.

n \ α	0.001	0.002	0.005	0.010	0.020	0.025	0.050	0.100
11	3.420	3.577	3.838	4.103	4.426	4.544	4.973	5.544
12	4.022	4.182	4.477	4.750	5.094	5.222	5.695	6.317
13	4.562	4.770	5.102	5.404	5.781	5.921	6.430	7.101
14	5.128	5.373	5.744	6.093	6.508	6.660	7.205	7.912
15	5.745	5.998	6.401	6.775	7.218	7.379	7.964	8.721
16	6.343	6.609	7.064	7.465	7.949	8.125	8.749	9.542
17	6.977	7.268	7.736	8.169	8.668	8.853	9.516	10.349
18	7.619	7.951	8.449	8.888	9.411	9.603	10.298	11.181
19	8.260	8.596	9.130	9.602	10.165	10.367	11.094	12.020
20	9.931	9.283	9.838	10.347	10.942	11.157	11.913	12.862
21	9.582	9.965	10.552	11.085	11.706	11.930	12.721	13.710
22	10.260	10.675	11.282	11.824	12.462	12.694	13.511	14.547
23	10.955	11.398	12.037	12.587	13.243	13.483	14.332	15.403
24	11.636	12.081	12.739	13.331	14.017	14.273	15.158	16.258
25	12.355	12.805	13.488	14.096	14.812	15.069	15.986	17.122
26	13.059	13.536	14.236	14.862	15.611	15.880	16.818	17.996
27	13.744	14.237	14.976	15.629	16.393	16.670	17.632	18.837
28	14.452	14.953	15.726	16.413	17.212	17.494	18.496	19.726
29	15.171	15.684	16.488	17.171	17.998	18.292	19.321	20.589
30	15.921	16.470	17.274	17.998	18.834	19.129	20.172	21.478
31	16.638	17.202	18.012	18.751	19.607	19.922	20.995	22.325
32	17.364	17.947	18.804	19.567	20.439	20.747	21.849	23.221
33	18.074	18.683	19.573	20.351	21.257	21.578	22.714	24.112
34	18.865	19.444	20.372	21.160	22.063	22.394	23.554	24.992
35	19.602	20.206	21.135	21.948	22.890	23.228	24.406	25.861
36	20.354	21.001	21.951	22.770	23.719	26.066	25.273	26.754
37	21.091	21.742	22.699	23.558	24.540	24.898	26.140	27.651
38	21.815	22.481	23.476	24.375	25.384	25.747	26.992	28.540
39	22.567	23.274	24.296	25.182	26.214	26.578	27.850	29.416
40	23.373	24.048	25.115	26.016	27.041	27.420	28.722	30.314
41	24.113	24.829	25.896	26.815	27.889	28.265	29.583	31.205
42	24.934	25.648	26.706	27.638	28.713	29.103	30.446	32.103

Table 5a (continued)

n \ α	0.001	0.002	0.005	0.010	0.020	0.025	0.050	0.100
43	25.644	26.375	27.493	28.459	29.562	29.945	31.321	33.001
44	26.404	27.160	28.306	29.292	30.421	30.816	32.213	33.915
45	27.169	27.942	29.106	30.090	31.237	31.650	33.068	34.801
46	27.977	28.764	29.930	30.919	32.089	32.503	33.946	35.705
47	28.697	29.513	30.728	31.768	32.944	33.355	34.803	36.575
48	29.577	30.342	31.557	32.580	33.765	34.189	35.677	37.488
49	30.321	31.167	32.378	32.438	34.652	35.080	36.561	38.408
50	31.061	31.903	33.161	34.229	35.478	35.924	37.447	39.302
51	31.817	32.655	33.969	35.093	36.347	36.788	38.323	40.225
52	32.703	33.579	34.842	35.937	37.213	37.660	39.212	41.130
53	33.473	34.311	35.621	36.787	38.055	38.505	40.106	42.036
54	34.290	35.152	36.454	37.594	38.912	39.365	40.990	42.931
55	35.042	35.927	37.271	38.422	39.765	40.229	41.861	43.836
56	35.902	38.816	38.152	39.316	40.649	41.114	42.752	44.753
57	36.689	37.567	38.933	40.125	41.472	41.961	43.633	45.653
58	37.479	38.363	39.744	40.975	42.353	42.841	44.526	46.580
59	38.226	39.173	40.595	41.817	43.200	43.689	45.408	47.492
60	39.062	40.019	41.462	42.683	44.093	44.578	46.309	48.411
61	39.895	40.858	42.299	43.531	44.948	45.454	47.201	49.316
62	40.689	41.641	43.103	44.389	45.835	46.360	48.120	50.247
63	41.527	42.503	43.950	45.236	36.686	47.198	48.982	51.143
64	42.338	43.349	44.821	46.086	47.553	48.062	49.859	52.054
65	43.088	44.146	45.617	46.944	48.420	48.951	50.786	52.997
66	43.911	44.933	46.471	47.812	49.305	49.834	51.665	53.910
67	44.707	45.737	47.286	48.621	50.149	50.684	52.568	54.806
68	45.522	46.572	48.149	49.485	51.028	51.568	53.454	55.738
69	46.286	47.337	49.000	50.356	51.904	52.448	54.350	56.646
70	47.141	48.250	49.845	51.227	52.774	53.326	55.259	57.572
71	47.959	49.079	50.644	52.069	53.658	54.221	56.153	58.506
72	48.686	49.844	51.503	52.936	54.526	55.082	57.050	59.408
73	49.567	50.738	52.397	53.827	55.426	56.000	59.972	60.348
74	50.389	51.540	53.197	54.661	56.301	56.890	58.870	61.273
75	51.204	52.371	54.112	55.553	57.196	57.785	59.788	62.188

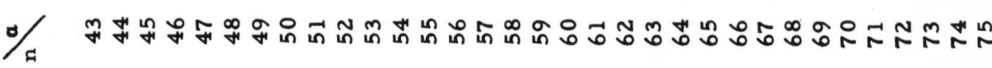

Table 5a (continued)

n \ α	0.001	0.002	0.005	0.010	0.020	0.025	0.050	0.100
76	52.009	53.204	54.955	56.395	58.060	58.642	60.655	63.098
77	52.953	54.089	55.797	57.281	58.924	59.511	61.552	64.031
78	53.766	54.891	56.643	58.167	59.858	60.442	62.485	64.958
79	54.473	55.702	57.487	59.007	60.722	61.323	63.405	65.898
80	55.422	56.614	58.379	59.883	61.586	62.197	64.282	66.793
81	56.234	57.398	59.206	60.741	62.468	63.090	65.203	67.744
82	57.053	58.257	60.082	61.623	63.354	63.981	66.108	68.666
83	57.857	59.084	60.907	62.472	64.232	64.850	67.001	69.579
84	58.647	59.959	61.771	63.335	65.113	65.738	67.895	70.515
85	59.485	60.753	62.652	64.243	66.042	66.675	68.823	71.448
86	60.428	61.646	63.504	65.103	66.911	67.544	69.738	72.387
87	61.328	62.538	64.415	65.986	67.811	68.447	70.625	73.298
88	61.969	63.309	65.218	66.844	68.662	69.319	71.550	74.217
89	63.033	64.298	66.177	67.771	69.608	70.253	72.484	75.168
90	63.765	65.068	66.946	68.563	70.432	71.073	73.351	76.078
91	64.632	65.917	67.880	69.548	71.380	72.039	74.287	77.026
92	65.437	66.803	68.764	70.419	72.306	72.953	75.213	77.965
93	66.212	67.517	69.523	71.195	73.127	73.782	76.102	78.872
94	67.038	68.426	70.450	72.132	74.032	74.693	77.022	79.797
95	67.902	69.272	71.222	72.972	74.913	75.587	77.923	80.726
96	68.761	70.168	72.168	73.881	75.815	76.491	78.845	81.675
97	69.652	71.016	73.046	74.755	76.713	77.394	79.756	82.602
98	70.473	71.954	73.937	75.691	77.611	78.314	80.688	83.538
99	71.307	72.649	74.748	76.560	78.524	79.204	81.592	84.477
100	72.184	73.547	75.644	77.401	79.376	80.096	82.497	85.412
101	73.098	74.435	76.537	78.307	80.327	81.030	83.425	86.361
102	73.953	75.368	77.466	79.192	81.204	81.911	84.347	87.279
103	74.666	76.172	78.259	80.062	82.084	82.814	85.277	88.220
104	75.527	77.053	79.142	80.958	82.985	83.700	86.172	89.152
105	76.472	77.844	79.979	81.808	83.874	84.598	87.102	90.086
106	77.181	78.698	80.855	82.660	84.725	85.453	87.985	90.990
107	78.162	79.569	81.748	83.586	85.661	86.404	88.930	91.950
108	79.072	80.470	82.618	84.494	86.590	87.330	89.851	92.883

Table 5a (continued)

n \ α	0.001	0.002	0.005	0.010	0.020	0.025	0.050	0.100
109	79.833	81.432	83.563	85.404	87.517	88.255	90.799	93.834
110	80.689	82.216	84.416	86.275	88.417	89.157	91.692	94.760
111	81.594	83.070	85.252	87.129	89.281	90.025	92.600	95.670
112	82.314	83.843	86.132	88.005	90.148	90.899	93.500	96.627
113	83.219	84.748	87.035	88.969	91.125	91.888	94.480	97.592
114	84.021	85.657	87.886	89.833	92.003	92.766	95.380	98.512
115	84.997	86.491	88.822	90.741	92.862	93.639	96.283	99.437
116	85.927	87.421	89.667	91.640	93.825	94.588	97.223	100.403
117	86.629	88.194	90.544	92.515	94.720	95.477	98.132	101.316
118	87.578	89.163	91.426	93.392	95.628	96.392	99.058	102.287
119	88.557	90.025	92.323	94.312	96.532	97.318	99.995	103.232
120	89.297	90.803	93.163	95.209	97.440	98.245	100.913	104.161
121	90.129	91.774	94.064	96.055	98.327	99.107	101.848	105.098
122	90.934	92.662	95.036	97.022	99.267	100.047	102.779	106.042
123	91.834	93.459	95.851	97.836	100.141	100.949	103.716	106.989
124	92.797	94.379	96.765	98.751	101.052	101.839	104.611	107.929
125	93.509	95.135	97.499	99.604	101.958	102.749	105.529	108.848
126	94.569	96.154	98.573	100.603	102.890	103.677	106.414	109.800
127	95.274	97.049	99.341	101.393	103.780	104.603	107.386	110.769
128	95.987	97.827	100.353	102.448	104.745	105.510	108.325	111.686
129	96.884	98.637	101.200	103.270	105.640	106.438	109.247	112.646
130	98.069	99.578	101.999	104.078	106.445	107.277	110.132	113.537
131	98.678	100.529	102.887	105.020	107.388	108.217	111.113	114.522
132	99.387	101.310	103.765	105.885	108.292	109.142	112.044	115.448
133	100.614	102.314	104.742	106.836	109.255	110.067	112.982	116.434
134	101.262	103.118	105.532	107.691	110.086	110.910	113.813	117.292
135	102.232	103.925	106.525	108.618	111.096	111.914	114.843	118.294
136	102.994	104.801	107.459	109.534	111.931	112.825	115.795	119.278
137	104.144	105.835	108.236	110.455	112.923	113.758	116.648	120.148
138	104.984	106.729	109.247	111.397	113.843	114.679	117.634	121.131
139	105.740	107.512	110.148	112.286	114.683	115.567	118.590	122.080
140	106.454	108.294	110.812	113.090	115.565	116.428	119.440	123.011
141	107.419	109.140	111.879	113.992	116.469	117.339	120.385	123.925

Table 5a (continued)

n \ α	0.001	0.002	0.005	0.010	0.020	0.025	0.050	0.100
142	108.289	110.004	112.585	114.862	117.352	118.219	121.275	124.861
143	109.129	110.994	113.670	115.962	118.477	119.316	122.255	125.851
144	109.934	111.772	114.408	116.757	119.277	120.198	123.207	126.814
145	111.030	112.784	115.432	117.666	120.235	121.088	124.148	127.728
146	111.955	113.879	116.365	118.549	121.098	121.919	124.970	128.601
147	112.789	114.518	117.257	119.519	122.045	122.907	125.923	129.592
148	113.394	115.214	117.985	120.287	122.890	123.783	126.888	130.517
149	114.219	116.064	118.875	121.272	123.878	124.753	127.854	131.503
150	115.489	117.314	119.902	122.068	124.707	125.624	128.790	132.470
151	116.494	118.225	120.883	123.109	125.627	126.512	129.670	133.402
152	117.159	119.094	121.650	123.951	126.495	127.422	130.626	134.370
153	118.105	119.859	122.468	124.816	127,446	128.345	131.491	135.300
154	118.908	120.648	123.515	125.900	128.475	129.431	132.565	136.260
155	119.609	121.507	124.155	126.628	129.260	130.178	133.368	137.146
156	120.529	122.442	125.351	127.669	130.262	131.194	134.371	138.133
157	121.129	123.138	126.040	128.500	131.264	132.198	135.346	139.137
158	122.257	124.135	126.949	129.413	132.109	133.017	136.207	140.013
159	123.070	124.927	127.830	130.188	132.880	133.842	137.065	140.922
160	124.138	126.080	128.904	131.262	133.912	134.852	138.047	141.930
161	124.854	126.812	129.586	132.040	134.735	135.712	138.799	142.850
162	125.600	127.590	130.552	132.970	135.717	136.683	139.935	143.835
163	126.705	128.633	131.512	133.948	136.631	137.571	140.886	144.785
164	127.434	129.419	132.315	134.798	137.513	138.417	141.783	145.659
165	128.367	130/397	133.279	135.744	138.470	139.415	142.732	146.662
166	129.174	131.269	134.283	136.694	139.413	140.412	143.656	147.590
167	130.038	132.163	135.235	137.698	140.467	141.420	144.736	148.652
168	131.105	133.214	135.995	138.439	141.221	142.244	145.560	149.511
169	131.994	134.099	137.078	139.448	142.231	143.198	146.528	150.547
170	133.059	135.000	137.762	140.304	143.126	144.115	147.434	151.430
171	133.484	135.515	138.494	141.001	143.858	144.869	148.282	152.386
172	134.482	136.586	139.654	142.127	144.920	145.932	149.323	153.284
173	135.444	137.608	140.580	143.120	145.890	146.900	150.296	154.265
174	136.124	138.349	141.322	143.910	146.756	147.755	151.142	155.174

Table 5a (continued)

n \ α	0.001	0.002	0.005	0.010	0.020	0.025	0.050	0.100
176	137.864	139.974	143.113	145.614	148.509	149.505	152.964	157.063
178	139.909	141.975	144.998	147.605	150.416	151.494	154.940	159.017
180	141.694	143.802	146.772	149.458	152.347	153.315	156.791	160.933
182	143.119	145.319	148.619	151.157	154.064	155.120	158.638	162.804
184	144.982	147.341	150.360	153.077	155.931	156.912	160.516	164.705
186	147.049	149.223	152.212	154.933	157.899	158.939	162.472	166.691
188	148.979	151.039	154.184	156.794	159.723	160.739	164.367	168.549
190	150.727	152.719	155.687	158.493	161.532	162.565	166.147	170.447
192	152.602	154.805	157.822	160.525	163.522	164.587	168.148	172.408
194	153.994	156.059	159.432	162.253	165.297	166.336	169.928	174.237
196	155.784	158.112	161.262	164.074	167.067	168.171	171.808	176.150
198	157.334	159.939	163.157	165.777	168.913	169.954	173.575	178.017
200	159.189	161.605	164.932	167.683	170.849	171.985	175.624	179.972
202	161.012	163.255	166.651	169.501	172.617	173.716	177.419	181.872
204	163.367	165.400	168.540	171.335	174.474	175.544	179.297	183.727
206	164.704	167.144	170.507	173.347	176.414	177.480	181.226	185.667
208	166.419	168.804	172.714	175.022	178.172	179.221	183.208	187.501
210	168.404	170.581	174.025	176.845	179.976	181.069	184.949	189.503
212	170.044	172.510	175.882	178.777	181.930	183.070	186.902	191.393
214	172.039	174.514	177.826	180.657	183.877	184.963	188.785	193.304
216	173.506	175.961	179.543	182.472	185.673	186.822	190.674	195.247
218	175.482	177.851	181.390	184.210	187.518	188.654	192.483	197.216
220	174.477	179.689	183.199	186.152	189.415	190.584	194.528	199.084
222	178.937	181.311	184.964	188.005	191.198	192.379	196.319	200.955
224	180.839	183.318	186.862	189.824	193.038	194.205	198.108	202.827
226	182.758	185.104	188.610	191.572	194.967	196.140	200.107	204.821
228	184.264	186.877	190.567	193.527	196.842	198.001	202.015	206.698
230	186.488	188.938	192.384	195.330	198.710	199.887	203.920	208.619
232	188.223	190.778	194.360	197.313	200.692	201.841	205.801	210.535
234	189.714	192.327	196.037	199.049	202.425	203.595	207.646	212.475
236	191.743	194.214	197.802	200.825	204.184	205.404	209.535	214.380
238	193.563	196.038	199.636	202.698	206.167	207.425	211.465	216.310
240	194.879	197.596	201.380	204.446	208.010	209.179	213.319	218.165

Table 5a (continued)

n \ α	0.001	0.002	0.005	0.010	0.020	0.025	0.050	0.100
242	197.149	199.657	203.317	206.371	209.856	211.075	215.218	220.098
244	198.899	201.414	205.104	208.277	211.726	212.944	217.061	221.991
246	200.684	203.179	206.855	210.011	213.522	214.724	219.013	223.977
248	202.569	205.142	208.751	211.893	215.431	216.685	220.928	225.871
250	204.340	206.793	210.595	213.738	217.271	218.539	222.767	227.764
260	213.089	216.009	219.858	223.088	226.635	227.879	232.256	237.362
270	222.284	225.204	229.115	232.392	236.085	237.340	241.757	246.938
280	231.417	234.164	238.311	241.665	245.477	246.777	251.258	256.542
290	240.524	243.280	247.353	250.866	254.788	256.153	260.712	266.161
300	249.624	252.492	256.742	260.401	264.325	265.674	270.302	275.749
310	258.822	261.849	265.954	269.611	273.645	275.031	279.821	285.535
320	267.529	270.795	270.082	278.889	282.960	284.430	289.285	295.052
330	277.024	280.054	284.407	288.259	292.361	293.871	298.821	304.760
340	285.969	289.020	293.694	297.388	301.807	303.332	308.330	314.267
350	295.237	298.431	303.217	307.113	311.461	312.946	318.059	323.820
360	304.370	307.942	312.535	316.439	320.832	322.384	327.558	333.694
370	313.444	316.742	321.478	325.573	330.063	331.669	337.037	343.293
380	322.771	326.247	331.159	335.201	339.695	341.259	346.639	352.944
390	332.058	335.445	340.430	344.523	349.176	350.808	356.245	362.655
400	342.008	345.144	349.998	354.048	358.617	360.244	365.823	372.039
410	350.337	354.005	359.159	363.385	368.144	369.816	375.418	381.978
420	359.649	363.362	368.521	372.793	377.509	379.203	384.935	391.625
430	368.295	372.207	377.747	382.061	387.027	388.688	394.517	401.248
440	378.406	381.902	387.270	391.811	396.699	398.374	404.180	411.059
450	387.456	391.227	396.591	401.112	406.130	407.943	413.824	420.746
460	396.860	400.667	406.048	410.656	415.701	417.474	423.420	430.449
470	405.999	409.851	415.416	419.928	425.150	426.947	432.949	440.098
480	415.551	419.528	424.972	429.647	434.677	436.478	442.649	449.859
490	425.026	428.828	434.349	438.962	444.257	446.048	452.234	459.603
500	433.958	438.162	443.569	448.426	453.752	455.630	461.983	469.272
900	810.188	815.762	823.689	830.504	837.768	840.367	848.989	859.155
2500	2349.389	2358.395	2372.212	2383.519	2396.386	2400.687	2415.540	2432.814
5000	4780.122	4794.958	4814.496	4831.732	4849.534	4856.118	4877.490	4902.398
9000	8693.341	8712.426	8738.532	8761.866	8787.454	8796.512	8826.573	8860.294

Table 5b. Estimated upper probability points, x_α, of the exponential score product-moment statistic, $R_2^{(2)}(n)$, from a Monte Carlo experiment with M = 1,000,000 repetitions, for series of length n = 11(1) 150; M = 500,000 for n = 151(1) 174, 176(2) 250, 260(10) 500, 900; M = 200,000 for n = 2, 500; M = 100,000 for n = 5,000; and M = 50,000 for n = 9,000.

n \ α	0.900	0.950	0.975	0.980	0.990	0.995	0.998	0.999
11	11.405	12.285	13.016	13.225	13.797	14.241	14.690	14.939
12	12.608	13.568	14.360	14.587	15.213	15.718	16.248	16.546
13	13.793	14.812	15.669	15.908	16.580	17.140	17.720	18.071
14	14.975	16.055	16.974	17.242	17.963	18.572	19.216	19.607
15	16.144	17.294	18.272	18.548	19.316	19.970	20.667	21.119
16	17.313	18.522	19.539	19.829	20.648	21.345	22.114	22.588
17	18.483	19.732	20.804	21.119	21.977	22.735	23.569	24.061
18	19.642	20.950	22.069	22.401	23.323	24.098	24.950	25.498
19	20.792	22.150	23.311	23.652	24.624	25.454	26.353	26.964
20	21.955	23.381	24.595	24.949	25.952	26.831	27.774	28.392
21	23.106	24.571	25.821	26.181	27.218	28.113	29.159	29.792
22	24.250	25.776	27.059	27.448	28.530	29.473	30.596	31.289
23	25.387	26.939	28.284	28.675	29.778	30.746	31.874	32.622
24	26.516	28.140	29.509	29.911	31.082	32.098	33.260	33.992
25	27.658	29.316	30.735	31.152	32.354	33.418	34.635	35.411
26	28.794	30.499	31.965	32.392	33.609	34.695	35.939	36.768
27	29.910	31.667	33.170	33.606	34.872	36.004	37.278	38.138
28	31.041	32.835	34.276	34.826	36.118	37.262	38.595	39.490
29	32.167	34.000	35.570	36.042	37.362	38.534	39.911	40.839
30	33.268	35.140	36.757	37.238	38.601	39.804	41.274	42.236
31	34.387	36.307	37.968	38.452	39.834	41.080	42.554	43.509
32	35.523	37.479	39.164	39.662	41.092	42.386	43.851	44.896
33	36.646	38.647	40.370	40.876	42.355	43.667	45.235	46.244
34	37.763	39.801	41.561	42.080	43.569	44.882	46.434	47.490
35	38.872	40.937	42.728	43.271	44.796	46.166	47.780	48.831
36	40.002	42.110	43.909	44.449	45.999	47.360	48.955	50.070
37	41.099	43.247	45.098	45.642	47.228	48.634	50.258	51.405
38	42.194	44.383	46.283	46.830	48.425	49.880	51.556	52.680
39	43.322	45.545	47.473	48.030	49.656	51.100	52.864	54.060
40	44.425	46.678	48.660	49.231	50.873	52.369	54.112	55.272
41	45.514	47.810	49.790	50.386	52.073	53.595	55.363	56.553
42	46.631	48.954	50.974	51.588	53.326	54.896	56.716	57.867

Table 5b (continued)

n \ α	0.900	0.950	0.975	0.980	0.990	0.995	0.998	0.999
43	47.716	50.100	52.151	52.753	54.535	56.113	57.936	59.245
44	48.839	51.240	53.297	53.912	55.683	57.273	59.209	60.451
45	49.939	52.358	54.464	55.082	56.887	58.509	60.455	61.759
46	51.031	53.486	55.602	56.244	58.086	59.804	61.822	63.208
47	52.129	54.622	56.792	57.420	59.308	61.004	62.982	64.393
48	53.229	55.764	57.945	58.592	60.484	62.195	64.202	65.535
49	54.338	56.903	59.141	59.789	61.705	63.398	65.437	66.771
50	55.412	58.016	60.261	60.929	62.840	64.590	66.630	68.065
51	56.497	59.113	61.420	62.084	64.076	65.855	67.959	69.379
52	57.617	60.262	62.559	63.257	65.255	67.054	69.151	70.644
53	58.691	61.374	63.706	64.386	66.409	68.211	70.375	71.802
54	59.790	62.499	64.810	65.519	67.552	69.408	71.597	73.041
55	60.875	63.591	65.976	66.689	68.745	70.587	72.857	74.370
56	61.948	64.720	67.118	67.840	69.916	71.808	74.060	75.598
57	63.060	65.856	68.284	69.009	71.098	72.984	75.255	76.873
58	64.146	66.954	69.445	70.165	72.280	74.116	76.425	77.946
59	65.223	68.065	70.558	71.294	73.482	75.401	77.677	79.262
60	66.335	69.205	71.720	72.467	74.643	76.588	78.907	80.486
61	67.421	70.300	72.821	73.593	75.791	77.717	80.136	81.835
62	68.497	71.415	73.948	74.709	76.930	78.926	81.363	83.066
63	69.583	72.528	75.100	75.869	78.069	80.057	82.462	84.102
64	70.647	73.624	76.206	76.981	79.246	81.266	83.779	85.537
65	71.738	74.734	77.392	78.182	80.493	82.569	85.113	86.761
66	72.809	75.858	78.498	79.293	81.581	83.675	86.145	87.902
67	73.882	76.971	79.652	80.441	82.761	84.846	87.407	89.199
68	74.987	78.058	80.740	81.567	83.927	85.977	88.562	90.406
69	76.056	79.192	81.906	82.706	85.062	87.208	89.766	91.462
70	77.169	80.286	83.021	83.828	86.235	88.374	90.959	92.827
71	78.219	81.397	84.157	84.991	87.352	89.577	92.210	94.012
72	79.290	82.471	85.246	86.069	88.560	90.717	93.339	95.147
73	80.387	83.595	86.384	87.225	89.652	91.853	94.522	96.363
74	81.481	84.699	87.529	88.361	90.828	93.030	95.726	97.652
75	82.545	85.815	88.693	89.540	92.027	94.247	96.962	98.708

Table 5b (continued)

n \ α	0.900	0.950	0.975	0.980	0.990	0.995	0.998	0.999
76	83.594	86.893	89.770	90.649	93.156	95.387	98.099	99.973
77	84.704	88.028	90.923	91.780	94.282	96.555	99.349	101.332
78	85.752	89.089	91.995	92.876	95.424	97.767	100.578	102.494
79	86.827	90.186	93.111	93.988	96.579	98.881	101.657	103.667
80	87.924	91.273	94.213	95.075	97.637	99.979	102.826	104.752
81	88.987	92.403	95.378	96.265	98.849	101.185	104.057	105.907
82	90.086	93.511	96.525	97.437	99.998	102.390	105.269	107.448
83	91.141	94.558	97.605	98.547	101.119	103.559	106.486	108.464
84	92.230	95.705	98.725	99.644	102.275	104.753	107.688	109.624
85	93.302	96.775	99.848	100.770	103.448	105.862	108.884	110.878
86	94.367	97.863	100.963	101.900	104.607	107.051	109.988	111.987
87	95.413	98.962	102.063	102.980	105.658	108.157	111.227	113.343
88	96.476	100.002	103.120	104.040	106.778	109.208	112.257	114.384
89	97.545	101.119	104.267	105.210	107.937	110.387	113.463	115.596
90	98.630	102.232	105.387	106.312	109.053	111.575	114.486	116.598
91	99.685	103.337	106.520	107.458	110.214	112.815	115.868	117.984
92	100.777	104.407	107.566	108.512	111.280	113.827	116.932	119.122
93	101.874	105.518	108.739	109.692	112.548	115.114	118.129	120.303
94	102.907	106.563	109.794	112.890	113.611	116.181	119.308	121.458
95	103.985	107.702	110.966	111.946	114.809	117.426	120.584	122.791
96	105.065	108.793	112.065	113.039	115.876	118.520	121.723	123.948
97	106.112	109.842	113.133	114.126	116.979	119.657	122.907	125.059
98	107.190	110.950	114.273	115.270	118.171	120.778	124.077	126.309
99	108.201	111.995	115.318	116.306	119.236	121.931	125.170	127.383
100	109.289	113.131	116.451	117.452	120.366	123.041	126.193	128.429
101	110.385	114.217	117.568	118.588	121.549	124.267	127.514	129.733
102	111.440	115.273	118.673	119.664	122.640	125.366	128.746	131.028
103	112.524	116.410	119.798	120.827	123.800	126.533	129.734	131.998
104	113.602	117.488	120.906	121.929	124.902	127.609	130.907	133.237
105	114.641	118.533	122.015	123.039	126.058	128.892	132.186	134.553
106	115.689	119.597	123.062	124.125	127.088	129.882	133.195	135.528
107	116.785	120.737	124.161	125.226	128.234	131.017	134.343	136.594
108	117.832	121.790	125.278	126.342	129.441	132.258	135.704	137.913

Table 5b (continued)

n \ α	0.900	0.950	0.975	0.980	0.990	0.995	0.998	0.999
109	118.893	122.886	126.404	127.460	130.525	133.234	136.631	139.074
110	119.984	123.999	127.525	128.563	131.610	134.370	137.807	140.108
111	121.010	125.018	128.580	129.666	132.757	135.534	138.992	141.394
112	122.098	126.163	129.737	130.770	133.856	136.685	140.188	142.694
113	123.179	127.277	130.855	131.895	135.051	137.902	141.463	143.904
114	124.203	128.279	131.882	132.960	136.092	138.939	142.524	144.938
115	125.271	129.353	132.945	134.008	137.167	140.087	143.656	146.163
116	126.345	130.451	134.064	135.150	138.296	141.231	144.798	147.267
117	127.397	131.532	135.178	136.278	139.472	142.437	146.024	148.504
118	128.460	132.622	136.293	137.413	140.543	143.448	146.884	149.468
119	129.500	133.685	137.345	138.458	141.682	144.678	148.187	150.698
120	130.568	134.756	138.439	139.544	142.731	145.703	149.231	151.689
121	131.653	135.860	139.601	140.705	143.979	146.940	150.532	153.047
122	132.690	136.932	140.651	141.752	144.941	147.896	151.522	154.047
123	133.769	138.038	141.757	142.893	146.157	149.268	152.982	155.680
124	134.837	139.115	142.872	144.000	147.270	150.270	153.902	156.456
125	135.898	140.165	143.945	145.081	148.417	151.366	154.896	157.445
126	136.974	141.252	145.026	146.101	149.302	152.290	155.947	158.514
127	137.993	142.315	146.141	147.369	150.700	153.825	157.468	160.019
128	139.016	143.260	147.150	148.240	151.724	154.812	158.374	160.845
129	140.166	144.540	148.300	149.427	152.725	155.761	159.673	162.435
130	141.119	145.513	149.496	150.603	154.017	156.900	160.817	163.550
131	142.219	146.638	150.522	151.671	155.029	158.230	161.889	164.903
132	143.241	147.654	151.555	152.718	156.039	159.314	163.205	165.755
133	144.277	148.710	152.611	153.808	157.175	160.295	164.209	167.240
134	145.285	149.755	153.694	154.846	158.359	161.499	165.294	167.668
135	146.500	150.925	154.842	156.000	159.367	162.690	166.393	169.070
136	147.433	151.902	155.814	156.986	160.514	163.846	167.727	170.168
137	148.590	153.135	157.126	158.330	161.840	165.065	168.922	171.333
138	149.594	154.023	158.112	159.286	162.681	165.870	169.648	172.311
139	150.734	155.313	159.279	160.514	164.032	167.222	170.869	174.033
140	151.703	156.208	160.180	161.355	164.968	168.283	172.134	174.837
141	152.797	157.316	161.409	162.610	166.174	169.297	173.133	175.815

Table 5b (continued)

n \ α	0.900	0.950	0.975	0.980	0.990	0.995	0.998	0.999
142	153.894	158.432	162.482	163.693	167.234	170.478	174.443	177.202
143	154.930	159.534	163.603	164.837	168.385	171.483	175.544	178.060
144	155.952	160.546	164.592	165.797	169.317	172.639	176.630	179.612
145	157.002	161.602	165.596	166.787	170.376	173.724	177.569	180.399
146	158.023	162.689	166.767	168.035	171.584	174.859	178.680	181.359
147	159.160	163.792	167.879	169.100	172.723	175.973	180.033	182.730
148	160.814	164.805	168.927	170.152	173.742	177.007	181.142	183.710
149	161.214	165.917	170.092	171.438	175.075	178.510	182.730	185.349
150	162.288	167.022	171.181	172.444	176.053	179.474	183.354	186.094
151	163.372	168.155	172.265	173.520	177.150	180.500	184.475	187.584
152	164.393	169.139	173.378	174.655	178.331	181.503	185.596	188.529
153	165.463	170.201	174.351	175.564	179.119	182.560	186.702	189.942
154	166.410	171.229	175.457	176.715	180.412	183.707	187.894	190.812
155	167.606	172.402	176.557	177.805	181.525	184.953	189.050	191.773
156	168.595	173.418	177.663	178.961	182.623	185.877	190.078	193.013
157	169.670	174.488	178.663	179.937	183.705	187.218	191.619	194.452
158	170.654	175.536	179.871	181.130	184.908	188.355	192.555	195.469
159	171.753	176.570	180.860	182.157	185.839	189.289	193.642	196.793
160	172.807	177.677	182.018	183.316	187.028	190.608	194.610	197.609
161	173.852	178.758	182.964	184.304	188.055	191.470	195.854	198.706
162	175.031	179.927	184.224	185.537	189.350	192.859	197.008	200.103
163	175.977	180.844	185.183	186.479	190.195	193.797	198.067	200.887
164	177.028	181.953	186.335	187.617	191.398	194.810	198.934	201.738
165	178.061	183.080	187.442	188.745	192.479	196.039	200.172	202.815
166	179.048	184.059	188.511	189.745	193.694	197.078	200.997	203.939
167	180.196	185.125	189.535	190.861	194.807	198.389	202.882	205.660
168	181.264	186.235	190.607	191.866	195.719	199.379	203.648	206.650
169	182.198	187.247	191.742	193.067	196.779	200.493	204.844	208.349
170	183.290	188.356	192.747	194.099	197.934	201.528	205.828	209.215
171	184.367	189.365	193.819	195.104	198.867	202.512	206.983	210.102
172	185.446	190.395	194.862	196.217	200.129	203.710	208.246	211.031
173	186.395	191.479	196.049	197.390	201.314	204.985	209.408	212.503
174	187.535	192.590	197.087	198.361	202.337	205.842	210.333	213.672

Table 5b (continued)

n \ α	0.900	0.950	0.975	0.980	0.990	0.995	0.998	0.999
176	189.566	194.681	199.072	200.363	204.372	207.968	212.114	215.422
178	191.787	196.888	201.411	202.707	206.547	210.040	214.529	218.039
180	193.807	198.977	203.492	204.872	208.920	212.661	217.009	220.130
182	195.915	201.109	205.664	206.998	211.053	214.892	219.444	222.457
184	198.000	203.207	207.753	209.117	213.008	216.563	221.358	224.480
186	200.021	205.341	209.877	211.287	215.337	219.087	223.502	226.634
188	202.232	207.484	212.173	213.593	217.717	221.412	225.000	229.093
190	204.219	209.519	214.202	215.599	219.735	223.625	228.139	231.647
192	206.410	211.741	216.370	217.803	221.843	225.672	230.347	233.704
194	208.456	213.815	218.657	220.089	224.229	228.199	232.523	235.700
196	210.578	216.010	220.802	222.212	226.424	230.255	234.783	238.105
198	212.725	218.160	222.982	224.403	228.495	232.305	237.181	240.513
200	214.755	220.224	225.027	226.487	230.722	234.700	239.447	243.035
202	216.894	222.393	227.165	228.615	232.842	236.627	241.297	244.708
204	218.936	224.503	229.404	230.829	235.142	238.884	243.574	247.020
206	221.007	226.553	231.451	232.932	237.252	241.375	245.930	249.014
208	223.185	228.717	233.663	235.231	239.527	243.425	248.159	251.287
210	225.197	230.779	235.741	237.263	241.683	246.042	250.821	254.196
212	227.318	232.887	237.879	239.465	243.900	247.929	252.668	255.998
214	229.338	234.976	240.074	241.612	246.126	250.150	255.046	258.588
216	231.543	237.192	242.168	243.691	248.099	252.164	256.101	260.622
218	233.556	239.276	244.269	245.762	250.205	254.178	259.362	262.667
220	235.681	241.441	246.444	247.916	252.466	256.607	261.691	265.315
222	237.738	243.514	248.585	250.064	254.508	258.600	263.614	266.275
224	239.849	245.591	250.727	252.226	256.685	260.688	265.891	269.452
226	241.979	247.803	252.910	254.441	258.881	263.166	268.053	271.712
228	244.014	249.795	254.897	256.509	260.959	265.095	270.003	273.563
230	246.142	251.982	257.091	258.642	263.230	267.385	272.584	276.342
232	248.213	254.100	259.222	260.786	265.229	269.421	274.642	277.985
234	250.299	256.189	261.360	262.907	267.455	271.697	276.803	280.664
236	252.353	258.286	263.482	265.001	269.754	273.976	278.934	282.754
238	254.441	260.330	265.590	267.161	271.767	276.029	281.552	285.182
240	256.554	262.516	267.762	269.442	274.068	278.350	283.743	287.294

Table 5b (continued)

n \ α	0.900	0.950	0.975	0.980	0.990	0.995	0.998	0.999
242	258.583	264.544	269.779	271.355	276.014	280.259	285.472	289.048
244	260.717	266.752	272.024	273.587	278.194	282.542	287.685	291.260
246	262.841	268.898	274.216	275.847	280.528	284.814	290.162	293.759
248	264.864	270.897	276.200	277.847	282.636	287.006	292.099	296.108
250	266.949	273.068	278.456	280.078	284.865	289.423	294.619	298.380
260	277.379	283.569	289.096	290.754	295.575	300.000	305.397	309.307
270	287.756	294.082	299.676	301.362	306.276	310.744	316.153	319.814
280	298.084	304.551	310.231	311.957	316.990	321.657	327.384	331.499
290	308.565	315.045	320.836	322.560	327.640	332.447	338.307	342.469
300	318.907	325.576	331.527	333.246	338.542	343.383	349.199	352.900
310	329.328	336.059	341.991	343.767	349.057	353.864	359.697	364.217
320	339.566	346.466	352.567	354.401	359.669	364.654	370.439	374.804
330	350.024	357.051	363.184	365.035	370.518	375.419	381.577	386.018
340	360.321	367.453	373.732	375.657	381.330	386.629	392.814	396.884
350	370.858	378.008	384.339	386.305	391.915	396.998	403.297	407.626
360	381.097	388.481	394.856	396.866	402.608	407.940	414.084	418.527
370	391.407	398.914	405.410	407.397	413.119	418.341	424.908	429.490
380	401.672	409.133	415.688	417.654	423.608	429.147	436.067	440.461
390	412.064	419.659	426.258	428.234	434.069	439.432	446.012	450.886
400	422.401	429.996	436.791	438.844	444.858	450.342	456.914	461.664
410	432.695	440.440	447.385	449.456	455.424	460.860	467.664	472.317
420	442.998	450.762	457.640	459.746	465.819	471.486	477.982	482.974
430	453.374	461.328	468.338	470.432	476.594	482.109	488.827	493.394
440	463.763	471.780	478.958	481.154	487.396	493.174	499.912	504.744
450	473.892	482.160	489.322	491.460	497.796	503.627	510.498	515.287
460	484.236	492.523	499.803	501.918	508.237	514.142	521.537	526.855
470	494.662	503.003	510.217	512.344	518.725	524.547	531.729	536.460
480	504.885	513.348	520.706	522.970	529.344	535.155	542.546	547.546
490	515.232	523.722	531.195	533.428	528.826	545.867	553.677	558.407
500	525.738	533.883	541.493	543.757	550.492	556.644	564.079	569.275
900	935.203	946.660	956.745	959.793	968.434	976.272	986.049	993.619
2500	2560.286	2579.185	2595.670	2600.628	2614.992	2628.438	2645.720	2658.694
5000	5082.762	5109.168	5131.880	5138.936	5158.894	5177.338	5199.086	5213.814
9000	9099.529	9135.356	9165.187	9176.736	9204.040	9229.760	9259.088	9281.365

Table 6a. Estimated lower probability points, x_α, of the exponential score product-moment statistic, $R_3^{(2)}(n)$, from a Monte Carlo experiment with M = 1,000,000 repetitions, for series of length n = 11 (1) 150; M = 500,000 for n = 151(1) 174, 176(2) 250, 260(10) 500, 900; M = 200,000 for n = 2, 500; M = 100,000 for n = 5,000; and M = 50,000 for n = 9,000.

n \ α	0.001	0.002	0.005	0.010	0.020	0.025	0.050	0.100
11	2.720	2.851	3.079	3.305	3.592	3.702	4.107	4.652
12	3.396	3.532	3.767	4.001	4.294	4.406	4.828	5.405
13	3.826	4.012	4.310	4.592	4.945	5.076	5.555	6.188
14	4.284	4.519	4.876	5.205	5.608	5.753	6.282	6.970
15	4.815	5.079	5.483	5.861	6.302	6.463	7.044	7.779
16	5.439	5.725	6.149	6.545	7.017	7.185	7.800	8.582
17	6.051	6.348	6.807	7.225	7.718	7.901	8.554	9.396
18	6.732	7.011	7.495	7.929	8.453	8.649	9.337	10.214
19	7.309	7.651	8.164	8.635	9.201	9.406	10.126	11.038
20	7.953	8.304	8.871	9.357	9.948	10.158	10.912	11.876
21	8.624	8.984	9.564	10.089	10.697	10.920	11.712	12.711
22	9.286	9.666	10.267	10.813	11.462	11.692	12.516	13.553
23	10.010	10.407	11.011	11.580	12.238	12.477	13.331	14.403
24	10.630	11.054	11.726	12.309	13.006	13.256	14.143	15.255
25	11.333	11.775	12.470	13.089	13.799	14.059	14.970	16.116
26	12.022	12.501	13.220	13.844	14.582	14.857	15.792	16.968
27	12.720	13.228	13.956	14.621	15.375	15.654	16.626	17.836
28	13.429	13.936	14.697	15.378	16.169	16.446	17.447	18.694
29	14.127	14.649	15.443	16.144	16.967	17.258	18.282	19.562
30	14.850	15.400	16.210	16.930	17.779	18.085	19.125	20.432
31	15.585	16.122	16.963	17.719	18.574	18.886	19.964	21.312
32	16.308	16.870	17.763	18.514	19.390	19.710	20.828	22.195
33	17.007	17.592	18.509	19.295	20.198	20.521	21.655	23.070
34	17.695	18.322	19.294	20.095	21.017	21.356	22.504	23.950
35	18.531	19.168	20.087	20.904	21.847	22.190	23.386	24.839
36	19.290	19.902	20.860	21.706	22.664	23.010	24.231	25.724
37	19.999	20.632	21.643	22.505	23.499	23.851	25.088	26.606
38	20.696	21.396	22.412	23.297	24.308	24.668	25.934	27.484
39	21.502	22.193	23.204	24.128	25.149	25.517	26.799	28.372
40	22.266	22.962	24.036	24.938	25.993	26.369	27.668	29.274
41	22.951	23.698	24.790	25.740	26.810	27.198	28.532	30.170
42	23.796	24.538	25.604	26.562	27.644	28.028	29.397	30.069

Table 6a (continued)

n \ α	0.001	0.002	0.005	0.010	0.020	0.025	0.050	0.100
43	24.517	25.284	26.402	27.380	28.478	28.876	30.262	31.962
44	25.351	26.127	27.241	28.217	29.335	29.738	31.148	32.855
45	26.096	26.895	28.027	29.033	30.186	30.592	32.010	33.760
46	26.828	27.640	28.822	29.839	31.016	31.435	32.884	34.654
47	27.699	28.477	29.669	30.697	31.882	32.297	33.767	35.553
48	28.475	29.244	30.436	31.500	32.722	33.151	34.631	36.454
49	29.185	30.024	31.269	32.347	33.566	33.998	35.513	37.365
50	30.002	30.834	32.073	33.166	34.414	34.858	36.388	38.260
51	30.799	31.626	32.922	34.000	35.253	35.705	37.264	39.162
52	31.546	32.405	33.715	34.834	36.136	36.590	38.151	40.075
53	32.313	33.213	34.545	35.669	36.977	37.441	39.033	40.976
54	33.146	34.075	35.408	36.534	37.850	38.322	39.934	41.898
55	33.910	34.849	36.209	37.344	38.697	39.168	40.801	42.792
56	34.721	35.642	37.003	38.176	39.532	40.016	41.686	43.715
57	35.478	36.481	37.866	39.057	40.420	40.912	42.601	44.631
58	36.353	37.311	38.705	39.906	41.300	41.785	43.467	45.522
59	37.187	38.148	39.539	40.757	42.143	42.649	44.369	46.452
60	37.803	38.822	40.341	41.583	42.998	43.496	45.252	47.373
61	38.694	39.714	41.200	42.438	43.879	44.387	46.142	48.279
62	39.551	40.561	42.039	43.307	44.752	45.253	47.011	49.155
63	40.426	41.475	42.903	44.189	45.629	46.139	47.935	50.108
64	41.158	42.197	43.719	45.011	46.491	47.010	48.824	51.016
65	42.030	43.021	44.564	45.864	47.345	47.872	49.695	51.916
66	42.853	43.891	45.421	46.725	48.211	48.749	50.589	52.843
67	43.551	44.619	46.207	47.578	49.113	49.657	51.498	53.764
68	44.447	45.452	47.063	48.392	49.956	50.510	52.400	54.666
69	45.149	46.240	47.841	49.268	50.821	51.384	53.297	55.605
70	46.059	47.141	48.740	50.124	51.734	52.281	54.188	56.530
71	46.882	48.011	49.591	50.999	52.590	53.155	55.085	57.436
72	47.689	48.793	50.441	51.838	53.444	54.014	55.981	58.362
73	48.519	49.631	51.295	52.747	54.357	54.935	56.900	59.298
74	49.269	50.438	52.129	53.590	55.229	55.796	57.796	60.219
75	50.147	51.323	53.015	54.446	56.085	56.687	58.698	61.128

Table 6a (continued)

n \ α	0.001	0.002	0.005	0.010	0.020	0.025	0.050	0.100
76	51.018	52.157	53.861	55.328	56.972	57.556	59.608	62.054
77	51.822	52.987	54.671	56.179	57.846	58.447	60.491	62.989
78	52.596	53.760	55.560	57.080	58.760	59.366	61.448	63.931
79	53.419	54.605	56.413	57.919	59.622	60.230	62.326	64.837
80	54.324	55.466	57.255	58.767	60.495	61.108	63.211	65.737
81	55.084	56.313	58.100	59.648	61.407	62.020	64.131	66.689
82	55.888	57.077	58.925	60.537	62.272	62.903	65.040	67.619
83	56.767	57.999	59.822	61.366	63.151	63.765	65.948	68.537
84	57.554	58.838	60.670	62.268	64.079	64.707	66.873	69.492
85	58.443	59.668	61.577	63.148	64.946	65.575	67.760	70.386
86	59.237	60.529	62.417	64.033	65.830	66.468	68.670	71.330
87	60.113	61.448	63.291	64.905	66.752	67.387	69.598	72.258
88	60.948	62.228	64.133	65.755	67.575	68.218	70.473	73.172
89	61.750	63.095	65.008	66.609	68.479	69.140	71.403	74.121
90	62.686	63.994	65.885	67.535	69.383	70.044	72.317	75.031
91	63.427	64.803	66.731	68.402	70.288	70.954	73.258	75.991
92	64.234	65.644	67.559	69.270	71.168	71.842	74.142	76.892
93	65.199	66.513	68.472	70.159	72.060	72.725	75.027	77.833
94	66.033	67.413	69.408	71.077	72.980	73.652	75.975	78.760
95	66.867	68.293	70.214	71.909	73.830	74.504	76.875	79.688
96	67.628	69.073	71.082	72.831	74.762	75.449	77.814	80.647
97	68.388	69.857	71.874	73.665	75.618	76.316	78.708	81.560
98	69.294	70.711	72.781	74.552	76.518	77.210	79.632	82.496
99	70.154	71.497	73.643	75.433	77.442	78.128	80.526	83.413
100	71.104	72.524	74.569	71.303	78.309	79.020	81.457	84.369
101	71.972	73.347	75.467	77.247	79.220	79.923	82.351	85.288
102	72.806	74.223	76.290	78.075	80.102	80.817	83.286	86.231
103	73.549	75.002	77.184	78.975	81.001	81.712	84.185	87.150
104	74.427	75.902	77.998	79.835	81.885	82.611	85.103	88.050
105	75.308	76.792	78.890	80.730	82.775	83.511	86.027	89.025
106	76.086	77.562	79.760	81.601	83.701	84.435	86.948	89.979
107	76.948	78.513	80.699	82.540	84.646	85.367	87.882	90.861
108	77.794	79.309	81.505	83.403	85.506	86.251	88.776	91.928

Table 6a (continued)

n \ α	0.001	0.002	0.005	0.010	0.020	0.025	0.050	0.100
109	78.713	80.259	82.431	84.280	86.406	87.175	89.720	92.776
110	79.622	81.094	83.296	85.185	87.339	88.088	90.656	93.735
111	80.452	81.939	84.131	86.040	88.202	88.961	91.533	94.648
112	81.218	82.767	85.018	86.918	89.100	89.845	92.465	95.567
113	82.051	83.680	85.885	87.805	90.013	90.777	93.383	96.527
114	82.913	84.487	86.753	88.707	90.930	91.703	94.313	97.470
115	83.792	85.378	87.615	89.612	91.805	92.565	95.257	98.393
116	84.744	86.291	88.582	90.540	92.755	93.530	96.186	99.363
117	85.687	87.159	89.475	91.459	93.681	94.447	97.103	100.315
118	86.509	88.040	90.361	92.283	94.527	95.311	98.008	101.211
119	87.314	88.890	91.191	93.208	95.430	96.224	98.923	102.162
120	88.229	89.767	92.080	94.090	96.347	97.148	99.858	103.117
121	88.946	90.546	92.960	94.952	97.260	98.048	100.798	104.058
122	89.761	91.404	93.836	95.842	98.114	98.933	101.689	104.971
123	90.671	92.372	94.771	96.795	99.083	99.895	102.646	105.933
124	91.489	93.195	95.625	97.635	99.967	100.775	103.586	106.881
125	92.431	94.044	96.499	98.561	100.887	101.692	104.458	107.788
126	93.265	94.893	97.253	99.320	101.677	102.510	105.394	108.746
127	94.254	96.028	98.317	100.365	102.681	103.457	106.335	109.679
128	95.045	96.797	99.282	101.282	103.596	104.428	107.275	110.683
129	95.549	97.447	100.056	102.167	104.468	105.326	108.164	111.533
130	96.733	98.627	101.048	103.065	105.430	106.262	109.087	112.543
131	97.604	99.314	101.912	103.962	106.364	107.170	110.021	113.446
132	98.449	100.244	102.757	104.972	107.183	108.049	110.940	114.396
133	99.424	100.983	103.492	105.694	108.107	108.965	111.938	115.346
134	100.264	101.910	104.411	106.578	109.056	109.923	112.788	116.275
135	101.159	102.738	105.223	107.505	109.971	110.813	113.748	117.227
136	101.889	103.658	106.209	108.375	110.817	111.687	114.661	118.143
137	102.635	104.454	107.137	109.359	111.832	112.720	115.635	119.098
138	103.502	105.294	107.875	110.091	112.617	113.541	116.520	120.038
139	104.502	106.135	108.840	110.958	113.511	114.378	117.403	120.939
140	105.722	107.504	109.931	112.082	114.567	115.441	118.432	121.984
141	106.285	108.079	110.667	112.899	115.416	116.265	119.255	122.883

Table 6a (continued)

n \ α	0.001	0.002	0.005	0.010	0.020	0.025	0.050	0.100
142	107.149	108.954	111.570	113.816	116.315	117.212	120.252	123.843
143	108.192	110.100	112.617	114.731	117.304	118.174	121.202	124.801
144	108.974	110.860	113.406	115.604	118.109	118.995	122.113	125.725
145	109.687	111.626	114.268	116.488	119.101	120.002	123.066	126.666
146	110.710	112.608	115.204	117.569	120.048	120.962	123.997	127.615
147	111.410	113.354	116.092	118.372	120.927	121.810	124.905	128.557
148	112.390	114.177	116.852	119.210	121.746	122.683	125.796	129.465
149	113.434	115.065	117.803	120.202	122.758	123.640	126.745	130.388
150	114.029	115.855	118.746	121.039	123.654	124.562	127.691	131.403
151	115.225	117.090	119.709	122.002	124.565	125.446	128.603	132.354
152	115.959	117.983	120.727	123.072	125.602	126.523	129.610	133.329
153	116.854	118.885	121.596	123.943	126.514	127.425	130.513	134.322
154	117.794	119.629	122.407	124.694	127.295	128.206	131.452	135.212
155	118.404	120.364	123.212	125.600	128.275	129.185	132.358	136.121
156	119.409	121.332	124.104	126.594	129.176	130.112	133.321	137.110
157	120.242	122.092	125.023	127.306	130.051	130.972	134.196	138.039
158	121.142	123.029	125.911	128.392	131.035	131.968	135.182	138.999
159	121.992	124.027	126.850	129.305	131.923	132.891	136.065	139.888
160	122.837	124.794	127.692	130.243	132.896	133.862	137.055	140.903
161	123.999	125.805	128.610	131.040	133.797	134.760	137.931	141.803
162	124.544	126.610	129.555	131.979	134.719	135.642	138.879	142.799
163	125.654	127.596	130.440	132.837	135.583	136.524	139.814	143.697
164	126.619	128.493	131.415	133.902	136.693	137.663	140.843	144.717
165	127.274	129.275	132.262	134.597	137.364	138.366	141.688	145.592
166	127.824	129.944	132.882	135.375	138.267	139.286	142.631	146.519
167	129.120	131.158	134.115	136.524	139.251	140.228	143.545	147.497
168	130.099	131.993	135.014	137.557	140.221	141.109	144.512	148.487
169	130.849	132.752	135.850	138.316	141.061	142.015	145.424	149.402
170	131.324	133.569	136.670	139.083	141.966	142.936	146.313	150.332
171	132.548	134.583	137.433	139.893	142.798	143.832	147.266	151.342
172	133.729	135.650	138.523	141.007	143.858	144.882	148.343	152.369
173	134.529	136.489	139.377	141.898	144.741	145.717	149.193	153.184
174	134.955	137.170	140.275	142.788	145.629	146.662	150.108	154.158

Table 6a (continued)

n \ α	0.001	0.002	0.005	0.010	0.020	0.025	0.050	0.100
176	136.764	139.003	142.129	144.580	147.547	148.555	151.956	156.045
178	138.534	140.760	143.748	146.436	149.311	150.382	153.878	157.988
180	140.554	142.700	145.719	148.285	151.221	152.249	155.741	159.889
182	142.389	144.493	147.636	150.348	153.218	154.206	157.720	161.882
184	143.834	145.988	149.304	151.947	154.934	155.955	159.512	163.690
186	145.895	148.070	151.227	153.768	156.770	157.802	161.415	165.615
188	147.649	149.905	153.069	155.691	158.610	159.663	163.218	167.395
190	149.554	151.589	154.727	157.427	160.439	161.509	165.133	169.412
192	151.149	153.295	156.497	159.356	162.342	163.390	166.968	171.265
194	152.972	155.173	158.406	161.192	164.263	165.300	168.854	173.202
196	154.812	157.047	160.280	162.979	166.108	167.149	170.728	175.040
198	156.734	158.950	162.146	164.838	167.899	168.981	172.731	177.060
200	158.449	160.672	163.964	166.528	169.706	170.779	174.429	178.832
202	159.979	162.121	165.551	168.378	171.506	172.612	176.351	180.800
204	161.863	164.059	167.404	170.304	173.430	174.549	179.327	182.700
206	163.640	166.053	169.310	172.058	175.256	176.386	180.205	184.722
208	165.399	167.582	171.320	174.146	177.159	178.285	182.070	186.618
210	166.944	169.317	172.762	175.792	179.052	180.168	183.913	188.432
212	169.255	171.327	174.732	177.535	180.944	182.095	185.909	190.475
214	171.007	173.349	176.778	179.594	182.695	183.820	187.666	192.244
216	172.667	174.971	178.437	181.369	184.664	185.782	189.642	194.185
218	174.322	176.761	180.294	183.181	186.449	187.607	191.508	196.210
220	176.166	178.633	182.203	185.012	188.255	189.449	193.360	198.028
222	177.859	180.404	183.865	186.771	190.185	191.353	195.276	199.928
224	180.842	182.185	185.686	188.603	192.069	192.231	197.186	201.855
226	181.518	183.867	187.476	190.561	193.920	195.084	199.033	203.760
228	183.299	185.774	189.398	192.479	195.885	197.029	201.017	205.733
230	185.217	187.679	191.367	194.375	197.726	198.867	202.847	207.631
232	187.054	189.364	193.003	196.064	199.432	200.616	204.626	209.464
234	188.919	191.244	194.909	197.963	201.316	202.555	206.612	211.426
236	190.714	193.223	196.927	199.902	203.267	204.462	208.498	213.363
238	192.419	194.852	198.596	201.659	205.075	206.261	210.388	215.236
240	194.149	196.694	200.395	203.359	206.879	208.087	212.261	217.164

Table 6a (continued)

n \ α	0.001	0.002	0.005	0.010	0.020	0.025	0.050	0.100
242	196.209	198.577	202.109	205.250	208.797	210.045	214.177	219.075
244	198.039	200.549	204.062	207.208	210.627	211.856	216.034	220.986
246	199.294	201.887	205.746	208.904	212.411	213.670	217.889	222.870
248	201.457	203.985	207.821	210.963	214.497	215.688	219.842	224.832
250	203.229	205.900	209.593	212.734	216.274	217.514	221.716	226.709
260	212.042	214.804	218.830	222.045	225.657	226.929	231.251	236.365
270	221.449	224.157	227.913	231.268	235.017	236.296	240.683	245.924
280	230.354	233.290	237.332	240.675	244.379	245.708	250.212	255.569
290	239.534	242.307	246.452	249.916	253.827	255.144	259.794	265.164
300	248.490	251.442	255.627	259.156	263.154	264.510	269.272	274.788
310	257.627	260.485	264.966	268.538	272.609	274.004	278.767	284.380
320	266.913	269.864	274.093	277.989	282.108	283.534	288.380	294.086
330	275.959	279.004	283.455	287.242	291.368	292.802	297.750	303.656
340	285.244	288.076	292.648	296.481	300.775	302.260	307.340	313.323
350	294.336	297.412	302.154	305.945	310.295	311.820	316.952	322.972
360	303.402	306.601	311.427	315.364	319.765	321.279	326.449	332.638
370	312.568	315.952	320.727	324.690	329.220	330.734	336.065	342.268
380	321.754	325.254	330.224	334.180	338.656	340.237	345.648	351.990
390	330.947	334.270	339.286	343.454	348.105	349.715	355.105	361.556
400	340.010	343.632	348.709	352.903	357.551	359.215	364.751	371.290
410	349.731	353.145	358.122	362.515	367.201	368.832	374.428	380.952
420	358.806	362.293	367.543	371.854	376.725	378.402	384.000	390.681
430	368.086	371.713	376.834	381.138	386.035	387.716	393.482	400.241
440	377.414	380.905	386.219	390.652	395.604	397.319	403.124	409.970
450	386.357	390.556	395.769	400.308	405.311	406.985	412.878	419.735
460	395.623	399.557	404.923	409.607	414.599	416.335	422.443	429.452
470	405.081	408.865	414.371	419.051	424.128	425.920	431.982	439.115
480	414.382	418.164	423.783	428.504	433.652	435.406	441.622	448.874
490	423.937	427.544	433.080	437.923	443.206	445.039	451.170	458.541
500	432.808	436.596	442.486	447.391	452.801	454.562	460.851	468.197
900	810.191	815.460	823.057	829.618	837.083	839.515	848.091	858.190
2500	2345.836	2356.233	2370.750	2382.468	2395.189	2399.709	2414.607	2432.139
5000	4777.262	4793.194	4813.944	4831.038	4849.082	4855.562	4876.774	4901.622
9000	8688.969	8715.480	8738.418	8762.201	8787.159	8796.126	8825.476	8858.706

Table 6b. Estimated upper probability points, x_α, of the exponential score product-moment statistic, $R_3^{(2)}(n)$, from a Monte Carlo experiment with M = 1, 000, 000 repetitions, fro series of length n = 11(1) 150; M = 500, 000 for n = 151(1) 174, 176(2) 250, 260(10) 500, 900; M = 200, 000 for n = 2, 500; M = 100, 000 for n = 5, 000; and M = 50, 000 for n = 9, 000.

n \ α	0.900	0.950	0.975	0.980	0.990	0.995	0.998	0.999
11	10.415	11.233	11.928	12.151	12.730	13.178	13.605	13.858
12	11.652	12.561	13.309	13.532	14.149	14.631	15.095	15.340
13	12.850	13.841	14.675	14.917	15.579	16.121	16.675	17.016
14	14.047	15.115	16.015	16.276	16.997	17.624	18.241	18.618
15	15.234	16.370	17.329	17.608	18.382	19.040	19.723	20.135
16	16.406	17.618	18.639	18.932	19.765	20.474	21.244	21.715
17	17.577	18.850	19.939	20.249	21.134	21.881	22.689	23.178
18	18.736	20.073	21.202	21.534	22.455	23.261	24.137	24.685
19	19.883	21.274	22.470	22.814	23.784	24.646	25.564	26.174
20	21.041	22.483	23.707	24.067	25.097	25.976	26.961	27.589
21	22.177	23.680	24.949	25.329	26.396	27.323	28.382	29.057
22	23.342	24.893	26.218	26.602	27.704	28.669	29.779	30.516
23	24.480	26.083	27.457	27.862	29.012	30.001	31.131	31.865
24	25.622	27.271	28.671	29.078	30.266	31.296	32.502	33.229
25	26.752	28.435	29.876	30.304	31.528	32.617	33.870	34.663
26	27.878	29.620	31.114	31.546	32.797	33.884	35.162	35.992
27	29.018	30.782	32.316	32.766	34.058	35.206	36.532	37.402
28	30.130	31.943	33.517	33.974	35.274	36.457	37.861	38.789
29	31.272	33.141	34.764	35.231	36.587	37.798	39.197	40.121
30	32.397	34.304	35.937	36.422	37.786	39.032	40.542	41.505
31	33.510	35.454	37.133	37.615	39.034	40.302	41.801	42.772
32	34.630	36.619	38.327	38.840	40.256	41.494	43.054	44.063
33	35.739	37.760	39.503	40.021	41.531	42.894	44.356	45.328
34	36.850	38.909	40.681	41.214	42.737	44.096	45.645	46.729
35	37.976	40.068	41.883	42.423	43.969	45.348	46.948	48.060
36	39.100	41.239	42.084	43.630	45.203	46.587	48.232	49.352
37	40.200	42.383	44.269	44.828	46.410	47.860	49.507	50.664
38	41.296	43.516	45.421	45.989	47.618	49.070	50.781	51.918
39	42.415	44.668	46.578	47.161	48.847	50.349	52.097	53.257
40	43.514	45.794	47.783	48.371	50.053	51.539	53.394	54.598
41	44.616	46.931	48.939	49.529	51.251	52.822	54.641	55.957
42	45.730	48.071	50.128	50.723	52.438	54.014	55.876	57.114

Table 6b (continued)

n \ α	0.900	0.950	0.975	0.980	0.990	0.995	0.998	0.999
43	46.839	49.223	51.273	51.876	53.686	55.294	57.131	58.413
44	47.925	50.345	52.432	53.046	54.856	56.451	58.354	59.733
45	49.025	51.461	53.583	54.222	56.029	57.652	59.654	60.953
46	50.124	52.614	54.776	55.406	57.253	58.908	60.941	62.232
47	51.220	53.736	55.940	56.602	58.436	60.133	62.162	63.514
48	52.332	54.869	57.057	57.714	59.646	61.377	63.393	64.820
49	53.414	55.995	58.249	58.918	60.852	62.618	64.671	66.075
50	54.497	57.088	59.358	60.052	61.997	63.776	65.856	67.313
51	55.613	58.260	60.541	61.240	63.201	64.962	67.142	68.627
52	56.702	59.363	61.700	62.402	64.389	66.220	68.401	69.894
53	57.777	60.459	62.785	63.485	65.528	67.358	69.516	70.962
54	58.883	61.616	64.003	64.709	66.712	68.617	70.855	72.370
55	59.950	62.719	65.146	65.851	67.906	69.796	72.017	73.551
56	61.066	63.847	66.268	66.987	69.037	70.937	73.132	74.695
57	62.176	64.984	67.469	68.220	70.320	72.206	74.489	76.077
58	63.241	66.073	68.542	69.272	71.353	73.301	75.601	77.152
59	64.310	67.166	69.670	70.415	72.589	74.539	76.950	78.601
60	65.392	68.266	70.804	71.553	73.762	75.775	78.102	79.673
61	66.486	69.409	71.976	72.739	74.961	77.013	79.378	81.061
62	67.589	70.532	73.080	73.857	76.059	78.071	80.589	82.255
63	68.677	71.629	74.220	74.997	77.245	79.298	81.837	83.545
64	69.737	72.730	75.356	76.150	78.426	80.468	82.922	84.657
65	70.802	73.832	76.509	77.300	79.566	81.628	84.081	85.774
66	71.913	74.969	77.660	78.469	80.754	82.865	85.359	87.090
67	72.979	76.067	78.762	79.568	81.927	84.039	86.511	88.194
68	74.089	77.180	79.838	80.648	82.968	85.104	87.694	89.369
69	75.161	78.292	81.023	81.844	84.210	86.338	88.911	90.797
70	76.207	79.353	82.079	82.902	85.316	87.475	90.073	92.011
71	77.277	80.447	83.262	84.520	86.582	88.799	91.374	93.242
72	78.384	81.591	84.396	85.234	87.643	89.890	92.499	94.428
73	79.457	82.694	85.538	86.378	88.784	91.013	93.689	95.532
74	80.525	83.758	86.612	87.454	89.929	92.233	94.958	96.719
75	81.623	84.899	87.793	88.661	91.158	93.404	96.169	97.942

Table 6b (continued)

n \ α	0.900	0.950	0.975	0.980	0.990	0.995	0.998	0.999
76	82.721	86.025	88.892	89.732	92.271	94.544	97.267	99.224
77	83.745	87.061	90.011	90.887	93.418	95.785	98.527	100.467
78	84.842	88.194	91.187	92.050	94.623	96.950	99.716	101.633
79	85.933	89.273	92.234	93.117	95.708	98.068	100.844	102.792
80	86.989	90.390	93.381	94.267	96.890	99.229	102.024	103.938
81	88.063	91.453	94.446	95.338	97.938	100.318	103.254	105.197
82	89.159	92.607	95.629	96.545	99.181	101.558	104.473	106.398
83	90.201	93.678	96.699	97.603	100.228	102.644	105.507	107.337
84	91.282	94.746	97.824	98.723	101.366	103.750	106.599	108.677
85	92.344	95.846	98.903	99.845	102.472	104.894	107.832	109.951
86	93.413	96.933	100.055	100.996	103.713	106.174	109.093	111.124
87	94.491	98.060	101.202	102.155	104.878	107.349	110.377	112.493
88	95.562	99.124	102.242	103.173	105.880	108.452	111.507	113.713
89	96.660	100.274	103.396	104.349	107.116	109.664	112.719	114.874
90	97.723	101.325	104.510	105.463	108.224	110.751	113.795	115.954
91	98.771	102.401	105.600	106.560	109.327	111.871	114.896	117.074
92	99.827	103.470	106.664	107.634	110.473	112.992	116.100	118.249
93	100.909	104.595	107.831	108.813	111.633	114.149	117.309	119.559
94	101.979	105.657	108.915	109.885	112.732	115.343	118.448	120.529
95	103.075	106.796	110.060	111.031	113.880	116.506	119.504	121.678
96	104.135	107.867	111.112	112.089	114.962	117.552	120.702	122.893
97	105.191	108.924	112.242	113.228	116.156	118.763	121.988	124.223
98	106.254	110.019	113.320	114.300	117.188	119.877	123.089	125.373
99	107.290	111.095	114.433	115.410	118.383	121.064	124.277	126.629
100	108.363	112.204	115.585	116.590	119.526	122.224	125.547	127.853
101	109.465	113.294	116.703	117.692	120.665	123.349	126.633	128.993
102	110.527	114.394	117.782	118.824	121.781	124.493	127.728	129.957
103	111.570	115.460	118.884	119.917	122.897	125.596	128.924	131.292
104	112.633	116.523	119.955	120.978	123.982	126.709	129.978	132.350
105	113.724	117.643	121.074	122.083	125.064	127.867	131.177	133.534
106	114.737	118.680	122.145	123.180	126.232	128.977	132.254	134.619
107	115.824	119.786	123.277	124.333	127.380	130.122	133.376	135.827
108	116.872	120.856	124.355	125.403	128.490	131.321	134.724	137.069

Table 6b (continued)

n \ α	0.900	0.950	0.975	0.980	0.990	0.995	0.998	0.999
109	117.946	121.931	125.443	126.500	129.561	132.393	135.868	139.293
110	118.988	123.027	126.563	127.635	130.723	133.519	137.092	139.477
111	120.102	124.133	127.701	128.784	131.937	134.758	138.205	140.760
112	121.152	125.235	128.764	129.829	132.955	135.815	139.288	141.627
113	122.223	126.260	129.876	130.943	134.080	136.942	140.477	142.942
114	123.274	127.360	130.958	132.035	135.185	138.114	141.629	144.089
115	124.358	128.490	132.073	133.162	136.297	139.244	142.986	145.237
116	125.392	129.535	133.207	134.281	137.500	140.447	143.984	146.563
117	126.433	130.565	134.240	135.350	138.573	141.523	145.072	147.588
118	127.504	131.684	135.343	136.464	139.657	142.588	146.163	148.668
119	128.592	132.782	136.474	137.577	140.805	143.678	147.222	149.837
120	129.665	133.859	137.526	138.626	141.870	144.836	148.562	151.048
121	130.719	134.945	138.649	139.759	143.038	146.083	149.623	152.105
122	131.753	136.006	139.759	140.853	144.087	147.108	150.734	153.295
123	132.795	137.077	140.859	141.999	145.238	148.261	151.994	154.573
124	133.933	138.199	141.989	143.126	146.409	149.385	153.103	155.607
125	134.926	139.234	143.033	144.183	147.535	150.609	154.354	156.912
126	136.017	140.338	144.168	145.299	148.574	151.673	155.180	157.574
127	137.098	141.384	145.201	146.322	149.630	152.589	156.479	158.918
128	138.152	142.518	146.413	147.598	150.957	154.030	157.727	160.374
129	139.138	143.471	147.347	148.547	151.899	154.954	158.638	161.217
130	140.202	144.633	148.530	149.692	153.075	156.124	160.079	162.544
131	141.266	145.692	149.610	150.752	154.282	157.255	161.168	163.789
132	142.423	146.802	150.663	151.841	155.195	158.334	162.030	164.634
133	143.416	147.850	151.725	152.883	156.294	159.570	163.279	165.873
134	144.496	148.941	152.836	154.031	157.352	160.514	164.320	166.640
135	145.615	149.957	153.911	155.062	158.483	161.569	165.458	168.054
136	146.492	150.932	154.909	156.037	159.501	162.707	166.529	169.078
137	147.639	152.112	156.139	157.292	160.633	163.947	167.737	170.477
138	148.688	153.316	157.282	158.486	161.965	165.252	169.197	172.058
139	149.720	154.252	158.297	159.559	163.080	166.285	170.340	172.960
140	150.737	155.298	159.431	160.669	164.248	167.465	171.414	174.189
141	151.825	156.358	160.502	161.768	165.414	168.812	172.549	175.382

Table 6b (continued)

n \ α	0.900	0.950	0.975	0.980	0.990	0.995	0.998	0.999
142	152.894	157.447	161.422	162.662	166.260	169.564	173.480	176.292
143	153.977	158.618	162.792	163.980	167.497	170.733	174.660	177.603
144	154.968	159.617	163.673	164.907	168.539	171.783	175.829	178.737
145	156.039	160.692	164.764	165.929	169.546	172.664	176.655	179.583
146	157.055	161.766	165.896	167.114	170.838	174.086	178.153	181.103
147	158.217	162.879	166.924	168.195	171.730	175.020	179.233	182.014
148	159.228	163.913	168.038	169.290	172.953	176.304	180.529	183.299
149	160.253	164.957	169.144	170.434	174.177	177.623	181.495	183.937
150	161.292	166.014	170.285	171.522	175.162	178.540	182.665	185.432
151	162.352	166.985	171.171	172.418	176.026	179.288	183.359	185.948
152	163.433	168.162	172.309	173.537	177.247	180.724	184.803	187.552
153	164.447	169.230	173.474	174.706	178.380	181.797	186.143	189.168
154	165.628	170.384	174.576	175.871	179.635	183.183	187.164	189.918
155	166.590	171.410	175.586	176.917	180.544	183.915	188.110	191.030
156	167.666	172.480	176.715	178.019	181.683	185.107	189.164	192.020
157	168.715	173.555	177.874	179.116	182.870	186.255	190.438	193.249
158	169.793	174.651	178.930	180.212	183.973	187.395	191.674	194.580
159	170.845	175.643	180.017	181.288	184.948	188.417	192.484	195.536
160	171.853	176.798	181.109	182.367	186.044	189.645	193.874	197.208
161	172.875	177.756	182.113	183.375	187.054	190.547	195.010	198.240
162	173.980	178.899	183.288	184.530	188.342	191.690	195.920	198.950
163	174.962	179.908	184.271	185.539	189.397	192.915	197.043	200.102
164	176.075	181.006	185.312	186.662	190.663	194.167	198.349	201.804
165	177.120	182.077	186.438	187.759	191.696	195.415	199.867	203.024
166	178.215	183.084	187.457	188.778	192.648	196.244	200.608	203.597
167	179.238	184.193	188.478	189.786	193.766	197.364	201.765	204.553
168	180.218	185.224	189.633	190.947	194.869	198.435	202.948	206.054
169	181.385	186.390	190.803	192.194	196.198	199.693	203.927	206.989
170	182.407	187.440	191.890	193.205	197.000	200.708	204.908	207.826
171	183.398	188.521	193.000	194.340	198.075	201.653	206.202	208.958
172	184.429	189.538	194.012	195.377	199.247	202.939	207.407	210.394
173	185.528	190.647	195.198	196.459	200.497	204.180	208.798	211.937
174	186.553	191.648	196.076	197.405	201.345	204.915	209.425	212.314

Table 6b (continued)

n \ α	0.900	0.950	0.975	0.980	0.990	0.995	0.998	0.999
176	188.683	193.770	198.349	199.768	203.645	207.337	211.657	214.939
178	190.715	195.807	200.381	201.702	205.822	209.585	213.963	217.432
180	192.923	198.103	202.499	203.818	207.891	211.590	215.980	219.045
182	195.046	200.168	204.752	206.182	210.151	213.883	218.428	221.614
184	197.090	202.362	207.059	208.495	212.563	216.353	220.559	224.998
186	199.123	204.388	209.051	210.506	214.443	218.128	222.740	225.893
188	201.253	206.595	211.308	212.695	216.644	220.167	224.464	227.755
190	203.336	208.636	213.365	214.719	218.823	222.689	227.284	230.758
192	205.384	210.763	215.380	216.784	220.937	224.848	229.813	232.744
194	207.478	212.948	217.668	219.079	223.387	227.157	231.914	235.568
196	209.637	215.025	219.831	221.256	225.378	229.079	233.575	236.892
198	211.709	217.027	221.829	223.264	227.375	231.395	235.925	239.332
200	213.939	219.451	224.304	225.820	230.020	233.903	238.519	241.569
202	215.860	221.346	226.222	227.681	231.948	235.714	240.200	243.700
204	217.949	223.437	228.202	229.667	234.065	238.050	242.720	246.219
206	220.118	225.733	230.707	232.187	236.344	240.323	245.040	248.455
208	222.212	227.724	232.607	234.104	238.502	242.535	247.365	250.882
210	224.342	229.892	234.803	236.295	240.577	244.509	249.615	252.994
212	226.403	232.017	237.023	238.523	242.960	247.022	251.675	255.163
214	228.512	234.201	239.290	240.737	245.037	249.100	254.154	257.654
216	230.594	236.268	241.278	242.771	247.047	251.212	256.410	259.741
218	232.565	238.269	243.314	245.857	249.244	253.199	258.267	261.733
220	234.737	240.507	245.613	247.099	251.520	255.723	260.602	264.089
222	236.801	242.571	247.646	249.201	253.597	257.732	262.701	266.187
224	238.831	244.560	249.624	251.149	255.674	259.741	264.811	268.846
226	240.943	246.718	251.778	253.363	257.893	262.019	266.977	270.583
228	243.137	248.995	254.166	255.606	260.229	264.557	269.883	273.275
230	245.184	251.064	256.223	257.752	262.197	266.414	271.484	274.964
232	247.196	253.041	258.200	259.736	264.226	268.605	273.838	277.184
234	249.319	255.300	260.552	262.067	266.712	270.907	276.202	279.912
236	251.399	257.359	262.591	264.167	268.748	272.939	277.989	281.882
238	253.591	259.536	264.770	266.342	270.977	275.359	280.475	284.064
240	255.645	261.617	266.924	268.474	273.141	277.412	282.583	286.257

Table 6B (continued)

n \ α	0.900	0.950	0.975	0.980	0.990	0.995	0.998	0.999
242	257.683	263.692	268.901	270.495	275.119	279.482	284.739	288.546
244	259.764	265.820	271.126	272.699	277.371	281.655	286.739	290.427
246	261.777	267.881	273.245	274.918	279.622	283.989	289.073	292.797
248	263.896	269.956	275.312	276.930	281.540	285.857	291.160	294.854
250	265.998	272.121	277.521	279.134	283.948	288.385	293.737	297.287
260	276.388	282.625	288.078	289.731	294.561	299.040	304.487	308.302
270	286.820	293.222	298.827	300.524	305.415	309.983	315.465	319.343
280	297.169	303.624	309.287	311.000	216.012	320.699	326.294	330.289
290	307.598	314.187	319.960	321.712	326.791	331.529	337.368	341.587
300	317.926	324.607	330.511	332.279	337.430	342.125	347.728	351.812
310	328.305	335.154	341.126	342.914	348.223	353.015	358.994	363.073
320	338.712	345.624	351.685	353.457	358.853	363.829	369.992	374.353
330	349.002	356.010	362.084	363.920	369.279	374.338	380.654	384.981
340	359.364	366.491	372.720	374.570	380.129	385.254	391.449	395.907
350	369.780	377.036	383.338	385.261	390.869	396.081	402.296	406.461
360	380.073	387.349	393.816	395.770	401.442	406.337	412.847	417.700
370	390.371	397.785	404.332	406.326	412.039	417.224	423.694	428.336
380	400.816	408.302	414.944	416.901	422.798	428.335	434.781	439.502
390	411.072	418.650	425.341	427.312	433.232	438.564	445.122	449.949
400	421.409	429.140	435.925	437.918	443.899	449.639	456.060	460.896
410	431.803	439.567	446.451	448.401	454.623	460.128	466.420	471.295
420	442.166	450.041	457.002	459.113	465.107	470.837	477.514	482.516
430	452.365	460.423	467.527	469.557	475.639	481.263	488.289	493.467
440	462.672	470.728	477.953	480.075	486.320	492.112	499.030	503.969
450	472.948	481.087	488.191	490.334	496.685	502.561	510.267	515.770
460	483.323	491.606	498.927	501.084	507.470	513.435	521.421	525.735
470	493.576	501.954	509.291	511.546	517.883	523.668	530.872	536.305
480	503.902	512.247	519.612	521.864	528.606	524.269	541.826	547.056
490	514.200	522.701	530.239	532.512	539.105	534.804	552.435	557.654
500	524.442	533.136	540.682	542.882	549.505	555.585	563.028	568.268
900	934.224	945.606	955.632	958.695	967.558	975.942	985.989	993.243
2500	2559.428	2578.375	2595.012	2599.842	2613.906	2626.557	2641.589	2652.126
5000	5083.250	5110.106	5132.531	5139.354	5158.258	5176.434	5201.378	5216.394
9000	9101.896	9135.575	9165.080	9174.648	9203.292	9227.576	9255.974	9278.135

Table 7 a. Estimated lower probability points, x_{α}, of the 1/2-Gamma score product-moment statistic, $R_1^{(3)}(n)$, from a Monte Carlo experiment with $\tilde{M} = 4,000,000$ repetitions, for series of length n = 10 to n = 40.

n \ α	Mean	0.001	0.002	0.005	0.010	0.020	0.025	0.050	0.100
10	1.9286	.4742	.5091	.5658	.6221	.6952	.7224	.8274	.9783
11	2.1688	.5533	.5928	.6594	.7235	.8077	.8395	.9588	1.1260
12	2.4095	.6336	.6811	.7578	.8313	.9246	.9605	1.0938	1.2782
13	2.6517	.7215	.7733	.8588	.9422	1.0465	1.0859	1.2333	1.4344
14	2.8957	.8123	.8699	.9665	1.0568	1.1708	1.2138	1.3752	1.5936
15	3.1389	.9060	.9701	1.0746	1.1756	1.3007	1.3474	1.5210	1.7570
16	3.3836	1.0030	1.0739	1.1885	1.2963	1.4310	1.4817	1.6690	1.9215
17	3.6278	1.1048	1.1807	1.3064	1.4225	1.5674	1.6211	1.8205	2.0888
18	3.8729	1.2058	1.2883	1.4219	1.5478	1.7039	1.7620	1.9750	2.2591
19	4.1180	1.3099	1.3992	1.5441	1.6778	1.8431	1.9049	2.1297	2.4306
20	4.3642	1.4163	1.5135	1.6684	1.8108	1.9855	2.0502	2.2877	2.6040
21	4.6098	1.5279	1.6329	1.7954	1.9468	2.1310	2.1989	2.4492	2.7796
22	4.8563	1.6406	1.7479	1.9238	2.0836	2.2773	2.3493	2.6122	2.9582
23	5.1054	1.7569	1.8721	2.0545	2.2222	2.4255	2.5016	2.7755	3.1390
24	5.3510	1.8756	1.9976	2.1879	2.3647	2.5788	2.6574	2.9430	3.3210
25	5.5980	1.9946	2.1201	2.3222	2.5068	2.7276	2.8103	3.1101	3.5052
26	5.8454	2.1146	2.2513	2.4614	2.6513	2.8834	2.9691	3.2785	3.6902
27	6.0940	2.2359	2.3766	2.5960	2.7967	3.0394	3.1281	3.4491	3.8680
28	6.3400	2.3662	2.5106	2.7363	2.9441	3.1941	3.2861	3.6195	4.0534
29	6.5898	2.4880	2.6406	2.8747	3.0936	3.3531	3.4481	3.7935	4.2418
30	6.8354	2.6147	2.7713	3.0209	3.2442	3.5137	3.6108	3.9681	4.4305
31	7.0859	2.7468	2.9103	3.1639	3.3977	3.6770	3.7787	4.1448	4.6203
32	7.3328	2.8835	3.0501	3.3135	3.5522	3.8380	3.9433	4.3220	4.8083
33	7.5827	3.0104	3.1844	3.4581	3.7064	4.0018	4.1090	4.4985	4.9993
34	7.8281	3.1478	3.3286	3.6077	3.8649	4.1686	4.2797	4.6769	5.1907
35	8.0797	3.2838	3.4659	3.7597	4.0201	4.3346	4.4484	4.8572	5.3867
36	8.3263	3.4106	3.6047	3.9071	4.1785	4.5013	4.6188	5.0406	5.5801
37	8.5750	3.5558	3.7523	4.0645	4.3401	4.6697	4.7897	5.2193	5.7756
38	8.8230	3.6895	3.8920	4.2117	4.4962	4.8354	4.9579	5.3989	5.9689
39	9.0701	3.8266	4.0406	4.3661	4.6585	5.0058	5.1330	5.5836	6.1699
40	9.3225	3.9779	4.1903	4.5241	4.8227	5.1797	5.3094	5.7719	6.3687

Table 7b. Estimated upper probability points, x_α, of the 1/2-Gamma score product-moment statistic, $R_1^{(3)}(n)$, from a Monte Carlo experiment with $M = 4,000,000$ repetitions, for series of length $n = 10$ to $n = 40$.

n \ α	0.900	0.950	0.975	0.980	0.990	0.995	0.998	0.999	σ
10	3.0429	3.3687	3.6768	3.7473	3.9216	4.0336	4.1418	4.2025	.7924
11	3.3968	3.7576	4.0928	4.1792	4.3984	4.5436	4.6784	4.7584	.8713
12	3.7430	4.1372	4.5018	4.6029	4.8481	5.0370	5.2025	5.3003	.9470
13	4.0888	4.5128	4.9046	5.0182	5.2965	5.5150	5.7232	5.8394	1.0198
14	4.4325	4.8897	5.0345	5.4293	5.7424	5.9886	6.2347	6.3707	1.0910
15	4.7711	5.2601	5.7000	5.8337	6.1815	6.4538	6.7368	6.8927	1.1594
16	5.1105	5.6325	6.0969	6.2393	6.6144	6.9151	7.2300	7.4147	1.2271
17	5.4439	5.9965	6.4813	6.6324	7.0418	7.3665	7.7135	7.7135	1.2914
18	5.7774	6.3613	6.8764	7.0343	7.4658	7.8141	8.1910	8.4203	1.3554
19	6.1100	6.7211	7.2564	7.4203	7.8778	8.2548	8.6602	8.9077	1.4173
20	6.4359	7.0761	7.6375	7.8075	8.2895	8.6935	9.1280	9.3938	1.4772
21	6.7634	7.4324	8.0204	8.1964	8.7006	9.1174	9.5835	9.8727	1.5364
22	7.0873	7.7843	8.3946	8.5785	9.1081	9.5544	10.0493	10.3609	1.5936
23	7.4151	8.1413	8.7717	8.9613	9.5063	9.9727	10.4882	10.8184	1.6515
24	7.7319	8.4862	9.1422	9.3410	9.9099	10.3999	10.9504	11.2989	1.7057
25	8.0535	8.8343	9.5127	9.7148	10.3037	10.8177	11.3883	11.7579	1.7605
26	8.3701	9.1813	9.8782	10.0878	10.6919	11.2218	11.8228	12.2072	1.8130
27	8.6890	9.5225	10.2432	10.4588	11.0806	11.6321	12.2522	12.6614	1.8658
28	9.0027	9.8602	10.6040	10.8279	11.4692	12.0461	12.6969	13.1165	1.9164
29	9.3228	10.2020	10.9655	11.1939	11.8596	12.4502	13.1198	13.5619	1.9678
30	9.6297	10.5384	11.3267	11.5603	12.2390	12.8416	13.5343	13.9927	2.0158
31	9.9470	10.8779	11.6860	11.9270	12.6287	13.2457	13.9573	14.4357	2.0655
32	10.2562	11.2087	12.0390	12.2857	13.0017	13.6428	14.3715	14.8555	2.1134
33	10.5660	11.5487	12.3952	12.6508	13.3926	14.0456	14.8103	15.3067	2.1617
34	10.8738	11.8738	12.7426	13.0012	13.7563	14.4302	15.1997	15.7183	2.2063
35	11.1872	12.2144	13.1016	13.3654	14.1357	14.8209	15.6317	16.1589	2.2534
36	11.4942	12.5380	13.4439	13.7148	14.5058	15.2124	16.0298	16.5728	2.2978
37	11.8006	12.8710	13.7969	14.0723	14.8678	15.5908	16.4268	16.9956	2.3431
38	12.1070	13.1999	14.1459	14.4271	15.2413	15.9734	16.8243	17.4007	2.3887
39	12.4102	13.5170	14.4840	14.7691	15.6046	16.3593	17.2457	17.8377	2.4307
40	12.7153	13.8467	14.8292	15.1235	15.9729	16.7445	17.6376	18.2498	2.4731

Table 8 a. Estimated lower probability points, x_α, of the 1/2 Gamma score product-moment statistic, $R_2^{(3)}(n)$, from a Monte Carlo experiment with $M = 4,000,000$ repetitions, for series of length $n = 10$ to $n = 40$.

n \ α	Mean	0.001	0.002	0.005	0.010	0.020	0.025	0.050	0.100
10	1.7136	.3149	.3442	.3991	.4545	.5257	.5527	.6522	.7876
11	1.9513	.3971	.4321	.4930	.5530	.6310	.6607	.7728	.9294
12	2.1900	.4868	.5254	.5898	.6555	.7410	.7743	.9007	1.0765
13	2.4314	.5627	.6098	.6867	.7623	.8590	.8959	1.0353	1.2278
14	2.6715	.6400	.6963	.7876	.8737	.9814	1.0219	1.1746	1.3848
15	2.9137	.7304	.7928	.8913	.9864	1.1058	1.1500	1.3173	1.5444
16	3.1572	.8267	.8922	1.0012	1.1048	1.2336	1.2825	1.4634	1.7065
17	3.4017	.9217	.9951	1.1144	1.2265	1.3657	1.4182	1.6124	1.8728
18	3.6451	1.0215	1.1017	1.2304	1.3513	1.5005	1.5567	1.7632	2.0408
19	3.8912	1.1243	1.2086	1.3480	1.4774	1.6380	1.6975	1.9171	2.2123
20	4.1345	1.2301	1.3231	1.4681	1.6072	1.7764	1.8402	2.0738	2.3830
21	4.3797	1.3369	1.4356	1.5944	1.7406	1.9197	1.9870	2.2319	2.5569
22	4.6259	1.4469	1.5538	1.7203	1.8753	2.0647	2.1351	2.3922	2.7328
23	4.8726	1.5629	1.6751	1.8497	2.0133	2.2118	2.2864	2.5563	2.9116
24	5.1199	1.6767	1.7933	1.9791	2.1526	2.3631	2.4410	2.7224	3.0921
25	5.3679	1.7940	1.9176	2.1133	2.2953	2.5137	2.5936	2.8881	3.2735
26	5.6107	1.9130	2.0432	2.2489	2.4380	2.6656	2.7490	3.0548	3.4556
27	5.8607	2.0338	2.1713	2.3857	2.5828	2.8214	2.9086	3.2245	3.6400
28	6.1054	2.1585	2.3063	2.5270	2.7293	2.9757	3.0653	3.3937	3.8232
29	6.3543	2.2831	2.4358	2.6675	2.8817	3.1367	3.2313	3.5691	4.0115
30	6.6002	2.4110	2.5662	2.8084	3.0280	3.2929	3.3901	3.7419	4.1971
31	6.8487	2.5413	2.7027	2.9519	3.1880	3.4517	3.5519	3.9164	4.3865
32	7.0983	2.6690	2.8385	3.0965	3.3312	3.6134	3.7167	4.0925	4.5770
33	7.3448	2.8024	2.9743	3.2422	3.4838	3.7773	3.8842	4.2700	4.7671
34	7.5934	2.9310	3.1115	3.3925	3.6436	3.9439	4.0524	4.4478	4.9581
35	7.8411	3.0662	3.2534	3.5406	3.7984	4.1068	4.2201	4.6265	5.1500
36	8.0886	3.1997	3.3947	3.6865	3.9544	4.2714	4.3884	4.8076	5.3446
37	8.3346	3.3435	3.5349	3.8443	4.1170	4.4435	4.5629	4.9897	5.5373
38	8.5861	3.4794	3.6789	3.9906	4.2732	4.6072	4.7300	5.1708	5.7332
39	8.8345	3.6126	3.8227	4.1466	4.4346	4.7782	4.9046	5.3516	5.9233
40	9.0806	3.7595	3.9759	4.3016	4.5967	4.9495	5.0766	5.5344	6.1163

Table 8b. **Estimated upper probability points, x_α, of the 1/2-Gamma score product moment statistics, $R_2^{(3)}(n)$, from a Monte Carlo experiment with $M = 4,000,000$ repetitions, for series of length $n = 10$ to $n = 40$.**

n \ α	0.900	0.950	0.975	0.980	0.990	0.995	0.998	0.999	σ
10	2.8319	3.1569	3.4458	3.5570	3.7612	3.8894	4.0109	4.0754	.7873
11	3.1840	3.5416	3.8688	3.9792	4.2156	4.3966	4.5419	4.6263	.8676
12	3.5287	3.9221	4.2792	4.4003	4.6717	4.8814	5.0722	5.1721	.9437
13	3.8764	4.3045	4.6879	4.8139	5.1168	5.3519	5.5864	5.7105	1.0189
14	4.2171	4.6791	5.0905	5.2232	5.5653	5.8227	6.0876	6.2311	1.0902
15	4.5566	5.0510	5.4868	5.6234	5.9979	6.2857	6.5861	6.7627	1.1595
16	4.8931	5.4195	5.8834	6.0289	6.427	6.7407	7.0846	7.2943	1.2270
17	5.2286	5.7865	6.2759	6.4261	6.8541	7.1904	7.5535	7.7671	1.2930
18	5.5594	6.1462	6.6617	6.8180	7.2661	7.6260	8.0244	8.2635	1.3556
19	5.8912	6.5104	7.0550	7.2196	7.6914	8.0798	8.5042	8.7655	1.4193
20	6.2143	6.8614	7.4304	7.6014	8.0964	8.5118	8.9594	9.2382	1.4787
21	6.5393	7.2146	7.8087	7.9858	8.5016	8.9361	9.4170	9.7219	1.5376
22	6.8651	7.5692	8.1861	8.3719	8.9123	9.3691	9.8768	10.1928	1.5959
23	7.1885	7.9208	8.5610	8.7526	9.3093	9.7887	10.3115	10.6518	1.6526
24	7.5100	8.2675	8.9352	9.1349	9.7106	10.2059	10.7707	11.1263	1.7082
25	7.8324	8.6202	9.3060	9.5139	10.1086	10.6271	11.2078	11.5890	1.7636
26	8.1409	8.9524	9.6593	9.8720	10.4867	11.0252	11.6363	12.0274	1.8138
27	8.4630	9.3029	10.0337	10.2529	10.8906	11.4524	12.0792	12.4818	1.8686
28	8.7726	9.6388	10.3888	10.6148	11.2648	11.8447	12.4985	12.9178	1.9179
29	9.0928	9.9819	10.7527	10.9833	11.6581	12.2533	12.9348	13.3805	1.9697
30	9.3995	10.3142	11.1045	11.3439	12.0349	12.6497	13.3508	13.8067	2.0181
31	9.7132	10.6514	11.4665	11.7086	12.4127	13.0404	13.7655	14.2517	2.0672
32	10.0248	10.9864	11.8223	12.0728	12.8004	13.4396	14.1915	14.6726	2.1154
33	10.3357	11.3205	12.1711	12.4254	13.1671	13.8391	14.6008	15.1085	2.1622
34	10.6452	11.6570	12.5290	12.7913	13.5528	14.2468	15.0326	15.5645	2.2556
35	10.9534	11.9818	12.8735	13.1380	13.9187	14.6116	15.4253	15.9753	2.2556
36	11.2583	12.3091	13.2253	13.4976	14.2875	14.9975	15.8375	16.3855	2.2997
37	11.5605	12.6348	13.5669	13.8454	14.6586	15.3873	16.2507	16.8297	2.3442
38	11.8722	12.9632	13.9172	14.2003	15.0179	15.7625	16.6405	17.2307	2.3883
39	12.1783	13.2941	14.2750	14.5682	15.4079	16.1703	17.0519	17.6576	2.4340
40	12.4802	13.6169	14.6033	14.8980	15.7539	16.5398	17.4293	18.0246	2,4750

Table 9 a. Estimated lower probability points, x_α, of the 1/2 Gamma score product-moment statistic, $R_3^{(3)}(n)$, from a Monte Carlo experiment with $\tilde{M} = 4,000,000$ repetitions, for series of length $n = 10$ to $n = 40$.

n \ α	Mean	0.001	0.002	0.005	0.010	0.020	0.025	0.050	0.100
10	1.5010	.2152	.2361	.2751	.3162	.3752	.2980	.4889	.6175
11	1.7349	.2967	.3243	.3725	.4211	.4862	.5119	.6113	.7537
12	1.9709	.4000	.4289	.4793	.5316	.6013	.6294	.7377	.8945
13	2.2098	.4514	.4920	.5589	.6245	.7090	.7418	.8669	1.0436
14	2.4487	.5126	.5619	.6435	.7215	.8205	.8583	1.0000	1.1959
15	2.6909	.5801	.6370	.7326	.8248	.9381	.9806	1.1382	1.3534
16	2.9325	.6712	.7372	.8398	.9380	1.0597	1.1064	1.2774	1.5110
17	3.1733	.7667	.8340	.9467	1.0541	1.1868	1.2363	1.4210	1.6716
18	3.4186	.8704	.9412	1.0613	1.1754	1.3181	1.3713	1.5703	1.8374
19	3.6611	.9694	1.0469	1.1764	1.2989	1.4510	1.5082	1.7189	2.0029
20	3.9045	1.0649	1.1531	1.2960	1.4274	1.5886	1.6495	1.8729	2.1743
21	4.1507	1.1689	1.2628	1.4146	1.5562	1.7295	1.7944	2.0306	2.3473
22	4.3960	1.2806	1.3793	1.5409	1.6891	1.8704	1.9388	2.1886	2.5220
23	4.6402	1.3877	1.4924	1.6635	1.8213	2.0156	2.0872	2.3494	2.6977
24	4.8861	1.5010	1.6121	1.7914	1.9588	2.1624	2.2384	2.5123	2.8754
25	5.1314	1.6149	1.7330	1.9233	2.0975	2.3112	2.3888	2.6755	3.0526
26	5.3797	1.7351	1.8577	2.0571	2.2406	2.4618	2.5445	2.8438	3.2364
27	5.6253	1.8512	1.9836	2.1913	2.3821	2.6155	2.7008	3.0130	3.4215
28	5.8709	1.9723	2.1082	2.3259	2.5250	2.7669	2.8561	3.1793	3.6009
29	6.1174	2.0934	2.2402	2.4461	2.6740	2.9251	3.0169	3.3515	3.7877
30	6.3639	2.2159	2.3675	2.6070	2.8224	3.0832	3.1781	3.5242	3.9730
31	6.6132	2.3523	2.5090	2.7539	2.9752	3.2413	3.3413	3.6980	4.1619
32	6.8596	2.4687	2.6302	2.8885	3.1206	3.4021	3.5041	3.8721	4.3493
33	7.1066	2.6014	2.7699	3.0321	3.2739	3.5609	3.6668	4.4071	4.5393
34	7.3542	2.7331	2.9099	3.1837	3.4306	3.7241	3.8325	4.2250	4.7307
35	7.6023	2.8656	3.0470	3.3272	3.5813	3.8886	4.0016	4.4036	4.9224
36	7.8510	2.9998	3.1868	3.4773	3.7390	4.0545	4.1692	4.5809	5.1137
37	8.0996	3.1316	3.3295	3.6287	3.8977	4.2205	4.3386	4.7641	5.3070
38	8.3475	3.2694	3.4751	3.7842	4.0634	4.3930	4.5139	4.9471	5.5020
39	8.5949	3.4138	3.6163	3.9326	4.2177	4.5569	4.6804	5.1250	5.6969
40	8.8448	3.5571	3.7645	4.0916	4.3831	4.7304	4.8568	5.3114	5.8928

Table 9b. Estimated upper probability points, x_α, of the 1/2 Gamma score product moment statistic, $R_3^{(3)}(n)$, from a Monte Carlo experiment with $M = 4,000,000$ repetitions, for series of length $n = 10$ to $n = 40$.

n \ α	0.900	0.950	0.975	0.980	0.990	0.995	0.998	0.999	σ
10	2.6047	2.8754	3.1231	3.2021	3.5283	3.6336	3.7590	3.8363	.7547
11	2.9583	3.2783	3.5708	3.6594	3.9935	4.1459	4.3140	4.3838	.8400
12	3.3058	3.6731	3.9973	4.0981	4.4443	4.6397	4.8243	4.9130	.9209
13	3.6499	4.0562	4.4127	4.5252	4.8863	5.1222	5.3525	5.4831	.9977
14	3.9897	4.4351	4.8291	4.9486	5.3288	5.5932	5.8633	6.0264	1.0721
15	4.3293	4.8131	5.2406	5.3710	5.7668	6.0654	6.3731	6.5490	1.1443
16	4.6651	5.1846	5.6409	5.7775	6.1932	6.5203	6.8619	7.0655	1.2133
17	4.9966	5.5470	6.0344	6.1819	6.6213	6.9717	7.3483	7.5711	1.2801
18	5.3304	5.9144	6.4316	6.5855	7.0452	7.4223	7.8254	8.0778	1.3456
19	5.6578	6.2726	6.8132	6.9776	7.4594	7.8609	8.2922	8.5607	1.4088
20	5.9819	6.6275	7.1915	7.3607	7.8585	8.2859	8.7418	9.0408	1.4688
21	6.3105	6.9879	7.5774	7.7561	8.2746	8.7176	9.2062	9.5244	1.5299
22	6.6306	7.3360	7.9550	8.1411	8.6844	9.1522	9.6674	9.9939	1.5880
23	6.9534	7.6876	8.3312	8.5245	9.0870	9.5774	10.1212	10.4634	1.6460
24	7.2717	8.0370	8.7002	8.8993	9.4843	9.9947	10.5687	10.9363	1.7012
25	7.5898	8.3841	9.0710	9.2758	9.8760	10.4035	10.9964	11.3868	1.7565
26	7.9128	8.7334	9.4405	9.6539	10.2745	10.8267	11.4402	11.8374	1.8115
27	8.2223	9.0697	9.8031	10.0224	10.6621	11.2289	11.8688	12.2829	1.8620
28	8.5401	9.4134	10.1650	10.3874	11.0472	11.6298	12.2877	12.7248	1.9147
29	8.8521	9.7500	10.5217	10.7541	11.4299	12.0384	12.7269	13.1855	1.9651
30	9.1593	10.0820	10.8793	11.1169	11.8120	12.4378	13.1405	13.6127	2.0140
31	9.4780	10.4238	11.2395	11.4817	12.1941	12.8263	13.5595	14.0537	2.0643
32	9.7784	10.7527	11.5862	11.8346	12.5657	13.2197	13.9812	14.4823	2.1120
33	10.0952	11.0880	11.9460	12.2007	12.9494	13.6255	14.4227	14.9278	2.1595
34	10.4025	11.4170	12.2970	12.5616	13.3265	14.0218	14.7968	15.3257	2.2061
35	10.7138	11.7498	12.6457	12.9089	13.6936	14.3993	15.2205	12.7625	2.2522
36	11.0263	12.0837	13.0000	13.2725	14.0675	14.7842	15.6255	16.1757	2.2997
37	11.3314	12.4132	13.3516	13.6312	14.4441	15.1874	16.0397	16.5945	2.3453
38	11.6333	12.7320	13.6845	13.9708	14.7977	15.5346	16.4128	16.9955	2.3859
39	11.9397	13.0614	14.0394	14.3298	15.1793	15.9537	16.8448	17.4367	2.4320
40	12.2483	13.3903	14.3792	14.6763	15.5412	16.3203	17.2237	17.8426	2.4752

TABLES TO FACILITATE THE USE OF ORTHOGONAL POLYNOMIALS FOR TWO TYPES OF ERROR STRUCTURES

Kirkland B. Stewart*

***Pacific Northwest Laboratory, Battelle Memorial Institute**

INTRODUCTION

When most of the work in fitting orthogonal polynomials was done on desk calculators the opportunity to work with evenly-spaced independent variables was always a welcome one. The amount of computational effort in using orthogonal polynomial data is small provided that it is not necessary to determine the coefficient estimates and the statistics associated with the model

$$y_a = a_0 + a_1\left(\frac{x_a - c}{d}\right) + \ldots + a_k\left(\frac{x_a - c}{d}\right)^k + e_a \qquad (1.$$

$$(x_a - x_{a-1} = d,\ a = 1, 2, \cdots, n$$

from those computed for the model

$$y_a = b_0' + b_1' \rho_{1,a} + \cdots + b_k' \rho_{k,a} + e_a$$

where $\rho_{i,a}$ is an orthogonal polynomial argument of the ith degree for the ath point. In most applications e_a, the error component, is assumed to have the structure

$$E(e_a) = 0, \qquad (a = 1, 2, \cdots, n)$$

$$\sigma_{e_a e_{a'}} = \delta_{aa'}\,\sigma^2$$

where $\delta_{aa'}$ is Kronecker's δ. There are many cases in applied problems, however, when the error structure is cumulative in the sense that

$$e_a = \sum_{i=1}^{a} \epsilon_i$$

where the ϵ_i random variables are statistically independent. These cases are called the type I and type II error models in this paper.

THE ADVANTAGES OF ORTHONORMAL ARGUMENTS

The emphasis in fitting polynomials by least squares has always been to use integral-valued arguments. The procedure of using orthonormal arguments $P_{i,a}$ instead of the integral-valued arguments $\rho_{i,a}$ has many advantages, however. If the model

$$y_a = b_0 P_{0,a} + b_1 P_{1,a} + \cdots + b_k P_{k,a} + e_a$$

is fitted by least squares, then

1. $\sigma^2_{\hat{b}_i} = \sigma^2$,

2. $\hat{\sigma}^2_{\hat{b}_i} = \hat{\sigma}^2$,

3. the reduction due to fitting b_i is $\hat{b}_i^2$,

4. $\hat{b}_i/\hat{\sigma} \sim t(n-k-1)$ under H_0: $b_i = 0$, if $e_a \sim \mathrm{NID}(0, \sigma^2)$,

5. If the elements of the transformation matrix T are available by which to effect the relationship

$$\hat{A} = T\hat{B}$$

ere $\hat{A}$ and $\hat{B}$ are column vectors of the $\hat{a}_i$ and $\hat{b}_i$ values, then $\Sigma_{\hat{a}}$, the covariance matrix of the a_i estimates, is given by

$$\Sigma_{\hat{a}} = \sigma^2 TT'.$$

t

$$\hat{y}_a = \sum_{i=0}^{k} \hat{b}_i P_{i,a}$$

6. Then $\hat{y}_a$ estimates $E(y_a)$ with a variance of

$$\sigma^2_{y_{a,L}} = \sigma^2 \sum_{i=0}^{k} P^2_{i,a} = \sigma^2 A^2_a.$$

7. $\hat{y}_a$ estimates a random y at x_a with a variance of

$$\sigma^2_{y_{a,s}} = \sigma^2 + \sigma^2_{y_{a,L}} = \sigma^2 B^2_a.$$

8. The residual $r_a = y_a - \hat{y}_a$ has variance

$$\sigma^2_{r_a} = \sigma^2 - \sigma^2_{y_{a,L}} = \sigma^2 C^2_a.$$

ables of the values T, TT', A_a, B_a, C_a, $t_{0.975}(n-k-1)A_a$, $t_{0.975}(n-k-1)B_a$, and $t_{0.975}(n-k-1)C_a$ are ovided in table Set I. $A_a = A_{a'}$, $B_a = B_{a'}$, $C_a = C_{a'}$ if $a' = n - a + 1$. These relationships are used to shorten the tables hen $n > 16$.

THE TYPE I ERROR MODEL

The most simple way of proceeding is as follows. Determine the values $\tilde{b}_i$ by

$$\tilde{b}_i = \sum_{a=1}^{n} \xi'_{i,a} y_a.$$

he values $\xi'_{i,a}$ are found in many reference sources such as [1], [2], [3] and [4]. Then compute $\hat{b}_i$ from

$$\hat{b}_i = \tilde{b}_i \left(\sum_{a=1}^{n} \xi'^2_{i,a} \right)^{-(1/2)}$$

here

$$\left(\sum_{a=1}^{n} \xi'^2_{i,a} \right)^{-(1/2)},$$

he orthonormalizing factors, are found in table set I.

For the type I error model it is convenient to set $c = x_0$ in equation (1). The transformation matrices T in table set were computed with this understanding. The tables extend from $n = 3$ to 20 and $k = 1$ to min $(n, 4)$.

Two examples will be shown. In the first example $\sigma^2 = 0$, so that the example provides an indication of the table ccuracies. In the second example a random error component is added, $e_a \sim N(0,1)$.

Example 1.

The model

$$y = 100 + 10x + x^2 + 0.1x^3$$

was used to generate the following y values

x	1	2	3	4	5	6	7
y	111.1	124.8	141.7	162.4	187.5	217.6	253.3

No error element was added to the y values. Then

$$\sum y = 1198.4 \quad \hat{b}_0 = 0.37796447\ (1198.4) = 452.9526209$$

$$\sum y\xi_1' = 658.0 \quad \hat{b}_1 = 0.18898224\ (658.0) = 124.3503139$$

$$\sum y\xi_2' = 184.8 \quad \hat{b}_2 = 0.10910895\ (184.8) = 20.16333396$$

$$\sum y\xi_3' = 3.6 \quad \hat{b}_3 = 0.40824829\ (3.6) = 1.469693844$$

$$\sum y^2 = 221{,}037.80.$$

The reduction due to fitting the model is

$$\sum_0^3 \hat{b}_1^2 = 221{,}037.7974$$

which would equal $\sum y^2$, of course, in the absence of rounding errors.

Values of the elements in $\hat{A}$ are obtained from $\hat{A} = T\hat{B}$, or

$$\begin{pmatrix} 99.99999549 \\ 10.000000160 \\ 1.000000200 \\ 0.1000000003 \end{pmatrix} = \begin{pmatrix} .37796447 & -.75592895 & 1.3093073 & -2.4494897 \\ & .18898224 & -.87287156 & 2.7896967 \\ & & .10910895 & -.81649658 \\ & & & .068041382 \end{pmatrix} \begin{pmatrix} 452.9526209 \\ 124.3503139 \\ 20.16333396 \\ 1.469693844 \end{pmatrix}$$

Example 2.

If the random errors $-0.5, -0.5, 1.4, -1.0, 1.4, 0.5, 0.7$, which are sampled from a random $N(0, 1)$ table, are added to the y values in the previous example, and the x values are transformed, one gets

x	154	158	162	166	170	174	178
y	110.6	124.3	143.1	161.4	188.9	218.1	254.0

for the second example.

re

$$\sum y = 1{,}200.4 \quad \hat{b}_i = \Sigma y \xi_i' \,(\Sigma \xi_i'^2)^{-(1/2)}$$

$$\sum y\xi_1' = 663.6 \quad \hat{b}_0 = 453.7085$$

$$\sum y\xi_2' = 181.4 \quad \hat{b}_1 = 125.4086$$

$$\sum y\xi_3' = 3.8 \quad \hat{b}_2 = 19.79236$$

$$\sum y^2 = 221{,}977.24 \quad \hat{b}_3 = 1.55134.$$

The reduction due to fitting the model is

$$\sum_0^3 \hat{b}_i^2 = 221{,}972.86$$

that the residual variance is estimated to be $\hat{\sigma}^2 = 1.460$, $\hat{\sigma} = 1.208$.

en

$$\hat{\sigma}^{-1}\,\hat{B} = \begin{pmatrix} 376. \\ 103.8 \\ 16.4 \\ 1.28 \end{pmatrix}$$

that $\hat{b}_3$ is the only parameter estimate which is consistent in a statistical sense with the hypotheses $H_0: b_i = 0, i = 0,$ 2, 3. From the relationship $\hat{A} = T\hat{B}$, one obtains

$$\hat{A} = \begin{pmatrix} 98.8000 \\ 10.7516 \\ 0.89286 \\ 0.10556 \end{pmatrix}.$$

The prediction equation is then

$$y = 98.80 + 10.75\left(\frac{x-150}{4}\right) + 0.8929\left(\frac{x-150}{4}\right)^2 + 0.1056\left(\frac{x-150}{4}\right)^3.$$

$_{\hat{a}}$, the covariance matrix of the $\hat{a}$ values, is estimated as $\hat{\sigma}^2\, T T'$, giving

$$\hat{\Sigma}_{\hat{a}} = 1.460\begin{pmatrix} 8.43 & -8.12 & 2.14 & -.167 \\ & 8.58 & -2.37 & 0.190 \\ & & 0.679 & -0.0556 \\ \text{symmetrical} & & & 0.004630 \end{pmatrix}.$$

The statistics pertinent to the individual observations are:

a	y_a	$\hat{y}_a$	r_a	$\hat{\sigma}_1$	$\hat{\sigma}_2$	$\hat{\sigma}_3$	t_1	t_2	t_3
1	110.6	110.6	0.0	1.2	1.7	0.3	3.7	5.3	1.0
2	124.3	124.7	−0.4	0.8	1.5	0.9	2.6	4.6	2.8
3	143.1	141.9	1.2	0.8	1.5	0.9	2.6	4.6	2.8
4	161.4	162.8	−1.4	0.7	1.4	1.0	2.2	4.4	3.1
5	188.9	188.1	0.8	0.8	1.5	0.9	2.6	4.6	2.8
6	218.1	218.3	−0.2	0.8	1.5	0.9	2.6	4.6	2.8
7	254.0	254.0	0.0	1.2	1.7	0.3	3.7	5.3	1.0

where

$$\hat{\sigma}_1 = \hat{\sigma} A_a \qquad t_1 = \hat{\sigma}\, t_{0.975}(3)\, A_a$$

$$\hat{\sigma}_2 = \hat{\sigma} B_a \qquad t_2 = \hat{\sigma}\, t_{0.975}(3)\, B_a$$

$$\hat{\sigma}_3 = \hat{\sigma} C_a \qquad t_3 = \hat{\sigma}\, t_{0.975}(3)\, C_a\,.$$

THE TYPE II ERROR MODEL

With the type II error model it is convenient to set $c = x_1$ in equation (1) and to calculate the transformation matri accordingly. The error model is cumulative with the structure

$$e_a = \sum_{j=1}^{a} \epsilon_j,$$

$$E(\epsilon_j) = 0,$$

$$\sigma^2_{\epsilon_1} = g\sigma^2,$$

$$\sigma^2_{\epsilon_j} = d\sigma^2, (j > 1)$$

$$\sigma_{\epsilon_j \epsilon_{j'}} = 0, (j \neq j').$$

The x_i values are monotone increasing and it is convenient to assume that $x_1 > 0$. The random errors ϵ_j have variance $d\sigma^2$ with the possible exception of ϵ_1. $\sigma^2_{\epsilon_1}$ depends on the circumstances, of course, but for the sake of expo sition assume that $g = x_1$, so that

$$\sigma^2_{\epsilon_1} = x_1 \sigma^2.$$

σ^2 is the random error in y associated with a unit increase in x. Since the variance of y_1 depends on the manner in whi the data were generated, the location of the y-axis in representing the curve is more of a question of meaningfulness or convenience than of necessity. Thus $x = 0$ is not necessarily the point at which the process of accumulating random errors in the dependent variable begins. For a detailed discussion of the process by which a cumulative error model occ see Mandel [5].

"A common situation in physical, chemical, and related experimentation involves the taking of measurements on essentially the same subject or unit of material at different stages of a given process. A simple example is provided by the testing of a specimen of plastic material for its resistance to abrasion by subjecting it for successive test periods to the action of an abrasive wheel and measuring the amount of wear at the end of each period. In chemical studies, the progress of a chemical reaction may be studied by taking samples of the reacting system at specified time intervals and subjecting the samples to chemical analysis. This procedure applies, of course, also to the control of large batches of reacting materials in chemical industrial processes." (Mandel [5], page 552.)

"At any rate it is clear that situations involving cumulative errors are quite frequent in laboratory experimentation." Mandel [7], page 303).

The ability to control the independent variable and to take measurements of the dependent variable when desired will in many of the experiments give rise to equally spaced independent variable situations. Within the region of experimentation the dependent or response variable is often a smooth, continuous, single-valued function. The polynomial, as an empirical model, will then give an adequate representation of the response.

The primary reason for using the cumulative error model under the appropriate circumstances is that the uncertainties associated with the fitting process will, in a statistical sense, be correctly adjudged. The use of the independent and equal variance error model, when in fact the cumulative model obtains, will give unbiased estimates of the parameters and often with great efficiency. The uncertainties estimated for the parameter estimates and the predicted values will tend to be optimistic, however.

Under the cumulative error structure, the variables

$$\begin{pmatrix} z_1' \\ z_2 \\ \cdots \\ z_n \end{pmatrix} = \begin{pmatrix} y_1\sqrt{d/x_1} \\ y_2 - y_1 \\ \cdots \\ y_n - y_{n-1} \end{pmatrix}$$

are independent with variance $d\sigma^2$. It is easier to work with $z_1 = y_1$ than with z_1'.

The model for the expected value of z_a is

$$E(z_a) = b_0\delta_{1,a} + \left(1 - \delta_{1,a}\right)\sum_{j=1}^{k} b_j W_{j,a}$$

where $\delta_{1,a}$ is Kronecker's δ, and

$$W_{j,a} = \left(\frac{x_a - x_1}{d}\right)^j - \left(\frac{x_{a-1} - x_1}{d}\right)^j$$

Let

$$U_{l,a} = \sum_{j=1}^{l} C_{l,j} W_{j,a}; \qquad (l = 1, \ldots, k).$$

where

$$U_{1,a} = (n-1)^{-(1/2)}, \qquad (a = 2, \ldots, n)$$

$$\sum_{a=2}^{n} U_{l,a} = 0, \qquad (l \neq 1)$$

$$\sum_{a=2}^{n} U_{l,a}^2 = 1,$$

$$\sum_{\substack{a=2 \\ l \neq l'}}^{n} U_{l,a} U_{l',a} = 0.$$

Then

$$(\delta_{1,a} \ (1 - \delta_{1,a}) \ U_{l,a}; l = 1, \cdots k)$$

are a set of orthonormal arguments over $a = 1, 2, \cdots, n$. Call these arguments

$$S_{j,a}; j = 0, 1, \cdots, k, \quad a = 1, 2, \cdots, n.$$

If the model is written as

$$z_a = \sum_{j=0}^{k} b_j S_{j,a} + \epsilon_a$$

then Q, the weighted sum of squares, is

$$Q = (z_1 - b_0)^2/(x_1\sigma^2) + \sum_{a=2}^{n} \left(z_a - \sum_{j=1}^{k} b_j S_{j,a} \right)^2 / (d\sigma^2),$$

which upon minimization yields in particular

$$\hat{b}_0 = z_1 = y_1$$

and in general

$$\hat{b}_j = \sum_{a=1}^{n} S_{j,a} z_a, \qquad j = 0, 1, \cdots, k.$$

The covariance matrix of the b_j estimates is given by the diagonal matrix

$$\Sigma_{\hat{b}} = \left(\begin{array}{c|ccc} \dfrac{1}{x_1 \sigma^2} & & & \\ \hline & \dfrac{1}{d\sigma^2} & & \\ & & \ddots & \\ & & & \dfrac{1}{d\sigma^2} \end{array} \right)^{-1} = \sigma^2 \left(\begin{array}{c|ccc} x_1 & & & \\ \hline & d & & \\ & & \ddots & \\ & & & d \end{array} \right).$$

The orthonormal arguments $U_{l,a}$ $(l = 1, \cdots, k; a = 2, \cdots, n)$ are equal to $P_{l-1,a-1}$ as defined in section 3. The transformation matrices are different, however. In matrix notation

$$\hat{b} = SZ = S(Y_2 - Y_1),$$

where the row a elements of the column vectors Y_2 and Y_1 respectively are y_a and y_{a-1}, $a > 1$, y_1 and 0, $a = 1$. Let H be the matrix such that

$$Y_1 = HY_2.$$

Then

$$\hat{b} = S(I - H)Y_2 = MY_2.$$

The elements in $\hat{b}$ can also be determined from

$$\hat{b}_0 = y_1.$$

$$\hat{b}_i = \left(\sum_{a=1}^{n-1} \xi'_{i-1,a} z_{a+1}\right)\left(\sum_{a=1}^{n-1} \xi'^2_{i-1,a}\right)^{-(1/2)}, (i > 1)$$

where the

$$\left(\sum_{a=1}^{n-1} \xi'^2_{i-1,a}\right)^{-(1/2)}$$

values are found in table set I.

The transformation matrix T is of the form

$$T = \left(\begin{array}{c|c} 1 & \emptyset_{12} \\ \hline \emptyset_{21} & W \end{array}\right)$$

where W is an upper triangular $(k \times k)$ matrix and the elements of $\emptyset_{12}$ and $\emptyset_{21}$ are zero. Then the covariance matrix of the a_i estimates, $(i = 1, 2, \cdots, k)$ is of the form

$$d\sigma^2 WW'.$$

The a_0 estimate has, according to our assumptions, a variance of $x_1\ \sigma^2$. The covariance matrix of the a_i estimates, $(i = 0, 1, \cdots, k)$ has the form

$$\Sigma_{\hat{a}} = \sigma^2 \left(\begin{array}{c|c} x_1 & \emptyset_{12} \\ \hline \emptyset_{21} & dWW' \end{array}\right)$$

Four types of tables are given in table set II. The first table gives the values of the matrix T by which the transformation

$$\hat{a} = T\hat{b}$$

is effected. The second type of table gives the matrix S such that

$$\hat{b} = SZ. .$$

This procedure requires the computation of the z_i values and is useful in the situation where the statistics associated with the fitting process are of interest.

The third type of table gives the matrix M by which the estimates

$$\hat{b} = MY_2$$

are obtained. This procedure eliminates the intermediate step of calculating the z_a values when the user is interested only in the parameter estimates.

The fourth kind of tables were generated as an aid in determining the predicted y_a values and the variance of the predicted y values as estimates of $E(y_a)$, the expected value of y_a. Suppose $\hat{y}_a$ is of interest. Then

$$\hat{y}_a = \hat{b}_0 + \hat{b}_1 \sum_{j=1}^{a} S_{1,j} + \cdots + \hat{b}_k \sum_{j=1}^{a} S_{k,j}.$$

The values $H(\mathrm{i}, a) = \sum_{j=1}^{a} S_{i,j}$ are given under the heading PREDICTION. $\hat{\sigma}^2_{\hat{y}_a}$ is easily determined by

$$\hat{\sigma}^2_{\hat{y}_a} = \hat{\sigma}^2 \left[x_1 + d\left(H^2(1,a) + \cdots + H^2(k,a)\right)\right]$$
$$= \hat{\sigma}^2 \left[x_1 + d\,G(k,a)\right].$$

The values $G(k, a)$ are given under the heading VARIANCE. The tables cover all situations from $n = 3$ to 20, $k = \min(n-1, 5)$. Their use will be shown by an example.

As an added comment, many of the statistics pertinent to the z_a values can be found from the tables in set I since the arguments $U_{l-1,a-1}$ are identical with $P_{l,a}$ as used in section 3.

Example

The dependent variables y_a were determined from the equation

$$y_a = 1000 + 100\left(\frac{x_a - 100}{5}\right) + 10\left(\frac{x_a - 100}{5}\right)^2 + \left(\frac{x_a - 100}{5}\right)^3 + e_a,$$

$$a = 1, \ldots, 7, e_a = \sum_{i=1}^{a} \epsilon_i.$$

The random errors $\eta_i, i = 1, \ldots, 7$ were sampled from an $N(0, 1)$ table. $\epsilon_i = \eta_i$ except that $\epsilon_1 = (\sqrt{100/5})\,\eta_1$. Then

i	1	2	3	4	5	6	7
ϵ_i	−3.2	−0.1	1.2	−0.2	2.0	0.2	0.8
$\sum_{i=1}^{a} \epsilon_i$	−3.2	−3.3	−2.1	−2.3	−0.3	−0.1	0.7

so that the relationships are

x_a	y_a	z_a
100	996.8	996.8
105	1107.7	110.9
110	1245.9	138.2
115	1414.7	168.8
120	1623.7	209.0
125	1874.9	251.2
130	2176.7	301.8

Parameter estimates can be obtained as follows

$\hat{b}_0 = y_1 = 996.8,$

$$\begin{pmatrix} \tilde{b}_1 \\ \tilde{b}_2 \\ \tilde{b}_3 \end{pmatrix} = \begin{pmatrix} 1 & -5 & 5 \\ 1 & -3 & -1 \\ 1 & -1 & -4 \\ 1 & 1 & 4 \\ 1 & 3 & -1 \\ 1 & 5 & 5 \end{pmatrix}' \begin{pmatrix} 110.9 \\ 138.2 \\ 168.8 \\ 209.0 \\ 251.2 \\ 301.8 \end{pmatrix} = \begin{pmatrix} 1179.9 \\ 1333.7 \\ 162.9 \end{pmatrix}$$

and using the orthonormalizing factors from table set I, one finds

$$\begin{pmatrix} \hat{b}_0 \\ \hat{b}_1 \\ \hat{b}_2 \\ \hat{b}_3 \end{pmatrix} = \begin{pmatrix} 996.8 \\ .408248\,(1179.9) \\ .119523\,(1333.7) \\ .109109\,(\ 162.9) \end{pmatrix} = \begin{pmatrix} 996.8 \\ 481.69 \\ 159.41 \\ 17.774 \end{pmatrix}$$

Equivalent results are obtained from

$$\hat{b} = SZ \text{ or } \hat{b} = MY_2.$$

Then

$$\hat{a} = T\hat{b}$$

or

$$\begin{vmatrix} \hat{a}_0 \\ \hat{a}_1 \\ \hat{a}_2 \\ \hat{a}_3 \end{vmatrix} = \begin{vmatrix} 996.8 \\ 99.784 \\ 10.3263 \\ 0.96965 \end{vmatrix}$$

so that the prediction equation is

$$y_a = 996.8 + 99.784 \left(\frac{x_a - 100}{5}\right) + 10.3263 \left(\frac{x_a - 100}{5}\right)^2 + 0.96965 \left(\frac{x_a - 100}{5}\right)^3$$

$$\sum_{a=1}^{7} z_a^2 = 1{,}251{,}367.410, \sum_{j=0}^{3} \hat{b}_j^2 = 1{,}251{,}364.279.$$

Then the residual sum of squares = 3.131.

The residual mean square = $d\hat{\sigma}^2 = 1.0437$

so that $\sqrt{d\hat{\sigma}^2} = 1.0216,$

and $\hat{\sigma}^2 = 0.2087$

so that $100\,\hat{\sigma}^2 = 20.87.$

Since

$$WW' = \begin{pmatrix} 1.645 & -0.568 & 0.0536 \\ -0.568 & 0.2554 & -0.0268 \\ 0.0536 & -0.0268 & 0.0030 \end{pmatrix}$$

the covariance matrix of the a_i estimates is estimated as

$$\hat{\Sigma}_{\hat{a}} = 1.0437 \begin{pmatrix} 20.00 & 0 & 0 & 0 \\ 0 & 1.645 & -0.568 & 0.0536 \\ 0 & -0.568 & 0.2554 & -0.0268 \\ 0 & 0.0536 & -0.0236 & 0.0030 \end{pmatrix}$$

Suppose that $\hat{y}_3$ is of interest. Then from the prediction equation

$$\hat{y}_3 = 996.8 + 99.784\,(2) + 10.3263\,(4) + 0.96965\,(8)$$

$$= 1245.4,$$

or from the fourth type of table in table set II,

$$\hat{y}_3 = 996.8 + 481.6921\ (.8165)$$
$$+ 159.4076\ (-0.9562) + 17.774\ (0.4364)$$
$$= 1245.4$$

or

$$\hat{y}_3 = \hat{z}_1 + \hat{z}_2 + \hat{z}_3$$
$$= 996.8 + 111.1 + 137.6$$
$$= 1245.4.$$

Let $X_3' = \begin{pmatrix} 1 \\ 2 \\ 4 \\ 8 \end{pmatrix}.$

Then

$$\hat{\sigma}^2_{y_3} = X_3 \hat{\sum}_{\hat{a}} X_3' = 22.5$$

or by recourse to the fourth kind of table

$$\hat{\sigma}^2_{\hat{y}_3} = \hat{\sigma}^2 \left[x_1 + d\,G\,(3, 3)\right]$$
$$= 0.2087 \left[100 + 5\ (1.581)\right]$$
$$= 22.5\ .$$

The statistics associated with z_3 are as follows:

$$z_3 = 138.2,$$
$$\hat{z}_3 = 481.69\ (0.4082482) + 159.41\ (-0.3585685) + 17.774\ (-0.1091089)$$
$$= 137.55,$$
$$r_3 = z_3 - \hat{z}_3 = 0.65$$
$$\hat{\sigma}_1 = 1.0216\ (0.554) = 0.566,$$
$$\hat{\sigma}_2 = 1.0216\ (1.143) = 1.168,$$
$$\hat{\sigma}_3 = 1.0216\ (0.832) = 0.850,$$

$$t_1 = 1.0216\,(1.764) = 1.802,$$

$$t_2 = 1.0216\,(3.638) = 3.717,$$

$$t_3 = 1.0216\,(2.649) = 2.706,$$

where $\hat{\sigma}_1, \hat{\sigma}_2, \hat{\sigma}_3, t_1, t_2$, and t_3 have the same meanings as in section 3 when applied to the statistics associated with z_3. The values 0.554, 1.143, 0.832, 1.764, 3.638, and 2.649 are obtained from table set I where $N = 6$, DEGREE = 2 and $I = 2$. t_1, t_2, and t_3 are meaningful, of course, only when $\epsilon_i \sim \text{NID}\,(0, d\sigma^2), 1 < i \leq n$.

Tables could also be developed to facilitate the calculation of the variance of the residual $R_a = y_a - \hat{y}_a$. It is felt, however, that these residuals would be too highly correlated to be of much help in understanding the pattern of results.

Missing Values

The author has also calculated tables to facilitate the fitting of orthogonal polynomials when the x values would be equidistant were it not for a missing observation. The tables extend from $n = 4$ to $n = 10$, and $k = \min\,(n - 1, 4)$. They are similar in form to the tables contained herein as they contain orthonormal arguments and transformation matrices. They are available upon request from the author.

They also contain a table whereby the "missing" observation can be calculated in terms of the other observations. This approach is in the style of the traditional missing value techniques where the value calculated for the missing observation is that which would be predicted by a least-squares fit of the original data. Then the least-squares fitting process is carried out in the usual way using the tables of orthogonal polynomial arguments. Adjustments must be made to take care of the diminished number of degrees of freedom, of course.

Reference [8] contains a simple and effective way of determining "missing" observations in the case of equidistant x values. This procedure has the advantage that only the ξ_i' tables are required. It can be used when more than one observation is missing and it is limited only by the range of the tabled ξ_i' values.

The advantages of the tables by the author are: the effect of adding an incremental degree in the model is easily assessed, in some cases the computations are easier than in [8], and they can be used in situations such as a one-way ANOVA with equally spaced levels where the data from one of the levels are missing, and it is desired to partition the sum of squares between levels into linear, quadratic, . . . , components.

Acknowledgment.

Programming assistance by Mrs. Rosalie Argo is gratefully acknowledged.

REFERENCES

Error Model I

1. Fisher, R. A. and Yates, F. (1957). *Statistical Tables for Biological, Agricultural and Medical Research,* 5th ed. Edinburgh: Oliver and Boyd.

2. Anderson, R. L., and Houseman, E. E. (1942). *Tables of Orthogonal Polynomial Values Extended to N = 104,* Research Bulletin No. 297, Agricultural Experiment Station, Iowa State College, Ames, Iowa.

3. DeLury, D. B. (1950). *Values and Integrals of Orthogonal Polynomials up to n = 26,* University of Toronto Press.

4. Natrella, M. G. (1963). *Experimental Statistics,* National Bureau of Standards Handbook 91, U. S. Government Printing Office, Washington, D. C.

Error Model II

5. Mandel, John (1957). "Fitting a straight line to certain types of cumulative data," *Journal of the American Statistical Association,* 52: 552–566.

6. Jaech, John L. (1964). "A note on the equivalence of two methods of fitting a straight line through cumulative data," *Journal of the American Statistical Association,* 59: 863–866.

7. Mandel, John (1964). *The Statistical Analysis of Experimental Data,* New York: Interscience Publishing, 295–309.

Missing Values

8. Hartley, H. O. (1951). "The fitting of polynomials to equidistant data with missing values," *Biometrika,* 38: 410–413.

TABLE I. THE INDEPENDENT ERROR MODEL

N = 3

DEGREE	NORMALIZATION FACTOR
0	5.7735027-001
1	7.0710678-001
2	4.0824829-001

TRANSFORMATION MATRIX

	0	1	2
0	5.7735027-001	-1.4142136+000	4.0824829+000
1		7.0710678-001	-4.8989795+000
2			1.2247449+000

AUXILIARY TABLES

DATA POINT STATISTICS

		STANDARD DEVIATIONS			95 PER CENT CONFIDENCE LIMITS		
DEGREE	I	MODEL	POINT	RESIDUAL	MODEL	POINT	RESIDUAL
1	1	.913	1.354	.408	11.599	17.204	5.187
	2	.577	1.155	.816	7.336	14.672	10.375
	3	.913	1.354	.408	11.599	17.204	5.187

COVARIANCE MATRIX OF PARAMETER ESTIMATES

	0	1
0	2.333+00	-1.000+00
1	-1.000+00	5.000-01

N = 4

DEGREE	NORMALIZATION FACTOR
0	5.0000000-001
1	2.2360680-001
2	5.0000000-001
3	2.2360680-001

TRANSFORMATION MATRIX

	0	1	2	3
0	5.0000000-001	-1.1180340+000	2.5000000+000	-7.8262379+000
1		4.4721360-001	-2.5000000+000	1.2447445+001
2			5.0000000-001	-5.5901699+000
3				7.4535599-001

AUXILIARY TABLES

DATA POINT STATISTICS

DEGREE	I	STANDARD DEVIATIONS MODEL	POINT	RESIDUAL	95 PER CENT CONFIDENCE LIMITS MODEL	POINT	RESIDUAL
1	1	.837	1.304	.548	3.600	5.610	2.357
	2	.548	1.140	.837	2.357	4.906	3.600
	3	.548	1.140	.837	2.357	4.906	3.600
	4	.837	1.304	.548	3.600	5.610	2.357
2	1	.975	1.396	.224	12.384	17.743	2.841
	2	.742	1.245	.671	9.423	15.819	8.524
	3	.742	1.245	.671	9.423	15.819	8.524
	4	.975	1.396	.224	12.384	17.743	2.841

COVARIANCE MATRIX OF PARAMETER ESTIMATES

		1
0	1.500+00	-5.000-01
1	-5.000-01	2.000-01

	0	1	2
0	7.750+00	-6.750+00	1.250+00
1	-6.750+00	6.450+00	-1.250+00
2	1.250+00	-1.250+00	2.500-01

N = 5

DEGREE	NORMALIZATION FACTOR
0	4.4721360-001
1	3.1622777-001
2	2.6726124-001
3	3.1622777-001
4	1.1952286-001

TRANSFORMATION MATRIX

	0	1	2	3	4
0	4.4721360-001	-9.4868330-001	1.8708287+000	-4.4271887+000	1.5059880+001
1		3.1622777-001	-1.6035675+000	6.2191461+000	-2.8386679+001
2			2.6726124-001	-2.3717082+000	1.7281013+001
3				2.6352314-001	-4.1833001+000
4					3.4860834-001

AUXILIARY TABLES

DATA POINT STATISTICS

DEGREE	I	STANDARD DEVIATIONS MODEL	STANDARD DEVIATIONS POINT	STANDARD DEVIATIONS RESIDUAL	95 PER CENT CONFIDENCE LIMITS MODEL	95 PER CENT CONFIDENCE LIMITS POINT	95 PER CENT CONFIDENCE LIMITS RESIDUAL
1	1	.775	1.265	.632	2.465	4.025	2.013
	2	.548	1.140	.837	1.743	3.628	2.663
	3	.447	1.095	.894	1.423	3.486	2.846
	4	.548	1.140	.837	1.743	3.628	2.663
	5	.775	1.265	.632	2.465	4.025	2.013
2	1	.941	1.373	.338	4.049	5.909	1.455
	2	.609	1.171	.793	2.622	5.039	3.411
	3	.697	1.219	.717	2.999	5.245	3.086
	4	.609	1.171	.793	2.622	5.039	3.411
	5	.941	1.373	.338	4.049	5.909	1.455
3	1	.993	1.409	.120	12.615	17.905	1.519
	2	.878	1.331	.478	11.160	16.911	6.075
	3	.697	1.219	.717	8.855	15.488	9.112
	4	.878	1.331	.478	11.160	16.911	6.075
	5	.993	1.409	.120	12.615	17.905	1.519

COVARIANCE MATRIX OF PARAMETER ESTIMATES

Degree 1:

	0	1
0	1.100+00	-3.000-01
1	-3.000-01	1.000-01

Degree 2:

	0	1	2
0	4.600+00	-3.300+00	5.000-01
1	-3.300+00	2.671+00	-4.286-01
2	5.000-01	-4.286-01	7.143-02

Degree 3:

	0	1	2	3
0	2.420+01	-3.083+01	1.100+01	-1.167+00
1	-3.083+01	4.135+01	-1.518+01	1.639+00
2	1.100+01	-1.518+01	5.696+00	-6.250-01
3	-1.167+00	1.639+00	-6.250-01	6.944-02

DEGREE	NORMALIZATION FACTOR
0	4.0824829-001
1	1.1952286-001
2	1.0910895-001
3	7.4535599-002
4	1.8898224-001

TRANSFORMATION MATRIX

	0	1	2	3	4
0	4.0824829-001	-8.3666003-001	1.5275252+000	-3.1304952+000	7.9372539+000
1		2.3904572-001	-1.1456439+000	3.9379642+000	-1.3669715+001
2			1.6366342-001	-1.3043730+000	7.3545587+000
3				1.2422600-001	-1.5433549+000
4					1.1023964-001

AUXILIARY TABLES

DATA POINT STATISTICS

DEGREE	I	STANDARD DEVIATIONS MODEL	POINT	RESIDUAL	95 PER CENT CONFIDENCE LIMITS MODEL	POINT	RESIDUAL
1	1	.724	1.234	.690	2.009	3.427	1.916
	2	.543	1.138	.840	1.509	3.160	2.331
	3	.425	1.087	.905	1.181	3.017	2.513
	4	.425	1.087	.905	1.181	3.017	2.513
	5	.543	1.138	.840	1.509	3.160	2.331
	6	.724	1.234	.690	2.009	3.427	1.916
2	1	.906	1.350	.423	2.884	4.295	1.345
	2	.554	1.143	.832	1.764	3.638	2.649
	3	.609	1.171	.793	1.940	3.727	2.523
	4	.609	1.171	.793	1.940	3.727	2.523
	5	.554	1.143	.832	1.764	3.638	2.649
	6	.906	1.350	.423	2.884	4.295	1.345
3	1	.980	1.400	.199	4.216	6.024	.857
	2	.761	1.257	.649	3.275	5.407	2.791
	3	.678	1.208	.735	2.919	5.200	3.161
	4	.678	1.208	.735	2.919	5.200	3.161
	5	.761	1.257	.649	3.275	5.407	2.791
	6	.980	1.400	.199	4.216	6.024	.857
4	1	.998	1.413	.063	12.681	17.951	.800
	2	.949	1.379	.315	12.059	17.518	4.002
	3	.777	1.266	.630	9.868	16.088	8.004
	4	.777	1.266	.630	9.868	16.088	8.004
	5	.949	1.379	.315	12.059	17.518	4.002
	6	.998	1.413	.063	12.681	17.951	.800

COVARIANCE MATRIX OF PARAMETER ESTIMATES

	0	1
0	8.667-01	-2.000-01
1	-2.000-01	5.714-02

	0	1	2
0	3.200+00	-1.950+00	2.500-01
1	-1.950+00	1.370+00	-1.875-01
2	2.500-01	-1.875-01	2.679-02

	0	1	2	3
0	1.300+01	-1.428+01	4.333+00	-3.889-01
1	-1.428+01	1.688+01	-5.324+00	4.892-01
2	4.333+00	-5.324+00	1.728+00	-1.620-01
3	-3.889-01	4.892-01	-1.620-01	1.543-02

	0	1	2	3	4
0	7.600+01	-1.228+02	6.271+01	-1.264+01	8.750-01
1	-1.228+02	2.037+02	-1.059+02	2.159+01	-1.507+00
2	6.271+01	-1.059+02	5.582+01	-1.151+01	8.108-01
3	-1.264+01	2.159+01	-1.151+01	2.397+00	-1.701-01
4	8.750-01	-1.507+00	8.108-01	-1.701-01	1.215-02

N = 7

DEGREE	NORMALIZATION FACTOR
0	3.7796447-001
1	1.8898224-001
2	1.0910895-001
3	4.0824829-001
4	8.0582296-002

TRANSFORMATION MATRIX

	0	1	2	3	4
0	3.7796447-001	-7.5592895-001	1.3093073+000	-2.4494897+000	5.3184316+000
1		1.8898224-001	-8.7287156-001	2.7896967+000	-8.4342804+000
2			1.0910895-001	-8.1649658-001	4.0626908+000
3				6.8041382-002	-7.5210143-001
4					4.7006340-002

AUXILIARY TABLES

DATA POINT STATISTICS

DEGREE	I	STANDARD DEVIATIONS MODEL	POINT	RESIDUAL	95 PER CENT CONFIDENCE LIMITS MODEL	POINT	RESIDUAL
1	1	.681	1.210	.732	1.752	3.111	1.881
	2	.535	1.134	.845	1.374	2.915	2.173
	3	.423	1.086	.906	1.086	2.791	2.330
	4	.378	1.069	.926	.972	2.748	2.380
	5	.423	1.086	.906	1.086	2.791	2.330
	6	.535	1.134	.845	1.374	2.915	2.173
	7	.681	1.210	.732	1.752	3.111	1.881
2	1	.873	1.327	.488	2.423	3.685	1.355
	2	.535	1.134	.845	1.484	3.148	2.346
	3	.535	1.134	.845	1.484	3.148	2.346
	4	.577	1.155	.816	1.603	3.206	2.267
	5	.535	1.134	.845	1.484	3.148	2.346
	6	.535	1.134	.845	1.484	3.148	2.346
	7	.873	1.327	.488	2.423	3.685	1.355
3	1	.964	1.389	.267	3.067	4.419	.851
	2	.673	1.205	.740	2.140	3.835	2.355
	3	.673	1.205	.740	2.140	3.835	2.355
	4	.577	1.155	.816	1.837	3.675	2.598
	5	.673	1.205	.740	2.140	3.835	2.355
	6	.673	1.205	.740	2.140	3.835	2.355
	7	.964	1.389	.267	3.067	4.419	.851

COVARIANCE MATRIX OF PARAMETER ESTIMATES

	0	1
0	7.143-01	-1.429-01
1	-1.429-01	3.571-02

	0	1	2
0	2.429+00	-1.286+00	1.429-01
1	-1.286+00	7.976-01	-9.524-02
2	1.429-01	-9.524-02	1.190-02

	0	1	2	3
0	8.429+00	-8.119+00	2.143+00	-1.667-01
1	-8.119+00	8.580+00	-2.373+00	1.898-01
2	2.143+00	-2.373+00	6.786-01	-5.556-02
3	-1.667-01	1.898-01	-5.556-02	4.630-03

4	1	.993	1.410	.114	4.275	6.065	.490
	2	.878	1.331	.479	3.777	5.725	2.061
	3	.677	1.208	.736	2.915	5.197	3.165
	4	.753	1.252	.658	3.240	5.386	2.831
	5	.677	1.208	.736	2.915	5.197	3.165
	6	.878	1.331	.479	3.777	5.725	2.061
	7	.993	1.410	.114	4.275	6.065	.490

	0	1	2	3	4
0	3.671+01	-5.298+01	2.375+01	-4.167+00	2.500-01
1	-5.298+01	7.972+01	-3.664+01	6.533+00	-3.965-01
2	2.375+01	-3.664+01	1.718+01	-3.111+00	1.910-01
3	-4.167+00	6.533+00	-3.111+00	5.703-01	-3.535-02
4	2.500-01	-3.965-01	1.910-01	-3.535-02	2.210-03

N = 8

DEGREE	NORMALIZATION FACTOR
0	3.5355339-001
1	7.7151675-002
2	7.7151675-002
3	6.1545745-002
4	4.0291148-002

TRANSFORMATION MATRIX

	0	1	2	3	4
0	3.5355339-001	-6.9436507-001	1.1572751+000	-2.0310096+000	3.9888237+000
1		1.5430335-001	-6.9436507-001	2.1130706+000	-5.8623621+000
2			7.7151675-002	-5.5391171-001	2.5551303+000
3				4.1030497-002	-4.2305706-001
4					2.3503170-002

AUXILIARY TABLES

DATA POINT STATISTICS

DEGREE	I	STANDARD DEVIATIONS MODEL	POINT	RESIDUAL	95 PER CENT CONFIDENCE LIMITS MODEL	POINT	RESIDUAL
1	1	.645	1.190	.764	1.579	2.912	1.869
	2	.523	1.129	.852	1.280	2.762	2.085
	3	.423	1.086	.906	1.034	2.656	2.218
	4	.362	1.063	.932	.885	2.602	2.281
	5	.362	1.063	.932	.885	2.602	2.281
	6	.423	1.086	.906	1.034	2.656	2.218
	7	.523	1.129	.852	1.280	2.762	2.085
	8	.645	1.190	.764	1.579	2.912	1.869
2	1	.842	1.307	.540	2.163	3.360	1.388
	2	.529	1.131	.849	1.360	2.908	2.182
	3	.482	1.110	.876	1.239	2.853	2.253
	4	.529	1.131	.849	1.360	2.908	2.182
	5	.529	1.131	.849	1.360	2.908	2.182
	6	.482	1.110	.876	1.239	2.853	2.253
	7	.529	1.131	.849	1.360	2.908	2.182
	8	.842	1.307	.540	2.163	3.360	1.388
3	1	.945	1.376	.326	2.625	3.821	.904
	2	.612	1.172	.791	1.699	3.255	2.196
	3	.646	1.191	.763	1.794	3.306	2.119
	4	.560	1.146	.828	1.555	3.182	2.300
	5	.560	1.146	.828	1.555	3.182	2.300

COVARIANCE MATRIX OF PARAMETER ESTIMATES

	0	1
0	6.071-01	-1.071-01
1	-1.071-01	2.381-02

	0	1	2
0	1.946+00	-9.107-01	8.929-02
1	-9.107-01	5.060-01	-5.357-02
2	8.929-02	-5.357-02	[illegible]

	0	1	2	3
0	6.071+00	-5.202+00	1.214+00	-8.333-02
1	-5.202+00	4.971+00	-1.224+00	8.670-02
2	1.214+00	-1.224+00	3.128-01	-2.273-02
3	-8.333-02	8.670-02	-2.273-02	1.684-03

	6	.646	1.191	.763	1.794	3.306	2.119
	7	.612	1.172	.791	1.699	3.255	2.196
	8	.945	1.376	.326	2.625	3.821	.904
4	1	.987	1.405	.163	3.140	4.471	.518
	2	.805	1.284	.593	2.563	4.086	1.886
	3	.658	1.197	.753	2.093	3.809	2.398
	4	.667	1.202	.745	2.124	3.826	2.370
	5	.667	1.202	.745	2.124	3.826	2.370
	6	.658	1.197	.753	2.093	3.809	2.398
	7	.805	1.284	.593	2.563	4.086	1.886
	8	.987	1.405	.163	3.140	4.471	.518

	0	1	2	3	
0	2.198+01	-2.859+01	1.[illegible]1+01	[illegible]771+00	9.3[illegible]02
1	-2.859+01	3.934+01	-1.[illegible]20+01	[illegible]567+00	-1.3[illegible]01
2	1.141+01	-1.620+01	6.[illegible]1+00	[illegible]104+00	6.0[illegible]2
3	-1.771+00	2.567+00	-1.[illegible]+00	[illegible]807-01	-9.94[illegible]3
4	9.375-02	-1.378-01	6.[illegible]-02	[illegible]943-03	5.52[illegible]

N = 9

DEGREE	NORMALIZATION FACTOR
0	3.3333333-001
1	1.2909944-001
2	1.8993429-002
3	3.1782086-002
4	2.2349508-002

TRANSFORMATION MATRIX

	0	1	2	3	4
0	3.3333333-001	-6.4549722-001	1.0446386+000	-1.7480147+000	3.1959796+000
1		1.2909944-001	-5.6980288-001	1.6738565+000	-4.3767786+000
2			5.6980288-002	-3.9727608-001	1.7413992+000
3				2.6485072-002	-2.6074426-001
4					1.3037213-002

AUXILIARY TABLES

DATA POINT STATISTICS

DEGREE	I	STANDARD DEVIATIONS MODEL	POINT	RESIDUAL	95 PER CENT CONFIDENCE LIMITS MODEL	POINT	RESIDUAL
1	1	.615	1.174	.789	1.453	2.776	1.865
	2	.511	1.123	.860	1.208	2.655	2.033
	3	.422	1.085	.907	.997	2.566	2.144
	4	.357	1.062	.934	.845	2.511	2.208
	5	.333	1.054	.943	.788	2.493	2.229
	6	.357	1.062	.934	.845	2.511	2.208
	7	.422	1.085	.907	.997	2.566	2.144
	8	.511	1.123	.860	1.208	2.655	2.033
	9	.615	1.174	.789	1.453	2.776	1.865
2	1	.813	1.289	.583	1.989	3.153	1.426
	2	.528	1.131	.849	1.292	2.767	2.078
	3	.448	1.096	.894	1.097	2.681	2.187
	4	.482	1.110	.876	1.179	2.716	2.144
	5	.505	1.120	.863	1.237	2.742	2.111
	6	.482	1.110	.876	1.179	2.716	2.144
	7	.448	1.096	.894	1.097	2.681	2.187
	8	.528	1.131	.849	1.292	2.767	2.078
	9	.813	1.289	.583	1.989	3.153	1.426

COVARIANCE MATRIX OF PARAMETER ESTIMATES

	0	1
0	5.278-01	-8.333-02
1	-8.333-02	1.667-02

	0	1	2
0	1.619+00	-6.786-01	5.952-02
1	-6.786-01	3.413-01	-3.247-02
2	5.952-02	-3.247-02	3.247-03

3	1	.927	1.363	.376	2.382	3.504	.967
	2	.573	1.153	.820	1.473	2.963	2.107
	3	.610	1.171	.793	1.567	3.011	2.038
	4	.560	1.146	.828	1.440	2.947	2.129
	5	.505	1.120	.863	1.299	2.880	2.218
	6	.560	1.146	.828	1.440	2.947	2.129
	7	.610	1.171	.793	1.567	3.011	2.038
	8	.573	1.153	.820	1.473	2.963	2.107
	9	.927	1.363	.376	2.382	3.504	.967

	0	1	2	3
0	4.675+00	-3.604+00	7.540-01	-4.630-02
1	-3.604+00	3.143+00	-6.975-01	4.433-02
2	7.540-01	-6.975-01	1.611-01	-1.052-02
3	-4.630-02	4.433-02	-1.052-02	7.015-04

4	1	.978	1.399	.209	2.715	3.883	.579
	2	.741	1.244	.672	2.056	3.455	1.865
	3	.657	1.197	.754	1.825	3.322	2.092
	4	.595	1.164	.804	1.653	3.231	2.231
	5	.646	1.190	.763	1.793	3.305	2.119
	6	.595	1.164	.804	1.653	3.231	2.231
	7	.657	1.197	.754	1.825	3.322	2.092
	8	.741	1.244	.672	2.056	3.455	1.865
	9	.978	1.399	.209	2.715	3.883	.579

	0	1	2	3	4
0	1.489+01	-1.759+01	6.319+00	-8.796-01	4.167-02
1	-1.759+01	2.230+01	-8.319+00	1.186+00	-5.706-02
2	6.319+00	-8.319+00	3.194+00	-4.646-01	2.270-02
3	-8.796-01	1.186+00	-4.646-01	6.869-02	-3.399-03
4	4.167-02	-5.706-02	2.270-02	-3.399-03	1.700-04

N = 10

DEGREE	NORMALIZATION FACTOR
0	3.1622777-001
1	5.5048188-002
2	8.7038828-002
3	1.0795838-002
4	1.8698940-002

TRANSFORMATION MATRIX

	0	1	2	3	4
0	3.1622777-001	-6.0553007-001	9.5742711-001	-1.5438048+000	2.6739484+000
1		1.1009638-001	-4.7871355-001	1.3692721+000	-3.4281390+000
2			4.3519414-002	-2.9688554-001	1.2543872+000
3				1.7993063-002	-1.7140695-001
4					7.7912249-003

AUXILIARY TABLES

DATA POINT STATISTICS

DEGREE	I	STANDARD DEVIATIONS MODEL	POINT	RESIDUAL	95 PER CENT CONFIDENCE LIMITS MODEL	POINT	RESIDUAL
1	1	.588	1.160	.809	1.355	2.675	1.866
	2	.498	1.117	.867	1.150	2.577	1.999
	3	.419	1.084	.908	.967	2.500	2.094
	4	.357	1.062	.934	.823	2.448	2.154
	5	.321	1.050	.947	.740	2.422	2.184
	6	.321	1.050	.947	.740	2.422	2.184
	7	.357	1.062	.934	.823	2.448	2.154
	8	.419	1.084	.908	.967	2.500	2.094
	9	.498	1.117	.867	1.150	2.577	1.999
	10	.588	1.160	.809	1.355	2.675	1.866
2	1	.786	1.272	.618	1.859	3.008	1.461
	2	.528	1.131	.849	1.249	2.674	2.008
	3	.428	1.088	.904	1.012	2.572	2.137
	4	.442	1.093	.897	1.045	2.585	2.121
	5	.474	1.106	.881	1.120	2.616	2.083
	6	.474	1.106	.881	1.120	2.616	2.083
	7	.442	1.093	.897	1.045	2.585	2.121
	8	.428	1.088	.904	1.012	2.572	2.137
	9	.528	1.131	.849	1.249	2.674	2.008
	10	.786	1.272	.618	1.859	3.008	1.461

COVARIANCE MATRIX OF PARAMETER ESTIMATES

Degree 1:

	0	1
0	4.667-01	-6.667-02
1	-6.667-02	1.212-02

Degree 2:

	0	1	2
0	1.383+00	-5.250-01	4.167-02
1	-5.250-01	2.413-01	-2.083-02
2	4.167-02	-2.083-02	1.894-03

3	1	•908	1.350	•420	2.221	3•304	1.027
	2	•549	1.141	•836	1.344	2•792	2.045
	3	•571	1.152	•821	1.397	2•818	2.009
	4	•554	1.143	•832	1.357	2•798	2.036
	5	•491	1.114	•871	1.201	2•726	2.132
	6	•491	1.114	•871	1.201	2•726	2.132
	7	•554	1.143	•832	1.357	2•798	2.036
	8	•571	1.152	•821	1.397	2•818	2.009
	9	•549	1.141	•836	1.344	2•792	2.045
	10	•908	1.350	•420	2.221	3•304	1.027

	0	1	2	3
0	3•767+00	-2•639+00	5•000-01	-2•778-02
1	-2•639+00	2•116+00	-4•274-01	2•464-02
2	5•000-01	-4•274-01	9•003-02	-5•342-03
3	-2•778-02	2•464-02	-5•342-03	3•238-04

4	1	•968	1.392	•251	2.488	3•578	.645
	2	•686	1.213	•727	1.764	3•118	1.870
	3	•654	1.195	•757	1.680	3•071	1.946
	4	•557	1.145	•830	1.433	2•943	2.134
	5	•595	1.164	•804	1.530	2•992	2.066
	6	•595	1.164	•804	1.530	2•992	2.066
	7	•557	1.145	•830	1.433	2•943	2.134
	8	•654	1.195	•757	1.680	3•071	1.946
	9	•686	1.213	•727	1.764	3•118	1.870
	10	•968	1.392	•251	2.488	3•578	.645

	0	1	2	3	4
0	1•092+01	-1•181+01	3•854+00	-4•861-01	2•083-02
1	-1•181+01	1•387+01	-4•728+00	6•122-01	-2•671-02
2	3•854+00	-4•728+00	1•664+00	-2•204-01	9•773-03
3	-4•861-01	6•122-01	-2•204-01	2•970-02	-1•335-03
4	2•083-02	-2•671-02	9•773-03	-1•335-03	6•070-05

N = 11

DEGREE	NORMALIZATION FACTOR
0	3.0151134-001
1	9.5346259-002
2	3.4139437-002
3	1.5267620-002
4	5.9131240-002

TRANSFORMATION MATRIX

	0	1	2	3	4
0	3.0151134-001	-5.7207755-001	8.8762536-001	-1.3893535+000	2.3061183+000
1		9.5346259-002	-4.0967325-001	1.1476161+000	-2.7791683+000
2			3.4139437-002	-2.2901431-001	9.4117223-001
3				1.2723017-002	-1.1826248-001
4					4.9276033-003

AUXILIARY TABLES

DATA POINT STATISTICS

		STANDARD DEVIATIONS			95 PER CENT CONFIDENCE LIMITS		
DEGREE	I	MODEL	POINT	RESIDUAL	MODEL	POINT	RESIDUAL
1	1	.564	1.148	.826	1.276	2.597	1.868
	2	.486	1.112	.874	1.100	2.515	1.977
	3	.416	1.083	.910	.940	2.450	2.058
	4	.357	1.062	.934	.807	2.402	2.113
	5	.316	1.049	.949	.715	2.373	2.146
	6	.302	1.044	.953	.682	2.363	2.157
	7	.316	1.049	.949	.715	2.373	2.146
	8	.357	1.062	.934	.807	2.402	2.113
	9	.416	1.083	.910	.940	2.450	2.058
	10	.486	1.112	.874	1.100	2.515	1.977
	11	.564	1.148	.826	1.276	2.597	1.868
2	1	.762	1.257	.648	1.757	2.899	1.494
	2	.528	1.131	.850	1.217	2.607	1.959
	3	.417	1.083	.909	.962	2.498	2.096
	4	.411	1.081	.911	.949	2.494	2.102
	5	.441	1.093	.898	1.017	2.520	2.070
	6	.455	1.099	.890	1.050	2.534	2.053
	7	.441	1.093	.898	1.017	2.520	2.070
	8	.411	1.081	.911	.949	2.494	2.102
	9	.417	1.083	.909	.962	2.498	2.096
	10	.528	1.131	.850	1.217	2.607	1.959

COVARIANCE MATRIX OF PARAMETER ESTIMATES

	0	1
0	4.182-01	-5.455-02
1	-5.455-02	9.091-03

	0	1	2
0	1.206+00	-4.182-01	3.030-02
1	-4.182-01	1.769-01	-1.399-02
2	3.030-02	-1.399-02	1.166-03

3	1	.889	1.338	.458	2.102	3.164	1.083
	2	.535	1.134	.845	1.266	2.682	1.997
	3	.535	1.134	.845	1.266	2.682	1.997
	4	.541	1.137	.841	1.279	2.688	1.989
	5	.490	1.114	.872	1.159	2.633	2.061
	6	.455	1.099	.890	1.077	2.598	2.105
	7	.490	1.114	.872	1.159	2.633	2.061
	8	.541	1.137	.841	1.279	2.688	1.989
	9	.535	1.134	.845	1.266	2.682	1.997
	10	.535	1.134	.845	1.266	2.682	1.997
	11	.889	1.338	.458	2.102	3.164	1.083

	0	1	2	3
0	3.136+00	-2.013+00	3.485-01	-1.768-02
1	-2.013+00	1.494+00	-2.768-01	1.460-02
2	3.485-01	-2.768-01	5.361-02	-2.914-03
3	-1.768-02	1.460-02	-2.914-03	1.619-04

4	1	.957	1.384	.290	2.342	3.387	.709
	2	.642	1.189	.766	1.572	2.908	1.875
	3	.642	1.189	.766	1.572	2.908	1.875
	4	.544	1.138	.839	1.331	2.786	2.053
	5	.544	1.138	.839	1.331	2.786	2.053
	6	.577	1.155	.816	1.413	2.825	1.998
	7	.544	1.138	.839	1.331	2.786	2.053
	8	.544	1.138	.839	1.331	2.786	2.053
	9	.642	1.189	.766	1.572	2.908	1.875
	10	.642	1.189	.766	1.572	2.908	1.875
	11	.957	1.384	.290	2.342	3.387	.709

	0	1	2	3	4
0	8.455+00	-8.422+00	2.519+00	-2.904-01	1.136-02
1	-8.422+00	9.218+00	-2.892+00	3.433-01	-1.369-02
2	2.519+00	-2.892+00	9.394-01	-1.142-01	4.638-03
3	-2.904-01	3.433-01	-1.142-01	1.415-02	-5.828-04
4	1.136-02	-1.369-02	4.638-03	-5.828-04	2.428-05

N = 12

DEGREE	NORMALIZATION FACTOR
0	2.8867513-001
1	4.1812100-002
2	9.1241484-003
3	1.3937367-002
4	1.1174754-002

TRANSFORMATION MATRIX

	0	1	2	3	4
0	2.8867513-001	-5.4355731-001	8.3029750-001	-1.2683004+000	2.0338052+000
1		8.3624201-002	-3.5584179-001	9.8026147-001	-2.3122428+000
2			2.7372445-002	-1.8118577-001	7.2868708-001
3				9.2915779-003	-8.4741884-002
4					3.2593032-003

AUXILIARY TABLES

DATA POINT STATISTICS

DEGREE	I	STANDARD DEVIATIONS MODEL	POINT	RESIDUAL	95 PER CENT CONFIDENCE LIMITS MODEL	POINT	RESIDUAL
1	1	.543	1.138	.840	1.210	2.535	1.871
	2	.474	1.107	.880	1.057	2.466	1.962
	3	.411	1.081	.912	.916	2.409	2.031
	4	.356	1.062	.934	.794	2.365	2.082
	5	.315	1.048	.949	.701	2.336	2.115
	6	.292	1.042	.957	.650	2.321	2.131
	7	.292	1.042	.957	.650	2.321	2.131
	8	.315	1.048	.949	.701	2.336	2.115
	9	.356	1.062	.934	.794	2.365	2.082
	10	.411	1.081	.912	.916	2.409	2.031
	11	.474	1.107	.880	1.057	2.466	1.962
	12	.543	1.138	.840	1.210	2.535	1.871
2	1	.739	1.244	.673	1.673	2.813	1.523
	2	.526	1.130	.850	1.191	2.556	1.924
	3	.411	1.081	.912	.930	2.446	2.062
	4	.389	1.073	.921	.879	2.427	2.084
	5	.411	1.081	.912	.930	2.446	2.062
	6	.433	1.090	.902	.978	2.465	2.040
	7	.433	1.090	.902	.978	2.465	2.040
	8	.411	1.081	.912	.930	2.446	2.062
	9	.389	1.073	.921	.879	2.427	2.084
	10	.411	1.081	.912	.930	2.446	2.062
	11	.526	1.130	.850	1.191	2.556	1.924

COVARIANCE MATRIX OF PARAMETER ESTIMATES

	0	1
0	3.788-01	-4.545-02
1	-4.545-02	6.993-03

	0	1	2
0	1.068+00	-3.409-01	2.273-02
1	-3.409-01	1.336-01	-9.740-03
2	2.273-02	-9.740-03	7.493-04

3	1	.871	1.326	.492	2.008	3.058	1.134
	2	.528	1.131	.849	1.217	2.608	1.958
	3	.505	1.120	.863	1.164	2.583	1.991
	4	.522	1.128	.853	1.204	2.601	1.967
	5	.489	1.113	.872	1.128	2.567	2.011
	6	.443	1.094	.896	1.022	2.522	2.067
	7	.443	1.094	.896	1.022	2.522	2.067
	8	.489	1.113	.872	1.128	2.567	2.011
	9	.522	1.128	.853	1.204	2.601	1.967
	10	.505	1.120	.863	1.164	2.583	1.991
	11	.528	1.131	.849	1.217	2.608	1.958
	12	.871	1.326	.492	2.008	3.058	1.134

	0	1	2	3
0	2.677+00	-1.584+00	2.525-01	-1.178-02
1	-1.584+00	1.095+00	-1.873-01	9.108-03
2	2.525-01	-1.873-01	3.358-02	-1.684-03
3	-1.178-02	9.108-03	-1.684-03	8.633-05

4	1	.946	1.376	.325	2.236	3.254	.769
	2	.608	1.170	.794	1.438	2.767	1.877
	3	.625	1.179	.781	1.478	2.789	1.846
	4	.542	1.137	.840	1.281	2.689	1.987
	5	.507	1.121	.862	1.199	2.651	2.038
	6	.543	1.138	.840	1.283	2.690	1.986
	7	.543	1.138	.840	1.283	2.690	1.986
	8	.507	1.121	.862	1.199	2.651	2.038
	9	.542	1.137	.840	1.281	2.689	1.987
	10	.625	1.179	.781	1.478	2.789	1.846
	11	.608	1.170	.794	1.438	2.767	1.877
	12	.946	1.376	.325	2.236	3.254	.769

	0	1	2	3	4
0	6.813+00	-6.287+00	1.735+00	-1.841-01	6.629-03
1	-6.287+00	6.441+00	-1.872+00	2.051-01	-7.536-03
2	1.735+00	-1.872+00	5.646-01	-6.343-02	2.375-03
3	-1.841-01	2.051-01	-6.343-02	7.268-03	-2.762-04
4	6.629-03	-7.536-03	2.375-03	-2.762-04	1.062-05

N = 13

DEGREE	NORMALIZATION FACTOR
0	2.7735010-001
1	7.4124932-002
2	2.2349508-002
3	4.1812100-002
4	3.8329090-003

TRANSFORMATION MATRIX

	0	1	2	3	4
0	2.7735010-001	-5.1887452-001	7.8223277-001	-1.1707388+000	1.8244647+000
1		7.4124932-002	-3.1289311-001	8.5017938-001	-1.9630882+000
2			2.2349508-002	-1.4634235-001	5.7844985-001
3				6.9686834-003	-6.2604180-002
4					2.2358636-003

AUXILIARY TABLES

DATA POINT STATISTICS

		STANDARD DEVIATIONS			95 PER CENT CONFIDENCE LIMITS		
DEGREE	I	MODEL	POINT	RESIDUAL	MODEL	POINT	RESIDUAL
1	1	.524	1.129	.852	1.154	2.485	1.874
	2	.463	1.102	.886	1.019	2.425	1.951
	3	.406	1.079	.914	.894	2.375	2.011
	4	.355	1.061	.935	.782	2.336	2.057
	5	.314	1.048	.949	.692	2.307	2.089
	6	.287	1.040	.958	.632	2.290	2.108
	7	.277	1.038	.961	.610	2.284	2.115
	8	.287	1.040	.958	.632	2.290	2.108
	9	.314	1.048	.949	.692	2.307	2.089
	10	.355	1.061	.935	.782	2.336	2.057
	11	.406	1.079	.914	.894	2.375	2.011
	12	.463	1.102	.886	1.019	2.425	1.951
	13	.524	1.129	.852	1.154	2.485	1.874
2	1	.719	1.231	.695	1.601	2.744	1.549
	2	.524	1.129	.852	1.168	2.516	1.898
	3	.408	1.080	.913	.910	2.407	2.034
	4	.373	1.067	.928	.830	2.378	2.068
	5	.386	1.072	.923	.860	2.388	2.056
	6	.408	1.080	.913	.910	2.407	2.034
	7	.418	1.084	.908	.932	2.415	2.024

COVARIANCE MATRIX OF PARAMETER ESTIMATES

Degree 1:

	0	1
0	3.462-01	-3.846-02
1	-3.846-02	5.495-03

Degree 2:

	0	1	2
0	9.580-01	-2.832-01	1.748-02
1	-2.832-01	1.034-01	-6.993-03
2	1.748-02	-6.993-03	4.995-04

	8	.408	1.080	.913	.910	2.407	2.034
	9	.386	1.072	.923	.860	2.388	2.056
	10	.373	1.067	.928	.830	2.378	2.068
	11	.408	1.080	.913	.910	2.407	2.034
	12	.524	1.129	.852	1.168	2.516	1.898
	13	.719	1.231	.695	1.601	2.744	1.549
3	1	.853	1.315	.522	1.930	2.974	1.180
	2	.524	1.129	.852	1.186	2.554	1.927
	3	.479	1.109	.878	1.084	2.509	1.985
	4	.501	1.118	.866	1.133	2.530	1.958
	5	.484	1.111	.875	1.096	2.514	1.979
	6	.441	1.093	.897	.998	2.473	2.030
	7	.418	1.084	.908	.946	2.452	2.055
	8	.441	1.093	.897	.998	2.473	2.030
	9	.484	1.111	.875	1.096	2.514	1.979
	10	.501	1.118	.866	1.133	2.530	1.958
	11	.479	1.109	.878	1.084	2.509	1.985
	12	.524	1.129	.852	1.186	2.554	1.927
	13	.853	1.315	.522	1.930	2.974	1.180

	0	1	2	3
0	2.329+00	-1.279+00	1.888-01	-8.159-03
1	-1.279+00	8.262-01	-1.314-01	5.925-03
2	1.888-01	-1.314-01	2.192-02	-1.020-03
3	-8.159-03	5.925-03	-1.020-03	4.856-05

4	1	.934	1.368	.358	2.153	3.155	.825
	2	.582	1.157	.813	1.342	2.668	1.875
	3	.604	1.168	.797	1.393	2.694	1.837
	4	.542	1.137	.840	1.249	2.623	1.938
	5	.486	1.112	.874	1.121	2.564	2.015
	6	.505	1.120	.863	1.164	2.583	1.990
	7	.528	1.131	.849	1.217	2.607	1.959
	8	.505	1.120	.863	1.164	2.583	1.990
	9	.486	1.112	.874	1.121	2.564	2.015
	10	.542	1.137	.840	1.249	2.623	1.938
	11	.604	1.168	.797	1.393	2.694	1.837
	12	.582	1.157	.813	1.342	2.668	1.875
	13	.934	1.368	.358	2.153	3.155	.825

	0	1	2	3	4
0	5.657+00	-4.860+00	1.244+00	-1.224-01	4.079-03
1	-4.860+00	4.680+00	-1.267+00	1.288-01	-4.389-03
2	1.244+00	-1.267+00	3.565-01	-3.723-02	1.293-03
3	-1.224-01	1.288-01	-3.723-02	3.968-03	-1.400-04
4	4.079-03	-4.389-03	1.293-03	-1.400-04	4.999-06

N = 14

DEGREE	NORMALIZATION FACTOR
0	2.6726124-001
1	3.3149677-002
2	3.7062466-002
3	3.2068417-003
4	2.7102759-003

TRANSFORMATION MATRIX

	0	1	2	3	4
0	2.6726124-001	-4.9724516-001	7.4124932-001	-1.0903262+000	1.6586889+000
1		6.6299354-002	-2.7796849-001	7.4665965-001	-1.6939225+000
2			1.8531233-002	-1.2025656-001	4.6865188-001
3				5.3447362-003	-4.7429829-002
4					1.5809943-003

AUXILIARY TABLES

DATA POINT STATISTICS

DEGREE	I	STANDARD DEVIATIONS MODEL	POINT	RESIDUAL	95 PER CENT CONFIDENCE LIMITS MODEL	POINT	RESIDUAL
1	1	.507	1.121	.862	1.105	2.443	1.878
	2	.452	1.097	.892	.985	2.391	1.943
	3	.401	1.077	.916	.873	2.347	1.996
	4	.354	1.061	.935	.771	2.311	2.038
	5	.314	1.048	.949	.685	2.284	2.068
	6	.285	1.040	.958	.621	2.266	2.088
	7	.269	1.036	.963	.587	2.256	2.098
	8	.269	1.036	.963	.587	2.256	2.098
	9	.285	1.040	.958	.621	2.266	2.088
	10	.314	1.048	.949	.685	2.284	2.068
	11	.354	1.061	.935	.771	2.311	2.038
	12	.401	1.077	.916	.873	2.347	1.996
	13	.452	1.097	.892	.985	2.391	1.943
	14	.507	1.121	.862	1.105	2.443	1.878
2	1	.699	1.220	.715	1.540	2.686	1.573
	2	.521	1.128	.853	1.147	2.482	1.878
	3	.407	1.080	.913	.897	2.377	2.010
	4	.362	1.063	.932	.796	2.340	2.052
	5	.365	1.065	.931	.803	2.343	2.049
	6	.386	1.072	.923	.849	2.359	2.031
	7	.401	1.077	.916	.882	2.371	2.017
	8	.401	1.077	.916	.882	2.371	2.017

COVARIANCE MATRIX OF PARAMETER ESTIMATES

	0	1
0	3.187-01	-3.297-02
1	-3.297-02	4.396-03

	0	1	2
0	8.681-01	-2.390-01	1.374-02
1	-2.390-01	8.166-02	-5.151-03
2	1.374-02	-5.151-03	3.434-04

	11	.362	1.063	.932	.796	2.340	2.052
	12	.407	1.080	.913	.897	2.377	2.010
	13	.521	1.128	.853	1.147	2.482	1.878
	14	.699	1.220	.715	1.540	2.686	1.573
3	1	.836	1.304	.548	1.864	2.905	1.221
	2	.522	1.128	.853	1.164	2.514	1.900
	3	.459	1.100	.888	1.023	2.452	1.979
	4	.479	1.109	.878	1.067	2.471	1.956
	5	.475	1.107	.880	1.059	2.467	1.960
	6	.441	1.093	.897	.983	2.435	1.999
	7	.408	1.080	.913	.909	2.406	2.034
	8	.408	1.080	.913	.909	2.406	2.034
	9	.441	1.093	.897	.983	2.435	1.999
	10	.475	1.107	.880	1.059	2.467	1.960
	11	.479	1.109	.878	1.067	2.471	1.956
	12	.459	1.100	.888	1.023	2.452	1.979
	13	.522	1.128	.853	1.164	2.514	1.900
	14	.836	1.304	.548	1.864	2.905	1.221
4	1	.922	1.360	.388	2.085	3.077	.877
	2	.563	1.147	.827	1.273	2.596	1.870
	3	.582	1.157	.813	1.317	2.617	1.840
	4	.540	1.137	.842	1.222	2.571	1.904
	5	.477	1.108	.879	1.079	2.506	1.989
	6	.473	1.106	.881	1.071	2.503	1.993
	7	.502	1.119	.865	1.136	2.531	1.956
	8	.502	1.119	.865	1.136	2.531	1.956
	9	.473	1.106	.881	1.071	2.503	1.993
	10	.477	1.108	.879	1.079	2.506	1.989
	11	.540	1.137	.842	1.222	2.571	1.904
	12	.582	1.157	.813	1.317	2.617	1.840
	13	.563	1.147	.827	1.273	2.596	1.870
	14	.922	1.360	.388	2.085	3.077	.877

	0	1	2	3
0	2.057+00	-1.053+00	1.449-01	-5.828-03
1	-1.053+00	6.392-01	-9.494-02	3.991-03
2	1.449-01	-9.494-02	1.481-02	-6.427-04
3	-5.828-03	3.991-03	-6.427-04	2.857-05

	0	1	2	3	4
0	4.808+00	-3.863+00	9.222-01	-8.450-02	2.622-03
1	-3.863+00	3.509+00	-8.888-01	8.433-02	-2.678-03
2	9.222-01	-8.888-01	2.344-01	-2.287-02	7.409-04
3	-8.450-02	8.433-02	-2.287-02	2.278-03	-7.499-05
4	2.622-03	-2.678-03	7.409-04	-7.499-05	2.500-06

N = 15

DEGREE	NORMALIZATION FACTOR
0	2.5819889-001
1	5.9761430-002
2	5.1897833-003
3	5.0138070-003
4	3.9324816-004

TRANSFORMATION MATRIX

	0	1	2	3	4
0	2.5819889-001	-4.7809144-001	7.0581053-001	-1.0228166+000	1.5242299+000
1		5.9761430-002	-2.4910960-001	6.6265816-001	-1.4812347+000
2			1.5569350-002	-1.0027614-001	3.8620247-001
3				4.1781725-003	-3.6703162-002
4					1.1469738-003

AUXILIARY TABLES

DATA POINT STATISTICS

		STANDARD DEVIATIONS			95 PER CENT CONFIDENCE LIMITS		
DEGREE	I	MODEL	POINT	RESIDUAL	MODEL	POINT	RESIDUAL
1	1	.492	1.114	.871	1.062	2.407	1.881
	2	.442	1.093	.897	.955	2.362	1.938
	3	.395	1.075	.919	.853	2.323	1.985
	4	.352	1.060	.936	.760	2.290	2.022
	5	.314	1.048	.949	.679	2.265	2.051
	6	.285	1.040	.959	.615	2.246	2.071
	7	.265	1.035	.964	.573	2.235	2.083
	8	.258	1.033	.966	.558	2.231	2.087
	9	.265	1.035	.964	.573	2.235	2.083
	10	.285	1.040	.959	.615	2.246	2.071
	11	.314	1.048	.949	.679	2.265	2.051
	12	.352	1.060	.936	.760	2.290	2.022
	13	.395	1.075	.919	.853	2.323	1.985
	14	.442	1.093	.897	.955	2.362	1.938
	15	.492	1.114	.871	1.062	2.407	1.881
2	1	.682	1.210	.732	1.485	2.637	1.594
	2	.518	1.126	.856	1.128	2.454	1.864
	3	.407	1.080	.913	.887	2.352	1.990
	4	.354	1.061	.935	.772	2.312	2.037
	5	.349	1.059	.937	.759	2.307	2.042
	6	.365	1.064	.931	.795	2.319	2.029
	7	.382	1.070	.924	.832	2.332	2.014
	8	.389	[illegible]	[illegible]	[illegible]	[illegible]	[illegible]

COVARIANCE MATRIX OF PARAMETER ESTIMATES

Degree 1:

	0	1
0	2.952-01	-2.857-02
1	-2.857-02	3.571-03

Degree 2:

	0	1	2
0	7.934-01	-2.044-01	1.099-02
1	-2.044-01	6.563-02	-3.878-03
2	1.099-02	-3.878-03	2.424-04

	10	.365	1.064	.931	.795	2.319	2.029
	11	.349	1.059	.937	.759	2.307	2.042
	12	.354	1.061	.935	.772	2.312	2.037
	13	.407	1.080	.913	.887	2.352	1.990
	14	.518	1.126	.856	1.128	2.454	1.864
	15	.682	1.210	.732	1.485	2.637	1.594
3	1	.820	1.293	.572	1.805	2.847	1.259
	2	.522	1.128	.853	1.149	2.483	1.878
	3	.443	1.094	.896	.976	2.408	1.973
	4	.458	1.100	.889	1.009	2.421	1.956
	5	.464	1.102	.886	1.021	2.426	1.950
	6	.440	1.092	.898	.968	2.404	1.977
	7	.405	1.079	.914	.892	2.375	2.012
	8	.389	1.073	.921	.856	2.361	2.028
	9	.405	1.079	.914	.892	2.375	2.012
	10	.440	1.092	.898	.968	2.404	1.977
	11	.464	1.102	.886	1.021	2.426	1.950
	12	.458	1.100	.889	1.009	2.421	1.956
	13	.443	1.094	.896	.976	2.408	1.973
	14	.522	1.128	.853	1.149	2.483	1.878
	15	.820	1.293	.572	1.805	2.847	1.259
4	1	.910	1.352	.415	2.027	3.012	.925
	2	.548	1.141	.836	1.222	2.541	1.863
	3	.560	1.146	.829	1.247	2.553	1.846
	4	.535	1.134	.845	1.193	2.527	1.882
	5	.474	1.107	.881	1.056	2.466	1.962
	6	.451	1.097	.893	1.004	2.444	1.989
	7	.473	1.106	.881	1.054	2.465	1.963
	8	.489	1.113	.872	1.090	2.481	1.943
	9	.473	1.106	.881	1.054	2.465	1.963
	10	.451	1.097	.893	1.004	2.444	1.989
	11	.474	1.107	.881	1.056	2.466	1.962
	12	.535	1.134	.845	1.193	2.527	1.882
	13	.560	1.146	.829	1.247	2.553	1.846
	14	.548	1.141	.836	1.222	2.541	1.863
	15	.910	1.352	.415	2.027	3.012	.925

	0	1	2	3
0	1.840+00	-8.822-01	1.136-01	-4.274-03
1	-8.822-01	5.047-01	-7.033-02	2.769-03
2	1.136-01	-7.033-02	1.030-02	-4.190-04
3	-4.274-03	2.769-03	-4.190-04	1.746-05

	0	1	2	3	4
0	4.163+00	-3.140+00	7.022-01	-6.022-02	1.748-03
1	-3.140+00	2.699+00	-6.424-01	5.713-02	-1.699-03
2	7.022-01	-6.424-01	1.595-01	-1.459-02	4.430-04
3	-6.022-02	5.713-02	-1.459-02	1.365-03	-4.210-05
4	1.748-03	-1.699-03	4.430-04	-4.210-05	1.316-06

N = 16

DEGREE	NORMALIZATION FACTOR
0	2.5000000-001
1	2.7116307-002
2	1.3231403-002
3	9.9614244-004
4	1.4582032-003

TRANSFORMATION MATRIX

	0	1	2	3	4
0	2.5000000-001	-4.6097722-001	6.7480156-001	-9.6526202-001	1.4129989+000
1		5.4232614-002	-2.2493385-001	5.9336884-001	-1.3097095+000
2			1.3231403-002	-8.4672107-002	3.2287050-001
3				3.3204748-003	-2.8921030-002
4					8.5061854-004

AUXILIARY TABLES

DATA POINT STATISTICS

DEGREE	I	STANDARD DEVIATIONS MODEL	POINT	RESIDUAL	95 PER CENT CONFIDENCE LIMITS MODEL	POINT	RESIDUAL
1	1	.477	1.108	.879	1.024	2.377	1.885
	2	.432	1.089	.902	.927	2.337	1.934
	3	.389	1.073	.921	.835	2.302	1.976
	4	.349	1.059	.937	.749	2.272	2.010
	5	.314	1.048	.949	.673	2.248	2.036
	6	.284	1.040	.959	.610	2.230	2.056
	7	.263	1.034	.965	.564	2.218	2.069
	8	.251	1.031	.968	.539	2.212	2.076
	9	.251	1.031	.968	.539	2.212	2.076
	10	.263	1.034	.965	.564	2.218	2.069
	11	.284	1.040	.959	.610	2.230	2.056
	12	.314	1.048	.949	.673	2.248	2.036
	13	.349	1.059	.937	.749	2.272	2.010
	14	.389	1.073	.921	.835	2.302	1.976
	15	.432	1.089	.902	.927	2.337	1.934
	16	.477	1.108	.879	1.024	2.377	1.885

COVARIANCE MATRIX OF PARAMETER ESTIMATES

	0	1
0	2.750-01	-2.500-02
1	-2.500-02	2.941-03

DEGREE	I	MODEL	POINT	RESIDUAL	MODEL	POINT	RESIDUAL
2	1	.665	1.201	.747	1.437	2.595	1.613
	2	.514	1.124	.858	1.110	2.429	1.853
	3	.407	1.080	.913	.879	2.332	1.973
	4	.350	1.059	.937	.755	2.289	2.024
	5	.336	1.055	.942	.725	2.279	2.035
	6	.347	1.058	.938	.749	2.287	2.026
	7	.364	1.064	.931	.786	2.299	2.012
	8	.375	1.068	.927	.810	2.307	2.003

	0	1	2
0	7.304-01	-1.768-01	8.929-03
1	-1.768-01	5.354-02	-2.976-03
2	8.929-03	-2.976-03	1.751-04

	9	.375	1.068	.927	.810	2.307	2.003
	10	.364	1.064	.931	.786	2.299	2.012
	11	.347	1.058	.938	.749	2.287	2.026
	12	.336	1.055	.942	.725	2.279	2.035
	13	.350	1.059	.937	.755	2.289	2.024
	14	.407	1.080	.913	.879	2.332	1.973
	15	.514	1.124	.858	1.110	2.429	1.853
	16	.665	1.201	.747	1.437	2.595	1.613
3	1	.805	1.284	.593	1.754	2.797	1.293
	2	.522	1.128	.853	1.137	2.457	1.859
	3	.431	1.089	.902	.940	2.373	1.966
	4	.439	1.092	.898	.957	2.380	1.957
	5	.450	1.097	.893	.981	2.389	1.946
	6	.436	1.091	.900	.950	2.377	1.961
	7	.405	1.079	.914	.883	2.351	1.992
	8	.380	1.070	.925	.828	2.331	2.015
	9	.380	1.070	.925	.828	2.331	2.015
	10	.405	1.079	.914	.883	2.351	1.992
	11	.436	1.091	.900	.950	2.377	1.961
	12	.450	1.097	.893	.981	2.389	1.946
	13	.439	1.092	.898	.957	2.380	1.957
	14	.431	1.089	.902	.940	2.373	1.966
	15	.522	1.128	.853	1.137	2.457	1.859
	16	.805	1.284	.593	1.754	2.797	1.293
4	1	.898	1.344	.440	1.976	2.958	.969
	2	.538	1.136	.843	1.185	2.500	1.855
	3	.538	1.136	.843	1.185	2.500	1.855
	4	.528	1.131	.849	1.162	2.489	1.869
	5	.474	1.106	.881	1.042	2.435	1.938
	6	.437	1.091	.899	.962	2.402	1.980
	7	.447	1.095	.895	.983	2.411	1.969
	8	.469	1.105	.883	1.033	2.431	1.943
	9	.469	1.105	.883	1.033	2.431	1.943
	10	.447	1.095	.895	.983	2.411	1.969
	11	.437	1.091	.899	.962	2.402	1.980
	12	.474	1.106	.881	1.042	2.435	1.938
	13	.528	1.131	.849	1.162	2.489	1.869
	14	.538	1.136	.843	1.185	2.500	1.855
	15	.538	1.136	.843	1.185	2.500	1.855
	16	.898	1.344	.440	1.976	2.958	.969

	0	1	2	3
0	1.662+00	-7.495-01	9.066-02	-3.205-03
1	-7.495-01	4.056-01	-5.322-02	1.970-03
2	9.066-02	-5.322-02	7.344-03	-2.812-04
3	-3.205-03	1.970-03	-2.812-04	1.103-05

	0	1	2	3	4
0	3.659+00	-2.600+00	5.469-01	-4.407-02	1.202-03
1	-2.600+00	2.121+00	-4.761-01	3.985-02	-1.114-03
2	5.469-01	-4.761-01	1.116-01	-9.619-03	2.746-04
3	-4.407-02	3.985-02	-9.619-03	8.475-04	-2.460-05
4	1.202-03	-1.114-03	2.746-04	-2.460-05	7.236-07

N = 17

DEGREE	NORMALIZATION FACTOR
0	2.4253563-001
1	4.9507377-002
2	1.1357771-002
3	1.6062314-002
4	7.7160861-003

TRANSFORMATION MATRIX

	0	1	2	3	4
0	2.4253563-001	-4.4556639-001	6.4739296-001	-9.1555191-001	1.3194507+000
1		4.9507377-002	-2.0443988-001	5.3541047-001	-1.1689870+000
2			1.1357771-002	-7.2280414-002	2.7327805-001
3				2.6770524-003	-2.3148258-002
4					6.4300718-004

AUXILIARY TABLES

DATA POINT STATISTICS

DEGREE	I	STANDARD DEVIATIONS MODEL	POINT	RESIDUAL	95 PER CENT CONFIDENCE LIMITS MODEL	POINT	RESIDUAL
1	1	.464	1.103	.886	.990	2.350	1.888
	2	.423	1.086	.906	.902	2.314	1.931
	3	.383	1.071	.924	.817	2.283	1.969
	4	.347	1.058	.938	.739	2.256	1.999
	5	.313	1.048	.950	.667	2.234	2.024
	6	.284	1.040	.959	.606	2.216	2.043
	7	.262	1.034	.965	.558	2.203	2.057
	8	.248	1.030	.969	.528	2.196	2.065
	9	.243	1.029	.970	.517	2.193	2.068
2	1	.650	1.193	.760	1.393	2.558	1.630
	2	.509	1.122	.860	1.093	2.407	1.846
	3	.407	1.080	.913	.873	2.316	1.959
	4	.347	1.058	.938	.744	2.270	2.012
	5	.326	1.052	.945	.699	2.256	2.028
	6	.332	1.054	.943	.711	2.260	2.024
	7	.347	1.058	.938	.744	2.270	2.012
	8	.360	1.063	.933	.772	2.279	2.001
	9	.365	1.064	.931	.783	2.283	1.997

COVARIANCE MATRIX OF PARAMETER ESTIMATES

	0	1
0	2.574-01	-2.206-02
1	-2.206-02	2.451-03

	0	1	2
0	6.765-01	-1.544-01	7.353-03
1	-1.544-01	4.425-02	-2.322-03
2	7.353-03	-2.322-03	1.290-04

3	1	.790	1.275	.613	1.707	2.753	1.324
	2	.522	1.128	.853	1.127	2.437	1.843
	3	.422	1.085	.906	.912	2.345	1.958
	4	.422	1.085	.906	.912	2.345	1.958
	5	.436	1.091	.900	.941	2.357	1.944
	6	.429	1.088	.903	.928	2.351	1.951
	7	.405	1.079	.914	.874	2.331	1.976
	8	.377	1.069	.926	.815	2.309	2.001
	9	.365	1.064	.931	.788	2.300	2.011

	0	1	2	3
0	1.515+00	-6.446-01	7.353-02	-2.451-03
1	-6.446-01	3.309-01	-4.102-02	1.433-03
2	7.353-02	-4.102-02	5.353-03	-1.935-04
3	-2.451-03	1.433-03	-1.935-04	7.167-06

4	1	.886	1.336	.463	1.931	2.911	1.009
	2	.531	1.132	.847	1.158	2.467	1.846
	3	.518	1.126	.855	1.130	2.454	1.863
	4	.518	1.126	.855	1.130	2.454	1.863
	5	.473	1.106	.881	1.032	2.411	1.919
	6	.430	1.089	.903	.937	2.372	1.967
	7	.425	1.087	.905	.927	2.368	1.972
	8	.447	1.095	.895	.973	2.386	1.950
	9	.459	1.100	.889	.999	2.397	1.936

	0	1	2	3	4
0	3.256+00	-2.187+00	4.341-01	-3.299-02	8.484-04
1	-2.187+00	1.697+00	-3.605-01	2.849-02	-7.517-04
2	4.341-01	-3.605-01	8.003-02	-6.519-03	1.757-04
3	-3.299-02	2.849-02	-6.519-03	5.430-04	-1.488-05
4	8.484-04	-7.517-04	1.757-04	-1.488-05	4.135-07

N = 18

DEGREE	NORMALIZATION FACTOR
0	2.3570226-001
1	2.2715543-002
2	6.5574123-003
3	6.5574123-003
4	5.9314026-003

TRANSFORMATION MATRIX

	0	1	2	3	4
0	2.3570226-001	-4.3159531-001	6.2295417-001	-8.7213584-001	1.2396631+000
1		4.5431085-002	-1.8688625-001	4.8634141-001	-1.0518354+000
2			9.8361184-003	-6.2295417-002	2.3379612-001
3				2.1858041-003	-1.8782775-002
4					4.9428355-004

AUXILIARY TABLES

DATA POINT STATISTICS

DEGREE	I	STANDARD DEVIATIONS MODEL	POINT	RESIDUAL	95 PER CENT CONFIDENCE LIMITS MODEL	POINT	RESIDUAL
1	1	.452	1.098	.892	.959	2.327	1.891
	2	.414	1.082	.910	.878	2.295	1.929
	3	.378	1.069	.926	.801	2.266	1.963
	4	.343	1.057	.939	.728	2.241	1.991
	5	.312	1.048	.950	.661	2.221	2.014
	6	.284	1.040	.959	.603	2.204	2.032
	7	.262	1.034	.965	.555	2.191	2.046
	8	.245	1.030	.969	.520	2.183	2.055
	9	.237	1.028	.972	.502	2.179	2.060
2	1	.635	1.185	.772	1.354	2.525	1.646
	2	.505	1.120	.863	1.076	2.388	1.840
	3	.407	1.080	.914	.867	2.301	1.947
	4	.345	1.058	.939	.735	2.255	2.001
	5	.319	1.050	.948	.680	2.237	2.020
	6	.319	1.050	.948	.680	2.237	2.020
	7	.331	1.053	.944	.706	2.245	2.011
	8	.345	1.058	.939	.735	2.255	2.001
	9	.353	1.061	.935	.753	2.261	1.994

COVARIANCE MATRIX OF PARAMETER ESTIMATES

Degree 1:

	0	1
0	2.418-01	-1.961-02
1	-1.961-02	2.064-03

Degree 2:

	0	1	2
0	6.299-01	-1.360-01	6.127-03
1	-1.360-01	3.699-02	-1.838-03
2	6.127-03	-1.838-03	9.675-05

3	1	.776	1.266	.631	1.665	2.715	1.353
	2	.522	1.128	.853	1.119	2.419	1.830
	3	.416	1.083	.910	.892	2.323	1.951
	4	.407	1.080	.913	.874	2.316	1.959
	5	.421	1.085	.907	.904	2.327	1.945
	6	.421	1.085	.907	.904	2.327	1.945
	7	.403	1.078	.915	.864	2.312	1.963
	8	.377	1.069	.926	.808	2.292	1.987
	9	.357	1.062	.934	.766	2.278	2.003

	0	1	2	3
0	1.391+00	-5.602-01	6.046-02	-1.906-03
1	-5.602-01	2.735-01	-3.214-02	1.063-03
2	6.046-02	-3.214-02	3.977-03	-1.362-04
3	-1.906-03	1.063-03	-1.362-04	4.778-06

4	1	.875	1.329	.485	1.890	2.870	1.047
	2	.526	1.130	.850	1.137	2.442	1.837
	3	.500	1.118	.866	1.081	2.416	1.870
	4	.507	1.121	.862	1.096	2.423	1.862
	5	.472	1.106	.881	1.020	2.389	1.904
	6	.427	1.087	.904	.923	2.349	1.953
	7	.410	1.081	.912	.887	2.335	1.970
	8	.424	1.086	.905	.917	2.347	1.956
	9	.442	1.093	.897	.956	2.362	1.937

	0	1	2	3	4
0	2.927+00	-1.864+00	3.503-01	-2.519-02	6.127-04
1	-1.864+00	1.380+00	-2.781-01	2.082-02	-5.199-04
2	3.503-01	-2.781-01	5.864-02	-4.528-03	1.156-04
3	-2.519-02	2.082-02	-4.528-03	3.576-04	-9.284-06
4	6.127-04	-5.199-04	1.156-04	-9.284-06	2.443-07

N = 19

DEGREE	NORMALIZATION FACTOR
0	2.2941573-001
1	4.1885391-002
2	8.5856681-003
3	2.1658420-003
4	6.6108829-004

TRANSFORMATION MATRIX

	0	1	2	3	4
0	2.2941573-001	-4.1885391-001	6.0099676-001	-8.3384917-001	1.1707874+000
1		4.1885391-002	-1.7171336-001	4.4435858-001	-9.5306895-001
2			8.5856681-003	-5.4146050-002	2.0190738-001
3				1.8048683-003	-1.5425393-002
4					3.8563483-004

AUXILIARY TABLES

DATA POINT STATISTICS

DEGREE	I	STANDARD DEVIATIONS MODEL	POINT	RESIDUAL	95 PER CENT CONFIDENCE LIMITS MODEL	POINT	RESIDUAL
1	1	.441	1.093	.897	.931	2.306	1.893
	2	.406	1.079	.914	.857	2.277	1.928
	3	.372	1.067	.928	.785	2.251	1.958
	4	.340	1.056	.940	.718	2.229	1.984
	5	.311	1.047	.951	.655	2.209	2.005
	6	.284	1.040	.959	.599	2.193	2.023
	7	.262	1.034	.965	.552	2.181	2.036
	8	.244	1.029	.970	.515	2.172	2.046
	9	.233	1.027	.972	.492	2.166	2.052
	10	.229	1.026	.973	.484	2.165	2.054
2	1	.622	1.177	.783	1.318	2.496	1.660
	2	.500	1.118	.866	1.060	2.370	1.836
	3	.406	1.079	.914	.862	2.288	1.937
	4	.344	1.058	.939	.730	2.242	1.990
	5	.314	1.048	.950	.665	2.222	2.013
	6	.308	1.046	.951	.654	2.218	2.017
	7	.318	1.049	.948	.673	2.224	2.010
	8	.331	1.053	.944	.701	2.233	2.000
	9	.341	1.057	.940	.723	2.240	1.993
	10	.345	1.058	.939	.731	2.242	1.990

COVARIANCE MATRIX OF PARAMETER ESTIMATES

	0	1
0	2.281-01	-1.754-02
1	-1.754-02	1.754-03

	0	1	2
0	5.893-01	-1.207-01	5.160-03
1	-1.207-01	3.124-02	-1.474-03
2	5.160-03	-1.474-03	7.371-05

3	1	.763	1.258	.647	1.626	2.681	1.379
	2	.521	1.128	.853	1.111	2.404	1.819
	3	.411	1.081	.912	.876	2.304	1.943
	4	.394	1.075	.919	.841	2.291	1.959
	5	.407	1.080	.913	.868	2.302	1.947
	6	.412	1.081	.911	.878	2.305	1.942
	7	.400	1.077	.917	.852	2.295	1.954
	8	.377	1.069	.926	.803	2.278	1.975
	9	.354	1.061	.935	.755	2.261	1.993
	10	.345	1.058	.939	.735	2.255	2.001

	0	1	2	3
0	1.285+00	-4.913-01	5.031-02	-1.505-03
1	-4.913-01	2.287-01	-2.553-02	8.020-04
2	5.031-02	-2.553-02	3.006-03	-9.773-05
3	-1.505-03	8.020-04	-9.773-05	3.258-06

4	1	.863	1.321	.505	1.852	2.834	1.082
	2	.523	1.129	.852	1.122	2.421	1.828
	3	.484	1.111	.875	1.039	2.383	1.876
	4	.495	1.116	.869	1.062	2.393	1.863
	5	.470	1.105	.883	1.007	2.370	1.893
	6	.427	1.087	.904	.915	2.332	1.940
	7	.401	1.077	.916	.859	2.311	1.965
	8	.405	1.079	.914	.869	2.314	1.961
	9	.424	1.086	.906	.909	2.329	1.943
	10	.433	1.090	.901	.929	2.337	1.933

	0	1	2	3	4
0	2.655+00	-1.607+00	2.867-01	-1.956-02	4.515-04
1	-1.607+00	1.137+00	-2.180-01	1.550-02	-3.675-04
2	2.867-01	-2.180-01	4.377-02	-3.212-03	7.786-05
3	-1.956-02	1.550-02	-3.212-03	2.412-04	-5.949-06
4	4.515-04	-3.675-04	7.786-05	-5.949-06	1.487-07

N = 20

DEGREE	NORMALIZATION FACTOR
0	2.2360680-001
1	1.9389168-002
2	7.5472235-003
3	4.5160928-004
4	2.0905447-004

TRANSFORMATION MATRIX

	0	1	2	3	4
0	2.2360680-001	-4.0717254-001	5.8113621-001	-7.9980003-001	1.1107064+000
1		3.8778337-002	-1.5849169-001	4.0810425-001	-8.6888265-001
2			7.5472235-003	-4.7418974-002	1.7582352-001
3				1.5053643-003	-1.2804586-002
4					3.0487111-004

AUXILIARY TABLES

DATA POINT STATISTICS

		STANDARD DEVIATIONS			95 PER CENT CONFIDENCE LIMITS		
DEGREE	I	MODEL	POINT	RESIDUAL	MODEL	POINT	RESIDUAL
1	1	.431	1.089	.902	.905	2.288	1.896
	2	.398	1.076	.917	.837	2.261	1.927
	3	.367	1.065	.930	.771	2.238	1.954
	4	.337	1.055	.942	.708	2.217	1.978
	5	.309	1.047	.951	.649	2.199	1.998
	6	.284	1.039	.959	.596	2.184	2.015
	7	.262	1.034	.965	.550	2.172	2.028
	8	.244	1.029	.970	.512	2.162	2.038
	9	.231	1.026	.973	.485	2.156	2.044
	10	.224	1.025	.974	.472	2.153	2.047
2	1	.609	1.171	.793	1.285	2.470	1.674
	2	.495	1.116	.869	1.045	2.354	1.833
	3	.406	1.079	.914	.856	2.277	1.928
	4	.344	1.057	.939	.725	2.231	1.981
	5	.310	1.047	.951	.654	2.209	2.006
	6	.300	1.044	.954	.633	2.203	2.013
	7	.306	1.046	.952	.645	2.206	2.009
	8	.318	1.049	.948	.670	2.214	2.001
	9	.329	1.053	.944	.694	2.221	1.992
	10	.335	1.055	.942	.707	2.225	1.988

COVARIANCE MATRIX OF PARAMETER ESTIMATES

	0	1
0	2.158-01	-1.579-02
1	-1.579-02	1.504-03

	0	1	2
0	5.535-01	-1.079-01	4.386-03
1	-1.079-01	2.662-02	-1.196-03
2	4.386-03	-1.196-03	5.696-05

3	1	•750	1.250	•662	1.590	2•650	1.403
	2	•521	1.128	•854	1.104	2•390	1.810
	3	•408	1.080	•913	.864	2•289	1.936
	4	•384	1.071	•924	.813	2•271	1.958
	5	•394	1.075	•919	.835	2•279	1.948
	6	•402	1.078	•916	.851	2•284	1.941
	7	•395	1.075	•919	.837	2•279	1.948
	8	•376	1.068	•927	.797	2•265	1.964
	9	•353	1.061	•935	.749	2•248	1.983
	10	•338	1.056	•941	.717	2•238	1.995

	0	1	2	3
0	1•193+00	-4•343-01	4•231-02	-1•204-03
1	-4•343-01	1•932-01	-2•055-02	6•143-04
2	4•231-02	-2•055-02	2•306-03	-7•138-05
3	-1•204-03	6•143-04	-7•138-05	2•266-06

4	1	•852	1.314	•523	1.817	2•801	1.115
	2	•521	1.128	•853	1.111	2•404	1.819
	3	•470	1.105	•882	1.003	2•355	1.881
	4	•483	1.110	•876	1.029	2•367	1.867
	5	•466	1.103	•885	.993	2•351	1.886
	6	•427	1.087	•904	.909	2•317	1.928
	7	•395	1.075	•919	.842	2•292	1.958
	8	•390	1.073	•921	.832	2•288	1.962
	9	•405	1.079	•914	.864	2•300	1.949
	10	•420	1.084	•908	.894	2•312	1.935

	0	1	2	3	4
0	2•427+00	-1•399+00	2•376-01	-1•543-02	3•386-04
1	-1•399+00	9•481-01	-1•733-01	1•174-02	-2•649-04
2	2•376-01	-1•733-01	3•322-02	-2•323-03	5•360-05
3	-1•543-02	1•174-02	-2•323-03	1•662-04	-3•904-06
4	3•386-04	-2•649-04	5•360-05	-3•904-06	9•295-08

TABLE II. THE CUMULATIVE ERROR MODEL

THERE ARE 3 OBSERVED Y VALUES
****

THE TRANSFORMATION MATRIX OF PARAMETER ESTIMATES
****

0	1	2
1.000000+000	0.000000	0.000000
	7.071068-001	-1.414214+000
		7.071068-001

THE ORTHONORMAL ARGUMENTS TO DETERMINE THE B(I) ESTIMATES FROM THE Z VALUES
****

	0	1	2
1	.0000000	.0000000	.0000000
2	.0000000	.7071068	-.7071068
3	.0000000	.7071068	.7071068

THE ARGUMENTS TO DETERMINE THE B(I) ESTIMATES FROM THE Y VALUES
****

	0	1	2
1	1.0000000	-.7071068	.7071068
2	.0000000	.0000000	-1.4142136
3	.0000000	.7071068	.7071068

TABLES TO AID IN FITTING THE DATA POINT STATISTICS
****

PREDICTION
****

	1	2
2	.7071	-.7071
3	1.4142	.0000

VARIANCE
****

1	2
.5000	1.0000
2.0000	2.0000

THERE ARE 4 OBSERVED Y VALUES

THE TRANSFORMATION MATRIX OF PARAMETER ESTIMATES

0	1	2	3
1.000000+000	0.000000	0.000000	0.000000
	5.773503-001	-1.060660+000	1.837117+000
		3.535534-001	-1.837117+000
			4.082483-001

THE ORTHONORMAL ARGUMENTS TO DETERMINE THE B(I) ESTIMATES FROM THE Z VALUES

	0	1	2	3
1	1.0000000	.0000000	.0000000	.0000000
2	.0000000	.5773503	-.7071068	.4082483
3	.0000000	.5773503	.0000000	-.8164966
4	.0000000	.5773503	.7071068	.4082483

THE ARGUMENTS TO DETERMINE THE B(I) ESTIMATES FROM THE Y VALUES

	0	1	2	3
1	1.0000000	-.5773503	.7071068	-.4082483
2	.0000000	.0000000	-.7071068	1.2247449
3	.0000000	.0000000	-.7071068	-1.2247449
4	.0000000	.5773503	.7071068	.4082483

TABLES TO AID IN FITTING THE DATA POINT STATISTICS

	PREDICTION			VARIANCE		
	1	2	3	1	2	3
2	.5774	-.7071	.4082	.3333	.8333	1.0000
3	1.1547	-.7071	-.4082	1.3333	1.8333	2.0000
4	1.7321	.0000	.0000	3.0000	3.0000	3.0000

```
THERE ARE  5 OBSERVED Y VALUES
**********************************

THE TRANSFORMATION MATRIX OF PARAMETER ESTIMATES
***************************************************

         0              1              2              3              4
   1.000000+000    0.000000       0.000000       0.000000       0.000000
                   5.000000-001  -8.944272-001   1.333333+000  -2.534210+000
                                  2.236068-001  -1.000000+000   3.614977+000
                                                 1.666667-001  -1.490712+000
                                                                1.863390-001


THE ORTHONORMAL ARGUMENTS TO DETERMINE THE B(I) ESTIMATES FROM THE Z VALUES
*************************************************************************

         0              1              2              3              4
1   1.0000000       .0000000       .0000000       .0000000       .0000000
2    .0000000       .5000000      -.6708204       .5000000      -.2236068
3    .0000000       .5000000      -.2236068      -.5000000       .6708204
4    .0000000       .5000000       .2236068      -.5000000      -.6708204
5    .0000000       .5000000       .6708204       .5000000       .2236068

THE ARGUMENTS TO DETERMINE THE B(I) ESTIMATES FROM THE Y VALUES
*************************************************************************

         0              1              2              3              4
1   1.0000000      -.5000000       .6708204      -.5000000       .2236068
2    .0000000       .0000000      -.4472136      1.0000000      -.8944272
3    .0000000       .0000000      -.4472136       .0000000      1.3416408
4    .0000000       .0000000      -.4472136     -1.0000000      -.8944272
5    .0000000       .5000000       .6708204       .5000000       .2236068
```

TABLES TO AID IN FITTING THE DATA POINT STATISTICS
****

PREDICTION
****

	1	2	3	4
2	.5000	-.6708	.5000	-.2236
3	1.0000	-.8944	.0000	.4472
4	1.5000	-.6708	-.5000	-.2236
5	2.0000	.0000	-.0000	-.0000

VARIANCE
****

1	2	3	4
.2500	.7000	.9500	1.0000
1.0000	1.8000	1.8000	2.0000
2.2500	2.7000	2.9500	3.0000
4.0000	4.0000	4.0000	4.0000

THERE ARE 6 OBSERVED Y VALUES
****

THE TRANSFORMATION MATRIX OF PARAMETER ESTIMATES
****

0	1	2	3	4	5
1.000000+000	0.000000	0.000000	0.000000	0.000000	0.000000
	4.472136-001	-7.905694-001	1.113589+000	-1.712900+000	3.735089+000
		1.581139-001	-6.681531-001	1.989600+000	-6.598658+000
			8.908708-002	-6.588078-001	3.784891+000
				6.588078-002	-8.715209-001
					6.972167-002

THE ORTHONORMAL ARGUMENTS TO DETERMINE THE B(I) ESTIMATES FROM THE Z VALUES
****

	0	1	2	3	4	5
1	1.0000000	.0000000	.0000000	.0000000	.0000000	.0000000
2	.0000000	.4472136	-.6324555	.5345225	-.3162278	.1195229
3	.0000000	.4472136	-.3162278	-.2672612	.6324555	-.4780914
4	.0000000	.4472136	.0000000	-.5345225	.0000000	.7171372
5	.0000000	.4472136	.3162278	-.2672612	-.6324555	-.4780914
6	.0000000	.4472136	.6324555	.5345225	.3162278	.1195229

THE ARGUMENTS TO DETERMINE THE B(I) ESTIMATES FROM THE Y VALUES
****

	0	1	2	3	4	5
1	1.0000000	-.4472136	.6324555	-.5345225	.3162278	-.1195229
2	.0000000	.0000000	-.3162278	.8017837	-.9486833	.5976143
3	.0000000	.0000000	-.3162278	.2672612	.6324555	-1.1952286
4	.0000000	.0000000	-.3162278	-.2672612	.6324555	1.1952286
5	.0000000	.0000000	-.3162278	-.8017837	-.9486833	-.5976143
6	.0000000	.4472136	.6324555	.5345225	.3162278	.1195229

TABLES TO AID IN FITTING THE DATA POINT STATISTICS
****

PREDICTION
****

	1	2	3	4	5
2	.4472	-.6325	.5345	-.3162	.1195
3	.8944	-.9487	.2673	.3162	-.3586
4	1.3416	-.9487	-.2673	.3162	.3586
5	1.7889	-.6325	-.5345	-.3162	-.1195
6	2.2361	.0000	.0000	.0000	.0000

VARIANCE
****

1	2	3	4	5
.2000	.6000	.8857	.9857	1.0000
.8000	1.7000	1.7714	1.8714	2.0000
1.8000	2.7000	2.7714	2.8714	3.0000
3.2000	3.6000	3.8857	3.9857	4.0000
5.0000	5.0000	5.0000	5.0000	5.0000

```
THERE ARE  7 OBSERVED Y VALUES
***********************************
```

```
THE TRANSFORMATION MATRIX OF PARAMETER ESTIMATES
*****************************************************
```

0	1	2	3	4	5
1.000000+000	0.000000	0.000000	0.000000	0.000000	0.000000
	4.082483-001	-7.171372-001	9.819805-001	-1.378909+000	2.324482+000
		1.195229-001	-4.909903-001	1.347852+000	-3.543417+000
			5.455447-002	-3.726780-001	1.716589+000
				3.105650-002	-3.307189-001
					2.204793-002

```
THE ORTHONORMAL ARGUMENTS TO DETERMINE THE B(I) ESTIMATES FROM THE Z VALUES
***************************************************************************
```

	0	1	2	3	4	5
1	1.0000000	.0000000	.0000000	.0000000	.0000000	.0000000
2	.0000000	.4082483	-.5976143	.5455447	-.3726780	.1889822
3	.0000000	.4082483	-.3585686	-.1091089	.5217492	-.5669467
4	.0000000	.4082483	-.1195229	-.4364358	.2981424	.3779645
5	.0000000	.4082483	.1195229	-.4364358	-.2981424	.3779645
6	.0000000	.4082483	.3585686	-.1091089	-.5217492	-.5669467
7	.0000000	.4082483	.5976143	.5455447	.3726780	.1889822

```
THE ARGUMENTS TO DETERMINE THE B(I) ESTIMATES FROM THE Y VALUES
***************************************************************************
```

	0	1	2	3	4	5
1	1.0000000	-.4082483	.5976143	-.5455447	.3726780	-.1889822
2	.0000000	.0000000	-.2390457	.6546537	-.8944272	.7559289
3	.0000000	.0000000	-.2390457	.3273268	.2236068	-.9449112
4	.0000000	.0000000	-.2390457	.0000000	.5962848	.0000000
5	.0000000	.0000000	-.2390457	-.3273268	.2236068	.9449112
6	.0000000	.0000000	-.2390457	-.6546537	-.8944272	-.7559289
7	.0000000	.4082483	.5976143	.5455447	.3726780	.1889822

TABLES TO AID IN FITTING THE DATA POINT STATISTICS
****

PREDICTION

****

	1	2	3	4	5
2	.4082	-.5976	.5455	-.3727	.1890
3	.8165	-.9562	.4364	.1491	-.3780
4	1.2247	-1.0757	-.0000	.4472	.0000
5	1.6330	-.9562	-.4364	.1491	.3780
6	2.0412	-.5976	-.5455	-.3727	-.1890
7	2.4495	.0000	-.0000	.0000	.0000

VARIANCE

****

1	2	3	4	5
.1667	.5238	.8214	.9603	.9960
.6667	1.5810	1.7714	1.7937	1.9365
1.5000	2.6571	2.6571	2.8571	2.8571
2.6667	3.5810	3.7714	3.7937	3.9365
4.1667	4.5238	4.8214	4.9603	4.9960
6.0000	6.0000	6.0000	6.0000	6.0000

THERE ARE 8 OBSERVED Y VALUES
****

THE TRANSFORMATION MATRIX OF PARAMETER ESTIMATES
****

0	1	2	3	4	5
1.000000+000	0.000000	0.000000	0.000000	0.000000	0.000000
	3.779645-001	-6.614378-001	8.910564-001	-1.190724+000	1.776840+000
		9.449112-002	-3.818813-001	1.003610+000	-2.373820+000
			3.636965-002	-2.381448-001	9.938483-001
				1.701035-002	-1.645222-001
					9.401268-003

THE ORTHONORMAL ARGUMENTS TO DETERMINE THE B(I) ESTIMATES FROM THE Z VALUES
****

	0	1	2	3	4	5
1	1.0000000	.0000000	.0000000	.0000000	.0000000	.0000000
2	.0000000	.3779645	-.5669467	.5455447	-.4082483	.2417469
3	.0000000	.3779645	-.3779645	.0000000	.4082483	-.5640761
4	.0000000	.3779645	-.1889822	-.3273268	.4082483	.0805823
5	.0000000	.3779645	.0000000	-.4364358	.0000000	.4834938
6	.0000000	.3779645	.1889822	-.3273268	-.4082483	.0805823
7	.0000000	.3779645	.3779645	.0000000	-.4082483	-.5640761
8	.0000000	.3779645	.5669467	.5455447	.4082483	.2417469

THE ARGUMENTS TO DETERMINE THE B(I) ESTIMATES FROM THE Y VALUES
****

	0	1	2	3	4	5
1	1.0000000	-.3779645	.5669467	-.5455447	.4082483	-.2417469
2	.0000000	.0000000	-.1889822	.5455447	-.8164966	.8058230
3	.0000000	.0000000	-.1889822	.3273268	.0000000	-.6446584
4	.0000000	.0000000	-.1889822	.1091089	.4082483	-.4029115
5	.0000000	.0000000	-.1889822	-.1091089	.4082483	.4029115
6	.0000000	.0000000	-.1889822	-.3273268	.0000000	.6446584
7	.0000000	.0000000	-.1889822	-.5455447	-.8164966	-.8058230
8	.0000000	.3779645	.5669467	.5455447	.4082483	.2417469

TABLES TO AID IN FITTING THE DATA POINT STATISTICS
****

PREDICTION
****

	1	2	3	4	5
2	.3780	-.5669	.5455	-.4082	.2417
3	.7559	-.9449	.5455	.0000	-.3223
4	1.1339	-1.1339	.2182	.4082	-.2417
5	1.5119	-1.1339	-.2182	.4082	.2417
6	1.8898	-.9449	-.5455	-.0000	.3223
7	2.2678	-.5669	-.5455	-.4082	-.2417
8	2.6458	.0000	-.0000	.0000	-.0000

VARIANCE
****

	1	2	3	4	5
2	.1429	.4643	.7619	.9286	.9870
3	.5714	1.4643	1.7619	1.7619	1.8658
4	1.2857	2.5714	2.6190	2.7857	2.8442
5	2.2857	3.5714	3.6190	3.7857	3.8442
6	3.5714	4.4643	4.7619	4.7619	4.8658
7	5.1429	5.4643	5.7619	5.9286	5.9870
8	7.0000	7.0000	7.0000	7.0000	7.0000

THERE ARE 9 OBSERVED Y VALUES
****

THE TRANSFORMATION MATRIX OF PARAMETER ESTIMATES
****

0	1	2	3	4	5
1.000000+000	0.000000	0.000000	0.000000	0.000000	0.000000
	3.535534-001	-6.172134-001	8.229512-001	-1.066793+000	1.482714+000
		7.715167-002	-3.086067-001	7.898371-001	-1.759380+000
			2.571722-002	-1.641220-001	6.480160-001
				1.025762-002	-9.401268-002
					4.700634-003

THE ORTHONORMAL ARGUMENTS TO DETERMINE THE B(I) ESTIMATES FROM THE Z VALUES
****

	0	1	2	3	4	5
1	1.0000000	.0000000	.0000000	.0000000	.0000000	.0000000
2	.0000000	.3535534	-.5400617	.5400617	-.4308202	.2820380
3	.0000000	.3535534	-.3857584	.0771517	.3077287	-.5237849
4	.0000000	.3535534	-.2314550	-.2314550	.4308202	-.1208734
5	.0000000	.3535534	-.0771517	-.3857584	.1846372	.3626203
6	.0000000	.3535534	.0771517	-.3857584	-.1846372	.3626203
7	.0000000	.3535534	.2314550	-.2314550	-.4308202	-.1208734
8	.0000000	.3535534	.3857584	.0771517	-.3077287	-.5237849
9	.0000000	.3535534	.5400617	.5400617	.4308202	.2820380

THE ARGUMENTS TO DETERMINE THE B(I) ESTIMATES FROM THE Y VALUES
****

	0	1	2	3	4	5
1	1.0000000	-.3535534	.5400617	-.5400617	.4308202	-.2820380
2	.0000000	.0000000	-.1543033	.4629100	-.7385489	.8058230
3	.0000000	.0000000	-.1543033	.3086067	-.1230915	-.4029115
4	.0000000	.0000000	-.1543033	.1543033	.2461830	-.4834938
5	.0000000	.0000000	-.1543033	.0000000	.3692745	.0000000
6	.0000000	.0000000	-.1543033	-.1543033	.2461830	.4834938
7	.0000000	.0000000	-.1543033	-.3086067	-.1230915	.4029115
8	.0000000	.0000000	-.1543033	-.4629100	-.7385489	-.8058230
9	.0000000	.3535534	.5400617	.5400617	.4308202	.2820380

TABLES TO AID IN FITTING THE DATA POINT STATISTICS
****

PREDICTION
****

	1	2	3	4	5
2	.3536	-.5401	.5401	-.4308	.2820
3	.7071	-.9258	.6172	-.1231	-.2417
4	1.0607	-1.1573	.3858	.3077	-.3626
5	1.4142	-1.2344	-.0000	.4924	.0000
6	1.7678	-1.1573	-.3858	.3077	.3626
7	2.1213	-.9258	-.6172	-.1231	.2417
8	2.4749	-.5401	-.5401	-.4308	-.2820
9	2.8284	.0000	-.0000	-.0000	.0000

VARIANCE
****

	1	2	3	4	5
2	.1250	.4167	.7083	.8939	.9735
3	.5000	1.3571	1.7381	1.7532	1.8117
4	1.1250	2.4643	2.6131	2.7078	2.8393
5	2.0000	3.5238	3.5238	3.7662	3.7662
6	3.1250	4.4643	4.6131	4.7078	4.8393
7	4.5000	5.3571	5.7381	5.7532	5.8117
8	6.1250	6.4167	6.7083	6.8939	6.9735
9	8.0000	8.0000	8.0000	8.0000	8.0000

THERE ARE 10 OBSERVED Y VALUES
****

THE TRANSFORMATION MATRIX OF PARAMETER ESTIMATES
****

0	1	2	3	4	5
1.000000+000	0.000000	0.000000	0.000000	0.000000	0.000000
	3.333333-001	-5.809475-001	7.692339-001	-9.772992-001	1.297389+000
		6.454972-002	-2.564113-001	6.449115-001	-1.382876+000
			1.899343-002	-1.191828-001	4.544400-001
				6.621268-003	-5.866746-002
					2.607443-003

THE ORTHONORMAL ARGUMENTS TO DETERMINE THE B(I) ESTIMATES FROM THE Z VALUES
****

	0	1	2	3	4	5
1	1.0000000	.0000000	.0000000	.0000000	.0000000	.0000000
2	.0000000	.3333333	-.5163978	.5318160	-.4449492	.3128931
3	.0000000	.3333333	-.3872983	.1329540	.2224746	-.4693397
4	.0000000	.3333333	-.2581989	-.1519474	.4131671	-.2458446
5	.0000000	.3333333	-.1290994	-.3228883	.2860388	.2011456
6	.0000000	.3333333	.0000000	-.3798686	.0000000	.4022911
7	.0000000	.3333333	.1290994	-.3228883	-.2860388	.2011456
8	.0000000	.3333333	.2581989	-.1519474	-.4131671	-.2458446
9	.0000000	.3333333	.3872983	.1329540	-.2224746	-.4693397
10	.0000000	.3333333	.5163978	.5318160	.4449492	.3128931

THE ARGUMENTS TO DETERMINE THE B(I) ESTIMATES FROM THE Y VALUES
****

	0	1	2	3	4	5
1	1.0000000	-.3333333	.5163978	-.5318160	.4449492	-.3128931
2	.0000000	.0000000	-.1290994	.3988620	-.6674238	.7822328
3	.0000000	.0000000	-.1290994	.2849014	-.1906925	-.2234951
4	.0000000	.0000000	-.1290994	.1709409	.1271283	-.4469902
5	.0000000	.0000000	-.1290994	.0569803	.2860388	-.2011456
6	.0000000	.0000000	-.1290994	-.0569803	.2860388	.2011456
7	.0000000	.0000000	-.1290994	-.1709409	.1271283	.4469902
8	.0000000	.0000000	-.1290994	-.2849014	-.1906925	.2234951
9	.0000000	.0000000	-.1290994	-.3988620	-.6674238	-.7822328
10	.0000000	.3333333	.5163978	.5318160	.4449492	.3128931

TABLES TO AID IN FITTING THE DATA POINT STATISTICS
****

PREDICTION	1	2	3	4	5	VARIANCE 1	2	3	4	5
2	.3333	-.5164	.5318	-.4449	.3129	.1111	.3778	.6606	.8586	.9565
3	.6667	-.9037	.6648	-.2225	-.1564	.4444	1.2611	1.7030	1.7525	1.7770
4	1.0000	-1.1619	.5128	.1907	-.4023	1.0000	2.3500	2.6130	2.6494	2.8112
5	1.3333	-1.2910	.1899	.4767	-.2011	1.7778	3.4444	3.4805	3.7078	3.7483
6	1.6667	-1.2910	-.1899	.4767	.2011	2.7778	4.4444	4.4805	4.7078	4.7483
7	2.0000	-1.1619	-.5128	.1907	.4023	4.0000	5.3500	5.6130	5.6494	5.8112
8	2.3333	-.9037	-.6648	-.2225	.1564	5.4444	6.2611	6.7030	6.7525	6.7770
9	2.6667	-.5164	-.5318	-.4449	-.3129	7.1111	7.3778	7.6606	7.8586	7.9565
10	3.0000	.0000	-.0000	.0000	-.0000	9.0000	9.0000	9.0000	9.0000	9.0000

THERE ARE 11 OBSERVED Y VALUES
****

THE TRANSFORMATION MATRIX OF PARAMETER ESTIMATES
****

0	1	2	3	4	5
1.000000+000	0.000000	0.000000	0.000000	0.000000	0.000000
	3.162278-001	-5.504819-001	7.253236-001	-9.086497-001	1.168684+000
		5.504819-002	-2.175971-001	5.406915-001	-1.129728+000
			1.450647-002	-8.996532-002	3.350227-001
				4.498266-003	-3.895612-002
					1.558245-003

THE ORTHONORMAL ARGUMENTS TO DETERMINE THE B(I) ESTIMATES FROM THE Z VALUES
****

	0	1	2	3	4	5
1	1.0000000	.0000000	.0000000	.0000000	.0000000	.0000000
2	.0000000	.3162278	-.4954337	.5222330	-.4534252	.3365809
3	.0000000	.3162278	-.3853373	.1740777	.1511417	-.4113767
4	.0000000	.3162278	-.2752409	-.0870388	.3778543	-.3178820
5	.0000000	.3162278	-.1651446	-.2611165	.3346710	.0560968
6	.0000000	.3162278	-.0550482	-.3481553	.1295501	.3365809
7	.0000000	.3162278	.0550482	-.3481553	-.1295501	.3365809
8	.0000000	.3162278	.1651446	-.2611165	-.3346710	.0560968
9	.0000000	.3162278	.2752409	-.0870388	-.3778543	-.3178820
10	.0000000	.3162278	.3853373	.1740777	-.1511417	-.4113767
11	.0000000	.3162278	.4954337	.5222330	.4534252	.3365809

THE ARGUMENTS TO DETERMINE THE B(I) ESTIMATES FROM THE Y VALUES
****

	0	1	2	3	4	5
1	1.0000000	-.3162278	.4954337	-.5222330	.4534252	-.3365809
2	.0000000	.0000000	-.1100964	.3481553	-.6045669	.7479576
3	.0000000	.0000000	-.1100964	.2611165	-.2267126	-.0934947
4	.0000000	.0000000	-.1100964	.1740777	.0431834	-.3739788
5	.0000000	.0000000	-.1100964	.0870388	.2051209	-.2804841
6	.0000000	.0000000	-.1100964	.0000000	.2591001	.0000000
7	.0000000	.0000000	-.1100964	-.0870388	.2051209	.2804841
8	.0000000	.0000000	-.1100964	-.1740777	.0431834	.3739788
9	.0000000	.0000000	-.1100964	-.2611165	-.2267126	.0934947
10	.0000000	.0000000	-.1100964	-.3481553	-.6045669	-.7479576

TABLES TO AID IN FITTING THE DATA POINT STATISTICS
****

PREDICTION
****

	1	2	3	4	5
2	.3162	-.4954	.5222	-.4534	.3366
3	.6325	-.8808	.6963	-.3023	-.0748
4	.9487	-1.1560	.6093	.0756	-.3927
5	1.2649	-1.3212	.3482	.4102	-.3366
6	1.5811	-1.3762	-.0000	.5398	.0000
7	1.8974	-1.3212	-.3482	.4102	.3366
8	2.2136	-1.1560	-.6093	.0756	.3927
9	2.5298	-.8808	-.6963	-.3023	.0748
10	2.8460	-.4954	-.5222	-.4534	-.3366
11	3.1623	.0000	.0000	.0000	-.0000

VARIANCE
****

	1	2	3	4	5
2	.1000	.3455	.6182	.8238	.9371
3	.4000	1.1758	1.6606	1.7520	1.7576
4	.9000	2.2364	2.6076	2.6133	2.7675
5	1.6000	3.3455	3.4667	3.6350	3.7483
6	2.5000	4.3939	4.3939	4.6853	4.6853
7	3.6000	5.3455	5.4667	5.6350	5.7483
8	4.9000	6.2364	6.6076	6.6133	6.7675
9	6.4000	7.1758	7.6606	7.7520	7.7576
10	8.1000	8.3455	8.6182	8.8238	8.9371
11	10.0000	10.0000	10.0000	10.0000	10.0000

THERE ARE 12 OBSERVED Y VALUES
****

THE TRANSFORMATION MATRIX OF PARAMETER ESTIMATES
****

0	1	2	3	4	5
1.000000+000	0.000000	0.000000	0.000000	0.000000	0.000000
	3.015113-001	-5.244044-001	6.884786-001	-8.537144-001	1.073232+000
		4.767313-002	-1.877669-001	4.624817-001	-9.485636-001
			1.137981-002	-6.997659-002	2.562354-001
				3.180754-003	-2.710182-002
					9.855207-004

THE ORTHONORMAL ARGUMENTS TO DETERMINE THE B(I) ESTIMATES FROM THE Z VALUES
****

	0	1	2	3	4	5
1	1.0000000	.0000000	.0000000	.0000000	.0000000	.0000000
2	.0000000	.3015113	-.4767313	.5120916	-.4580286	.3547874
3	.0000000	.3015113	-.3813850	.2048366	.0916057	-.3547874
4	.0000000	.3015113	-.2860388	-.0341394	.3358876	-.3547874
5	.0000000	.3015113	-.1906925	-.2048366	.3511553	-.0591312
6	.0000000	.3015113	-.0953463	-.3072549	.2137467	.2365250
7	.0000000	.3015113	.0000000	-.3413944	.0000000	.3547874
8	.0000000	.3015113	.0953463	-.3072549	-.2137467	.2365250
9	.0000000	.3015113	.1906925	-.2048366	-.3511553	-.0591312
10	.0000000	.3015113	.2860388	-.0341394	-.3358876	-.3547874
11	.0000000	.3015113	.3813850	.2048366	-.0916057	-.3547874
12	.0000000	.3015113	.4767313	.5120916	.4580286	.3547874

THE ARGUMENTS TO DETERMINE THE B(I) ESTIMATES FROM THE Y VALUES
****

	0	1	2	3	4	5
1	1.0000000	-.3015113	.4767313	-.5120916	.4580286	-.3547874
2	.0000000	.0000000	-.0953463	.3072549	-.5496343	.7095749
3	.0000000	.0000000	-.0953463	.2389761	-.2442819	.0000000
4	.0000000	.0000000	-.0953463	.1706972	-.0152676	-.2956562
5	.0000000	.0000000	-.0953463	.1024183	.1374086	-.2956562
6	.0000000	.0000000	-.0953463	.0341394	.2137467	-.1182625
7	.0000000	.0000000	-.0953463	-.0341394	.2137467	.1182625

9	.0000000	.0000000	-.0953463	-.1706972	-.0152676	.2956562
10	.0000000	.0000000	-.0953463	-.2389761	-.2442819	.0000000
11	.0000000	.0000000	-.0953463	-.3072549	-.5496343	-.7095749
12	.0000000	.3015113	.4767313	.5120916	.4580286	.3547874

TABLES TO AID IN FITTING THE DATA POINT STATISTICS
****

PREDICTION
****

	1	2	3	4	5
2	.3015	-.4767	.5121	-.4580	.3548
3	.6030	-.8581	.7169	-.3664	-.0000
4	.9045	-1.1442	.6828	-.0305	-.3548
5	1.2060	-1.3348	.4780	.3206	-.4139
6	1.5076	-1.4302	.1707	.5344	-.1774
7	1.8091	-1.4302	-.1707	.5344	.1774
8	2.1106	-1.3348	-.4780	.3206	.4139
9	2.4121	-1.1442	-.6828	-.0305	.3548
10	2.7136	-.8581	-.7169	-.3664	-.0000
11	3.0151	-.4767	-.5121	-.4580	-.3548
12	3.3166	.0000	.0000	-.0000	-.0000

VARIANCE
****

1	2	3	4	5
.0909	.3182	.5804	.7902	.9161
.3636	1.1000	1.6140	1.7483	1.7483
.8182	2.1273	2.5935	2.5944	2.7203
1.4545	3.2364	3.4648	3.5676	3.7389
2.2727	4.3182	4.3473	4.6329	4.6643
3.2727	5.3182	5.3473	5.6329	5.6643
4.4545	6.2364	6.4648	6.5676	6.7389
5.8182	7.1273	7.5935	7.5944	7.7203
7.3636	8.1000	8.6140	8.7483	8.7483
9.0909	9.3182	9.5804	9.7902	9.9161
11.0000	11.0000	11.0000	11.0000	11.0000

THERE ARE 13 OBSERVED Y VALUES
****

THE TRANSFORMATION MATRIX OF PARAMETER ESTIMATES
****

0	1	2	3	4	5
1.000000+000	0.000000	0.000000	0.000000	0.000000	0.000000
	2.886751-001	-5.017452-001	6.569387-001	-8.083673-001	9.990230-001
		4.181210-002	-1.642347-001	4.018607-001	-8.129633-001
			9.124148-003	-5.574947-002	2.016112-001
				2.322894-003	-1.955582-002
					6.518606-004

THE ORTHONORMAL ARGUMENTS TO DETERMINE THE B(I) ESTIMATES FROM THE Z VALUES
****

	0	1	2	3	4	5
1	1.0000000	.0000000	.0000000	.0000000	.0000000	.0000000
2	.0000000	.2886751	-.4599331	.5018282	-.4599331	.3687669
3	.0000000	.2886751	-.3763089	.2281037	.0418121	-.3017184
4	.0000000	.2886751	-.2926847	.0091241	.2926847	-.3687669
5	.0000000	.2886751	-.2090605	-.1551105	.3484342	-.1452718
6	.0000000	.2886751	-.1254363	-.2646003	.2648100	.1340970
7	.0000000	.2886751	-.0418121	-.3193452	.0975616	.3128931
8	.0000000	.2886751	.0418121	-.3193452	-.0975616	.3128931
9	.0000000	.2886751	.1254363	-.2646003	-.2648100	.1340970
10	.0000000	.2886751	.2090605	-.1551105	-.3484342	-.1452718
11	.0000000	.2886751	.2926847	.0091241	-.2926847	-.3687669
12	.0000000	.2886751	.3763089	.2281037	-.0418121	-.3017184
13	.0000000	.2886751	.4599331	.5018282	.4599331	.3687669

THE ARGUMENTS TO DETERMINE THE B(I) ESTIMATES FROM THE Y VALUES
****

	0	1	2	3	4	5
1	1.0000000	-.2886751	.4599331	-.5018282	.4599331	-.3687669
2	.0000000	.0000000	-.0836242	.2737245	-.5017452	.6704852
3	.0000000	.0000000	-.0836242	.2189796	-.2508726	.0670485
4	.0000000	.0000000	-.0836242	.1642347	-.0557495	-.2234951
5	.0000000	.0000000	-.0836242	.1094898	.0836242	-.2793688
6	.0000000	.0000000	-.0836242	.0547449	.1672484	-.1787961

8	.0000000	.0000000	-.0836242	-.0547449	.1672484	.1787961
9	.0000000	.0000000	-.0836242	-.1094898	.0836242	.2793688
10	.0000000	.0000000	-.0836242	-.1642347	-.0557495	.2234951
11	.0000000	.0000000	-.0836242	-.2189796	-.2508726	-.0670485
12	.0000000	.0000000	-.0836242	-.2737245	-.5017452	-.6704852
13	.0000000	.2886751	.4599331	.5018282	.4599331	.3687669

TABLES TO AID IN FITTING THE DATA POINT STATISTICS
****

PREDICTION
****

	1	2	3	4	5
2	.2887	-.4599	.5018	-.4599	.3688
3	.5774	-.8362	.7299	-.4181	.0670
4	.8660	-1.1289	.7391	-.1254	-.3017
5	1.1547	-1.3380	.5839	.2230	-.4470
6	1.4434	-1.4634	.3193	.4878	-.3129
7	1.7321	-1.5052	-.0000	.5854	.0000
8	2.0207	-1.4634	-.3193	.4878	.3129
9	2.3094	-1.3380	-.5839	.2230	.4470
10	2.5981	-1.1289	-.7391	-.1254	.3017
11	2.8868	-.8362	-.7299	-.4181	-.0670
12	3.1754	-.4599	-.5018	-.4599	-.3688
13	3.4641	.0000	.0000	.0000	-.0000

VARIANCE
****

1	2	3	4	5
.0833	.2949	.5467	.7582	.8942
.3333	1.0326	1.5654	1.7403	1.7448
.7500	2.0245	2.5707	2.5864	2.6774
1.3333	3.1235	3.4645	3.5143	3.7141
2.0833	4.2249	4.3269	4.5649	4.6628
3.0000	5.2657	5.2657	5.6084	5.6084
4.0833	6.2249	6.3269	6.5649	6.6628
5.3333	7.1235	7.4645	7.5143	7.7141
6.7500	8.0245	8.5707	8.5864	8.6774
8.3333	9.0326	9.5654	9.7403	9.7448
10.0833	10.2949	10.5467	10.7582	10.8942
12.0000	12.0000	12.0000	12.0000	12.0000

THERE ARE 14 OBSERVED Y VALUES
****

THE TRANSFORMATION MATRIX OF PARAMETER ESTIMATES
****

0	1	2	3	4	5
1.000000+000	0.000000	0.000000	0.000000	0.000000	0.000000
	2.773501-001	-4.818121-001	6.295111-001	-7.700395-001	9.392543-001
		3.706247-002	-1.452718-001	3.536607-001	-7.079702-001
			7.449836-003	-4.529644-002	1.622598-001
				1.742171-003	-1.453311-002
					4.471727-004

THE ORTHONORMAL ARGUMENTS TO DETERMINE THE B(I) ESTIMATES FROM THE Z VALUES
****

	0	1	2	3	4	5
1	1.0000000	.0000000	.0000000	.0000000	.0000000	.0000000
2	.0000000	.2773501	-.4447496	.4916892	-.4599331	.3794580
3	.0000000	.2773501	-.3706247	.2458446	.0000000	-.2529720
4	.0000000	.2773501	-.2964997	.0446990	.2508726	-.3679593
5	.0000000	.2773501	-.2223748	-.1117475	.3344968	-.2069771
6	.0000000	.2773501	-.1482499	-.2234951	.2926847	.0421620
7	.0000000	.2773501	-.0741249	-.2905436	.1672484	.2453062
8	.0000000	.2773501	.0000000	-.3128931	.0000000	.3219644
9	.0000000	.2773501	.0741249	-.2905436	-.1672484	.2453062
10	.0000000	.2773501	.1482499	-.2234951	-.2926847	.0421620
11	.0000000	.2773501	.2223748	-.1117475	-.3344968	-.2069771
12	.0000000	.2773501	.2964997	.0446990	-.2508726	-.3679593
13	.0000000	.2773501	.3706247	.2458446	.0000000	-.2529720
14	.0000000	.2773501	.4447496	.4916892	.4599331	.3794580

THE ARGUMENTS TO DETERMINE THE B(I) ESTIMATES FROM THE Y VALUES
****

	0	1	2	3	4	5
1	1.0000000	-.2773501	.4447496	-.4916892	.4599331	-.3794580
2	.0000000	.0000000	-.0741249	.2458446	-.4599331	.6324300
3	.0000000	.0000000	-.0741249	.2011456	-.2508726	.1149873
4	.0000000	.0000000	-.0741249	.1564466	-.0836242	-.1609822
5	.0000000	.0000000	-.0741249	.1117475	.0418121	-.2491391
6	.0000000	.0000000	-.0741249	.0670485	.1254363	-.2031442

7	.0000000	.0000000	-.0741249	.0223495	.1672484	-.0766582
8	.0000000	.0000000	-.0741249	-.0223495	.1672484	.0766582
9	.0000000	.0000000	-.0741249	-.0670485	.1254363	.2031442
10	.0000000	.0000000	-.0741249	-.1117475	.0418121	.2491391
11	.0000000	.0000000	-.0741249	-.1564466	-.0836242	.1609822
12	.0000000	.0000000	-.0741249	-.2011456	-.2508726	-.1149873
13	.0000000	.0000000	-.0741249	-.2458446	-.4599331	-.6324300
14	.0000000	.2773501	.4447496	.4916892	.4599331	.3794580

TABLES TO AID IN FITTING THE DATA POINT STATISTICS
****

PREDICTION
****

	1	2	3	4	5
2	.2774	-.4447	.4917	-.4599	.3795
3	.5547	-.8154	.7375	-.4599	.1265
4	.8321	-1.1119	.7822	-.2091	-.2415
5	1.1094	-1.3342	.6705	.1254	-.4485
6	1.3868	-1.4825	.4470	.4181	-.4063
7	1.6641	-1.5566	.1564	.5854	-.1610
8	1.9415	-1.5566	-.1564	.5854	.1610
9	2.2188	-1.4825	-.4470	.4181	.4063
10	2.4962	-1.3342	-.6705	.1254	.4485
11	2.7735	-1.1119	-.7822	-.2091	.2415
12	3.0509	-.8154	-.7375	-.4599	-.1265
13	3.3282	-.4447	-.4917	-.4599	-.3795
14	3.6056	.0000	.0000	-.0000	.0000

VARIANCE
****

1	2	3	4	5
.0769	.2747	.5165	.7280	.8720
.3077	.9725	1.5165	1.7280	1.7440
.6923	1.9286	2.5405	2.5842	2.6425
1.2308	3.0110	3.4605	3.4763	3.6774
1.9231	4.1209	4.3207	4.4955	4.6606
2.7692	5.1923	5.2168	5.5594	5.5854
3.7692	6.1923	6.2168	6.5594	6.5854
4.9231	7.1209	7.3207	7.4955	7.6606
6.2308	8.0110	8.4605	8.4763	8.6774
7.6923	8.9286	9.5405	9.5842	9.6425
9.3077	9.9725	10.5165	10.7280	10.7440
11.0769	11.2747	11.5165	11.7280	11.8720
13.0000	13.0000	13.0000	13.0000	13.0000

```
THERE ARE 15 OBSERVED Y VALUES
**********************************

THE TRANSFORMATION MATRIX OF PARAMETER ESTIMATES
**************************************************
```

0	1	2	3	4	5
1.000000+000	0.000000	0.000000	0.000000	0.000000	0.000000
	2.672612-001	-4.640955-001	6.053536-001	-7.370391-001	8.897836-001
		3.314968-002	-1.297186-001	3.145377-001	-6.244927-001
			6.177078-003	-3.741315-002	1.330294-001
				1.336184-003	-1.106696-002
					3.161989-004

```
THE ORTHONORMAL ARGUMENTS TO DETERMINE THE B(I) ESTIMATES FROM THE Z VALUES
****************************************************************************
```

	0	1	2	3	4	5
1	1.0000000	.0000000	.0000000	.0000000	.0000000	.0000000
2	.0000000	.2672612	-.4309458	.4818121	-.4585784	.3875695
3	.0000000	.2672612	-.3646464	.2594373	-.0352753	-.2086912
4	.0000000	.2672612	-.2983471	.0741249	.2116516	-.3577564
5	.0000000	.2672612	-.2320477	-.0741249	.3142705	-.2493454
6	.0000000	.2672612	-.1657484	-.1853123	.3046500	-.0352336
7	.0000000	.2672612	-.0994490	-.2594373	.2148584	.1707474
8	.0000000	.2672612	-.0331497	-.2964997	.0769642	.2927098
9	.0000000	.2672612	.0331497	-.2964997	-.0769642	.2927098
10	.0000000	.2672612	.0994490	-.2594373	-.2148584	.1707474
11	.0000000	.2672612	.1657484	-.1853123	-.3046500	-.0352336
12	.0000000	.2672612	.2320477	-.0741249	-.3142705	-.2493454
13	.0000000	.2672612	.2983471	.0741249	-.2116516	-.3577564
14	.0000000	.2672612	.3646464	.2594373	.0352753	-.2086912
15	.0000000	.2672612	.4309458	.4818121	.4585784	.3875695

```
THE ARGUMENTS TO DETERMINE THE B(I) ESTIMATES FROM THE Y VALUES
****************************************************************
```

	0	1	2	3	4	5
1	1.0000000	-.2672612	.4309458	-.4818121	.4585784	-.3875695
2	.0000000	.0000000	-.0662994	.2223748	-.4233031	.5962607
3	.0000000	.0000000	-.0662994	.1853123	-.2469268	.1490652
4	.0000000	.0000000	-.0662994	.1482499	-.1026189	-.1084110
5	.0000000	.0000000	-.0662994	.1111874	.0096205	-.2141118

6	.0000000	.0000000	-.0662994	.0741249	.0897916	-.2059810
7	.0000000	.0000000	-.0662994	.0370625	.1378942	-.1219624
8	.0000000	.0000000	-.0662994	.0000000	.1539284	.0000000
9	.0000000	.0000000	-.0662994	-.0370625	.1378942	.1219624
10	.0000000	.0000000	-.0662994	-.0741249	.0897916	.2059810
11	.0000000	.0000000	-.0662994	-.1111874	.0096205	.2141118
12	.0000000	.0000000	-.0662994	-.1482499	-.1026189	.1084110
13	.0000000	.0000000	-.0662994	-.1853123	-.2469268	-.1490652
14	.0000000	.0000000	-.0662994	-.2223748	-.4233031	-.5962607
15	.0000000	.2672612	.4309458	.4818121	.4585784	.3875695

TABLES TO AID IN FITTING THE DATA POINT STATISTICS
****

PREDICTION
****

	1	2	3	4	5
2	.2673	-.4309	.4818	-.4586	.3876
3	.5345	-.7956	.7412	-.4939	.1789
4	.8018	-1.0939	.8154	-.2822	-.1789
5	1.0690	-1.3260	.7412	.0321	-.4282
6	1.3363	-1.4917	.5559	.3367	-.4635
7	1.6036	-1.5912	.2965	.5516	-.2927
8	1.8708	-1.6243	-.0000	.6285	.0000
9	2.1381	-1.5912	-.2965	.5516	.2927
10	2.4054	-1.4917	-.5559	.3367	.4635
11	2.6726	-1.3260	-.7412	.0321	.4282
12	2.9399	-1.0939	-.8154	-.2822	.1789
13	3.2071	-.7956	-.7412	-.4939	-.1789
14	3.4744	-.4309	-.4818	-.4586	-.3876
15	3.7417	.0000	.0000	-.0000	-.0000

VARIANCE
****

1	2	3	4	5
.0714	.2571	.4893	.6996	.8498
.2857	.9187	1.4681	1.7120	1.7440
.6429	1.8396	2.5044	2.5840	2.6160
1.1429	2.9011	3.4505	3.4516	3.6350
1.7857	4.0110	4.3201	4.4334	4.6482
2.5714	5.1033	5.1912	5.4954	5.5811
3.5000	6.1385	6.1385	6.5335	6.5335
4.5714	7.1033	7.1912	7.4954	7.5811
5.7857	8.0110	8.3201	8.4334	8.6482
7.1429	8.9011	9.4505	9.4516	9.6350
8.6429	9.8396	10.5044	10.5840	10.6160
10.2857	10.9187	11.4681	11.7120	11.7440
12.0714	12.2571	12.4893	12.6996	12.8498
14.0000	14.0000	14.0000	14.0000	14.0000

THERE ARE 16 OBSERVED Y VALUES
****

THE TRANSFORMATION MATRIX OF PARAMETER ESTIMATES
****

0	1	2	3	4	5
1.000000+000	0.000000	0.000000	0.000000	0.000000	0.000000
	2.581989-001	-4.482107-001	5.838506-001	-7.082002-001	8.479414-001
		2.988072-002	-1.167701-001	2.822356-001	-5.566919-001
			5.189783-003	-3.133629-002	1.107649-001
				1.044543-003	-8.602304-003
					2.293948-004

THE ORTHONORMAL ARGUMENTS TO DETERMINE THE B(I) ESTIMATES FROM THE Z VALUES
****

	0	1	2	3	4	5
1	1.0000000	.0000000	.0000000	.0000000	.0000000	.0000000
2	.0000000	.2581989	-.4183300	.4722703	-.4562564	.3936414
3	.0000000	.2581989	-.3585686	.2698687	-.0651795	-.1687035
4	.0000000	.2581989	-.2988072	.0986059	.1754832	-.3417327
5	.0000000	.2581989	-.2390457	-.0415183	.2908008	-.2768467
6	.0000000	.2581989	-.1792843	-.1505037	.3058422	-.0979188
7	.0000000	.2581989	-.1195229	-.2283505	.2456765	.0987053
8	.0000000	.2581989	-.0597614	-.2750585	.1353728	.2442071
9	.0000000	.2581989	.0000000	-.2906279	.0000000	.2972956
10	.0000000	.2581989	.0597614	-.2750585	-.1353728	.2442071
11	.0000000	.2581989	.1195229	-.2283505	-.2456765	.0987053
12	.0000000	.2581989	.1792843	-.1505037	-.3058422	-.0979188
13	.0000000	.2581989	.2390457	-.0415183	-.2908008	-.2768467
14	.0000000	.2581989	.2988072	.0986059	-.1754832	-.3417327
15	.0000000	.2581989	.3585686	.2698687	.0651795	-.1687035
16	.0000000	.2581989	.4183300	.4722703	.4562564	.3936414

THE ARGUMENTS TO DETERMINE THE B(I) ESTIMATES FROM THE Y VALUES
****

	0	1	2	3	4	5
1	1.0000000	-.2581989	.4183300	-.4722703	.4562564	-.3936414
2	.0000000	.0000000	-.0597614	.2024015	-.3910769	.5623449
3	.0000000	.0000000	-.0597614	.1712628	-.2406627	.1730292
4	.0000000	.0000000	-.0597614	.1401241	-.1153176	-.0648859

5	.0000000	.0000000	-.0597614	.1089854	-.0150414	-.1789279
6	.0000000	.0000000	-.0597614	.0778467	.0601657	-.1966241
7	.0000000	.0000000	-.0597614	.0467080	.1103038	-.1455018
8	.0000000	.0000000	-.0597614	.0155693	.1353728	-.0530885
9	.0000000	.0000000	-.0597614	-.0155693	.1353728	.0530885
10	.0000000	.0000000	-.0597614	-.0467080	.1103038	.1455018
11	.0000000	.0000000	-.0597614	-.0778467	.0601657	.1966241
12	.0000000	.0000000	-.0597614	-.1089854	-.0150414	.1789279
13	.0000000	.0000000	-.0597614	-.1401241	-.1153176	.0648859
14	.0000000	.0000000	-.0597614	-.1712628	-.2406627	-.1730292
15	.0000000	.0000000	-.0597614	-.2024015	-.3910769	-.5623449
16	.0000000	.2581989	.4183300	.4722703	.4562564	.3936414

TABLES TO AID IN FITTING THE DATA POINT STATISTICS
****

PREDICTION
****

	1	2	3	4	5
2	.2582	-.4183	.4723	-.4563	.3936
3	.5164	-.7769	.7421	-.5214	.2249
4	.7746	-1.0757	.8407	-.3460	-.1168
5	1.0328	-1.3148	.7992	-.0552	-.3936
6	1.2910	-1.4940	.6487	.2507	-.4916
7	1.5492	-1.6136	.4204	.4964	-.3929
8	1.8074	-1.6733	.1453	.6317	-.1486
9	2.0656	-1.6733	-.1453	.6317	.1486
10	2.3238	-1.6136	-.4204	.4964	.3929
11	2.5820	-1.4940	-.6487	.2507	.4916
12	2.8402	-1.3148	-.7992	-.0552	.3936
13	3.0984	-1.0757	-.8407	-.3460	.1168
14	3.3566	-.7769	-.7421	-.5214	-.2249
15	3.6148	-.4183	-.4723	-.4563	-.3936
16	3.8730	.0000	.0000	.0000	.0000

VARIANCE
****

1	2	3	4	5
.0667	.2417	.4647	.6729	.8278
.2667	.8702	1.4210	1.6929	1.7435
.6000	1.7571	2.4640	2.5837	2.5973
1.0667	2.7952	3.4340	3.4370	3.5920
1.6667	3.8988	4.3197	4.3825	4.6241
2.4000	5.0036	5.1803	5.4267	5.5810
3.2667	6.0667	6.0878	6.4869	6.5090
4.2667	7.0667	7.0878	7.4869	7.5090
5.4000	8.0036	8.1803	8.4267	8.5810
6.6667	8.8988	9.3197	9.3825	9.6241
8.0667	9.7952	10.4340	10.4370	10.5920
9.6000	10.7571	11.4640	11.5837	11.5973
11.2667	11.8702	12.4210	12.6929	12.7435
13.0667	13.2417	13.4647	13.6729	13.8278
15.0000	15.0000	15.0000	15.0000	15.0000

THERE ARE 17 OBSERVED Y VALUES
****

THE TRANSFORMATION MATRIX OF PARAMETER ESTIMATES
****

0	1	2	3	4	5
1.000000+000	0.000000	0.000000	0.000000	0.000000	0.000000
	2.500000-001	-4.338609-001	5.645399-001	-6.826896-001	8.119275-001
		2.711631-002	-1.058512-001	2.551785-001	-5.006498-001
			4.410468-003	-2.656380-002	9.344652-002
				8.301187-004	-6.804948-003
					1.701237-004

THE ORTHONORMAL ARGUMENTS TO DETERMINE THE B(I) ESTIMATES FROM THE Z VALUES
****

	0	1	2	3	4	5
1	1.0000000	.0000000	.0000000	.0000000	.0000000	.0000000
2	.0000000	.2500000	-.4067446	.4630991	-.4532448	.3980895
3	.0000000	.2500000	-.3525120	.2778595	-.0906490	-.1326965
4	.0000000	.2500000	-.2982794	.1190826	.1424484	-.3222629
5	.0000000	.2500000	-.2440468	-.0132314	.2659700	-.2930988
6	.0000000	.2500000	-.1898142	-.1190826	.2998389	-.1472785
7	.0000000	.2500000	-.1355815	-.1984710	.2639777	.0335387
8	.0000000	.2500000	-.0813489	-.2513967	.1783095	.1881082
9	.0000000	.2500000	-.0271163	-.2778595	.0627570	.2756004
10	.0000000	.2500000	.0271163	-.2778595	-.0627570	.2756004
11	.0000000	.2500000	.0813489	-.2513967	-.1783095	.1881082
12	.0000000	.2500000	.1355815	-.1984710	-.2639777	.0335387
13	.0000000	.2500000	.1898142	-.1190826	-.2998389	-.1472785
14	.0000000	.2500000	.2440468	-.0132314	-.2659700	-.2930988
15	.0000000	.2500000	.2982794	.1190826	-.1424484	-.3222629
16	.0000000	.2500000	.3525120	.2778595	.0906490	-.1326965
17	.0000000	.2500000	.4067446	.4630991	.4532448	.3980895

THE ARGUMENTS TO DETERMINE THE B(I) ESTIMATES FROM THE Y VALUES
****

	0	1	2	3	4	5
1	1.0000000	-.2500000	.4067446	-.4630991	.4532448	-.3980895
2	.0000000	.0000000	-.0542326	.1852396	-.3625958	.5307860
3	.0000000	.0000000	-.0542326	.1587768	-.2330973	.1895664

4	.0000000	.0000000	-.0542326	.1323140	-.1235217	-.0291641
5	.0000000	.0000000	-.0542326	.1058512	-.0338688	-.1458203
6	.0000000	.0000000	-.0542326	.0793884	.0358611	-.1808172
7	.0000000	.0000000	-.0542326	.0529256	.0856682	-.1545695
8	.0000000	.0000000	-.0542326	.0264628	.1155525	-.0874922
9	.0000000	.0000000	-.0542326	.0000000	.1255139	.0000000
10	.0000000	.0000000	-.0542326	-.0264628	.1155525	.0874922
11	.0000000	.0000000	-.0542326	-.0529256	.0856682	.1545695
12	.0000000	.0000000	-.0542326	-.0793884	.0358611	.1808172
13	.0000000	.0000000	-.0542326	-.1058512	-.0338688	.1458203
14	.0000000	.0000000	-.0542326	-.1323140	-.1235217	.0291641
15	.0000000	.0000000	-.0542326	-.1587768	-.2330973	-.1895664
16	.0000000	.0000000	-.0542326	-.1852396	-.3625958	-.5307860
17	.0000000	.2500000	.4067446	.4630991	.4532448	.3980895

TABLES TO AID IN FITTING THE DATA POINT STATISTICS
****

	PREDICTION					VARIANCE				
	1	2	3	4	5	1	2	3	4	5
2	.2500	-.4067	.4631	-.4532	.3981	.0625	.2279	.4424	.6478	.8063
3	.5000	-.7593	.7410	-.5439	.2654	.2500	.8265	1.3755	1.6713	1.7417
4	.7500	-1.0575	.8600	-.4014	-.0569	.5625	1.6809	2.4206	2.5817	2.5849
5	1.0000	-1.3016	.8468	-.1355	-.3500	1.0000	2.6941	3.4112	3.4296	3.5520
6	1.2500	-1.4914	.7277	.1644	-.4972	1.5625	3.7868	4.3164	4.3434	4.5906
7	1.5000	-1.6270	.5293	.4283	-.4637	2.2500	4.8971	5.1772	5.3606	5.5757
8	1.7500	-1.7083	.2779	.6067	-.2756	3.0625	5.9809	6.0581	6.4261	6.5021
9	2.0000	-1.7354	-.0000	.6694	.0000	4.0000	7.0118	7.0118	7.4599	7.4599
10	2.2500	-1.7083	-.2779	.6067	.2756	5.0625	7.9809	8.0581	8.4261	8.5021
11	2.5000	-1.6270	-.5293	.4283	.4637	6.2500	8.8971	9.1772	9.3606	9.5757
12	2.7500	-1.4914	-.7277	.1644	.4972	7.5625	9.7868	10.3164	10.3434	10.5906
13	3.0000	-1.3016	-.8468	-.1355	.3500	9.0000	10.6941	11.4112	11.4296	11.5520
14	3.2500	-1.0575	-.8600	-.4014	.0569	10.5625	11.6809	12.4206	12.5817	12.5849
15	3.5000	-.7593	-.7410	-.5439	-.2654	12.2500	12.8265	13.3755	13.6713	13.7417
16	3.7500	-.4067	-.4631	-.4532	-.3981	14.0625	14.2279	14.4424	14.6478	14.8063
17	4.0000	.0000	.0000	.0000	.0000	16.0000	16.0000	16.0000	16.0000	16.0000

THERE ARE 18 OBSERVED Y VALUES
****

THE TRANSFORMATION MATRIX OF PARAMETER ESTIMATES
****

0	1	2	3	4	5
1.000000+000	0.000000	0.000000	0.000000	0.000000	0.000000
	2.425356-001	-4.208127-001	5.470660-001	-6.598934-001	7.804821-001
		2.475369-002	-9.654106-002	2.322343-001	-4.536416-001
			3.785924-003	-2.275495-002	7.973289-002
				6.692631-004	-5.465561-003
					1.286014-004

THE ORTHONORMAL ARGUMENTS TO DETERMINE THE B(I) ESTIMATES FROM THE Z VALUES
****

	0	1	2	3	4	5
1	1.0000000	.0000000	.0000000	.0000000	.0000000	.0000000
2	.0000000	.2425356	-.3960590	.4543109	-.4497448	.4012365
3	.0000000	.2425356	-.3465516	.2839443	-.1124362	-.1003091
4	.0000000	.2425356	-.2970443	.1362933	.1124362	-.3009274
5	.0000000	.2425356	-.2475369	.0113578	.2409347	-.3009274
6	.0000000	.2425356	-.1980295	-.0908622	.2891217	-.1851861
7	.0000000	.2425356	-.1485221	-.1703666	.2730593	-.0231483
8	.0000000	.2425356	-.0990148	-.2271554	.2088101	.1311735
9	.0000000	.2425356	-.0495074	-.2612287	.1124362	.2391987
10	.0000000	.2425356	.0000000	-.2725865	.0000000	.2777791
11	.0000000	.2425356	.0495074	-.2612287	-.1124362	.2391987
12	.0000000	.2425356	.0990148	-.2271554	-.2088101	.1311735
13	.0000000	.2425356	.1485221	-.1703666	-.2730593	-.0231483
14	.0000000	.2425356	.1980295	-.0908622	-.2891217	-.1851861
15	.0000000	.2425356	.2475369	.0113578	-.2409347	-.3009274
16	.0000000	.2425356	.2970443	.1362933	-.1124362	-.3009274
17	.0000000	.2425356	.3465516	.2839443	.1124362	-.1003091
18	.0000000	.2425356	.3960590	.4543109	.4497448	.4012365

THE ARGUMENTS TO DETERMINE THE B(I) ESTIMATES FROM THE Y VALUES
****

	0	1	2	3	4	5
1	1.0000000	-.2425356	.3960590	.4543109	.4497448	-.4012365
2	.0000000	.0000000	-.0495074	.1703666	-.3373086	.5015456

3	.0000000	.0000000	-.0495074	.1476510	-.2248724	.2006182
4	.0000000	.0000000	-.0495074	.1249355	-.1284985	.0000000
5	.0000000	.0000000	-.0495074	.1022199	-.0481869	-.1157413
6	.0000000	.0000000	-.0495074	.0795044	.0160623	-.1620378
7	.0000000	.0000000	-.0495074	.0567889	.0642493	-.1543217
8	.0000000	.0000000	-.0495074	.0340733	.0963739	-.1080252
9	.0000000	.0000000	-.0495074	.0113578	.1124362	-.0385804
10	.0000000	.0000000	-.0495074	-.0113578	.1124362	.0385804
11	.0000000	.0000000	-.0495074	-.0340733	.0963739	.1080252
12	.0000000	.0000000	-.0495074	-.0567889	.0642493	.1543217
13	.0000000	.0000000	-.0495074	-.0795044	.0160623	.1620378
14	.0000000	.0000000	-.0495074	-.1022199	-.0481869	.1157413
15	.0000000	.0000000	-.0495074	-.1249355	-.1284985	.0000000
16	.0000000	.0000000	-.0495074	-.1476510	-.2248724	-.2006182
17	.0000000	.0000000	-.0495074	-.1703666	-.3373086	-.5015456
18	.0000000	.2425356	.3960590	.4543109	.4497448	.4012365

TABLES TO AID IN FITTING THE DATA POINT STATISTICS
****

PREDICTION
****

	1	2	3	4	5
2	.2425	-.3961	.4543	-.4497	.4012
3	.4851	-.7426	.7383	-.5622	.3009
4	.7276	-1.0397	.8745	-.4497	-.0000
5	.9701	-1.2872	.8859	-.2088	-.3009
6	1.2127	-1.4852	.7950	.0803	-.4861
7	1.4552	-1.6337	.6247	.3534	-.5093
8	1.6977	-1.7328	.3975	.5622	-.3781
9	1.9403	-1.7823	.1363	.6746	-.1389
10	2.1828	-1.7823	-.1363	.6746	.1389
11	2.4254	-1.7328	-.3975	.5622	.3781
12	2.6679	-1.6337	-.6247	.3534	.5093
13	2.9104	-1.4852	-.7950	.0803	.4861
14	3.1530	-1.2872	-.8859	-.2088	.3009
15	3.3955	-1.0397	-.8745	-.4497	-.0000
16	3.6380	-.7426	-.7383	-.5622	-.3009
17	3.8806	-.3961	-.4543	-.4497	-.4012
18	4.1231	.0000	.0000	.0000	-.0000

VARIANCE
****

1	2	3	4	5
.0588	.2157	.4221	.6244	.7853
.2353	.7868	1.3318	1.6478	1.7384
.5294	1.6103	2.3751	2.5774	2.5774
.9412	2.5980	3.3829	3.4265	3.5170
1.4706	3.6765	4.3086	4.3150	4.5513
2.1176	4.7868	5.1770	5.3019	5.5612
2.8824	5.8848	6.0428	6.3589	6.5018
3.7647	6.9412	6.9598	7.4149	7.4342
4.7647	7.9412	7.9598	8.4149	8.4342
5.8824	8.8848	9.0428	9.3589	9.5018
7.1176	9.7868	10.1770	10.3019	10.5612
8.4706	10.6765	11.3086	11.3150	11.5513
9.9412	11.5980	12.3829	12.4265	12.5170
11.5294	12.6103	13.3751	13.5774	13.5774
13.2353	13.7868	14.3318	14.6478	14.7384
15.0588	15.2157	15.4221	15.6244	15.7853
17.0000	17.0000	17.0000	17.0000	17.0000

THERE ARE 19 OBSERVED Y VALUES
****

THE TRANSFORMATION MATRIX OF PARAMETER ESTIMATES
****

0	1	2	3	4	5
1.000000+000	0.000000	0.000000	0.000000	0.000000	0.000000
	2.357023-001	-4.088796-001	5.311504-001	-6.393477-001	7.526950-001
		2.271554-002	-8.852507-002	2.125694-001	-4.137153-001
			3.278706-003	-1.967224-002	6.870541-002
				5.464510-004	-4.448552-003
					9.885671-005

THE ORTHONORMAL ARGUMENTS TO DETERMINE THE B(I) ESTIMATES FROM THE Z VALUES
****

	0	1	2	3	4	5
1	1.0000000	.0000000	.0000000	.0000000	.0000000	.0000000
2	.0000000	.2357023	-.3861642	.4459040	-.4459040	.4033354
3	.0000000	.2357023	-.3407331	.2885261	-.1311482	-.0711768
4	.0000000	.2357023	-.2953021	.1508205	.0852464	-.2787759
5	.0000000	.2357023	-.2498710	.0327871	.2163946	-.3025015
6	.0000000	.2357023	-.2044399	-.0655741	.2754113	-.2135305
7	.0000000	.2357023	-.1590088	-.1442631	.2754113	-.0711768
8	.0000000	.2357023	-.1135777	-.2032798	.2295094	.0771082
9	.0000000	.2357023	-.0681466	-.2426243	.1508205	.1957363
10	.0000000	.2357023	-.0227155	-.2622965	.0524593	.2609817
11	.0000000	.2357023	.0227155	-.2622965	-.0524593	.2609817
12	.0000000	.2357023	.0681466	-.2426243	-.1508205	.1957363
13	.0000000	.2357023	.1135777	-.2032798	-.2295094	.0771082
14	.0000000	.2357023	.1590088	-.1442631	-.2754113	-.0711768
15	.0000000	.2357023	.2044399	-.0655741	-.2754113	-.2135305
16	.0000000	.2357023	.2498710	.0327871	-.2163946	-.3025015
17	.0000000	.2357023	.2953021	.1508205	-.0852464	-.2787759
18	.0000000	.2357023	.3407331	.2885261	.1311482	-.0711768
19	.0000000	.2357023	.3861642	.4459040	.4459040	.4033354

THE ARGUMENTS TO DETERMINE THE B(I) ESTIMATES FROM THE Y VALUES
****

	0	1	2	3	4	5
1	1.0000000	-.2357023	.3861642	-.4459040	.4459040	-.4033354

2	.0000000	.0000000	-.0454311	.1573779	-.3147558	.4745122
3	.0000000	.0000000	-.0454311	.1377057	-.2163946	.2075991
4	.0000000	.0000000	-.0454311	.1180334	-.1311482	.0237256
5	.0000000	.0000000	-.0454311	.0983612	-.0590167	-.0889710
6	.0000000	.0000000	-.0454311	.0786889	.0000000	-.1423537
7	.0000000	.0000000	-.0454311	.0590167	.0459019	-.1482851
8	.0000000	.0000000	-.0454311	.0393445	.0786889	-.1186281
9	.0000000	.0000000	-.0454311	.0196722	.0983612	-.0652454
10	.0000000	.0000000	-.0454311	.0000000	.1049186	.0000000
11	.0000000	.0000000	-.0454311	-.0196722	.0983612	.0652454
12	.0000000	.0000000	-.0454311	-.0393445	.0786889	.1186281
13	.0000000	.0000000	-.0454311	-.0590167	.0459019	.1482851
14	.0000000	.0000000	-.0454311	-.0786889	.0000000	.1423537
15	.0000000	.0000000	-.0454311	-.0983612	-.0590167	.0889710
16	.0000000	.0000000	-.0454311	-.1180334	-.1311482	-.0237256
17	.0000000	.0000000	-.0454311	-.1377057	-.2163946	-.2075991
18	.0000000	.0000000	-.0454311	-.1573779	-.3147558	-.4745122
19	.0000000	.2357023	.3861642	.4459040	.4459040	.4033354

TABLES TO AID IN FITTING THE DATA POINT STATISTICS
****

PREDICTION
****

	1	2	3	4	5
2	.2357	-.3862	.4459	-.4459	.4033
3	.4714	-.7269	.7344	-.5771	.3322
4	.7071	-1.0222	.8853	-.4918	.0534
5	.9428	-1.2721	.9180	-.2754	-.2491
6	1.1785	-1.4765	.8525	.0000	-.4626
7	1.4142	-1.6355	.7082	.2754	-.5338
8	1.6499	-1.7491	.5049	.5049	-.4567
9	1.8856	-1.8172	.2623	.6557	-.2610
10	2.1213	-1.8400	-.0000	.7062	.0000
11	2.3570	-1.8172	-.2623	.6557	.2610
12	2.5927	-1.7491	-.5049	.5049	.4567
13	2.8284	-1.6355	-.7082	.2754	.5338
14	3.0641	-1.4765	-.8525	-.0000	.4626
15	3.2998	-1.2721	-.9180	-.2754	.2491
16	3.5355	-1.0222	-.8853	-.4918	-.0534
17	3.7712	-.7269	-.7344	-.5771	-.3322
18	4.0069	-.3862	-.4459	-.4459	-.4033
19	4.2426	.0000	.0000	-.0000	-.0000

VARIANCE
****

1	2	3	4	5
.0556	.2047	.4035	.6023	.7650
.2222	.7506	1.2900	1.6230	1.7333
.5000	1.5449	2.3286	2.5704	2.5733
.8889	2.5071	3.3498	3.4257	3.4878
1.3889	3.5690	4.2957	4.2957	4.5097
2.0000	4.6749	5.1765	5.2523	5.5373
2.7222	5.7816	6.0365	6.2915	6.5000
3.5556	6.8579	6.9267	7.3567	7.4248
4.5000	7.8854	7.8854	8.3870	8.3870
5.5556	8.8579	8.9267	9.3567	9.4248
6.7222	9.7816	10.0365	10.2915	10.5000
8.0000	10.6749	11.1765	11.2523	11.5373
9.3889	11.5690	12.2957	12.2957	12.5097
10.8889	12.5071	13.3498	13.4257	13.4878
12.5000	13.5449	14.3286	14.5704	14.5733
14.2222	14.7506	15.2900	15.6230	15.7333
16.0556	16.2047	16.4035	16.6023	16.7650
18.0000	18.0000	18.0000	18.0000	18.0000

THERE ARE 20 OBSERVED Y VALUES
****

THE TRANSFORMATION MATRIX OF PARAMETER ESTIMATES
****

0	1	2	3	4	5
1.000000+000	0.000000	0.000000	0.000000	0.000000	0.000000
	2.294157-001	-3.979112-001	5.165710-001	-6.206942-001	7.278913-001
		2.094270-002	-8.156385-002	1.955575-001	-3.794371-001
			2.861889-003	-1.714625-002	5.971831-002
				4.512171-004	-3.663531-003
					7.712697-005

THE ORTHONORMAL ARGUMENTS TO DETERMINE THE B(I) ESTIMATES FROM THE Z VALUES
****

	0	1	2	3	4	5
1	1.0000000	.0000000	.0000000	.0000000	.0000000	.0000000
2	.0000000	.2294157	-.3769685	.4378691	-.4418318	.4045860
3	.0000000	.2294157	-.3350831	.2919127	-.1472773	-.0449540
4	.0000000	.2294157	-.2931977	.1631277	.0606436	-.2565023
5	.0000000	.2294157	-.2513123	.0515140	.1927599	-.2994730
6	.0000000	.2294157	-.2094270	-.0429283	.2599010	-.2340253
7	.0000000	.2294157	-.1675416	-.1201994	.2728961	-.1110628
8	.0000000	.2294157	-.1256562	-.1802990	.2425743	.0277657
9	.0000000	.2294157	-.0837708	-.2232274	.1797649	.1500670
10	.0000000	.2294157	-.0418854	-.2489844	.0952970	.2327031
11	.0000000	.2294157	.0000000	-.2575700	.0000000	.2617910
12	.0000000	.2294157	.0418854	-.2489844	-.0952970	.2327031
13	.0000000	.2294157	.0837708	-.2232274	-.1797649	.1500670
14	.0000000	.2294157	.1256562	-.1802990	-.2425743	.0277657
15	.0000000	.2294157	.1675416	-.1201994	-.2728961	-.1110628
16	.0000000	.2294157	.2094270	-.0429283	-.2599010	-.2340253
17	.0000000	.2294157	.2513123	.0515140	-.1927599	-.2994730
18	.0000000	.2294157	.2931977	.1631277	-.0606436	-.2565023
19	.0000000	.2294157	.3350831	.2919127	.1472773	-.0449540
20	.0000000	.2294157	.3769685	.4378691	.4418318	.4045860

THE ARGUMENTS TO DETERMINE THE B(I) ESTIMATES FROM THE Y VALUES
****

0	1	2	3	4	5

1	1.0000000	-.2294157	.3769685	-.4378691	.4418318	-.4045860
2	.0000000	.0000000	-.0418854	.1459564	-.2945545	.4495400
3	.0000000	.0000000	-.0418854	.1287850	-.2079208	.2115483
4	.0000000	.0000000	-.0418854	.1116137	-.1321164	.0429707
5	.0000000	.0000000	-.0418854	.0944423	-.0671411	-.0654477
6	.0000000	.0000000	-.0418854	.0772710	-.0129951	-.1229624
7	.0000000	.0000000	-.0418854	.0600997	.0303218	-.1388285
8	.0000000	.0000000	-.0418854	.0429283	.0628094	-.1223013
9	.0000000	.0000000	-.0418854	.0257570	.0844678	-.0826360
10	.0000000	.0000000	-.0418854	.0085857	.0952970	-.0290879
11	.0000000	.0000000	-.0418854	-.0085857	.0952970	.0290879
12	.0000000	.0000000	-.0418854	-.0257570	.0844678	.0826360
13	.0000000	.0000000	-.0418854	-.0429283	.0628094	.1223013
14	.0000000	.0000000	-.0418854	-.0600997	.0303218	.1388285
15	.0000000	.0000000	-.0418854	-.0772710	-.0129951	.1229624
16	.0000000	.0000000	-.0418854	-.0944423	-.0671411	.0654477
17	.0000000	.0000000	-.0418854	-.1116137	-.1321164	-.0429707
18	.0000000	.0000000	-.0418854	-.1287850	-.2079208	-.2115483
19	.0000000	.0000000	-.0418854	-.1459564	-.2945545	-.4495400
20	.0000000	.2294157	.3769685	.4378691	.4418318	.4045860

TABLES TO AID IN FITTING THE DATA POINT STATISTICS
****

PREDICTION						VARIANCE				
**********						********				
	1	2	3	4	5	1	2	3	4	5
2	.2294	-.3770	.4379	-.4418	.4046	.0526	.1947	.3865	.5817	.7454
3	.4588	-.7121	.7298	-.5891	.3596	.2105	.7175	1.2501	1.5972	1.7265
4	.6882	-1.0052	.8929	-.5285	.1031	.4737	1.4842	2.2815	2.5608	2.5714
5	.9177	-1.2566	.9444	-.3357	-.1963	.8421	2.4211	3.3130	3.4257	3.4642
6	1.1471	-1.4660	.9015	-.0758	-.4304	1.3158	3.4649	4.2776	4.2834	4.4686
7	1.3765	-1.6335	.7813	.1971	-.5414	1.8947	4.5632	5.1736	5.2124	5.5056
8	1.6059	-1.7592	.6010	.4397	-.5137	2.5789	5.6737	6.0349	6.2282	6.4920
9	1.8353	-1.8430	.3778	.6194	-.3636	3.3684	6.7649	6.9076	7.2913	7.4235
10	2.0647	-1.8848	.1288	.7147	-.1309	4.2632	7.8158	7.8324	8.3432	8.3603
11	2.2942	-1.8848	-.1288	.7147	.1309	5.2632	8.8158	8.8324	9.3432	9.3603
12	2.5236	-1.8430	-.3778	.6194	.3636	6.3684	9.7649	9.9076	10.2913	10.4235
13	2.7530	-1.7592	-.6010	.4397	.5137	7.5789	10.6737	11.0349	11.2282	11.4920
14	2.9824	-1.6335	-.7813	.1971	.5414	8.8947	11.5632	12.1736	12.2124	12.5056
15	3.2118	-1.4660	-.9015	-.0758	.4304	10.3158	12.4649	13.2776	13.2834	13.4686
16	3.4412	-1.2566	-.9444	-.3357	.1963	11.8421	13.4211	14.3130	14.4257	14.4642
17	3.6707	-1.0052	-.8929	-.5285	-.1031	13.4737	14.4842	15.2815	15.5608	15.5714
18	3.9001	-.7121	-.7298	-.5891	-.3596	15.2105	15.7175	16.2501	16.5972	16.7265
19	4.1295	-.3770	-.4379	-.4418	-.4046	17.0526	17.1947	17.3865	17.5817	17.7454
20	4.3589	.0000	-.0000	.0000	-.0000	19.0000	19.0000	19.0000	19.0000	19.0000